ARITHMETIC AND ALGEBRA

THE PRINDLE, WEBER, AND SCHMIDT SERIES IN MATHEMATICS

Althoen and Bumcrot, *Introduction to Discrete Mathematics*
Brown and Sherbert, *Introductory Linear Algebra with Applications*
Buchthal and Cameron, *Modern Abstract Algebra*
Burden and Faires, *Numerical Analysis*, Fourth Edition
Cullen, *Linear Algebra and Differential Equations*
Cullen, *Mathematics for the Biosciences*
Eves, *In Mathematical Circles*
Eves, *Mathematical Circles Adieu*
Eves, *Mathematical Circles Revisited*
Eves, *Mathematical Circles Squared*
Eves, *Return to Mathematical Circles*
Fletcher and Patty, *Foundations of Higher Mathematics*
Geltner and Peterson, *Geometry for College Students*
Gilbert and Gilbert, *Elements of Modern Algebra*, Second Edition
Gobran, *Beginning Algebra*, Fourth Edition
Gobran, *College Algebra*
Gobran, *Intermediate Algebra*, Fourth Edition
Gordon, *Calculus and the Computer*
Hall, *Algebra for College Students*
Hall and Bennett, *College Algebra with Applications*, Second Edition
Hartfiel and Hobbs, *Elementary Linear Algebra*
Hunkins and Mugridge, *Applied Finite Mathematics*, Second Edition
Kaufmann, *Algebra for College Students*, Third Edition
Kaufmann, *Algebra with Trigonometry for College Students*, Second Edition
Kaufmann, *College Algebra*
Kaufmann, *College Algebra and Trigonometry*
Kaufmann, *Elementary Algebra for College Students*, Third Edition
Kaufmann, *Intermediate Algebra for College Students*, Third Edition
Kaufmann, *Precalculus*
Kaufmann, *Trigonometry*
Keisler, *Elementary Calculus: An Infinitesimal Approach*, Second Edition
Kirkwood, *Introduction to Real Analysis*
Laufer, *Discrete Mathematics and Applied Modern Algebra*
Nicholson, *Linear Algebra with Applications*
Pasahow, *Mathematics for Electronics*

Powers, *Elementary Differential Equations*
Powers, *Elementary Differential Equations with Boundary-Value Problems*
Powers, *Elementary Differential Equations with Linear Algebra*
Proga, *Arithmetic and Algebra*, Second Edition
Proga, *Basic Mathematics*, Second Edition
Radford, Vavra, and Rychlicki, *Introduction to Technical Mathematics*
Radford, Vavra, and Rychlicki, *Technical Mathematics with Calculus*
Rice and Strange, *Calculus and Analytic Geometry for Engineering Technology*
Rice and Strange, *Plane Trigonometry*, Fifth Edition
Rice and Strange, *Technical Mathematics*
Rice and Strange, *Technical Mathematics and Calculus*
Schelin and Bange, *Mathematical Analysis for Business and Economics*, Second Edition
Strnad, *Introductory Algebra*
Swokowski, *Algebra and Trigonometry with Analytic Geometry*, Seventh Edition
Swokowski, *Calculus with Analytic Geometry*, Second Alternate Edition
Swokowski, *Calculus with Analytic Geometry*, Fourth Edition
Swokowski, *Fundamentals of Algebra and Trigonometry*, Seventh Edition
Swokowski, *Fundamentals of College Algebra*, Seventh Edition
Swokowski, *Fundamentals of Trigonometry*, Seventh Edition
Swokowski, *Precalculus: Functions and Graphs*, Fifth Edition
Tan, *Applied Calculus*
Tan, *Applied Finite Mathematics*, Second Edition
Tan, *Calculus for the Managerial, Life, and Social Sciences*
Tan, *College Mathematics*, Second Edition
Venit and Bishop, *Elementary Linear Algebra*, Third Edition
Venit and Bishop, *Elementary Linear Algebra*, Alternate Second Edition
Willard, *Calculus and Its Applications*, Second Edition
Wood and Capell, *Arithmetic*
Wood, Capell, and Hall, *Developmental Mathematics*, Third Edition
Wood, Capell, and Hall, *Intermediate Algebra*
Wood, Capell, and Hall, *Introductory Algebra*
Zill, *A First Course in Differential Equations with Applications*, Fourth Edition
Zill, *Calculus with Analytic Geometry*, Second Edition
Zill, *Differential Equations with Boundary-Value Problems*, Second Edition

ARITHMETIC AND ALGEBRA
second edition

Rosanne Proga

PWS-KENT Publishing Company
Boston

PWS-KENT
Publishing Company

20 Park Plaza
Boston, Massachusetts 02116

PWS-KENT Publishing Company is a division of Wadsworth, Inc.

Library of Congress Cataloging-in-Publication Data

Proga, Rosanne.
Arithmetic and algebra.

Includes index.
1. Arithmetic—1961- . 2. Algebra.
I. Title.
QA107.P74 1989 513′.12 88–19643
ISBN 0-534-91596-5

Printed in the United States of America

90 91 92 93 — 10 9 8 7 6 5 4 3

Production Editor: Susan M. C. Caffey
Manufacturing Coordinator: Margaret Sullivan Higgins
Text and Cover Designer: Susan M. C. Caffey
Interior Illustration: Jan Rapp
Typesetting: Polyglot Pte Ltd.
Cover Printing: New England Book Components
Printing and Binding: The Murray Printing Company

Cover Art: "Nebula I" by Marko Spalatin, a serigraph used with the permission of the artist. Please note that this is a horizontal image used vertically.

PREFACE

Arithmetic and Algebra, Second Edition, is a practical approach to the fundamentals of arithmetic and elementary algebra for students who elect non-mathematical areas of study. The approach to the subject matter is mature, yet easily understandable, and assumes no prior knowledge of mathematics. Various mathematical techniques are thoroughly explained because knowing *why* a technique works is an aid to remembering *how* to use it.

A main feature of this book is its flexibility. This book is especially suited for a combination arithmetic-elementary algebra course. Both the arithmetic and algebra portions of the book are complete in themselves and could be covered independently in separate arithmetic or elementary algebra courses. A large problem that students confront in an elementary algebra course is that they might have forgotten many arithmetic skills that need to be applied in order to master algebraic techniques. In this text, examples from arithmetic are used to introduce new algebraic techniques in order to illustrate the operations being performed. These examples are accompanied by references to sections of the book for the student who might need a more thorough review of a topic. In response to reviewers' comments, the following revisions have been made in the second edition:

- Estimation techniques have been added to Chapter 3 (Decimals).
- Additional problems on converting units of measurement have been added to Section 4.1 (Ratio).
- The topics presented in Chapter 8 (Measurement in the English and Metric Systems) and Chapter 9 (Introduction to Geometry) are covered earlier.
- Section 10.6 (Formulas) has been added to Chapter 10 (Introduction to Algebra).
- Chapter 13 (Factoring) has been expanded to include factoring by grouping, factoring the difference of two cubes, factoring the sum of two cubes, and a general factoring strategy.
- Additional word problems have been added to Chapter 16 (Rational Expressions).

Topics

The highlights of this book include the following:

- Techniques for estimating answers are presented in Chapter 1 (Whole Numbers) so that students are better able to spot errors in their calculations.
- In Chapter 3 (Fractions), mixed numbers are presented early to achieve an integrated approach to the discussion of basic operations with fractions.
- Though algebraic concepts are introduced in Chapter 4 (Ratio and Proportion) and are used to solve equations in Chapter 5 (Percent), Chapter 10 (Introduction to Algebra) assumes no prior knowledge of algebra.
- Word problems emphasizing consumer applications appear throughout the text and predominate Section 5.4 (Applications of Percent) and Section 5.6 (Reading Graphs).
- Chapter 6 (Positive and Negative Numbers) includes examples of various arithmetic operations on signed integers, decimals, and fractions. This chapter may serve as a starting point for an elementary algebra course.
- Procedures for simplifying square roots are introduced in Chapter 7 (Roots of Numbers) and expanded upon to apply to algebraic expressions in Chapter 17 (Radical Expressions)
- Methods for approximating conversions between the English and Metric systems

appear in Chapter 8 (Measurement in the English and Metrix Systems) to develop the student's ability to "think metric."

• A thorough treatment of elementary geometry appears in Chapter 9 (Introduction to Geometry).

• Word problems that can be solved using quadratic equations are covered in Chapter 13 (Factoring) in anticipation that some instructors may not be able to cover Chapter 18 (Quadratic Equations).

• A thorough presentation of graphing linear equations and inequalities is presented in Chapter 14 (Graphing), graphing systems of linear equations and inequalities in Chapter 15 (Systems of Linear Equations), and graphing quadratic equations in Chapter 18 (Quadratic Equations).

• Chapter 16 (Rational Expressions) briefly reviews the concepts of mixed numbers and the basic operations with fractions in order to perform the same operations on algebraic expressions. Section 16.3 (Lowest Common Denominator) is designed to pave the way for students to be able to add and subtract rational expressions having different denominators, a skill that is often difficult to master.

Features

• *Word problems* reinforce each newly learned skill and emphasize applications of mathematics to the real world.

• *Quick Quizzes* appear at the end of each section enabling students to check their progress while reading the text.

• *Calculator problems* are designated by the symbol , and a brief description of how to use a calculator appears in the appendix. This book may be used with or without a pocket calculator.

• *Warnings* designated by the symbol alert the student to common algebra mistakes.

• *Chapter Summaries* highlight important definitions and calculations and are followed by review exercises that reinforce the material learned. All items are keyed to the appropriate section of the text for quick referencing.

Supplements

• *The Answer Key* includes answers to even-numbered exercises. The answers to the odd-numbered exercises are found in the back of the book.

• *The Test Bank* includes problems similar to those in the book. These can be used to supplement exercises in the text and/or as quizzes or self-testing tools. Answers are provided.

• *Computerized Test Generator for IBM*

Acknowledgments

I would like to express my appreciation to the many people who contributed to the development of this book. I am especially grateful to the following reviewers for their helpful suggestions:

Robert Wendling, *Ashland College*; Paul Pontius, *Pan American University*; Dorothy Schwellenbach, *Hartnell College*; Norma Alexander, *Ohlone College*; Kathleen Tomine, *Western Wyoming College*; James Choike, *Oklahoma State University*; John Hansen, *Armstrong State College*; Faye Hendrix Thames, *La Mar University*; Sandra Wager, *Pennsylvania State University*; Jerry Gilpin, *Pan American University*; Virginia Kaiser, *Morraine Valley Community College*; Dick Holliday, *Rogue Community College*; Froylan Tiscareño, *Mt. San Antonio Community College*; and Lovenia De Conge-Watson, *Southern University*.

My editors at PWS-KENT have each made invaluable contributions during various stages of the project. Tom Orowan did an excellent job of typing the manuscript. Finally, my special thanks to my father for his thoughtful comments and diligent proofreading, and to Peter Marzuk for his continued support and encouragement.

CONTENTS

1 WHOLE NUMBERS

1.1 The Base Ten System

When we begin to study any new field of knowledge, it is essential that we learn the vocabulary that is used to communicate the fundamental ideas of the subject. Thus, in order to study mathematics, we must familiarize ourselves with the language of mathematics, whose alphabet consists not only of letters, but also of numbers.

In this chapter, we will consider the numbers most frequently encountered—the **whole numbers**. These include {0, 1, 2, 3, 4, 5, 6, 7, 8, 9, 10, 11, 12, ...}. The braces { } that enclose the list indicate that the whole numbers are considered a set. A *set* is simply a collection or group of objects. The symbol ... indicates that it is impossible to list all the whole numbers. Since we can add the number 1 to any whole number and obtain another number that is larger, there is no end to the list of whole numbers. We express this idea by saying that the set of whole numbers is *infinite*.

The whole numbers can be represented by a diagram called a number line, shown in Figure 1.1. Notice that larger numbers are located to the right of smaller numbers. The arrow on the number line indicates that there are an infinite number of whole numbers.

figure 1.1

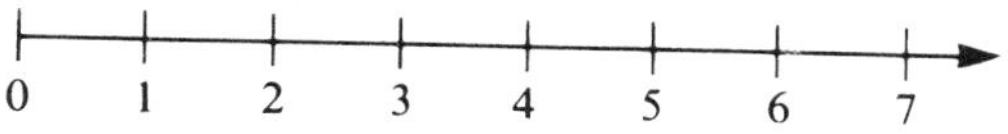

The number system we most frequently use is called the **base ten system**. In this system, only ten digits (0, 1, 2, 3, 4, 5, 6, 7, 8, 9) are used to express all numbers. The location of each digit, or **place value**, determines the value of the number.

definition

The *place value* of a digit is the name of the location of that digit.

The place value increases by a factor of ten as we move from right to left. The rightmost location is the ones place, the position immediately to the left of it is the tens place, the next place is the hundreds place, and so on, as shown in Table 1.1.

We name numbers according to the location of their digits. For example, the number one thousand has a 1 in the thousands place followed by three zeros: 1,000. After every group of three digits, moving from right to left, we place a comma to make it easier for us to read the number. Each of these three-digit groups is called a **period**. A chart indicating the place values of the first 15 locations included in the first five periods is

table 1.1
Place Values

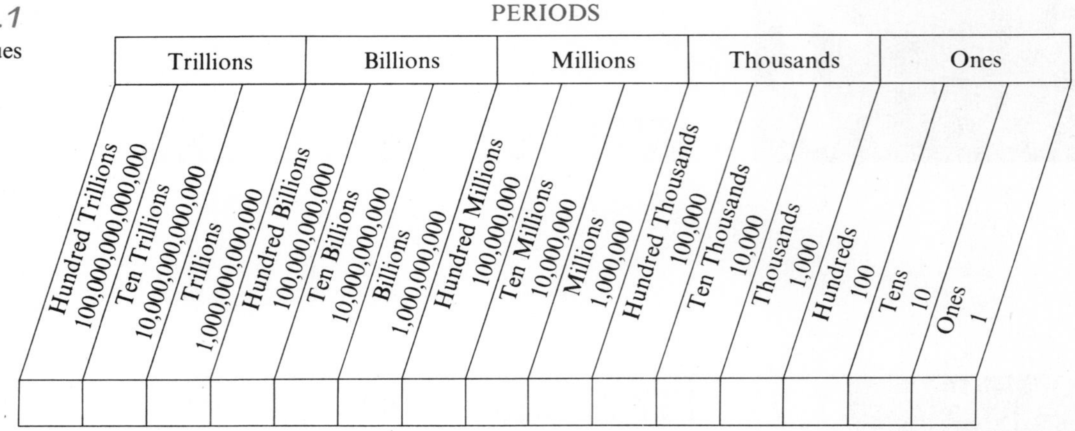

shown in Table 1.1. The place value names the location of the digit 1 in each numerical representation given.

The place value of each digit in the number

8,416,923

can be determined using the chart as illustrated, beginning with the ones place and moving from right to left.

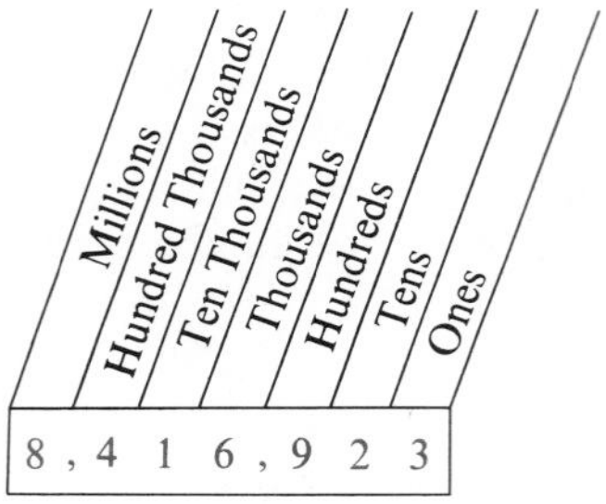

Once we have identified the place value of each digit, we can write the number in words as follows:

eight million, four hundred sixteen thousand, nine hundred twenty-three

Now let us look at some examples that illustrate how place value is used to assign names to numbers.

example 1

For each of the following numbers, give the place value of each digit and rewrite the number in words.

(a) 325

3: hundreds place
2: tens place
5: ones place
three hundred twenty-five

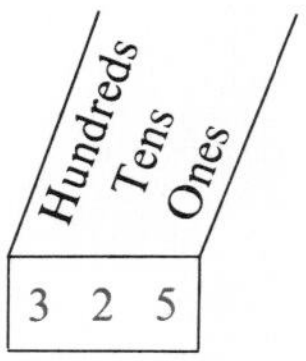

(b) 8,016

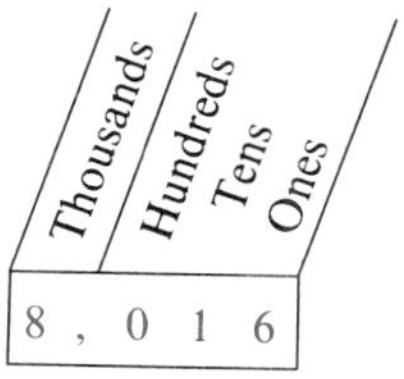

8: thousands place
0: hundreds place
1: tens place
6: ones place
eight thousand, sixteen

(c) 423,701

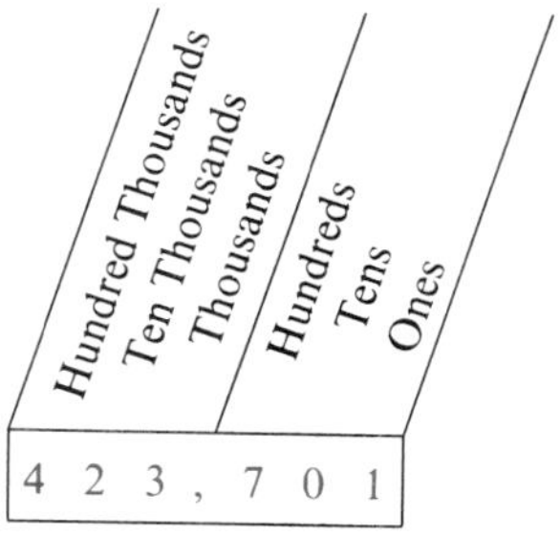

4: hundred thousands place
2: ten thousands place
3: thousands place
7: hundreds place
0: tens place
1: ones place
four hundred twenty-three thousand, seven hundred one

(d) 61,004,072

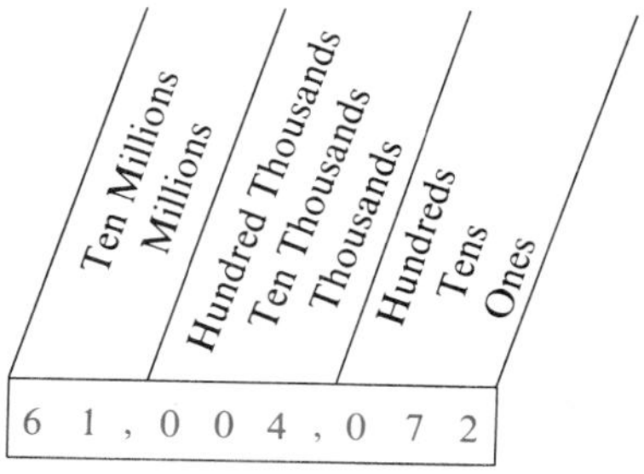

6: ten millions place
1: millions place
0: hundred thousands place
0: ten thousands place
4: thousands place
0: hundreds place
7: tens place
2: ones place
sixty-one million, four thousand, seventy-two

(e) 7,325,018,640

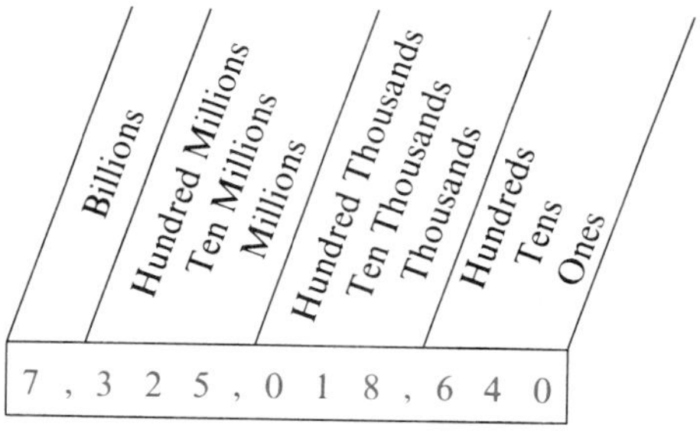

7: billions place
3: hundred millions place
2: ten millions place
5: millions place
0: hundred thousands place
1: ten thousands place
8: thousands place
6: hundreds place
4: tens place
0: ones place
seven billion, three hundred twenty-five million, eighteen thousand, six hundred forty

example 2

Write each of the following numbers using digits.

(a) three thousand, eight
3,008

(b) two hundred fifty-seven thousand, seven hundred twenty-two

257,722

(c) six million, three thousand, one

6,003,001 ●

We can also use the concept of place value to rewrite any number in **expanded notation**. This notation will be particularly useful to us when learning how to perform the various arithmetic operations on whole numbers (see box). We put parentheses around each quantity obtained in step 2 to indicate that the multiplication is to be performed before the addition (step 3).

To Write a Number in Expanded Notation

1. Identify the place value of each digit.
2. Multiply each digit by its place value.
3. Add the results found in step 2 together.

For example, the number 736 can be written in expanded notation as follows:

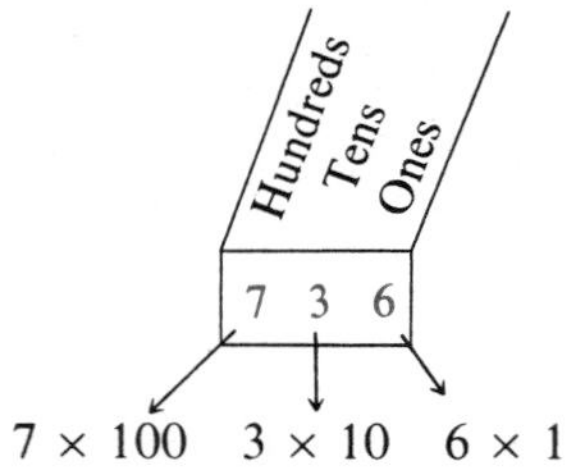

$$736 = 7 \text{ hundreds} + 3 \text{ tens} + 6 \text{ ones}$$
$$= (7 \times 100) + (3 \times 10) + (6 \times 1)$$

Remember that the parentheses are used to indicate that multiplication is performed before addition.

In some forms of expanded notation, the results in parentheses are simplified further as follows:

$$736 = (7 \times 100) + (3 \times 10) + (6 \times 1)$$
$$= 700 + 30 + 6$$

We will not perform this extra step when we write numbers in expanded notation.

example 3

Rewrite each of the following numbers in expanded notation.

(a) 85

$85 = 8 \text{ tens} + 5 \text{ ones} = (8 \times 10) + (5 \times 1)$

(b) 409

$$409 = 4 \text{ hundreds} + 0 \text{ tens} + 9 \text{ ones}$$
$$= (4 \times 100) + (0 \times 10) + (9 \times 1)$$

(c) 350,081

$$350{,}081 = 3 \text{ hundred thousands} + 5 \text{ ten thousands} + 0 \text{ thousands}$$
$$+ 0 \text{ hundreds} + 8 \text{ tens} + 1 \text{ one}$$
$$= (3 \times 100{,}000) + (5 \times 10{,}000) + (8 \times 10) + (1 \times 1)$$

●

Certain symbols are used in mathematics to represent the various mathematical operations and to compare numbers. The four basic arithmetic operations are indicated by the symbols in the accompanying box. The letters a and b are used to represent any two whole numbers.

Four Basic Arithmetic Operations and Their Symbols

Addition:	$a + b$	means to add a and b.
Subtraction:	$a - b$	means to subtract b from a.
Multiplication:	$a \times b$	means to multiply a and b.
Division:	$a \div b$	means to divide a by b.

example 4

State what arithmetic operation is indicated by each of the following expressions.

(a) 2×100

Multiply 2 and 100.

(b) $8 - 3$

Subtract 3 from 8.

(c) $4 + b$

Add 4 and b.

(d) $x \div y$

Divide x by y.

A list of symbols used to compare two numbers is shown in another box. The letters a and b are again used to represent any two whole numbers.

Symbols Used to Compare Numbers

Equality:	$a = b$	means a is equal to b.
Inequality:	$a \neq b$	means a is not equal to b.
Less than:	$a < b$	means a is less than b.
Greater than:	$a > b$	means a is greater than b.

Notice that in using the $>$ or $<$ symbol, the wide end of the symbol faces the larger number. Also note the respective locations of a and b on the following number lines.

$a < b$

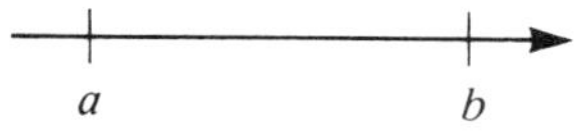

a is less than b indicates that a is to the left of b on a number line.

$a > b$

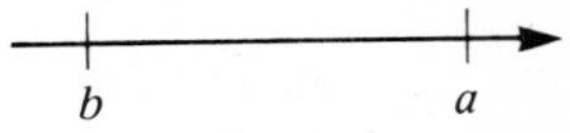

a is greater than b indicates that a is to the right of b on a number line.

example 5

Compare each of the following pairs of numbers represented on this number line. Use the $=$, $<$, or $>$ sign to indicate the relationship between them.

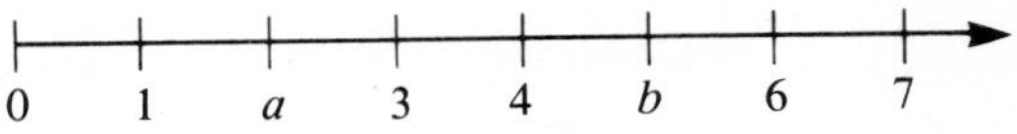

(a) $3 ? 6$

Since 3 is to the left of 6 on the number line, 3 is less than 6.

$3 < 6$

(b) $b ? 5$

Since b is at the same location as 5 on the number line, b is equal to 5.

$b = 5$

(c) $4 ? a$

Since 4 is to the right of a, 4 is greater than a.

$4 > a$

(d) $a ? b$

Since a is to the left of b, a is less than b.

$a < b$

example 6

For each of the following expressions, use the $=$ or $\neq$ sign to express the appropriate relationship between the two quantities. For quantities that are unequal, use the greater than ($>$) or less than ($<$) sign to further define the relationship.

(a) $6 \ ? \ 16$

$6 \neq 16$ — 6 is not equal to 16.

$6 < 16$ — 6 is less than 16.

(b) $5{,}028 \ ? \ (5 \times 1{,}000) + (2 \times 10) + (8 \times 1)$

$5{,}028 = (5 \times 1{,}000) + (2 \times 10) + (8 \times 1)$

Both sides are equal since the right-hand side is simply 5,028 written in expanded notation.

(c) $82 \ ? \ 28$

$82 \neq 28$ — 82 is not equal to 28.

$82 > 28$ — 82 is greater than 28.

(d) $9 \times 10 \ ? \ 10 \times 9$

$9 \times 10 = 10 \times 9$

Since both sides express the number 90, they are equal.

(e) $(6 \times 100) + (2 \times 10) \ ? \ (6 \times 100) + (2 \times 1)$

6 hundreds + 2 tens $\neq$ 6 hundreds + 2 ones

$600 + 20 > 600 + 2$

$620 > 602$

By rewriting in words, it is apparent that the two sides are unequal. Simplifying, we can see that $620 > 602$.

(f) If $a > b$, $b \ ? \ a$.

$b \neq a$. If a is greater than b, they cannot be equal. Draw a number line to show the relationship between a and b. a is greater than b means that a is to the right of b on a number line. Therefore, b is to the left of a, meaning that b is less than a.

$b < a$.

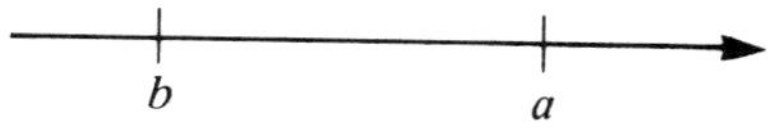

QUICK QUIZ

1. Write 3,061 in words.
2. Write 405,790 in expanded notation.
3. Write sixty thousand, five hundred twelve using digits.

ANSWERS

1. three thousand, sixty-one
2. $(4 \times 100{,}000) + (5 \times 1{,}000) + (7 \times 100) + (9 \times 10)$
3. 60,512

1.1 Exercises

Indicate which of the following statements are true and which are false. For those that are false, change the italic expression to make the statement true.

1. The set of whole numbers *does not* include the number 0.
2. The set of whole numbers is *finite*.
3. The number seven thousand, eight is written *7,080*.
4. Our number system, which uses ten digits, is called the *number ten system*.

For each of the following numbers, give the place value of the digit in parentheses and rewrite the number in words.

5. 58 (5)	**6.** 29 (9)	**7.** 806 (8)	**8.** 321 (2)
9. 4,030 (4)	**10.** 7,514 (5)	**11.** 21,009 (2)	**12.** 39,456 (9)
13. 500,001 (5)	**14.** 407,200 (7)	**15.** 1,496,628 (4)	**16.** 7,820,413 (2)
17. 38,002,540 (8)	**18.** 718,243,659 (7)	**19.** 715,151,515 (7)	**20.** 1,601,070,402 (7)
21. 89,641,372,055 (9)	**22.** 526,828,424,121 (5)	**23.** 7,000,000,486,532 (7)	**24.** 603,030,222,111,300 (6)

Write each of the following numbers using digits.

25. forty-seven
26. ninety-nine
27. six hundred two
28. five hundred seventy-eight

29. one thousand, sixty-six
30. three thousand, four hundred seventeen
31. eight thousand, eight
32. seventy-two thousand, four hundred fifty-nine
33. fifty thousand, six hundred
34. two hundred thousand, two
35. nine hundred forty-four thousand, twenty-six
36. three hundred seven thousand, six hundred sixty-seven
37. seven million, eight
38. three million, nine hundred two thousand, seven hundred fifty-three
39. eleven million, sixty-four thousand, two hundred eighteen
40. sixty-seven million, four hundred seventy thousand, seven
41. two billion, fifty-three million, one hundred three thousand, four hundred eighty-one
42. seven billion, five hundred five million, twenty thousand, four
43. fifty billion, nine million, three hundred thousand, five hundred
44. six hundred eighty billion, two hundred sixty-six million, forty thousand, twenty-eight
45. five trillion, seven hundred five billion, six million, four hundred seventy-four thousand, three hundred two
46. nine hundred trillion, nine hundred thousand, ninety

Write each of the following numbers using expanded notation.

47. 33
48. 51
49. 605
50. 777
51. 4,002
52. 8,115
53. 10,040
54. 63,874
55. 638,518
56. 500,005
57. 30,041,003
58. 48,673,851
59. 999,999,999
60. 348,627,841
61. 5,038,602,124
62. 88,777,666,555
63. 300,002,404,100
64. 5,937,628,416,572
65. 44,290,053,777,119
66. 600,000,007,000,000

State what is meant by each of the following mathematical expressions.

67. 7×3
68. $4 \neq 9$
69. $18 \div 2$
70. $2 < 7$
71. $a = 10$
72. $9 - x$
73. $w > t$
74. $c + d$

Compare each of the following pairs of numbers represented on this number line. Use the =, >, or < sign to indicate the relationship between them.

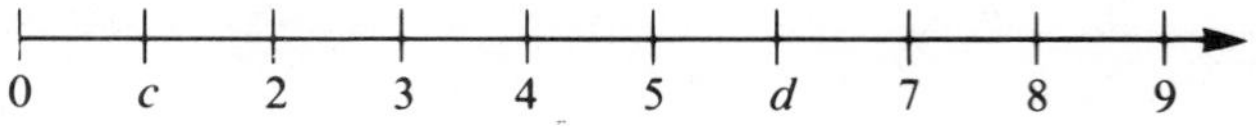

75. $0 ? 3$
76. $8 ? 2$
77. $d ? 7$
78. $4 ? c$
79. $8 ? c$
80. $c ? 1$
81. $6 ? d$
82. $d ? c$

For each of the following expressions, use the = or ≠ sign to express the appropriate relationship between the two quantities. For quantities that are unequal, use the greater than (>) or less than (<) sign to further define the relationship.

83. $45 \ ? \ 54$
84. $860 \ ? \ 8{,}060$
85. $(3 \times 10) + (2 \times 1) \ ? \ 32$
86. $637 \ ? \ (6 \times 100) + (3 \times 10) + (7 \times 1)$
87. $4 \times 100 \ ? \ 100 \times 4$
88. $0 \times 10 \ ? \ 0 \times 100$
89. $1 \times 10 \ ? \ 1 \times 100$
90. $(5 \times 1{,}000) + (6 \times 100) \ ? \ (5 \times 1{,}000) + (6 \times 10) + (9 \times 1)$
91. If $a = b$, then $b \ ? \ a$.
92. If $a < b < c$, then $a \ ? \ c$.

Answer the following questions.

93. The Pacific Ocean covers sixty-four million square miles. How is this number expressed using digits?

94. The area of the Caspian Sea, which is between the USSR and Iran, is 152,239 square miles. What is the place value of 5 in this number?

95. A light year is the distance light travels in one year. It is equivalent to 5,880,000,000,000 miles. How is this number written in words?

96. The Grand Canyon National Park covers an area of 1,218,375 acres. How is this number written in words?

97. The planet Pluto is 3,675,270,000 miles from the sun. What is the place value of the digit 6 in this number?

98. The U.S. gross national product was approximately one trillion, five hundred billion dollars in 1975. How is this number expressed using digits?

99. According to the 1980 census, the population of New York City was 6,808,370. How is this number expressed in words?

100. In 1970, the national debt was $370,094,000,000. How is this number written in words?

1.2 Addition of Whole Numbers

We will now consider how to perform the four basic arithmetic operations—addition, subtraction, multiplication, and division—on whole numbers. Let us begin our discussion by defining some of the terminology and properties that apply to addition.

definition

An *equation* is a mathematical expression that states that two quantities are equal.

Consider the following example. If there are five apples on one table and three apples on another, the total number of apples on both tables is eight. This can be represented by the equation

$$3 + 5 = 8$$

definition

To add two numbers a and b, we can express the operation performed by the equation $a + b = c$, where the numbers a and b are called *addends* and the result, c, is called the *sum*.

For example, in the equation $1 + 2 = 3$, the numbers 1 and 2 are the addends, and the number 3 is the sum.

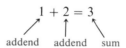

$$1 + 2 = 3$$

addend ↗ addend ↑ sum ↖

For small numbers, we can determine the result of an addition operation by simply counting the total number of items, using such devices as our fingers or tally marks. The Chinese invented an instrument called the abacus, which consists of sliding beads on parallel wires, which enables them to add even very large numbers very rapidly by counting. However, since we use addition so frequently, certain addition facts should be memorized, thus allowing us to build speed and accuracy in performing our calculations.

Table 1.2 gives important addition facts that must be memorized. The numbers to be added, or addends, are chosen from the numbers in the top row and from those in the leftmost column. The sum is obtained by moving your right index finger down the column containing the addend in the top row and moving your left index finger across the row containing the other addend in the leftmost column. The block where your fingers meet, which is the intersection of this row and column, contains the sum.

For example, to add $5 + 7$ we find the number 5 in the top row and the number 7 in the leftmost column. Move one finger down the column containing 5 and another across the row containing 7. Your two fingers should meet at the block containing 12. Thus,

$$5 + 7 = 12$$

Some sums that may be calculated using Table 1.2 are:

$3 + 4 = 7 \qquad 7 + 9 = 16 \qquad 6 + 6 = 12 \qquad 0 + 5 = 5$

table 1.2
Addition Table

+	0	1	2	3	4	5	6	7	8	9
0	0	1	2	3	4	**5**	6	7	8	9
1	1	2	3	4	5	**6**	7	8	9	10
2	2	3	4	5	6	**7**	8	9	10	11
3	3	4	5	6	7	**8**	9	10	11	12
4	4	5	6	7	8	**9**	10	11	12	13
5	5	6	7	8	9	**10**	11	12	13	14
6	6	7	8	9	10	**11**	12	13	14	15
7	**7**	**8**	**9**	**10**	**11**	**12**	13	14	15	16
8	8	9	10	11	12	13	14	15	16	17
9	9	10	11	12	13	14	15	16	17	18

Notice that whenever any number is added to 0, the resulting sum is the same as that number. Thus,

property

The number 0 is said to be the *identity element* of addition, since for any number a, $a + 0 = a$ and $0 + a = a$.

It is also important to note that when adding any two numbers, the same sum is obtained regardless of the *order* of the addends. For example, since

$$3 + 8 = 11 \quad \text{and} \quad 8 + 3 = 11$$

then

$$3 + 8 = 8 + 3$$

This property is called the **commutative property** of addition and may be stated as follows:

property

For any numbers a and b, $a + b = b + a$. Thus, according to the *commutative property*, we can change the *order* of the addends without affecting the result.

For example, $1 + 5 = 5 + 1$.

In order to add more than two numbers, we first add any two together, and then we add that result to the remaining numbers to obtain the total sum. For example, to add $5 + 3 + 2$, we first add $5 + 3$ to obtain 8, and then we add $8 + 2$ to obtain a total sum of 10. We can indicate this process by using parentheses, which group together the numbers to be added together first:

$$5 + 3 + 2 = (5 + 3) + 2 = 8 + 2 = 10$$

Notice that we obtain the same result if we first add 3 + 2 to obtain 5 and then add 5 + 5 to obtain 10. By using parentheses to group the numbers that are added first, we obtain:

$$5 + 3 + 2 = 5 + (3 + 2) = 5 + 5 = 10$$

This example illustrates the **associative property** of addition, which may be stated as follows:

property

For any numbers a, b, and c, $(a + b) + c = a + (b + c)$. Thus, according to the *associative property*, we can change the way in which we *group* the addends without affecting the result.

For example, (4 + 2) + 7 = 4 + (2 + 7).

example 1

State whether the commutative or associative property is illustrated by each of the following equations.

(a) 8 + 7 = 7 + 8

Commutative property, since the order is changed.

(b) 2 + (6 + 4) = (2 + 6) + 4

Associative property, since 6 and 4 are grouped together on the left, and 2 and 6 are grouped together on the right.

(c) 5 + (1 + 3) = 5 + (3 + 1)

Commutative property, since the order of 1 and 3 is reversed. Note that this is not the associative property since the same numbers are grouped together on both sides of the equation.

(d) (4 + 1) + 3 = 4 + (3 + 1)

Both the commutative and associative properties, since the order and the grouping are changed. ●

In order to add numbers greater than nine, it is easiest to write the problem vertically, lining up the digits with the same place values. We then add the digits in the individual columns separately, moving from right to left and keeping in mind the place value of the digits in each column.

example 2

Add each of the following.

(a) 82 + 56

addend	82 =	8 tens + 2 ones	Write the problem vertically and indicate the place value of each digit.
addend	+ 56 =	5 tens + 6 ones	Add the numbers in the ones column: 2 + 6 = 8.
sum		13 tens + 8 ones	Add the numbers in the tens column: 8 + 5 = 13.

When the sum of the numbers in any place is greater than nine, the result can be rewritten using the next higher place value. Thus, in Example 2, 13 tens = 1 hundred + 3 tens, so the final answer is:

1 hundred + 3 tens + 8 ones = 138.

(b) 385 + 267

$$\begin{array}{lrl}
\text{addend} & 385 = & 3 \text{ hundreds} + 8 \text{ tens} + 5 \text{ ones} \\
\text{addend} & +\,267 = & 2 \text{ hundreds} + 6 \text{ tens} + 7 \text{ ones} \\ \hline
\text{sum} & & 5 \text{ hundreds} + 14 \text{ tens} + 12 \text{ ones}
\end{array}$$

Since 12 ones = 1 ten + 2 ones, and 14 tens = 1 hundred + 4 tens, the final sum is:

$$\begin{aligned}
&5 \text{ hundreds} + 1 \text{ hundred} + 4 \text{ tens} + 1 \text{ ten} + 2 \text{ ones} \\
&\quad = 6 \text{ hundreds} + 5 \text{ tens} + 2 \text{ ones} = 652.
\end{aligned}$$

●

example 3

Add 527 + 68 + 245.

solution

$$\begin{array}{lrl}
\text{addend} & 527 = & 5(100) + 2(10) + 7(1) \\
\text{addend} & 68 = & 6(10) + 8(1) \\
\text{addend} & 245 = & 2(100) + 4(10) + 5(1) \\ \hline
& & 7(100) + 12(10) + 20(1) \\
& = & 700 + 120 + 20 \\
& = & 840
\end{array}$$

●

A short cut for adding numbers in this fashion is called *carrying*. The procedure involves the three steps listed in the box.

To Add Whole Numbers

1. Write the problem vertically, and line up the numbers with the same place value.
2. Add the numbers in each column separately, moving from right to left.
3. If the sum of any column is greater than nine, put down the appropriate digit in the ones place and add, or **carry**, the other digit to the next column immediately to the left.

If we use carrying to add the numbers in Example 2(b), we obtain the following result.

$$\begin{array}{r}
{}^{1} \\
38\boxed{5} \\
+\,26\boxed{7} \\ \hline
\boxed{2}
\end{array}$$

5 + 7 = 12. Put down 2. Carry 1 to the tens place.

$$\begin{array}{r}
{}^{1}\ {}^{1} \\
3\,\boxed{8}\,5 \\
+\,2\,\boxed{6}\,7 \\ \hline
\boxed{5}\,2
\end{array}$$

1 + 8 + 6 = 15. Put down 5. Carry 1 to the hundreds place.

$$\begin{array}{r|r} & 1 \\ & 3\ 85 \\ + & 2\ 67 \\ \hline & 6\ 52 \end{array}$$

1 + 3 + 2 = 6. Put down 6.

The sum is 652.

Let us now look at some more examples of addition that illustrate carrying.

example 4

Add 39 + 28 + 83.

solution

$$\begin{array}{r} {}^{12} \\ 39 \\ 28 \\ +\ 83 \\ \hline 150 \end{array}$$

9 + 8 + 3 = 20. Put down 0. Carry 2 to the tens place.

2 + 3 + 2 + 8 = 15. Put down 5. Carry 1 to the hundreds place. The sum is 150. ●

example 5

Add 507 + 1036.

solution

$$\begin{array}{r} {}^{1} \\ 507 \\ +\ 1{,}036 \\ \hline 1{,}543 \end{array}$$

7 + 6 = 13. Put down 3. Carry 1.
1 + 0 + 3 = 4.
5 + 0 = 5.
The sum is 1,543. ●

Notice that because of the commutative property, we can reverse the addends to check our answer:

$$\begin{array}{r} {}^{1} \\ 1{,}036 \\ +\ 507 \\ \hline 1{,}543 \end{array}$$

example 6

Add 3,694,862 + 8,581,077. Check your answer by reversing the addends.

solution

$$\begin{array}{r} {}^{1\,1\quad 1} \\ 3{,}694{,}862 \\ +\ 8{,}581{,}077 \\ \hline 12{,}275{,}939 \end{array}$$

CHECK

$$\begin{array}{r} {}^{1\,1\quad 1} \\ 8{,}581{,}077 \\ +\ 3{,}694{,}862 \\ \hline 12{,}275{,}939 \checkmark \end{array}$$

●

Estimation

When adding a column of numbers, it is important to be able to predict the answer roughly so that you can recognize an unreasonable answer caused by a careless mistake. You can do this by approximating each of the addends and mentally adding them up. This process is called *estimation.*

It is extremely important to estimate answers whenever you use a pocket calculator to do a calculation. Since most mistakes in using a calculator are caused by pressing the wrong buttons, you should always estimate the answer before using a calculator, and then compare the calculator's answer to your estimate.

example 7

Estimate 293 + 512. Then determine the exact answer.

solution

293 is between 200 and 300 on the number line, but much closer to 300. Therefore 293 ≈ 300. The symbol ≈ means "is approximately equal to." Similarly, 512 ≈ 500. Adding 300 + 500 provides us with an estimate of 800. Now we will determine the exact answer:

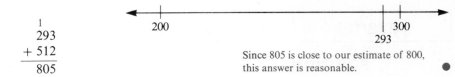

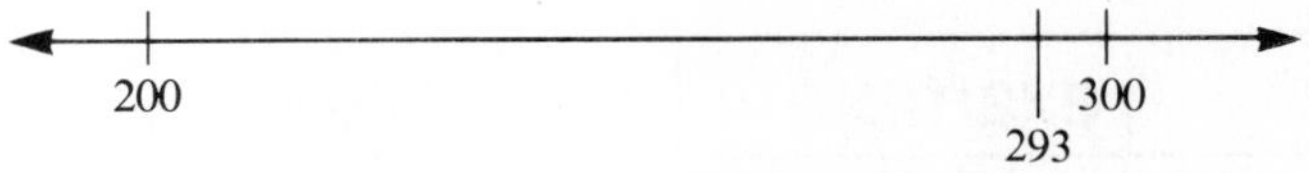

$$\begin{array}{r} {}^{1} \\ 293 \\ +\ 512 \\ \hline 805 \end{array}$$

Since 805 is close to our estimate of 800, this answer is reasonable. ●

example 8

Estimate 597 + 83 + 681 + 3,469. Then determine the exact answer.

solution

Approximate each addend:

$$\begin{array}{rcr} 597 & \approx & 600 \\ 83 & \approx & 100 \\ 681 & \approx & 700 \\ 3{,}469 & \approx & 3{,}500 \\ \cline{3-3} & & 4{,}900 \end{array}$$

Adding these approximations provides us with an estimate of 4900. Now we will determine the exact answer:

$$\begin{array}{r} {}^{1\,3\,2} \\ 597 \\ 83 \\ 681 \\ +\ 3{,}469 \\ \hline 4{,}830 \end{array}$$

Since 4,830 is close to our estimate of 4,900, this answer is reasonable. ●

Solving Word Problems Using Addition

Addition is probably the most frequently used arithmetic operation. It is used to solve a wide variety of word problems encountered in daily life. Certain words appearing in word problems such as *add, sum, plus, total, increased by,* and *more than* are a clue that addition may be used to solve the problem. Table 1.3 illustrates how words can be translated into arithmetic expressions using addition.

table 1.3
Expressing Addition

USING WORDS	USING SYMBOLS
The sum of 5 and 6 is 11.	$5 + 6 = 11$
The total of 6, 9, 3, and 5.	$6 + 9 + 3 + 5$
The number 8 increased by 7.	$8 + 7$
3 more than x is 14.	$x + 3 = 14$
a plus b equals c.	$a + b = c$

example 9

On a certain day the following transactions were processed at a Boston bank: 106 cash deposits, 392 check deposits, 281 cash withdrawals, 434 checks cashed, 27 money order purchases, 15 currency exchanges, and 213 other miscellaneous transactions. What were the total number of transactions conducted on that day?

solution Add 106 + 392 + 281 + 434 + 27 + 15 + 213. Let us first estimate the answer by approximating each addend and adding the approximations.

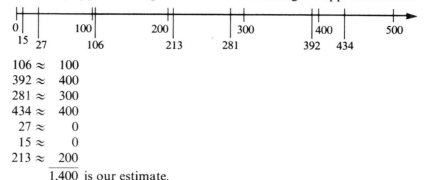

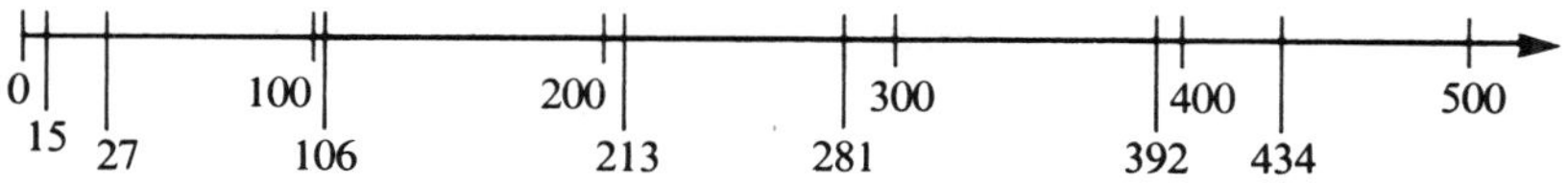

$$\begin{array}{rr} 106 \approx & 100 \\ 392 \approx & 400 \\ 281 \approx & 300 \\ 434 \approx & 400 \\ 27 \approx & 0 \\ 15 \approx & 0 \\ 213 \approx & \underline{200} \\ & 1{,}400 \end{array}$$

1,400 is our estimate.

Now let us determine the exact answer.

$$\begin{array}{r} 106 \\ 392 \\ 281 \\ 434 \\ 27 \\ 15 \\ +\ 213 \\ \hline 1{,}468 \end{array}$$

Since 1,468 is close to our estimate of 1,400, this answer is reasonable.

Thus, 1,468 transactions were processed. ●

QUICK QUIZ

Add each of the following groups of numbers.

1. 37 + 89

2. 456 + 797

3. 3,682 + 56 + 876

ANSWERS

1. 126

2. 1,253

3. 4,614

1.2 Exercises

The problems marked with the symbol ▦ are calculator problems. The procedure for using a calculator is discussed in the appendix.

Indicate which of the following statements are true and which are false. For those that are false, change the italic expression to make the statement true.

1. In the equation 5 + 7 = 12, the number 12 is called the *sum.*

2. According to the *associative* property of addition, the order of the addends can be changed without changing the sum.

3. When adding numbers that have more than one digit, we add the numbers in the *leftmost* column first.

4. We perform a carry when the sum of the numbers in any column is *greater* than 9.

5. When adding two whole numbers, each addend is always *greater* than or equal to the sum.

6. The number 0 is called the *auxiliary* element of addition.

Perform each of the following additions from memory without referring to the addition table.

7. $0 + 4$ **8.** $7 + 9$ **9.** $5 + 9$ **10.** $4 + 7$ **11.** $6 + 0$
12. $6 + 9$ **13.** $9 + 4$ **14.** $4 + 8$ **15.** $8 + 9$ **16.** $7 + 6$
17. $8 + 7$ **18.** $9 + 9$ **19.** $6 + 3 + 8$ **20.** $7 + 0 + 2$ **21.** $5 + 2 + 9$
22. $9 + 1 + 4$ **23.** $8 + 7 + 3$ **24.** $9 + 4 + 7$ **25.** $4 + 8 + 7$ **26.** $3 + 9 + 6$

State which property, commutative or associative (or both), is illustrated by each of the following equations.

27. $7 + 5 = 5 + 7$ **28.** $3 + (5 + 2) = 3 + (2 + 5)$ **29.** $8 + (1 + 7) = (8 + 1) + 7$
30. $2 + 9 = 9 + 2$ **31.** $(6 + 3) + 1 = 6 + (3 + 1)$ **32.** $2 + (7 + 4) = (2 + 7) + 4$
33. $8 + (3 + 2) = 8 + (2 + 3)$ **34.** $(9 + 2) + 5 = 5 + (9 + 2)$ **35.** $6 + (4 + 2) = (4 + 6) + 2$
36. $8 + (1 + 6) = (8 + 6) + 1$

Estimate the following sums and then determine the exact answers.

37. $18 + 54$ **38.** $24 + 81$ **39.** $127 + 521$ **40.** $97 + 212$
41. $62 + 13 + 59$ **42.** $8 + 32 + 77$ **43.** $323 + 410 + 168$ **44.** $125 + 684 + 537$
45. $1{,}021 + 317 + 88$ **46.** $112 + 3{,}107 + 498$

Add each of the following groups of numbers.

47. $18 + 31$ **48.** $53 + 26$
49. $87 + 46$ **50.** $75 + 39$
51. $862 + 509$ **52.** $380 + 477$
53. $537 + 248$ **54.** $199 + 801$
55. $37 + 86 + 23$ **56.** $52 + 94 + 18$
57. $687 + 23 + 15$ **58.** $97 + 324 + 62$
59. $5{,}029 + 8{,}653$ **60.** $8{,}967 + 4{,}036$
61. $39 + 62 + 384 + 7$ **62.** $12 + 284 + 6 + 39$
63. $57{,}656 + 40{,}932$ **64.** $61{,}486 + 29{,}513$
65. $567 + 84 + 67 + 8{,}248 + 76 + 319$ **66.** $52 + 928 + 3{,}516 + 88 + 547 + 43$
67. $738{,}927 + 39{,}724 + 6{,}421{,}507$ **68.** $5{,}459{,}638 + 80{,}167 + 279{,}156$
69. $3{,}648 + 294 + 6{,}527 + 14{,}687 + 388 + 18{,}493$ **70.** $52{,}468 + 3{,}284 + 957 + 62{,}748 + 54{,}682 + 728$

Translate each of the following statements into an arithmetic expression using addition.

71. The sum of 4 and 9 is 13. **72.** The number 2 increased by 6. **73.** 8 more than 3.
74. The sum of 7 and x. **75.** 5 increased by x is 9. **76.** x more than 8 is 17.
77. The sum of x and y. **78.** The total of a, b, and c is 20.

Solve each of the following word problems.

79. Last March, a salesman drove 213 miles from New York to Boston, 457 miles from Boston to Buffalo, 192 miles from Buffalo to Cleveland, and 175 miles from Cleveland to Detroit, calling on a number of customers along the way. What is the total number of miles he put on his car while traveling this route?

80. During the 1977–1978 academic year in the state of Ohio, there were 1,424,000 students enrolled in public elementary schools and 717,000 students enrolled in public secondary schools. How many students were enrolled in Ohio public schools during the 1977–1978 school year?

81. A student budgeted the following amounts to cover her expenses during her freshman year in college: tuition, \$5,500; room and board, \$3,500; books, \$250; transportation, \$500; miscellaneous expenses, \$400. What did she expect the total cost of her freshman year to be?

82. The area of Greenland is 839,999 square miles, and the area of Iceland is 39,768 square miles. What is the combined area of Greenland and Iceland?

83. A movie theater sold 82 tickets to the 2:00 P.M. show, 53 to the 5:30 P.M. show, 158 to the 8:00 P.M. show, and 144 to the 10:30 P.M. show. What was the total number of tickets sold to all shows?

84. In 1976, it was estimated that the number of eligible voters in the state of California could be broken down into the following age brackets: 18–24 years, 2,918; 25–44 years, 5,856; 45–64 years, 4,400; 65 years and older, 2,119. What was the total number of eligible voters in California in 1976?

85. In 1978, the number of passenger cars made by the leading American automobile manufacturers was as follows:

American Motors	164,351
Chrysler Corporation	1,082,274
Ford Motor Company	2,511,888
General Motors	5,316,616

What was the total number of cars manufactured by these four companies in 1978?

86. In 1960, the populations of five Midwestern states were as follows:

Ohio	9,706,397
Indiana	4,662,498
Illinois	10,081,158
Michigan	7,823,194
Wisconsin	3,951,777

What was the total population of these five states in 1960?

1.3 Subtraction of Whole Numbers

The operation **subtraction** may be defined as the *inverse* of addition. For example, consider the following problem. If ten people are in a room, and three of them leave, how many remain? Using your knowledge of addition, you might solve this problem by asking yourself what number must be added to three to obtain a total of ten.

$$? + 3 = 10$$

It is often easier, however, to pose the problem as ten reduced by three yields what result? Written as an equation this becomes:

$$10 - 3 = ?$$

In both equations the unknown number is 7:

$$7 + 3 = 10$$
$$10 - 3 = 7$$

The second equation illustrates the operation subtraction, and it is the reverse process of the first equation, which illustrates addition.

definition

To *subtract* b from a, we express the operation performed by the equation $a - b = c$, where a is the *minuend*, b is the *subtrahend*, and c is the *difference*.

For example, in the equation $7 - 2 = 5$, 7 is the minuend, 2 is the subtrahend, and 5 is the difference.

minuend ↗ 7 − 2 = 5 ↖ difference
↑
subtrahend

Whenever we do a subtraction problem, we can check our answer by adding the difference to the subtrahend to obtain the minuend: $c + b = a$.

definition

Addition and subtraction are said to be *inverse* operations, since if $a - b = c$, then $c + b = a$.

For example, if $9 - 3 = 6$, then $6 + 3 = 9$.

example 1

Subtract each of the following pairs of numbers and check your answers using addition.

(a) $13 - 6$

$13 - 6 = 7$ Since $7 + 6 = 13$.

(b) $9 - 4$

$9 - 4 = 5$ Since $5 + 4 = 9$. ●

Notice that whenever we perform a subtraction operation, the minuend and the subtrahend cannot be reversed without changing the answer to the problem. Thus,

$$11 - 5 \neq 5 - 11$$
$$6 \neq -6$$

The right-hand side yields a negative number that does not belong to the set of whole numbers. On a number line, numbers located to the *left of zero* are called *negative numbers*, and numbers located to the *right of zero* are called *positive numbers*, as shown in Figure 1.2. Whenever the subtrahend is greater than the minuend, the result is a negative number. (Negative numbers will be discussed in detail in Chapter 6.) Since the order of the minuend and subtrahend cannot be changed without creating an inequality, the *commutative* property does *not* hold true for subtraction.

figure 1.2

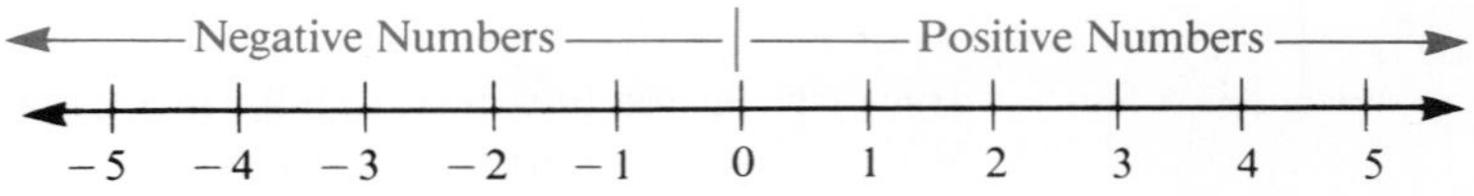

Let us now determine whether the associative property holds true for subtraction by looking at the following example.

Does $(8 - 4) - 2 = 8 - (4 - 2)$?

$$4 - 2 \; ? \; 8 - 2$$
$$2 \neq 6$$

Thus, the *associative* property does *not* hold true for subtraction.

When more than one subtraction operation appears in a problem, we perform each subtraction in order, moving from left to right, unless otherwise indicated by parentheses.

example 2

Perform the following subtraction operations.

(a) $7 - 3 - 2$

Do the leftmost subtraction first.

$(7 - 3) - 2 = 4 - 2 = 2$

(b) $12 - 6 - 3$

Do the leftmost subtraction first.

$(12 - 6) - 3 = 6 - 3 = 3$

In order to subtract numbers greater than nine, we write the problem vertically, lining up digits with the same place value. We then subtract the numbers in each column, moving from right to left. If a digit in the ones column of the subtrahend is larger than a digit in the ones column of the minuend, we must rewrite the minuend. Since 1 ten = 10 ones, we can do this by subtracting 1 from the digit in the tens column and adding 10 to the digit in the ones column.

example 3 Subtract $52 - 27$.

solution

minuend	52	$= 5 \text{ tens} + 2 \text{ ones}$	$= 4 \text{ tens} + 12 \text{ ones}$
subtrahend	-27	$= 2 \text{ tens} + 7 \text{ ones}$	$= 2 \text{ tens} + 7 \text{ ones}$
difference			$2 \text{ tens} + 5 \text{ ones} = 25$

1. Write the problem vertically and indicate the place value of each digit.
2. Since in the ones column, $2 < 7$, we must rewrite the minuend. Since 1 ten = 10 ones, subtract 1 from the digit in the tens column, and add 10 to the digit in the ones column to obtain 4 tens + 12 ones.
3. Subtract the numbers in the ones column: $12 - 7 = 5$.
4. Subtract the numbers in the tens column: $4 - 2 = 2$. The result is 25.

A short cut for subtracting numbers in this fashion is called *borrowing*. The borrowing procedure is outlined in the box.

To Subtract Whole Numbers

1. Write the number vertically and line up numbers with the same place value.
2. Subtract the numbers in each column separately, moving from right to left.
3. If a digit in the minuend is less than the digit that has the same place value in the subtrahend, rewrite the minuend by subtracting, or borrowing, 1 from the digit immediately to the left of the smaller digit, and adding 10 to the smaller digit.

If we use borrowing to subtract the numbers in Example 3 we obtain the following:

$$\begin{array}{r} \overset{4}{\cancel{5}}\,{}^{1}2 \\ -\;2\;7 \\ \hline \end{array}$$

Since in the ones column $2 < 7$, we must borrow. Since 1 ten = 10 ones, borrow 1 from the 5 in the tens column to get 4 tens, and add 10 to the 2 in the ones column to get 12 ones.

$$\begin{array}{r} \overset{4}{\cancel{5}}\,{}^{1}2 \\ -\;2\;7 \\ \hline 5 \end{array}$$

Subtract $12 - 7 = 5$.

$$\begin{array}{r} {}^{4} \\ \not{5}\,{}^{1}2 \\ -\,2\,7 \\ \hline 2\,5 \end{array}$$

Subtract 4 − 2 = 2. The result is 25.

Let us now look at some more examples of subtraction that illustrate borrowing.

example 4

Perform each of the following subtractions and check your answers by using addition.

(a) 516 − 42

$$\begin{array}{r} {}^{4} \\ \not{5}\,{}^{1}16 \\ -\quad 42 \\ \hline 474 \end{array}$$

Subtract 6 − 2 = 4.

Since in the tens column 1 < 4, we must borrow. Since 1 hundred = 10 tens, borrow 1 from the 5 in the hundreds column to get 4 hundreds, and add 10 to the 1 in the tens column to get 11 tens.

Subtract 11 − 4 = 7.

Subtract 4 − 0 = 4. The result is 474.

CHECK

$$\begin{array}{r} {}^{1} \\ 474 \\ +\ \ 42 \\ \hline 516 \end{array} \ \checkmark$$

(b) 602 − 89

$$\begin{array}{r} {}^{5} \\ \not{6}\,{}^{1}0\,2 \\ -\quad 8\,9 \\ \hline \end{array}$$

Since 2 < 9, we try to borrow from the tens column, but nothing is there. So we borrow 1 from the hundreds column to get 5 hundreds, and add 10 to the tens column to get 10 tens.

$$\begin{array}{r} {}^{5\ 9} \\ \not{6}\,{}^{1}\not{0}\,{}^{1}2 \\ -\quad 8\,9 \\ \hline 5\,1\,3 \end{array}$$

Now borrow 1 from the tens column to get 9 tens, and add 10 to the ones column to get 12 ones.

12 − 9 = 3. 9 − 8 = 1. 5 − 0 = 5.
The result is 513.

CHECK

$$\begin{array}{r} {}^{1\,1} \\ 513 \\ +\ \ 89 \\ \hline 602 \end{array} \ \checkmark$$

●

Estimation

It is important to be able to estimate answers when performing subtraction, just as it was when doing addition. Example 5 illustrates how estimation is used to approximate answers to subtraction problems.

example 5

Estimate the following differences. Then determine the exact answers. Check your answers using addition.

(a) 5,182 − 2,328

Approximate the minuend and subtrahend and subtract:

$$\begin{array}{rl} 5{,}182 \approx & 5{,}000 \\ 2{,}328 \approx & 2{,}000 \\ \cline{2-2} & 3{,}000 \text{ is our estimate.} \end{array}$$

Now determine the exact answer:

$$\begin{array}{r} {}^{4}\ {}^{7}\ \ \\ \not{5},{}^{1}\not{1}\not{8}{}^{1}2 \\ -2,32\ 8 \\ \hline 2,85\ 4 \end{array}$$

Since $2 < 8$, borrow 1 from 8 to get 7 tens, and add 10 to 2 to get 12 ones.

$12 - 8 = 4$. $7 - 2 = 5$.

Since $1 < 3$, borrow 1 from 5 to get 4 thousands, and add 10 to 1 to get 11 hundreds.

$11 - 3 = 8$. $4 - 2 = 2$. The result is 2,854.

CHECK

$$\begin{array}{r} {}^{1}\ \ {}^{1}\ \ \\ 2,854 \\ +2,328 \\ \hline 5,182 \end{array} \checkmark$$

Since 2,854 is close to our estimate of 3,000, this answer is reasonable.

(b) 1,000 − 327

Approximate the minuend and subtrahend and subtract:

$$\begin{array}{rr} 1,000 \approx & 1,000 \\ 327 \approx & 300 \\ \hline & 700 \end{array}$$ is our estimate.

Now determine the exact answer:

$$\begin{array}{r} {}^{0}\ \ \ \ \ \\ \not{1},{}^{1}000 \\ -\ \ 327 \\ \hline \end{array}$$

Since $0 < 7$, we try to borrow from the tens column, but nothing is there. Since nothing is in the hundreds column either, we borrow 1 from the thousands column to get 0 thousands, and add 10 to the hundreds column to get 10 hundreds.

$$\begin{array}{r} {}^{0}\ {}^{9}\ \ \ \ \\ \not{1},{}^{1}\not{0}{}^{1}00 \\ -\ \ 3\ 27 \\ \hline \end{array}$$

Borrow 1 from the hundreds column to get 9 hundreds, and add 10 to the tens column to get 10 tens.

$$\begin{array}{r} {}^{0}\ {}^{9}\ {}^{9}\ \ \\ \not{1},{}^{1}\not{0}{}^{1}\not{0}{}^{1}0 \\ -\ \ 3\ 2\ 7 \\ \hline \end{array}$$

Borrow 1 from the tens column to get 9 tens, and add 10 to the ones column to get 10 ones.

$$\begin{array}{r} {}^{0}\ {}^{9}\ {}^{9}\ \ \\ \not{1},{}^{1}\not{0}{}^{1}\not{0}{}^{1}0 \\ -\ \ 3\ 2\ 7 \\ \hline 6\ 7\ 3 \end{array}$$

$10 - 7 = 3$. $9 - 2 = 7$. $9 - 3 = 6$. The result is 673.

Since 673 is close to our estimate of 700, this answer is reasonable.

CHECK

$$\begin{array}{r} {}^{1}{}^{1}\ \\ 673 \\ +\ \ 327 \\ \hline 1,000 \end{array} \checkmark$$

Solving Word Problems Using Subtraction

Word problems that contain words such as *subtract*, *difference*, *minus*, *from*, *decreased by*, and *less than* can usually be solved using subtraction. Table 1.4 illustrates how words can be translated into arithmetic expressions using subtraction.

table *1.4*
Expressing Subtraction

USING WORDS	USING SYMBOLS
The difference between 8 and 3 is 5.	$8 - 3 = 5$
11 decreased by 4.	$11 - 4$
5 from 12 is 7.	$12 - 5 = 7$
2 less than x is 9.	$x - 2 = 9$
Subtract a from b.	$b - a$

example 6

In 1940 there were approximately 186,500 bachelor's degrees awarded in the United States, and in 1975 there were approximately 944,000 bachelor's degrees awarded. How many more people earned bachelor's degrees in 1975 than in 1940?

solution

Let us first estimate the answer by approximating the minuend and subtrahend and subtracting:

$$\begin{array}{rl} 944{,}000 \approx & 900{,}000 \\ 186{,}500 \approx & \underline{200{,}000} \\ & 700{,}000 \text{ is our estimate.} \end{array}$$

Now let us find the exact answer:

$$\begin{array}{r} {\scriptstyle 8\,13\,13} \\ \not{9}\,^{1}\not{4}\,^{1}\not{4},{}^{1}000 \\ -\,1\;8\;6,500 \\ \hline 7\;5\;7,500 \end{array}$$

Since 757,500 is close to our estimate of 700,000, this answer is reasonable.

CHECK

$$\begin{array}{r} {\scriptstyle 1\,1\,1} \\ 757{,}500 \\ +\;186{,}500 \\ \hline 944{,}000 \end{array} \;\checkmark$$

Since 757,500 is close to our estimate of 700,000, this answer is reasonable. ●

example 7

On the first day of class, 272 students were registered to take Biology I. Since then, 79 students have dropped the course, and 28 students who did not register originally have added the course. How many students are now registered to take Biology I?

solution

Subtract 79 from 272. Then add 28 to the result.

$$\begin{array}{r} {\scriptstyle 1\,16} \\ \not{2}\,^{1}\not{7}\,^{1}2 \\ -\quad 7\;9 \\ \hline 1\;9\;3 \end{array} \qquad \begin{array}{r} {\scriptstyle 1\,1} \\ 193 \\ +\;\;28 \\ \hline 221 \end{array}$$

Thus, 221 students are now registered for Biology I. ●

QUICK QUIZ
Subtract each of the following pairs of numbers. **1.** 72 − 49 **2.** 827 − 658 **3.** 5,000 − 1,436

ANSWERS
1. 23 **2.** 169 **3.** 3,564

1.3 Exercises

Indicate which of the following statements are true and which are false. For those that are false, change the italic expression to make the statement true.

1. The *associative* property does not hold true for subtraction.
2. Whenever the subtrahend exceeds the minuend, the difference is *positive*.
3. You can check the result of a subtraction problem by adding the difference to the *subtrahend*.
4. When subtracting numbers that have more than one digit, we begin at the *rightmost* column.

Perform each of the following subtractions from memory.

5. 15 − 6
6. 12 − 7
7. 11 − 3
8. 16 − 9
9. 13 − 6
10. 17 − 8
11. 14 − 6
12. 11 − 7
13. 14 − 9
14. 16 − 7
15. 17 − 8
16. 15 − 9
17. 12 − 3 − 6
18. 15 − 7 − 0
19. 18 − 5 − 4
20. 13 − 9 − 2
21. 19 − 0 − 9
22. 14 − 8 − 6

Estimate the following differences. Then determine the exact answers.

23. 821 − 285
24. 679 − 312
25. 1,029 − 506
26. 8,172 − 987
27. 7,178 − 2,954
28. 4,938 − 2,068
29. 32,197 − 13,465
30. 59,238 − 41,069

Perform each of the following subtractions. Check your answers using addition.

31. 58 − 13
32. 85 − 42
33. 31 − 18
34. 83 − 27
35. 138 − 92
36. 415 − 73
37. 812 − 167
38. 652 − 283
39. 206 − 89
40. 520 − 122
41. 713 − 487
42. 824 − 398
43. 5,203 − 584
44. 6,046 − 967
45. 4,731 − 2,847
46. 8,485 − 7,942
47. 5,013 − 847
48. 6,403 − 586
49. 36,004 − 8,276
50. 40,027 − 5,473
51. 8,326,412 − 2,718,637
52. 5,000,000 − 1,841,652
53. 67,362,107 − 35,684,793
54. 62,158,235 − 48,473,829

Translate each of the following statements into an arithmetic expression using subtraction.

55. 5 less than 12.
56. Subtract 2 from 9.
57. 15 decreased by 7 is 8.
58. 4 less than x.
59. The difference between 10 and 3 is 7.
60. 6 decreased by x.
61. The difference between x and y.
62. a less than b is 9.

Solve each of the following word problems.

63. A stereo receiver that retails for $420 is marked down $59. What is the sale price of this receiver?

64. The Nile River is 4,180 miles long, and the Mississippi River is 2,348 miles long. How much longer is the Nile than the Mississippi?

65. If the population of the United States was 23,261,000 in 1850 and 204,879,000 in 1970, by how much did the population increase during these years?

66. The distance from the sun to Venus is 67,270,000 miles, and the distance from the sun to Saturn is 887,140,000 miles. How much farther from the sun is Saturn than Venus?

67. If a couple bought a house for $62,500 and spent $19,000 on home improvements before selling it for $85,200, how much money did they make on the home?

68. A woman's checking account contained $903 at the beginning of the month. During the month she wrote checks for $37, $129, $425, $62, and $26, and she also deposited $150. What is her balance at the end of the month?

69. If the IRS collected a total of $195,772,096,000 in 1970 and a total of $358,139,416,000 in 1977, how much more money did the IRS collect in 1977 than in 1970?

70. Between 1829 and 1978, 36,200,005 persons immigrated from Europe to the United States, and 2,859,940 persons immigrated from Asia to the United States. How many more European immigrants were there than Asian between 1829 and 1978?

1.4 Multiplication of Whole Numbers

The operation **multiplication** may be defined as repeated addition. For example, if we add five twos together, we obtain:

$$2 + 2 + 2 + 2 + 2 = 10$$

This could also be expressed by saying that the sum of five twos is 10, or 5 times 2 equals 10:

$$5 \times 2 = 10$$

The "times" symbol, ×, indicates the operation multiplication. The number 2 in our example, which is the number being multiplied, is called the **multiplicand**. The number 5, which is the number of times the multiplicand is being counted, is called the **multiplier**. The number 10, which is the result, is called the **product**.

When a multiplication problem is written horizontally, the multiplier appears first. When it is written vertically, the multiplicand appears on top:

$5 \times 2 = 10$ (5: Multiplier, 2: Multiplicand, 10: Product)

Multiplicand	2
Multiplier	× 5
Product	10

The multiplicand and multiplier are also called **factors**.

definition

When multiplying two numbers *a* and *b*, we express the operation performed by the equation $a \times b = c$, where *a* is called the *multiplier*, *b* is called the *multiplicand*, and *c* is called the *product*. *a* and *b* are also referred to as *factors* of the product *c*.

For example, in the expression $3 \times 2 = 6$, 3 is the multiplier, 2 is the multiplicand, and 6 is the product. The numbers 2 and 3 are also called factors of the number 6.

In addition to using a × to express multiplication, we can use a raised dot, · , or put parentheses around some or all of the factors. Thus,

$$3 \times 2 = 3 \cdot 2 = (3)2 = 3(2) = (3)(2)$$

For small numbers, we can determine the results of a multiplication operation by rewriting the problem as repeated addition and calculating the sum. For example,

$$3 \times 8 = 8 + 8 + 8 = 24$$

However, this process can become very cumbersome and time consuming. It is essential, therefore, to memorize the products of all single-digit numbers. These simple multiplications are shown in Table 1.5. Some products that may be calculated using this table are:

$3 \times 0 = 0$ $\quad$ $8 \times 1 = 8$

$4 \times 5 = 20$ $\quad$ $7 \times 5 = 35$

table 1.5
Multiplication Table

×	**0**	**1**	**2**	**3**	**4**	**5**	**6**	**7**	**8**	**9**
0	0	0	0	0	0	0	0	0	0	0
1	0	1	2	3	4	5	6	7	8	9
2	0	2	4	6	8	10	12	14	16	18
3	0	3	6	9	12	15	18	21	24	27
4	0	4	8	12	16	20	24	28	32	36
5	0	5	10	15	20	25	30	35	40	45
6	0	6	12	18	24	30	36	42	48	54
7	0	7	14	21	28	35	42	49	56	63
8	0	8	16	24	32	40	48	56	64	72
9	0	9	18	27	36	45	54	63	72	81

Notice that whenever 0 is multiplied by any number, the resulting product is 0. Also recognize that products with 5 as a factor end in 0 or 5. Finally, note that whenever a number is multiplied by 1, the resulting product is the same as that number.

property

The number 1 is said to be the *identity element* of multiplication, since for any number a, $a \cdot 1 = a$ and $1 \cdot a = a$.

Let us now examine some of the other properties that apply to multiplication. To illustrate the commutative property, let us compare both sides of this equation:

Does $3 \times 6 = 6 \times 3$?

Using repeated addition we obtain:

$$6 + 6 + 6 = 3 + 3 + 3 + 3 + 3 + 3$$
$$18 = 18$$

Thus, the order of the factors may be changed without changing the result.

property

The *commutative property* of multiplication is expressed as $a \cdot b = b \cdot a$ for any numbers a and b.

For example, $7 \cdot 4 = 4 \cdot 7$.

In order to multiply more than two numbers together, we can multiply any two together, and then multiply that result by the remaining factor(s) to obtain the final product. For example,

$$(4 \times 2) \times 3 = 4 \times (2 \times 3)$$
$$8 \times 3 = 4 \times 6$$
$$24 = 24$$

Thus, according to the associative property, we can change the grouping of the factors without affecting the resulting product.

property

The *associative property* of multiplication is expressed as $a \cdot (b \cdot c) = (a \cdot b) \cdot c$ for any numbers a, b, and c.

For example, $5 \cdot (6 \cdot 2) = (5 \cdot 6) \cdot 2$.

The commutative and associative properties may sometimes be used to simplify a multiplication problem that contains three or more factors. For example, to multiply

$$5 \times 7 \times 6$$

we might think

$$(5 \times 6) \times 7 = 30 \times 7 = 210$$

The next property to be learned applies to problems that involve both addition and multiplication, as illustrated by this example.

$$3(7 + 2) = 3(7) + 3(2)$$
$$3(9) = 21 + 6$$
$$27 = 27$$

This property is called the **distributive property** of multiplication over addition. It is used whenever the sum of two or more numbers is to be multiplied by another number that appears outside the parentheses that group together the numbers to be added. In the example just cited, the factor 3 is distributed over each of the addends in parentheses:

$$3(7 + 2) = 3(7) + 3(2)$$

property

The *distributive property* of multiplication over addition is expressed as $a(b + c) = a(b) + a(c)$ for any numbers a, b, and c.

For example, $2(1 + 3) = 2(1) + 2(3)$.

The distributive property may also be used to multiply the difference of two numbers by another number. For example,

$$4(9 - 2) = 4(9) - 4(2)$$
$$4(7) = 36 - 8$$
$$28 = 28$$

If two sums are multiplied together, the distributive property can be applied twice:

$$(a + b)(c + d) = a(c + d) + b(c + d)$$
$$= a(c) + a(d) + b(c) + b(d)$$

For example,

$$(3 + 5)(2 + 7) = 3(2 + 7) + 5(2 + 7)$$
$$= 3(2) + 3(7) + 5(2) + 5(7)$$

example 1

State which property, commutative, associative, or distributive, is illustrated by each of the following equations.

(a) $3 \cdot (7 \cdot 2) = (3 \cdot 7) \cdot 2$

Associative, since 7 and 2 are grouped together on the left, and 3 and 7 are grouped together on the right.

(b) $4(5 + 3) = 4(5) + 4(3)$

Distributive, since the factor 4 is distributed over each of the addends in parentheses.

(c) $8 \cdot (3 \cdot 7) = 8 \cdot (7 \cdot 3)$

Commutative, since the 3 and 7 are reversed.

(d) $(2 + 5)(3 + 4) = 2(3 + 4) + 5(3 + 4) = 2(3) + 2(4) + 5(3) + 5(4)$

Distributive, applied twice to the product of two sums.

(e) $(3 \cdot 9) \cdot 2 = 3 \cdot (2 \cdot 9)$

Commutative and associative, since both the order and the grouping of the factors is changed. ●

example 2

Use the distributive property to rewrite each of the following multiplications and then simplify.

(a) $5(2 + 7)$

$$\begin{aligned} 5(2 + 7) &= 5(2) + 5(7) \\ &= 10 + 35 \\ &= 45 \end{aligned}$$

(b) $6(13) + 6(7)$

$$\begin{aligned} 6(13) + 6(7) &= 6(13 + 7) \\ &= 6(20) \\ &= 120 \end{aligned}$$

(c) $8(62) - 8(2)$

$$\begin{aligned} 8(62) - 8(2) &= 8(62 - 2) \\ &= 8(60) \\ &= 480 \end{aligned}$$ ●

In order to multiply numbers that have more than one digit, we can write the numbers to be multiplied in expanded notation and apply the distributive property.

example 3

Multiply 36×4.

solution

$36 \times 4 = (30 + 6) \times 4$	Write 36 in expanded notation.
$= 4(30 + 6)$	By the commutative property.
$= 4(30) + 4(6)$	By the distributive property.
$= 120 + 24 = 144$	Add to obtain the result. ●

We can write the same problem vertically and obtain:

36	
$\times\ \ 4$	
24	$6 \times 4 = 24$
120	$30 \times 4 = 120$
144	Add to obtain the result of 144.

A short cut for the same process is:

	2	
Multiplicand	36	Write the problem vertically.
Multiplier	$\times$ 4	Multiply $6 \times 4 = 24$. Put down 4, carry 2 (two tens) to the tens column.
Product	144	Multiply $3 \times 4 = 12$. Add the 2 that was carried to 12. $12 + 2 = 14$. Put down 14. ●

If both numbers to be multiplied have more than one digit, we apply the distributive property twice, as shown in Example 4.

example 4 Multiply 87 × 32.

solution

$$\begin{aligned} 87 \times 32 &= (80 + 7) \cdot (30 + 2) \\ &= (80 + 7)(30) + (80 + 7)(2) \\ &= (80)(30) + (7)(30) + (80)(2) + (7)(2) \\ &= 2{,}400 + 210 + 160 + 14 \\ &= 2{,}784 \end{aligned}$$

Now let us do Example 4 using the short cut.

```
    2
    1
   87
 × 32
 ----
  174
 2 61
 ----
2,784
```

First multiply 87 by 2. 7 × 2 = 14. Put down 4, carry 1. 8 × 2 = 16. 16 + 1 = 17. Put down 17.

Now multiply 87 by 30. Think of multiplying 87 by 3 and place the result under the first product, beginning in the tens column. 7 × 3 = 21. Put down 1, carry 2. 8 × 3 = 24. 24 + 2 = 26. Put down 26.

Add the two products to obtain 2,784.

CHECK

```
    1
    1
   32
 × 87
 ----
  224
 2 56
 ----
2,784 ✓
```

Notice that because of the commutative property, we can reverse the multiplicand and the multiplier to check our answer.

The steps for multiplying numbers that have more than one digit are outlined in the box.

To Multiply Whole Numbers

1. Write the problem vertically and place the number with the larger number of digits on top and the smaller number below it.
2. Multiply each digit of the top number (multiplicand), by the ones digit in the bottom number (multiplier), moving from right to left.
3. For a product that exceeds nine, write down the rightmost digit of the product and carry the other digit to the next column on the left and write it above the multiplicand. Calculate the next product and be sure to add to that product the digit that was carried.
4. Multiply each digit in the multiplicand by the next digit to the left in the multiplier. Place each product under the previous one calculated, but displaced one column to the left.
5. Repeat step 4 for all remaining digits in the multiplier.
6. Add the products to obtain the final answer.

example 5

Multiply 65 × 328.

solution

```
   14
   14
  328
×  65
-----
 111
 1 640
19 68
------
21,320
```

Write the problem vertically with 328 on top.

Multiply 328 by 5. 8 × 5 = 40. Put down 0, carry 4. 2 × 5 = 10. 10 + 4 = 14. Put down 4, carry 1. 3 × 5 = 15. 15 + 1 = 16. Put down 16.

Multiply 328 by 6. Place result under first product, but displaced one column to the left. 8 × 6 = 48. Put down 8, carry 4. 2 × 6 = 12. 12 + 4 = 16. Put down 6, carry 1. 3 × 6 = 18. 18 + 1 = 19. Put down 19.

Add the products to obtain 21,320. ●

QUICK QUIZ

Multiply each of the following pairs of numbers.

1. 89 × 4
2. 27 × 58
3. 396 × 712

ANSWERS

1. 356
2. 1,566
3. 281,952

Multiplication with Zeros

Numbers that end in zeros can be multiplied very quickly. Observe the pattern shown in the following examples.

$$3 \times 6 = 18$$
$$3 \times 60 = 3 \times 6 \times 10 = (3 \times 6) \times 10 = 18 \times 10 = 180$$
$$3 \times 600 = 3 \times 6 \times 100 = (3 \times 6) \times 100 = 18 \times 100 = 1{,}800$$
$$3 \times 6{,}000 = 3 \times 6 \times 1{,}000 = (3 \times 6) \times 1{,}000 = 18 \times 1{,}000 = 18{,}000$$
$$30 \times 60 = 3 \times 10 \times 6 \times 10 = (3 \times 6) \times (10 \times 10) = 18 \times 100 = 1{,}800$$
$$30 \times 600 = 3 \times 10 \times 6 \times 100 = (3 \times 6) \times (10 \times 100) = 18 \times 1{,}000 = 18{,}000$$
$$300 \times 600 = 3 \times 100 \times 6 \times 100 = (3 \times 6) \times (100 \times 100) = 18 \times 10{,}000 = 180{,}000$$

Notice that in each case, the product can be obtained by first multiplying 3 × 6 to get 18, and then attaching to the right of 18 the appropriate number of zeros, which is equal to the sum of the ending zeros in all numbers being multiplied. We have thus established a procedure for multiplying whole numbers ending in zeros (see box).

To Multiply Whole Numbers Ending in Zero(s)

1. Ignore the string of consecutive zeros at the end of each number being multiplied. Find the product of the remaining digits of the numbers being multiplied.
2. Find the sum of the ending zeros in all numbers being multiplied. Attach that number of zeros to the right end of the product calculated in step 1.

example 6

Multiply each of the following pairs of numbers.

(a) 3×200

$3 \times 2 = 6$ — Multiply $3 \times 2 = 6$.

$3 \times 200 = 600$ — Attach two zeros to the right of 6.

(b) 40×60

$4 \times 6 = 24$ — Multiply $4 \times 6 = 24$.

$40 \times 60 = 2{,}400$ — Attach two zeros to the right of 24.

(c) $700 \times 3{,}000$

$7 \times 3 = 21$ — Multiply $7 \times 3 = 21$.

$700 \times 3{,}000 = 2{,}100{,}000$ — Attach five zeros to the right of 21.

(d) $4{,}200 \times 90$

```
   1
  42
×  9
----
 378
```

Multiply $42 \times 9 = 378$.

$4{,}200 \times 90 = 378{,}000$ — Attach three zeros to the right of 378.

(e) $680{,}000 \times 5{,}100$

```
    4
   68
×  51
-----
   68
 3 40
-----
3,468
```

Multiply $68 \times 51 = 3{,}468$

$680{,}000 \times 5{,}100 = 3{,}468{,}000{,}000$ — Attach six zeros to the right of 3,468. ●

Let us now establish a technique for multiplying numbers that have zeros between the nonzero digits of the number.

example 7

Multiply 132×308.

solution

```
    21
   132
 × 308
------
  1 056
  0 00
 39 6
------
 40,656
```

Notice that we can omit the second row of zeros if we begin writing the product of 132×3 under the hundreds place as shown on the right. The result is 40,656.

```
    21
   132
 × 308
------
  1 056
 39 6
------
 40,656
```

●

example 8

Multiply 6,127 × 4,001.

solution

```
     12
   6,127
 × 4,001
   6 127
  00 00
 000 0
24 508
24,514,127
```

We can omit the two rows of zeros if we begin writing the product of 6,127 × 4 under the thousand place as shown on the right. The result is 24,514,127.

```
     12
   6 127
 × 4,001
   6 127
24 508
24,514,127
```

The two previous examples illustrate that every zero that appears in the multiplier causes the product of the multiplicand and the next nonzero digit in the multiplier to be displaced one additional column to the left.

example 9

Multiply 2,187 × 50,030.

solution

```
     43
     22
   2,187
× 50,030
  65 610
109 35
109,415,610
```

Write the product of 2,187 × 3 displaced one column to the left; put a 0 in the ones place as a placeholder.

Write the product of 2,187 × 5 displaced two additional columns to the left.

Add the two products to obtain a result of 109,415,610.

Estimation

It is important to be able to estimate answers to multiplication problems. If you have a rough idea of the result, then you can recognize an unreasonable answer. The following examples illustrate the use of estimation when performing multiplication operations.

example 10

Estimate the following products. Then determine the exact answers.

(a) 52 × 27

Approximate each factor and then multiply:

$$52 \approx 50$$
$$27 \approx 30$$

1,500 is our estimate.

Now determine the exact answer:

```
   1
  52
× 27
 364
104
1,404
```

Since 1,404 is close to our estimate of 1,500, this answer is reasonable.

Notice that the estimate of a product is not as close to the actual answer as is the estimate of a sum or a difference.

(b) 285×893

Approximate each factor and then multiply:

$$\begin{array}{rr} 285 \approx & 300 \\ 893 \approx & 900 \\ \hline & 270{,}000 \end{array}$$ is our estimate.

Now determine the exact answer:

$$\begin{array}{rr} & \scriptstyle 64 \\ & \scriptstyle 74 \\ & \scriptstyle 21 \\ & 285 \\ \times & 893 \\ \hline & 855 \\ & 25\,65 \\ & 228\,0 \\ \hline & 254{,}505 \end{array}$$

Since 254,505 is close to our estimate of 270,000, this answer is reasonable. ●

Solving Word Problems Using Multiplication

Word problems containing the words *multiply*, *product*, and *times* can usually be solved using multiplication. Table 1.6 illustrates how words can be translated into arithmetic expressions using multiplication.

table *1.6*
Expressing Multiplication

USING WORDS	USING SYMBOLS
Multiply 2 by 3.	2×3
8 times 7 is 56.	$8 \times 7 = 56$
The product of a and b equals 18.	$a \cdot b = 18$

Many word problems that are solved using multiplication, however, do not contain any of these "clue" words. Consider the following example.

example *11*

An engineer earns a weekly salary of \$632. What is the engineer's yearly salary?

solution

Since there are 52 weeks in a year, multiply 632×52. Let us first estimate the answer by approximating each factor and then multiplying.

$$\begin{array}{rr} 632 \approx & 600 \\ 52 \approx & 50 \\ \hline & 30{,}000 \end{array}$$ is our estimate.

Now let us find the exact answer.

$$\begin{array}{rr} & 632 \\ \times & 52 \\ \hline & 1\,264 \\ & 31\,60 \\ \hline & 32{,}864 \end{array}$$

Since 32,864 is close to our estimate of 30,000, this answer is reasonable.

The engineer's yearly salary is \$32,864. ●

QUICK QUIZ

Multiply each of the following pairs of numbers.

1. 700 × 50 **2.** 135 × 40 **3.** 726 × 407

ANSWERS

1. 35,000 **2.** 5,400 **3.** 295,482

1.4 Exercises

Indicate which of the following statements are true and which are false. For those that are false, change the italic expression to make the statement true.

1. Multiplication is repeated *subtraction.*
2. The identity element for multiplication is *0.*
3. When multiplying numbers that have more than one digit, we use the *distributive* property.
4. When multiplying three or more numbers, the way in which we group factors *does not* affect the product.

Perform each of the following multiplications from memory without referring to the multiplication table.

5. 7 × 6	**6.** 3 × 8	**7.** 7 × 7	**8.** 6 × 9
9. 9 × 5	**10.** 4 × 6	**11.** 9 × 7	**12.** 6 × 6
13. 9 × 9	**14.** 5 × 4	**15.** 7 × 3	**16.** 4 × 9
17. 4 × 2 × 9	**18.** 2 × 7 × 5	**19.** 3 × 7 × 3	**20.** 6 × 8 × 5

State which property, commutative, associative, or distributive, is illustrated by each of the following equations.

21. $6 \cdot (3 \cdot 5) = (6 \cdot 3) \cdot 5$	**22.** $3 \cdot 9 = 9 \cdot 3$
23. $8(5 + 3) = 8(5) + 8(3)$	**24.** $(7 \cdot 2) \cdot 4 = (2 \cdot 7) \cdot 4$
25. $2(8 - 3) = 2(8) - 2(3)$	**26.** $5(3 + 6) = (3 + 6)5$
27. $(4 + 2)(6 + 3) = 4(6 + 3) + 2(6 + 3)$	**28.** $(8 \cdot 9) \cdot 7 = 8 \cdot (9 \cdot 7)$

Use the distributive property to rewrite each of the following expressions and then simplify.

29. 3(6 + 2)	**30.** 4(7 + 5)	**31.** 8(7) + 8(3)	**32.** 7(8) + 7(2)	**33.** 5(7 − 2)
34. 4(9 − 3)	**35.** (9 + 8)3	**36.** (6 + 7)5	**37.** 6(12) + 6(8)	**38.** 8(13) + 8(7)
39. 7(9) − 7(4)	**40.** 6(8) − 6(5)	**41.** 8(40 + 5)	**42.** 4(70 + 5)	**43.** 5(73) + 5(7)
44. 3(94) − 3(4)				

Multiply each of the following pairs of numbers.

45. 53 × 6	**46.** 84 × 4	**47.** 63 × 9	**48.** 27 × 5
49. 98 × 3	**50.** 77 × 6	**51.** 41 × 32	**52.** 27 × 53
53. 66 × 44	**54.** 33 × 55	**55.** 213 × 47	**56.** 421 × 36
57. 854 × 69	**58.** 722 × 38	**59.** 513 × 282	**60.** 467 × 674
61. 3,135 × 832	**62.** 2,716 × 741	**63.** 8,629 × 3,468	**64.** 2,738 × 4,572

Multiply each of the following pairs of numbers.

65. 50 × 30	**66.** 40 × 70	**67.** 500 × 60	**68.** 200 × 90
69. 806 × 93	**70.** 530 × 78	**71.** 800 × 800	**72.** 400 × 900
73. 859 × 506	**74.** 703 × 487	**75.** 8,000 × 600	**76.** 3,000 × 700
77. 8,602 × 535	**78.** 7,094 × 396	**79.** 2,000 × 70,000	**80.** 5,000 × 50,000
81. 7,248 × 3,007	**82.** 2,073 × 4,603		

Estimate each of the following products. Then determine the exact answers.

83. 27×51 **84.** 12×79 **85.** 283×41 **86.** 669×31 **87.** 703×821
88. 393×715 **89.** $5{,}262 \times 291$ **90.** $6{,}884 \times 913$

Translate each of the following statements into an arithmetic expression using multiplication.

91. The product of 5 and 4 is 20. **92.** Multiply 6 by 7. **93.** 3 times a.
94. The product of x and y is z.

Solve each of the following word problems.

95. If a car can travel 28 miles on one gallon of gas, how far can it travel on 18 gallons of gas?

96. If a student studies 26 hours a week, and there are 14 weeks in a semester, how many hours does he study during an entire semester?

97. A stockholder buys 115 shares of a certain stock that sells for \$34 a share. What is her total investment?

98. Lake Huron is 13 times the size of the Great Salt Lake, which covers 1,800 square miles. How large is Lake Huron?

99. If 181,000 cars were sold in the United States in 1910, and 37 times that number were sold in 1975, how many cars were sold in the United States in 1975?

100. If the population of California in 1960 was 69 times that of Alaska, and 226,167 people were living in Alaska that year, what was the population of California in 1960?

101. How many minutes are there in a year?

102. If 1 acre = 43,560 square feet, how many square feet are in 1,386 acres?

1.5 Division of Whole Numbers

The operation **division** may be defined as repeated subtraction. For example, if we were asked to determine how many fives go into 15, we could proceed as follows.

$$\begin{array}{r l}
15 & \text{Subtract 5 from 15.} \\
-\ 5 & \\
\hline
10 & \text{Subtract 5 from the result.} \\
-\ 5 & \\
\hline
5 & \text{Subtract 5 from the next result.} \\
-\ 5 & \\
\hline
0 & \text{Stop when the number remaining is less than 5.}
\end{array}$$

Since we were able to subtract three fives before the remaining number was less than five, we can say that five goes into 15 three times. This can be expressed by the equation $15 \div 5 = 3$, where the symbol $\div$ indicates the operation division. This same equation, which states that 15 divided by 5 equals 3, can be written in three different ways.

$$15 \div 5 = 3 \quad \text{or} \quad \frac{15}{5} = 3 \quad \text{or} \quad 5\,\overset{3}{\overline{)15}}$$

definition

***Division* of a number *a* by another number *b*, is expressed by the equation**

$$a \div b = c \quad \text{or} \quad \frac{a}{b} = c \quad \text{or} \quad b\,\overset{c}{\overline{)a}}$$

where *a* is the *dividend*, *b* is the *divisor*, and *c* is the *quotient*.

For example, in the equation $16 \div 2 = 8$, 16 is the dividend, 2 is the divisor, and 8 is the quotient.

example 1

Perform each of the following divisions using repeated subtraction.

(a) $18 \div 9$

$$\begin{array}{r} 18 \\ -\ 9 \\ \hline 9 \\ -\ 9 \\ \hline 0 \end{array}$$

Subtract 9 from 18.
Subtract 9 from the result.
The number remaining is 0.

Since we were able to subtract 9 from 18 twice before the number remaining was less than 9, then $18 \div 9 = 2$. The number that remains after the procedure of repeated subtraction is completed is called the **remainder**. Note that *the remainder must always be a whole number that is less than the divisor*. In this example, the remainder is 0. Dividends are said to be *evenly divisible* by a given divisor if the remainder is 0. Thus, in this example, 18 is evenly divisible by 9.

(b) $17 \div 4$

$$\begin{array}{r} 17 \\ -\ 4 \\ \hline 13 \\ -\ 4 \\ \hline 9 \\ -\ 4 \\ \hline 5 \\ -\ 4 \\ \hline 1 \end{array}$$

Subtract 4 from 17.
Subtract 4 from the result.
Subtract 4 from the result.
Subtract 4 from the result.
Since $1 < 4$, the remainder is 1.

Since we were able to subtract 4 from 17 four times before obtaining a remainder of 1, we write the answer as 4 R1, which is read 4 remainder 1.

(c) $2 \div 7$

$$\begin{array}{r} 2 \\ -7 \\ \hline -5 \end{array}$$

If we subtract 7 from 2, the result is not a whole number, but a negative number.

The established procedure is to subtract the divisor from the dividend and from the ensuing results until the remaining result is a *whole number* that is less than the divisor. In this example, the dividend is less than the divisor, and the result of the subtraction is not a whole number. Therefore, the subtraction need not have been done. This result can be interpreted to mean that 7 goes into 2 zero times. If we do no subtraction, the remainder is the same as the dividend, 2. Thus, the quotient is 0 with a remainder of 2, which is written 0 R2. Therefore, when dividing whole numbers, if the dividend is less than the divisor, the quotient is 0, and the remainder is equal to the dividend. ●

We will now use this process of repeated subtraction to illustrate a few principles regarding division.

Any number divided by 1 is equal to the number itself. For any number *a*,

$a \div 1 = a$. As an example, let us consider $3 \div 1$:

$$\left.\begin{array}{r} 3 \\ -\,1 \\ \hline 2 \\ -\,1 \\ \hline 1 \\ -\,1 \\ \hline 0 \end{array}\right\} \text{3 times}$$

Notice that the number 1 can be subtracted three times from 3. Thus, $3 \div 1 = 3$.

Any nonzero number divided by itself is equal to 1. For any number $a \neq 0$, $a \div a = 1$. As an example, let us consider $8 \div 8$:

$$\left.\begin{array}{r} 8 \\ -\,8 \\ \hline 0 \end{array}\right\} \text{1 time}$$

Notice that the number 8 can be subtracted one time from 8. Thus, $8 \div 8 = 1$.

Zero divided by any nonzero number is equal to zero. For any number $a \neq 0$, $0 \div a = 0$. As an example, let us consider $0 \div 6$. This problem is similar to Example 1(c) since the dividend, 0, is less than the divisor, 6. The quotient is therefore 0 with a remainder equal to the dividend, which is 0. Thus, $0 \div 6 = 0$.

Division by zero is impossible. As an example, let us consider $5 \div 0$.

$$\begin{array}{r} 5 \\ -\,0 \\ \hline 5 \\ -\,0 \\ \hline 5 \\ -\,0 \\ \hline 5 \end{array}$$

Notice that we could continue this process indefinitely and still be no closer to arriving at a solution. No matter how many times we subtract 0 from 5, it is impossible to end up with a remainder other than 5. We express this idea by saying that $5 \div 0$ is **undefined**.

We will now look at some examples that illustrate the different notations that can be used to express division.

example 2

Rewrite each of the following division problems using the other two notations, translate the equation into words, and identify the dividend, divisor, and quotient.

(a) $\dfrac{36}{9} = 4$

$36 \div 9 = 4$ $\qquad 9\overline{)36}$ with quotient 4

Thirty-six divided by nine equals four.

36, dividend; 9, divisor; 4, quotient

Notice that $4 \cdot 9 = 36$.

(b) $56 \div 7 = 8$

$\dfrac{56}{7} = 8$ $\qquad 7\overline{)56}$ with quotient 8

Fifty-six divided by seven equals eight.

56, dividend; 7, divisor; 8, quotient

Notice that $8 \cdot 7 = 56$.

(c) $6\overline{)54}$ with quotient 9

$54 \div 6 = 9$ $\qquad \dfrac{54}{6} = 9$

Fifty-four divided by six equals nine.

54, dividend; 6, divisor; 9, quotient

Notice that $9 \cdot 6 = 54$.

(d) $11 \div 5 = 2\ \text{R}1$

$\dfrac{11}{5} = 2\ \text{R}1$ $\qquad 5\overline{)11}$ with quotient 2 R1

Eleven divided by five equals two remainder one.

11, dividend; 5, divisor;

2, quotient; 1, remainder

Notice that $(2 \cdot 5) + 1 = 11$. ●

The problems in Example 2 lead us to believe that there is a relationship between multiplication and division. In fact, division is the **inverse** of multiplication. For example,

$$24 \div 3 = 8 \quad \text{since} \quad 8 \cdot 3 = 24$$
$$42 \div 6 = 7 \quad \text{since} \quad 7 \cdot 6 = 42$$
$$63 \div 7 = 9 \quad \text{since} \quad 9 \cdot 7 = 63$$

Notice that the product of the quotient and the divisor is equal to the dividend. Thus, multiplication can be used to check the result of a division problem.

definition

Multiplication and division are said to be *inverse* operations, since if $a \div b = c$, then $c \cdot b = a$.

For example, if $36 \div 9 = 4$, then $4 \cdot 9 = 36$. Since multiplication is repeated addition, division is repeated subtraction, and addition and subtraction are inverse operations, it makes sense that multiplication and division are also inverse operations.

Notice that whenever we perform a division operation, the dividend and divisor cannot be reversed without changing the answer to the problem. Thus,

$$20 \div 4 \neq 4 \div 20$$
$$5 \neq 0\ \text{R}4$$

Therefore, the *commutative* property does *not* hold true for division.

Let us now determine whether the associative property holds true for division by looking at the following example:

Does $\quad (18 \div 6) \div 3 = 18 \div (6 \div 3)$?

$$3 \div 3 \ ?\ 18 \div 2$$
$$1 \neq 9$$

Thus, the *associative* property does *not* hold true for division.

When more than one division operation appears in a problem, we perform each division in order, moving from left to right, unless otherwise indicated by parentheses.

example 3

Perform each of the following divisions.

(a) $28 \div 7 \div 2$

Do the leftmost division first:

$$(28 \div 7) \div 2 = 4 \div 2 = 2$$

(b) $48 \div 8 \div 2$

Do the leftmost division first:

$$(48 \div 8) \div 2 = 6 \div 2 = 3$$

Let us now consider a technique that uses multiplication instead of repeated subtraction to solve a division problem. This procedure is outlined in the box.

To Divide Whole Numbers

1. Write the problem using divisor $\overline{)\text{dividend}}$ notation.
2. Determine the largest number that can be multiplied by the divisor to obtain a product that is less than or equal to the dividend. Write that number above the dividend.
3. Multiply that number by the divisor.
4. Subtract the product obtained in step 3 from the dividend. The result is the remainder.

To check the result of a division problem, we can follow these steps:

1. Multiply the quotient by the divisor.
2. Add the remainder.
3. The result should equal the dividend.

example 4

Divide $29 \div 4$ and check your answer.

solution

$$\begin{array}{r} 7\text{ R}1 \\ 4\overline{)\ 29} \\ -\ 28 \\ \hline 1 \end{array}$$

Write the problem as shown on the left.

Ask yourself what number multiplied by 4 produces a product less than or equal to 29.

Write 7 in the space for the quotient.

Multiply $7 \times 4 = 28$.

Subtract $29 - 28 = 1$, the remainder. The quotient is 7 R1.

CHECK

$$\begin{array}{r} 7 \\ \times\ 4 \\ \hline 28 \\ +\ 1 \\ \hline 29\ \checkmark \end{array}$$

Multiply $7 \times 4 = 28$.

Add the remainder: $28 + 1 = 29$.

Since 29 is the dividend, the answer is correct.

example 5

Perform each of the following divisions and check your answers.

(a) $53 \div 7$

$$\begin{array}{r} 7\ \text{R4} \\ 7\overline{)53} \\ 49 \\ \hline 4 \end{array}$$

7 goes into 53 seven times. Write down 7.

Multiply $7 \times 7 = 49$.

Subtract $53 - 49 = 4$.

The quotient is 7 R4.

CHECK

$$\begin{array}{r} 7 \\ \times\ 7 \\ \hline 49 \\ +\ 4 \\ \hline 53\ \checkmark \end{array}$$

Multiply quotient and divisor.

Add the remainder.

Always be certain that the remainder is less than the divisor. If it isn't, the number chosen for the quotient is not large enough.

(b) $153 \div 9$

In this problem, 9 goes into 153 more than 10 times. To determine exactly how many times, we will use the same procedure more than once.

$$\begin{array}{r} 17 \\ 9\overline{)153} \\ 90 \\ \hline 63 \\ 63 \\ \hline 0 \end{array}$$

9 goes into 150 ten times. (It is often easier to mentally approximate the dividend by omitting the ones digit.) Put the 1 over the tens place of the dividend.

$9 \times 10 = 90$. $153 - 90 = 63$.

9 goes into 63 seven times. Put the 7 over the ones place of the dividend.

$9 \times 7 = 63$. $63 - 63 = 0$.

The quotient is 17.

CHECK

$$\begin{array}{r} 17 \\ \times\ 9 \\ \hline 153\ \checkmark \end{array}$$

(c) $486 \div 7$

$$\begin{array}{r} 69\ \text{R3} \\ 7\overline{)486} \\ 420 \\ \hline 66 \\ 63 \\ \hline 3 \end{array}$$

7 goes into 480 about 60 times. Put the 6 over the tens place of the dividend.

$7 \times 60 = 420$. $486 - 420 = 66$.

7 goes into 66 nine times. Put the 9 over the ones place of the dividend.

$7 \times 9 = 63$. $66 - 63 = 3$.

CHECK

$$\begin{array}{r} 69 \\ \times \quad 7 \\ \hline 483 \\ + \quad 3 \\ \hline 486 \end{array} \checkmark$$

The quotient is 69 R3.

(d) 2,695 ÷ 4

$$\begin{array}{r} 673 \text{ R3} \\ 4\overline{)2{,}695} \\ 2{,}400 \\ \hline 295 \\ 280 \\ \hline 15 \\ 12 \\ \hline 3 \end{array}$$

4 goes into 2,600 six hundred times. Put the 6 over the hundreds place of the dividend.

4 × 600 = 2,400. 2,695 − 2,400 = 295.

4 goes into 290 seventy times. Put the 7 over the tens place of the dividend.

4 × 70 = 280. 295 − 280 = 15.

4 goes into 15 three times. Put the 3 over the ones place of the dividend.

4 × 3 = 12. 15 − 12 = 3.

The quotient is 673 R3.

CHECK

$$\begin{array}{r} 673 \\ \times \quad 4 \\ \hline 2{,}692 \\ + \quad 3 \\ \hline 2{,}695 \end{array} \checkmark$$

(e) 38,527 ÷ 6

We can save ourselves even more work by omitting zeros in the numbers we subtract, as long as we remember to line up numbers that have the same place value.

$$\begin{array}{r} 6{,}421 \text{ R1} \\ 6\overline{)38{,}527} \\ 36 \\ \hline 2\,5 \\ 2\,4 \\ \hline 12 \\ 12 \\ \hline 7 \\ 6 \\ \hline 1 \end{array}$$

6 goes into 38 six times. Put the 6 over the 8 in the dividend.
6 × 6 = 36. 38 − 36 = 2. Bring down 5, the next number in the dividend.
6 goes into 25 four times. Put the 4 over the 5 in the dividend.
6 × 4 = 24. 25 − 24 = 1. Bring down the 2 from the dividend.
6 goes into 12 two times. Place the 2 over the 2 in the dividend.
6 × 2 = 12. 12 − 12 = 0.
Bring down 7.
6 goes into 7 one time. Put 1 over the 7.
6 × 1 = 6. 7 − 6 = 1.
The quotient is 6,421 R1.

CHECK

$$\begin{array}{r} 6{,}421 \\ \times \quad 6 \\ \hline 38{,}526 \\ + \quad 1 \\ \hline 38{,}527 \end{array} \checkmark$$

●

When the divisor contains more than one digit, we can mentally approximate it by considering only the first digit. However, the numbers we subtract from the dividend must be the exact products of the quotients and the divisor.

example 6

Perform each of the following divisions and check your answers.

(a) $284 \div 32$

$$\begin{array}{r|l} & 8 \text{ R28} \\ \cline{2-2} 32 & 284 \\ & 256 \\ \cline{2-2} & 28 \end{array}$$

30 goes into 280 about 9 times.

$$\begin{array}{r} 32 \\ \times 9 \\ \hline 288 > 284 \end{array}$$

This guess is too big. Let us try 8.

$$\begin{array}{r} 32 \\ \times 8 \\ \hline 256 < 284 \end{array}$$

$$\begin{array}{r} 284 \\ -\ 256 \\ \hline 28 < 32 \end{array}$$

Since the remainder is less than the divisor, this guess is correct.

CHECK

$$\begin{array}{r} 32 \\ \times 8 \\ \hline 256 \\ +\ 28 \\ \hline 284 \ \checkmark \end{array}$$

(b) $8{,}467 \div 53$

$$\begin{array}{r|l} & 159 \text{ R40} \\ \cline{2-2} 53 & 8{,}467 \\ & 5\,3 \\ \cline{2-2} & 3\,16 \\ & 2\,65 \\ \cline{2-2} & 517 \\ & 477 \\ \cline{2-2} & 40 \end{array}$$

50 goes into 80 about 1 time.

$$\begin{array}{r} 53 \\ \times 1 \\ \hline 53 < 84 \end{array} \qquad \begin{array}{r} 84 \\ \times 53 \\ \hline 31 < 53 \end{array}$$

Put 1 over the 4. Bring down the 6.

50 goes into 300 about 6 times.

$$\begin{array}{r} 53 \\ \times 6 \\ \hline 318 > 316 \end{array}$$

This guess is too big. Try 5.

$$\begin{array}{r} 53 \\ \times 5 \\ \hline 265 < 316 \end{array} \qquad \begin{array}{r} 316 \\ -\ 265 \\ \hline 51 < 53 \end{array}$$

Put 5 over the 6. Bring down the 7.

50 goes into 517 about 9 times.

$$\begin{array}{r} 53 \\ \times 9 \\ \hline 477 < 517 \end{array} \qquad \begin{array}{r} 517 \\ -\ 477 \\ \hline 40 < 53 \end{array}$$

Put 9 over the 7. The remainder is 40.

CHECK

$$\begin{array}{r} 159 \\ \times 53 \\ \hline 477 \\ 795 \\ \hline 8{,}427 \\ +\ 40 \\ \hline 8{,}467 \ \checkmark \end{array}$$

(c) $68{,}921 \div 213$

$$\begin{array}{r|l} & 323 \text{ R122} \\ \cline{2-2} 213 & 68{,}921 \\ & 63\,9 \\ \cline{2-2} & 5\,02 \\ & 4\,26 \\ \cline{2-2} & 761 \\ & 639 \\ \cline{2-2} & 122 \end{array}$$

200 goes into 600 about 3 times.

$$\begin{array}{r} 213 \\ \times 3 \\ \hline 639 < 689 \end{array} \qquad \begin{array}{r} 689 \\ -\ 639 \\ \hline 50 < 213 \end{array}$$

Put 3 over the 9. Bring down the 2.

200 goes into 500 about 2 times.

$$\begin{array}{r} 213 \\ \times 2 \\ \hline 426 < 502 \end{array} \qquad \begin{array}{r} 502 \\ -\ 426 \\ \hline 76 < 213 \end{array}$$

Put 2 over the 2. Bring down the 1.

200 goes into 700 about 3 times.

CHECK

$$\begin{array}{r} 323 \\ \times \quad 213 \\ \hline 969 \\ 3\,23 \\ 64\,6 \\ \hline 68{,}799 \\ + \quad 122 \\ \hline 68{,}921 \end{array} \checkmark$$

$$\begin{array}{r} 213 \\ \times \quad 3 \\ \hline 639 < 761 \end{array} \qquad \begin{array}{r} 761 \\ -\ 639 \\ \hline 122 \text{ remainder} \end{array}$$

Put 3 over the 1.

Solving Word Problems Using Division

Word problems containing the words *divide* and *quotient* can usually be solved using division. Table 1.7 illustrates how words can be translated into arithmetic expressions using division.

table 1.7
Expressing Division

USING WORDS	USING SYMBOLS
The quotient of 15 and 3.	$15 \div 3$
Divide 28 by 4.	$28 \div 4$
6 divided by x is 2.	$6 \div x = 2$

example 7

A company has a total of $21,350 in bonus money to be divided equally among 14 employees. How much should each person receive?

solution

Divide $21{,}350 \div 14$.

$$\begin{array}{r} 1{,}525 \\ 14\,\overline{)\,21{,}350} \\ 14\phantom{{,}350} \\ \hline 73\phantom{{,}50} \\ 70\phantom{{,}50} \\ \hline 35 \\ 28 \\ \hline 70 \\ 70 \\ \hline 0 \end{array}$$

CHECK

$$\begin{array}{r} 1{,}525 \\ \times \quad 14 \\ \hline 6\,100 \\ 15\,25 \\ \hline 21{,}350 \end{array} \checkmark$$

Each person should receive $1,525.

QUICK QUIZ

Divide each of the following pairs of numbers.

1. $2{,}610 \div 9$
2. $400 \div 25$
3. $5{,}638 \div 107$

ANSWERS

1. 290
2. 16
3. 52 R74

1.5 Exercises

Indicate which of the following statements are true and which are false. For those that are false, change the italic expression to make the statement true.

1. In the equation $18 \div 3 = 6$, the number 18 is called the *divisor.*
2. A number is said to be evenly divisible by another number if the remainder is equal to *zero.*
3. Division may be defined as repeated *substitution.*
4. To check the result of a division problem, we multiply the quotient by the divisor and *add* the remainder.
5. In a division problem, the remainder must be *greater* than the divisor.
6. Division by *one* is undefined.

Perform each of the following divisions from memory.

7. $36 \div 4$
8. $15 \div 5$
9. $0 \div 3$
10. $8 \div 0$
11. $64 \div 8$
12. $42 \div 7$
13. $21 \div 7$
14. $72 \div 9$
15. $28 \div 4$
16. $54 \div 6$
17. $56 \div 7$
18. $81 \div 9$
19. $16 \div 4$
20. $30 \div 6$
21. $15 \div 0 \div 3$
22. $0 \div 8 \div 2$
23. $72 \div 8 \div 3$
24. $32 \div 4 \div 2$
25. $42 \div 7 \div 6$
26. $64 \div 8 \div 4$

Perform each of the following divisions. Check your answers.

27. $52 \div 7$
28. $83 \div 9$
29. $34 \div 6$
30. $71 \div 8$
31. $25 \div 4$
32. $43 \div 5$
33. $108 \div 6$
34. $72 \div 3$
35. $208 \div 4$
36. $155 \div 5$
37. $287 \div 7$
38. $576 \div 8$
39. $113 \div 3$
40. $309 \div 5$
41. $614 \div 4$
42. $293 \div 9$
43. $4{,}381 \div 3$
44. $8{,}652 \div 7$
45. $5{,}106 \div 4$
46. $2{,}790 \div 8$
47. $4{,}102 \div 7$
48. $2{,}346 \div 6$
49. $50{,}813 \div 8$
50. $76{,}042 \div 5$
51. $49{,}303 \div 3$
52. $38{,}412 \div 7$
53. $8{,}627{,}327 \div 4$
54. $2{,}918{,}327 \div 2$
55. $4{,}000{,}000 \div 3$
56. $7{,}000{,}000 \div 9$
57. $86 \div 21$
58. $93 \div 15$
59. $62 \div 38$
60. $76 \div 27$
61. $154 \div 13$
62. $294 \div 81$
63. $367 \div 62$
64. $843 \div 47$
65. $4{,}263 \div 23$
66. $7{,}916 \div 92$
67. $28{,}368 \div 76$
68. $36{,}941 \div 84$
69. $9{,}246 \div 356$
70. $5{,}481 \div 127$
71. $7{,}304 \div 281$
72. $8{,}015 \div 673$
73. $36{,}843 \div 502$
74. $13{,}817 \div 709$
75. $62{,}302 \div 671$
76. $70{,}149 \div 946$
77. $280{,}490 \div 8{,}014$
78. $31{,}668 \div 2{,}639$
79. $1{,}118{,}042 \div 4{,}678$
80. $4{,}627{,}389 \div 9{,}127$

Translate each of the following statements into an arithmetic expression using division.

81. The quotient of 36 and 9.
82. 14 divided by 2 equals 7.
83. x divided by y.
84. The quotient of 20 and a is 4.

Solve each of the following word problems.

85. If an inheritance of $7,740 (after taxes) is to be divided equally among nine relatives of the deceased, how much money will each person receive?

86. If 79,143 people voted in a certain election, how many votes would each candidate receive if the votes were split evenly among the three candidates?

87. At the beginning of a card game, a deck of 52 cards is dealt out evenly among three players, and the remaining cards are put in a pile. How many cards does each player receive? How many cards are put in the pile?

88. If 159 marbles are to be divided evenly among twelve children, how many marbles will each child get? How many marbles will be left over?

89. Thirty-two bricks weight 128 pounds. How much does a single brick weigh?

90. If a station wagon travels 368 miles on 16 gallons of gas, how many miles per gallon does the car get?

91. If a computer programmer earns $33,540 a year, what is her weekly salary?

92. If eight dozen calculus books cost $1,632, what is the price of a single book?

93. An auditorium has a total of 1,176 seats. If there are 84 rows of seats, how many seats are in each row?

94. A businessman bought 57 shares of stock for a total cost of $2,166. What was the price of a single share?

95. A machine can fill 2,728 bottles with milk in 1 hour. How long will it take for this machine to fill 35,464 bottles?

96. A wholesale distributor purchased 6,782 radios for a price of $393,356. How much did he pay for one radio?

1.6 Exponents

Lengthy arithmetic expressions that occur frequently in mathematics can be very cumbersome. Therefore, useful shorthand notations have been developed to make writing these expressions less clumsy. For example, we can express the product

$$\underbrace{2 \cdot 2 \cdot 2 \cdot 2 \cdot 2 \cdot 2 \cdot 2}_{\text{7 twos}} \quad \text{as } 2^7$$

which is read "two to the seventh power." The shorthand notation 2^7 is an exponential expression. Two is called the **base** and 7 is called the **exponent**. Exponents are used to express repeated multiplication. The exponent, 7, tells you that the base, 2, appears as a factor seven times.

definition

The expression b^n means that b appears as a factor n times. b is called the *base* and n is called the *exponent*, or *power*.

For example, 5^3 means that 5 appears as a factor three times. The base is 5, and the exponent, or power, is 3.

$$\text{base} \longrightarrow 5^3 \longleftarrow \text{exponent}$$

example 1

Rewrite each of the following expressions using exponents.

(a) $3 \cdot 3 \cdot 3 \cdot 3$

$$\underbrace{3 \cdot 3 \cdot 3 \cdot 3}_{\text{4 threes}} = 3^4$$

3 appears as a factor 4 times.

(b) $5 \cdot 5 \cdot 4 \cdot 4 \cdot 4$

$$\underbrace{5 \cdot 5}_{\text{2 fives}} \cdot \underbrace{4 \cdot 4 \cdot 4}_{\text{3 fours}} = 5^2 \cdot 4^3$$

5 appears as a factor 2 times, and 4 appears as a factor 3 times.

(c) $9 \cdot 9 \cdot 8 \cdot 7 \cdot 7 \cdot 7$

$$\underbrace{9 \cdot 9}_{\text{2 nines}} \cdot \underbrace{8}_{\text{1 eight}} \cdot \underbrace{7 \cdot 7 \cdot 7}_{\text{3 sevens}} = 9^2 \cdot 8^1 \cdot 7^3$$

Notice that since 8 appears as a factor once, it may be rewritten as 8^1, or "8 to the first power." ●

example 2

Rewrite each of the following numbers as an exponential expression with a base of 2 or 3.

(a) 9

$9 = 3^2$ since $3 \cdot 3 = 9$.

3^2 is read as "3 to the second power" or "3 squared."

(b) 8

$8 = 2^3$ since $2 \cdot 2 \cdot 2 = 8$.

2^3 is read as "2 to the third power" or "2 cubed." ●

definition

For any base b, $b^1 = b$. In other words, any number raised to the first power is the number itself.

For example, $2^1 = 2$, $15^1 = 15$, $304^1 = 304$, and so on.

The next definition is an important fact to remember about the meaning of zero as an exponent.

definition

For any base $b \neq 0$, $b^0 = 1$. In other words, any nonzero number raised to the zero power equals one.

For example, $5^0 = 1$, $12^0 = 1$, $153^0 = 1$, and so on.

example 3

Express the following exponential expressions as products, and then simplify to whole numbers.

(a) $2^3 \cdot 3^2$

$$\begin{aligned} 2^3 \cdot 3^2 &= (2 \cdot 2 \cdot 2) \cdot (3 \cdot 3) \\ &= 8 \cdot 9 = 72 \end{aligned}$$

(b) $2^2 \cdot 5^2 \cdot 3^3$

$$\begin{aligned} 2^2 \cdot 5^2 \cdot 3^3 &= (2 \cdot 2) \cdot (5 \cdot 5) \cdot (3 \cdot 3 \cdot 3) \\ &= 4 \cdot 25 \cdot 27 = 100 \cdot 27 = 2{,}700 \end{aligned}$$

(c) $4^2 \cdot 8^0 \cdot 10^4$

$$4^2 \cdot 8^0 \cdot 10^4 = (4 \cdot 4) \cdot 1 \cdot (10 \cdot 10 \cdot 10 \cdot 10)$$
$$= 16 \cdot 10{,}000 = 160{,}000$$

Notice that in Example 3(c), the product $16 \cdot 10^4 = 160{,}000$, which is the number 16 followed by four zeros.

> Any number multiplied by 10^n is that number followed by n zeros.

example 4

Simplify each of the following exponential expressions.

(a) $6^2 \cdot 10^2$

$$6^2 \cdot 10^2 = (6 \cdot 6) \cdot 10^2$$
$$= 36 \cdot 10^2$$
$$= 3{,}600 \qquad \text{36 followed by two zeros}$$

(b) $3^3 \cdot 10^5$

$$3^3 \cdot 10^5 = (3 \cdot 3 \cdot 3) \cdot 10^5$$
$$= 27 \cdot 10^5$$
$$= 2{,}700{,}000 \qquad \text{27 followed by five zeros}$$

(c) $2^5 \cdot 10^4$

$$2^5 \cdot 10^4 = (2 \cdot 2 \cdot 2 \cdot 2 \cdot 2) \cdot 10^4$$
$$= 32 \cdot 10^4$$
$$= 320{,}000 \qquad \text{32 followed by four zeros}$$

Every place value in the base ten system can be rewritten as a **power of ten**, as shown in Table 1.8. Every whole number can thus be rewritten in expanded notation using powers of ten. For example,

$$5{,}823 = (5 \times 1{,}000) + (8 \times 100) + (2 \times 10) + (3 \times 1)$$
$$= (5 \times 10^3) + (8 \times 10^2) + (2 \times 10^1) + (3 \times 10^0)$$

table *1.8*
Place Values as Powers of Ten

Ones	$10^0 = 1$	Hundred thousands	$10^5 = 100{,}000$
Tens	$10^1 = 10$	Millions	$10^6 = 1{,}000{,}000$
Hundreds	$10^2 = 100$	Ten millions	$10^7 = 10{,}000{,}000$
Thousands	$10^3 = 1{,}000$	Hundred millions	$10^8 = 100{,}000{,}000$
Ten thousands	$10^4 = 10{,}000$	Billions	$10^9 = 1{,}000{,}000{,}000$

example 5

Rewrite each of the following numbers in expanded notation using powers of ten.

(a) 742

$$742 = (7 \times 10^2) + (4 \times 10^1) + (2 \times 10^0)$$

(b) 30,469

$$30{,}469 = (3 \times 10^4) + (0 \times 10^3) + (4 \times 10^2) + (6 \times 10^1) + (9 \times 10^0)$$
$$= (3 \times 10^4) + (4 \times 10^2) + (6 \times 10^1) + (9 \times 10^0)$$

When using exponents, we can detect certain properties that illustrate rules that enable us to simplify our calculations.

Let us first look at the following examples.

$$2^2 \cdot 2^3 = \underbrace{(2 \cdot 2) \cdot (2 \cdot 2 \cdot 2)}_{\text{5 twos}} = 2^5$$

$$3^3 \cdot 3^4 = \underbrace{(3 \cdot 3 \cdot 3) \cdot (3 \cdot 3 \cdot 3 \cdot 3)}_{\text{7 threes}} = 3^7$$

$$5^2 \cdot 5^3 \cdot 5^4 = \underbrace{(5 \cdot 5) \cdot (5 \cdot 5 \cdot 5) \cdot (5 \cdot 5 \cdot 5 \cdot 5)}_{\text{9 fives}} = 5^9$$

In each of these examples, we multiplied powers of the same base. Notice the relationship between the exponents of the numbers being multiplied and the exponent of the result. In each case, the result could have been found by raising the common base to an exponent that is the sum of all the exponents in the expressions being multiplied. Thus, we have illustrated the following rule.

rule 1

To multiply exponential expressions of the same base, keep the base the same and add the exponents.

Using this rule to simplify the previous examples, we obtain:

$$2^2 \cdot 2^3 = 2^{2+3} = 2^5$$
$$3^3 \cdot 3^4 = 3^{3+4} = 3^7$$
$$5^2 \cdot 5^3 \cdot 5^4 = 5^{2+3+4} = 5^9$$

When multiplying very large numbers that end in zeros, it is often convenient to first rewrite each factor using powers of ten, and then use Rule 1 to calculate the product.

example 6

Calculate the following products by first rewriting each factor using powers to ten.

(a) 30×600

$$\begin{aligned} 30 \times 600 &= (3 \times 10^1) \times (6 \times 10^2) \\ &= (3 \times 6) \times (10^1 \times 10^2) \\ &= 18 \times 10^3 \end{aligned}$$

Rewrite using powers of ten. By commutative and associative properties. Multiply powers of the same base.

(b) $8{,}000 \times 5{,}000$

$$\begin{aligned} 8{,}000 \times 5{,}000 &= (8 \times 10^3) \times (5 \times 10^3) \\ &= (8 \times 5) \times (10^3 \times 10^3) \\ &= 40 \times 10^6 \\ &= 4 \times 10^1 \times 10^6 \\ &= 4 \times 10^7 \end{aligned}$$

(c) $70{,}000 \times 200{,}000$

$$\begin{aligned} 70{,}000 \times 200{,}000 &= (7 \times 10^4) \times (2 \times 10^5) \\ &= (7 \times 2) \times (10^4 \times 10^5) \\ &= 14 \times 10^9 \end{aligned}$$

Notice that 14×10^9 is much easier to read than 14,000,000,000. For this reason, very large numbers ending in zeros are almost always expressed using powers of ten.

(d) $120{,}000 \times 3{,}000{,}000$

$$\begin{aligned} 120{,}000 \times 3{,}000{,}000 &= (12 \times 10^4) \times (3 \times 10^6) \\ &= (12 \times 3) \times (10^4 \times 10^6) \\ &= 36 \times 10^{10} \end{aligned}$$

Since 120,000 is 12 followed by four zeros, it can be rewritten as 12×10^4.

Let us now look at the following examples, which illustrate what happens when we raise an exponential expression to a power.

$$(2^2)^3 = (2^2)(2^2)(2^2) = 2^6$$
$$(5^4)^2 = (5^4)(5^4) = 5^8$$
$$(9^5)^4 = (9^5)(9^5)(9^5)(9^5) = 9^{20}$$

Notice that in each case, the base remains unchanged. Also notice that the exponent of each result is the product of the exponents in the original expression. Thus, we have illustrated another rule for simplifying exponential expressions.

rule 2

To raise an exponential expression to a power, keep the base the same and multiply the exponents.

Using this rule to simplify the previous examples, we obtain:

$$(2^2)^3 = 2^{2 \cdot 3} = 2^6$$
$$(5^4)^2 = 5^{4 \cdot 2} = 5^8$$
$$(9^5)^4 = 9^{5 \cdot 4} = 9^{20}$$

example 7

Simplify each of the following operations using the rules for exponents.

(a) $6^3 \cdot 6^5$

$$6^3 \cdot 6^5 = 6^{3+5} = 6^8$$

Here we are multiplying powers of the same base, so we add exponents.

(b) $4^3 \cdot 4^5 \cdot 4^7$

$$4^3 \cdot 4^5 \cdot 4^7 = 4^{3+5+7} = 4^{15}$$

Here, again, we are multiplying powers of the same base, so we add exponents.

(c) $3 \cdot 10^4 \cdot 5 \cdot 10^6$

$$\begin{aligned} 3 \cdot 10^4 \cdot 5 \cdot 10^6 &= (3 \cdot 5) \cdot (10^4 \cdot 10^6) \\ &= 15 \cdot 10^{4+6} \\ &= 15 \cdot 10^{10} \end{aligned}$$

By commutative and associative properties.

(d) $(7^4)^2$

$$(7^4)^2 = 7^{4 \cdot 2} = 7^8$$

Here we are raising an exponential expression to a power, so we multiply exponents.

(e) $(8^3)^4 \cdot (5^2)^3$

$$(8^3)^4 \cdot (5^2)^3 = 8^{3 \cdot 4} \cdot 5^{2 \cdot 3} = 8^{12} \cdot 5^6$$

Here we are raising two exponential expressions to powers, so we multiply exponents in both cases.

(f) $2^5 \cdot 2^3 \cdot (2^3)^4$

$$2^5 \cdot 2^3 \cdot (2^3)^4 = 2^{5+3} \cdot 2^{3 \cdot 4} = 2^8 \cdot 2^{12} = 2^{8+12} = 2^{20}$$

Here we use both rules. We must add exponents and multiply exponents.

QUICK QUIZ

1. Rewrite $7 \cdot 7 \cdot 7 \cdot 2 \cdot 2 \cdot 2 \cdot 2 \cdot 2$ using exponents.
2. Express $4^2 \cdot 5^0 \cdot 3^1$ as a whole number.
3. Simplify $6^2 \cdot 6^7$.
4. Simplify $(3^8)^2$.

ANSWERS

1. $7^3 \cdot 2^5$
2. 48
3. 6^9
4. 3^{16}

1.6 Exercises

Indicate which of the following statements are true and which are false. For those that are false, change the italic word to make the statement true.

1. In the expression b^n, b is called the *exponent*.
2. An exponential expression indicates repeated *addition*.
3. Any nonzero number raised to the zero power equals *one*.
4. When multiplying powers of the same base, keep the base the same and *multiply* the exponents.

Rewrite each of the following expressions using exponents.

5. $3 \cdot 3 \cdot 3 \cdot 8 \cdot 8$ **6.** $5 \cdot 5 \cdot 9 \cdot 9 \cdot 9 \cdot 9 \cdot 9$ **7.** $2 \cdot 2 \cdot 2 \cdot 2 \cdot 2 \cdot 2 \cdot 6 \cdot 7 \cdot 7$

8. $3 \cdot 3 \cdot 4 \cdot 4 \cdot 5 \cdot 5 \cdot 9$ **9.** $5 \cdot 5 \cdot 5 \cdot 8 \cdot 8 + 5 \cdot 5 \cdot 7 \cdot 7 \cdot 7$

10. $4 \cdot 4 \cdot 4 \cdot 9 \cdot 9 \cdot 9 \cdot 9 + 2 \cdot 2 \cdot 4 \cdot 6 \cdot 6$

Rewrite each of the following numbers as an exponential expression with a base of 2, 3, 5, or 7.

11. 4 **12.** 27 **13.** 25 **14.** 49 **15.** 32

16. 125 **17.** 81 **18.** 16 **19.** 625 **20.** 64

Rewrite the following exponential expressions as whole numbers.

21. $2^2 \cdot 5^2$ **22.** $2^3 \cdot 4^2$ **23.** $6^0 \cdot 3^2$ **24.** $7^2 \cdot 2^1$

25. $3^3 \cdot 5^0 \cdot 8^1$ **26.** $2^2 \cdot 5^3 \cdot 7^0$ **27.** $5^1 \cdot 2^2 \cdot 6^2$ **28.** $2^5 \cdot 1^3 \cdot 3^2$

29. $7^2 \cdot 10^3$ **30.** $4^3 \cdot 10^2$ **31.** $2^4 \cdot 10^4$ **32.** $3^3 \cdot 10^3$

33. $6^7 \cdot 11^2$ **34.** $3^9 \cdot 4^5$

Write each of the following numbers in expanded notation using powers of ten.

35. 538
36. 914
37. 602
38. 830
39. 7,436
40. 2,951
41. 41,783
42. 20,841
43. 74,062
44. 80,027
45. 340,005
46. 804,070
47. 6,947,089
48. 3,078,614
49. 5,201,840,693
50. 7,003,438,020

Calculate the following products by first rewriting each factor using powers of ten.

51. 70×400
52. 20×900
53. $600 \times 8{,}000$
54. $9{,}000 \times 300$
55. $1{,}100 \times 7{,}000$
56. $6{,}000 \times 12{,}000$
57. $40{,}000 \times 50{,}000$
58. $60{,}000 \times 500{,}000$
59. $700{,}000 \times 300{,}000$
60. $5{,}000{,}000 \times 9{,}000{,}000$
61. $250{,}000 \times 200{,}000$
62. $15{,}000{,}000 \times 4{,}000{,}000$

Simplify each of the following exponential expressions using the rules for exponents.

63. $3^4 \cdot 3^5$
64. $8^4 \cdot 8^9$
65. $4^2 \cdot 4^4$
66. $5^5 \cdot 5^7$
67. $2^3 \cdot 2^5 \cdot 2^6$
68. $6^3 \cdot 6^0 \cdot 6^9$
69. $5^2 \cdot 5^4 \cdot 5^5$
70. $4^8 \cdot 4^3 \cdot 4^0$
71. $(3^4)^3$
72. $(8^5)^3$
73. $(4^2)^5$
74. $(7^3)^6$
75. $(7^2)^3 \cdot 7^4$
76. $(9^4)^6 \cdot 9^7$
77. $5^3 \cdot (5^4)^2$
78. $4^3 \cdot (4^5)^4$
79. $8^2 \cdot 8^4 \cdot 6^2 \cdot 6^5$
80. $3^5 \cdot 3^3 \cdot 9^4 \cdot 9^7$
81. $2^3 \cdot 10^5 \cdot 2^6 \cdot 10^3$
82. $8^2 \cdot 10^4 \cdot 8^5 \cdot 10^7$
83. $7^8 \cdot 8^6 \cdot 7^5 \cdot 8^2$
84. $5^4 \cdot 6^3 \cdot 5^9 \cdot 6^7$
85. $(6^4)^2 \cdot (6^3)^4$
86. $(8^2)^4 \cdot (8^5)^2$
87. $(4^8)^3 \cdot (4^7)^2$
88. $(3^5)^3 \cdot (3^4)^2$

1.7 Order of Operations

Now that we have learned how to perform a variety of arithmetic operations, it is important to learn how to solve problems that involve more than one operation. Parentheses always indicate which operation is to be performed first. However, problems that contain no parentheses present us with the possibility of different answers depending on the order in which we perform the various operations. For example, if we attempt to simplify $3 + 5 \cdot 4$, we obtain two different answers, depending on which operation we perform first. Performing addition first we obtain

$$3 + 5 \cdot 4 = (3 + 5) \cdot 4 = 8 \cdot 4 = 32$$

Performing multiplication first we obtain

$$3 + 5 \cdot 4 = 3 + (5 \cdot 4) = 3 + 20 = 23$$

To eliminate this confusion, the order of operations rules have been developed. They establish a conventional order that must be followed when simplifying arithmetic expressions (see box). According to the order of operations rules, to solve the previous example correctly, we perform multiplication before addition.

$$3 + 5 \cdot 4 = 3 + (5 \cdot 4) = 3 + 20 = 23$$

The Order of Operations Rule

1. Perform all operations in parentheses first.
2. Simplify all exponential expressions.
3. Do multiplication and division as they occur, moving from left to right.
4. Do addition and subtraction as they occur, moving from left to right.

example 1

Simplify each of the following expressions using the order of operations rules.

(a) $8 \cdot (4 + 6)$

$$8 \cdot (4 + 6) = 8 \cdot 10 = 80$$

We do addition first since it is enclosed in parentheses.

(b) $9 - 2 \cdot 3$

$$9 - 2 \cdot 3 = 9 - 6 = 3$$

Do multiplication before subtraction.

(c) $48 \div 8 \cdot 5 \div 2$

$$48 \div 8 \cdot 5 \div 2 = 6 \cdot 5 \div 2 = 30 \div 2 = 15$$

Do multiplication and division moving from left to right.

(d) $2^3 \cdot 3^2$

$$2^3 \cdot 3^2 = 8 \cdot 9 = 72$$

Simplify exponential expressions and then multiply.

(e) $5^2(9 - 3) + 2(8 + 7)$

$5^2(9 - 3) + 2(8 + 7)$

$= 5^2(6) + 2(15)$ — Do operations in ( ) first.

$= 25(6) + 2(15)$ — Simplify exponential expressions.

$= 150 + 30$ — Do multiplication.

$= 180$ — Do addition.

(f) $[4(3 + 2) - 3(9 - 7)] \div 7$

$[4(3 + 2) - 3(9 - 7)] \div 7$

$= [4(5) - 3(2)] \div 7$ — Simplify innermost ( ) first.

$= [20 - 6] \div 7$ — In [ ] multiply before subtracting.

$= 14 \div 7$ — Do subtraction.

$= 2$ — Do division.

Notice that when parentheses ( ) are nested inside brackets [ ], we simplify the innermost parentheses first and work outwards.

(g) $4[3(5 - 3) + (6 - 4)^2]$

$4[3(5 - 3) + (6 - 4)^2]$ — Simplify innermost ( ) first.

$= 4[3(2) + 2^2]$ — In [ ] raise to powers, then multiply.

$= 4[6 + 4]$ — In [ ] do addition.

$= 4[10]$ — Do multiplication.

$= 40$

(h) $5^2 \cdot 10^3 + 3^3 \cdot 10^2 + 7^2 \cdot 10^1$

$5^2 \cdot 10^3 + 3^3 \cdot 10^2 + 7^2 \cdot 10^1$

$= 25 \cdot 1{,}000 + 27 \cdot 100 + 49 \cdot 10$ — First raise to powers.

$= 25{,}000 + 2{,}700 + 490$ — Next do multiplication.

$= 28{,}190$ — Finally, do addition.

(i) $3^2 \cdot 3^3 + 3^6$

$$3^2 \cdot 3^3 + 3^6 = 3^{2+3} + 3^6 = 3^5 + 3^6$$

WARNING: Do not attempt to simplify further. We have not established a rule for adding powers of the same base. ●

The wording of a particular problem often indicates which operation should be performed first. When translating words into mathematical expressions, we must remember to use parentheses to enclose those operations that are to be performed first.

example 2

Translate the following words into mathematical expressions and simplify.

(a) 4 times the sum of 7 and 2.

$$4 \cdot (7 + 2) = 4 \cdot 9 = 36$$

(b) 6 more than 3 squared.

$$6 + 3^2 = 6 + 9 = 15$$

Since we simplify powers before multiplying, no ( ) are needed.

(c) 5 less than the product of 3 and 4.

$$3 \cdot 4 - 5 = 12 - 5 = 7$$

Since multiplication is performed before subtraction, no ( ) are needed.

(d) The sum of the product of 5 and 4 and the quotient of 18 and 2.

$$(5 \cdot 4) + (18 \div 2) = 20 + 9 = 29$$

Parentheses are not needed but they do simplify translation.

(e) The difference between the sum of 9 and 8 and the product of 3 and 5.

$$(9 + 8) - (3 \cdot 5) = 17 - 15 = 2$$

●

QUICK QUIZ

Simplify each of the following expressions.

1. $7 + 4 \cdot 5$

2. $8 \div 2 + 6 \cdot 2 - 9 \div 3$

3. $[(6 - 2)^2 + (9 - 7)^3] \div 4$

ANSWERS

1. 27

2. 13

3. 6

1.7 Exercises

Indicate which of the following statements are true and which are false. For those that are false, change the italic expression to make the statement true.

1. Parentheses enclose the operation to be performed *last.*

2. Unless otherwise indicated by parentheses, we perform addition *before* multiplication.

3. Multiplication and division are performed in *the same* step moving from left to right.

4. If parentheses are nested inside brackets, we evaluate the *innermost* expression first.

Simplify each of the following expressions using the order of operations rules.

5. $5 + 3 \cdot 4$

6. $11 - 2 \cdot 4$

7. $5^2 \cdot 2^2$

8. $3^2 \cdot 2^3$

9. $18 \div 3 + 7$

10. $7 - 21 \div 3$

11. $32 \div 8 + 6 \cdot 2$

12. $5 \cdot 4 - 24 \div 8$

13. $2 \cdot 5^2 + 3^3$

14. $8^0 \cdot 6 - 2^2$

15. $72 \div 8 + 63 \div 9$
16. $56 \div 7 - 24 \div 8$
17. $8(2 + 3)$
18. $5(6 + 2)$
19. $2 \cdot 4^2 + 7^0 \cdot 3$
20. $3^2 \cdot 7 - 2^3 \div 4$
21. $81 \div (13 - 4)$
22. $66 \div (5 + 6)$
23. $56 \div 7 \div 4$
24. $72 \div 8 \div 3$
25. $5(6 - 3)^3$
26. $(5 + 1)^2 \div 4$
27. $(6 \cdot 3 - 5 + 2) \div 3$
28. $(4 \cdot 7 - 6 + 5) \div 9$
29. $6^2 \cdot 10^2 + 5^2 \cdot 10^1$
30. $4^2 \cdot 10^3 + 3^2 \cdot 10^2$
31. $2^3 \cdot 10^3 + 2^3 \cdot 10^2 + 2^3 \cdot 10^1 + 2^3 \cdot 10^0$
32. $3^2 \cdot 10^3 + 3^2 \cdot 10^2 + 3^2 \cdot 10^1 + 3^2 \cdot 10^0$
33. $63 \div 7 + 5 \cdot 2 - 6 \div 3$
34. $8 \cdot 6 - 4 \cdot 3 + 14 \div 2$
35. $3(4 + 5) - 8(6 - 3)$
36. $7(8 - 4) + 3(4 + 3)$
37. $3(8 - 5)^2 - 36 \div (13 - 4)$
38. $48 \div (17 - 9) + 4(7 - 5)^3$
39. $[6(2 + 3) - 4(7 - 2)] \div 5$
40. $[8(13 - 8) - 3(11 - 5)] \div 2$
41. $2[(5 - 3)^2 + 4(10 - 7)]$
42. $5[6(11 - 8) - (12 - 8)^2]$
43. $7^5 + 7^3 \cdot 7^4$
44. $6^4 \cdot 6^5 + 6^2$
45. $5^2 \cdot 5^8 + (5^3)^2$
46. $2^9 \cdot 2^2 + (2^4)^3$
47. $85{,}008 \div 184 + 67 \cdot 23$
48. $347 \cdot 216 \div (4{,}086 - 3{,}978)$
49. $[(13)^5 - (28)^3] \div 3$
50. $(34)(19)^3 + (55)(6)^5$

Find the mistake in each of the following equations and determine the correct answer.

51. $8 + 3 \cdot 2 = 11 \cdot 2 = 22$
52. $4 + 2 \cdot 5 = 6 \cdot 5 = 30$
53. $6 - 2 \div 2 = 4 \div 2 = 2$
54. $9 - 3 \div 3 = 6 \div 3 = 2$
55. $2 \cdot 3^2 = 6^2 = 36$
56. $5 \cdot 2^2 = 10^2 = 100$
57. $8 - 6 + 2 = 8 - 8 = 0$
58. $9 - 5 - 3 = 9 - 2 = 7$
59. $(4 + 5)^2 = 4 + 25 = 29$
60. $(9 - 2)^2 = 9 - 4 = 5$
61. $3(8 - 6) = 24 - 6 = 18$
62. $5(4 + 3) = 20 + 3 = 23$

Translate each of the following word expressions into mathematical expressions and simplify.

63. Twice the sum of 6 and 2.
64. 4 times the difference between 8 and 3.
65. 2 less than the product of 5 and 4.
66. 5 more than the quotient of 15 and 3.
67. 6 more than 2 cubed.
68. 9 less than 5 squared.
69. The sum of the product of 3 and 4 and the quotient of 10 and 2.
70. The difference between the product of 8 and 3 and the product of 6 and 2.
71. The product of the sum of 2 and 5 and the sum of 6 and 3.
72. The product of the difference between 9 and 2 and the sum of 5 and 4.

1.8 Averages

One of the most frequently used applications of division is to calculate an **average**. Our daily living experiences include many references to averages, such as batting averages, average test grades, average income, average speed, and the Dow Jones Industrial Average.

definition

The *average* of a set of numbers is a number that is used to represent the entire set. To calculate an average, first add all the numbers in the set, and then divide that sum by the number of numbers that were added together.

In mathematics, the average is often referred to as the *mean*.

example 1

Calculate the average of the set of numbers {28, 17, 34, 53, 21, 38, 19}.

solution

Add the seven numbers. Then divide the total by 7.

$$\begin{array}{r} {}^{4}\\ 28\\ 17\\ 34\\ 53\\ 21\\ 38\\ 19\\ \hline 210 \end{array} \qquad \begin{array}{r} 30\\ 7\overline{)210}\\ 21\\ \hline 0 \end{array}$$

The average is 30. ●

Notice that the average of a given set of numbers is always greater than or equal to the smallest number in the set and less than or equal to the largest number in the set. For instance, in Example 1, $17 < 30 < 53$.

example 2

A student received grades of 83, 67, 75, 72, and 88 on mathematics quizzes. What is this student's average quiz grade?

solution

Add the five quiz grades. Then divide that sum by 5.

$$\begin{array}{r} 83\\ 67\\ 75\\ 72\\ 88\\ \hline 385 \end{array} \qquad \begin{array}{r} 77\\ 5\overline{)385}\\ 35\\ \hline 35\\ 35\\ \hline 0 \end{array}$$

The student's average quiz grade is 77. ●

example 3

A stockbroker purchased a variety of stocks for a client at the following prices: six shares at \$35 a share, twenty shares at \$18 a share, four shares at \$40 a share, and ten shares at \$27 a share. What was the average price per share of all the stocks purchased?

solution

Calculate the total amount of money spent by multiplying the number of shares purchased at each price by the price per share and then adding those results. Then divide that sum by the total number of shares purchased.

PRICE PER SHARE	×	NUMBER OF SHARES	
\$35	×	6	= \$ 210
\$18	×	20	= \$ 360
\$40	×	4	= \$ 160
\$27	×	10	= \$ 270
	Total:	40	\$1,000

Forty shares were purchased for a total of $1,000. To calculate the average price per share, divide $1,000 by 40:

$$\begin{array}{r} 25 \\ 40\overline{)1{,}000} \\ 80 \\ \hline 200 \\ 200 \\ \hline 0 \end{array}$$

The average price per share was $25. ●

When you are asked to calculate an average, it is not always necessary to add up numbers before dividing, because the total often is given in the problem.

example 4

During a particular week at a computer firm, 518 employees worked a total of 19,684 hours. What was the average number of hours worked by each employee that week?

solution

Divide 19,684 ÷ 518.

$$\begin{array}{r} 38 \\ 518\overline{)19{,}684} \\ 15\ 54 \\ \hline 4\ 144 \\ 4\ 144 \\ \hline 0 \end{array} \qquad \begin{array}{r} 518 \\ \times\quad 3 \\ \hline 1{,}554 \end{array} < 1{,}968 \qquad \begin{array}{r} 518 \\ \times\quad 8 \\ \hline 4{,}144 \end{array} \qquad \begin{array}{r} 1{,}968 \\ -\ 1{,}554 \\ \hline 414 \end{array} < 518$$

On the average, each employee worked 38 hours that week. ●

QUICK QUIZ

Find the average of each of the following sets of numbers.

1. {9, 13, 6, 20}
2. {216, 462, 381}

ANSWERS

1. 12
2. 353

1.8 Exercises

Indicate which of the following statements are true and which are false. For those that are false, change the italic expression to make the statement true.

1. When calculating an average, we perform the operation addition *before* division.
2. The *mean* is another name for the average.
3. The average of a set of whole numbers is always *greater than* the largest number in the set.
4. To calculate the average of three numbers, we divide the sum of the numbers by *4*.

Calculate the average of each of the following sets of numbers.

5. {54, 32, 71, 27}
6. {89, 65, 38, 41, 72}
7. {314, 281, 437}
8. {652, 587, 736, 801}
9. {52,679, 47,325, 26,839, 39,461}
10. {857, 799, 934, 687, 492, 865, 528, 753, 412, 683, 536}

Solve each of the following word problems.

11. A ten-year-old practiced his piano lessons for 35 minutes on Monday, 27 minutes on Tuesday, 24 minutes on Wednesday, 42 minutes on Thursday, 18 minutes on Friday, 31 minutes on Saturday, and 26 minutes on Sunday. What was the average length of a daily practice session that week?
12. The heights of the players on the starting lineup of a basketball team are 78 inches, 81 inches, 84 inches, 83 inches, and 79 inches. What is the average height of a player on that team?
13. A secretary types 868 words in 14 minutes. What is his average typing speed in words per minute?
14. A plane flew 4,284 miles in 12 hours. What was its average speed in miles per hour?
15. A salesman drove 235 miles on Monday, 318 miles on Tuesday, 203 miles on Wednesday, 188 miles on Thursday, and 126 miles on Friday. What was the average distance he traveled daily that week?
16. The members of a bowling team bowled the following scores: 198, 160, 214, 231, and 187. What was the average score for that team?
17. On a math quiz worth 10 points, eight students scored 10, twelve students scored 9, five students scored 8, four students scored 5, one student scored 4, and two students scored 2. What was the average score on that quiz?
18. The following bets were made on the outcome of a football game: six people bet $1, three people bet $5, four bet $10, one bet $15, and three bet $20. What was the size of the average bet?
19. A contest is advertising that one $7,500 prize, four $2,500 prizes, ten $1,000 prizes, and twenty-five $500 prizes will be awarded to winners. What will be the average amount awarded to a winner of this contest?
20. The members of a track team ran the following races: five ran the 100-meter race, two ran the 200-meter race, three ran the 400-meter, two ran the 1,000-meter, four ran the 5,000-meter, and two ran the 10,000-meter. What was the average distance run by a member of that track team?
21. If 51,100 people died in traffic accidents in 1978, what was the average number of deaths per day related to motor vehicles in 1978?
22. If 6,135 students at a certain university bought a total of 79,755 books in one semester, what was the average number of books bought by a student for one semester of study at that university?
23. The areas in square miles of five western European countries are France, 212,973; Spain, 194,884; Italy, 116,304; West Germany, 95,815; and the United Kingdom, 94,249. What is the average area?
24. The number of marriages in the United States for the years 1970–1977 were: 1970: 2,158,802; 1971: 2,190,481; 1972: 2,282,154; 1973: 2,284,108; 1974: 2,229,667; 1975: 2,152,662; 1976: 2,154,807; and 1977: 2,178,367. What was the average number of marriages per year for the period 1970–1977?

1.9 Short Division and Tests for Divisibility

Short Division

When dividing by a one-digit number, we can save ourselves some work by using a technique called *short division.* Instead of showing the subtraction to find the remainder for each step as we do in long division, we mentally subtract and write the remainder in front of the next digit in the dividend. Example 1 illustrates the difference between the technique we have been using, long division, and short division.

example 1

Divide 4,524 ÷ 6.

solution

Long Division

```
    754
 6|4,524
   4 2
   ---
    32
    30
   ---
     24
     24
    ---
      0
```

The answer is 754.

Short Division

$$\begin{array}{r} 7\ \ 5\ \ 4 \\ 6\overline{)4\ \ 5\ {}^{3}2\ {}^{2}4} \end{array}$$

6 goes into 45, 7 times with a remainder of 3. Put 7 over the dividend. Write 3 in front of the next digit.

6 goes into 32, 5 times with a remainder of 2. Put 5 over the dividend. Write 2 in front of the next digit.

6 goes into 24, 4 R0. Put 4 over the dividend. ●

example 2

Divide each of the following numbers using short division.

(a) 31,825 ÷ 5

$$\begin{array}{r} 6,\ 3\ \ 6\ \ 5 \\ 5\overline{)3\ \ 1,{}^{1}8\ {}^{3}2\ {}^{2}5} \end{array}$$

5 goes into 31, 6 R1. Put 6 over the dividend. Write 1 in front of the next digit.

5 goes into 18, 3 R3. Put 3 over the dividend. Write 3 in front of the next digit.

5 goes into 32, 6 R2. Put 6 over the dividend. Write 2 in front of the next digit.

5 goes into 25, 5 R0. Write 5 over the dividend.

The answer is 6,365.

(b) 9,063 ÷ 7

$$\begin{array}{r} 1,\ 2\ \ 9\ \ 4\ \text{R5} \\ 7\overline{)9,{}^{2}0\ {}^{6}6\ {}^{3}3\ \ \ \ \ \ } \end{array}$$

7 goes into 9, 1 R2. Put 1 over the dividend. Write 2 in front of the next digit.

7 goes into 20, 2 R6. Put 2 over the dividend. Write 6 in front of the next digit.

7 goes into 66, 9 R3. Put 9 over the dividend. Write 3 in front of the next digit.

7 goes into 33, 4 R5. Put 4 over the dividend, and indicate the remainder as in long division.

The answer is 1,294 R5. ●

Tests for Divisibility

When working with whole numbers, it is often quite useful to be able to determine quickly if a certain number can be evenly divided by another, without actually going through the division process. We will now learn techniques for determining whether or not a number is **divisible** by a given divisor.

definition

A number is *divisible* by a given divisor if the division process yields a whole number quotient and a remainder of zero.

Divisibility by 2

A number is divisible by 2 if its ones digit is 0, 2, 4, 6, or 8.

Numbers divisible by 2 are called *even* numbers. Thus, numbers with a ones digit of 0, 2, 4, 6, or 8 are even. Numbers not divisible by 2 are called *odd* numbers. Thus, numbers with a ones digit of 1, 3, 5, 7, or 9 are odd.

example 3

Determine whether or not each of the following numbers is divisible by 2.

(a) 54

54 is divisible by 2. — Its ones digit is 4.
54 is an even number.

(b) 221

221 is not divisible by 2. — Its ones digit is 1.
221 is an odd number.

(c) 2,000

2,000 is divisible by 2. — Its ones digit is 0.
2,000 is an even number.

Divisibility by 3

A number is divisible by 3 if the sum of its digits is divisible by 3.

example 4

Determine whether or not each of the following numbers is divisible by 3.

(a) 265

The sum of the digits is $2 + 6 + 5 = 13$. — 13 is not divisible by 3.
Therefore, 265 is not divisible by 3.

(b) 5,187

The sum of the digits is
$5 + 1 + 8 + 7 = 21$. — 21 is divisible by 3.
Therefore, 5,187 is divisible by 3.

(c) 468,293

The sum of the digits is
$4 + 6 + 8 + 2 + 9 + 3 = 32$. — 32 is not divisible by 3.
Therefore, 468,293 is not divisible by 3.

Divisibility by 4

A number is divisible by 4 if the last two digits form a number divisible by 4.

example 5

Determine whether or not each of the following numbers is divisible by 4.

(a) 536

536: The last two digits are 36. — 36 is divisible by 4.

Therefore, 536 is divisible by 4.

(b) 1,749

1,749: The last two digits are 49. — 49 is not divisible by 4.

Therefore, 1,749 is not divisible by 4.

(c) 69,304

69,304: The last two digits are 04. — 04 is divisible by 4.

Therefore, 69,304 is divisible by 4.

Divisibility by 5

A number is divisible by 5 if its ones digit is 0 or 5.

example 6

Determine whether or not each of the following numbers is divisible by 5.

(a) 6,051

6,051 is not divisible by 5 since its ones digit is 1.

(b) 936,275

936,275 is divisible by 5 since its ones digit is 5.

Divisibility by 6

A number is divisible by 6 if it is divisible by both 2 and 3.

example 7

Determine whether or not each of the following numbers is divisible by 6.

(a) 582

582 is divisible by 2. — Its ones digit is 2.
The sum of the digits is $5 + 8 + 2 = 15$. — 15 is divisible by 3.
582 is divisible by 3.
Therefore, 582 is divisible by 6.

(b) 1,438

1,438 is divisible by 2. — Its ones digit is 8.
The sum of the digits is
$$1 + 4 + 3 + 8 = 16.$$
— 16 is not divisible by 3.
1,438 is not divisible by 3.
Therefore, 1,438 is not divisible by 6.

(c) 61,275

61,275 is not divisible by 2. — Its ones digit is 5.
Therefore, 61,275 is not divisible by 6.

There is no simple test for divisibility by 7. ●

Divisibility by 8

A number is divisible by 8 if the last three digits form a number divisible by 8.

example 8

Determine whether or not each of the following numbers is divisible by 8.

(a) 3,265

3,265: The last three digits are 265.

$$\begin{array}{r} 3\ \ 3\ \text{R}1 \\ 8\overline{)2\ \ 6^{2}5} \end{array}$$

265 is not divisible by 8.

Therefore, 3,265 is not divisible by 8.

(b) 78,544

78,544: The last three digits are 544.

$$\begin{array}{r} 6\ \ 8 \\ 8\overline{)5\ \ 4^{6}4} \end{array}$$

544 is divisible by 8.

Therefore, 78,544 is divisible by 8. ●

Divisibility by 9

A number is divisible by 9 if the sum of its digits is divisible by 9.

example 9

Determine whether or not each of the following numbers is divisible by 9.

(a) 309

The sum of the digits is $3 + 0 + 9 = 12$. — 12 is not divisible by 9.
Therefore, 309 is not divisible by 9.

(b) 418,374

The sum of the digits is

$$4 + 1 + 8 + 3 + 7 + 4 = 27.$$

27 is divisible by 9.

Therefore, 418,374 is divisible by 9. ●

Divisibility by 10

A number is divisible by 10 if its ones digit is 0.

example 10

Determine whether or not the number 4,005 is divisible by 10.

solution

4,005 is not divisible by 10. — Its ones digit is 5.

example 11

The number 4,51_ has a missing ones digit. Fill in the appropriate digit(s) so that the result is divisible by the following numbers.

(a) Divisible by 2.

The ones digit must be 0, 2, 4, 6, or 8. Therefore, $4{,}51\underline{0}$, $4{,}51\underline{2}$, $4{,}51\underline{4}$, $4{,}51\underline{6}$, and $4{,}51\underline{8}$, are all divisible by 2.

(b) Divisible by 3.

The sum of the digits must be divisible by 3.

$$4{,}51\_: \quad \underbrace{4 + 5 + 1}_{10} + \_ = \text{a number divisible by 3}$$

$$10 + \_ = \text{a number divisible by 3}$$

$10 + \underline{2} = 12$ — Divisible by 3.

$10 + \underline{5} = 15$ — Divisible by 3.

$10 + \underline{8} = 18$ — Divisible by 3.

Therefore, $4{,}51\underline{2}$, $4{,}51\underline{5}$, and $4{,}51\underline{8}$ are all divisible by 3.

(c) Divisible by 6.

The number must be divisible by both 2 and 3.
The numbers divisible by 3 are $4{,}51\underline{2}$, $4{,}51\underline{5}$, and $4{,}51\underline{8}$. — Example 11(b)
Of these, only $4{,}51\underline{2}$ and $4{,}51\underline{8}$ are also divisible by 2. — Example 11(a)
Therefore, $4{,}51\underline{2}$ and $4{,}51\underline{8}$ are divisible by 6.

example 12

The number 81,53_ has a missing ones digit. Fill in the appropriate digit(s) so that the result is divisible by each of the following numbers.

(a) Divisible by 4.

The last two digits must form a number divisible by 4. $3\underline{2}$ and $3\underline{6}$ are both divisible by 4.
Therefore, $81{,}53\underline{2}$ and $81{,}53\underline{6}$ are both divisible by 4.

(b) Divisible by 8.

The last three digits must make a number divisible by 8.

$$\begin{array}{r} 6\ 7 \\ 8\overline{)53\,{}^{5}\underline{6}} \end{array}$$

8 goes into 53, 6 R5. Fill in the ones digit so 5 is divisible by 8. Since 56 is divisible by 8, the ones digit is 6.

Thus, $81{,}53\underline{6}$ is divisible by 8.

(c) Divisible by 9.

The sum of the digits must be divisible by 9.

$$81{,}53\_: \underbrace{8+1+5+3}_{17} + \_ = \text{a number divisible by 9}$$

$$17 + \_ = \text{a number divisible by 9}$$

$$17 + \underline{1} = 18 \quad \text{Divisible by 9.}$$

Therefore, 81,53$\underline{1}$ is divisible by 9.

QUICK QUIZ

Use the divisibility tests to answer each of the following questions.

1. Is 458 divisible by 3?
2. Is 1,740 divisible by 6?
3. Is 93,264 divisible by 8?

ANSWERS

1. No
2. Yes
3. Yes

1.9 Exercises

Indicate which of the following statements are true and which are false. For those that are false, change the italic expression to make the statement true.

1. A number that is divisible by 10 is *always* divisible by 5.
2. Numbers divisible by 2 are called *odd*.
3. A number is divisible by 8 if the last *four* digits form a number divisible by 8.
4. If a number is divisible by 9, it is also divisible by *2*.

Use short division to find each of the following quotients.

5. 436 ÷ 4	**6.** 216 ÷ 3	**7.** 919 ÷ 7	**8.** 813 ÷ 5	**9.** 1,344 ÷ 2
10. 6,284 ÷ 6	**11.** 3,719 ÷ 8	**12.** 4,602 ÷ 9	**13.** 73,841 ÷ 3	**14.** 82,036 ÷ 4
15. 37,245 ÷ 5	**16.** 20,493 ÷ 8	**17.** 1,003,456 ÷ 7	**18.** 3,842,461 ÷ 2	

Use the divisibility tests to determine whether or not each of the following numbers is divisible by 2, 3, 4, 5, 6, 8, 9, and 10.

19. 182	**20.** 215	**21.** 600	**22.** 416	**23.** 552
24. 738	**25.** 1,407	**26.** 1,233	**27.** 2,385	**28.** 1,230
29. 4,356	**30.** 7,206	**31.** 8,469	**32.** 6,210	**33.** 3,824
34. 5,308	**35.** 23,186	**36.** 37,239	**37.** 86,044	**38.** 79,600
39. 56,328	**40.** 63,747	**41.** 438,816	**42.** 524,090	**43.** 360,045
44. 473,284	**45.** 5,392,074	**46.** 9,005,635	**47.** 20,625,080	**48.** 549,627,088

The number 2,60_ has a missing ones digit. Fill in the appropriate digit(s) so that the result is divisible by the following numbers.

49. Divisible by 2.	**50.** Divisible by 3.	**51.** Divisible by 4.	**52.** Divisible by 5.
53. Divisible by 6.	**54.** Divisible by 8.	**55.** Divisible by 9.	**56.** Divisible by 10.

The number 54, 89_ has a missing ones digit. Fill in the appropriate digit(s) so that the result is divisible by the following numbers.

57. Divisible by 2.
58. Divisible by 3.
59. Divisible by 4.
60. Divisible by 5.
61. Divisible by 6.
62. Divisible by 8.
63. Divisible by 9.
64. Divisible by 10.

A number is divisible by 12 if it is divisible by both 3 and 4. Determine whether or not each of the following numbers is divisible by 12.

65. 288
66. 356
67. 1,536
68. 5,124
69. 2,444
70. 4,698
71. 14,355
72. 92,184

A number is divisible by 15 if it is divisible by both 3 and 5. Determine whether or not each of the following numbers is divisible by 15.

73. 230
74. 555
75. 8,095
76. 6,345
77. 2,961
78. 40,280
79. 10,335
80. 163,550

A number is divisible by 18 if it is divisible by both 2 and 9. Determine whether or not each of the following numbers is divisible by 18.

81. 216
82. 908
83. 4,982
84. 27,846
85. 41,756
86. 328,464
87. 1,029,114
88. 30,478,162

1.10 Prime Numbers and Factoring

When we factor a number, we rewrite it as a product. Sometimes it is possible to factor a number in more than one way. For example, the number 12 may be factored in each of the following ways.

$$12 = 2 \cdot 6$$
$$12 = 3 \cdot 4$$
$$12 = 2 \cdot 2 \cdot 3$$

The last factorization of 12 is the most complete, since 2 cannot be rewritten as a product that does not contain 2, and 3 cannot be rewritten as a product that does not contain 3. The numbers 2 and 3 are therefore called **prime numbers**.

definition

A *prime number* is a whole number greater than 1 that is divisible only by 1 and itself.

definition

A *composite number* is a whole number greater than 1 that is not prime.

Notice that by definition, the numbers 0 and 1 are considered neither prime nor composite.

In order to determine which whole numbers are prime, we can use a device called the Sieve of Eratosthenes, developed by the Greek mathematician Eratosthenes. This technique requires us to find all the **multiples** of each successive prime number to eliminate whole numbers that are not prime.

definition

A *multiple* of a given whole number is the product of that number and another whole number.

To find all the multiples of a given whole number, just multiply that number by every whole number. For example, the multiples of 2 can be determined by multiplying every whole number (0, 1, 2, 3, 4, 5, . . .) by 2 to obtain the following.

Multiples of 2: 0, 2, 4, 6, 8, 10, . . .

Notice that 0 is a multiple of every whole number.

example 1

List the multiples of each of the following numbers.

(a) 3

Multiply every whole number by 3.
Multiples of 3: 0, 3, 6, 9, 12, 15, . . .

(b) 5

Multiply every whole number by 5.
Multiples of 5: 0, 5, 10, 15, 20, 25, . . .

(c) 12

Multiply every whole number by 12.
Multiples of 12: 0, 12, 24, 36, 48, 60, . . .

To find all the prime numbers less than 100, we use the Sieve of Eratosthenes to perform the following steps. The result of this process is shown in Table 1.9.

1. List all the whole numbers starting with 2 that are less than 100.
2. Circle the number 2 and then cancel out all multiples of 2.
3. Circle the lowest number that is not canceled, 3. Cancel out all multiples of 3.
4. Circle the lowest uncanceled number, 5. Cancel all multiples of 5.
5. Continue in this manner. Circle the lowest uncanceled number, and then cancel all of its multiples.

table ***1.9***
The Sieve of Eratosthenes

		2	3	~~4~~	5	~~6~~	7	~~8~~	~~9~~
~~10~~	11	~~12~~	13	~~14~~	~~15~~	~~16~~	17	~~18~~	19
~~20~~	~~21~~	~~22~~	23	~~24~~	~~25~~	~~26~~	~~27~~	~~28~~	29
~~30~~	31	~~32~~	~~33~~	~~34~~	~~35~~	~~36~~	37	~~38~~	~~39~~
~~40~~	41	~~42~~	43	~~44~~	~~45~~	~~46~~	47	~~48~~	~~49~~
~~50~~	~~51~~	~~52~~	53	~~54~~	~~55~~	~~56~~	~~57~~	~~58~~	59
~~60~~	61	~~62~~	~~63~~	~~64~~	~~65~~	~~66~~	67	~~68~~	~~69~~
~~70~~	71	~~72~~	73	~~74~~	~~75~~	~~76~~	~~77~~	~~78~~	79
~~80~~	~~81~~	~~82~~	83	~~84~~	~~85~~	~~86~~	~~87~~	~~88~~	89
~~90~~	~~91~~	~~92~~	~~93~~	~~94~~	~~95~~	~~96~~	97	~~98~~	~~99~~

6. When all the numbers have been circled or canceled, you are finished. The circled numbers are prime, and the canceled numbers are composite. Notice that the number 2 is the only even prime number.

Now that we are familiar with prime numbers, we can factor any number completely by rewriting it as the product of only prime numbers. **Prime factorization** will be a very useful technique for working with fractions in Chapter 3.

definition

The *complete*, or *prime*, *factorization* of a number is the unique representation of that number as the product of its prime factors.

To completely factor a number, we can use a device called a *factor tree*. First we rewrite the number as the product of any two numbers. Then we rewrite all composite factors as the product of two other numbers. We continue to rewrite all composite factors in this manner, until all factors of the original number are prime. For example, the number 72 can be completely factored as follows:

$72 = 8 \quad 9$	Rewrite 72 as $8 \cdot 9$.
$= 2 \cdot 4 \cdot 3 \cdot 3$	Factor 8 and 9.
$= 2 \cdot 2 \cdot 2 \cdot 3 \cdot 3$	Factor 4.

The prime factorization of 72 is, therefore,

$$72 = 2 \cdot 2 \cdot 2 \cdot 3 \cdot 3 = 2^3 \cdot 3^2$$

which can be written in shorthand notation using exponents.

Let us now consider some further examples of complete factorization.

example 2

Completely factor each of the following numbers.

(a) 81

$$81 = 9 \cdot 9$$
$$= 3 \cdot 3 \cdot 3 \cdot 3$$

$$81 = 3 \cdot 3 \cdot 3 \cdot 3 = 3^4$$

(b) 48

$$48 = 6 \cdot 8$$
$$= 2 \cdot 3 \cdot 2 \cdot 4$$
$$= 2 \cdot 3 \cdot 2 \cdot 2 \cdot 2$$

$$48 = 2 \cdot 2 \cdot 2 \cdot 2 \cdot 3 = 2^4 \cdot 3$$

(c) 60

$$60 = 6 \quad 10$$
$$= 2 \cdot 3 \cdot 2 \cdot 5$$

$$60 = 2 \cdot 2 \cdot 3 \cdot 5 = 2^2 \cdot 3 \cdot 5$$

(d) 216

$$216 = 3 \cdot 72$$
$$= 3 \cdot 8 \cdot 9$$
$$= 3 \cdot 2 \cdot 4 \cdot 3 \cdot 3$$
$$= 3 \cdot 2 \cdot 2 \cdot 2 \cdot 3 \cdot 3$$
$$216 = 2 \cdot 2 \cdot 2 \cdot 3 \cdot 3 \cdot 3 = 2^3 \cdot 3^3$$

QUICK QUIZ

Completely factor each of the following numbers.

1. 40

2. 66

3. 168

ANSWERS

1. $2^3 \cdot 5$

2. $2 \cdot 3 \cdot 11$

3. $2^3 \cdot 3 \cdot 7$

1.10 Exercises

Indicate which of the following statements are true and which are false. For those that are false, change the italic word to make the statement true.

1. A *composite* number is evenly divisible only by 1 and itself.

2. A nonzero multiple of a number is always *greater* than or equal to that number.

3. The prime factorization of any composite number *is* unique.

4. The number 2 is considered to be *prime.*

State whether each of the following numbers is prime or composite.

5. 9	**6.** 5	**7.** 17	**8.** 27	**9.** 31	**10.** 2
11. 15	**12.** 47	**13.** 53	**14.** 67	**15.** 81	**16.** 87
17. 93	**18.** 73	**19.** 49	**20.** 59	**21.** 79	**22.** 69
23. 23	**24.** 63	**25.** 121	**26.** 97	**27.** 101	**28.** 89
29. 105	**30.** 207	**31.** 145	**32.** 113	**33.** 333	**34.** 276

List the first five multiples of each of the following numbers.

35. 4	**36.** 6	**37.** 7	**38.** 9	**39.** 10
40. 13	**41.** 15	**42.** 16	**43.** 20	**44.** 25

Completely factor each of the following numbers.

45. 16	**46.** 36	**47.** 54	**48.** 18	**49.** 42	**50.** 63
51. 32	**52.** 80	**53.** 78	**54.** 45	**55.** 56	**56.** 99
57. 64	**58.** 24	**59.** 60	**60.** 84	**61.** 108	**62.** 144
63. 225	**64.** 169	**65.** 150	**66.** 210	**67.** 178	**68.** 256

69. 360 **70.** 425 **71.** 690 **72.** 512 **73.** 700 **74.** 388
75. 950 **76.** 1,000 **77.** 2,500 **78.** 4,000 **79.** 6,250 **80.** 7,500
81. 10,000 **82.** 35,000

1.11 Greatest Common Factor (GCF) and Least Common Multiple (LCM)

In this section, we will learn two more techniques that will be helpful for our work with fractions. The first is a procedure to find the **greatest common factor (GCF)** of two or more numbers.

definition

The *greatest common factor*, or *GCF*, of a set of numbers is the largest number by which each number in the set is evenly divisible.

To Find the Greatest Common Factor of a Set of Numbers

1. Completely factor each number in the set.
2. Find the prime factors common to every number in the set.
3. The GCF is the product of these common prime factors.

example 1

Find the GCF of 12 and 18.

solution

$12 = 4 \cdot 3 = \boxed{2} \cdot 2 \cdot \boxed{3}$

$18 = 6 \cdot 3 = \boxed{2} \cdot 3 \cdot \boxed{3}$

Factor 12 and 18 completely.

The prime numbers 2 and 3 are factors of both 12 and 18.

$\text{GCF} = 2 \cdot 3 = 6$

The GCF is the product of 2 and 3.

example 2

Find the GCF of each of the following sets of numbers.

(a) $\{28, 32\}$

$28 = 4 \cdot 7 = \boxed{2} \cdot \boxed{2} \cdot 7$

$32 = 4 \cdot 8 = \boxed{2} \cdot \boxed{2} \cdot 2 \cdot 2 \cdot 2$

$\text{GCF} = 2 \cdot 2 = 4$

(b) $\{64, 96\}$

$64 = 8 \cdot 8 = 2 \cdot \boxed{2} \cdot \boxed{2} \cdot \boxed{2} \cdot \boxed{2} \cdot \boxed{2}$

$96 = 3 \cdot 32 = 3 \cdot \boxed{2} \cdot \boxed{2} \cdot \boxed{2} \cdot \boxed{2} \cdot \boxed{2}$

$\text{GCF} = 2 \cdot 2 \cdot 2 \cdot 2 \cdot 2 = 32$

(c) $\{36, 90\}$

$36 = 6 \cdot 6 = 2 \cdot 3 \cdot 2 \cdot 3 = 2 \cdot \boxed{2} \cdot \boxed{3} \cdot \boxed{3}$

$90 = 9 \cdot 10 = 3 \cdot 3 \cdot 2 \cdot 5 = \boxed{2} \cdot \boxed{3} \cdot \boxed{3} \cdot 5$

$\text{GCF} = 2 \cdot 3 \cdot 3 = 18$

(d) $\{24, 60, 108\}$

$$24 = 3 \cdot 8 = 3 \cdot 2 \cdot 2 \cdot 2 = \boxed{2} \cdot \boxed{2} \cdot \boxed{3} \cdot 2$$
$$60 = 6 \cdot 10 = 2 \cdot 3 \cdot 2 \cdot 5 = \boxed{2} \cdot \boxed{2} \cdot \boxed{3} \cdot 5$$
$$108 = 12 \cdot 9 = 4 \cdot 3 \cdot 3 \cdot 3 = \boxed{2} \cdot \boxed{2} \cdot \boxed{3} \cdot 3 \cdot 3$$
$$\text{GCF} = 2 \cdot 2 \cdot 3 = 12$$

The second technique we will learn is a procedure to find the **least common multiple (LCM)** of two or more numbers.

definition

The *least common multiple*, or *LCM*, of a set of numbers is the smallest nonzero multiple common to every number in the set.

example 3

Find four common multiples of 2 and 6, and determine the LCM.

solution

Multiples of 2: 0, 2, 4, 6, 8, 10, 12, 14, 16, 18,...
Multiples of 6: 0, 6, 12, 18, 24, 30, 36,...

Four multiples common to both 2 and 6 are 0, 6, 12, and 18.
Since the LCM is the smallest nonzero common multiple, the LCM of 2 and 6 is 6.

Notice that the LCM of a set of numbers is the smallest number that is evenly divisible by every number in the set.

To Find the Least Common Multiple of a Set of Numbers

1. Completely factor each number in the set and write each factorization in exponential notation.
2. List every unique prime factor that appears in any of the factorizations.
3. Raise each of the prime factors to the highest exponent it is raised to in any of the factorizations.
4. The LCM is the product of these exponential expressions.

example 4

Find the LCM of 24 and 32.

solution

$24 = 6 \cdot 4 = 2 \cdot 3 \cdot 2 \cdot 2 = 2^3 \cdot 3$ — Factor each number completely and write in exponential notation.
$32 = 4 \cdot 8 = 2 \cdot 2 \cdot 2 \cdot 2 \cdot 2 = 2^5$

The unique prime factors that appear in the factorizations are 2 and 3.

The highest power 2 is raised to is 5: 2^5

The highest power 3 is raised to is 1: 3^1.

$\text{LCM} = 2^5 \cdot 3^1 = 32 \cdot 3 = 96$ — The LCM is the product $2^5 \cdot 3^1$.

example 5 Find the LCM of {12, 42, 56}.

solution

$12 = 3 \cdot 4 = 3 \cdot 2 \cdot 2 = 2^2 \cdot 3$

$42 = 6 \cdot 7 = 2 \cdot 3 \cdot 7$

$56 = 7 \cdot 8 = 7 \cdot 2 \cdot 2 \cdot 2 = 2^3 \cdot 7$

Factor each number completely and write it in exponential notation.

The unique prime factors appearing in the factorizations are 2, 3, and 7.

The highest exponent 2 is raised to is 3: 2^3.

The highest exponent 3 is raised to is 1: 3^1.

The highest exponent 7 is raised to is 1: 7^1.

$\text{LCM} = 2^3 \cdot 3^1 \cdot 7 = 8 \cdot 3 \cdot 7 = 168$

The LCM is the product $2^3 \cdot 3^1 \cdot 7^1$.

QUICK QUIZ

1. Find the GCF of {18, 63}.
2. Find the LCM of {16, 36}.
3. Find the GCF and LCM of {12, 24, 66}.

ANSWERS

1. GCF = 9
2. LCM = 144
3. GCF = 6; LCM = 264.

1.11 Exercises

Indicate which of the following statements are true and which are false. For those that are false, change the italic expression to make the statement true.

1. The GCF of a set of numbers is the *sum* of all prime factors common to every number in the set.
2. The LCM of a set of numbers can never equal *zero*.
3. The GCF of a set of numbers is always *greater than* the largest number in the set.
4. The LCM of a set of numbers is always *greater than* the smallest number in the set.

Find the greatest common factor (GCF) of each of the following sets of numbers.

5. {18, 30}	**6.** {42, 28}	**7.** {64, 40}	**8.** {15, 35}
9. {60, 36}	**10.** {81, 27}	**11.** {45, 80}	**12.** {54, 66}
13. {96, 36}	**14.** {48, 72}	**15.** {86, 42}	**16.** {49, 112}
17. {48, 136}	**18.** {108, 81}	**19.** {65, 169}	**20.** {144, 72}
21. {36, 24, 48}	**22.** {27, 12, 42}	**23.** {84, 28, 42}	**24.** {96, 32, 48}
25. {64, 48, 120}	**26.** {18, 42, 96}	**27.** {15, 45, 90, 75}	**28.** {24, 84, 132, 36}

Find the least common multiple (LCM) of each of the following sets of numbers.

29. {4, 6}	**30.** {3, 9}	**31.** {10, 15}	**32.** {12, 16}
33. {20, 8}	**34.** {24, 18}	**35.** {36, 27}	**36.** {15, 35}
37. {12, 32}	**38.** {48, 21}	**39.** {56, 24}	**40.** {30, 55}
41. {8, 4, 12}	**42.** {6, 9, 15}	**43.** {10, 8, 25}	**44.** {24, 16, 12}
45. {21, 18, 14}	**46.** {36, 9, 12}	**47.** {20, 36, 48}	**48.** {27, 42, 54}
49. {60, 49, 84}	**50.** {24, 42, 96}	**51.** {12, 18, 24, 32}	**52.** {15, 45, 60, 75}

Find the GCF and LCM of each of the following sets of numbers.

53. {21, 36}
54. {18, 54}
55. {6, 30, 54}
56. {9, 21, 36}
57. {14, 42, 49}
58. {18, 45, 63}
59. {24, 84, 108}
60. {32, 48, 80}
61. {16, 28, 36, 48}
62. {12, 18, 21, 39}

Summary and Review

Key Terms

[1.1] The set of **whole numbers** includes $\{0, 1, 2, 3, 4, \ldots\}$.

Our number system is called the **base ten system** since it includes ten digits: 0, 1, 2, 3, 4, 5, 6, 7, 8, 9.

The **place value** of a digit is the name of the location of that digit.

A **period** is the name given to every group of three digits in a whole number moving from right to left.

[1.2] An **equation** is a mathematical expression that states that two quantities are equal.

Addition is expressed by the equation $a + b = c$, where a and b are the **addends** and c is the **sum**.

Zero is the **identity element of addition**, since for any number a, $a + 0 = a$ and $0 + a = a$.

According to the **commutative property of addition**, we can change the **order** of the addends without affecting the result.

According to the **associative property of addition**, we can change the **grouping** of the addends without affecting the result.

[1.3] **Subtraction** is expressed by the equation $a - b = c$, where a is the **minuend**, b is the **subtrahend**, and c is the **difference**.

Addition and subtraction are **inverse operations**.

[1.4] **Multiplication** is repeated addition. Multiplication is expressed by the equation $a \cdot b = c$, where a is the **multiplier**, b is the **multiplicand**, and c is the **product**. a and b are also called **factors** of the product c.

One is the **identity element of multiplication** since for any number a, $a \cdot 1 = a$ and $1 \cdot a = a$.

According to the **commutative property of multiplication**, we can change the **order** of the factors without affecting the result.

According to the **associative property of multiplication**, we can change the **grouping** of the factors without affecting the result.

The **distributive property of multiplication over addition** states that for any numbers a, b, and c, $a(b + c) = a(b) + a(c)$.

[1.5] **Division** is repeated subtraction. Division is expressed by the equation $a \div b = c$, where a is the **dividend**, b is the **divisor**, and c is the **quotient**.

The **remainder** is a whole number less than the divisor that is left over after the process of repeated subtraction has been completed.

Division by zero is **undefined** since it is impossible to divide a number by zero.

Multiplication and division are **inverse** operations.

[1.6] **Exponents** express repeated multiplication.

The exponential expression b^n means that b appears as a factor n times. b is the **base**, and n is the **exponent**, or **power**.

For any base b, $b^1 = b$. If $b \neq 0$, $b^0 = 1$.

Powers of ten represent place values in the base ten system.

[1.9] A number is **divisible** by a given divisor if the division process yields a whole number quotient and a remainder of zero.

[1.10] A **prime number** is a whole number greater than 1, that is divisible only by 1 and itself.

A **composite number** is a whole number greater than 1 that is not prime.

A **multiple** of a given whole number is the product of that number and another whole number.

The **complete**, or **prime, factorization** of a number is the unique representation of that number as the product of its prime factors.

[1.11] The **greatest common factor**, or **GCF**, of a set of numbers is the largest number by which every number in the set is evenly divisible.

The **least common multiple**, or **LCM**, of a set of numbers is the smallest nonzero multiple common to every number in the set.

Calculations

[1.1] **To write a whole number in expanded notation**, each digit is multiplied by its corresponding place value, and the results are then added together.

[1.2] **To add whole numbers** when the sum of any column is greater than nine, put down the ones digit of the sum and *carry* the other digit to the next column immediately to the left.

[1.3] **To subtract whole numbers** when a digit in the minuend is less than the digit that has the same place value in the subtrahend, rewrite the minuend by *borrowing* 1 from the digit immediately to the left of the smaller digit, and add 10 to the smaller digit.

[1.4] **To multiply whole numbers** that have more than one digit, multiply each digit in the multiplicand by each digit in the multiplier and add the products.

[1.5] **To divide whole numbers**, determine the largest whole number that can be multiplied by the divisor to obtain a product less than or equal to the dividend. Subtract that product from the dividend. Repeat the process until the result of the subtraction is less than the divisor.

To check the answer to a division problem, multiply the quotient by the divisor and add the remainder. The result should equal the dividend.

[1.6] **To multiply a number by 10^n**, where n is a whole number, write the number followed by n zeros.

To multiply exponential expressions of the same base, keep the base the same and add the exponents.

To raise an exponential expression to a power, keep the base the same and multiply the exponents.

[1.8] **To calculate the average** of a set of numbers, first add all the numbers in the set, and then divide that sum by the number of numbers that were added together.

[1.10] **To find all the multiples of a given whole number**, multiply that number by every whole number.

[1.11] **To find the greatest common factor (GCF)** of a set of numbers, completely factor each number and calculate the product of all prime factors common to every number in the set.

To find the least common multiple (LCM) of a set of numbers, completely factor each number and write each factorization in exponential notation. Calculate the product of all unique prime factors, each of which is raised to the highest exponent that appears in any one of the factorizations.

Order of Operations Rules

[1.7] 1. Perform all operations in parentheses first.
2. Simplify all exponential expressions.
3. Do multiplication and division as they occur, moving from left to right.
4. Do addition and subtraction as they occur, moving from left to right.

Tests for Divisibility

[1.9] **A number is divisible by 2** if its ones digit is 0, 2, 4, 6, or 8.
A number is divisible by 3 if the sum of its digits is divisible by 3.
A number is divisible by 4 if the last two digits make a number divisible by 4.
A number is divisible by 5 if its ones digit is 0 or 5.
A number is divisible by 6 if it is divisible by both 2 and 3.
A number is divisible by 8 if the last three digits form a number divisible by 8.
A number is divisible by 9 if the sum of its digits is divisible by 9.
A number is divisible by 10 if its ones digit is 0.

Chapter 1 Review Exercises

Indicate which of the following statements are true and which are false. For those that are false, change the italic expression to make the statement true.

1. The associative property *does* hold true for division.
2. The number 1 is the identity element for *addition.*
3. The commutative property does not hold true for subtraction and *multiplication.*
4. To raise an exponential expression to a power, keep the base the same and *multiply* the exponents.
5. For a given set of unique numbers, the GCF is always *greater than* the LCM.
6. Unless otherwise indicated by parentheses, we do subtraction *before* division.

[1.1] Write each of the following numbers in words and in expanded notation.

7. 907 8. 4,892 9. 13,069
10. 890,036 11. 4,079,581 12. 3,000,000,000,001

[1.1] Compare each of the following pairs of numbers represented on the number line. Use the $=$, $>$, or $<$ sign to indicate the relationship between them.

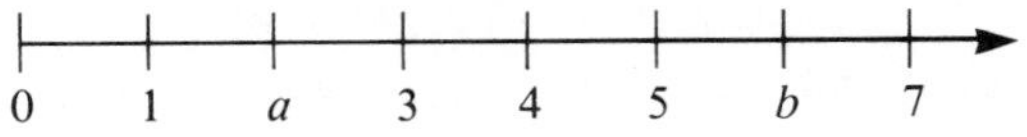

13. 4 ? 0 14. *a* ? 2 15. 6 ? *b* 16. 5 ? *b* 17. *a* ? 0 18. *a* ? *b*

[1.2, 1.4] State which property, the commutative or associative property of addition, the commutative or associative property of multiplication, or the distributive property, is illustrated by each of the following expressions.

19. $3 \cdot 9 = 9 \cdot 3$
20. $8 + (3 + 7) = 8 + (7 + 3)$
21. $3(2 + 6) = 3(2) + 3(6)$
22. $(3 \cdot 6) \cdot 2 = 3 \cdot (6 \cdot 2)$
23. $8 \cdot (5 \cdot 4) = 8 \cdot (4 \cdot 5)$
24. $(5 + 2)(8 + 3) = 5(8 + 3) + 2(8 + 3)$
25. $6(9 - 2) = 6(9) - 6(2)$
26. $(3 + 4) + 7 = 3 + (4 + 7)$
27. $(5 \cdot 3) \cdot 9 = (3 \cdot 5) \cdot 9$
28. $4(13) - 4(5) = 4(13 - 5)$
29. $8 + 4(5) = 8 + 5(4)$
30. $(7 + 3) \cdot 2 = 2 \cdot (7 + 3)$

[1.2–1.5] Perform each of the operations indicated.

31. $897 + 362$
32. $705 - 29$
33. $47 \cdot 62$
34. $149 \cdot 16$
35. $169 \div 5$
36. $1{,}507 + 838$
37. $300 - 53$
38. $96 \div 7$
39. $18 + 294 + 39$
40. $140 + 6{,}073 + 89$
41. $1{,}734 \div 11$
42. $8{,}369 \div 24$
43. $234 \cdot 407$
44. $678 \cdot 44$
45. $14{,}003 - 7{,}945$
46. $53{,}861 + 92{,}749$
47. $1{,}836 \cdot 493$
48. $83{,}492 \div 687$
49. $1{,}000{,}000 - 473{,}817$
50. $5{,}003 \cdot 7{,}020$
51. $62{,}859{,}310 - 38{,}672{,}197$
52. $84{,}824 + 62{,}527 + 41{,}689 + 57{,}730$
53. $24{,}862 \cdot 1{,}897$
54. $51{,}228{,}684 \div 6{,}987$

[1.2–1.5] Translate each of the following statements into mathematical expressions.

55. 2 less than 5.
56. w more than 3.
57. The quotient of 33 and 3.
58. The difference between 15 and 6.
59. The product of 2 and 8 is 16.
60. The total of x, y, and z.

[1.6] Simplify each of the following expressions using the rules for exponents.

61. $5^7 \cdot 5^2$
62. $4^4 \cdot 4^9$
63. $2^5 \cdot 2^0 \cdot 2^3$
64. $8^7 \cdot 8^3 \cdot 8^1$
65. $(7^4)^5$
66. $(9^2)^6$
67. $(2^3)^4 \cdot (2^5)^3$
68. $(3^7)^2 \cdot (3^2)^5$

[1.7] Perform each of the following operations indicated, and express all answers as whole numbers.

69. $6^2 + 2^3$
70. $3^3 - 4^2$
71. $2^3 \cdot 3^4$
72. $4^2 \cdot 7^0$
73. $(9 - 3)^2$
74. $(7 + 3)^5$
75. $4^2 \cdot (3^0)^2$
76. $(2^2)^3 \cdot 5^2$
77. $6 \cdot 3 - 15 \div 5$
78. $12 \div 2 + 4 \cdot 6$
79. $(8 + 1)^2 \div 3^3$
80. $2^5 \div (6 - 2)^2$
81. $72 \div 3^2 + 4$
82. $64 \div 2^3 - 5$
83. $4(6 - 3) + 3(9 - 7)^2$
84. $2(1 + 4)^2 - 5(2 + 6)$
85. $5 + 7 \cdot 3 - 12 \div 4$
86. $2 \cdot 3 + 4 - 9 \div 3$
87. $(8 \div 2^2 \cdot 6 + 7 - 1) \div 6$
88. $48 \div (4 + 6 \cdot 3 - 4^2)$
89. $3[(8 - 3)^2 + 5(11 - 7)]$
90. $[6(9 - 4) + (2 + 3)^3] \div 5$
91. $36{,}801 \div 423 + (614)(96)$
92. $[(872)(128) + 1196] \div 476$

[1.9] Use the divisibility tests to determine whether or not each of the following numbers is divisible by 2, 3, 4, 5, 6, 8, 9, and 10.

93. 204 **94.** 545 **95.** 2,137 **96.** 4,698 **97.** 71,265 **98.** 43,056

[1.10] State whether each of the following numbers is prime or composite.

99. 65 **100.** 29 **101.** 47 **102.** 93 **103.** 115 **104.** 201

[1.10] Completely factor each of the following numbers.

105. 68 **106.** 96 **107.** 120 **108.** 164 **109.** 1,500 **110.** 4,000

[1.11] Find the greatest common factor (GCF) and least common multiple (LCM) of each of the following sets of numbers.

111. $\{32, 56\}$
112. $\{81, 21\}$
113. $\{144, 54\}$
114. $\{72, 104\}$
115. $\{12, 32, 48\}$
116. $\{54, 63, 36\}$
117. $\{14, 21, 42, 56\}$
118. $\{24, 32, 36, 48\}$

[1.7] Translate each of the following expressions into mathematical expressions and simplify.

119. 8 less than the product of 6 and 9.
120. 6 times the sum of 4 and 9.
121. The sum of the product of 3 and 9 and the quotient of 20 and 4.
122. The square of the difference between 17 and 3.
123. 3 less than the quotient of 63 and 7.
124. 8 less than the product of 3 and 9.

Solve each of the following word problems.

125. In 1978, the United States admitted 6,739 immigrants from Germany, 12,076 from England, 21,315 from China, and 92,367 from Mexico. What was the total number of immigrants admitted from these countries in 1978?

126. At the beginning of the month, a lawyer's checking account had a balance of $418. During the month she wrote checks for $132, $173, $29, $156, $87, $43, and $62, and she also made deposits of $250 and $367. What was her balance at the end of the month?

127. A couple made a down payment of $90 on a refrigerator that cost $495, and they agreed to pay off the rest in 90 days to avoid interest charges. What will be their monthly payment?

128. During the time in which a motorist used 17 gallons of gas, her odometer reading changed from 13,979 miles to 14,438 miles. How many miles per gallon did she get?

129. The distance from New York to San Francisco is 2,571 miles, from San Francisco to Honolulu is 2,393 miles, from Honolulu to Tokyo is 3,853 miles, from Tokyo to Hong Kong is 1,794 miles, from Hong Kong to Calcutta is 1,648 miles, and from Calcutta to Moscow is 3,321 miles. What is the total distance traveled by a plane that flies this route?

130. A teacher's yearly gross salary is $12,000. Every month the following amounts are deducted from his paycheck: $177 for federal taxes, $68 for social security, $42 for state taxes, $39 for retirement fund, and $19 for health insurance. What is his monthly take-home pay?

131. If the population of London was 2,043,000 more than that of Bangkok in 1978, and the population of Bangkok was 4,875,000, what was the combined population of Bangkok and London in 1978?

132. During her freshman year, a student completed six four-credit courses, three two-credit courses, and two one-credit courses. How many credits did she earn that year?

133. During a football game, a team scored three touchdowns, four field goals, and two extra points. What was their final score? (A touchdown is worth six points, and a field goal is worth three points.)

134. During a certain semester at a small college, 258 students took general chemistry, and twice the number of students who took chemistry took calculus. Three times the number of students who took calculus took English composition. What was the total number of students enrolled in these three courses for that semester?

135. A woman is deciding whether to buy a car that costs $6,972 or another that costs $6,144. She plans to make a down payment of $1,500 and pay off the balance over a three-year period. How much more would her monthly payments be (ignoring interest) if she bought the more expensive car?

136. How many seconds are there in a week?

137. A student obtained the scores of 61, 78, 83, 65, and 73 on five examinations. What was his average grade?

138. A basketball player scored the following points during a six-game series: 18, 29, 24, 20, 31, and 22. What was her scoring average for that series?

139. In 1970, the state of New Hampshire had a population of 738,000. If New Hampshire covers an area of roughly 9,000 square miles, what was its average population density per square mile in 1970?

140. In 1981, the annual salaries of seven state governors were Massachusetts, $60,000; New York, $85,000; Ohio, $50,000; California, $49,000; Alaska, $70,000; New Hampshire, $45,000; and Tennessee, $68,000. What was the average salary?

141. In 1979, 47,842,375 passengers flew into or out of O'Hare International Airport in Chicago. What was the average number of passengers served daily by the airport that year?

142. The diameters of the nine planets of our solar system are Mercury, 3,100 miles; Venus, 7,700 miles; Earth, 7,950 miles; Mars, 4,200 miles; Jupiter, 88,700 miles; Saturn, 75,100 miles; Uranus, 3,200 miles; Neptune, 27,700 miles; and Pluto, 1,500 miles. What is the average diameter of a planet in our solar system?

2 FRACTIONS

2.1 Definitions

This chapter will deal with numbers called **fractions**. We will begin our discussion by introducing four different ways in which we use fractions to represent ideas that cannot be adequately conveyed by using only whole numbers.

In the first example, shown in Figure 2.1, a circle is divided into four equal pieces. If we want to focus our attention on only one of those pieces, we can shade one of the four equal pieces of the circle. This one piece can be represented mathematically by the fraction $\frac{1}{4}$, which is read, "one-fourth." The fraction $\frac{1}{4}$ means that we are interested in one piece of a whole circle that consists of four equal pieces. Thus, one use of fractions is to represent the parts of a whole.

figure 2.1

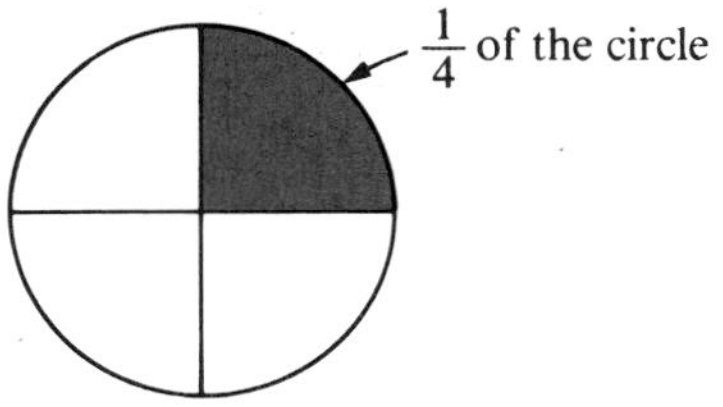

As a second example, let us suppose that there are three boxes lined up in front of us, as shown in Figure 2.2, and that a ball has been put inside one of them without our knowledge. If we are asked to guess which box contains the ball, our chances are one out of three that we will guess correctly. Expressed as a fraction, our chances are represented by $\frac{1}{3}$, or one-third. If two balls are placed inside two different boxes, and we are asked to pick a box containing a ball, our chances of picking correctly are two out of three, or $\frac{2}{3}$ (two-thirds). Thus, a fraction also can mean the number of correct choices out of the total number of choices available.

figure 2.2
"Which box contains the ball? Our chances are one out of three that we guess correctly."

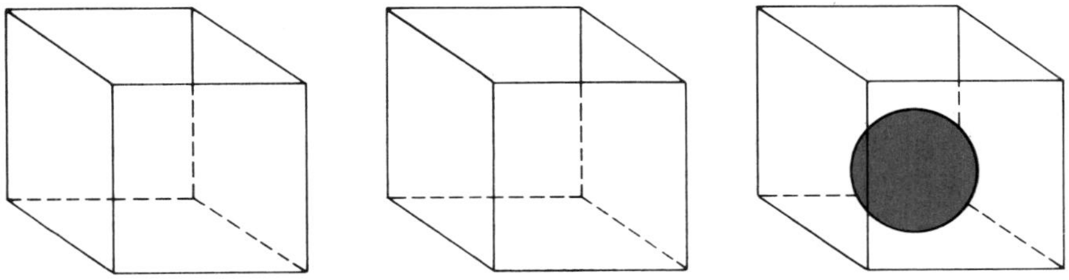

As a third example, let us suppose that the distance between your home and your place of employment is exactly 1 mile. To illustrate this, let us use a number line labeled from 0 (home) to 1 (work), as shown in Figure 2.3. In order to pinpoint any location between work and home, we can use fractions to represent the distance away from your home. Suppose your car breaks down one morning on your way to work at the place marked with an X in Figure 2.3(a). If we divide the mile into two equal segments, we can estimate your location as approximately $\frac{1}{2}$ mile away from home. If we divide the mile into three equal parts, as in Figure 2.3(b), we obtain a better approximation of $\frac{1}{3}$ mile away. If we divide the mile into four equal parts, as in Figure 2.3(c), we see that the location of your car is between $\frac{1}{4}$ and $\frac{2}{4}$ mile away. Figure 2.3(d) illustrates that the closest approximation is $\frac{2}{5}$ mile away from home. Notice that by dividing the mile into a greater number of parts, we are able to indicate locations with greater accuracy.

Finally, fractions can also be used simply as another way to represent division. For example, the fraction $\frac{2}{3}$ means $2 \div 3$. We can write $20 \div 5$ as $\frac{20}{5}$, which in simplest form is the number 4.

Now let us apply our understanding of fractions to some problems.

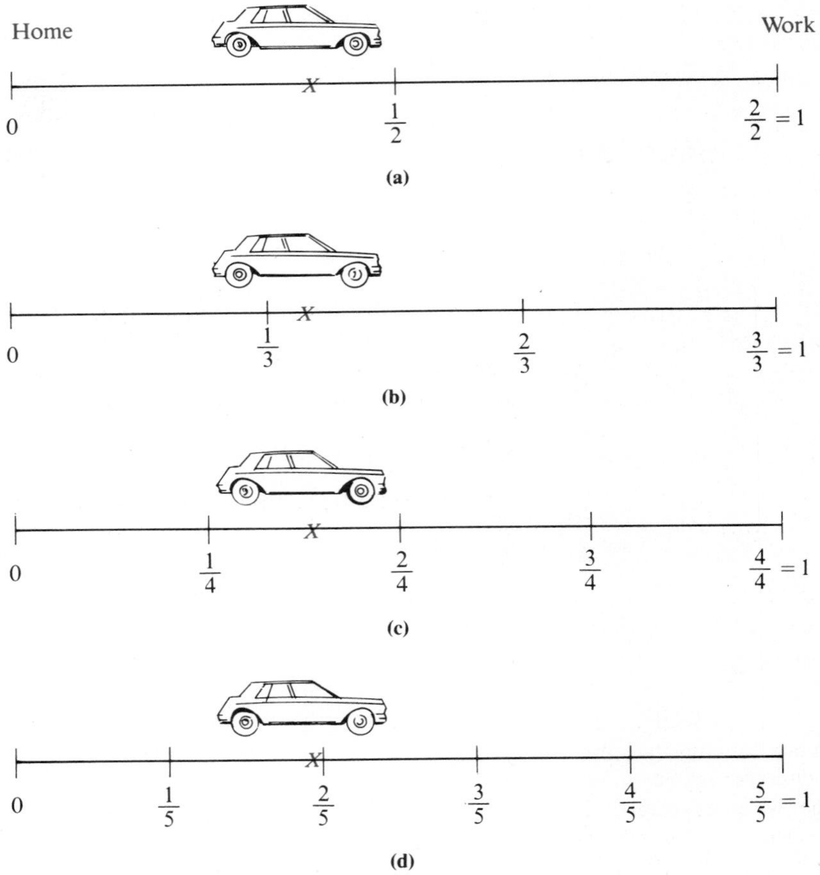

figure 2.3

example 1

On a coat that is supposed to have eight buttons, three buttons are missing. What fraction of the buttons are missing?

solution

Since three out of eight buttons are missing, the fraction that represents the missing buttons is $\frac{3}{8}$. ●

example 2

A kennel has nine dogs and five cats. What fraction of the population consists of cats?

solution

A fraction represents the number of parts of the whole. We must determine the size of the whole population of the kennel, which is $9 + 5 = 14$ animals. Thus, the fraction of cats is 5 out of 14, or $\frac{5}{14}$. ●

example 3

If a single card is drawn at random from a regular deck of 52 cards, what are the chances that the card is a king?

solution

There are four kings in a deck of 52 cards. Thus, there are four desirable choices out of a total of 52 possible choices. The chances that the card drawn is a king are, therefore, 4 out of 52, or $\frac{4}{52}$. ●

Now that we are familiar with some of the uses of fractions, it is time to learn some formal vocabulary and notation.

definition

A *fraction* is any number that can be written in the form $\frac{a}{b}$, where a and b are numbers such that b is not equal to zero. Since $\frac{a}{b}$ means $a \div b$, b cannot be zero since division by zero is undefined. The *terms* of the fraction are a and b. The term a is called the *numerator* and b is called the *denominator*.

For example, in the fraction $\frac{2}{7}$, 2 and 7 are the **terms** of the fraction. The **numerator** is 2, and the **denominator** is 7.

$$\frac{2}{7}\begin{array}{l}\longleftarrow \text{numerator}\\ \longleftarrow \text{denominator}\end{array}$$

definition

A *proper fraction* is one in which the numerator is less than the denominator.

Examples of proper fractions are $\frac{1}{5}$, $\frac{3}{41}$, and $\frac{19}{120}$.

definition

An *improper fraction* is one in which the numerator is greater than or equal to the denominator.

Examples of improper fractions are $\frac{7}{3}$, $\frac{99}{99}$, and $\frac{287}{103}$.

Characteristics of Fractions

We will now investigate some of the characteristics of fractions by looking at the division problems represented by various fractions. Consider the following examples.

$$\frac{5}{1} = 5 \div 1 = 5$$

$$\frac{33}{1} = 33 \div 1 = 33$$

$$\frac{168}{1} = 168 \div 1 = 168$$

We have illustrated the following.

Any fraction whose denominator is 1 is equal to its numerator. Likewise, any whole number can be expressed as a fraction whose denominator is 1.

For example, $7 = \frac{7}{1}$, $22 = \frac{22}{1}$, and $300 = \frac{300}{1}$.

Let us now consider these examples:

$$\frac{8}{8} = 8 \div 8 = 1$$

$$\frac{19}{19} = 19 \div 19 = 1$$

$$\frac{153}{153} = 153 \div 153 = 1$$

We have shown the following.

Any fraction with a nonzero denominator that is equal to its numerator is equal to 1.

Observe what happens when the numerator is zero:

$$\frac{0}{6} = 0 \div 6 = 0$$

$$\frac{0}{13} = 0 \div 13 = 0$$

$$\frac{0}{257} = 0 \div 257 = 0$$

These examples show the following.

> Any fraction with a numerator of zero and a nonzero denominator is equal to zero.

Recall from Section 1.5 that division by zero is undefined. It is important to keep that fact in mind when working with fractions since fractions represent division. Therefore, $\frac{9}{0}$, $\frac{31}{0}$, $\frac{0}{0}$, and $\frac{405}{0}$ are all examples of fractions (divisions) that are undefined.

> Since division by zero is undefined, any fraction with a denominator of zero is undefined.

Mixed Numbers

Improper fractions represent numbers that are greater than or equal to one, since their numerators are greater than or equal to their denominators. Those improper fractions that represent numbers greater than one also can be expressed as **mixed numbers**.

definition

A *mixed number* is the sum of a whole number and a proper fraction.

We can gain some understanding of mixed numbers by looking at the names of various locations on the line segment shown in Figure 2.4. In this drawing, a line segment between the values of 0 and 3 is divided into 12 equal pieces. The improper fraction equivalent to each mixed number is listed below the mixed number.

figure *2.4*

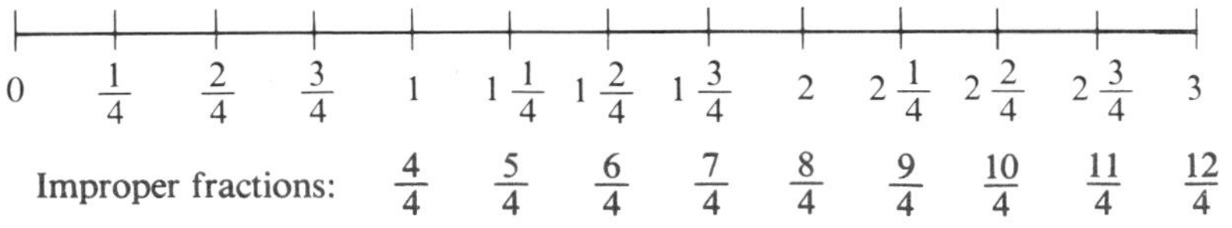

A procedure for changing improper fractions to mixed numbers is shown in the box.

To Convert an Improper Fraction to a Mixed Number

1. Divide the numerator by the denominator.
2. The quotient is the whole number portion of the mixed number.
3. The remainder is the numerator of the fractional portion of the mixed number. The denominator of the fractional portion is the same as the denominator of the original improper fraction.
4. The mixed number is the sum of the whole number found in step 2 and the fraction found in step 3.

example 4

Convert $\frac{83}{4}$ to a mixed number.

solution

$\frac{83}{4} = 83 \div 4 = 20 \text{ R}3$

Divide 83 by 4:

$$\begin{array}{r} 20 \text{ R}3 \\ 4\overline{)83} \\ \underline{80} \\ 3 \end{array}$$

The whole number portion is the quotient, 20.

The fractional portion is the remainder over the divisor, $\frac{3}{4}$.

The mixed number is $20 + \frac{3}{4} = 20\frac{3}{4}$.

Even though the addition sign between the whole number and fractional portion is omitted, it is understood to be there.

Therefore, $\frac{83}{4} = 20\frac{3}{4}$

$20\frac{3}{4}$ is read, "twenty and three-fourths." ●

example 5

Convert each of the following improper fractions to mixed numbers.

(a) $\frac{123}{7}$

$123 \div 7 = 17 \text{ R}4$

Divide 123 by 7:

$$\begin{array}{r} 1\ 7 \text{ R}4 \\ 7\overline{)1\ 2^5 3} \end{array}$$

The whole number portion is the quotient, 17. The fractional portion is $\frac{4}{7}$.

Thus, $\frac{123}{7} = 17 + \frac{4}{7} = 17\frac{4}{7}$

(b) $\frac{1{,}024}{11}$

Divide 1,024 by 11:

$$\begin{array}{r} 93 \text{ R}1 \\ 11\overline{)1{,}024} \\ \underline{99} \\ 34 \\ \underline{33} \\ 1 \end{array}$$

$1{,}024 \div 11 = 93 \text{ R}1$

The whole number portion is 93. The fractional portion is $\frac{1}{11}$.

Thus, $\frac{1{,}024}{11} = 93 + \frac{1}{11} = 93\frac{1}{11}$ ●

We can think of the problem of changing an improper fraction to a mixed number as a division problem. The numerator is the dividend, and the denominator is the divisor. The mixed number equivalent to the improper fraction is the answer to the division problem. The mixed number is expressed as the sum of the quotient, a whole number, and a fraction whose numerator is the remainder and whose denominator is the divisor.

To convert a mixed number to an improper fraction, we reverse the process that we have just learned. That is, we find the dividend of a division problem, knowing the quotient, remainder, and divisor. The divisor is the denominator of the fractional part of the mixed number, and it is also the denominator of the improper fraction. The numerator of the improper fraction is the dividend, which can be calculated in the same way we check a division problem. Multiply the whole number (quotient) by the denominator (divisor) and add the numerator (remainder). This procedure is outlined in the box.

To Convert a Mixed Number to an Improper Fraction

1. Multiply the whole number part of the mixed number by the denominator of the fractional part.
2. Add the numerator of the fractional part to the result of step 1.
3. The numerator of the improper fraction is the result of step 2. The denominator of the improper fraction is the same as the denominator of the fractional part of the mixed number.

example 6

Convert $8\frac{2}{5}$ to an improper fraction.

solution

$8 \cdot 5 = 40$ Multiply the whole number by the denominator of the fractional part.

$40 + 2 = 42$ Add the numerator of the fractional part to the product of the whole number and the denominator.

$8\frac{2}{5} = \frac{42}{5}$ The numerator of the improper fraction is 42.

The denominator of the improper fraction is 5. ●

Notice that the process used to find the numerator of an improper fraction is the same as that used to check the division problem represented by an improper fraction. For example,

$$\frac{42}{5} = 42 \div 5 = 8 \text{ R}2$$

$$\begin{array}{r} 8 \text{ R}2 \\ 5\overline{)42} \\ \underline{40} \\ 2 \end{array}$$

CHECK

$$\begin{array}{rl} 8 & \text{quotient} \\ \times\ 5 & \text{divisor} \\ \hline 40 & \\ +\ 2 & \text{remainder} \\ \hline 42 & \text{dividend} \checkmark \end{array}$$

example 7

Convert each of the following mixed numbers to improper fractions.

(a) $13\frac{7}{10}$

$13 \cdot 10 = 130$ Multiply whole number by denominator.

$130 + 7 = 137$ Add product and numerator.

$13\frac{7}{10} = \frac{137}{10}$ The numerator of the improper fraction is 137.

The denominator of the improper fraction is 10.

We can also solve this problem by setting up our work as follows:

$$13\frac{7}{10} = \frac{(13 \cdot 10) + 7}{10} = \frac{130 + 7}{10} = \frac{137}{10}$$

(b) $53\frac{2}{11}$

$$53\frac{2}{11} = \frac{(53 \cdot 11) + 2}{11} = \frac{583 + 2}{11} = \frac{585}{11}$$

example 8

A train trip from Boston to Philadelphia takes 7 hours 13 minutes. Express this time in hours as a mixed number and as an improper fraction.

solution

$$7 \text{ hr } 13 \text{ min} = 7 \text{ hr} + \frac{13}{60} \text{ hr}$$ Since 60 min = 1 hr.

$$= 7\frac{13}{60} \text{ hr}$$

$$7\frac{13}{60} = \frac{(7 \cdot 60) + 13}{60} = \frac{420 + 13}{60} = \frac{433}{60}$$ Now convert $7\frac{13}{60}$ to an improper fraction.

$$7 \text{ hr } 13 \text{ min} = 7\frac{13}{60} \text{ hr} = \frac{433}{60} \text{ hr}$$

QUICK QUIZ

1. Express $3 \div 8$ as a fraction.
2. Change $\frac{47}{5}$ to a mixed number.
3. Change $6\frac{3}{8}$ to an improper fraction.

ANSWERS

1. $\frac{3}{8}$
2. $9\frac{1}{5}$
3. $\frac{51}{8}$

2.1 Exercises

Indicate which of the following statements are true and which are false. For those that are false, change the italic expression to make the statement true.

1. In the fraction $\frac{7}{8}$, 7 is the *numerator*.
2. The fraction $\frac{12}{12}$ is equal to *12*.
3. The fraction $\frac{5}{0}$ is *undefined*.

4. The number $2\frac{3}{4}$ is a *mixed number.*

5. The fraction $\frac{111}{6}$ is called an *invalid* fraction.

6. A proper fraction represents a number *less than* one.

7. The fraction $\frac{5}{20}$ means $20 \div 5$.

8. The *numerator* of a fraction can never be zero.

9. The fraction $\frac{9}{1}$ is equal to 9.

10. In the mixed number $7\frac{3}{5}$, the number 3 represents the *quotient* of a division problem.

Use a fraction to represent the parts of a whole, division, or the chances of a favorable outcome.

11. The shaded portion of this circle.

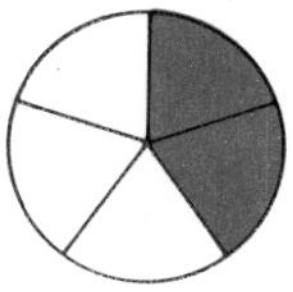

12. The shaded portion of this box.

13. $39 \div 8$ **14.** $3 \div 7$ **15.** $1 \div 11$ **16.** $28 \div 1$ **17.** 6 **18.** 25

19. The chance of picking a spade at random from a regular deck of 52 cards.

20. The chance of obtaining a head when flipping a fair coin.

Identify each of the following fractions as proper or improper.

21. $\frac{8}{2}$ **22.** $\frac{37}{89}$ **23.** $\frac{3}{3}$ **24.** $\frac{5}{1}$ **25.** $\frac{3}{200}$

26. $\frac{0}{50}$ **27.** $\frac{52}{7}$ **28.** $\frac{63}{63}$ **29.** $\frac{300}{299}$ **30.** $\frac{645}{700}$

Rewrite each of the following fractions as a division problem and simplify if possible.

31. $\frac{54}{6}$ **32.** $\frac{8}{1}$ **33.** $\frac{0}{6}$ **34.** $\frac{39}{3}$ **35.** $\frac{21}{7}$ **36.** $\frac{4}{0}$

Use the following diagram, in which a line segment from 0 to 4 is divided into 12 equal pieces, to determine the answers to the following problems.

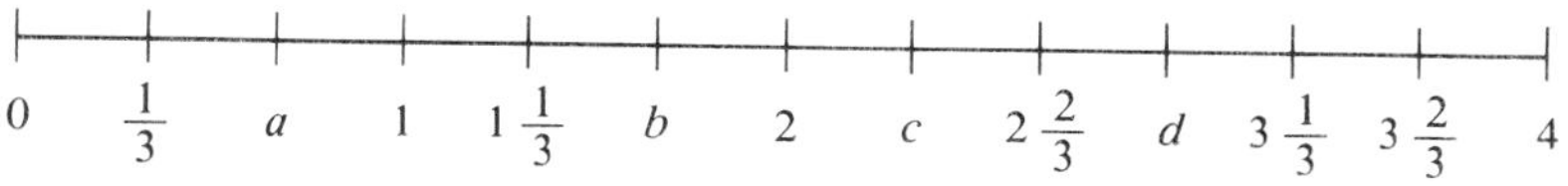

37. Express location a as a fraction.

38. Express location d as a whole number.

39. Express location b as a mixed number and as an improper fraction.

40. Express location c as a mixed number and as an improper fraction.

Convert the following improper fractions to mixed numbers.

41. $\frac{7}{4}$ **42.** $\frac{9}{2}$ **43.** $\frac{19}{3}$ **44.** $\frac{23}{6}$ **45.** $\frac{41}{5}$ **46.** $\frac{52}{7}$

47. $\frac{82}{3}$ **48.** $\frac{71}{2}$ **49.** $\frac{101}{7}$ **50.** $\frac{181}{9}$ **51.** $\frac{320}{9}$ **52.** $\frac{213}{5}$

53. $\frac{543}{7}$ **54.** $\frac{905}{8}$ **55.** $\frac{115}{12}$ **56.** $\frac{200}{11}$ **57.** $\frac{853}{21}$ **58.** $\frac{687}{31}$

Convert the following mixed numbers to improper fractions.

59. $3\frac{1}{5}$ **60.** $6\frac{2}{3}$ **61.** $8\frac{2}{9}$ **62.** $4\frac{5}{8}$ **63.** $4\frac{3}{7}$ **64.** $6\frac{1}{4}$

65. $10\frac{5}{8}$ **66.** $12\frac{2}{5}$ **67.** $12\frac{3}{8}$ **68.** $20\frac{7}{9}$ **69.** $21\frac{5}{6}$ **70.** $32\frac{4}{5}$

71. $7\frac{11}{20}$ **72.** $9\frac{2}{15}$ **73.** $35\frac{9}{10}$ **74.** $67\frac{4}{10}$ **75.** $52\frac{4}{13}$ **76.** $78\frac{5}{11}$

77. $1\frac{100}{101}$ **78.** $2\frac{300}{301}$ **79.** $69\frac{73}{94}$ **80.** $82\frac{51}{135}$ **81.** $137\frac{17}{23}$ **82.** $389\frac{22}{41}$

Answer each of the following questions with an appropriate fraction.

83. In a bus with 32 windows, three are broken. What fraction of the windows are broken?

84. In a class of 27 students, five failed their first test. What fraction of the students failed?

85. In a regular deck of 52 cards, what fraction of the cards are hearts?

86. If you roll a single die, what is your chance of rolling a 2?

87. In a pack of soda containing five root beers and seven colas, what fraction of the bottles are colas?

88. A bag contains eight hard candies and 13 chocolate candies. What fraction of the candies are chocolate?

89. If there are 11 adults in a room and seven of them are men, what fraction of the people are men? What fraction are women?

90. A fruit basket contains two bananas, four apples, six oranges, and three pears. What portion of the fruit consists of apples?

91. A runner finished a race in 7 minutes 21 seconds. Express this time in minutes as a mixed number and as an improper fraction.

92. A woman is 5 feet 7 inches tall. Express her height in feet as a mixed number and as an improper fraction.

93. A bag of peanuts weighs 1 pound 5 ounces. Express this weight in pounds as a mixed number and as an improper fraction.

94. It is estimated that American men between the ages of 25 and 54 watch 28 hours 46 minutes of television on the average per week. Express this amount of time in hours as a mixed number and as an improper fraction.

2.2 Reducing to Lowest Terms

Equivalent Fractions

There is often more than one fraction that can be used to represent the same value. For example,

$$\frac{8}{8} = \frac{3}{3} = \frac{17}{17} = \frac{100}{100} = 1$$

definition

Fractions that have the same numerical value are called *equivalent fractions.*

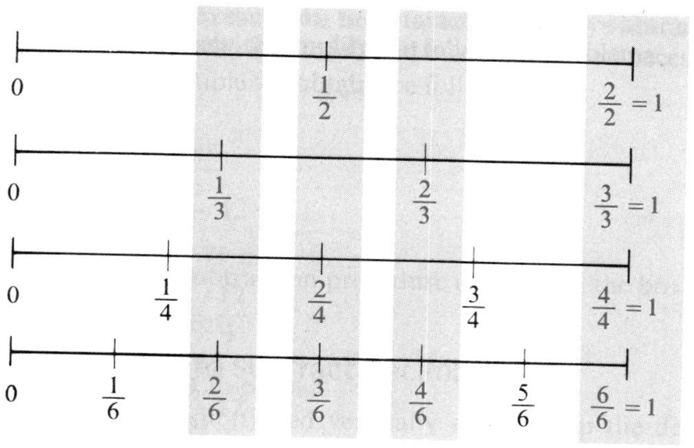

figure 2.5

Figure 2.5 illustrates more examples of equivalent fractions. Four line segments of length one are divided into two, three, four, and six pieces, respectively. From this diagram we can see that the fractions $\frac{1}{2}$, $\frac{2}{4}$, and $\frac{3}{6}$ all represent the same distance. These are, therefore, equivalent fractions. We represent this by writing

$$\frac{1}{2} = \frac{2}{4} = \frac{3}{6}$$

We can also see from Figure 2.5 that

$$\frac{1}{3} = \frac{2}{6}, \quad \frac{2}{3} = \frac{4}{6}, \quad \text{and} \quad \frac{2}{2} = \frac{3}{3} = \frac{4}{4} = \frac{6}{6} = 1$$

We will now look at some examples that show us how to find equivalent fractions.

> To find an equivalent fraction of one that is given, multiply or divide the numerator and denominator of the given fraction by the same nonzero number.

example 1 Find an equivalent fraction of $\frac{3}{5}$ with a denominator of 10.

solution

$$\frac{3}{5} = \frac{?}{10}$$

To obtain a denominator of 10, we had to multiply 5 by 2.

$$\frac{3 \cdot 2}{5 \cdot 2} = \frac{6}{10}$$

Therefore, we must multiply the numerator 3 by 2. ●

Notice that since $\frac{2}{2} = 1$, multiplying both numerator and denominator by 2 in Example 1 is the same as multiplying the original fraction by 1. That is, $\frac{3}{5} \cdot \frac{2}{2} = \frac{3}{5} \cdot 1$. Hence, $\frac{6}{10}$ has the same value as $\frac{3}{5}$.

example 2

Find an equivalent fraction of $\frac{15}{20}$ with a numerator of 3.

solution $\frac{15}{20} = \frac{3}{?}$

$\frac{15 \div 5}{20 \div 5} = \frac{3}{4}$ Divide both numerator and denominator by 5.

example 3

Find two equivalent fractions of $\frac{4}{12}$ with denominators of 3 and 27, respectively.

solution $\frac{4}{12} = \frac{?}{3}$

$\frac{4 \div 4}{12 \div 4} = \frac{1}{3}$ Divide both numerator and denominator by 4.

$\frac{4}{12} = \frac{?}{27}$ There is no whole number we can multiply 12 by to get 27, so we begin with the fraction equivalent to $\frac{4}{12}$ that we just calculated, which is $\frac{1}{3}$.

$\frac{4}{12} = \frac{1}{3} = \frac{?}{27}$

$\frac{1 \cdot 9}{3 \cdot 9} = \frac{9}{27}$ Multiply numerator and denominator by 9.

example 4

Find an equivalent fraction of $4\frac{2}{5}$ with a denominator of 15.

solution $4\frac{2}{5} = \frac{22}{5}$ Change $4\frac{2}{5}$ to an improper fraction.

$\frac{22}{5} = \frac{?}{15} = \frac{22 \cdot 3}{5 \cdot 3} = \frac{66}{15}$ Multiply numerator and denominator by 3.

Lowest Terms

Fractions are often easier to understand when the numerator and denominator are as small as possible, or when they are in *lowest terms.*

definition

A fraction is in *lowest terms*, or *simplest terms*, if the only common factor of both the numerator and denominator is the number one.

If a fraction is not in lowest terms, we can simplify it by **reducing** to lowest terms.

To **reduce** a fraction to lowest terms, divide both the numerator and denominator by their greatest common factor.

example 5

Reduce $\frac{15}{60}$ to lowest terms.

solution

$15 = \boxed{3 \cdot 5}$

$60 = 2 \cdot 2 \cdot \boxed{3 \cdot 5}$

$\text{GCF} = 3 \cdot 5 = 15$

Find the greatest common factor of 15 and 60.

$$\frac{15 \div 15}{60 \div 15} = \frac{1}{4}$$

Now divide both numerator and denominator by 15. ●

Another way to reduce a fraction to lowest terms is to completely factor both the numerator and denominator, and then cross out the common factors that appear in both. This technique of crossing out like terms is sometimes called *canceling.*

example 6

Reduce $\frac{12}{18}$.

solution

$$\frac{12}{18} = \frac{2 \cdot 2 \cdot 3}{2 \cdot 3 \cdot 3} = \frac{\cancel{2} \cdot 2 \cdot \cancel{3}}{\cancel{2} \cdot 3 \cdot \cancel{3}} = \frac{2}{3}$$

Completely factor the numerator and denominator. Cross out their common factors.

$$\frac{12}{18} = \frac{2}{2} \cdot \frac{2}{3} \cdot \frac{3}{3} = 1 \cdot \frac{2}{3} \cdot 1 = \frac{2}{3}$$

Notice that since $\frac{2}{2}$ and $\frac{3}{3}$ both equal 1, crossing out common factors is the same as eliminating factors of 1. Because of the commutative property of multiplication, it does not matter which two 3s we cancel, as long as one is in the numerator and the other is in the denominator. ●

Often, it is easiest to divide numerator and denominator by any common factor we can think of and then repeat the process until the fraction is in simplest terms.

example 7

Reduce $\frac{72}{54}$ to lowest terms.

solution

$$\frac{72}{54} = \frac{72 \div 2}{54 \div 2} = \frac{36 \div 3}{27 \div 3} = \frac{12 \div 3}{9 \div 3} = \frac{4}{3}$$

Notice that improper fractions reduced to lowest terms are still improper fractions. ●

Sometimes we will show division by a common factor by canceling the original numerator and denominator and writing the results next to them.

example 8

Reduce $\frac{42}{252}$ to lowest terms.

solution

$$\frac{42}{252} = \frac{\overset{21}{\cancel{42}}}{\underset{126}{\cancel{252}}}$$

Divide numerator and denominator by 2.

$$\frac{21}{126} = \frac{\overset{7}{\cancel{21}}}{\underset{42}{\cancel{126}}}$$

Divide numerator and denominator by 3.

$$\frac{7}{42} = \frac{\overset{1}{\cancel{7}}}{\underset{6}{\cancel{42}}} = \frac{1}{6}$$

Divide numerator and denominator by 7. Notice that the numerator of the reduced fraction is 1 and not 0 since $7 \div 7 = 1$. ●

Instead of showing each individual step, as in Example 8, we usually reduce in the following manner.

$$\frac{42}{252} = \frac{\overset{\overset{\overset{1}{\cancel{7}}}{\cancel{21}}}{\cancel{42}}}{\underset{\underset{\underset{6}{\cancel{42}}}{\cancel{126}}}{\cancel{252}}}$$

- 7 ← Divide by 7.
- 21 ← Divide by 3.
- 42 ← Divide by 2.
- 252 ← Divide by 2.
- 126 ← Divide by 3.
- 42 ← Divide by 7.

$$\frac{42}{252} = \frac{1}{6}$$

example 9

There are 5,500 students at a major university, 220 of whom are history majors. Express the fraction of history majors in the student body as a fraction reduced to lowest terms.

solution

$$\frac{220}{5{,}500}$$

Since 220 out of a total of 5,500 students are history majors.

$$\frac{220}{5{,}500} = \frac{\overset{\overset{\overset{1}{\cancel{2}}}{\cancel{22}}}{\cancel{220}}}{\underset{\underset{\underset{25}{\cancel{50}}}{\cancel{550}}}{\cancel{5{,}500}}}$$

- 2 ← Divide by 2.
- 22 ← Divide by 11.
- 220 ← Divide by 10.
- 5,500 ← Divide by 10.
- 550 ← Divide by 11.
- 50 ← Divide by 2.

Reduce to lowest terms.

$$\frac{220}{5{,}500} = \frac{1}{25}$$

Therefore, $\frac{1}{25}$ of the student body are history majors. ●

QUICK QUIZ

Reduce each of the following fractions to lowest terms.

1. $\frac{9}{21}$
2. $\frac{24}{72}$
3. $\frac{132}{55}$

ANSWERS

1. $\frac{3}{7}$
2. $\frac{1}{3}$
3. $\frac{12}{5}$

2.2 Exercises

Indicate which of the following statements are true and which are false. For those that are false, change the italic word to make the statement true.

1. An equivalent fraction is obtained by multiplying or *adding* the same number to both numerator and denominator of a given fraction.
2. The fraction $\frac{8}{12}$ *has* been reduced to lowest terms.
3. To reduce a fraction to lowest terms, the operation *subtraction* must be performed on both numerator and denominator.
4. Multiplying both numerator and denominator of a fraction by the same number is the same as multiplying the fraction by *one*.

Change each of the following fractions to equivalent fractions by performing the indicated operation on *both the numerator and the denominator*.

5. $\frac{3}{7}$ Multiply by 6.
6. $\frac{4}{5}$ Multiply by 2.
7. $\frac{21}{49}$ Divide by 7.
8. $\frac{32}{48}$ Divide by 8.
9. $\frac{8}{13}$ Multiply by 2.
10. $\frac{4}{21}$ Multiply by 4.
11. $\frac{15}{27}$ Divide by 3.
12. $\frac{35}{75}$ Divide by 5.
13. $\frac{5}{4}$ Multiply by 7.
14. $\frac{6}{5}$ Multiply by 6.
15. $\frac{36}{42}$ Divide by 6.
16. $\frac{72}{81}$ Divide by 9.
17. $\frac{11}{22}$ Divide by 11.
18. $\frac{36}{60}$ Divide by 12.
19. $\frac{105}{20}$ Divide by 5.
20. $\frac{130}{30}$ Divide by 10.
21. $\frac{66}{84}$ Divide by 6.
22. $\frac{49}{91}$ Divide by 7.
23. $\frac{66}{55}$ Divide by 11.
24. $\frac{39}{26}$ Divide by 13.
25. $\frac{150}{80}$ Divide by 10.
26. $\frac{144}{96}$ Divide by 12.
27. $\frac{153}{218}$ Multiply by 49.
28. $\frac{327}{289}$ Multiply by 34.
29. $\frac{702}{4{,}374}$ Divide by 54.
30. $\frac{1{,}073}{1{,}769}$ Divide by 29.

Fill in the missing number in each of these equivalent fractions.

31. $\frac{3}{6} = \frac{?}{2}$
32. $\frac{9}{27} = \frac{?}{3}$
33. $\frac{5}{8} = \frac{25}{?}$
34. $\frac{4}{16} = \frac{1}{?}$
35. $\frac{3}{4} = \frac{?}{20}$
36. $\frac{5}{9} = \frac{25}{?}$
37. $\frac{9}{36} = \frac{3}{?}$
38. $\frac{8}{44} = \frac{2}{?}$
39. $\frac{3}{16} = \frac{9}{?}$
40. $\frac{5}{21} = \frac{10}{?}$
41. $\frac{6}{24} = \frac{1}{?}$
42. $\frac{12}{16} = \frac{3}{?}$
43. $\frac{12}{48} = \frac{?}{12}$
44. $\frac{15}{60} = \frac{1}{?}$
45. $\frac{9}{18} = \frac{?}{2}$
46. $\frac{3}{27} = \frac{?}{9}$
47. $\frac{42}{6} = \frac{?}{2}$
48. $\frac{32}{8} = \frac{4}{?}$
49. $\frac{72}{45} = \frac{8}{?}$
50. $\frac{48}{16} = \frac{6}{?}$
51. $\frac{9}{12} = \frac{?}{144}$
52. $\frac{3}{16} = \frac{24}{?}$
53. $\frac{102}{11} = \frac{?}{121}$
54. $\frac{2}{13} = \frac{?}{182}$
55. $\frac{30}{225} = \frac{?}{15}$

56. $\frac{26}{169} = \frac{2}{?}$ 57. $7 = \frac{35}{?}$ 58. $9 = \frac{63}{?}$ 59. $3\frac{1}{8} = \frac{?}{48}$ 60. $5\frac{3}{4} = \frac{?}{12}$

61. $6\frac{2}{3} = \frac{60}{?}$ 62. $4\frac{1}{8} = \frac{99}{?}$ 63. $\frac{5}{9} = \frac{?}{6{,}039}$ 64. $\frac{7}{8} = \frac{2{,}492}{?}$ 65. $\frac{2{,}202}{2{,}936} = \frac{?}{4}$

66. $\frac{3{,}315}{4{,}641} = \frac{5}{?}$

Reduce each of the following fractions to lowest terms.

67. $\frac{4}{10}$ 68. $\frac{6}{8}$ 69. $\frac{3}{9}$ 70. $\frac{7}{28}$ 71. $\frac{9}{12}$ 72. $\frac{8}{24}$

73. $\frac{15}{20}$ 74. $\frac{21}{56}$ 75. $\frac{18}{3}$ 76. $\frac{32}{4}$ 77. $\frac{16}{28}$ 78. $\frac{27}{63}$

79. $\frac{11}{66}$ 80. $\frac{15}{75}$ 81. $\frac{25}{10}$ 82. $\frac{48}{18}$ 83. $\frac{18}{36}$ 84. $\frac{9}{54}$

85. $\frac{14}{49}$ 86. $\frac{63}{70}$ 87. $\frac{70}{90}$ 88. $\frac{55}{88}$ 89. $\frac{42}{88}$ 90. $\frac{16}{64}$

91. $\frac{55}{90}$ 92. $\frac{28}{40}$ 93. $\frac{24}{144}$ 94. $\frac{39}{169}$ 95. $\frac{120}{270}$ 96. $\frac{450}{750}$

97. $\frac{125}{35}$ 98. $\frac{180}{150}$ 99. $\frac{121}{132}$ 100. $\frac{50}{625}$ 101. $\frac{48}{256}$ 102. $\frac{36}{432}$

103. $\frac{1{,}024}{48}$ 104. $\frac{1{,}125}{250}$

Answer each of the following questions with a fraction reduced to lowest terms.

105. In a class of 132 students, 22 are over 6 feet tall. What fraction of the total class are more than 6 feet tall?

106. In a recent survey, 63 out of 91 people responded that they regularly use the same brand of toothpaste. What fraction of the people surveyed are loyal to a single brand of toothpaste?

107. A zoologist who makes a monthly salary of $1,250 pays $550 a month for rent. What fraction of her monthly salary does she pay in rent?

108. An editorial assistant who makes a yearly salary of $14,500 pays $4,350 in federal income tax. What fraction of his salary does he pay in federal income tax?

109. An eye clinic was visited by 104 nearsighted people and 13 farsighted people in one day. What fraction of the patients were nearsighted?

110. A photographer took 48 black-and-white pictures and 112 color pictures. What fraction of his photos were black-and-white?

111. Eight hours is equivalent to what fraction of a week?

112. Nine inches is equivalent to what fraction of a yard?

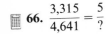

2.3 Addition of Fractions

In Chapter 1 we learned how to perform the four basic arithmetic operations on whole numbers. Now we will establish some rules for arithmetic with fractions, beginning with addition. The following examples will give us an intuitive understanding of how these rules were developed.

figure *2.6*

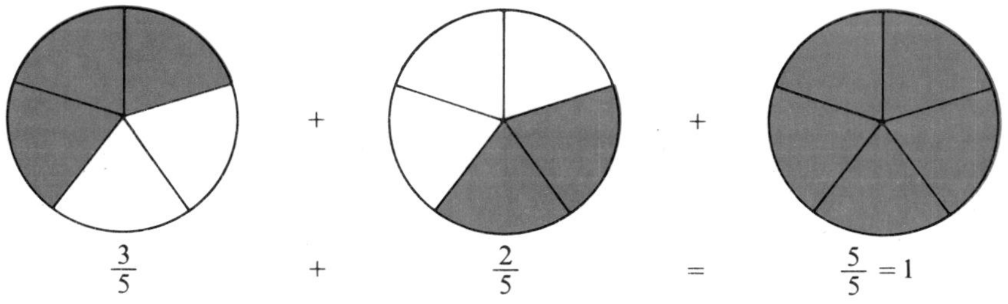

In the first example, let us imagine that you baked two identical pies for a party and that you divided each pie into five equal pieces, as shown in Figure 2.6. Suppose your guests ate three pieces of the first pie and two pieces of the second pie. Would one pie have been sufficient for the guests?

Looking at Figure 2.6, we can see that the guests ate $\frac{3}{5}$ of the first pie and $\frac{2}{5}$ of the second pie. Adding these two fractions together visually, we can see that the guests consumed a total of $\frac{5}{5}$ of a pie, or one entire *pie*. So one pie would have been enough.

In the second example, imagine that you baked two identical pies, but you did not cut them into the same number of pieces. At the end of the party you noticed that your guests ate $\frac{1}{2}$ of the first pie and $\frac{1}{3}$ of the second pie. How can you determine the total amount of pie that they ate.

Looking at Figure 2.7 we see that this is not an easy problem to do visually because both pies are not cut into the same number of equal pieces. Therefore, let us slice each pie into an appropriate number of equal pieces, that is, a number that is evenly divisible by both 2 and 3, such as the number 6. As shown in Figure 2.8, $\frac{1}{2}$ of the first pie is equivalent to $\frac{3}{6}$, and $\frac{1}{3}$ of the second pie is equivalent to $\frac{2}{6}$. Adding the two visually, we obtain a total of $\frac{5}{6}$ of a pie.

figure *2.7*

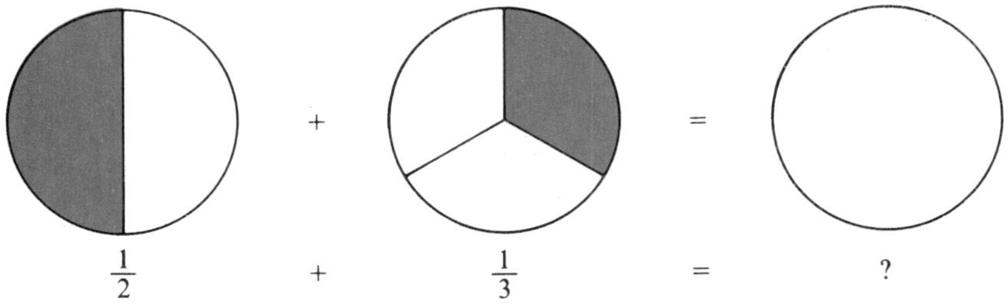

figure *2.8*

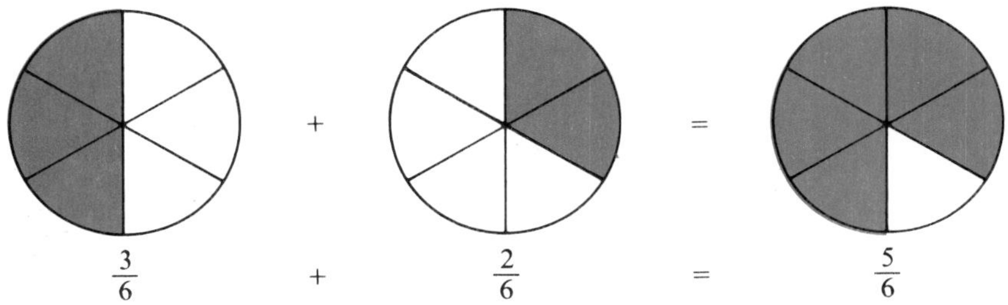

Once the two pies are cut into the same number of equal pieces, the denominators of the fractions we wish to add are the same, and the sum can be obtained by adding the numerators. Notice that the common denominator must be evenly divisible by each of the denominators of the fractions to be added. Thus, we have illustrated the procedure outlined in the box.

To Add Fractions

1. Find a common denominator.
2. Change each fraction to an equivalent fraction with the common denominator.
3. Add the numerators and write the result over the common denominator.
4. If necessary, reduce the answer to lowest terms.

example 1 Add $\frac{2}{5} + \frac{1}{4}$.

solution

$\frac{2}{5} + \frac{1}{4} = \frac{2 \cdot 4}{5 \cdot 4} + \frac{1 \cdot 5}{4 \cdot 5}$ A common denominator is $5 \cdot 4 = 20$.

$= \frac{8}{20} + \frac{5}{20}$ Change each fraction to an equivalent fraction with a denominator of 20.

$= \frac{8 + 5}{20}$ Add the numerators and write the result over the common denominator.

$= \frac{13}{20}$ ●

example 2 Add $\frac{3}{8} + \frac{5}{12}$. A common denominator is $8 \cdot 9 = 72$.

solution

$\frac{3}{8} + \frac{5}{12} = \frac{3 \cdot 9}{8 \cdot 9} + \frac{5 \cdot 6}{12 \cdot 6}$ Change each fraction to an equivalent fraction with a denominator of 72.

$= \frac{27}{72} + \frac{30}{72}$ Add the numerators and write the result over the common denominator.

$= \frac{27 + 30}{72}$

$= \frac{57}{72}$

$= \frac{57 \div 3}{72 \div 3} = \frac{19}{24}$ Reduce to lowest terms. ●

Finding the Lowest Common Denominator (LCD)

In order to keep our calculations as simple as possible, we will now introduce a procedure for finding the **lowest common denominator**, or **LCD**, of two or more fractions.

definition

The *lowest common denominator*, or *LCD*, is the smallest number that is divisible by the denominators of all fractions being considered.

To find the LCD of two or more fractions, find the least common multiple (LCM) of their denominators.

The technique for finding the LCM of two or more numbers was discussed in Section 1.11. Example 3 will help us to review the procedure.

example 3

Find the LCD of $\frac{5}{18}$ and $\frac{11}{48}$.

solution

$18 = 3 \cdot 6 = 3 \cdot 2 \cdot 3 = 2 \cdot 3^2$

$48 = 6 \cdot 8 = 2 \cdot 3 \cdot 2 \cdot 2 \cdot 2 = 2^4 \cdot 3$

Factor each denominator completely and write in exponential notation.

The highest exponent 2 is raised to is 4: 2^4.

The highest exponent 3 is raised to is 2: 3^2.

The unique prime factors in the factorizations are 2 and 3.

$\text{LCD} = 2^4 \cdot 3^2 = 16 \cdot 9 = 144$

The LCD is the product $2^4 \cdot 3^2$. ●

Once the LCD is found, it is important to be able to identify what number each of the original denominators must be multiplied by to obtain the LCD. This number can be found by comparing the prime factorizations of each denominator with the prime factorization of the LCD. Cancel the factors common to the denominator and the LCD to obtain the answer. In Example 3, we can determine what number each denominator must be multiplied by to obtain the LCD as follows:

$\text{LCD} = 2^4 \cdot 3^2 = 2 \cdot 2 \cdot 2 \cdot 2 \cdot 3 \cdot 3$

$18 = 2 \cdot 3 \cdot 3$

$\text{LCD} = \boxed{2 \cdot 2 \cdot 2} \cdot \cancel{2} \cdot \cancel{3} \cdot \cancel{3}$ Multiply by 8.

$48 = 2 \cdot 2 \cdot 2 \cdot 2 \cdot 3$

$\text{LCD} = \cancel{2} \cdot \cancel{2} \cdot \cancel{2} \cdot \cancel{2} \cdot \cancel{3} \cdot \boxed{3}$ Multiply by 3.

We can see, therefore, that we must multiply 18 by 8 and we must multiply 48 by 3 to obtain the LCD.

example 4

Find the LCD of $\frac{7}{90}$ and $\frac{5}{24}$. Then determine what number each denominator must be multiplied by to obtain the LCD.

solution

$90 = 10 \cdot 9 = 2 \cdot 5 \cdot 3 \cdot 3 = 2 \cdot 3^2 \cdot 5$

$24 = 6 \cdot 4 = 2 \cdot 3 \cdot 2 \cdot 2 = 2^3 \cdot 3$

$\text{LCD} = 2^3 \cdot 3^2 \cdot 5 = 8 \cdot 9 \cdot 5$

$= 72 \cdot 5 = 360$

The unique prime factors are 2, 3, and 5.

The highest exponent 2 is raised to is 3: 2^3.
The highest exponent 3 is raised to is 2: 3^2.
The highest exponent 5 is raised to is 1: 5^1.

$\text{LCD} = 2^3 \cdot 3^2 \cdot 5 = 2 \cdot 2 \cdot 2 \cdot 3 \cdot 3 \cdot 5$

$90 = 2 \cdot 3 \cdot 3 \cdot 5$

$\text{LCD} = \boxed{2 \cdot 2} \cdot \not{2} \cdot \not{3} \cdot \not{3} \cdot \not{5}$

Multiply by 4.

$24 = 2 \cdot 2 \cdot 2 \cdot 3$

$\text{LCD} = \not{2} \cdot \not{2} \cdot \not{2} \cdot \not{3} \cdot \boxed{3 \cdot 5}$

Multiply by 15.

To determine what number each denominator must be multiplied by to obtain the LCD, we compare prime factorizations.

Thus, to obtain the LCD we must multiply 90 by 4 and we must multiply 24 by 15. ●

Let us now add the two fractions in Example 2 by first finding the LCD.

example 5

Add $\frac{3}{8} + \frac{5}{12}$.

solution

$8 = 2 \cdot 2 \cdot 2 = 2^3$

$12 = 2 \cdot 2 \cdot 3 = 2^2 \cdot 3$

$\text{LCD} = 2^3 \cdot 3 = 8 \cdot 3 = 24$

Find the LCD. Compare the prime factorizations. To obtain the LCD we must multiply 8 by 3 and 12 by 2.

$$\frac{3}{8} + \frac{5}{12} = \frac{3 \cdot 3}{8 \cdot 3} + \frac{5 \cdot 2}{12 \cdot 2}$$

Change each fraction to an equivalent fraction with the same LCD.

$$= \frac{9}{24} + \frac{10}{24}$$

$$= \frac{9 + 10}{24}$$

Add the numerators and place the result over the LCD.

$$= \frac{19}{24}$$ ●

Notice that the answer in Example 5 is the same answer as in Example 2. However, since we used 24 instead of 72 as our common denominator, we were able to work with smaller numbers in Example 5, which simplified our calculations.

The following examples illustrate addition fractions using a vertical format.

example 6

Add $\frac{7}{30} + \frac{5}{42}$.

solution

$30 = 5 \cdot 6 = 5 \cdot 2 \cdot 3$
$42 = 6 \cdot 7 = 2 \cdot 3 \cdot 7$
$\text{LCD} = 2 \cdot 3 \cdot 5 \cdot 7 = 210$

Find the LCD. Compare the prime factorizations. To obtain the LCD we must multiply 30 by 7 and 42 by 5.

$$\frac{7}{30} \cdot \frac{7}{7} = \frac{49}{210}$$
$$+\frac{5}{42} \cdot \frac{5}{5} = \frac{25}{210}$$

Change each fraction to an equivalent fraction with the same LCD.

$$\frac{74}{210}$$

Add the numerators and place the result over the LCD.

$$\frac{74}{210} = \frac{37}{105}$$

Reduce to lowest terms. ●

example 7 Add $\frac{7}{16} + \frac{11}{12} + \frac{3}{20}$.

solution

$16 = 2 \cdot 2 \cdot 2 \cdot 2 = 2^4$
$12 = 2 \cdot 2 \cdot 3 = 2^2 \cdot 3$
$20 = 2 \cdot 2 \cdot 5 = 2^2 \cdot 5$
$\text{LCD} = 2^4 \cdot 3 \cdot 5 = 16 \cdot 15 = 240$

Find the LCD. To obtain the LCD we must multiply 16 by 15, 12 by 20, and 20 by 12.

$$\frac{7}{16} \cdot \frac{15}{15} = \frac{105}{240}$$
$$\frac{11}{12} \cdot \frac{20}{20} = \frac{220}{240}$$
$$+\frac{3}{20} \cdot \frac{12}{12} = \frac{36}{240}$$

Change each fraction to an equivalent fraction with a denominator of 240.

$$\frac{361}{240}$$

Add the numerators and place the result over the LCD. ●

Another Look at Mixed Numbers

Now that we have learned how to add fractions, we can examine more closely the meaning of a mixed number. For example, we have already established that $5\frac{2}{3}$ means $5 + \frac{2}{3}$. Since any whole number can be represented by a fraction with a denominator of one, we can rewrite 5 as $\frac{5}{1}$ and add the two fractions as follows:

$$5\frac{2}{3} = 5 + \frac{2}{3} = \frac{5}{1} + \frac{2}{3}$$

The LCD is 3.

$$= \frac{5 \cdot 3}{1 \cdot 3} + \frac{2}{3}$$

Change $\frac{5}{1}$ to an equivalent fraction with a denominator of 3.

$$= \frac{15}{3} + \frac{2}{3} = \frac{17}{3}$$

Add the numerators.

The result of our addition is the same as that which we would obtain by converting $5\frac{2}{3}$ to an improper fraction using the technique introduced in Section 2.1.

$$5\frac{2}{3} = \frac{(5 \cdot 3) + 2}{3} = \frac{17}{3}$$

example 8

Use addition of fractions to convert $2\frac{9}{11}$ to an improper fraction.

solution

$$2\frac{9}{11} = 2 + \frac{9}{11} = \frac{2}{1} + \frac{9}{11}$$

The LCD is 11.

$$= \frac{2 \cdot 11}{1 \cdot 11} + \frac{9}{11}$$

Change $\frac{2}{1}$ to an equivalent fraction with a denominator of 11.

$$= \frac{22}{11} + \frac{9}{11} = \frac{31}{11}$$

Add the numerators.

example 9

On three consecutive days, a jogger ran $\frac{1}{4}$ mile, $\frac{3}{8}$ mile, and $\frac{2}{5}$ mile. What was the total distance he ran on those three days?

solution

Add $\frac{1}{4} + \frac{3}{8} + \frac{2}{5}$.

$4 = 2 \cdot 2 = 2^2$

$8 = 2 \cdot 2 \cdot 2 = 2^3$

$5 = 5$

$\text{LCD} = 2^3 \cdot 5 = 40$

Find the LCD. To obtain the LCD we must multiply 4 by 10, 8 by 5, and 5 by 8.

$$\frac{1}{4} \cdot \frac{10}{10} = \frac{10}{40}$$
$$\frac{3}{8} \cdot \frac{5}{5} = \frac{15}{40}$$
$$+\frac{2}{5} \cdot \frac{8}{8} = \frac{16}{40}$$
$$\overline{\frac{41}{40}}$$

Change each fraction to an equivalent fraction with a denominator of 40.

Add the numerators and place the result over the LCD.

$$\frac{41}{40} = 41 \div 40 = 1\frac{1}{40}$$

Convert the answer to a mixed number, which is the way we express distances.

He ran a total of $1\frac{1}{40}$ miles.

QUICK QUIZ

Add each of the following fractions.

1. $\frac{1}{2} + \frac{1}{3}$
2. $\frac{3}{16} + \frac{5}{12}$
3. $\frac{2}{9} + \frac{7}{24} + \frac{5}{18}$

ANSWERS

1. $\frac{5}{6}$
2. $\frac{29}{48}$
3. $\frac{19}{24}$

2.3 Exercises

Indicate which of the following statements are true and which are false. For those that are false, change the italic word to make the statement true.

1. In order to add two fractions, you must first find a common *denominator*.
2. The LCD of two fractions is the *largest* number that is evenly divisible by both denominators.
3. When adding two fractions, each greater than zero, the sum is always *greater* than either of the two addends.
4. The sum of two fractions, each greater than zero and less than one, is *always* less than one.

Add each of the following fractions. Reduce all answers to lowest terms.

5. $\frac{3}{7}+\frac{1}{7}$
6. $\frac{4}{9}+\frac{2}{9}$
7. $\frac{5}{8}+\frac{3}{8}$
8. $\frac{3}{4}+\frac{1}{4}$
9. $\frac{3}{10}+\frac{5}{10}$
10. $\frac{4}{11}+\frac{5}{11}$
11. $\frac{8}{17}+\frac{5}{17}+\frac{2}{17}$
12. $\frac{4}{21}+\frac{8}{21}+\frac{5}{21}$
13. $\frac{7}{100}+\frac{42}{100}+\frac{66}{100}$
14. $\frac{21}{50}+\frac{16}{50}+\frac{33}{50}$
15. $\frac{1}{3}+\frac{3}{4}$
16. $\frac{3}{7}+\frac{2}{5}$
17. $\frac{7}{12}+\frac{2}{3}$
18. $\frac{3}{5}+\frac{8}{15}$
19. $\frac{1}{11}+\frac{2}{3}$
20. $\frac{3}{4}+\frac{2}{13}$
21. $\frac{7}{10}+\frac{3}{25}$
22. $\frac{3}{28}+\frac{5}{42}$
23. $\frac{3}{56}+\frac{1}{8}$
24. $\frac{1}{6}+\frac{11}{48}$
25. $\frac{12}{7}+\frac{9}{5}$
26. $\frac{21}{4}+\frac{8}{3}$
27. $\frac{9}{10}+\frac{37}{100}$
28. $\frac{81}{100}+\frac{7}{10}$
29. $\frac{5}{48}+\frac{5}{36}$
30. $\frac{8}{49}+\frac{8}{21}$
31. $\frac{4}{15}+\frac{3}{50}$
32. $\frac{7}{40}+\frac{3}{16}$
33. $\frac{5}{18}+\frac{4}{27}$
34. $\frac{1}{40}+\frac{5}{24}$
35. $\frac{4}{21}+\frac{5}{63}$
36. $\frac{7}{52}+\frac{8}{13}$
37. $\frac{3}{11}+\frac{5}{12}$
38. $\frac{9}{10}+\frac{12}{13}$
39. $\frac{15}{4}+\frac{27}{5}$
40. $\frac{24}{7}+\frac{14}{9}$
41. $\frac{37}{5}+\frac{19}{10}$
42. $\frac{22}{3}+\frac{17}{24}$
43. $\frac{11}{49}+\frac{9}{14}$
44. $\frac{7}{54}+\frac{13}{36}$
45. $\frac{1}{2}+\frac{1}{3}+\frac{1}{5}$
46. $\frac{1}{4}+\frac{1}{6}+\frac{1}{9}$
47. $\frac{3}{10}+\frac{23}{100}+\frac{561}{1{,}000}$
48. $\frac{7}{10}+\frac{69}{100}+\frac{103}{1{,}000}$
49. $\frac{3}{25}+\frac{2}{15}+\frac{1}{5}$
50. $\frac{2}{3}+\frac{7}{36}+\frac{5}{27}$
51. $\frac{3}{8}+\frac{5}{12}+\frac{1}{4}$
52. $\frac{1}{48}+\frac{3}{16}+\frac{5}{18}$
53. $\frac{5}{16}+\frac{3}{8}+\frac{5}{14}$
54. $\frac{6}{49}+\frac{3}{56}+\frac{5}{14}$
55. $\frac{51}{639}+\frac{87}{213}$

56. $\frac{13}{182} + \frac{47}{728}$

57. $\frac{15}{29} + \frac{25}{49}$

58. $\frac{11}{37} + \frac{31}{53}$

Use addition of fractions to convert each of the following mixed numbers to an improper fraction.

59. $2\frac{2}{7}$

60. $6\frac{3}{4}$

61. $3\frac{1}{17}$

62. $4\frac{4}{13}$

63. $13\frac{7}{8}$

64. $16\frac{3}{4}$

Solve each of the following word problems.

65. On Saturday a student studied $\frac{3}{5}$ hour, and on Sunday she studied an additional $\frac{5}{6}$ hour. What is the total amount of time she spent studying that weekend?

66. A recipe calls for $\frac{2}{3}$ cup of sugar and $\frac{3}{4}$ cup of flour. What is the total amount of flour and sugar needed?

67. Of the people attending a conference, $\frac{7}{12}$ have brown eyes, and $\frac{2}{5}$ have blue eyes. What fraction of the people have brown eyes or blue eyes?

68. A businessman invested $\frac{3}{10}$ of an inheritance in the stock market and $\frac{5}{8}$ of it in real estate. What fraction of the inheritance did he invest in the stock market or real estate?

69. A gourmet cook bought $\frac{1}{2}$ pound of swiss cheese, $\frac{3}{8}$ pound of cheddar cheese, and $\frac{1}{4}$ pound of blue cheese. What is the total amount of cheese that he bought?

70. A salesclerk at a fabric store cut a remnant of fabric into three pieces that were of lengths $\frac{3}{4}$ yard, $\frac{5}{8}$ yard, and $\frac{2}{5}$ yard. How long was the original remnant before she cut it?

2.4 Subtraction of Fractions

In this section we will find the difference between two fractions. As in the case with whole numbers, we will only work with problems in which the minuend is greater than the subtrahend so that our results will always be greater than zero. In Chapter 6 we will consider problems in which the results are less than zero.

Comparing Fractions

Before we begin our discussion of subtraction of fractions, let us establish a procedure for determining the larger of two fractions. If the two denominators of the fractions being compared are the same, the procedure is relatively simple. For example, in Figure 2.9 we can see easily that $\frac{5}{8}$ is greater than $\frac{3}{8}$. Using mathematical notation we write $\frac{5}{8} > \frac{3}{8}$. Thus, if two fractions have the same denominator, the larger fraction has the larger numerator.

Now let us compare two fractions that have different denominators. Look at Figure 2.10 and determine which is larger, $\frac{3}{5}$ or $\frac{2}{3}$. Even a very close inspection of the diagrams leaves us with some doubt, because the shaded areas appear to be roughly the same size. To answer the question with certainty, we need to divide both circles into an equal number of pieces; that is, we need to find a common denominator for the two

figure 2.9

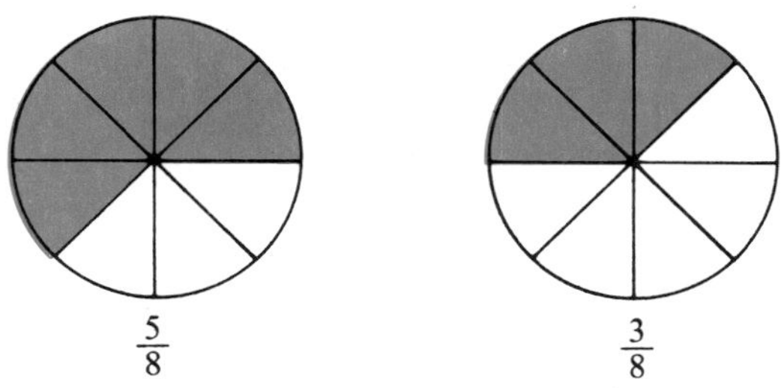

figure 2.10

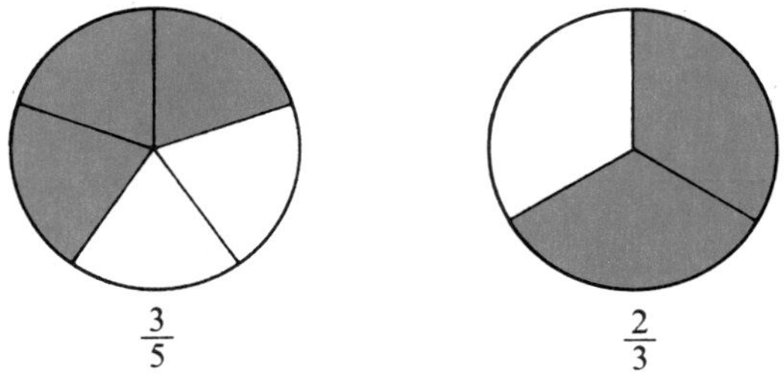

fractions that we wish to compare. The LCD of 3 and 5 is 15. Converting to equivalent fractions we obtain

$$\frac{3}{5} = \frac{3 \cdot 3}{5 \cdot 3} = \frac{9}{15} \quad \text{and} \quad \frac{2}{3} = \frac{2 \cdot 5}{3 \cdot 5} = \frac{10}{15}$$

Since $\frac{9}{15}$ is less than $\frac{10}{15}$, we conclude that $\frac{3}{5}$ is less than $\frac{2}{3}$. This is written as $\frac{3}{5} < \frac{2}{3}$. Thus, $\frac{2}{3}$ is the larger fraction.

We have illustrated a procedure for comparing two fractions, as outlined in the box.

To Find the Larger of Two Fractions

1. Find the LCD.
2. Change each fraction to an equivalent fraction with a denominator equal to the LCD.
3. Compare the two fractions with the same denominator. The one with the larger numerator is the larger fraction.

example 1

Determine which fraction is larger, $\frac{4}{7}$ or $\frac{6}{11}$, and indicate this by placing a $>$ or $<$ sign between them.

solution

$\frac{4}{7} ? \frac{6}{11}$

LCD $= 7 \cdot 11 = 77$. Find the LCD.

$$\frac{4}{7} \cdot \frac{11}{11} = \frac{44}{77}$$

Change each fraction to an equivalent fraction with a denominator of 77.

$$\frac{6}{11} \cdot \frac{7}{7} = \frac{42}{77}$$

$$\frac{44}{77} > \frac{42}{77}$$

Compare the equivalent fractions.

Therefore, $\frac{4}{7} > \frac{6}{11}$

example 2

For each of the following pairs of fractions, indicate which is larger by replacing the ? with a > or < sign.

(a) $\frac{7}{10} ? \frac{11}{15}$

$10 = 2 \cdot 5$

$15 = 3 \cdot 5$

$\text{LCD} = 2 \cdot 3 \cdot 5 = 30$

Find the LCD.

$$\frac{7}{10} \cdot \frac{3}{3} = \frac{21}{30}$$

Change each fraction to an equivalent fraction with a denominator of 30.

$$\frac{11}{15} \cdot \frac{2}{2} = \frac{22}{30}$$

$$\frac{21}{30} < \frac{22}{30}$$

Compare the equivalent fractions.

Therefore, $\frac{7}{10} < \frac{11}{15}$

(b) $\frac{7}{24} ? \frac{5}{18}$

$24 = 4 \cdot 6 = 2 \cdot 2 \cdot 2 \cdot 3 = 2^3 \cdot 3$

$18 = 3 \cdot 6 = 3 \cdot 2 \cdot 3 = 2 \cdot 3^2$

$\text{LCD} = 2^3 \cdot 3^2 = 8 \cdot 9 = 72$

Find the LCD.

$$\frac{7}{24} \cdot \frac{3}{3} = \frac{21}{72}$$

Change each fraction to an equivalent fraction with a denominator of 72.

$$\frac{5}{18} \cdot \frac{4}{4} = \frac{20}{72}$$

$$\frac{21}{72} > \frac{20}{72}$$

Compare the equivalent fractions.

Therefore, $\frac{7}{24} > \frac{5}{18}$

Subtracting Fractions

Let us now begin our discussion of subtraction of fractions with an example. Suppose you have $\frac{1}{2}$ of a pie and you would like to save a piece $\frac{1}{3}$ the size of the whole pie. How much of the pie can you eat and still have $\frac{1}{3}$ of the pie left?

To solve this problem, we need to calculate the difference

$$\frac{1}{2} - \frac{1}{3}$$

Look at Figure 2.11, which illustrates both $\frac{1}{2}$ and $\frac{1}{3}$ of a pie. Both pie pictures must be divided into the same number of equal pieces before they can be compared and subtracted, which is the same as finding the LCD of the two fractions $\frac{1}{2}$ and $\frac{1}{3}$.

$$\text{LCD} = 2 \cdot 3 = 6$$

figure 2.11

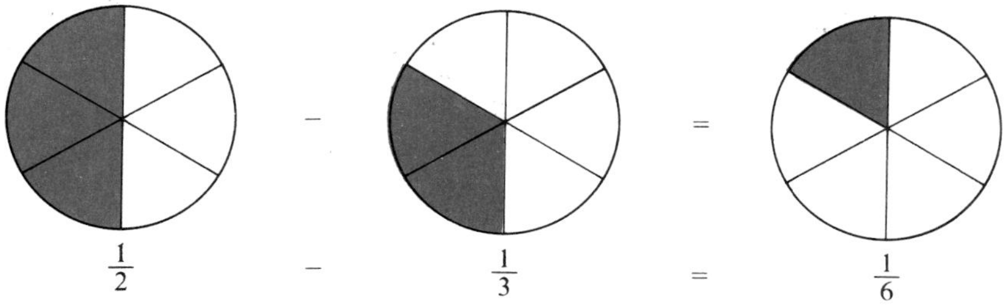

Now change each fraction to an equivalent fraction with a denominator of 6.

$$\frac{1}{2} = \frac{1}{2} \cdot \frac{3}{3} = \frac{3}{6} \qquad \frac{1}{3} = \frac{1}{3} \cdot \frac{2}{2} = \frac{2}{6}$$

Subtract the numerators and write the result over the LCD.

$$\frac{3}{6} - \frac{2}{6} = \frac{3-2}{6} = \frac{1}{6}$$

Thus, you can eat $\frac{1}{6}$ of the pie and still have $\frac{1}{3}$ of the pie left.

The procedure that we have illustrated for subtracting two fractions is outlined in the box.

To Subtract Two Fractions

1. Find the LCD.
2. Change each fraction to an equivalent fraction with a denominator equal to the LCD.
3. Find the difference between the two numerators, and write the result over the common denominator.

example 3

Subtract $\frac{3}{4} - \frac{5}{14}$.

solution

$4 = 2 \cdot 2 = 2^2$

$14 = 2 \cdot 7$

$\text{LCD} = 2^2 \cdot 7 = 28$

Find the LCD. To obtain the LCD we must multiply 4 by 7 and 14 by 2.

$$\frac{3}{4} - \frac{5}{14} = \frac{3 \cdot 7}{4 \cdot 7} - \frac{5 \cdot 2}{14 \cdot 2}$$

Change each fraction to an equivalent fraction with a denominator of 28.

$$= \frac{21}{28} - \frac{10}{28}$$

$$= \frac{21 - 10}{28}$$

Subtract the numerators and place the result over the LCD.

$$= \frac{11}{28}$$

Now that we know the technique for subtracting fractions, we will use the vertical format for setting up problems.

example 4

Subtract $\frac{9}{28} - \frac{2}{21}$.

solution

$28 = 2 \cdot 2 \cdot 7 = 2^2 \cdot 7$

$21 = 3 \cdot 7$

$\text{LCD} = 2^2 \cdot 3 \cdot 7 = 84$

Find the LCD. To obtain the LCD we must multiply 28 by 3 and 21 by 4.

$$\begin{array}{r} \frac{9}{28} \cdot \frac{3}{3} = \frac{27}{84} \\ -\frac{2}{21} \cdot \frac{4}{4} = \frac{8}{84} \\ \hline \frac{19}{84} \end{array}$$

Change each fraction to an equivalent fraction with a denominator of 84.

Subtract the numerators and place the result over the LCD.

example 5

Subtract $5 - \frac{2}{9}$.

solution

$$5 - \frac{2}{9} = \frac{5}{1} - \frac{2}{9}$$

$$= \frac{5 \cdot 9}{1 \cdot 9} - \frac{2}{9}$$

$$= \frac{45}{9} - \frac{2}{9} = \frac{43}{9}$$

Rewrite 5 as the improper fraction $\frac{5}{1}$ and proceed as usual. The LCD is 9. Change $\frac{5}{1}$ to an equivalent fraction with a denominator of 9. Subtract the numerators.

Let us now look at an example that combines both addition and subtraction.

example 6

Calculate $\frac{3}{8} + \frac{1}{3} - \frac{1}{2}$.

solution

$$\begin{aligned} \frac{3}{8}\cdot\frac{3}{3} &= \frac{9}{24} \\ +\frac{1}{3}\cdot\frac{8}{8} &= \frac{8}{24} \\ \hline &\quad \frac{17}{24} \end{aligned}$$

First add $\frac{3}{8}+\frac{1}{3}$.

$$\begin{aligned} \frac{17}{24}\cdot 1 &= \frac{17}{24} \\ -\frac{1}{2}\cdot\frac{12}{12} &= \frac{12}{24} \\ \hline &\quad \frac{5}{24} \end{aligned}$$

Then subtract $\frac{17}{24}-\frac{1}{2}$.

The result is $\frac{5}{24}$.

example 7

A banker owns $\frac{7}{8}$ of an acre of land, and his son owns $\frac{1}{6}$ acre less than he does. How much land does his son own?

solution

Subtract $\frac{7}{8}-\frac{1}{6}$.

$8 = 2\cdot 2\cdot 2 = 2^3$

$6 = 2\cdot 3$

$\text{LCD} = 2^3\cdot 3 = 24$

Find the LCD.

$$\begin{aligned} \frac{7}{8}\cdot\frac{3}{3} &= \frac{21}{24} \\ -\frac{1}{6}\cdot\frac{4}{4} &= \frac{4}{24} \\ \hline &\quad \frac{17}{24} \end{aligned}$$

Change to equivalent fractions and subtract.

His son owns $\frac{17}{24}$ acre of land.

QUICK QUIZ

Perform each of the operations indicated.

1. $\frac{3}{5}-\frac{1}{4}$
2. $\frac{5}{18}-\frac{5}{24}$
3. $\frac{8}{9}-\frac{1}{2}+\frac{5}{6}$

ANSWERS

1. $\frac{7}{20}$
2. $\frac{5}{72}$
3. $\frac{11}{9}$

2.4 Exercises

Indicate which of the following statements are true and which are false. For those that are false, change the italic expression to make the statement true.

1. When comparing two fractions with the same denominator, the larger fraction *always* has the larger numerator.
2. To obtain a result *greater* than zero, we subtract the smaller fraction from the larger fraction.
3. When comparing two fractions with the same numerator, the larger fraction has the smaller denominator.
4. In order to subtract a fraction from a whole number, we must first rewrite the whole number as an *improper fraction.*

For each of the following pairs of fractions, indicate which is larger by replacing the ? by a > or < sign.

5. $\frac{3}{5} ? \frac{2}{3}$ **6.** $\frac{7}{12} ? \frac{3}{4}$ **7.** $\frac{11}{18} ? \frac{2}{3}$ **8.** $\frac{1}{5} ? \frac{2}{7}$

9. $\frac{5}{16} ? \frac{3}{8}$ **10.** $\frac{5}{8} ? \frac{7}{12}$ **11.** $\frac{2}{9} ? \frac{3}{13}$ **12.** $\frac{4}{11} ? \frac{1}{3}$

13. $\frac{7}{18} ? \frac{2}{5}$ **14.** $\frac{2}{7} ? \frac{3}{10}$ **15.** $\frac{1}{3} ? \frac{6}{19}$ **16.** $\frac{2}{5} ? \frac{7}{16}$

17. $\frac{11}{12} ? \frac{12}{13}$ **18.** $\frac{15}{16} ? \frac{14}{15}$ **19.** $\frac{10}{3} ? \frac{16}{5}$ **20.** $\frac{18}{5} ? \frac{7}{2}$

21. $\frac{9}{42} ? \frac{4}{21}$ **22.** $\frac{9}{49} ? \frac{4}{21}$ **23.** $\frac{3}{81} ? \frac{2}{45}$ **24.** $\frac{40}{121} ? \frac{10}{35}$

Subtract each of the following fractions.

25. $\frac{6}{7} - \frac{2}{7}$ **26.** $\frac{8}{11} - \frac{5}{11}$ **27.** $\frac{4}{9} - \frac{1}{3}$ **28.** $\frac{3}{4} - \frac{7}{12}$

29. $\frac{3}{8} - \frac{2}{9}$ **30.** $\frac{5}{7} - \frac{1}{6}$ **31.** $\frac{5}{6} - \frac{1}{9}$ **32.** $\frac{7}{8} - \frac{5}{12}$

33. $\frac{13}{30} - \frac{4}{45}$ **34.** $\frac{11}{24} - \frac{7}{36}$ **35.** $\frac{7}{18} - \frac{7}{24}$ **36.** $\frac{5}{36} - \frac{5}{48}$

37. $\frac{8}{13} - \frac{2}{39}$ **38.** $\frac{17}{60} - \frac{4}{15}$ **39.** $\frac{11}{16} - \frac{5}{24}$ **40.** $\frac{9}{16} - \frac{5}{64}$

41. $\frac{9}{25} - \frac{3}{55}$ **42.** $\frac{5}{54} - \frac{2}{45}$ **43.** $\frac{5}{24} - \frac{5}{84}$ **44.** $\frac{7}{72} - \frac{1}{45}$

45. $5 - \frac{5}{6}$ **46.** $8 - \frac{2}{3}$ **47.** $7 - \frac{5}{12}$ **48.** $3 - \frac{9}{16}$

49. $\frac{38}{294} - \frac{13}{588}$ **50.** $\frac{54}{471} - \frac{16}{157}$ **51.** $\frac{13}{37} - \frac{17}{59}$ **52.** $\frac{52}{67} - \frac{23}{41}$

Perform each of the operations indicated.

53. $\frac{1}{6} - \frac{1}{24} + \frac{7}{18}$ **54.** $\frac{6}{7} - \frac{2}{3} + \frac{5}{21}$ **55.** $\frac{2}{5} + \frac{7}{10} - \frac{4}{15}$ **56.** $\frac{8}{9} + \frac{2}{3} - \frac{5}{6}$ **57.** $\frac{3}{14} - \frac{2}{21} + \frac{4}{7}$

58. $\frac{7}{8} + \frac{3}{12} - \frac{1}{6}$ **59.** $\frac{4}{5} - \frac{1}{4} - \frac{1}{3}$ **60.** $\frac{7}{9} - \frac{1}{2} - \frac{1}{7}$ **61.** $\frac{11}{24} + \frac{7}{36} - \frac{5}{48}$ **62.** $\frac{9}{56} + \frac{11}{16} - \frac{3}{28}$

Solve each of the following word problems.

63. A recipe calls for $\frac{3}{4}$ cup brown sugar and $\frac{2}{3}$ cup rolled oats. How much more sugar than oats does the recipe require?

64. A tailor purchased $\frac{5}{8}$ yard of wool fabric and $\frac{3}{4}$ yard of corduroy. How much more corduroy than wool did he buy?

65. The captain of a swim team swam $\frac{3}{8}$ mile, and her teammate swam $\frac{2}{5}$ mile. Who swam farther, the captain or her teammate? By how much?

66. A high school football player completed his math homework in $\frac{4}{5}$ hour, and a basketball player completed his in $\frac{5}{6}$ hour. Who worked faster? By how much?

67. In a suburb of a large midwestern city, $\frac{5}{12}$ of the voters are registered Democrats, and $\frac{7}{18}$ are registered Republicans. Are there more registered Democrats or Republicans?

68. If $\frac{17}{36}$ of the students in a math class received a grade of B or better, and $\frac{11}{24}$ of the students received a grade of C or below, did more students do well or poorly?

2.5 Addition and Subtraction of Mixed Numbers

Since most applications of fractions in daily life involve mixed numbers, we will now investigate some examples that illustrate how to add and subtract mixed numbers.

example 1 Add $3\frac{7}{8} + 5\frac{2}{9}$.

solution

$$3\frac{7}{8} + 5\frac{2}{9} = \left(3 + \frac{7}{8}\right) + \left(5 + \frac{2}{9}\right)$$ Since a mixed number is the sum of a whole number and a fraction.

$$= (3 + 5) + \left(\frac{7}{8} + \frac{2}{9}\right)$$ By the commutative and associative properties.

$$= 8 + \left(\frac{7}{8} + \frac{2}{9}\right)$$ Add the whole number parts first.

$$= 8 + \left(\frac{7 \cdot 9}{8 \cdot 9} + \frac{2 \cdot 8}{9 \cdot 8}\right)$$ Now add the fractional parts. Convert to equivalent fractions with denominators of 72.

$$= 8 + \left(\frac{63}{72} + \frac{16}{72}\right)$$

$$= 8 + \frac{79}{72}$$ Add the numerators.

$$= 8 + 1 + \frac{7}{72}$$ Convert the result to a mixed number.

$$= 9 + \frac{7}{72}$$ Combine the whole numbers to obtain the final answer.

$$= 9\frac{7}{72}$$

Example 1 illustrates the procedure for adding mixed numbers, as outlined in the box.

To Add Mixed Numbers

1. Add the whole number parts and fractional parts separately.
2. Change any resulting improper fraction to a mixed number.
3. Add both sums and express the final answer as a mixed number.

Let us now look at Example 2, in which we use the vertical format.

example 2

Add $5\frac{3}{4} + 7\frac{5}{12} + 2\frac{1}{8}$.

solution

$4 = 2 \cdot 2 = 2^2$ — Find the LCD of the fractional parts.

$12 = 3 \cdot 2 \cdot 2 = 2^2 \cdot 3$

$8 = 2 \cdot 2 \cdot 2 = 2^3$

$\text{LCD} = 2^3 \cdot 3 = 8 \cdot 3 = 24$

$$\begin{aligned} 5\frac{3}{4} &= 5 + \frac{3}{4} \cdot \frac{6}{6} = 5 + \frac{18}{24} \\ 7\frac{5}{12} &= 7 + \frac{5}{12} \cdot \frac{2}{2} = 7 + \frac{10}{24} \\ +\,2\frac{1}{8} &= 2 + \frac{1}{8} \cdot \frac{3}{3} = 2 + \frac{3}{24} \\ \hline & \qquad\qquad\quad 14 + \frac{31}{24} \end{aligned}$$

Add the whole number parts and fractional parts separately.

$$14 + \frac{31}{24} = 14 + 1 + \frac{7}{24}$$

Change $\frac{31}{24}$ to a mixed number.

$$= 15 + \frac{7}{24}$$

Add both sums to simplify.

$$= 15\frac{7}{24}$$

Example 3 illustrates how to subtract mixed numbers.

example 3

Subtract $5\frac{2}{7} - 3\frac{4}{7}$.

solution

Minuend	$5\frac{2}{7} = 5 + \frac{2}{7}$
Subtrahend	$-\,3\frac{4}{7} = 3 + \frac{4}{7}$

After writing the problem vertically and attempting to subtract the whole number parts and fractional parts, we see that the fractional part of the minuend is less than the fractional part of the subtrahend: $\frac{2}{7} < \frac{4}{7}$. We therefore need to rewrite the minuend by borrowing one from the whole number part, and adding one in the form of $\frac{7}{7}$ to the fractional part:

$$5 + \frac{2}{7} = 4 + 1 + \frac{2}{7} = 4 + \frac{7}{7} + \frac{2}{7} = 4 + \frac{9}{7}$$

$$\begin{array}{r} 5 + \frac{2}{7} = 4 + \frac{9}{7} \\ -\ 3 + \frac{4}{7} = 3 + \frac{4}{7} \\ \hline 1 + \frac{5}{7} = 1\frac{5}{7} \end{array}$$

Now subtract the whole number parts and fractional parts.

●

We have thus illustrated the procedure for subtracting mixed numbers, as outlined in the box. We will use this procedure in Example 4 to solve another problem.

To Subtract Mixed Numbers

1. Change the fractional parts to equivalent fractions with the same LCD.
2. If the fractional part of the minuend is less than that of the subtrahend, borrow one from the whole number part of the minuend, and add one to its fractional part.
3. Subtract the whole number parts and fractional parts separately.

example 4

Subtract $7\frac{1}{6} - 2\frac{8}{15}$.

solution

$6 = 2 \cdot 3$

$15 = 3 \cdot 5$

$\text{LCD} = 2 \cdot 3 \cdot 5 = 30$

Change the fractional parts to equivalent fractions with the same LCD.

$$\begin{array}{r} 7\frac{1}{6} = 7 + \frac{1}{6} \cdot \frac{5}{5} = 7 + \frac{5}{30} \\ -\ 2\frac{8}{15} = 2 + \frac{8}{15} \cdot \frac{2}{2} = 2 + \frac{16}{30} \\ \hline \end{array}$$

Since $\frac{5}{30} < \frac{16}{30}$, we must borrow 1 from the whole number part of the minuend and add $1 = \frac{30}{30}$ to its fractional part.

$$7 + \frac{5}{30} = 6 + 1 + \frac{5}{30} = 6 + \frac{30}{30} + \frac{5}{30} = 6 + \frac{35}{30}$$

$$\begin{array}{r} 7 + \frac{5}{30} = 6 + \frac{35}{30} \\ -\ 2 + \frac{16}{30} = 2 + \frac{16}{30} \\ \hline 4 + \frac{19}{30} = 4\frac{19}{30} \end{array}$$

Subtract the whole number parts and fractional parts separately.

●

Observe what happens in Example 5 when we subtract a mixed number from a whole number.

example 5

Subtract $7 - 2\frac{5}{8}$.

solution

$7 = 6 + 1 = 6 + \frac{8}{8}$

Since there is no fractional part to the minuend, we rewrite 7 as a mixed number.

$$\begin{array}{rl} 7 & = 6 + 1 = 6 + \frac{8}{8} \\ -\, 2\frac{5}{8} & = 2 + \frac{5}{8} = 2 + \frac{5}{8} \\ \hline & \quad 4 + \frac{3}{8} = 4\frac{3}{8} \end{array}$$

Now we subtract the whole number parts and fractional parts.

Let us now look at an example that combines both addition and subtraction.

example 6

Calculate $5\frac{1}{8} - 2\frac{3}{4} + 3\frac{5}{6}$.

solution

First subtract $5\frac{1}{8} - 2\frac{3}{4}$.

$$\begin{array}{rl} 5\frac{1}{8} & = 5 + \frac{1}{8} \qquad = 4 + \frac{9}{8} \\ -\, 2\frac{3}{4} & = 2 + \frac{3}{4}\cdot\frac{2}{2} = 2 + \frac{6}{8} \\ \hline & \qquad 2 + \frac{3}{8} = 2\frac{3}{8} \end{array}$$

Now add $2\frac{3}{8} + 3\frac{5}{6}$.

$8 = 2 \cdot 2 \cdot 2 = 2^3$

$6 = 2 \cdot 3$

LCD $= 2^3 \cdot 3 = 24$

$$\begin{array}{rl} 2\frac{3}{8} & = 2 + \frac{3}{8}\cdot\frac{3}{3} = 2 + \frac{9}{24} \\ +\, 3\frac{5}{6} & = 3 + \frac{5}{6}\cdot\frac{4}{4} = 3 + \frac{20}{24} \\ \hline & \qquad 5 + \frac{29}{24} = 5 + 1 + \frac{5}{24} \\ & \qquad = 6 + \frac{5}{24} = 6\frac{5}{24} \end{array}$$

We will now solve some word problems that involve addition and subtraction of mixed numbers.

example 7

A flight from Los Angeles to New York that stops in Chicago takes 3 hours 5 minutes to go from Los Angeles to Chicago and 2 hours 40 minutes to go from Chicago to New York. Express the flight time for each part of the trip and the total number of hours of flight time as mixed fractions.

solution

Los Angeles to Chicago: 3 hr 5 min

$$= 3 \text{ hr} + \frac{5}{60} \text{ hr} = 3\frac{1}{12} \text{ hr}$$

Since 1 hr = 60 min.

Chicago to New York: 2 hr 40 min

$$= 2 \text{ hr} + \frac{40}{60} \text{ hr} = 2\frac{2}{3} \text{ hr}$$

$$\begin{array}{rl} 3\frac{1}{12} = 3 + \frac{1}{12} & = 3 + \frac{1}{12} \\ + 2\frac{2}{3} = 2 + \frac{2}{3} \cdot \frac{4}{4} & = 2 + \frac{8}{12} \\ \hline & 5 + \frac{9}{12} = 5 + \frac{3}{4} = 5\frac{3}{4} \end{array}$$

To find the total flight time, add $3\frac{1}{12}$ hr + $2\frac{2}{3}$ hr.

The total flight time is $5\frac{3}{4}$ hr.

example 8

For a distance of $\frac{1}{4}$ mile, the maximum speed for a greyhound is $39\frac{7}{20}$ mph, and that for a human is $27\frac{22}{25}$ mph. How much faster can a greyhound run than a human?

solution

Subtract $39\frac{7}{20} - 27\frac{22}{25}$.

$$\begin{array}{rl} 39\frac{7}{20} = 39 + \frac{7}{20} \cdot \frac{5}{5} = 39 + \frac{35}{100} & = 38 + \frac{135}{100} \\ - 27\frac{22}{25} = 27 + \frac{22}{25} \cdot \frac{4}{4} = 27 + \frac{88}{100} & = 27 + \frac{88}{100} \\ \hline & 11 + \frac{47}{100} = 11\frac{47}{100} \end{array}$$

$20 = 2^2 \cdot 5$

$25 = 5^2$

$\text{LCD} = 2^2 \cdot 5^2 = 100$

A greyhound can run $11\frac{47}{100}$ mph faster than a human.

QUICK QUIZ

Perform each of the following calculations.

1. $6\frac{3}{4} + 2\frac{5}{6}$
2. $7\frac{3}{8} - 3\frac{1}{2}$
3. $8\frac{1}{4} - 2\frac{2}{3} + 3\frac{1}{6}$

ANSWERS

1. $9\frac{7}{12}$
2. $3\frac{7}{8}$
3. $8\frac{3}{4}$

2.5 Exercises

Perform each of the following additions. Express your answers as mixed numbers reduced to lowest terms.

1. $3\frac{1}{8} + 4\frac{5}{8}$

2. $6\frac{2}{9} + 2\frac{4}{9}$

3. $2\frac{1}{4} + 5\frac{3}{8}$

4. $7\frac{2}{3} + 8\frac{1}{6}$

5. $3\frac{3}{4} + 6\frac{7}{8}$

6. $1\frac{3}{5} + 8\frac{9}{10}$

7. $88\frac{1}{2} + 47\frac{3}{8}$

8. $59\frac{2}{9} + 71\frac{1}{3}$

9. $27\frac{5}{16} + 19\frac{3}{4}$

10. $11\frac{4}{5} + 24\frac{7}{20}$

11. $5\frac{1}{3} + 11\frac{1}{4}$

12. $9\frac{1}{2} + 3\frac{2}{7}$

13. $6\frac{7}{8} + 10\frac{5}{6}$

14. $12\frac{8}{9} + 2\frac{7}{12}$

15. $7\frac{3}{16} + 9\frac{9}{20}$

16. $6\frac{9}{10} + 3\frac{8}{15}$

17. $19\frac{5}{12} + 7\frac{7}{18}$

18. $8\frac{4}{21} + 15\frac{9}{14}$

19. $15\frac{5}{7} + 3\frac{3}{8}$

20. $19\frac{3}{4} + 4\frac{7}{9}$

21. $93\frac{3}{5} + 47\frac{4}{25}$

22. $62\frac{5}{49} + 88\frac{2}{7}$

23. $2\frac{22}{45} + 7\frac{11}{20}$

24. $5\frac{25}{36} + 1\frac{13}{16}$

25. $13\frac{5}{48} + 4\frac{7}{36}$

26. $7\frac{4}{63} + 16\frac{8}{27}$

27. $53\frac{7}{18} + 31\frac{5}{54}$

28. $62\frac{5}{48} + 27\frac{9}{16}$

29. $2\frac{3}{5} + 7\frac{1}{5} + 4\frac{4}{5}$

30. $6\frac{7}{8} + 2\frac{5}{8} + 9\frac{3}{8}$

31. $13\frac{3}{4} + 5\frac{1}{6} + 27\frac{5}{12}$

32. $2\frac{1}{2} + 16\frac{5}{16} + 33\frac{3}{8}$

33. $2\frac{5}{6} + 8\frac{2}{3} + 4\frac{2}{27}$

34. $6\frac{5}{12} + 1\frac{2}{9} + 3\frac{4}{15}$

35. $6\frac{2}{3} + \frac{5}{12} + 2\frac{5}{6}$

36. $4\frac{3}{26} + \frac{5}{13} + 1\frac{8}{39}$

37. $2\frac{3}{8} + 1\frac{7}{16} + 7\frac{5}{24}$

38. $5\frac{2}{21} + 2\frac{3}{7} + 6\frac{8}{35}$

39. $3\frac{5}{48} + 8\frac{7}{24} + 7\frac{5}{12}$

40. $7\frac{9}{14} + 4\frac{3}{56} + 6\frac{5}{28}$

41. $85\frac{35}{817} + 167\frac{17}{43}$

42. $284\frac{13}{37} + 58\frac{28}{777}$

43. $473\frac{15}{26} + 519\frac{22}{35}$

44. $267\frac{38}{41} + 734\frac{18}{19}$

Perform each of the following subtractions. Express your answers as mixed numbers reduced to lowest terms.

45. $8\frac{3}{4} - 2\frac{1}{4}$

46. $7\frac{5}{7} - 3\frac{2}{7}$

47. $9\frac{3}{4} - 4\frac{1}{8}$

48. $8\frac{4}{5} - 1\frac{3}{10}$

49. $4\frac{5}{6} - 2\frac{5}{12}$

50. $6\frac{3}{8} - 3\frac{3}{16}$

51. $36\frac{2}{3} - 27\frac{1}{27}$

52. $25\frac{3}{4} - 18\frac{2}{5}$

53. $56\frac{3}{5} - 29\frac{2}{15}$

54. $95\frac{5}{24} - 47\frac{1}{6}$

55. $5\frac{1}{3} - 1\frac{2}{3}$

56. $7\frac{2}{5} - 3\frac{4}{5}$

57. $4\frac{2}{9} - 2\frac{8}{9}$

58. $9\frac{3}{10} - 3\frac{7}{10}$

59. $6\frac{3}{14} - 2\frac{5}{7}$

60. $4\frac{2}{9} - 1\frac{11}{18}$

61. $32\frac{4}{27} - 16\frac{5}{9}$

62. $55\frac{1}{8} - 29\frac{9}{32}$

63. $85\frac{5}{9} - 72\frac{1}{2}$

64. $55\frac{2}{3} - 43\frac{7}{8}$

65. $4\frac{1}{6} - 3\frac{5}{8}$

66. $7\frac{3}{8} - 6\frac{11}{12}$

67. $8 - 5\frac{3}{7}$

68. $9 - 4\frac{2}{9}$

69. $15 - 7\frac{5}{18}$

70. $23 - 16\frac{9}{32}$

71. $54 - 29\frac{5}{8}$

72. $61 - 43\frac{3}{4}$

73. $8\frac{5}{42} - 4\frac{5}{14}$

74. $5\frac{5}{54} - 1\frac{5}{18}$

75. $11\frac{3}{56} - 4\frac{4}{7}$

76. $22\frac{6}{45} - 8\frac{5}{9}$

77. $4\frac{7}{36} - 2\frac{3}{48}$

78. $6\frac{5}{42} - 3\frac{2}{63}$

79. $11\frac{5}{6} - 6\frac{3}{8} - 3\frac{1}{24}$

80. $15\frac{11}{12} - 7\frac{1}{4} - 2\frac{1}{6}$

81. $22\frac{9}{20} - 10\frac{3}{5} - 4\frac{2}{15}$

82. $31\frac{5}{6} - 12\frac{7}{18} - 9\frac{11}{36}$

83. $18\frac{7}{20} - 3\frac{3}{32} - 7\frac{9}{16}$

84. $25\frac{8}{9} - 5\frac{7}{18} - 10\frac{4}{27}$

85. $16\frac{8}{15} - 5\frac{4}{25} - 2\frac{2}{75}$

86. $19\frac{5}{40} - 7\frac{7}{24} - 4\frac{9}{16}$

87. $12\frac{71}{288} - 3\frac{5}{16}$

88. $19\frac{7}{24} - 8\frac{59}{360}$

89. $8\frac{9}{14} - 3\frac{5}{27}$

90. $7\frac{11}{36} - 4\frac{8}{19}$

Perform each of the operations indicated.

91. $6\frac{2}{5} - 2\frac{3}{5} + 4\frac{1}{5}$

92. $7\frac{2}{7} + 3\frac{1}{7} - 5\frac{5}{7}$

93. $8 - 2\frac{3}{4} + 9\frac{4}{5}$

94. $5 - 3\frac{5}{6} + 1\frac{1}{7}$

95. $53\frac{5}{21} - 25\frac{2}{7} + 19\frac{3}{14}$

96. $18\frac{8}{15} + 34\frac{3}{10} - 27\frac{7}{30}$

97. $8\frac{3}{10} + 9\frac{7}{100} - 4\frac{9}{1{,}000}$

98. $5\frac{7}{1{,}000} + 3\frac{1}{10} - 2\frac{3}{100}$

99. $8\frac{3}{16} - 6\frac{5}{12} + 3\frac{9}{32}$

100. $5\frac{5}{18} - 2\frac{7}{36} + 7\frac{4}{27}$

Solve each of the following word problems.

101. The Mont Royal Tunnel in Montreal is about $3\frac{1}{5}$ miles long, and the Holland Tunnel between New York and New Jersey is about $1\frac{7}{11}$ miles long. What is the combined length of these two tunnels?

102. A recipe calls for $\frac{1}{4}$ teaspoon ginger, $\frac{7}{8}$ teaspoon salt, $1\frac{1}{2}$ teaspoons cornstarch, and $1\frac{1}{4}$ teaspoons sugar. What is the total amount of these dry ingredients?

103. A swimming pool with a capacity of $25\frac{3}{20}$ gallons contains $8\frac{5}{6}$ gallons of water. How much more water is needed to fill the pool?

104. For city driving, cars average $14\frac{3}{50}$ miles per gallon, and buses average $5\frac{13}{20}$ miles per gallon. How much farther can a car go than a bus on a gallon of gas?

105. A dress that has a $\frac{5}{8}$-inch hem needs to be shortened $1\frac{3}{4}$ inches. What will be the length of the hem after the dress is shortened?

106. A stock opened on the American Stock Exchange at $30\frac{1}{8}$ and closed at $28\frac{3}{4}$. How much did the stock go down that day?

107. A musician practiced 5 hours 20 minutes on Saturday and 5 hours 35 minutes on Sunday. Express the number of hours she practiced each day and the total number of hours she practiced that weekend as mixed fractions.

108. A piece of ham weighs 2 pounds 7 ounces, and a piece of beef weighs 4 pounds 2 ounces. Express the weight of each in pounds as mixed fractions. How many more pounds does the beef weigh than the ham.

109. A dresser is $4\frac{11}{24}$ feet long and a desk is $3\frac{7}{16}$ feet long. If they are placed end to end along a $14\frac{1}{2}$-foot wall, how much space along the wall will remain open?

110. A bookshelf $2\frac{5}{8}$ feet long and another $1\frac{7}{12}$ feet long are both filled with books. If the books on both shelves are transferred to a shelf 6 feet long, how much space will remain on that bookshelf for additional books?

2.6 Multiplication of Fractions

To begin our discussion of multiplication of fractions, let us consider a variety of situations in which we have different amounts of pie. For each situation, let us indicate $\frac{1}{2}$ *of* the total amount of pie available, as shown in Figure 2.12.

Notice that in each situation, the result can be obtained by converting the two numbers separated by the word *of* to fractions, and multiplying the numerators and the denominators. We can conclude, therefore, that the word *of* means to *multiply*.

$$\frac{1}{2} \text{ of } 4 \quad \text{means} \quad \frac{1}{2}\cdot 4 = \frac{1}{2}\cdot\frac{4}{1} = \frac{1\cdot 4}{2\cdot 1} = \frac{4}{2} = 2$$

$$\frac{1}{2} \text{ of } 3 \quad \text{means} \quad \frac{1}{2}\cdot 3 = \frac{1}{2}\cdot\frac{3}{1} = \frac{1\cdot 3}{2\cdot 1} = \frac{3}{2} = 1\frac{1}{2}$$

$$\frac{1}{2} \text{ of } 2 \quad \text{means} \quad \frac{1}{2}\cdot 2 = \frac{1}{2}\cdot\frac{2}{1} = \frac{1\cdot 2}{2\cdot 1} = \frac{2}{2} = 1$$

$$\frac{1}{2} \text{ of } 1 \quad \text{means} \quad \frac{1}{2}\cdot 1 = \frac{1}{2}\cdot\frac{1}{1} = \frac{1\cdot 1}{2\cdot 1} = \frac{1}{2}$$

$$\frac{1}{2} \text{ of } \frac{1}{2} \quad \text{means} \quad \frac{1}{2}\cdot\frac{1}{2} = \frac{1\cdot 1}{2\cdot 2} = \frac{1}{4}$$

$$\frac{1}{2} \text{ of } \frac{1}{3} \quad \text{means} \quad \frac{1}{2}\cdot\frac{1}{3} = \frac{1\cdot 1}{2\cdot 3} = \frac{1}{6}$$

$$\frac{1}{2} \text{ of } \frac{1}{4} \quad \text{means} \quad \frac{1}{2}\cdot\frac{1}{4} = \frac{1\cdot 1}{2\cdot 4} = \frac{1}{8}$$

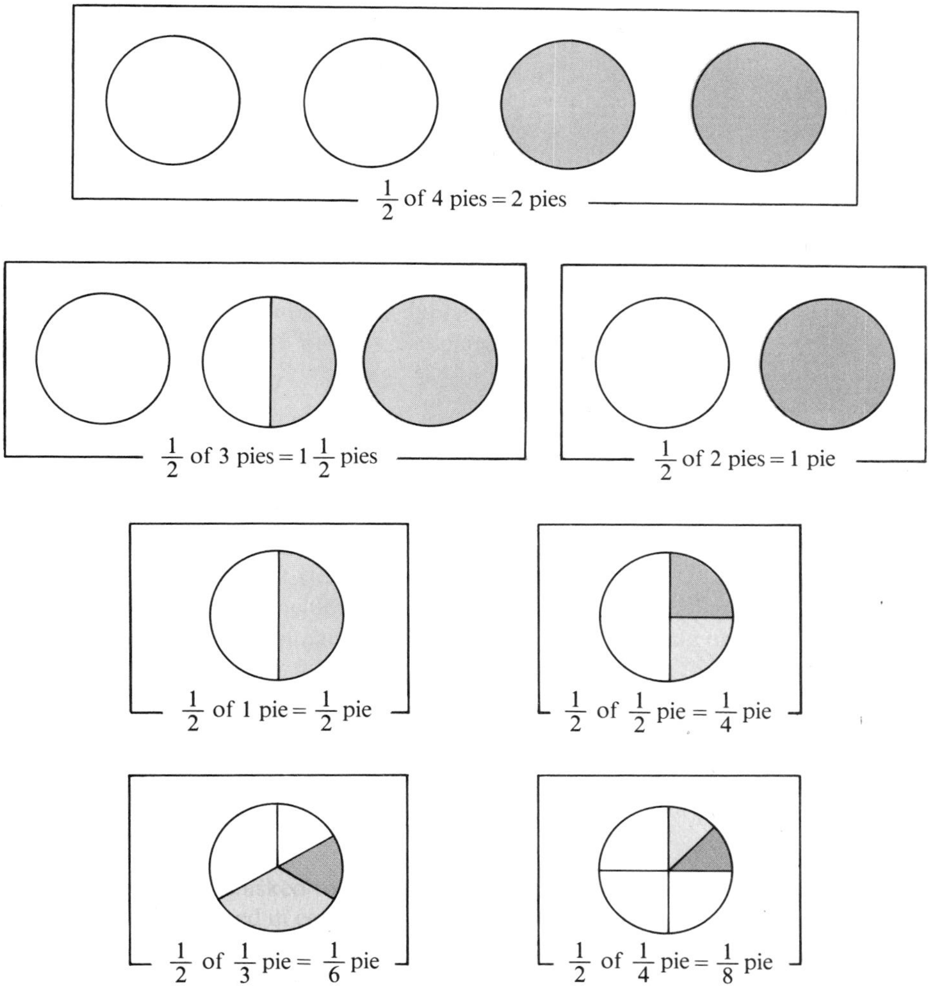

figure 2.12

As another example, let us consider the box shown in Figure 2.13. One-half of the box is lightly shaded. One-third of this shaded area, or $\frac{1}{3}$ of $\frac{1}{2}$, is then shaded again. The area shaded twice is $\frac{1}{6}$ of the total area of the box. Thus,

$$\frac{1}{3} \text{ of } \frac{1}{2} \quad \text{means} \quad \frac{1}{3} \cdot \frac{1}{2} = \frac{1 \cdot 1}{3 \cdot 2} = \frac{1}{6}$$

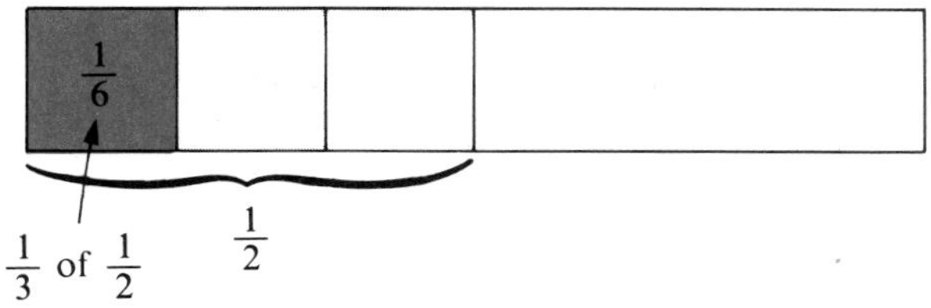

figure 2.13

We have therefore illustrated a procedure for multiplying fractions, as outlined in the box.

To Multiply Fractions

1. Multiply the numerators.
2. Place the result over the product of the denominators.
3. If necessary, reduce the answer to lowest terms.

example 1

Find each of the following products.

(a) $\frac{3}{7} \cdot \frac{2}{5}$

$$\frac{3}{7} \cdot \frac{2}{5} = \frac{3 \cdot 2}{7 \cdot 5} = \frac{6}{35}$$

Multiply the numerators. Place the result over the product of the denominators.

(b) $\frac{2}{9} \cdot \frac{5}{8}$

$$\frac{2}{9} \cdot \frac{5}{8} = \frac{2 \cdot 5}{9 \cdot 8} = \frac{10}{72}$$

Multiply numerators and place result over product of denominators.

$$\frac{10}{72} = \frac{5}{36}$$

Reduce answer to lowest terms.

(c) $\frac{3}{4}$ of 5

$$\frac{3}{4} \text{ of } 5 = \frac{3}{4} \cdot 5$$

Remember that *of* means to multiply.

$$= \frac{3}{4} \cdot \frac{5}{1}$$

Change 5 to an improper fraction.

$$= \frac{3 \cdot 5}{4 \cdot 1} = \frac{15}{4}$$

(d) $\frac{3}{7} \cdot \frac{1}{4} \cdot \frac{2}{5}$

$$\frac{3}{7} \cdot \frac{1}{4} \cdot \frac{2}{5} = \frac{3 \cdot 1 \cdot 2}{7 \cdot 4 \cdot 5} = \frac{6}{140} = \frac{3}{70}$$

(e) $\left(\frac{2}{3}\right)^3$

$$\left(\frac{2}{3}\right)^3 = \frac{2}{3} \cdot \frac{2}{3} \cdot \frac{2}{3}$$

Remember that to cube a number, we must use it as a factor three times.

$$= \frac{2 \cdot 2 \cdot 2}{3 \cdot 3 \cdot 3} = \frac{8}{27}$$

The procedure for multiplying mixed numbers is listed in the box. Example 2 illustrates how to apply the procedure.

To Multiply Mixed Numbers

1. Change each mixed number to an improper fraction.
2. Multiply the numerators.
3. Place the result over the product of the denominators.
4. Express the answer as a mixed number or as a proper fraction reduced to lowest terms.

example 2

Multiply each of the following.

(a) $2\frac{1}{3} \cdot 5\frac{3}{4}$

$$2\frac{1}{3} \cdot 5\frac{3}{4} = \frac{7}{3} \cdot \frac{23}{4}$$

Change each mixed number to an improper fraction.

$$= \frac{7 \cdot 23}{3 \cdot 4}$$

$$= \frac{161}{12} = 13\frac{5}{12}$$

Express the answer as a mixed number.

(b) $\frac{1}{8}$ of $7\frac{2}{3}$

$$\frac{1}{8} \cdot 7\frac{2}{3} = \frac{1}{8} \cdot \frac{23}{3} = \frac{23}{24}$$

To avoid obtaining an answer that needs to be reduced to lowest terms, we divide both numerator and denominator by their common factors before doing any multiplication.

example 3

Perform each of the following multiplications.

(a) $\frac{2}{3} \cdot \frac{3}{5}$

$$\frac{2}{3} \cdot \frac{3}{5} = \frac{2 \cdot \cancel{3}}{\cancel{3} \cdot 5} = \frac{2}{5}$$

Since $\frac{2}{5} \cdot \frac{3}{3} = \frac{2}{5} \cdot 1 = \frac{2}{5}$.

(b) $\frac{5}{36} \cdot \frac{8}{15}$

$$\frac{5}{36} \cdot \frac{8}{15} = \frac{5 \cdot 8}{36 \cdot 15} = \frac{\cancel{5} \cdot \cancel{2} \cdot \cancel{2} \cdot 2}{\cancel{2} \cdot 3 \cdot \cancel{2} \cdot 3 \cdot 3 \cdot \cancel{5}} = \frac{2}{3 \cdot 3 \cdot 3} = \frac{2}{27}$$

Cancel common factors.

(c) $\frac{7}{30} \cdot \frac{12}{21}$

$$\frac{7}{30}\cdot\frac{12}{21}=\frac{\overset{1}{\cancel{7}}\cdot\overset{2}{\cancel{12}}}{\underset{5}{\cancel{30}}\cdot\underset{3}{\cancel{21}}}=\frac{1\cdot 2}{5\cdot 3}=\frac{2}{15}$$

Instead of completely factoring numerator and denominator, simply divide both by a series of common factors.

(d) $\frac{3}{8}\cdot\frac{4}{5}\cdot\frac{2}{9}$

$$\frac{3}{8}\cdot\frac{4}{5}\cdot\frac{2}{9}=\frac{\overset{1}{\cancel{3}}\cdot\overset{1}{\cancel{4}}\cdot\overset{1}{\cancel{2}}}{\underset{4}{\cancel{8}}\cdot 5\cdot\underset{3}{\cancel{9}}}=\frac{1}{5\cdot 3}=\frac{1}{15}$$

Notice that when all the numbers in the numerator are canceled, the numerator becomes 1 and not 0.

(e) $1\frac{2}{5}\cdot\frac{4}{7}\cdot 3\frac{1}{8}$

$$1\frac{2}{5}\cdot\frac{4}{7}\cdot 3\frac{1}{8}=\frac{7}{5}\cdot\frac{4}{7}\cdot\frac{25}{8}=\frac{\overset{1}{\cancel{7}}\cdot\overset{1}{\cancel{4}}\cdot\overset{5}{\cancel{25}}}{\underset{1}{\cancel{5}}\cdot\underset{1}{\cancel{7}}\cdot\underset{2}{\cancel{8}}}=\frac{5}{2}=2\frac{1}{2}$$

Now let us review some of the order of operations rules.

example 4

Perform each of the following operations.

(a) $\frac{2}{7}\left(\frac{1}{3}+\frac{3}{4}\right)$

$$\frac{2}{7}\left(\frac{1}{3}+\frac{3}{4}\right)=\frac{2}{7}\left(\frac{1\cdot 4}{3\cdot 4}+\frac{3\cdot 3}{4\cdot 3}\right)$$

Remember to do what is in the parentheses first.

$$=\frac{2}{7}\left(\frac{4}{12}+\frac{9}{12}\right)$$

$$=\frac{2}{7}\left(\frac{13}{12}\right)=\frac{\cancel{2}\cdot 13}{7\cdot\underset{6}{\cancel{12}}}=\frac{13}{7\cdot 6}=\frac{13}{42}$$

(b) $\frac{3}{4}\cdot\frac{2}{9}+\frac{1}{2}$

$$\frac{3}{4}\cdot\frac{2}{9}+\frac{1}{2}=\frac{3\cdot 2}{4\cdot 9}+\frac{1}{2}=\frac{\cancel{3}\cdot\cancel{2}}{\underset{2}{\cancel{4}}\cdot\underset{3}{\cancel{9}}}+\frac{1}{2}$$

Remember to do multiplication before addition.

$$=\frac{1}{2\cdot 3}+\frac{1}{2}=\frac{1}{6}+\frac{1}{2}$$

$$=\frac{1}{6}+\frac{1\cdot 3}{2\cdot 3}=\frac{1}{6}+\frac{3}{6}=\frac{4}{6}=\frac{2}{3}$$

The distributive property will often make it easier for us to simplify a problem, as Example 4 illustrates.

example 5

Use the distributive property to simplify the following problems.

(a) $12\left(\frac{5}{6} - \frac{3}{4}\right)$

$$12\left(\frac{5}{6} - \frac{3}{4}\right) = 12 \cdot \frac{5}{6} - 12 \cdot \frac{3}{4}$$

$$= \frac{12}{1} \cdot \frac{5}{6} - \frac{12}{1} \cdot \frac{3}{4}$$

$$= \frac{\overset{2}{\cancel{12}} \cdot 5}{1 \cdot \cancel{6}} - \frac{\overset{3}{\cancel{12}} \cdot 3}{1 \cdot \cancel{4}}$$

$$= \frac{10}{1} - \frac{9}{1}$$

$$= 10 - 9 = 1$$

(b) $\frac{2}{3}\left(\frac{9}{8} + \frac{3}{14}\right)$

$$\frac{2}{3}\left(\frac{9}{8} + \frac{3}{14}\right) = \frac{2}{3} \cdot \frac{9}{8} + \frac{2}{3} \cdot \frac{3}{14}$$

$$= \frac{\cancel{2} \cdot \overset{3}{\cancel{9}}}{\cancel{3} \cdot \underset{4}{\cancel{8}}} + \frac{\cancel{2} \cdot \cancel{3}}{\cancel{3} \cdot \underset{7}{\cancel{14}}}$$

$$= \frac{3}{4} + \frac{1}{7}$$

$$= \frac{3 \cdot 7}{4 \cdot 7} + \frac{1 \cdot 4}{7 \cdot 4}$$

$$= \frac{21}{28} + \frac{4}{28} = \frac{25}{28}$$

example 6

A father is 5 feet 8 inches tall, and his daughter is $\frac{3}{4}$ of his height. How tall is his daughter?

solution

$$5 \text{ ft } 8 \text{ in.} = 5\frac{8}{12} \text{ ft} = 5\frac{2}{3} \text{ ft}$$

$$\frac{3}{4} \text{ of } 5\frac{2}{3} = \frac{3}{4} \cdot 5\frac{2}{3} = \frac{3}{4} \cdot \frac{17}{3} = \frac{\cancel{3} \cdot 17}{4 \cdot \cancel{3}} = \frac{17}{4} = 4\frac{1}{4} \text{ ft}$$

Find $\frac{3}{4}$ of $5\frac{2}{3}$.

His daughter is $4\frac{1}{4}$ feet tall, which can also be expressed as 4 feet 3 inches.

Notice that $4\frac{1}{4}$ ft $= 4\frac{3}{12}$ ft $= 4$ ft 3 in.

QUICK QUIZ

Perform each of the following multiplications.

1. $\frac{3}{8} \cdot \frac{5}{7}$
2. $\frac{4}{9} \cdot \frac{6}{7} \cdot \frac{3}{4}$
3. $5\frac{1}{2} \cdot 2\frac{2}{3}$

ANSWERS

1. $\frac{15}{56}$
2. $\frac{2}{7}$
3. $14\frac{2}{3}$

2.6 Exercises

Indicate which of the following statements are true and which are false. For those that are false, change the italic expression to make the statement true.

1. The expression $\frac{1}{3}$ of 2 means $\frac{1}{3}$ *divided by* 2.
2. To multiply two fractions, multiply the denominators and *add* the numerators.
3. The product $5 \cdot \frac{3}{8}$ is equal to $5\frac{3}{8}$.
4. To multiply mixed numbers, first change each mixed number to an *improper fraction.*

Perform each of the following multiplications.

5. $\frac{3}{7} \cdot \frac{2}{9}$
6. $\frac{1}{5} \cdot \frac{3}{4}$
7. $\frac{5}{8} \cdot \frac{1}{4}$
8. $\frac{6}{7} \cdot \frac{2}{5}$
9. $\frac{6}{7} \cdot \frac{2}{11}$
10. $\frac{1}{2} \cdot \frac{5}{12}$
11. $\frac{2}{3} \cdot \frac{5}{9}$
12. $\frac{3}{8} \cdot \frac{7}{10}$
13. $\frac{8}{9} \cdot \frac{3}{4}$
14. $\frac{7}{8} \cdot \frac{4}{9}$
15. $\frac{7}{16} \cdot \frac{2}{49}$
16. $\frac{2}{15} \cdot \frac{5}{12}$
17. $\frac{10}{3} \cdot \frac{15}{8}$
18. $\frac{12}{5} \cdot \frac{20}{3}$
19. $\frac{3}{11} \cdot \frac{5}{18}$
20. $\frac{7}{24} \cdot \frac{3}{16}$
21. $\frac{7}{12} \cdot \frac{16}{49}$
22. $\frac{8}{15} \cdot \frac{35}{63}$
23. $\frac{5}{64} \cdot \frac{8}{3}$
24. $\frac{7}{72} \cdot \frac{9}{2}$
25. $\frac{3}{4} \cdot \frac{1}{7} \cdot \frac{5}{8}$
26. $\frac{5}{9} \cdot \frac{3}{8} \cdot \frac{4}{15}$
27. $\frac{7}{36} \cdot \frac{28}{3} \cdot \frac{6}{7}$
28. $\frac{5}{6} \cdot \frac{54}{7} \cdot \frac{3}{25}$
29. $\frac{18}{121} \cdot \frac{11}{36} \cdot \frac{12}{33}$
30. $\frac{21}{32} \cdot \frac{3}{56} \cdot \frac{64}{15}$
31. $8 \cdot \frac{7}{8}$
32. $9 \cdot \frac{5}{6}$
33. $\frac{2}{7} \cdot 6$
34. $\frac{3}{5} \cdot 4$
35. $4 \cdot 8\frac{3}{5}$
36. $8 \cdot 9\frac{1}{4}$

37. $\frac{1}{2} \cdot 2\frac{1}{8}$

38. $\frac{1}{9} \cdot 6\frac{2}{3}$

39. $9\frac{3}{8} \cdot 5\frac{3}{5}$

40. $4\frac{2}{3} \cdot 1\frac{1}{2}$

41. $2\frac{1}{3} \cdot 3\frac{1}{4}$

42. $4\frac{2}{7} \cdot 1\frac{3}{4}$

43. $5\frac{1}{5} \cdot 6\frac{1}{4}$

44. $2\frac{1}{3} \cdot 4\frac{7}{8}$

45. $5\frac{1}{3} \cdot 6\frac{1}{4}$

46. $7\frac{1}{9} \cdot 3\frac{3}{4}$

47. $6\frac{7}{8} \cdot 5\frac{3}{5}$

48. $3\frac{3}{4} \cdot 8\frac{2}{3}$

49. $7\frac{3}{4} \cdot 2\frac{1}{9}$

50. $7\frac{1}{7} \cdot 8\frac{2}{5}$

51. $4\frac{2}{7} \cdot \frac{5}{6} \cdot 3$

52. $7\frac{1}{5} \cdot \frac{5}{8} \cdot 4$

53. $3\frac{1}{8} \cdot 4\frac{2}{5} \cdot 5\frac{1}{3}$

54. $5\frac{2}{5} \cdot 1\frac{2}{3} \cdot 6\frac{1}{4}$

55. $\frac{37}{52} \cdot \frac{65}{81}$

56. $\frac{63}{98} \cdot \frac{27}{48}$

57. $82\frac{14}{29} \cdot 19\frac{18}{31}$

58. $58\frac{37}{41} \cdot 21\frac{73}{85}$

Perform each of the operations indicated.

59. $\frac{3}{4}\left(\frac{1}{2} - \frac{2}{9}\right)$

60. $\frac{2}{3}\left(\frac{5}{8} + \frac{7}{12}\right)$

61. $\frac{1}{8} \cdot \frac{7}{10} + \frac{4}{5}$

62. $\frac{3}{5} \cdot \frac{1}{3} + \frac{7}{8}$

63. $\frac{1}{2} - \frac{3}{5} \cdot \frac{1}{4}$

64. $\frac{7}{12} + \frac{3}{4} \cdot \frac{5}{6}$

65. $\left(\frac{1}{3}\right)^3 \cdot \left(\frac{1}{2}\right)^2$

66. $\left(\frac{4}{9}\right)^2 \cdot \left(\frac{3}{8}\right)^3$

67. $\frac{55}{72} \cdot \frac{18}{25} + \frac{32}{61}$

68. $\frac{23}{47}\left(\frac{18}{29} - \frac{31}{84}\right)$

Use the distributive property to simplify each of the following problems.

69. $24\left(\frac{5}{8} + \frac{2}{3}\right)$

70. $35\left(\frac{5}{7} - \frac{2}{5}\right)$

71. $6\left(\frac{2}{15} + \frac{1}{6}\right)$

72. $8\left(\frac{3}{4} + \frac{7}{12}\right)$

73. $\frac{1}{4}\left(\frac{4}{7} + \frac{8}{21}\right)$

74. $\frac{3}{7}\left(\frac{14}{3} - \frac{7}{9}\right)$

75. $\frac{3}{11}\left(\frac{22}{5} - \frac{11}{30}\right)$

76. $\frac{5}{12}\left(\frac{24}{7} - \frac{9}{14}\right)$

Solve each of the following word problems.

77. In the United States, $\frac{4}{5}$ of the people have vision problems that require corrective lenses. Of these people, $\frac{2}{7}$ wear contact lenses. What fraction represents the people in the United States who wear contacts?

78. At a PTA meeting $\frac{3}{4}$ of the people have brown eyes and $\frac{1}{5}$ have blue eyes. If $\frac{1}{3}$ of each group is wearing glasses, what fraction represents the total number of blue-eyed and brown-eyed people wearing glasses?

79. A woman swam eight laps of a pool that measures $16\frac{2}{3}$ yards in length. How far did she swim?

80. A cookie recipe calls for $3\frac{3}{4}$ cups of flour. How much flour would you need to make $\frac{1}{3}$ of this recipe?

81. If a rug measures $8\frac{3}{4}$ feet in length, and a room is $2\frac{1}{2}$ times as long as this rug, how long is the room?

82. A car can travel $21\frac{5}{8}$ miles on 1 gallon of gas. How far can it travel on $\frac{1}{4}$ gallon of gas?

2.7 Division of Fractions

Before we begin our discussion of division of fractions, let us look at some interesting results of multiplication of fractions.

$$\frac{1}{3} \cdot 3 = \frac{1}{3} \cdot \frac{3}{1} = \frac{3}{3} = 1 \qquad 17 \cdot \frac{1}{17} = \frac{17}{1} \cdot \frac{1}{17} = \frac{17}{17} = 1$$

$$\frac{9}{7} \cdot \frac{7}{9} = \frac{63}{63} = 1 \qquad 2\frac{1}{4} \cdot \frac{4}{5} = \frac{5}{4} \cdot \frac{4}{5} = \frac{20}{20} = 1$$

The numbers 3 and $\frac{1}{3}$ are said to be **reciprocals** since their product is one. The fraction $\frac{1}{17}$ is the reciprocal of 17 since $17 \cdot \frac{1}{17} = 1$. Likewise, 17 is the reciprocal of $\frac{1}{17}$.

definition

The *reciprocal* of a number is what you must multiply that number by to obtain a product of one. For any fraction not equal to zero, the reciprocal is obtained by inverting the numerator and denominator.

example 1

Give the reciprocal of each of the following numbers.

(a) $\frac{2}{3}$

The reciprocal is $\frac{3}{2}$.

Since $\frac{2}{3} \cdot \frac{3}{2} = \frac{6}{6} = 1$.

(b) 100

The reciprocal is $\frac{1}{100}$.

Since $\frac{100}{1} \cdot \frac{1}{100} = \frac{100}{100} = 1$.

(c) $9\frac{1}{5}$

$9\frac{1}{5} = \frac{46}{5}$

First convert $9\frac{1}{5}$ to an improper fraction, and then invert the numerator and denominator.

The reciprocal is $\frac{5}{46}$.

Since $\frac{46}{5} \cdot \frac{5}{46} = 1$.

(d) $\frac{0}{5}$

Since $\frac{0}{5} = 0$, it has no reciprocal. When zero is multiplied by any number, the product is always zero. ●

To illustrate division of fractions, let us consider this example. If we divide $\frac{1}{4}$ of a pie equally among three people, how large of a piece will each person receive? As we can see in Figure 2.14, each person will receive $\frac{1}{12}$ of the whole pie. Mathematically this can be written as

$$\frac{1}{4} \div 3 = \frac{1}{12}$$

figure 2.14

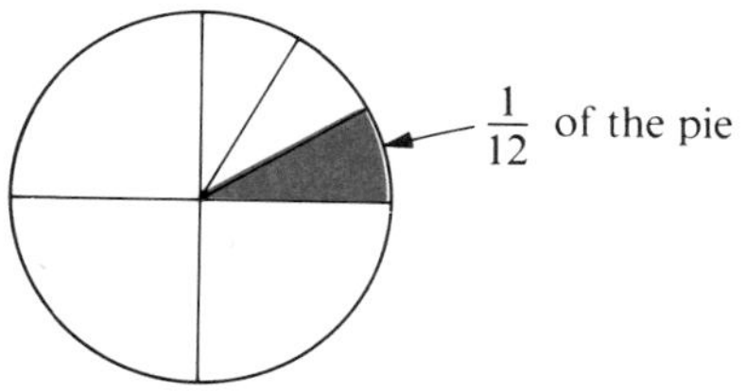

As another example, suppose that we want to determine how many sixths go into $\frac{1}{3}$. In other words, what is $\frac{1}{3} \div \frac{1}{6}$? This problem can be illustrated by a box divided into six equal pieces, as shown in Figure 2.15. Each piece represents $\frac{1}{6}$ of the total box. We can

figure 2.15

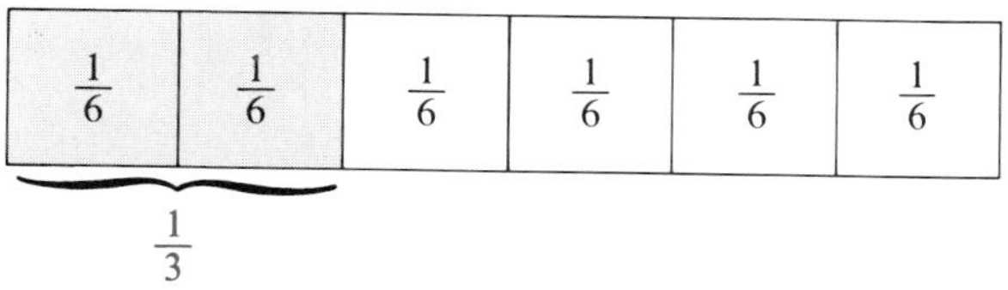

see that $\frac{1}{3}$ of the box (shaded area) contains two of these pieces. Thus, $\frac{1}{6}$ goes into $\frac{1}{3}$ exactly 2 times, or

$$\frac{1}{3} \div \frac{1}{6} = 2$$

If we closely examine these two examples, we can see that when a fraction is divided by a number, the fraction is multiplied by the reciprocal of the number.

$$\frac{1}{4} \div 3 = \frac{1}{4} \cdot \frac{1}{3} = \frac{1}{12}$$

$$\frac{1}{3} \div \frac{1}{6} = \frac{1}{3} \cdot \frac{6}{1} = \frac{6}{3} = 2$$

We can show *why* this technique works by rewriting each division problem as one large fraction.

$$\frac{1}{4} \div 3 \quad \text{means} \quad \frac{\frac{1}{4}}{3}$$

This number is called a **complex fraction**.

definition

A *complex fraction* is a fraction whose numerator and/or denominator contains fractions.

We can multiply both numerator and denominator of the complex fraction by the reciprocal of the denominator.

$$\frac{\frac{1}{4}}{3} = \frac{\frac{1}{4} \cdot \frac{1}{3}}{3 \cdot \frac{1}{3}} = \frac{\frac{1}{12}}{1} = \frac{1}{12}$$

Since we get a 1 in the denominator, the answer is the fraction in the numerator.

This technique also works for problems in which fractions are divided by fractions.

example 2

Divide $\frac{3}{7} \div \frac{2}{5}$.

solution

$$\frac{3}{7} \div \frac{2}{5} = \frac{\frac{3}{7}}{\frac{2}{5}} = \frac{\frac{3}{7} \cdot \frac{5}{2}}{\frac{2}{5} \cdot \frac{5}{2}} = \frac{\frac{15}{14}}{1} = \frac{15}{14}$$

Notice that the answer in Example 2 is the same as the result we would obtain if we simply multiplied the first fraction (dividend) by the reciprocal of the second (divisor).

$$\frac{3}{7} \div \frac{2}{5} = \frac{3}{7} \cdot \frac{5}{2} = \frac{15}{14}$$

We have thus established a procedure for division of fractions, as outlined in the box.

To Divide Fractions

1. Find the reciprocal of the divisor.
2. Multiply the dividend by the reciprocal of the divisor.
3. If necessary, reduce answer to lowest terms.

example 3

Perform each of the following divisions.

(a) $\frac{7}{9} \div 5$

The reciprocal of 5 is $\frac{1}{5}$.

$$\frac{7}{9} \div 5 = \frac{7}{9} \cdot \frac{1}{5} = \frac{7}{45}$$

Multiply $\frac{7}{9}$ by $\frac{1}{5}$.

(b) $\frac{5}{8} \div \frac{2}{3}$

The reciprocal of $\frac{2}{3}$ is $\frac{3}{2}$.

$$\frac{5}{8} \div \frac{2}{3} = \frac{5}{8} \cdot \frac{3}{2} = \frac{15}{16}$$ Multiply $\frac{5}{8}$ by $\frac{3}{2}$.

(c) $8 \div \frac{3}{8}$ The reciprocal of $\frac{3}{8}$ is $\frac{8}{3}$.

$$8 \div \frac{3}{8} = \frac{8}{1} \cdot \frac{8}{3} = \frac{64}{3}$$ Multiply 8 by $\frac{8}{3}$. ●

The next examples illustrate how to divide mixed numbers. The procedure used in the examples is outlined in the box.

To Divide Mixed Numbers

1. Change each mixed number to an improper fraction.
2. Find the reciprocal of the divisor.
3. Multiply the dividend by the reciprocal of the divisor.
4. Express the answer as a mixed number or as a proper fraction reduced to lowest terms.

example 4

Perform each of the following divisions.

(a) $3\frac{2}{7} \div 4$

$$3\frac{2}{7} \div 4 = \frac{23}{7} \div 4$$ Change $3\frac{2}{7}$ to an improper fraction.

$$= \frac{23}{7} \cdot \frac{1}{4}$$ The reciprocal of 4 is $\frac{1}{4}$.

$$= \frac{23}{28}$$

(b) $6\frac{2}{9} \div 5\frac{5}{6}$

$$6\frac{2}{9} \div 5\frac{5}{6} = \frac{56}{9} \div \frac{35}{6}$$ Change mixed numbers to improper fractions.

$$= \frac{56}{9} \cdot \frac{6}{35} = \frac{\overset{8}{\cancel{56}} \cdot \overset{2}{\cancel{6}}}{\underset{3}{\cancel{9}} \cdot \underset{5}{\cancel{35}}} = \frac{16}{15} = 1\frac{1}{15}$$ Multiply by the reciprocal and simplify the answer. ●

example 5

Use the order of operations rules to simplify the following problems.

(a) $\frac{3}{5} \div \frac{9}{10} \cdot \frac{15}{22}$

$$\frac{3}{5} \div \frac{9}{10} \cdot \frac{15}{22} = \left(\frac{3}{5} \div \frac{9}{10}\right) \cdot \frac{15}{22}$$

Multiplication and division are performed in the same step, moving from left to right.

$$= \left(\frac{3}{5} \cdot \frac{10}{9}\right) \cdot \frac{15}{22}$$

$$= \frac{\cancel{3} \cdot \overset{\cancel{5}}{\cancel{10}} \cdot \overset{5}{\cancel{15}}}{\cancel{5} \cdot \underset{3}{\cancel{9}} \cdot \underset{11}{\cancel{22}}} = \frac{5}{11}$$

(b) $\frac{7}{8} - \frac{5}{12} \div \frac{10}{3}$

$$\frac{7}{8} - \frac{5}{12} \div \frac{10}{3} = \frac{7}{8} - \left(\frac{5}{12} \div \frac{10}{3}\right)$$

Perform division before subtraction.

$$= \frac{7}{8} - \left(\frac{5}{12} \cdot \frac{3}{10}\right) = \frac{7}{8} - \frac{\cancel{5} \cdot \cancel{3}}{\underset{4}{\cancel{12}} \cdot \underset{2}{\cancel{10}}}$$

$$= \frac{7}{8} - \frac{1}{4 \cdot 2} = \frac{7}{8} - \frac{1}{8}$$

$$= \frac{6}{8} = \frac{3}{4}$$

(c) $\left(\frac{1}{2}\right)^3 \div \left(\frac{2}{5}\right)^2$

$$\left(\frac{1}{2}\right)^3 \div \left(\frac{2}{5}\right)^2 = \left(\frac{1}{2} \cdot \frac{1}{2} \cdot \frac{1}{2}\right) \div \left(\frac{2}{5} \cdot \frac{2}{5}\right)$$

Raise to powers before dividing.

$$= \frac{1}{8} \div \frac{4}{25}$$

$$= \frac{1}{8} \cdot \frac{25}{4} = \frac{25}{32}$$

example 6

A piece of ribbon that is $7\frac{7}{8}$ yards long is cut into three equal pieces. How long is each piece?

Divide $7\frac{7}{8} \div 3$.

$$7\frac{7}{8} \div 3 = \frac{63}{8} \div 3 = \frac{63}{8} \cdot \frac{1}{3}$$

$$= \frac{\overset{21}{\cancel{63}} \cdot 1}{8 \cdot \cancel{3}} = \frac{21}{8} = 2\frac{5}{8} \text{ yd}$$

Each piece is $2\frac{5}{8}$ yards long.

QUICK QUIZ

Perform each of the following divisions.

1. $\frac{5}{9} \div 7$

2. $\frac{3}{8} \div \frac{27}{32}$

3. $6\frac{2}{3} \div 3\frac{3}{4}$

ANSWERS

1. $\frac{5}{63}$

2. $\frac{4}{9}$

3. $1\frac{7}{9}$

2.7 Exercises

Indicate which of the following statements are true and which are false. For those that are false, change the italic expression to make the statement true.

1. To divide two fractions, multiply the dividend by the *reciprocal* of the divisor.

2. When you *add* a number and its reciprocal, you obtain a result of one.

3. Three divided by one-third is equal to *one-ninth.*

4. Dividing by $\frac{1}{7}$ is the same as multiplying by *7.*

Find the reciprocal of each of the following numbers.

5. $\frac{3}{8}$ **6.** $\frac{7}{12}$ **7.** $\frac{1}{7}$ **8.** $\frac{18}{57}$ **9.** 11 **10.** 29

11. 1 **12.** 0 **13.** $7\frac{3}{4}$ **14.** $15\frac{2}{3}$

Perform each of the following divisions.

15. $\frac{2}{3} \div 9$ **16.** $\frac{5}{8} \div 7$ **17.** $\frac{13}{7} \div 4$ **18.** $\frac{18}{5} \div 5$

19. $\frac{3}{4} \div \frac{1}{8}$ **20.** $\frac{5}{9} \div \frac{1}{4}$ **21.** $\frac{2}{5} \div \frac{3}{7}$ **22.** $\frac{3}{8} \div \frac{2}{3}$

23. $\frac{2}{7} \div \frac{3}{5}$ **24.** $\frac{7}{9} \div \frac{3}{4}$ **25.** $\frac{14}{15} \div \frac{7}{9}$ **26.** $\frac{7}{12} \div \frac{5}{8}$

27. $\frac{5}{36} \div \frac{7}{18}$ **28.** $\frac{3}{16} \div \frac{5}{24}$ **29.** $\frac{3}{11} \div \frac{5}{22}$ **30.** $\frac{5}{36} \div \frac{7}{12}$

31. $\frac{14}{3} \div \frac{7}{2}$ **32.** $\frac{18}{5} \div \frac{9}{4}$ **33.** $\frac{20}{49} \div \frac{15}{14}$ **34.** $\frac{5}{12} \div \frac{25}{18}$

35. $\frac{21}{10} \div \frac{6}{25}$ **36.** $\frac{45}{28} \div \frac{5}{16}$ **37.** $8 \div \frac{4}{7}$ **38.** $5 \div \frac{3}{8}$

39. $\frac{11}{12} \div \frac{2}{3} \div \frac{33}{2}$ **40.** $\frac{5}{3} \div \frac{1}{2} \div \frac{5}{6}$ **41.** $\frac{9}{16} \div \frac{1}{12} \div \frac{3}{4}$ **42.** $\frac{18}{25} \div \frac{9}{10} \div \frac{1}{3}$

43. $\frac{3}{5} \div 1\frac{1}{2}$ **44.** $4\frac{7}{8} \div \frac{3}{8}$ **45.** $5\frac{1}{3} \div \frac{4}{5}$ **46.** $\frac{1}{8} \div 3\frac{5}{8}$

47. $1\frac{2}{5} \div 3\frac{1}{2}$ **48.** $5\frac{1}{3} \div 7\frac{5}{7}$ **49.** $6\frac{3}{4} \div 4\frac{2}{5}$ **50.** $5\frac{2}{3} \div 2\frac{3}{7}$

51. $7\frac{1}{3} \div 2\frac{1}{2}$ **52.** $8\frac{1}{2} \div 1\frac{5}{6}$ **53.** $2\frac{1}{12} \div 3\frac{3}{4}$ **54.** $5\frac{1}{3} \div 1\frac{5}{9}$

55. $6\frac{1}{8} \div 1\frac{3}{4}$ **56.** $3\frac{1}{3} \div 2\frac{6}{7}$ **57.** $2\frac{2}{9} \div 5\frac{7}{11}$ **58.** $4\frac{3}{5} \div 8\frac{2}{3}$

59. $\frac{53}{81} \div \frac{24}{37}$ **60.** $\frac{16}{49} \div \frac{61}{72}$ **61.** $26\frac{12}{31} \div 11\frac{26}{59}$ **62.** $84\frac{23}{62} \div 29\frac{37}{81}$

Perform each of the operations indicated.

63. $\frac{1}{3} \cdot \frac{8}{15} \div \frac{4}{5}$ **64.** $\frac{7}{9} \cdot \frac{5}{6} \cdot \frac{3}{10}$ **65.** $\frac{1}{9} \div \frac{5}{9} \cdot \frac{15}{49} \div \frac{3}{14}$ **66.** $\frac{1}{7} \cdot \frac{2}{9} \div \frac{4}{27} \cdot \frac{7}{2}$

67. $\left(\frac{2}{7} + \frac{3}{5}\right) \div \frac{1}{5}$ **68.** $\left(\frac{3}{4} - \frac{5}{9}\right) \div \frac{2}{3}$ **69.** $\frac{1}{3} \div \frac{1}{4} + \frac{5}{6}$ **70.** $\frac{2}{3} + \frac{5}{8} \div \frac{15}{4}$

71. $\left(\frac{51}{67} + \frac{11}{19}\right) \div \frac{15}{43}$ **72.** $\frac{43}{77} \div \frac{18}{53} \cdot \frac{27}{31}$

Solve each of the following word problems.

73. One lap of an indoor track is $\frac{1}{8}$ mile long. How many laps must a jogger run to go $\frac{2}{3}$ mile?

74. A car went $\frac{11}{12}$ mile on $\frac{1}{30}$ gallon of gas. How many miles per gallon did the car get?

75. A total of $3\frac{1}{8}$ pounds of chocolate is to be divided evenly among ten children. How much chocolate should each child receive?

76. An eight-story building is $90\frac{2}{3}$ feet high. What is the height of each story?

77. How many boxes can be filled with $7\frac{2}{9}$ pounds of crackers if each box has a capacity of $1\frac{2}{3}$ pounds?

78. A drinking glass can hold $1\frac{1}{3}$ cups of liquid. How many glasses of this size can be filled from a pitcher that contains $14\frac{2}{3}$ cups of water?

79. The planet Jupiter makes one rotation on its axis in $9\frac{5}{6}$ hours. How many times will it rotate in $44\frac{1}{4}$ hours?

80. A bus traveled $65\frac{7}{8}$ miles in $1\frac{5}{12}$ hours. What was its average speed in miles per hour?

2.8 Simplifying Complex Fractions

We have already introduced one technique to simplify complex fractions, that is, multiplying numerator and denominator by the reciprocal of the denominator. For example,

$$\frac{\frac{2}{9}}{\frac{3}{5}} = \frac{\frac{2}{9} \cdot \frac{5}{3}}{\frac{3}{5} \cdot \frac{5}{3}} = \frac{\frac{10}{27}}{1} = \frac{10}{27}$$

Notice that the same result can be obtained by multiplying both numerator and denominator by the LCD of the fractions $\frac{2}{9}$ and $\frac{3}{5}$, which is 45.

$$\frac{\frac{2}{9}}{\frac{3}{5}} = \frac{\frac{2}{9} \cdot 45}{\frac{3}{5} \cdot 45} = \frac{\frac{2}{\not{9}} \cdot \frac{\not{45}^{5}}{1}}{\frac{3}{\not{5}} \cdot \frac{\not{45}^{9}}{1}} = \frac{2 \cdot 5}{3 \cdot 9} = \frac{10}{27}$$

The second technique produces a result in which there is a whole number in both the numerator and denominator. This procedure is very useful when evaluating complicated complex fractions.

example 1

Simplify the following complex fraction.

$$\frac{\frac{1}{3} + \frac{1}{2}}{\frac{5}{6}}$$

solution

$$\frac{\frac{1}{3} + \frac{1}{2}}{\frac{5}{6}}$$

Find the LCD of all fractions that appear in the numerator and denominator. The LCD of $\frac{1}{3}$, $\frac{1}{2}$, and $\frac{1}{6}$ is 6.

$$\frac{\left(\frac{1}{3} + \frac{1}{2}\right) \cdot 6}{\frac{5}{6} \cdot 6} = \frac{\frac{1}{3} \cdot \frac{6}{1} + \frac{1}{2} \cdot \frac{6}{1}}{\frac{5}{6} \cdot \frac{6}{1}}$$

Multiply both numerator and denominator of the complex fraction by the LCD. Evaluate the numerator using the distributive property.

$$= \frac{2 + 3}{5} = \frac{5}{5} = 1$$

●

In general, complex fractions that contain arithmetic operations in the numerator and/or denominator can be simplified in one of two ways. Both methods are listed in the box.

To Simplify Complex Fractions

Simplify the numerator and denominator separately. Then multiply the numerator by the reciprocal of the denominator.

OR

Find the LCD of all fractions that appear in the numerator and denominator. Then multiply both numerator and denominator by the LCD, using the distributive property.

The following example illustrates both techniques.

example 2

Simplify the following complex fraction.

$$\frac{\frac{5}{8}-\frac{1}{3}}{\frac{1}{4}+\frac{7}{12}}$$

(a) $$\frac{\frac{5}{8}-\frac{1}{3}}{\frac{1}{4}+\frac{7}{12}} = \frac{\frac{5\cdot 3}{8\cdot 3}-\frac{1\cdot 8}{3\cdot 8}}{\frac{1\cdot 3}{4\cdot 3}+\frac{7}{12}} = \frac{\frac{7}{24}}{\frac{10}{12}}$$

Simplify numerator and denominator separately.

$$= \frac{7}{24} \div \frac{10}{12} = \frac{7}{24}\cdot\frac{12}{10}$$

Now multiply the numerator by the reciprocal of the denominator.

$$= \frac{7\cdot \cancel{12}}{\underset{2}{\cancel{24}}\cdot 10} = \frac{7}{2\cdot 10} = \frac{7}{20}$$

(b) $8 = 2^3$

$3 = 3$

$4 = 2^3$

$12 = 2^2 \cdot 3$

$\text{LCD} = 2^3 \cdot 3 = 24$

Find the LCD of $\frac{5}{8}, \frac{1}{3}, \frac{1}{4}$, and $\frac{7}{12}$.

$$\frac{\left(\frac{5}{8}-\frac{1}{3}\right)\cdot 24}{\left(\frac{1}{4}+\frac{7}{12}\right)\cdot 24} = \frac{\left(\frac{5}{8}\cdot 24\right)-\left(\frac{1}{3}\cdot 24\right)}{\left(\frac{1}{4}\cdot 24\right)+\left(\frac{7}{12}\cdot 24\right)}$$

Multiply numerator and denominator by 24, using the distributive property.

$$= \frac{\left(\frac{5}{\cancel{8}}\cdot\frac{\overset{3}{\cancel{24}}}{1}\right)-\left(\frac{1}{\cancel{3}}\cdot\frac{\overset{8}{\cancel{24}}}{1}\right)}{\left(\frac{1}{\cancel{4}}\cdot\frac{\overset{6}{\cancel{24}}}{1}\right)+\left(\frac{7}{\cancel{12}}\cdot\frac{\overset{2}{\cancel{24}}}{1}\right)}$$

$$= \frac{(5\cdot 3)-8}{6+(7\cdot 2)}$$

$$= \frac{15-8}{6+14} = \frac{7}{20}$$

example 3

Simplify the following complex fraction.

$$\frac{5\frac{7}{18}+2\frac{1}{2}}{9\frac{3}{4}-4\frac{5}{12}}$$

solution

$18 = 2 \cdot 3^2$

$2 = 2$ Find the LCD of all fractions.

$4 = 2^2$

$12 = 2^2 \cdot 3$

$\text{LCD} = 2^2 \cdot 3^2 = 36$

$$\frac{5\frac{7}{18} + 2\frac{1}{2}}{9\frac{3}{4} - 4\frac{5}{12}} = \frac{\left(\frac{97}{18} + \frac{5}{2}\right) \cdot 36}{\left(\frac{39}{4} - \frac{53}{12}\right) \cdot 36}$$

Change each mixed number to an improper fraction and multiply numerator and denominator by 36.

$$= \frac{\left(\frac{97}{18} \cdot 36\right) + \left(\frac{5}{2} \cdot 36\right)}{\left(\frac{39}{4} \cdot 36\right) - \left(\frac{53}{12} \cdot 36\right)}$$

$$= \frac{\left(\frac{97}{\cancel{18}} \cdot \frac{\overset{2}{\cancel{36}}}{1}\right) + \left(\frac{5}{\cancel{2}} \cdot \frac{\overset{18}{\cancel{36}}}{1}\right)}{\left(\frac{39}{\cancel{4}} \cdot \frac{\overset{9}{\cancel{36}}}{1}\right) - \left(\frac{53}{\cancel{12}} \cdot \frac{\overset{3}{\cancel{36}}}{1}\right)}$$

$$= \frac{(97 \cdot 2) + (5 \cdot 18)}{(39 \cdot 9) - (53 \cdot 3)} = \frac{194 + 90}{351 - 159} = \frac{284}{192} = \frac{71}{48} = 1\frac{23}{48}$$

●

QUICK QUIZ

Simplify each of the following complex fractions.

1. $\dfrac{\frac{5}{6} - \frac{3}{4}}{\frac{2}{3}}$

2. $\dfrac{2\frac{1}{3}}{\frac{1}{2} + \frac{4}{9}}$

ANSWERS

1. $\frac{1}{8}$

2. $2\frac{8}{17}$

2.8 Exercises

Simplify each of the following complex fractions.

1. $\dfrac{\frac{3}{8}}{\frac{5}{6}}$ 2. $\dfrac{\frac{2}{5}}{\frac{7}{9}}$ 3. $\dfrac{\frac{8}{11}}{\frac{5}{22}}$ 4. $\dfrac{\frac{7}{18}}{\frac{13}{27}}$ 5. $\dfrac{\frac{2}{5} + \frac{4}{3}}{\frac{8}{9}}$ 6. $\dfrac{\frac{3}{4} + \frac{5}{7}}{\frac{1}{2}}$

7. $\dfrac{\frac{3}{4}}{\frac{9}{16}-\frac{1}{2}}$
8. $\dfrac{\frac{4}{5}}{\frac{11}{15}-\frac{2}{3}}$
9. $\dfrac{\frac{2}{9}+\frac{5}{6}}{\frac{1}{2}+\frac{2}{3}}$
10. $\dfrac{\frac{5}{12}+\frac{1}{4}}{\frac{2}{3}+\frac{5}{6}}$
11. $\dfrac{\frac{5}{24}+\frac{3}{8}}{\frac{7}{12}-\frac{1}{6}}$
12. $\dfrac{\frac{4}{5}+\frac{6}{15}}{\frac{2}{3}-\frac{7}{30}}$

13. $\dfrac{\frac{5}{7}-\frac{1}{4}}{\frac{1}{2}+\frac{5}{8}}$
14. $\dfrac{\frac{1}{2}-\frac{2}{5}}{\frac{3}{10}+\frac{6}{7}}$
15. $\dfrac{\frac{11}{15}-\frac{7}{12}}{\frac{3}{4}-\frac{1}{30}}$
16. $\dfrac{\frac{11}{21}-\frac{2}{9}}{\frac{6}{7}-\frac{1}{3}}$
17. $\dfrac{\frac{5}{8}+\frac{1}{2}}{4\frac{3}{4}}$
18. $\dfrac{\frac{7}{9}+\frac{2}{3}}{5\frac{1}{2}}$

19. $\dfrac{3\frac{5}{8}}{\frac{3}{4}-\frac{7}{12}}$
20. $\dfrac{1\frac{7}{9}}{\frac{2}{3}-\frac{5}{18}}$
21. $\dfrac{\frac{3}{10}}{2\frac{1}{5}+4\frac{7}{15}}$
22. $\dfrac{\frac{1}{6}}{5\frac{7}{18}+1\frac{5}{12}}$
23. $\dfrac{7\frac{4}{5}-3\frac{1}{4}}{\frac{2}{3}}$
24. $\dfrac{3\frac{4}{7}-1\frac{2}{21}}{\frac{5}{6}}$

25. $\dfrac{3\frac{5}{12}-1\frac{2}{15}}{\frac{7}{30}+1\frac{3}{20}}$
26. $\dfrac{\frac{8}{21}+3\frac{2}{3}}{4\frac{1}{2}-2\frac{3}{7}}$
27. $\dfrac{3\frac{2}{5}-2\frac{1}{6}}{7\frac{1}{2}+\frac{4}{5}}$
28. $\dfrac{3\frac{2}{9}+2\frac{5}{18}}{5\frac{5}{6}-\frac{7}{27}}$
29. $\dfrac{3\frac{2}{9}+4\frac{1}{2}}{1\frac{5}{18}+6\frac{2}{3}}$
30. $\dfrac{3\frac{5}{6}+5\frac{3}{4}}{2\frac{7}{12}+6\frac{1}{3}}$

31. $\dfrac{5\frac{3}{10}-2\frac{5}{6}}{7\frac{4}{5}-3\frac{1}{2}}$
32. $\dfrac{6\frac{7}{12}-2\frac{1}{6}}{3\frac{5}{8}-1\frac{3}{4}}$

Summary and Review

Key Terms

[2.1] A **fraction** is any number that can be written in the form $\frac{a}{b}$, where a and b are numbers such that b is not equal to zero. The **terms** of the fraction are a and b. The term a is called the **numerator**, and b is called the **denominator**.

A **proper fraction** is one in which the numerator is less than the denominator.

An **improper fraction** is one in which the numerator is greater than or equal to the denominator.

A **mixed number** is the sum of a whole number and a proper fraction.

[2.2] **Equivalent fractions** are fractions that have the same numerical value.

A fraction is in **lowest terms**, or **simplest terms**, if the only common factor of both the numerator and denominator is the number one.

[2.3] The **lowest common denominator**, or **LCD**, is the smallest number that is divisible by the denominators of all fractions being considered.

[2.7] The **reciprocal** of a number is what you must multiply that number by to obtain a product of one.

A **complex fraction** is a fraction whose numerator and/or denominator contains fractions.

Calculations

[2.1] **To convert an improper fraction to a mixed number**, divide the numerator by the denominator. The quotient is the whole number portion of the mixed number, and the fractional portion of the mixed number is the remainder divided by the divisor.

To convert a mixed number to an improper fraction, multiply the whole number part by the denominator, add the product to the numerator, and put the result over the same denominator.

[2.2] **To find an equivalent fraction of one that is given**, multiply or divide the numerator and denominator of the given fraction by the same number.

To reduce a fraction to lowest terms, divide both numerator and denominator by their greatest common factor.

[2.3] **To find the lowest common denominator (LCD) of two or more fractions**, completely factor each denominator, and calculate the product of all unique prime factors, each of which is raised to the highest exponent that appears in the complete factorization of any denominator. In other words, find the LCM of the denominators.

To add (or subtract) fractions, find the LCD and change each fraction to an equivalent fraction with a denominator equal to the LCD. Add (or subtract) the numerators and place the result over the LCD.

[2.4] **To find the larger of two fractions**, change each to an equivalent fraction with a denominator equal to the LCD and compare the numerators. The larger fraction has the larger numerator.

[2.5] **To add mixed numbers**, add the whole number parts and fractional parts separately, change any resulting improper fraction to a mixed number, and add the sums together.

To subtract mixed numbers, change the fractional parts to equivalent fractions with the same LCD. If the fractional part of the minuend is less than that of the subtrahend, borrow one from the whole number part of the minuend, and add one to its fractional part. Then subtract the whole number and fractional parts separately.

[2.6] **To multiply fractions**, multiply the numerators and place the result over the product of the denominators.

To multiply mixed numbers, change each mixed number to an improper fraction. Then multiply the numerators and place the result over the product of the denominators.

[2.7] **To find the reciprocal of a nonzero fraction**, invert the numerator and the denominator.

To divide two fractions, multiply the first (dividend) by the reciprocal of the second (divisor).

To divide mixed numbers, change each mixed number to an improper fraction, and multiply the dividend by the reciprocal of the divisor.

[2.8] **To simplify a complex fraction**, multiply the numerator and denominator by the LCD of all fractions that appear in the numerator and denominator.

Chapter 2 Review Exercises

Indicate which of the following statements are true and which are false. For those that are false, change the italic word to make the statement true.

1. A mixed number is the *product* of a whole number and a fraction.
2. The commutative property holds true for addition and *subtraction* of fractions.

3. Fractions in which the numerator is greater than or equal to the denominator are called *improper* fractions.
4. To convert a mixed number to an improper fraction, you multiply the whole number by the *denominator* and add the numerator.
5. A *reduced* fraction is obtained by multiplying the numerator and denominator by the same whole number.
6. Any fraction that has a *one* in the denominator is equal to the numerator.

State which of the following problems are correct and which are incorrect. For those that are incorrect, replace the underlined answer with the correct one.

7. $3 \cdot \frac{5}{8} = \underline{3\frac{5}{8}}$ **8.** $\frac{1}{2} + \frac{3}{3} = \underline{\frac{4}{5}}$ **9.** $\frac{0}{9}$ is undefined. **10.** $\frac{1}{3} \cdot \frac{9}{5} + \frac{1}{5} = \underline{\frac{2}{3}}$

11. $2 - \frac{1}{6} = \underline{1\frac{5}{6}}$ **12.** $\frac{132}{144} = \underline{\frac{11}{12}}$ **13.** $\frac{5}{8}$ is less than $\frac{3}{5}$. **14.** The reciprocal of $1\frac{1}{8}$ is $\underline{8}$.

[2.3–2.8] Perform each of the following operations indicated. Be sure to reduce all answers to lowest terms.

15. $\frac{3}{8} + \frac{1}{4}$ **16.** $\frac{5}{9} + \frac{2}{3}$ **17.** $\frac{5}{12} - \frac{1}{8}$ **18.** $\frac{5}{6} - \frac{2}{15}$

19. $\frac{8}{11} \cdot \frac{5}{3}$ **20.** $\frac{3}{8} \cdot \frac{7}{12}$ **21.** $\frac{7}{16} \div \frac{1}{20}$ **22.** $\frac{42}{9} \div \frac{10}{6}$

23. $\frac{3}{18} + \frac{5}{27}$ **24.** $\frac{7}{16} + \frac{11}{24}$ **25.** $6\frac{3}{4} \cdot \frac{2}{9}$ **26.** $\frac{3}{5} \cdot 2\frac{7}{8}$

27. $7 \div \frac{1}{3}$ **28.** $\frac{3}{8} \div 9$ **29.** $\frac{9}{28} + \frac{1}{21}$ **30.** $\frac{5}{36} + \frac{7}{48}$

31. $5\frac{3}{4} - \frac{5}{12}$ **32.** $8\frac{2}{3} - \frac{7}{15}$ **33.** $5\frac{1}{8} + 6\frac{2}{3}$ **34.** $3\frac{5}{9} + 2\frac{7}{12}$

35. $7\frac{5}{6} - 2\frac{7}{8}$ **36.** $9\frac{1}{3} - 5\frac{8}{15}$ **37.** $\frac{17}{72} \div 3\frac{7}{9}$ **38.** $5\frac{2}{7} \div \frac{3}{4}$

39. $3\frac{2}{25} \cdot 8\frac{4}{7}$ **40.** $6\frac{5}{9} \cdot 5\frac{3}{4}$ **41.** $7\frac{5}{6} \div 2\frac{3}{4}$ **42.** $5\frac{1}{4} \div 2\frac{5}{8}$

43. $\frac{7}{15} \cdot \frac{35}{18} \div \frac{49}{12}$ **44.** $\frac{3}{32} \div \frac{5}{8} \cdot \frac{5}{16}$ **45.** $8\frac{4}{9} + 3\frac{5}{18} - 2\frac{2}{27}$ **46.** $7\frac{3}{5} - 2\frac{7}{20} + 9\frac{3}{4}$

47. $\frac{5}{3}\left(\frac{12}{5} - \frac{1}{2}\right)$ **48.** $\frac{2}{5}\left(\frac{8}{9} + \frac{1}{4}\right)$ **49.** $\frac{6}{7} \cdot \frac{5}{9} + \frac{2}{3}$ **50.** $\frac{5}{7} - \frac{1}{3} \cdot \frac{1}{4}$

51. $\left(\frac{1}{3}\right)^2 + \frac{5}{12}$ **52.** $\frac{7}{9} - \left(\frac{1}{2}\right)^3$ **53.** $\left(\frac{3}{4}\right)^3 - \left(\frac{1}{8}\right)^2$ **54.** $\left(\frac{2}{3}\right)^3 + \left(\frac{5}{6}\right)^2$

55. $\dfrac{\frac{5}{13}}{\frac{7}{39}}$ **56.** $\dfrac{\frac{7}{55}}{\frac{3}{11}}$ **57.** $\dfrac{8\frac{1}{6}}{18\frac{2}{3}}$ **58.** $\dfrac{5\frac{1}{8}}{20\frac{3}{4}}$

59. $\dfrac{\frac{7}{9} + \frac{5}{6}}{\frac{11}{12}}$ **60.** $\dfrac{\frac{8}{15} - \frac{1}{3}}{\frac{2}{5}}$ **61.** $\dfrac{\frac{21}{25}}{\frac{4}{5} + \frac{2}{15}}$ **62.** $\dfrac{\frac{7}{36}}{\frac{8}{9} - \frac{5}{12}}$

63. $\dfrac{\frac{2}{3}+\frac{1}{2}}{\frac{3}{4}+\frac{5}{12}}$

64. $\dfrac{\frac{3}{16}+\frac{1}{2}}{\frac{3}{4}+\frac{5}{12}}$

65. $\dfrac{\frac{5}{9}+2\frac{3}{4}}{\frac{7}{12}-\frac{1}{8}}$

66. $\dfrac{5\frac{1}{6}-\frac{3}{4}}{\frac{2}{9}+\frac{5}{18}}$

67. $\dfrac{5\frac{2}{7}-3\frac{1}{3}}{2\frac{5}{6}+1\frac{9}{42}}$

68. $\dfrac{3\frac{2}{3}-1\frac{4}{63}}{5\frac{3}{7}-2\frac{1}{9}}$

69. $\dfrac{4\frac{3}{4}+5\frac{2}{3}}{9\frac{5}{6}-6\frac{1}{2}}$

70. $\dfrac{3\frac{3}{4}+1\frac{1}{2}}{5\frac{4}{5}-2\frac{7}{10}}$

Solve each of the following word problems.

71. In a small town in New Hampshire, $\frac{1}{12}$ of the people are between 6 feet and $6\frac{1}{2}$ feet tall. Of the town's population, $\frac{3}{64}$ are over $6\frac{1}{2}$ feet tall. What fraction of the people in this town are over 6 feet tall?

72. In 1975, the average expected lifetime for a man living in the United States was $68\frac{7}{9}$ years, and that for a man living in Denmark was $71\frac{1}{12}$ years. On the average, how much greater was the male life expectancy for a Dane than for an American in 1975?

73. A cookie recipe calls for $2\frac{2}{3}$ cups of sugar and $3\frac{1}{2}$ cups of flour. How much flour and sugar would you need to make a double recipe?

74. A race car driver drove 200 miles in $1\frac{1}{5}$ hours. What was the car's average speed in miles per hour?

75. If $\frac{5}{12}$ of the people attending a convention are Democrats and $\frac{4}{9}$ are Republicans, are there more Democrats or Republicans at that convention? What fraction of the people are affiliated with neither party?

76. If $\frac{2}{9}$ of the students in a history class received a grade above C, and $\frac{8}{21}$ received a grade below C, what fraction of the students received a grade of C?

77. A company produces soda bottles that have a capacity of $1\frac{1}{8}$ pints each. How many bottles can be filled with $20\frac{1}{4}$ pints of soda?

78. At a well-known university, there are 2,550 freshmen. Two-thirds of the freshmen take mathematics, and $\frac{1}{5}$ of those taking mathematics also take economics. How many freshmen at this school study both mathematics and economics?

79. On a certain day, $\frac{3}{14}$ of a bank's customers made check deposits, $\frac{1}{6}$ made cash deposits, and $\frac{4}{7}$ made withdrawals. If the bank had 546 customers on that day, how many more people made withdrawals than deposits?

80. Of the people who work at a particular company, $\frac{3}{4}$ drive 10 miles or less to work, and $\frac{1}{6}$ drive more than 10 miles to work. Of all those who drive to work, $\frac{4}{9}$ car-pool. What fraction of the people who work at this company car-pool?

3 DECIMALS

3.1 Decimals and Place Value

In this chapter, we will increase our knowledge of place value by discussing the values of digits located to the right of the ones place. These are called *decimal places* and are represented by powers of ten that appear in the denominator.

The decimal places are named from left to right beginning with the tenths place, which is represented by the fraction $\frac{1}{10}$ (one tenth). The value of the next place to the right is obtained by multiplying the denominator of one-tenth by 10. Thus, to the right of the tenths place is the hundredths place, represented by the fraction $\frac{1}{100}$ (one hundredth). A chart showing the first six decimal places to the right of the ones place, along with the six corresponding place values to the left of the ones place, is given in Figure 3.1.

Notice that the place values located to the left of the ones place all end in *s*: ten*s*, hundred*s*, thousand*s*, and so on. The decimal places, located to the right of the ones place, all end in *ths*: ten*ths*, hundred*ths*, thousand*ths*, and so on. The fractions that

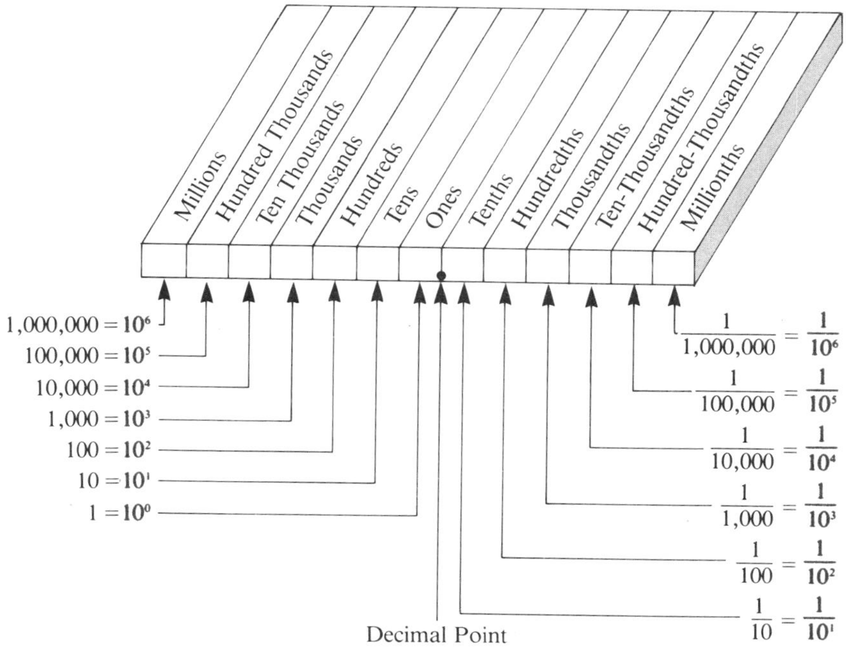

figure 3.1
Decimal Places

represent the decimal places, such as $\frac{1}{10}$, $\frac{1}{100}$, and $\frac{1}{1{,}000}$, are examples of **decimal fractions**.

definition

A *decimal fraction* is a fraction whose denominator can be expressed as a power of ten.

If the numerator of a decimal fraction is a whole number, the name of the fraction will tell us the place value of the rightmost digit in the numerator.

example 1

Write the name of each of the following decimal fractions in words, and give the place value of each digit in the numerator.

(a) $\frac{3}{10}$

three tenths
3 is in the tenths place.

(b) $\frac{8}{1{,}000}$

eight thousandths
8 is in the thousandths place.

(c) $\frac{16}{100}$

sixteen hundredths
6 is in the hundredths place.
1 is in the tenths place.

One place to the left of the hundredths place is the tenths place.

(d) $\frac{4{,}853}{10{,}000}$

four thousand, eight hundred fifty-three ten-thousandths
3 is in the ten-thousandths place.
5 is in the thousandths place.
8 is in the hundredths place.
4 is in the tenths place.

(e) $\frac{521}{10}$

five hundred twenty-one tenths
1 is in the tenths place.
2 is in the ones place.
5 is in the tens place.

Notice that $\frac{521}{10}$ is an improper fraction that can be rewritten as $52\frac{1}{10}$ (fifty-two and one tenth).

Numbers that can be expressed as decimal fractions are called **decimals.**

definition

A *decimal* is a number that can be represented as a fraction whose denominator is a power of ten greater than or equal to one.

Writing decimals as decimal fractions can be very awkward, so it is more common to express decimals in *decimal notation.* When a number is written in decimal notation, the digits in the various decimal places are written to the right of the ones place, separated from the ones place by a *decimal point*, a symbol that looks like a period.

As an example, let us look at how the following numbers are written as decimal fractions and in decimal notation.

seven tenths	$\frac{7}{10}$	$= .7$
seven hundredths	$\frac{7}{100}$	$= .07$
seven thousandths	$\frac{7}{1{,}000}$	$= .007$
seven ten-thousandths	$\frac{7}{10{,}000}$	$= .0007$

The number of decimal places is equal to the number of zeros in the denominator of the decimal fraction. Zeros that appear between the decimal point and the first nonzero digit are necessary placeholders. To determine the number of zeros between the decimal point and the first nonzero digit, subtract the number of digits in the numerator of the decimal fraction from the number of zeros in the denominator. For example,

2 digits
4 zeros
$$\frac{61}{10{,}000} = .0061 \qquad 4 - 2 = 2 \text{ zeros}$$

This number is read as sixty-one ten-thousandths. The rightmost digit appears in the ten-thousandths place.

We can also write .0061 as 0.0061. Even though the first zero is not a necessary placeholder in this case, it is often included to make the number easier to read.

example 2

Write each of the following decimal fractions in decimal notation and translate it into words.

(a) $\frac{4}{100}$

$$\frac{4}{100} = 0.04$$

four hundredths

(b) $\frac{81}{1{,}000}$

$$\frac{81}{1{,}000} = 0.081$$

eighty-one thousandths

(c) $\dfrac{747}{100{,}000}$

$$\frac{747}{100{,}000} = 0.00747$$

seven hundred forty-seven hundred-thousandths ●

When a number greater than one is written in decimal notation, we read the whole number and decimal portions separately, replacing the decimal point by the word *and*. For example,

548.372

is read as

five hundred forty-eight *and* three hundred seventy-two thousandths

The place value of each digit can be indicated as follows:

hundreds	tens	ones		tenths	hundredths	thousandths
5	4	8	.	3	7	2
			↑ and			

In decimal notation, we know that zeros to the right of the decimal point between the decimal point and the first nonzero digit are essential. Others frequently are not. For example, the value of a number is not affected by zeros to the right of the decimal point that appear after the *last* nonzero digit, or by zeros to the left of the decimal point that appear in front of the first nonzero digit. Zeros that appear between the decimal point and the first nonzero digit on *either side* of the decimal point are necessary placeholders. For example, in the number

0400.020

the first and last zeros are optional, but the middle three zeros are necessary placeholders.

0400.020 = 400.02

example 3

Indicate the place value of each nonzero digit in the following decimals, and translate the decimal into words.

(a) 45.8

4	5	.	8
↗	↑		↖
tens	ones		tenths

forty-five and eight tenths

(b) 6.02

6	.	0	2
↗			↖
ones			hundredths

six and two hundredths

(c) 89.6305

8	9	.	6	3	0	5
↗	↗		↗	↖		↖
tens	ones		tenths	hundredths		ten-thousandths

eighty-nine and six thousand, three hundred five ten-thousandths

(d) 500.007

5	0	0	.	0	0	7
↗						↖
hundreds						thousandths

five hundred and seven thousandths

(e) 2,002.0202

2	,	0	0	2	.	0	2	0	2
↗				↗			↖		↖
thousands				ones			hundredths		ten-thousandths

two thousand, two and two hundred two ten-thousandths ●

A decimal fraction whose numerator is greater than or equal to its denominator has a value that is greater than or equal to one, and is sometimes called an improper decimal fraction. To write an improper decimal fraction in decimal notation, we determine the place value of the last digit in the numerator and put the decimal point in the appropriate location. For example, in the decimal fraction

$$\frac{527}{10} \qquad \text{five hundred twenty-seven tenths}$$

7 is in the tenths place. Therefore, the decimal point is placed between the 2 and the 7:

$$\frac{527}{10} = 52.7 \qquad \text{fifty-two and seven tenths}$$

example 4

Write each of the following improper decimal fractions in decimal notation and translate it into words.

(a) $\frac{71}{10}$

7.1

seven and one tenth — 1 is in the tenths place.

(b) $\frac{8,352}{100}$

83.52

eighty-three and fifty-two hundredths — 2 is in the hundredths place.

(c) $\frac{738,461}{1,000}$

738.461 — 1 is in the thousandths place.

seven hundred thirty-eight and four hundred sixty-one thousandths ●

Now that we are familiar with decimal notation, we can use our knowledge of place value to rewrite decimals in expanded notation. For example, to express the number

49.613 in expanded notation, we write

$$49.613 = (4 \times 10) + (9 \times 1) + \left(6 \times \frac{1}{10}\right) + \left(1 \times \frac{1}{100}\right) + \left(3 \times \frac{1}{1{,}000}\right)$$

Using powers of ten,

$$49.613 = (4 \times 10^1) + (9 \times 10^0) + \left(6 \times \frac{1}{10^1}\right) + \left(1 \times \frac{1}{10^2}\right) + \left(3 \times \frac{1}{10^3}\right)$$

example 5

Translate each of the following numbers into decimal notation and then rewrite it in expanded notation.

(a) six and three hundredths

$$6.03 = (6 \times 1) + \left(3 \times \frac{1}{100}\right)$$
$$= (6 \times 10^0) + \left(3 \times \frac{1}{10^2}\right)$$

(b) twenty-nine and sixty-two thousandths

$$29.062 = (2 \times 10) + (9 \times 1) + \left(6 \times \frac{1}{100}\right) + \left(2 \times \frac{1}{1{,}000}\right)$$
$$= (2 \times 10^1) + (9 \times 10^0) + \left(6 \times \frac{1}{10^2}\right) + \left(2 \times \frac{1}{10^3}\right)$$

(c) eight hundred fifteen and seven hundred forty-two thousandths

$$815.742 = (8 \times 100) + (1 \times 10) + (5 \times 1) + \left(7 \times \frac{1}{10}\right)$$
$$+ \left(4 \times \frac{1}{100}\right) + \left(2 \times \frac{1}{1{,}000}\right)$$
$$= (8 \times 10^2) + (1 \times 10^1) + (5 \times 10^0) + \left(7 \times \frac{1}{10^1}\right)$$
$$+ \left(4 \times \frac{1}{10^2}\right) + \left(2 \times \frac{1}{10^3}\right)$$

(d) six thousand, two and nine hundred-thousandths

$$6{,}002.00009 = (6 \times 1{,}000) + (2 \times 1) + \left(9 \times \frac{1}{100{,}000}\right)$$
$$= (6 \times 10^3) + (2 \times 10^0) + \left(9 \times \frac{1}{10^5}\right)$$

●

One of the most common uses of decimal notation is our monetary system. The smallest unit of money is one cent, and 100 cents = 1 dollar. The dollar is the monetary unit most frequently used, and since 1 cent $= \frac{1}{100}$ dollar, any amount of money can be expressed in dollars as a decimal with two places to the right of the decimal point—the tenths and hundredths place. For example, 450 cents (¢) is equivalent to $\frac{450}{100}$ dollars ($), which, expressed as a decimal, is 4.50 dollars. Using monetary notation this becomes \$4.50, which is read as "four dollars and fifty cents." Notice that even though $4.50 = 4.5$, we leave the zero in the hundredths place when expressing units of money to make them easier to read.

It is important to be able to write monetary units in words for the purpose of writing checks. All dollar amounts are written out in words on checks, and the cents are written as the numerator of a decimal fraction with a denominator of 100, as shown in Figure 3.2.

figure 3.2

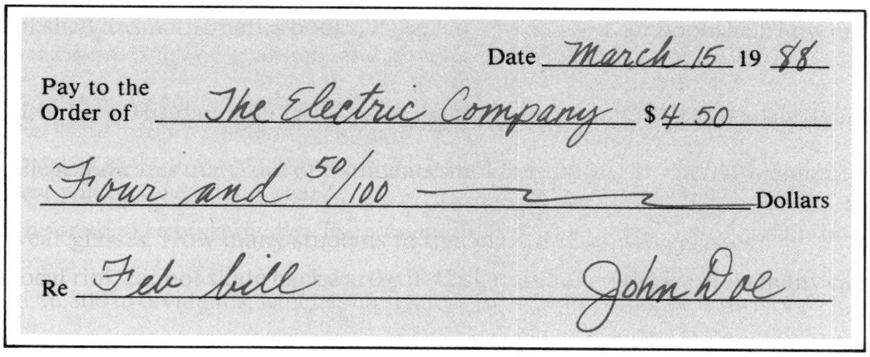

example 6

Write the following amounts of money as decimal fractions and as decimals in monetary notation. Rewrite the amount in words as it would appear on a check.

(a) 33¢

$$33 \text{ cents} = \frac{33}{100} \text{ dollars} = \$0.33$$

Zero and 33/100 dollars

(b) 4,700¢

$$4{,}700 \text{ cents} = \frac{4{,}700}{100} \text{ dollars} = \$47.00,$$

or \$47

If there are zeros in both the tenths place and hundredths place, these places are sometimes omitted.

Forty-seven and 00/100 dollars

(c) 203¢

$$203 \text{ cents} = \frac{203}{100} \text{ dollars} = \$2.03$$

Two and 3/100 dollars

(d) 8 dollars and 630 cents

$$8 + \frac{630}{100} \text{ dollars} = 8 + 6 + \frac{30}{100} \text{ dollars} = \$14.30$$

Fourteen and 30/100 dollars

(e) 561,004¢

$$561{,}004 \text{ cents} = \frac{561{,}004}{100} \text{ dollars} = \$5{,}610.04$$

Five thousand, six hundred ten and 4/100 dollars

QUICK QUIZ	ANSWERS
1. Write 8.27 in words.	**1.** eight and twenty-seven hundredths
2. Write sixty-two and five thousandths in decimal notation.	**2.** 62.005
3. Write 9.031 in expanded notation.	**3.** $(9 \times 10^0) + \left(3 \times \frac{1}{10^2}\right) + \left(1 \times \frac{1}{10^3}\right)$

3.1 Exercises

Indicate which of the following statements are true and which are false. For those that are false, change the italic word to make the statement true.

1. The place value two places to the right of the decimal point is called the *tenths* place.

2. One cent is equivalent to *one-tenth* of a dollar.

3. Decimal fractions have a power of ten in the *numerator*.

4. Zeros that appear between the decimal point and the first nonzero digit are *necessary*.

Write each of the following decimal fractions in decimal notation.

5. $\frac{3}{10}$ **6.** $\frac{5}{10}$ **7.** $\frac{8}{100}$ **8.** $\frac{32}{100}$ **9.** $\frac{5}{1{,}000}$ **10.** $\frac{19}{1{,}000}$

11. $\frac{529}{1{,}000}$ **12.** $\frac{8}{10{,}000}$ **13.** $\frac{21}{10{,}000}$ **14.** $\frac{621}{10{,}000}$ **15.** $\frac{3}{100{,}000}$ **16.** $\frac{48}{100{,}000}$

17. $\frac{127}{100{,}000}$ **18.** $\frac{69}{1{,}000{,}000}$ **19.** $\frac{6{,}589}{1{,}000{,}000}$ **20.** $\frac{973{,}642}{1{,}000{,}000}$ **21.** $\frac{73}{10}$ **22.** $\frac{804}{10}$

23. $\frac{519}{100}$ **24.** $\frac{1{,}362}{100}$ **25.** $\frac{2{,}837}{1{,}000}$ **26.** $\frac{50{,}505}{1{,}000}$ **27.** $\frac{70{,}803}{10{,}000}$ **28.** $\frac{6{,}328{,}174}{10{,}000}$

For each of the following decimals, give the place value of the digit in parentheses, and rewrite the number in words.

29. 0.3 (3) **30.** 0.04 (4) **31.** 0.53 (3)
32. 0.835 (5) **33.** 5.9 (9) **34.** 7.06 (0)
35. 47.8 (8) **36.** 60.02 (2) **37.** 38.974 (4)
38. 82.7091 (9) **39.** 800.65327 (7) **40.** 302.305301 (1)
41. 3,603.46 (4) **42.** 4,000.0073 (7) **43.** 500.4689 (9)
44. 302.00005 (5) **45.** 8,067.0004 (4) **46.** 80,305.0702 (2)
47. 362,006.200008 (8) **48.** 6,002,003.008007 (7)

Write each of the following numbers in decimal notation.

49. four tenths
50. five hundredths
51. six thousandths
52. eighty-seven thousandths
53. four ten-thousandths
54. seventy-eight ten-thousandths
55. three hundred eighty-six ten-thousandths
56. fifty-three hundred-thousandths
57. six millionths
58. five hundred six millionths
59. two thousand, fifty-nine millionths
60. sixty-two thousand, four hundred thirty-eight millionths
61. two and six tenths
62. seven and fifty-four hundredths
63. three hundred eight and twelve hundredths
64. six thousand and six thousandths
65. eleven thousand and fifty-eight hundred-thousandths
66. five million, six hundred and forty-three millionths

Write each of the following decimals in expanded notation.

67. 5.2
68. 8.06
69. 36.46
70. 89.364
71. 230.005
72. 5,627.364
73. 30,005.6278
74. 200,002.00002
75. 50,083,070.0406
76. 4,070,801.050602

Write the following amounts of money as decimals in monetary notation, and rewrite them in words as they would appear on a check.

77. 16¢
78. 89¢
79. 374¢
80. 620¢
81. 4,648¢
82. 7,923¢
83. 80,567¢
84. 368,902¢
85. 5 dollars and 214 cents
86. 27 dollars and 480 cents

Answer each of the following questions.

87. Jupiter is 5.203 astronomical units away from the sun. What is the place value of the digit 3 in this number?

88. The cost of a new refrigerator is $659.00. How would this amount be written out on a check?

89. The moon makes one revolution around the earth in twenty-seven and thirty-two hundredths days. How is this number written in decimal notation?

90. The average annual rainfall in Providence, Rhode Island, is 42.75 inches. How is this number written out in words?

3.2 Rounding Off

It is often very awkward to work with numbers that have a large number of digits. For example, if the computerized scale at a supermarket indicates that a bag of tomatoes weighs 1.51827 pounds, the grocer will probably tell you that it weighs 1.5 pounds. This approximation was calculated by **rounding off** the number 1.51827 to the nearest tenth to obtain 1.5.

definition

The procedure used to estimate a number to a given degree of precision is called *rounding off.*

When rounding off a number, we state the place value of the last digit of our result to indicate the precision of the approximation. For example, the number 1.51827 can be rounded off in the following ways.

To the nearest tenth	1.5
To the nearest hundredth	1.52
To the nearest thousandth	1.518
To the nearest ten-thousandth	1.5183

Notice that in rounding off the number 1.51827 to the nearest hundredth, 1.52 is a better approximation than 1.51, so we added one to the digit in the hundredths place. This process is called *rounding up.* When rounding off 5.1827 to the nearest thousandth, 5.18 is a better approximation than 5.19, so the digit in the thousandths place stayed the same. This process is called *rounding down.*

To Round Off a Number

1. Examine the digit immediately to the right of the place value to which you are rounding off (the place value of the desired precision).
2. If the digit is 5 or greater, round up. That is, add one to the digit in the place value of the desired precision.
3. If the digit is less than 5, round down. That is, retain the digit in the place value of the desired precision.

example 1

Round off the following numbers to the nearest hundredth.

(a) 3.6248

2 is in the hundredths place. 4 is on its right.

$4 < 5$ — Round down: 2 stays the same.

$3.6248 \approx 3.62$ — The symbol $\approx$ means "is approximately equal to."

(b) 52.475

7 is in the hundredths place. 5 is on its right.

$5 = 5$ — Round up: $7 + 1 = 8$.

$53.475 \approx 53.48$

(c) 0.297

9 is in the hundredths place. 7 is on its right.

$7 > 5$ — Round up: $9 + 1 = 10$. Write 0 in the hundredths place and carry 1 to the tenths place to obtain .30.

$0.297 \approx 0.30$ — Notice that we leave the zero in the hundredths place to indicate precision to the nearest hundredth. ●

example 2

Round off the number 7,439.855 as indicated.

(a) To the nearest thousand.

7,439.855

7,000

4 < 5, so round down. Notice that the zeros are necessary placeholders.

(b) To the nearest hundred.

7,439.855

7,400

3 < 5, so round down.

(c) To the nearest ten.

7,439.855

7,440

9 > 5, so round up.

(d) To the nearest whole number.

7,439.855

7,440

8 > 5, so round up.

(e) To the nearest tenth.

7,439.855

7,439.9

5 = 5, so round up.

(f) To the nearest hundredth.

7,439.855

7,439.86

5 = 5, so round up. ●

You may be asked to round off a number expressed in dollars to a degree of precision that is expressed in cents. It is helpful, therefore, to know the place values of the locations to the right of the decimal point in terms of both dollars and cents. As an example, let us determine the place value of each digit in the number \$6.52814. Remember that since 1 dollar = 100 cents, \$6.52814 = 652.814¢. Table 3.1 gives the place value of digit in terms of both dollars and cents.

table 3.1
Place Values in Dollars and Cents

DIGIT	PLACE VALUE \$6.52814 = 652.814¢ DOLLARS	CENTS
6	ones	hundreds
5	tenths	tens
2	hundredths	ones
8	thousandths	tenths
1	ten-thousandths	hundredths
4	hundred-thousandths	thousandths

example 3

Round off \$20.59748 to the places indicated.

(a) To the nearest dollar.

\$20.59748

\$21

5 = 5, so round up.

(b) To the nearest tenth of a dollar.

\$20.59748

\$20.6 — 9 > 5, so round up.

(c) To the nearest cent.

\$20.59748

\$20.60 — 7 > 5, so round up.

(d) To the nearest tenth of a cent.

\$20.59748

\$20.597 — 4 < 5, so round down.

(e) To the nearest hundredth of a cent.

\$20.59748

\$20.5975 — 8 > 5, so round up.

QUICK QUIZ

Round off each of the following numbers as indicated.

1. 4.857 (nearest tenth)
2. 267.416 (nearest hundred)
3. \$85.95 (nearest dollar)

ANSWERS

1. 4.9
2. 300
3. \$86

3.2 Exercises

Indicate which of the following statements are true and which are false. For those that are false, change the italic expression to make the statement true.

1. When rounding off, we examine the digit immediately to the *right* of the place value that indicates the desired precision.
2. When the digit examined is *5 or less*, we round down.
3. Rounding off to the nearest *hundredth* of a dollar is the same as rounding off to the nearest tenth of a cent.
4. To round up, the digit with the same place value as the desired precision *is increased by one*.

Round off each of the following decimals to the nearest tenth, the nearest hundredth, and the nearest thousandth.

5. 3.2785 **6.** 0.1063 **7.** 13.9628 **8.** 82.3974 **9.** 7.0058 **10.** 3.2109
11. 0.5020 **12.** 8.0305 **13.** 62.8299 **14.** 27.3906 **15.** 38.0705 **16.** 52.5060
17. 4.7979 **18.** 5.9898 **19.** 67.9999 **20.** 11.9999

Round off each of the following decimals to the nearest hundred, the nearest ten, and the nearest whole number.

21. 6,846.7 **22.** 9,320.3 **23.** 7,045.8 **24.** 340.69 **25.** 709.38 **26.** 281.97
27. 695.7 **28.** 219.5 **29.** 7,080.4 **30.** 3,020.7 **31.** 499.6 **32.** 529.9
33. 985.9 **34.** 976.5 **35.** 60,709.92 **36.** 34,050.99

Round off each of the following numbers as indicated.

37. 325,873 (nearest ten thousand) **38.** 738,359 (nearest hundred thousand)
39. 6.05038 (nearest ten-thousandth) **40.** 8.29731 (nearest hundredth)

41. 2.5874095 (nearest millionth)
42. 95,601,382 (nearest million)
43. 69.9583 (nearest tenth)
44. 499.71 (nearest whole number)
45. 9,999.99 (nearest tenth)
46. 909.999 (nearest hundredth)

Round off each of the following dollar amounts to the values indicated.

47. \$367.69 (nearest dollar)
48. \$204.27 (nearest dollar)
49. \$32.845 (nearest cent)
50. \$65.602 (nearest cent)
51. \$88.59 (nearest tenth of a dollar)
52. \$12.47 (nearest tenth of a dollar)
53. \$5.0305 (nearest tenth of a cent)
54. \$8.4297 (nearest tenth of a cent)
55. \$6.83402 (nearest hundredth of a cent)
56. \$3.76995 (nearest hundredth of a cent)

Answer each of the following questions.

57. The Panama Canal is 50.7 miles long. What is its length rounded off to the nearest mile?

58. Saturn is 9.539 astronomical units away from the sun. What is this distance rounded off to the nearest hundredth of an astronomical unit?

59. A runner completed a race in 53.684 seconds. What was the runner's time rounded off to the nearest tenth of a second?

60. Hudson Bay covers 475,800 square miles. What is its area rounded off to the nearest thousand square miles?

61. A color television is on sale for \$499.99. What is its price rounded off to the nearest dollar?

62. A gallon of gas sells for \$1.599. What is the price per gallon rounded off to the nearest cent?

3.3 Addition and Subtraction of Decimals

We will now illustrate procedures for performing the operations addition and subtraction on decimals. As an example, let us consider how to add 3.6 + 9.5 by changing each decimal to a mixed number as follows:

$$\begin{array}{rr} 3.6 = & 3\frac{6}{10} \\ +\ 9.5 = & 9\frac{5}{10} \\ \hline & 12\frac{11}{10} \end{array}$$

Add the whole number and fractional parts separately.

$$= 12 + 1 + \frac{1}{10}$$

Convert: $\frac{11}{10} = 1\frac{1}{10}$

$$= 13\frac{1}{10}$$

Add the whole numbers.

$$= 13.1$$

Convert to decimal notation.

Notice that this same result can be obtained by simply adding the digits with the same place value, moving from right to left and carrying when necessary. Using this procedure for the previous example we obtain

$$\begin{array}{lr} & \overset{1}{3.6} \\ \text{Addend} & \\ \text{Addend} & +\ 9.5 \\ \hline \text{Sum} & 13.1 \end{array}$$

Thus, we have illustrated the addition procedure outlined in the box.

To Add Decimals

1. Write the numbers to be added vertically and line up the decimal points.
2. Add all digits with the same place value, beginning with the rightmost column. Carry to the next column when the sum of the digits in any column is greater than 9.
3. Be sure to place a decimal point in the sum in the correct location. Check that all its digits are in line with the digits in the addends with the same place value.

Example 1 provides some additional illustrations of adding decimals.

example 1

Add each of the following numbers.

(a) 18.36 + 40.962

```
  1 1
  18.36
+ 40.962
--------
  59.322
```

(b) 83.76 + 0.0673 + 9.73

```
  11 1
  83.7600
   0.0673
+  9.7300
---------
  93.5573
```

Sometimes we add zeros to the right of the digits following the decimal point to help us line up the addends correctly.

(c) 3067.04 + 0.3678 + 402.071 + 2.14095

```
   1  2 1
  3067.04
     0.3678
   402.0710
+    2.14095
------------
  3471.61975
```

(d) $4.03 + $0.67 + $995.84

```
      111 1
  $     4.03
        0.67
+     995.84
------------
  $1,000.54
```

When adding monetary amounts, the dollar sign usually appears in the first addend and in the sum.

Now let us consider how to subtract 8.2 − 3.8 by converting each decimal to a mixed number as follows:

$$\begin{array}{r} 8.2 = 8\frac{2}{10} = 7\frac{12}{10} \\ -\ 3.8 = 3\frac{8}{10} = 3\frac{8}{10} \\ \hline 4\frac{4}{10} = 4.4 \end{array}$$

Notice that the same result can be obtained by simply subtracting digits with the same place value, moving from right to left and borrowing when necessary. Using this procedure for the previous example we obtain the following.

	7
Minuend	8.2
Subtrahend	− 3.8
Difference	4.4

Thus, we have established the subtraction procedure outlined in the box.

To Subtract Decimals

1. Write the numbers to be subtracted vertically and line up the decimal points.
2. If the number of decimal places in the subtrahend exceeds that in the minuend, annex the necessary number of zeros to the right of the last decimal place in the minuend.
3. Subtract all digits with the same place value, beginning in the rightmost column and borrowing when necessary.
4. Insert a decimal point in the correct location in the difference.

Now we will look at some additional examples of subtracting decimals.

example 2

Subtract each of the following numbers.

(a) 27.4 − 12.9

$$\begin{array}{r} {\scriptstyle 6} \\ 27.4 \\ -\ 12.9 \\ \hline 14.5 \end{array}$$

(b) 102.3 − 8.62

$$\begin{array}{r} {\scriptstyle 9\,11\,12} \\ 102.30 \\ -\ 8.62 \\ \hline 93.68 \end{array}$$

Add a zero to the right of the last decimal place in the minuend.

(c) 462 − 26.528

$$\begin{array}{r} {\scriptstyle 5\,11\ \ 9\ 9} \\ 462.000 \\ -\ 26.528 \\ \hline 435.472 \end{array}$$

When adding zeros to the right of a whole number, be certain to insert a decimal point to the right of the ones place.

(d) \$5,200 − \$92.95

$$\begin{array}{r} {\scriptstyle 1\ 9\ 9\ 9} \\ \$5{,}200.00 \\ -\ 92.95 \\ \hline \$5{,}107.05 \end{array}$$

●

We will now use the estimation process introduced in Chapter 1 to approximate the answers to problems involving addition and subtraction of decimals.

example 3

Estimate each of the following sums and differences, then determine the exact answers.

(a) $58.93 + 129.67 + 70.14$

$$\begin{array}{rcr} 58.93 & \approx & 60 \\ 129.67 & \approx & 130 \\ 70.14 & \approx & \underline{70} \\ & & 260 \end{array}$$

Approximate each addend by rounding off to the nearest ten.

Our estimate is 260.

Now determine the exact answer.

$$\begin{array}{r} 58.93 \\ 129.67 \\ \underline{70.14} \\ 258.74 \end{array}$$

Since 258.74 is close to our estimate of 260, this answer is reasonable.

(b) $3797.3 - 412.5$

$$\begin{array}{rcr} 3797.3 & \approx & 3800 \\ 412.5 & \approx & \underline{400} \\ & & 3400 \end{array}$$

Approximate the minuend and subtrahend by rounding them off to the nearest hundred.
Our estimate is 3400.

Now determine the exact answer.

$$\begin{array}{r} 3797.3 \\ \underline{412.5} \\ 3384.8 \end{array}$$

Since 3384.8 is close to our estimate of 3400, this answer is reasonable.

example 4

At the beginning of the month, an architect had a balance of \$482.76 in his checking account. During the month, he wrote checks for \$53.62, \$137.00, \$86.99, \$398.12, and \$78.26. He also made deposits of \$55.89, \$125.00, and \$738.64. What was his balance at the end of the month?

solution

$$\begin{array}{r} {\scriptstyle 222\ 1} \\ \$\ 482.76 \\ 55.89 \\ 125.00 \\ \underline{738.64} \\ \$1{,}402.29 \end{array}$$

Add the deposits to the original balance.

$$\begin{array}{r} {\scriptstyle 331\ 1} \\ \$\ 53.62 \\ 137.00 \\ 86.99 \\ 398.12 \\ \underline{78.26} \\ \$753.99 \end{array}$$

Add the amounts of the checks written.

$$\begin{array}{r} {\scriptstyle 13\,9\,11} \\ \$1{,}402.29 \\ \underline{-\quad 753.99} \\ \$\ 648.30 \end{array}$$

Then subtract the total of the check amounts.

His balance at the end of the month was \$648.30.

QUICK QUIZ

Perform each of the operations indicated.

1. 6.389 + 29.47

2. 61.43 − 47.85

3. 87.13 + 5.62 − 13.7

ANSWERS

1. 35.859

2. 13.58

3. 79.05

3.3 Exercises

Indicate which of the following statements are true and which are false. For those that are false, change the italic expression to make the statement true.

1. To add or subtract decimals, write the problem *horizontally* and line up the decimal points.

2. When adding and subtracting decimals, we follow *the same* rules for borrowing and carrying as apply to adding and subtracting whole numbers.

3. When adding decimals, we add the digits with the same place value, moving from *left to right*.

4. When subtracting decimals, the difference is *always* greater than the subtrahend.

Add each of the following decimals.

5. 5.3 + 6.8

6. 4.6 + 9.7

7. 14.28 + 32.67

8. 27.84 + 16.29

9. 3.864 + 12.7

10. 35.8 + 2.169

11. \$45.83 + \$0.92

12. \$0.57 + \$86.75

13. 0.6783 + 153.6

14. 289.9 + 0.0481

15. 5.3 + 20.82 + 0.067

16. 93.6 + 4.863 + 0.39

17. 1.863 + 17.8 + 1289.6 + 0.4519

18. 0.6793 + 4382.7 + 12.7 + 2.1869

19. \$362.94 + \$487.02 + \$13.62

20. \$576.03 + \$0.79 + \$93.67

21. 37.289 + 0.86897 + 2.9935

22. 0.05916 + 8.9472 + 67.00847

Subtract each of the following decimals.

23. 7.4 − 2.8

24. 9.2 − 1.7

25. \$13.07 − \$4.95

26. \$81.31 − \$3.86

27. 53.1 − 7.432

28. 71.3 − 0.8516

29. \$163.84 − \$4.78

30. \$781.36 − \$5.07

31. 6.073 − 0.6354

32. 4.602 − 0.8307

33. 3 − 0.2774

34. 7 − 0.5382

35. 420 − 9.934

36. 809 − 5.6214

37. 701.3 − 69.4795

38. 2,000 − 0.728

Perform each of the operations indicated.

39. 8.9 − 6.2 + 3.7

40. 4.1 + 5.2 − 7.6

41. 18.47 + 406.049

42. 502.06 − 289.74 + 21.48

43. \$0.37 + \$82.03 − \$5.95

44. \$560 − \$92.37 + \$25.62

45. 8,000 − 2.6384 + 647.29

46. 6,002.999 + 211.001 − 152.78

47. \$3,000 − \$15.95 + \$167.82

48. \$55,000 − \$7,999.99 + \$72.88

49. 0.24 + 34 − 0.489

50. 17 − 11 + 3.2

Estimate each of the following sums and differences, then determine the exact answers.

51. 48.41 + 121.73

52. 61.07 + 299.34

53. 60.2 + 259.3 + 29.7

54. 28.7 + 819.6 + 41.2

55. 840.7 − 19.8

56. 469.1 − 30.4

57. 5693.9 − 301.5

58. 8802.3 − 598.7

Solve each of the following word problems.

59. A jogger ran the following distances in a single week: 3.25 miles, 2.8 miles, 3.5 miles, 2.75 miles, 3.1 miles, 2.9 miles, and 2.5 miles. What was the total distance she ran that week?

60. If a $498 stereo receiver is marked down $29.99, what is its sale price?

61. A man bought items at a grocery store that cost $1.89, $0.43, $3.89, $0.65, $0.39, $1.29, $5.62, $0.59, and $1.19. He had coupons for some of these items worth $0.15, $0.40, and $0.25. What was his grocery bill?

62. If it takes the earth 365.256 days to revolve about the sun, and it takes Mercury 87.969 days to revolve about the sun, how much longer does it take the earth than Mercury to revolve about the sun?

63. If a student bought a notebook for $1.57, a pen for $0.49, typing paper for $1.89, and a chocolate bar for $0.35, how much change would he receive from a $10 bill?

64. If the St. Lawrence Seaway is 2,400 miles long, and the Panama Canal is 50.7 miles long, how much longer is the St. Lawrence Seaway than the Panama Canal?

65. If a secretary's monthly take-home pay is $889.45, and he spends $465 for rent, $15.86 for electricity, $38.17 for gas, $52.43 for phone, and $132 for car payments, how much does he have left for other expenses?

66. A student's checking account showed a balance of $255.39 at the beginning of the month. During the month she wrote checks for $15.23, $128.75, $62.45, $89.36, and $12.71, and she made a deposit of $193.67. What was her balance at the end of the month?

3.4 Multiplication of Decimals

In order to establish a procedure for multiplying decimals, we can convert each decimal to an equivalent fraction, and then apply the rules for multiplying fractions. As an example, let us consider the following products.

$$0.2 \times 3 = \frac{2}{10} \times \frac{3}{1} = \frac{6}{10} = 0.6$$

$$0.02 \times 3 = \frac{2}{100} \times \frac{3}{1} = \frac{6}{100} = 0.06$$

$$0.2 \times 0.3 = \frac{2}{10} \times \frac{3}{10} = \frac{6}{100} = 0.06$$

$$0.002 \times 3 = \frac{2}{1{,}000} \times \frac{3}{1} = \frac{6}{1{,}000} = 0.006$$

$$0.02 \times 0.3 = \frac{2}{100} \times \frac{3}{10} = \frac{6}{1{,}000} = 0.006$$

$$0.002 \times 0.3 = \frac{2}{1{,}000} \times \frac{3}{10} = \frac{6}{10{,}000} = 0.0006$$

$$0.02 \times 0.03 = \frac{2}{100} \times \frac{3}{100} = \frac{6}{10{,}000} = 0.0006$$

Notice that each product can be determined by multiplying 2×3 and inserting the appropriate number of zeros between the decimal point and the number 6. In each example, the number of decimal places in the product is equal to the sum of the number of decimal places in all numbers being multiplied. We have thus illustrated a procedure for multiplying decimals, as outlined in the box.

To Multiply Decimals

1. Calculate the product as usual.
2. Put a decimal point in the appropriate location so that the number of decimal places in the product equals the sum of the number of decimal places in all factors.

Let us look at some additional examples of multiplying decimals.

example 1

Multiply each of the following decimals.

(a) 0.36×0.81

$$\begin{array}{r} {}^{4} \\ 0.36 \\ \times\ 0.81 \\ \hline 36 \\ 288 \\ \hline 0.2916 \end{array}$$

Two decimal places.
Two decimal places.

The number of decimal places in the product is equal to $2 + 2 = 4$.

(b) 0.052×8.6

$$\begin{array}{r} {}^{1\,1} \\ 0.052 \\ \times\ 8.6 \\ \hline 312 \\ 416 \\ \hline 0.4472 \end{array}$$

Three decimal places.
One decimal place.

The number of decimal places in the product is equal to $3 + 1 = 4$.

(c) 43.7×0.00035

$$\begin{array}{r} {}^{1\,2} \\ {}^{1\,3} \\ 43.7 \\ \times\ 0.00035 \\ \hline 2185 \\ 1311 \\ \hline 0.015295 \end{array}$$

The number of decimal places in the product is equal to $1 + 5 = 6$.

(d) $\$59.95 \times 60$

$$\begin{array}{r} {}^{5\,5\,\,3} \\ \$\ \ 59.95 \\ \times\ \ 60 \\ \hline \$3{,}597.00 \end{array}$$

The number of decimal places in the product is equal to $2 + 0 = 2$. ●

We will now investigate how to multiply decimals by powers of ten. For example,

$$2.5 \times 10^1 = 2.5 \times 10 = 25$$

$$2.5 \times 10^2 = 2.5 \times 100 = 250$$

$$2.5 \times 10^3 = 2.5 \times 1{,}000 = 2{,}500$$

Notice that in each case, the product can be obtained by moving the decimal point to the right the number of places that is indicated by the power of ten (see box).

> To multiply a decimal by 10^n, where n is a whole number, move the decimal point n places to the right.

example 2

Find each of the following products.

(a) 0.3×10^4

$0.3 \times 10^4 = 0.3000_{\wedge} = 3{,}000$ — Move the decimal point four places to the right.

(b) 82.73×10^3

$82.73 \times 10^3 = 82.730_{\wedge} = 82{,}730$ — Move the decimal point three places to the right.

(c) 6.41896×10^5

$6.41896 \times 10^5 = 6.41896_{\wedge} = 641{,}896$ — Move the decimal point five places to the right.

(d) 25.3×10^9

$25.3 \times 10^9 = 25.300000000_{\wedge}$ — Move the decimal point nine places to the right.

$= 25{,}300{,}000{,}000$

Now let us look at some examples of multiplying decimals by whole numbers that are multiples of powers of ten.

example 3

Multiply each of the following numbers.

(a) 1.5×300

$1.5 \times 300 = 1.5 \times 3 \times 100$	Rewrite 300 in expanded notation.
$= (1.5 \times 3) \times 10^2$	Express 100 as a power of 10.
$= 4.5 \times 10^2$	Multiply using associative property.
$= 450$	Multiply result by power of 10.

(b) $7.2 \times 20{,}000$

$7.2 \times 20{,}000 = 7.2 \times 2 \times 10{,}000$	Rewrite 20,000 in expanded notation.
$= (7.2 \times 2) \times 10^4$	Express 10,000 as a power of 10.
$= 14.4 \times 10^4$	Multiply using associative property.
$= 144{,}000$	Multiply result by power of 10.

(c) $0.012 \times 80{,}000$

$0.012 \times 80{,}000$

$= \dfrac{12}{1{,}000} \times 8 \times 10{,}000$ Rewrite factors.

$= (12 \times 8) \times \left(\dfrac{1}{1{,}000} \times 10{,}000\right)$ By the commutative and associative properties.

$= 96 \times \dfrac{10{,}000}{1{,}000}$ Multiply.

$= 96 \times 10$ Reduce to lowest terms.

$= 960$ Multiply.

(d) $150{,}000{,}000 \times 0.0003$

$150{,}000{,}000 \times 0.0003$

$= 15 \times 10{,}000{,}000 \times \dfrac{3}{10{,}000}$ Rewrite factors.

$= (15 \times 3) \times \dfrac{10{,}000{,}000}{10{,}000}$ By the commutative and associative properties.

$= 45 \times 1{,}000$ Multiply and reduce.

$= 45{,}000$ Multiply. ●

Decimals are used often as part of a shorthand notation for very large or very small numbers.

example 4

Rewrite the following numbers.

(a) 3.2 million

$$3.2 \text{ million} = 3.2 \times 1{,}000{,}000 = 3.2 \times 10^6 = 3{,}200{,}000$$

(b) 1.8 thousandths

$$1.8 \text{ thousandths} = 1.8 \times \frac{1}{1{,}000} = \frac{1.8}{1{,}000} = \frac{1.8}{1{,}000} \times \frac{10}{10} = \frac{18}{10{,}000} = 0.0018$$

●

When working with decimals, we sometimes encounter arithmetic expressions that contain more than one operation. To simplify these expressions, we follow the same order of operations rules that apply when working with whole numbers.

example 5

Simplify each of the following expressions.

(a) $3(1.8 + 2.4)$

$$3(1.8 + 2.4) = 3(4.2) = 12.6$$

Perform the operation in ( ) first. Then multiply the sum by 3.

$$\begin{array}{r} 1.8 \\ +2.4 \\ \hline 4.2 \end{array} \qquad \begin{array}{r} 4.2 \\ \times \quad 3 \\ \hline 12.6 \end{array}$$

(b) $(0.05)(1.2)^2$

$$(0.05)(1.2)^2 = (0.05)(1.44) = 0.0720$$

Square 1.2. Then multiply result by 0.05.

$$\begin{array}{r} 1.2 \\ \times \ 1.2 \\ \hline 24 \\ 12\ \ \\ \hline 1.44 \end{array} \qquad \begin{array}{r} 1.44 \\ \times \ 0.05 \\ \hline 0.0720 \end{array}$$

(c) $[5.8 - 4(0.7)]^3$

$$[5.8 - 4(0.7)]^3 = [5.8 - 2.8]^3 = (3.0)^3 = 27$$

In [ ], do multiplication before subtraction. Then cube result.

example 6

The planet Mercury is 0.387 astronomical units away from the sun. If one astronomical unit is equivalent to 93 million miles, how far away in miles is Mercury from the sun?

solution

$$\begin{aligned} 0.387 \times 93 \text{ million} &= 0.387 \times 93 \times 1{,}000{,}000 \\ &= (0.387 \times 93) \times 10^6 \\ &= 35.991 \times 10^6 \\ &= 35{,}991{,}000 \text{ miles} \end{aligned}$$

Multiply.

$$\begin{array}{r} 0.387 \\ \times \quad 93 \\ \hline 1161 \\ 3483\ \ \\ \hline 35.991 \end{array}$$

QUICK QUIZ

Multiply each of the following decimals.

1. 0.8×0.09
2. 7.3×0.21
3. 1.48×60.7

ANSWERS

1. 0.072
2. 1.533
3. 89.836

3.4 Exercises

Indicate which of the following statements are true and which are false. For those that are false, change the italic expression to make the statement true.

1. When multiplying decimals, the number of decimal places in the product is equal to the *sum* of the number of decimal places in all factors.
2. To multiply a decimal by 10^n, where n is a whole number, move the decimal point n places to the *left*.
3. The number 6.9 million is equal to the *sum* of 6.9 and 1,000,000.
4. If two decimals, each greater than zero and less than one, are multiplied together, the product is *less than* either factor.

Multiply each of the following decimals.

5. 0.07×0.2
6. 0.5×0.06
7. 0.28×0.008
8. 0.04×0.096
9. 3.8×0.81
10. 0.62×9.7
11. 0.534×2.7
12. 6.32×8.09
13. $\$6.28 \times 7.4$
14. $\$18.09 \times 6.2$
15. 0.0071×0.038
16. 0.6704×0.605
17. 15.07×3.012
18. 7.0063×217.14
19. 382.68×0.00079
20. 0.000603×579.28
21. $7{,}006.08 \times 0.0731$
22. $0.2053 \times 5{,}607.39$
23. $\$2{,}999 \times 12$
24. $\$5{,}085.75 \times 52$
25. 3.285×0.4176
26. 728.8×456.3
27. 562.7×329.4
28. 0.8394×6.196

Find each of the following products.

29. 0.8×10^5
30. 2.93×10^4
31. 0.0031×10^2
32. 2.0084×10^2
33. 0.3602005×10^4
34. 0.0370059×10^3
35. 5.7×10^8
36. 6.2×10^7
37. 0.02803×10^7
38. 0.30705×10^8
39. $4.8 \times 3{,}000$
40. $7{,}000 \times 0.25$
41. 0.003×9.4
42. 1.84×400
43. $60{,}000 \times 0.379$
44. $0.807 \times 50{,}000$
45. $3{,}000 \times 0.09$
46. 0.005×800
47. $70{,}000 \times 0.002$
48. $0.0004 \times 6{,}000$

Rewrite the following numbers using digits.

49. 4.4 million
50. 7.2 million
51. 8.9 hundredths
52. 5.3 thousandths
53. 64 thousandths
54. 814 hundredths
55. 1.75 thousand
56. 2.5 billion

Simplify each of the following expressions.

57. $7(3.4 + 9.6)$
58. $8(18.4 - 11.4)$
59. $(0.05)(0.02)^2$
60. $(0.01)^3(0.08)$
61. $2[1.5 + (30)(0.03)]$
62. $3[4.2 + (0.25)(60)]$
63. $(1.5 + 3.3)(7.2 - 5.6)$
64. $(9.4 - 8.5)(6.8 + 5.2)$
65. $(7.2 + 9.8)^2$
66. $(0.018 + 0.002)^3$
67. $5[(3.8 + 6.9) - 4.7]$
68. $3[18.8 - (9.3 + 7.5)]$
69. $[(0.5)(0.2)]^3(16.8 - 9.3 + 1.5)^2$
70. $[5.6 + (4.5)(3.2)][(0.6)(0.5)]^3$

Solve each of the following word problems.

71. If 1 meter is equal to 39.37 inches, how many inches are there in 27 meters?
72. If salami costs $1.29 per pound, and ham costs $2.35 per pound, what is the cost of 1.5 pounds of salami and 2.25 pounds of ham? Round off your answer to the nearest cent.
73. In 1980, the population of Finland was 4,770,000, and the population of Denmark was 1.08 times that of Finland. What was the population of Denmark in 1980?
74. If a mechanic makes $442.33 a week, what is his yearly salary?

75. An electrician earns $12.65 an hour and is paid time-and-a-half for overtime (1.5 times her hourly wage for every hour above 40 hours worked in a single week). What would her pay be for a 47.5-hour work week? Round off the answer to the nearest cent.

76. A welder worked the following hours each day during a single week: 7.8, 8.1, 4.2, 8.0, and 7.9. If he earns $10.45 an hour, what were his gross earnings that week?

77. A psychiatrist bought a new car and made a down payment of $2,500. If he pays off the remainder in 36 monthly payments of $186.25 each, what is the total amount he will pay for the car?

78. A notebook costs 98¢, a biology book costs 12.5 times the price of the notebook, and a mathematics book costs 1.2 times the price of the biology book. What is the price of the mathematics book?

79. To determine state income tax, a couple filing jointly must subtract from their gross salary $2,000 for a spouse's exemption and $700 for each dependent, and then multiply that result by 0.05375. How much state income tax will a couple with three children have to pay if they file jointly and earn a total of $46,530.00? Round your answer to the nearest cent.

80. A house that costs $80,000.00 is purchased with a down payment of $15,000.00 and a 30-year mortgage that specifies monthly payments of $521.67. How much more than the original selling price do the buyers agree to pay for the house by taking out a mortgage?

3.5 Division of Decimals

In order to establish a procedure for dividing decimals, we can rewrite any division problem as a fraction, and then reduce it to lowest terms. For example,

$$0.8 \div 0.4 = \frac{0.8}{0.4}$$

$$= \frac{0.8}{0.4} \times \frac{10}{10} = \frac{8}{4} = 2$$

We will now investigate some additional examples of dividing decimals in this manner.

example 1

Divide each of the following decimals.

(a) $0.9 \div 0.03$

$$0.9 \div 0.03 = \frac{0.9}{0.03}$$ Rewrite the division as a fraction.

$$= \frac{0.9}{0.03} \times \frac{100}{100}$$

$$= \frac{90}{3} = 30$$ Reduce to lowest terms.

(b) $0.36 \div 0.004$

$$0.36 \div 0.004 = \frac{0.36}{0.004}$$ Rewrite as a fraction.

$$= \frac{0.36}{0.004} \times \frac{1{,}000}{1{,}000}$$

$$= \frac{360}{4} = 90$$ Reduce.

(c) $0.042 \div 0.0007$

$$0.042 \div 0.007 = \frac{0.042}{0.0007}$$

$$= \frac{0.042}{0.0007} \times \frac{10{,}000}{10{,}000}$$

$$= \frac{420}{7} = 60$$

●

Notice that for each division in Example 1, we first multiplied numerator and denominator by the lowest power of ten necessary to make the denominator a whole number. The same result can be achieved simply by moving the decimal point in the numerator and denominator the same number of places to the right.

For example, to divide

$$0.0072 \div 0.0009 = \frac{0.0072}{0.0009}$$

move the decimal point four places to the right in both the numerator and denominator and then perform the division:

$$\frac{0.0072_{\curvearrowright}}{0.0009_{\curvearrowright}} = \frac{72}{9} = 8$$

If we write this same problem using the alternative notation for division,

$$0.0009_{\curvearrowright}\overline{)0.0072_{\curvearrowright}}$$

we indicate that the decimal point has been moved by placing the caret symbol ($\wedge$) in the new location before proceeding with the division. The decimal point in the quotient must be placed immediately above the caret in the dividend as shown:

$$\begin{array}{r} 8. \\ 0.0009_{\wedge}\overline{)0.0072_{\wedge}} \end{array}$$

We can now summarize the procedure for dividing decimals, as outlined in the box.

To Divide Decimals

1. Move the decimal point in the divisor the necessary number of places to the right to make it a whole number, and place a caret in its new location.
2. Move the decimal point in the dividend the same number of places to the right, and place a caret in its new location.
3. Put a decimal point in the quotient immediately above the caret in the dividend.
4. Proceed with the technique for dividing whole numbers, noting the location of the decimal point in the quotient.

Let us now look at some additional examples of dividing decimals.

example 2

Divide each of the following decimals.

(a) $0.02904 \div 0.04$

```
               . 726
0.04 ⌃ |0.02 ⌃ 904
           2 8
         -----
            10
             8
           ---
            24
            24
           ---
             0
```

Move the decimal point two places to the right.

CHECK

```
   0.726
×   0.04
--------
 0.02904 √
```

(b) $0.040678 \div 0.473$

```
                  . 086
0.473 ⌃ |0.040 ⌃ 678
             37  84
             ------
              2838
              2838
              ----
                 0
```

Move the decimal point three places to the right. Notice that zero is a necessary placeholder in the quotient.

CHECK

```
    0.473
×   0.086
---------
     2838
    3784
---------
 0.040678 √
```

(c) $114 \div 0.06$

```
           1900.
0.06 ⌃ |114.00 ⌃
         6
        --
         54
         54
        --
          0
```

Notice that it is necessary to insert two zeros before moving the decimal point in the dividend, and that zeros are necessary placeholders in the quotient.

CHECK

```
   1900
× 0.06
-------
 114.00 √
```

●

In problems where the division process does not yield a remainder of zero, we are often asked to round off the quotient to a specified place value. In order to do so, it is necessary to carry out the division procedure until the quotient contains a digit that is one place value to the right of that to which we intend to round off. For example, to round off to the nearest hundredth, we must carry out the division to the thousandths place.

example 3

Calculate the following quotients and round off as indicated.

(a) $1.3682 \div 0.017$ (nearest hundredth)

```
               80 . 482 ≈ 80.48
0.017^ )1.368^ 200
         136
         ---
           8 2
           6 8
           ---
           1 40
           1 36
           ----
              40
              34
              --
               6
```

Carry out division to the thousandths place.

CHECK

```
     80.482
  ×   0.017
  ---------
     563374
     80482
  ---------
    1368194
  +       6
  ---------
   1.368200 ✓
```

(b) $0.436 \div 3.7$ (nearest thousandth)

```
          . 1178 ≈ 0.118
3.7^ )0.4^ 3600
        37
        --
         66
         37
         ---
         290
         259
         ---
          310
          296
          ---
           14
```

Carry out division to the ten-thousandths place.

CHECK

```
     0.1178
  ×     3.7
  ---------
       8246
      3534
  ---------
      43586
  +      14
  ---------
    0.43600 ✓
```

(c) $259 ÷ 12 (nearest cent)

```
        $ 21.583 ≈ $21.58
   12 )$259.000
        24
         19
         12
          70
          60
          100
           96
           40
           36
            4
```

Carry out division to the nearest tenth of a cent or nearest thousandth of a dollar.

CHECK

```
     21.583
  ×      12
  ---------
      43166
     21583
  ---------
     258996
  +       4
  ---------
    259.000  √
```

Now we will learn how to divide decimals by powers of ten.

$$3.2 \div 10^1 = \frac{3.2}{10} = \frac{3.2}{10.0} = \frac{.32}{1.0} = 0.32$$

$$3.2 \div 10^2 = \frac{3.2}{100} = \frac{03.2}{100.0} = \frac{.032}{1.0} = 0.032$$

$$3.2 \div 10^3 = \frac{3.2}{1{,}000} = \frac{003.2}{1000.0} = \frac{.0032}{1.0} = 0.0032$$

Notice that in each case, the quotient can be obtained by moving the decimal point to the left the number of places that is indicated by the power of ten.

> To divide a decimal by 10^n, where n is a whole number, move the decimal point n places to the left.

example 4

Find each of the following quotients.

(a) $0.83 \div 10^2$

$0.83 \div 10^2 = 00.83 = 0.0083$

Move the decimal point 2 places to the left.

(b) $5.29 \div 100{,}000$

$100{,}000 = 10^5$

$5.29 \div 10^5 = 00005.29 = 0.0000529$

Move the decimal point five places to the left.

(c) $\dfrac{8{,}678.432}{10^3}$

$8{,}678.432 \div 10^3 = 8{,}678.432 = 8.678432$ — Move the decimal point three places to the left. ●

When division is expressed by fractional notation, we use the order of operations rules to *simplify the numerator and denominator first*, before doing the division. For example, let us simplify the following expression.

$$\frac{7.3 + 2.7}{(0.5)^2} = \frac{10.0}{0.25} = \frac{1{,}000}{25} = 40$$

Notice that this expression could be rewritten with parentheses to indicate the operation performed first, as follows:

$$\frac{7.3 + 2.7}{(0.5)^2} = (7.3 + 2.7) \div (0.5)^2 = 10 \div 0.25 = 40$$

example 5

Simplify each of the following divisions in fractional notation.

(a) $\dfrac{15.6 - 8.4}{0.07 + 0.05}$

$$\frac{15.6 - 8.4}{0.07 + 0.05} = \frac{7.2}{0.12} = \frac{720}{12} = 60$$

Perform the operations in the numerator and the denominator. Reduce to lowest terms.

(b) $\dfrac{(0.03)^3}{5.81 - 5.801}$

$$\frac{(0.03)^3}{5.81 - 5.801} = \frac{0.000027}{0.009} = \frac{0.027}{9} = 0.003$$

(c) $\dfrac{(0.3)(0.2)^3}{(0.02)(0.5)^2}$

$$\frac{(0.3)(0.2)^3}{(0.02)(0.5)^2} = \frac{(0.3)(0.008)}{(0.02)(0.25)} = \frac{0.0024}{0.0050} = \frac{24}{50}$$

$$= \frac{48}{100} = 0.48$$

●

The computation of averages is an important application of division of decimals. Remember that to calculate an average, we first add all the numbers in the set, and then we divide that sum by the number of numbers that were added together.

example 6

During the fall semester, a student obtained the following grades on his math quizzes: 78, 62, 83, 77, and 86. Each quiz was worth 100 points. What was his average quiz grade for that semester?

solution

```
  78
  62
  83
  77
  86
 ---
 386
```

Add the five scores.

```
     77.2
  5 )386.0
     35
     --
      36
      35
      --
       10
       10
       --
        0
```

Divide the total by 5.

His average grade is 77.2.

QUICK QUIZ

Calculate each of the following quotients.

1. $0.63 \div 0.009$
2. $0.3648 \div 0.08$
3. $0.4053 \div 2.1$

ANSWERS

1. 70
2. 4.56
3. 0.193

3.5 Exercises

Indicate which of the following statements are true and which are false. For those that are false, change the italic word to make the statement true.

1. To divide a decimal by 10^n, where n is a whole number, move the decimal point n places to the *right*.
2. A *caret* indicates the new location of a decimal point.
3. When simplifying a fractional expression, we perform the arithmetic operations in the numerator and denominator *before* dividing the numerator by the denominator.
4. To round off a quotient, carry out the division until the quotient has a digit in the location immediately to the *left* of the place value to which you intend to round off.

Calculate the following quotients.

5. $0.81 \div 0.09$	**6.** $0.56 \div 0.8$	**7.** $0.0049 \div 0.7$	**8.** $2.4 \div 0.003$
9. $16.9 \div 0.0013$	**10.** $1.08 \div 0.0012$	**11.** $0.0221 \div 1.7$	**12.** $0.00225 \div 1.5$
13. $0.00252 \div 0.018$	**14.** $3.78 \div 0.0021$	**15.** $73.6 \div 0.0032$	**16.** $148.5 \div 4.5$
17. $5.096 \div 9.8$	**18.** $38.64 \div 8.4$	**19.** $0.4088 \div 0.0073$	**20.** $0.1624 \div 0.0056$
21. $17{,}860 \div 0.038$	**22.** $59{,}520 \div 0.064$		

Calculate the following quotients and round off as indicated.

23. $4.2 \div 0.9$ (nearest ten-thousandth)
24. $6.1 \div 0.7$ (nearest hundredth)
25. $168 \div 0.9$ (nearest tenth)
26. $771 \div 0.7$ (nearest whole number)
27. $20.8 \div 0.03$ (nearest thousandth)
28. $19.7 \div 0.07$ (nearest hundredth)
29. \$17.89 ÷ 6 (nearest cent)
30. \$52.37 ÷ 3 (nearest tenth of a cent)
31. $689 \div 0.27$ (nearest hundredth)
32. $457 \div 0.65$ (nearest thousandth)
33. $946 \div 0.31$ (nearest whole number)
34. $673 \div 0.84$ (nearest tenth)
35. \$254.99 ÷ 18 (nearest tenth of a dollar)
36. \$567.45 ÷ 14 (nearest tenth of a dollar)
37. $53.485 \div 0.002154$ (nearest whole number)
38. $10.005 \div 24.3$ (nearest ten-thousandth)
39. $0.6284 \div 9.217$ (nearest hundred-thousandth)
40. $82{,}547 \div 0.00764$ (nearest hundred)

Find each of the following quotients.

41. $7.83 \div 10^4$
42. $92{,}174 \div 10^7$
43. $0.006005 \div 10$
44. $0.0638 \div 1{,}000$
45. $90.009 \div 10^2$
46. $3{,}070 \div 10^5$
47. $\frac{281.7}{10^6}$
48. $\frac{80.4}{10^3}$
49. $\frac{0.5386}{100}$
50. $\frac{3{,}684}{10{,}000}$

Simplify each of the following expressions.

51. $\frac{5.8 + 9.2}{0.003}$
52. $\frac{13.6 - 2.6}{0.022}$
53. $\frac{(0.2)^3}{(0.05)^2}$
54. $\frac{(0.3)^2}{(0.001)^3}$
55. $\frac{7.3 + 12.7}{20 - 19.5}$
56. $\frac{23.6 - 8.1}{13.8 + 1.7}$
57. $\frac{(0.2)^3(0.3)^2}{(0.5)^2}$
58. $\frac{(0.06)^2}{(0.1)^2(0.2)^2}$
59. $\frac{(0.04)^3}{5.01 - (0.1)^2}$
60. $\frac{8.04 - (0.2)^2}{(4)(0.05)^2}$
61. $\frac{(0.3)^2}{4.61 - 4.601}$
62. $\frac{(0.6)^2}{7.31 - 7.301}$
63. $4.2 \div 0.7 + (0.2)(0.6)$
64. $0.36 \div 0.06 + (0.8)(0.3)$
65. $[3.8 + (0.4)(3)] \div (0.5)^2$
66. $[(6)(0.4) + 15.6] \div (0.3)^2$
67. $(0.2)^4 + 0.048 \div 0.6$
68. $(0.9)^2 - 0.56 \div 0.8$
69. $6.4 \div 0.08 \div (0.2)^2$
70. $(0.2)^5 \div 0.004 \div 0.8$

Solve each of the following word problems.

71. If an engineer's yearly salary is \$37,000, what is her weekly salary? Round off to the nearest cent.
72. If a car travels 513 miles on 17.2 gallons of gas, how many miles per gallon (correct to the nearest tenth) does the car get?
73. If chicken costs 79¢ per pound, how many pounds of chicken (correct to the nearest tenth) could you buy for \$5.00?
74. A batting average is calculated by dividing the number of hits by the number of times at bat and rounding off that result to the nearest thousandth. If a player made 23 hits in 89 times at bat, what was his batting average?
75. Iceland has an area of 39,702 square miles and a population of 230,000. What is the average population density per square mile? Round your answer to the nearest thousandth.
76. During a practice session, an olympic swimmer completed a 100-meter course in the following times: 50.45 seconds, 50.53 seconds, 49.89 seconds, 51.06 seconds, and 50.93 seconds. What was his average time during that practice session?

77. During a particular academic year, a student received one A, two B's, four C's, and one D. The numerical equivalents for these grades are A = 4.0, B = 3.0, C = 2.0, and D = 1.0. What was the student's GPA (grade point average) for that year? Round your answer to the nearest tenth.

78. A jogger can run 4 miles in 38 minutes 6 seconds. What is the jogger's average speed in miles per hour? Round off to the nearest tenth.

79. A race car driver finished a 500-mile race in 3 hours 6 minutes 27 seconds. What was the driver's average speed in miles per hour? Round your answer to the nearest tenth.

80. A woman bought a condominium that cost $65,000. She made a down payment of $20,000, and obtained a 15-year mortgage totaling $92,574 to pay off the remainder. What are her monthly payments?

3.6 Fractions and Decimals

Since any decimal can also be expressed as a fraction, it is often useful to convert decimals to fractions and fractions to decimals. To convert a decimal to a fraction, we simply rewrite the number as a decimal fraction and reduce to lowest terms. For example,

$$0.35 = \frac{35}{100} = \frac{7}{20}$$

example 1

Convert each of the following decimals to fractions.

(a) 0.06

$$0.06 = \frac{6}{100} = \frac{3}{50}$$

Rewrite as a decimal fraction and reduce to lowest terms.

(b) 0.125

$$0.125 = \frac{125}{1,000} = \frac{5}{40} = \frac{1}{8}$$

Rewrite as a decimal fraction and reduce.

(c) 0.0032

$$0.0032 = \frac{32}{10,000} = \frac{8}{2,500} = \frac{4}{1,250} = \frac{2}{625}$$

Rewrite and reduce.

(d) 8.25

$$8.25 = 8\frac{25}{100} = \frac{825}{100} = \frac{165}{20} = \frac{33}{4}$$

Recall that fractional notation expresses the operation division. Therefore, to convert a fraction to a decimal, simply divide the numerator by the denominator. For example,

$$\frac{3}{4} = 3 \div 4 = 0.75$$

```
     .75
  4 |3.00
     2 8
     ---
      20
      20
      --
       0
```

These two procedures are outlined in the box.

> To convert a decimal to a fraction, rewrite the decimal as a decimal fraction and reduce it to lowest terms.
> To convert a fraction to a decimal, divide the numerator by the denominator.

Because we obtained a remainder of zero when calculating the decimal representation for $\frac{3}{4}$, 0.75 is called a **terminating decimal**.

definition

A *terminating decimal* is the exact representation of a fraction. That is, a terminating decimal is the quotient obtained by dividing the numerator of a fraction by the denominator when the remainder is zero.

Sometimes, when converting a fraction to its decimal representation, the long division process will never yield a remainder of zero, regardless of how far we carry out the division. For example, to convert $\frac{1}{3}$ to a decimal we must divide 1 by 3, as illustrated.

$$\frac{1}{3} = 1 \div 3 = 0.333\ldots$$

```
    .333
3 |1.000
    9
    --
    10
     9
    --
     10
      9
     --
      1
```

The decimal we obtained as our answer is called a **repeating decimal**. The symbol ... indicates that the pattern of digits repeats itself indefinitely.

definition

A *repeating decimal* is the representation of a fraction in which a pattern of digits repeats itself indefinitely.

We can also write the decimal representation for $\frac{1}{3}$ as

$$\frac{1}{3} = 0.\overline{3}$$

The bar over the 3 indicates that the digit 3 is repeated indefinitely.

example 2

Convert each of the following fractions to decimals.

(a) $\frac{5}{8}$

$$\frac{5}{8} = 5 \div 8 = 0.625$$

```
    .625
8 |5.000
   4 8
   ---
    20
    16
    --
     40
     40
     --
      0
```

The result is a terminating decimal.

(b) $\frac{4}{11}$

$$\frac{4}{11} = 4 \div 11 = 0.3636\ldots = 0.\overline{36}$$

$$\begin{array}{r} .3636 \\ 11\overline{)4.0000} \\ 3\,3 \\ \hline 70 \\ 66 \\ \hline 40 \\ 33 \\ \hline 70 \\ 66 \\ \hline 4 \end{array}$$

Notice that the digits 36 will repeat themselves indefinitely. We place a bar over both digits that repeat.

The result is a repeating decimal.

(c) $\frac{38}{25}$

$$\frac{38}{25} = 1.52$$

$$\begin{array}{r} 1.52 \\ 25\overline{)38.00} \\ 25 \\ \hline 13\,0 \\ 12\,5 \\ \hline 50 \\ 50 \\ \hline 0 \end{array}$$

The result is a terminating decimal. ●

Any fraction whose numerator is a whole number, and whose denominator is a whole number other than zero, can be represented as a terminating or repeating decimal. Sometimes, however, when a decimal does not terminate, a pattern of repeating digits does not become apparent, even after carrying out the division four or five places. In such cases, instead of carrying out the division further, we can approximate the answer by rounding off.

example 3

Convert $\frac{4}{13}$ to a decimal rounded off to the nearest ten-thousandth.

solution

$$\frac{4}{13} = 4 \div 13 \approx 0.30769 \approx 0.3077$$

Carry out the division to the hundred-thousandths place and round off.

$$\begin{array}{r} .30769 \\ 13\overline{)4.00000} \\ 3\,9 \\ \hline 100 \\ 91 \\ \hline 90 \\ 78 \\ \hline 120 \\ 117 \\ \hline 3 \end{array}$$

●

QUICK QUIZ

1. Convert 0.075 to a fraction.
2. Convert $\frac{8}{9}$ to a decimal.
3. Convert $\frac{3}{17}$ to a decimal correct to the nearest thousandth.

ANSWERS

1. $\frac{3}{40}$
2. $0.\overline{8}$
3. 0.176

3.6 Exercises

Indicate which of the following statements are true and which are false. For those that are false, change the italic expression to make the statement true.

1. A *continuing* decimal is the representation of a fraction in which a pattern of digits repeats itself indefinitely.
2. To convert a fraction to a decimal, divide the *denominator by the numerator*.
3. If the long division process yields a remainder of zero, the fraction can be represented as a *terminating* decimal.
4. Fractions whose numerators are less than their denominators can be rewritten as decimals *greater* than one.

Convert each of the following decimals to fractions reduced to lowest terms.

5. 0.08	**6.** 0.46	**7.** 0.16	**8.** 0.36	**9.** 0.024	**10.** 0.125
11. 0.0075	**12.** 0.0028	**13.** 0.3125	**14.** 0.625	**15.** 2.5	**16.** 6.4
17. 12.8	**18.** 5.75	**19.** 15.45	**20.** 70.75	**21.** 40.15	**22.** 20.05
23. 4.0375	**24.** 8.2025				

Convert each of the following fractions to decimals. Carry out the division until the remainder is zero or until a pattern of repeating digits becomes apparent.

25. $\frac{1}{4}$	**26.** $\frac{1}{3}$	**27.** $\frac{2}{5}$	**28.** $\frac{7}{8}$	**29.** $\frac{5}{11}$	**30.** $\frac{3}{8}$
31. $\frac{5}{32}$	**32.** $\frac{5}{6}$	**33.** $\frac{4}{15}$	**34.** $\frac{5}{16}$	**35.** $\frac{4}{9}$	**36.** $\frac{3}{25}$
37. $\frac{7}{18}$	**38.** $\frac{11}{15}$	**39.** $\frac{3}{4}$	**40.** $\frac{1}{8}$	**41.** $\frac{2}{9}$	**42.** $\frac{2}{55}$
43. $\frac{8}{3}$	**44.** $\frac{7}{4}$	**45.** $\frac{13}{9}$	**46.** $\frac{15}{11}$	**47.** $\frac{19}{16}$	**48.** $\frac{26}{15}$
49. $\frac{367}{12}$	**50.** $\frac{461}{15}$	**51.** $\frac{727}{3}$	**52.** $\frac{555}{9}$	**53.** $\frac{2}{21}$	**54.** $\frac{6}{7}$
55. $\frac{678}{999}$	**56.** $\frac{281}{999}$				

Find a decimal approximation to each of the following fractions correct to the place value indicated.

57. $\frac{3}{19}$ (nearest thousandth)

58. $\frac{6}{21}$ (nearest thousandth)

59. $\frac{7}{31}$ (nearest hundredth)

60. $\frac{6}{17}$ (nearest thousandth)

61. $\frac{13}{15}$ (nearest ten-thousandth)

62. $\frac{5}{9}$ (nearest ten-thousandth)

63. $\frac{6}{11}$ (nearest hundred-thousandth)

64. $\frac{5}{14}$ (nearest hundredth)

65. $\frac{2}{7}$ (nearest thousandth)

66. $\frac{9}{14}$ (nearest ten-thousandth)

67. $\frac{18}{7}$ (nearest hundredth)

68. $\frac{29}{13}$ (nearest thousandth)

69. $\frac{368}{9}$ (nearest hundredth)

70. $\frac{555}{19}$ (nearest thousandth)

71. $\frac{521}{683}$ (nearest thousandth)

72. $\frac{37}{749}$ (nearest ten-thousandth)

Summary and Review

Key Terms

[3.1] A **decimal** is a number that can be represented as a fraction whose denominator is a power of ten greater than or equal to one.

A **decimal fraction** is a fraction whose denominator can be expressed as a power of ten.

[3.2] **Rounding off** is the procedure used to estimate a number to a given degree of precision.

[3.6] A **terminating decimal** is the exact representation of a fraction. It is the quotient obtained by dividing the numerator of the fraction by the denominator when the remainder is zero.

A **repeating decimal** is the representation of a fraction in which a pattern of digits repeats itself indefinitely.

Calculations

[3.1] **To write a number in decimal notation**, write the digits in the various decimal places to the right of the ones place, separated from the ones place by a *decimal point*.

[3.2] **To round off a number**, examine the digit immediately to the right of the place value to which you are rounding off. If that digit is 5 or greater, add one to the digit in the place value of the desired precision (*round up*). If the digit to the right is less than 5, retain the digit in the place value of desired precision (*round down*).

[3.3] **To add (or subtract) decimals**, write the numbers vertically and line up the decimal points. Add (or subtract) the digits in each column separately, moving from right to left, and bring down the decimal point.

[3.4] **To multiply decimals**, calculate the product as usual, and put a decimal point in the appropriate location so that the number of decimal places in the product equals the sum of the number of decimal places in all factors.

To multiply a decimal by 10^n, where n is a whole number, move the decimal point n places to the right.

[3.5] **To divide decimals**, move the decimal point in the divisor and in the dividend the same number of places to the right in order to make the divisor a whole number. Place a caret in the new location of each decimal point. Put a decimal point in the quotient immediately above the caret in the dividend, and apply the technique for dividing whole numbers.

To divide a decimal by 10^n, where n is a whole number, move the decimal point n places to the left.

[3.6] **To convert a decimal to a fraction**, rewrite the number as a decimal fraction and reduce it to lowest terms.

To convert a fraction to a decimal, divide the numerator by the denominator.

Chapter 3 Review Exercises

Indicate which of the following statements are true and which are false. For those that are false, change the italic expression to make the statement true.

1. The decimal places are located to the *left* of the decimal point.
2. *Rounding off* is a procedure for estimating numbers.
3. To find the average of a set of numbers, we use the operations addition and *subtraction.*
4. *All* fractions whose numerators and denominators are nonzero whole numbers can be represented as either terminating or repeating decimals.
5. When adding and *multiplying* decimals, we write the problem vertically and line up the decimal points.
6. If the numerator of a fraction is evenly divisible by the denominator, the fraction is equivalent to a *repeating* decimal.

[3.1] Write each of the following decimals in words and in expanded notation.

7. 8.7
8. 6.2
9. 5.87
10. 12.04
11. 43.628
12. 70.296
13. 806.5614
14. 4,083.0609

[3.2] Round off each of the following decimals to the places indicated.

15. 4.537 (nearest tenth)
16. 8.984 (nearest hundredth)
17. 17.86 (nearest whole number)
18. 8.2786 (nearest thousandth)
19. 55.44 (nearest ten)
20. 467.329 (nearest hundredth)
21. 953.628 (nearest hundred)
22. 74.4893 (nearest thousandth)
23. $5.0477 (nearest tenth of a cent)
24. $27.568 (nearest tenth of a dollar)

[3.3–3.5] Perform each of the operations indicated.

25. 623.56 + 4.872
26. 6.3784 + 297.69
27. 741.36 − 3.89
28. 83 − 2.743
29. 6.42 × 0.03
30. 13.93 × 0.005
31. 142.8 ÷ 0.06
32. 872.4 ÷ 0.4
33. 5.468 + 12.82 − 8.9273
34. 48.1 + 362.76 − 7.9857
35. 8.534 × 0.38
36. 31.746 × 0.26
37. 5.6079 ÷ 0.67
38. 3.9312 ÷ 0.72
39. $632.50 + $3,028.99
40. $1,813.45 − $546.89
41. $89.99 × 28
42. $220.02 ÷ 57
43. 58.473 − 9.5841 + 127.8 − 88.63
44. 17.357 − 9.461 + 752.3 − 68.598
45. 7.13 × 0.04 × 5.2
46. 8.5 × 0.26 × 0.009
47. 5.67 ÷ 27 ÷ 0.03
48. 4.82 ÷ 0.004 ÷ 0.25

49. $7.2 \times 400{,}000$
50. 0.009×0.0005
51. $(3.2 + 6.3) \div 0.05$
52. $(2.6 + 1.6) \div 0.06$
53. $(0.6)^2 - (0.4)^3$
54. $(1.1)^2 - (0.2)(0.8)$
55. $0.63 \div 30{,}000$
56. $432 \div 0.0002$
57. $(1.2)^2 \div (3.2 - 2.8)$
58. $(1.4)^2 \div (0.2)^2$
59. $(4.7 + 3.7) \div (7.3 - 6.6)$
60. $(9.3 - 1.8) \div (4.2 + 1.8)$
61. $[8.143 + 0.007 - (0.3)(0.5)]^2$
62. $[5.219 + 0.081 - 3.3]^3$
63. $5.4309 + 621.847 + 0.5649$
64. $7{,}004 - 5.3682 - 29.637$
65. 3.218×0.4527
66. $62.2836 \div 0.657$

[3.4, 3.5] Find each of the following products and quotients.

67. 0.0053×10^3
68. 862×10^5
69. $\frac{0.078}{10^6}$
70. $\frac{3.892}{10^7}$
71. 16.894×10^0
72. $\frac{4.5831}{10^0}$
73. $9{,}209 \times 1{,}000$
74. $\frac{260{,}047}{10{,}000}$

[3.6] Convert each of the following decimals to fractions reduced to lowest terms.

75. 0.12
76. 0.08
77. 0.025
78. 0.004
79. 0.00016
80. 0.00068
81. 0.00225
82. 0.00144
83. 4.4
84. 7.5
85. 20.45
86. 247.5

[3.6] Convert each of the following fractions to decimals. Round off only when indicated.

87. $\frac{2}{3}$
88. $\frac{3}{8}$
89. $\frac{7}{16}$
90. $\frac{5}{6}$
91. $\frac{8}{11}$
92. $\frac{9}{25}$
93. $\frac{5}{7}$ (nearest thousandth)
94. $\frac{9}{21}$ (nearest ten-thousandth)
95. $\frac{12}{5}$
96. $\frac{23}{8}$
97. $\frac{52}{11}$
98. $\frac{86}{13}$ (nearest thousandth)

[3.3–3.5] Solve each of the following word problems.

99. If a baseball player made 17 hits in 69 times at bat, what was his batting average? Round off to the nearest thousandth.

100. A carpenter earns $13.65 an hour and gets time-and-a-half for overtime (more than 40 hours worked in a single week). What would be the carpenter's salary for a 47.25-hour work week?

101. Before a salesman began a two-day trip, the odometer on his car read 29,086.9 miles. At the end of the first day, the reading was 29,586.7 miles, and at the end of the second day, the reading was 30,070.2 miles. Which day did he travel the longer distance? What was the total distance that he traveled on these two days combined?

102. If seven umbrellas cost $96.95, what is the cost of two umbrellas?

103. During a student's undergraduate years at a state university, she received the following grades: five A's, eleven B's, fourteen C's, two D's, and one F. If all courses are worth the same number of credit hours, what was the student's GPA (grade point average)? (Recall that A = 4.0, B = 3.0, C = 2.0, D = 1.0, and F = 0.0.) Round your answer to the nearest tenth.

104. The scale on a map indicates that 1 inch represents 12.8 miles. If the distance between two cities on that map is 4.5 inches, what is the actual distance between them?

105. A retailer buys scarves at a cost of \$51.96 a dozen. He sells the scarves for \$7.98 apiece. How much profit will he make by selling five scarves?

106. A nurse took a patient's temperature every hour and recorded these temperatures (in °F): 100.2, 99.8, 98.7, 99.2, 99.4, 100.4, and 101.1. What was the patient's average temperature for this 7-hour period? Round your answer to the nearest tenth.

107. If you bought three identical notebooks and received \$4.06 in change from a \$10 bill, what was the price of a single notebook?

108. During a 585-mile trip, a car averaged 32.5 miles per gallon and used a full tank of gas. How much would it cost to refill the tank with gas that costs \$1.545 a gallon?

4 RATIO AND PROPORTION

4.1 Ratio

In this chapter we will learn how fractions can be used to express a relationship between two quantities. For example, if a class of students contains 13 men and 27 women, we can say that the ratio of men to women is 13 to 27, or 13:27. Ratios are usually written as fractions in which the numerator is the first quantity being compared, and the denominator is the second. Thus, the ratio of men to women in this case can also be written as $\frac{13}{27}$.

definition

A *ratio* is a fraction that expresses a relationship between two quantities. The ratio of a to b can be written in the form $a:b$ or $\frac{a}{b}$. The first quantity being compared is placed in the numerator, and the second quantity is placed in the denominator.

example 1

In a basket of fruit, there are 5 apples, 6 oranges, and 3 bananas.

(a) What is the ratio of oranges to apples?

Since there are 6 oranges and 5 apples, the ratio of oranges to apples is 6 : 5 or $\frac{6}{5}$.

(b) What is the ratio of bananas to oranges?

Since there are 3 bananas and 6 oranges, the ratio of bananas to oranges is 3 : 6 or $\frac{3}{6}$.

$\frac{3}{6}$ reduces to $\frac{1}{2}$, so the ratio expressed in lowest terms is $\frac{1}{2}$ or 1 : 2.

(c) What fraction of the fruit is oranges?

First add the number of apples, oranges, and bananas to find the total amount of fruit.

$$5 + 6 + 3 = 14 \text{ fruit}$$

Since there are 6 oranges and 14 pieces of fruit, the ratio of oranges to fruit is $6:14$ or $\frac{6}{14}$.

$\frac{6}{14}$ reduces to $\frac{3}{7}$, so $\frac{3}{7}$ of the fruit is oranges. ●

When using ratios, it is important to keep in mind the units in which the two quantities we are comparing are expressed. For example, suppose one pole vaulter cleared 8 feet and a second vaulter cleared 92 inches. Which pole vaulter jumped higher?

In order to compare these two heights in a meaningful way, they must be expressed in the same units. To convert from one unit of measurement to another, we can multiply the quantity we wish to convert by a fraction equivalent to one. Since 12 inches = 1 foot, the fractions

$$\frac{1 \text{ ft}}{12 \text{ in.}} \quad \text{and} \quad \frac{12 \text{ in.}}{1 \text{ ft}} \quad \text{both are equal to 1.}$$

This can be shown by rewriting the numerators as follows:

$$\frac{1 \text{ ft}}{12 \text{ in.}} = \frac{12 \text{ in.}}{12 \text{ in.}} = 1$$

$$\frac{12 \text{ in.}}{1 \text{ ft}} = \frac{1 \text{ ft}}{1 \text{ ft}} = 1$$

Fractions such as this are sometimes called **unit fractions**.

definition

A *unit fraction* is a fraction equivalent to one.

We can convert 92 inches to feet by multiplying 92 inches by the unit fraction $\frac{1 \text{ ft}}{12 \text{ in.}}$.

$$\begin{aligned} 92 \text{ in.} &= 92 \text{ in.} \times 1 \\ &= \frac{92 \text{ in.}}{1} \times \frac{1 \text{ ft}}{12 \text{ in.}} \\ &= \frac{92}{12} \text{ ft} \\ &= \frac{23}{3} \text{ ft} \\ &= 7\frac{2}{3} \text{ ft} \end{aligned}$$

Notice that we can cancel out common units in the numerator and denominator just as we do with common factors.

We determine the appropriate unit fraction to use for a conversion by looking at the units we need to cancel for the conversion to take place. If we had tried to multiply

$$92 \text{ in.} \times \frac{12 \text{ in.}}{1 \text{ ft}}$$

the inches would not have canceled, and our attempt at conversion would not have worked.

In order to convert 8 feet to inches, we must multiply by a unit fraction that has feet in the denominator, $\frac{12 \text{ in.}}{1 \text{ ft}}$, so that the feet cancel.

$$\begin{aligned} 8 \text{ ft} &= 8 \text{ ft} \times 1 \\ &= \frac{8 \cancel{\text{ft}}}{1} \times \frac{12 \text{ in.}}{1 \cancel{\text{ft}}} \\ &= 96 \text{ in.} \end{aligned}$$

We can now compare the heights cleared by the two pole vaulters expressed in either feet or inches, because we have calculated both distances in both units.

$$\frac{\text{1st vaulter}}{\text{2nd vaulter}} = \frac{8 \cancel{\text{ft}}}{7\frac{2}{3} \cancel{\text{ft}}} = \frac{8 \cdot 3}{\frac{23}{3} \cdot 3} = \frac{24}{23} \qquad \frac{\text{1st vaulter}}{\text{2nd vaulter}} = \frac{96 \cancel{\text{in.}}}{92 \cancel{\text{in.}}} = \frac{96 \cdot \frac{1}{4}}{92 \cdot \frac{1}{4}} = \frac{24}{23}$$

Notice that in both ratios, the common units cancel, and both reduce to the same fraction. Thus, the ratio of the height cleared by the first vaulter to that cleared by the second is 24:23. We can therefore conclude that the first vaulter jumped higher.

We will now summarize the procedure for converting units of measurement.

To Convert from One Unit of Measurement to Another

Multiply by the unit fraction that has the old unit in the denominator and the new unit in the numerator.

For example, to convert 24 ounces to pounds, multiply 24 ounces by the unit fraction $\frac{1 \text{ lb}}{16 \text{ oz}}$:

$$\begin{aligned} 24 \text{ oz} &= \frac{24 \text{ oz}}{1} \times \frac{1 \text{ lb}}{16 \text{ oz}} \\ &= \frac{24}{16} \text{ lb} = \frac{3}{2} \text{ lb} = 1\frac{1}{2} \text{ lb} \end{aligned}$$

example 2

Convert each of the following units of measurement.

(a) 3 gallons to quarts

$$\begin{aligned} 3 \text{ gal} &= \frac{3 \text{ gal}}{1} \times \frac{4 \text{ qt}}{1 \text{ gal}} \\ &= \frac{3 \cancel{\text{gal}}}{1} \times \frac{4 \text{ qt}}{1 \cancel{\text{gal}}} \\ &= 12 \text{ qt} \end{aligned}$$

Multiply by the unit fraction that has qt in the numerator and gal in the denominator.

(b) 25 minutes to hours

$$25 \text{ min} = \frac{25 \text{ min}}{1} \times \frac{1 \text{ hr}}{60 \text{ min}}$$

Multiply by the unit fraction that has hours in the numerator and minutes in the denominator.

$$= \frac{25 \cancel{\text{min}}}{1} \times \frac{1 \text{ hr}}{60 \cancel{\text{min}}}$$

$$= \frac{25}{60} \text{ hr}$$

$$= \frac{5}{12} \text{ hr}$$

Divide numerator and denominator by 5.

In order to compare two quantities, we will now look at some examples that require us to convert units of measurement.

example 3

Express each of the following comparisons as a ratio with the same units in the numerator and denominator. Reduce to lowest terms.

(a) 8 nickels to 3 dimes

$$8 \text{ nickels} = 8 \text{ nickels} \times 1$$

$$= \frac{8 \cancel{\text{nickels}}}{1} \times \frac{1 \text{ dime}}{2 \cancel{\text{nickels}}}$$

$$= \frac{8}{2} \text{ dimes}$$

$$= 4 \text{ dimes}$$

We can convert 8 nickels to dimes by multiplying 8 nickels by a unit fraction with nickels in the denominator (so that nickels cancel) and dimes in the numerator. Since 2 nickels = 1 dime, this unit fraction is $\frac{1 \text{ dime}}{2 \text{ nickels}}$.

Thus, 8 nickels : 3 dimes

$$= 4 \text{ dimes} : 3 \text{ dimes}$$

Express each quantity in the same units.

$$= \frac{4 \cancel{\text{dimes}}}{3 \cancel{\text{dimes}}}$$

Rewrite as a fraction and simplify.

$$= \frac{4}{3}$$

The ratio of 8 nickels to 3 dimes is $\frac{4}{3}$.

(b) 15 minutes to 5 hours

$$15 \text{ min} = 15 \text{ min} \times 1$$

$$= \frac{15 \cancel{\text{min}}}{1} \times \frac{1 \text{ hr}}{60 \cancel{\text{min}}}$$

$$= \frac{15}{60} \text{ hr}$$

$$= \frac{1}{4} \text{ hr}$$

We can convert 15 min to hours by multiplying 15 min by a unit fraction with minutes in the denominator (so minutes cancel) and hours in the numerator. Since 1 hr = 60 min, this unit fraction is $\frac{1 \text{ hr}}{60 \text{ min}}$.

Thus, 15 min : 5 hr $= \frac{1}{4}$ hr : 5 hr

Express each quantity in the same units.

$= \dfrac{\frac{1}{4}\ \cancel{\text{hr}}}{5\ \cancel{\text{hr}}}$

Rewrite as a fraction.

$= \dfrac{\frac{1}{4} \times 4}{5 \times 4}$

Simplify.

$= \dfrac{1}{20}$

The ratio of 15 minutes to 5 hours is $\frac{1}{20}$.

(c) 8 pints to $5\frac{1}{2}$ quarts

$5\frac{1}{2}\text{ qt} = 5\frac{1}{2}\text{ qt} \times 1$

$= \dfrac{11}{2}\ \cancel{\text{qt}} \times \dfrac{2\text{ pt}}{1\ \cancel{\text{qt}}}$

$= \dfrac{22}{2}\text{ pt}$

$= 11\text{ pt}$

We can convert $5\frac{1}{2}$ qt to pints by multiplying $5\frac{1}{2}$ qt by a unit fraction with quarts in the denominator (so that quarts cancel) and pints in the numerator. Since 2 pt = 1 qt, this unit fraction is $\frac{2\text{ pt}}{1\text{ qt}}$.

Thus, 8 pt : $5\frac{1}{2}$ qt = 8 pt : 11 pt

$= \dfrac{8\ \cancel{\text{pt}}}{11\ \cancel{\text{pt}}} = \dfrac{8}{11}$

Express each quantity in the same units and simplify.

The ratio of 8 pints to $5\frac{1}{2}$ quarts is $\frac{8}{11}$.

(d) 4 yards to 200 inches

It is not always necessary to use one of the units given as the common unit to compare the two quantities. In this problem we will convert both units to feet.

$4\text{ yd} = 4\text{ yd} \times 1$

$= 4\ \cancel{\text{yd}} \times \dfrac{3\text{ ft}}{1\ \cancel{\text{yd}}}$

$= 12\text{ ft}$

Since 3 ft = 1 yd, we can convert 4 yd to feet.

$200\text{ in.} = 200\text{ in.} \times 1$

$= 200\ \cancel{\text{in.}} \times \dfrac{1\text{ ft}}{12\ \cancel{\text{in.}}}$

$= \dfrac{200}{12}\text{ ft} = \dfrac{50}{3}\text{ ft}$

Since 12 in. = 1 ft, we can convert 200 in. to feet.

Thus, 4 yd : 200 in. $= 12 \text{ ft} : \frac{50}{3} \text{ ft}$ — Express the ratio of 4 yd to 200 in. in feet.

$$= \frac{12 \cancel{\text{ft}}}{\frac{50}{3} \cancel{\text{ft}}}$$ Express the ratio as a fraction.

$$= \frac{12 \times 3}{\frac{50}{3} \times 3}$$ Simplify.

$$= \frac{36}{50}$$ Reduce to lowest terms.

$$= \frac{18}{25}$$

The ratio of 4 yards to 200 inches is $\frac{18}{25}$. ●

One of the most common examples of a ratio is a **rate**. Rates express the relationship between two unlike quantities.

definition

A *rate* is a ratio that compares two different quantities and has a denominator equal to one.

For example, if you traveled 150 miles in 3 hours, your rate of speed could be calculated by rewriting the ratio of miles to hours as an equivalent ratio with a denominator of one. The ratio of miles to hours is

$$\frac{150 \text{ mi}}{3 \text{ hr}} = \frac{150 \text{ mi} \times \frac{1}{3}}{3 \text{ hr} \times \frac{1}{3}}$$

$$= \frac{50 \text{ mi}}{1 \text{ hr}}$$

Your rate of speed is $\frac{50 \text{ mi}}{1 \text{ hr}}$, which is more commonly written as 50 mi/hr or 50 miles per hour. Notice that the / is read as "per."

example 4

Before lunch, a secretary typed 9 pages in 50 minutes. After lunch, she typed 15 pages in $1\frac{1}{4}$ hours. If each page typed was approximately the same length, did she type faster before or after lunch?

solution

Determine her rate before lunch:

$$\frac{9 \text{ pp}}{50 \text{ min}} = 0.18 \text{ pp/min}$$ Before lunch, the ratio of pages to minutes was 9 to 50.

Determine her rate after lunch in the same units as her rate before lunch:

$$1\frac{1}{4}\text{ hr} = 1\frac{1}{4}\text{ hr} \times 1$$

$$= \frac{5}{4}\cancel{\text{hr}} \times \frac{60\text{ min}}{1\ \cancel{\text{hr}}}$$

$$= \frac{5(60)}{4}\text{ min} = 75\text{ min}$$

We must convert $1\frac{1}{4}$ hr to minutes. Since 1 hr = 60 min, we can multiply $1\frac{1}{4}$ hr by the unit fraction $\frac{60\text{ min}}{1\text{ hr}}$.

So, she typed 15 pp in 75 min.

Her rate after lunch was, therefore,

$$\frac{15\text{ pp}}{75\text{ min}} = \frac{1\text{ p}}{5\text{ min}} = 0.20\text{ pp/min}$$

Thus, she typed faster after lunch.

$0.18\text{ pp/min} < 0.20\text{ pp/min}$. ●

Ratios are especially useful for determining *unit prices*; that is, for finding the price of a single item given the price of a larger quantity of the items. For example, if soup is priced at 5 cans for a dollar, the price of one can could be determined by examining the ratio of dollars to cans, which is 1:5 or $\frac{1\text{ dollar}}{5\text{ cans}}$. Changing $\frac{1}{5}$ to a decimal so that the denominator is equal to one, $\frac{1}{5} = \frac{0.20}{1}$, we obtain the unit price of \$0.20/can. We want the denominator of the ratio to be equal to one, because unit pricing is a rate.

example 5

Determine the unit price of each of the following items with the given prices.

(a) Soap—3 bars for \$0.99

$$\frac{0.99\text{ dollars}}{3\text{ bars}}$$

Find the ratio of dollars to bars.

$$\frac{0.99}{3} = \frac{0.33}{1}$$

Unit price is \$0.33/bar.

(b) Socks—5 pairs for \$7.29

$$\frac{7.29\text{ dollars}}{5\text{ pairs}}$$

Find the ratio of dollars to pairs.

$$\frac{7.29}{5} = \frac{1.458}{1}$$

Unit price is \$1.458/pair.

(c) Orange juice—16 fluid ounces for 52¢

$$\frac{52\text{ cents}}{16\text{ fl oz}}$$

Find the ratio of cents to fluid ounces.

$$\frac{52}{16} = \frac{3.25}{1}$$

Unit price is 3.25¢/fl oz. ●

example 6

A 16-fluid ounce bottle of vegetable oil costs \$0.94, and a 24-fluid ounce bottle costs \$1.29. Determine which size is the better buy.

solution

$$\frac{0.94 \text{ dollars}}{16 \text{ fl oz}} = \frac{0.94}{16} = \frac{0.05875}{1}$$

Find the ratio of dollars to fluid ounces for the smaller size.

Unit price is \$0.05875/fl oz for the 16-fl oz bottle.

$$\frac{1.29 \text{ dollars}}{24 \text{ fl oz}} = \frac{1.29}{24} = \frac{0.05375}{1}$$

Find the ratio of dollars to fluid ounces for the larger size.

Unit price is \$0.05375/fl oz for the 24-fl oz bottle.
Since the unit price of the 24-fl oz size is less, it is the better buy. ●

QUICK QUIZ

Express each of the following ratios as a fraction without units. Reduce to lowest terms.

1. 9 to 36
2. $3\frac{1}{2}$ to $5\frac{1}{4}$
3. 3 yards to 8 feet

ANSWERS

1. $\frac{1}{4}$
2. $\frac{2}{3}$
3. $\frac{9}{8}$

4.1 Exercises

Indicate which of the following statements are true and which are false. For those that are false, change the italic expression to make the statement true.

1. The ratio $x:y$ can be written as a fraction in which x is the *denominator*.
2. A ratio expresses the operation *subtraction*.
3. If the units in the numerator and denominator are *the same*, they cancel.
4. Changing the order of the two quantities being compared *does not* affect the value of the ratio.

Express each of the following ratios as a fraction reduced to lowest terms.

5. 3:5 **6.** 7:9 **7.** 8:2 **8.** 11:1 **9.** 5 to 8
10. 12 to 3 **11.** 88:4 **12.** 6:100 **13.** 4:56 **14.** 63:3
15. $\frac{1}{4}:3$ **16.** $6:\frac{2}{3}$ **17.** $\frac{5}{8}:\frac{2}{3}$ **18.** $\frac{2}{9}:\frac{3}{5}$ **19.** $3\frac{1}{2}:8\frac{1}{4}$
20. $6\frac{2}{3}:1\frac{3}{4}$ **21.** 2.7 to 0.9 **22.** 0.0032 to 0.04

23. Convert 4 yards to feet.
24. Convert 3 hours to minutes.
25. Convert 20 nickels to quarters.
26. Convert 35 days to weeks.
27. Convert 5 quarts to pints.
28. Convert 9 pints to fluid ounces.
29. Convert 18 inches to feet.
30. Convert 36 ounces to pounds.
31. Convert $\frac{5}{8}$ pound to ounces.
32. Convert $1\frac{3}{4}$ gallons to quarts.

Express each of the following comparisons as a ratio with the same units in the numerator and denominator. Reduce to lowest terms.

33. 3 weeks to 18 days
34. 16 months to 2 years
35. 19 nickels to 5 quarters
36. 14 dimes to 29 nickels
37. 64 items to 4 dozen
38. 6 dozen to 99 items
39. 42 inches to 4 feet
40. 10 feet to 100 inches
41. 6 dollars to 15 quarters
42. 81 nickels to 8 dollars
43. 7 pints to 3 quarts
44. 5 gallons to 35 quarts
45. 12 ounces to 2 pounds
46. $\frac{3}{4}$ pound to 8 ounces
47. 12 fluid ounces to 2 pints
48. 3 tablespoons to 8 teaspoons
49. 9 months to $\frac{1}{2}$ year
50. $3\frac{1}{4}$ days to 100 hours
51. $9\frac{1}{3}$ dozen to 400 items
52. 1,000 pounds to $2\frac{1}{2}$ tons
53. 4.25 dollars to 75 nickels
54. 150 dimes to 9.35 dollars
55. 3 gallons to 18 pints
56. 30 cups to 7 quarts
57. 8 yards to 48 inches
58. 2 miles to 176 yards
59. 720 seconds to $\frac{1}{2}$ hour
60. $1\frac{3}{4}$ hours to 6,000 seconds

Answer each of the following questions.

61. In a math class, 4 students have blue eyes, 2 students have green eyes, and 14 students have brown eyes.
  a. What is the ratio of blue-eyed to brown-eyed students?
  b. What is the ratio of brown-eyed to green-eyed students?
  c. What fraction of the students have blue eyes?
  d. What fraction of the students have blue or green eyes?
  e. What fraction of the students do not have green eyes?

62. A bag contains 9 red marbles, 12 blue marbles, 8 white marbles, and 6 green marbles.
  a. What is the ratio of red marbles to green marbles?
  b. What is the ratio of red marbles to blue marbles?
  c. What fraction of the marbles in the bag are green?
  d. What fraction of the marbles in the bag are either blue or white?
  e. What fraction of the marbles in the bag are not red or white?

63. A serving of oatmeal has the following nutritional value.

Protein	5 grams
Carbohydrate	18 grams
Fat	2 grams
Potassium	0.045 grams
Sodium	0.005 grams

  a. What is the ratio of fat to carbohydrate?
  b. What is the ratio of potassium to sodium?
  c. What is the ratio of potassium to protein?
  d. What fraction of the total nutritional value contains protein?
  e. What fraction of the total nutritional value contains carbohydrate or fat?

64. A lawyer has the following monthly budget.

Rent	\$575	Transportation	\$120
Utilities	\$145	Clothing	\$170
Food	\$230	Miscellaneous	\$260

a. What is the ratio of her allowance for transportation to that for clothing?
b. What is the ratio of her allowance for food to that for rent?
c. What is the ratio of her rent to the amount budgeted for utilities?
d. What fraction of her total budget does she allocate for miscellaneous expenses?
e. What fraction of the total budget does she allocate for food and clothing?

Solve each of the following word problems.

65. A regular deck of 52 cards contains 13 hearts. What ratio represents the fraction of cards in a regular deck that are hearts?

66. An orange has 75 calories and an apple has 115 calories. What is the ratio of calories in an apple to calories in an orange?

67. A station wagon traveled 435 miles on 15 gallons of gas. A sedan traveled 124 miles on 4 gallons of gas. Which car got better mileage?

68. In a single day, an umbrella factory produced 12,000 umbrellas and, of these, 72 were defective. What ratio represents the fraction of umbrellas produced that were defective?

69. A 16-ounce box of rice sells for 88¢. What is the unit price of 1 ounce of rice?

70. A 13-ounce can of tuna costs \$2.34. What is the unit price of 1 ounce of tuna?

71. An acid solution contains 5 fluid ounces of hydrochloric acid and 95 fluid ounces of water. What is the ratio of acid to water?

72. If a cake recipe requires $2\frac{1}{4}$ cups of flour and $\frac{2}{3}$ cup of sugar, what is the ratio of flour to sugar?

73. The area of Cambodia is 70,000 square miles, and the area of Ireland is 26,600 square miles. What is the ratio of the area of Ireland to that of Cambodia?

74. In 1980, the population of Brazil was 122,000,000, and the population of Haiti was 5,000,000. What is the ratio of the population of Brazil to that of Haiti?

75. A can of frozen orange juice concentrate that makes 24 fluid ounces sells for 57¢. A quart of ready-made orange juice sells for 88¢. Determine the unit price of each per fluid ounce. (1 qt = 32 fl oz.) Which is the better buy?

76. An 8-ounce package of cheese that contains 12 slices costs \$1.29. What is the unit price per slice? What is the unit price per ounce?

77. A jogger ran for $\frac{3}{8}$ hour on Monday and for 20 minutes on Tuesday. Express the ratio of these two times as a fraction with the same units in the numerator and denominator. On which day was the jogger's running time longer?

78. A football player drank $\frac{2}{3}$ pint of water followed by 12 fluid ounces of orange juice. Express the ratio of water to orange juice as a fraction with the same units in the numerator and denominator. Did he drink more water or more orange juice?

79. A sports car traveled 217 miles in $3\frac{1}{2}$ hours. What was its average rate of speed?

80. A jogger can run $3\frac{1}{5}$ miles in 40 minutes. Another jogger can run 5.4 miles in 1 hour. Find the rate of each jogger and determine who runs faster.

81. In 1975, the population of Virginia was approximately 5,000,000. If the area of the state is almost 40,000 square miles, what was the average population density per square mile of Virginia in 1975?

82. The Yukon River is 1,900 miles long and the Mekong River is 2,500 miles long. What is the ratio of the length of the Yukon to that of the Mekong?

83. A carton with 12 eggs sells for \$1.05, and another carton containing 8 eggs sells for 68¢. Determine the unit price for each carton. Which is the better buy?

84. If $\frac{3}{4}$ pound of freshly ground coffee costs \$1.95, what is the unit price per pound?

85. A seamstress purchased 20 inches of blue ribbon and $\frac{5}{8}$ yard of red ribbon. Express the ratio of blue ribbon to red ribbon as a fraction with the same units in the numerator and denominator. Which piece of ribbon is longer?

86. A secretary spoke on the phone to a client for 27 minutes, and then her boss spoke to the client for 0.4 hour. Express the ratio of these two times as a fraction with the same units in the numerator and denominator. Who spoke longer to the client?

87. In 1980, the population of Canada was 23,850,000. If Canada covers an area of 3,851,809 square miles, what was the average population density per square mile of Canada in 1980? Round your answer to the nearest hundredth.

88. A 13-ounce can of coffee sells for \$2.19. What is the unit price per ounce? Round your answer to the nearest tenth of a cent.

89. A runner completed a 6.2-mile race in 1 hour 8 minutes. What was her average rate of speed in miles per hour? Round your answer to the nearest hundredth.

90. A toaster factory produced 13,457 toasters, of which 83 were defective. Find a decimal that represents the fraction of toasters that were defective. Round your answer to the nearest hundred-thousandth.

4.2 Proportion

As was illustrated in the previous section, more than one equivalent ratio is often used to express the same relationship between two quantities. For example, if a purse contains 4 nickels and 12 dimes, the ratio of nickels to dimes is 4 to 12, or 1 to 3. This can be written mathematically as

$$4:12::1:3$$

and is read, "4 is to 12 as 1 is to 3."

This statement of equality between two ratios is called a **proportion**, and it most frequently appears in fractional notation. That is,

$$\frac{4}{12}=\frac{1}{3}\ .$$

definition

A *proportion* is a statement of equality between two ratios. It can be written in the form

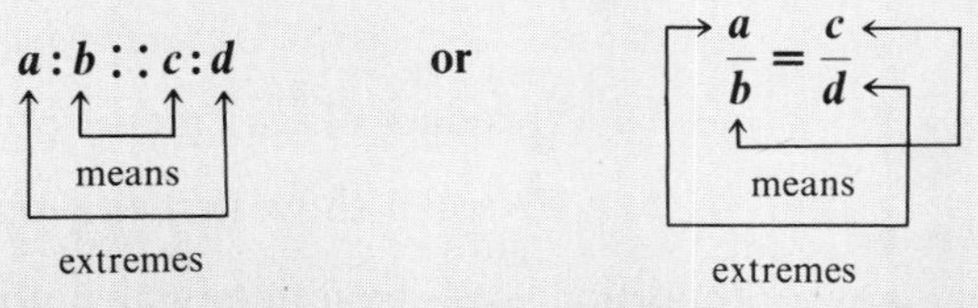

The numbers *a*, *b*, *c*, and *d* are called the *terms* of the proportion. The terms *a* and *d* are called the *extremes*. The terms *b* and *c* are called the *means*.

In our example of the ratio of 4 nickels to 12 dimes, the numbers 4 and 3 are the *extremes*, and the numbers 12 and 1 are the *means*.

$$4:12::1:3$$

means (12 and 1); extremes (4 and 3)

example 1

For each of the following proportions, identify the extremes and the means.

(a) $2:9::6:27$

Extremes are 2 and 27.
Means are 6 and 9.

(b) $\frac{3}{24} = \frac{2}{16}$

Extremes are 3 and 16.
Means are 24 and 2. ●

An important property of proportions can be illustrated by comparing the product of the means to the product of the extremes. For example, in the proportion $\frac{1}{2} = \frac{3}{6}$, the product of the extremes, 1×6, is equal to the product of the means, 2×3, since both products are equal to 6. This property is called the **fundamental property of proportions**.

The Fundamental Property of Proportions

In any proportion, the product of the extremes is equal to the product of the means.

If $\quad \frac{a}{b} = \frac{c}{d}$

then $\quad a \cdot d = b \cdot c$

The products $a \cdot d$ and $b \cdot c$ are sometimes called **cross products**, because they can be obtained by cross multiplying.

$$\frac{a}{b} \times \frac{c}{d}$$

$$ad = bc$$

Notice that when we use letters to represent numbers, we often omit the symbol $\times$ or $\cdot$ used to express the operation multiplication.

product of extremes = product of means

We can show that the fundamental property of proportions is true for the proportion $\frac{a}{b} = \frac{c}{d}$ by multiplying both sides by the LCD of the two ratios, which is bd.

$$(bd)\frac{a}{b} = \frac{c}{d}(bd)$$

Now, cancel common factors in the numerator and denominator.

$$\frac{\cancel{b}da}{\cancel{b}} = \frac{cb\cancel{d}}{\cancel{d}}$$

$$ad = bc$$

example 2

Use the fundamental property of proportions to determine whether each of the following statements is true or false.

(a) $\frac{4}{16} = \frac{5}{20}$

Does $(4)(20) = (16)(5)$?

$$80 = 80$$

Cross multiply to determine the products of the extremes and the means.

$\frac{4}{16} = \frac{5}{20}$ is a true statement.

(b) $\frac{6}{11} = \frac{7}{12}$

Does $(6)(12) = (11)(7)$?

$$72 \neq 77$$

Cross multiply.

$\frac{6}{11} = \frac{7}{12}$ is a false statement.

(c) $\frac{\frac{2}{3}}{8} = \frac{\frac{6}{9}}{8}$

Does $(2)\left(\frac{9}{8}\right) = \left(\frac{3}{8}\right)(6)$?

$$\frac{18}{8} = \frac{18}{8}$$

Cross multiply.

$\frac{\frac{2}{3}}{8} = \frac{\frac{6}{9}}{8}$ is a true statement. ●

We can use a similar technique to determine the larger of two ratios whose numerators and denominators are all greater than zero. Suppose that

$$\frac{a}{b} > \frac{c}{d}$$

We can multiply both sides of the inequality by bd to obtain

$$(bd)\frac{a}{b} > \frac{c}{d}(bd)$$

$$\frac{\cancel{b}da}{\cancel{b}} > \frac{cb\cancel{d}}{\cancel{d}}$$

$$ad > bc$$

The extremes are greater than the means.

We have therefore shown that

If the product of the extremes is greater than the product of the means, the first ratio is greater than the second.

Similarly, if

$$\frac{a}{b} < \frac{c}{d}$$

we can multiply both sides of the inequality by bd to obtain

$$(bd)\frac{a}{b} < \frac{c}{d}(bd)$$

$$\frac{\not{b}da}{\not{b}} < \frac{cb\not{d}}{\not{d}}$$

$$ad < bc$$

We have now also shown that — The extremes are less than the means.

If the product of the extremes is less than the product of the means, the first ratio is less than the second.

example 3

Determine which of the following ratios is greater, and place the appropriate sign (> or <) between them.

(a) $\frac{3}{8} \;?\; \frac{1}{3}$

$3(3) \;?\; 8(1)$ — Cross multiply.

$9 > 8$ — The extremes are greater than the means.

$\frac{3}{8} > \frac{1}{3}$ — The first ratio is greater than the second.

(b) $\frac{4}{3} \;?\; \frac{7}{5}$

$4(5) \;?\; 3(7)$ — Cross multiply.

$20 < 21$ — The extremes are less than the means.

$\frac{4}{3} < \frac{7}{5}$ — The first ratio is less than the second.

(c) $\dfrac{\frac{2}{3}}{\frac{3}{4}} \;?\; \dfrac{\frac{6}{11}}{\frac{7}{2}}$

$$\frac{2}{3}\left(\frac{7}{2}\right) \; ? \; 4\left(\frac{6}{11}\right)$$ Cross multiply.

$$\frac{14}{6} \; ? \; \frac{24}{11}$$ Simplify.

$$\frac{7}{3} \; ? \; \frac{24}{11}$$ Reduce to lowest terms.

$$7(11) \; ? \; 3(24)$$ Since it is still not clear which ratio is greater, we cross multiply again.

$$77 > 72$$ The extremes are greater than the means.

$$\frac{\frac{2}{3}}{4} > \frac{\frac{6}{11}}{\frac{7}{2}}$$ The first ratio is larger than the second.

Notice that the same result can be obtained by simplifying the complex fractions in the original ratios before cross multiplying.

$$\frac{\frac{2}{3}\cdot 3}{4\cdot 3} = \frac{2}{12} = \frac{1}{6}$$ Simplify $\frac{\frac{2}{3}}{4}$.

$$\frac{\frac{6}{11}\cdot 22}{\frac{7}{2}\cdot 22} = \frac{12}{77}$$ Simplify $\frac{\frac{6}{11}}{\frac{7}{2}}$.

$$\frac{1}{6} \; ? \; \frac{12}{77}$$ Restate in simplified terms and cross multiply.

$$77 > 72$$ The extremes are greater than the means.

$$\frac{\frac{2}{3}}{4} > \frac{\frac{6}{11}}{\frac{7}{2}}$$ The first ratio is greater than the second.

Example 4 illustrates how a word problem can be solved by using this same technique to compare ratios.

example 4

A Japanese car can travel 20 miles on 0.75 gallons of gas, and an American car can travel 160 miles on 6.2 gallons of gas. Which car gets the better gas mileage?

solution

Japanese car American car

$$\frac{20}{0.75} \; ? \; \frac{160}{6.2}$$ Set up two ratios that express the mileage for each car, and determine which is greater.

$(20)(6.2) \ ? \ (0.75)(160)$ — Cross multiply.

$124 > 120$ — The extremes are greater than the means.

Japanese car > American car — The first ratio is larger.

Thus, the Japanese car gets better mileage.

Note that since multiplication by a two-digit number is usually easier than division by a two-digit number, this method of comparing ratios is easier than determining the actual mileage for each car.

QUICK QUIZ

Determine which of the following ratios is larger and put the appropriate sign (> or <) between them.

1. $\frac{3}{5} \ ? \ \frac{5}{8}$

2. $\frac{2\frac{2}{5}}{5\frac{1}{4}} \ ? \ \frac{1\frac{3}{7}}{3\frac{1}{3}}$

3. $\frac{1.1}{0.03} \ ? \ \frac{0.7}{0.02}$

ANSWERS

1. $3(8) < 5(5)$

   $\frac{3}{5} < \frac{5}{8}$

2. $\left(\frac{12}{5}\right)\left(\frac{10}{3}\right) > \left(\frac{21}{4}\right)\left(\frac{10}{7}\right)$

   $\frac{2\frac{2}{5}}{5\frac{1}{4}} > \frac{1\frac{3}{7}}{3\frac{1}{3}}$

3. $(1.1)(0.02) > (0.03)(0.7)$

   $\frac{1.1}{0.03} > \frac{0.7}{0.02}$

4.2 Exercises

Indicate which of the following statements are true and which are false. For those that are false, change the italic or underlined expression to make the statement true.

1. A *proportion* is a statement of equality between two ratios.
2. In the proportion, $a:b::c:d$, the terms <u>a and b</u> are the means.
3. When comparing two ratios, if the product of the means exceeds the product of the extremes, the first ratio is *greater* than the second.
4. If two ratios are equal, the cross products are *equal*.

For each of the following proportions, identify the extremes and the means, and rewrite the expression in words.

5. $6:9::2:3$
6. $1:7::3:21$
7. $\frac{5}{2} = \frac{20}{8}$
8. $\frac{24}{6} = \frac{4}{1}$

9. $\frac{1}{2}:11::1:22$

10. $4:9::\frac{4}{3}:3$

11. $\frac{2\frac{1}{3}}{8} = \frac{7}{24}$

12. $\frac{7}{6\frac{2}{5}} = \frac{5\frac{1}{4}}{4\frac{4}{5}}$

Use the fundamental property of proportions to determine whether each of the following statements is true or false.

13. $4:9::2:3$

14. $6:18::2:6$

15. $\frac{1}{3} = \frac{8}{24}$

16. $\frac{9}{2} = \frac{27}{6}$

17. $5:3 = 7:4$

18. $2:7 = 6:21$

19. $\frac{5}{21} = \frac{6}{20}$

20. $\frac{9}{36} = \frac{2}{8}$

21. $\frac{13}{2} = \frac{17}{3}$

22. $\frac{8}{27} = \frac{16}{54}$

23. $\frac{\frac{3}{4}}{\frac{5}{8}} = \frac{\frac{8}{15}}{\frac{4}{9}}$

24. $\frac{\frac{1}{6}}{\frac{3}{11}} = \frac{\frac{2}{3}}{\frac{9}{10}}$

25. $\frac{\frac{6}{11}}{\frac{5}{22}} = \frac{\frac{4}{15}}{\frac{1}{9}}$

26. $\frac{\frac{3}{5}}{\frac{7}{15}} = \frac{\frac{4}{7}}{\frac{5}{9}}$

27. $\frac{3\frac{1}{4}}{\frac{2}{3}} = \frac{8}{1\frac{1}{3}}$

28. $\frac{2\frac{7}{8}}{3\frac{1}{2}} = \frac{4\frac{2}{3}}{6\frac{2}{9}}$

29. $\frac{2.1}{0.7} = \frac{3.3}{1.1}$

30. $\frac{8.4}{0.42} = \frac{6.0}{0.3}$

31. $\frac{3.2}{1.2} = \frac{5.6}{2.1}$

32. $\frac{8.6}{4.2} = \frac{7.3}{3.6}$

33. $\frac{527}{61} = \frac{242}{28}$

34. $\frac{184}{95} = \frac{473}{245}$

35. $\frac{0.0417}{8.62} = \frac{0.254}{52.6}$

36. $\frac{0.32}{1.2} = \frac{0.0128}{0.048}$

Determine which of the following ratios is greater, and put the appropriate sign (> or <) between them.

37. $\frac{4}{9} ? \frac{5}{11}$

38. $\frac{2}{3} ? \frac{5}{8}$

39. $\frac{7}{16} ? \frac{3}{7}$

40. $\frac{5}{8} ? \frac{2}{7}$

41. $\frac{\frac{3}{8}}{\frac{5}{6}} ? \frac{\frac{1}{4}}{\frac{4}{9}}$

42. $\frac{\frac{1}{3}}{\frac{7}{12}} ? \frac{\frac{3}{8}}{\frac{9}{14}}$

43. $\frac{\frac{3}{7}}{\frac{3}{44}} ? \frac{\frac{4}{5}}{\frac{1}{8}}$

44. $\frac{\frac{3}{8}}{\frac{2}{9}} ? \frac{\frac{1}{3}}{\frac{4}{19}}$

45. $\frac{5\frac{5}{8}}{4\frac{1}{2}} ? \frac{1\frac{2}{3}}{1\frac{1}{5}}$

46. $\frac{3\frac{7}{8}}{6\frac{1}{4}} ? \frac{3\frac{1}{5}}{5\frac{5}{7}}$

47. $\frac{1.2}{3.5} ? \frac{0.15}{0.5}$

48. $\frac{0.43}{0.04} ? \frac{0.21}{0.02}$

49. $\frac{5.2}{0.13} ? \frac{0.11}{0.003}$

50. $\frac{0.05}{0.72} ? \frac{0.44}{6.2}$

51. $\frac{20.7}{0.039} ? \frac{7.93}{0.015}$

52. $\frac{0.0067}{4.25} ? \frac{0.501}{318}$

53. $\frac{849}{0.616} ? \frac{23.4}{0.017}$

54. $\frac{0.0042}{0.119} ? \frac{0.031}{0.879}$

Use the techniques for comparing ratios presented in this section to solve each of the following word problems.

55. A baseball player for the Boston Red Sox made 15 hits in 39 times at bat. A player for the Chicago White Sox made 28 hits in 71 times at bat. Which player has the higher batting average?

56. A sports car can travel 300 miles on 18 gallons of gas, and a luxury car can travel 400 miles on 25 gallons of gas. Which car gets better mileage?

57. Deodorant soap costs 99¢ for 4 bars, and facial soap costs $1.49 for 6 bars. If all bars are the same size, which type is the better buy?

58. If a 7-ounce tube of toothpaste costs $1.29, and a 12-ounce tube costs $2.25, which size is the better buy?

59. A Russian olympic skater finished a 500-meter race in 37 seconds, and a U.S. skater finished a 1,000-meter race in 1-minute 13 seconds. Who skated faster?

60. Chicken wings cost $1.76 for 3.2 pounds, and chicken legs cost $2.22 for 4.1 pounds. Which is more expensive, chicken legs or wings?

61. A 32-ounce bottle of dishwashing detergent sells for $2.19. An 18.5-ounce bottle sells for $1.25. Which size is the better buy?

62. In May, 17.2 gallons of gas sold for $32.49. In August, 12.9 gallons sold for $23.87. During which month was the price of gas higher?

63. Alabama has 148 hospitals and admits 752,371 patients per year. Florida admits 1,608,248 patients yearly and has a total of 245 hospitals. Which state has the higher ratio of hospitals to yearly admissions?

64. In 1946, there were 2,291,045 marriages and 610,000 divorces. In 1960, there were 393,000 divorces and 1,523,000 marriages. Was the ratio of divorces to marriages higher in 1946 or 1960?

4.3 Solving Proportions

Thus far, we have looked only at proportions whose terms are all numbers. Proportions often contain an unknown term that is represented by a letter. The letter that is used to represent an unknown quantity is called a **variable**.

We will encounter many mathematical expressions that contain a combination of numbers and variables. For example, in the expression $5 \cdot x$, the number 5 is multiplied by the variable x. The number 5 is often called the **coefficient** of x. The **coefficient** of a variable is the number that is multiplied by the variable. When expressing the product of a number and a variable, the multiplication sign is usually omitted.

$$5 \cdot x = 5x$$

example 1

Identify the variables and their coefficients in each of the following mathematical expressions.

(a) $12y$

y is a variable.

12 is the coefficient of y.

(b) $\frac{3}{4}w$

w is a variable.

$\frac{3}{4}$ is the coefficient of w.

(c) t

t is a variable.

Since $t = 1 \cdot t$, the coefficient of t is 1.

(d) $(8)(3)x$

$(8)(3)x = 24x$

x is a variable.

24 is the coefficient of x. ●

The fundamental property of proportions, which states that the product of the extremes is equal to the product of the means, is used to find the value of the unknown term in a proportion. For example, to find the value of x in the proportion

$$\frac{3}{7} = \frac{9}{x}$$

we first cross multiply:

$$3 \cdot x = 7 \cdot 9$$

Since our goal is to solve for x, we need to perform some mathematical operation on both sides of the equation that will isolate the variable x on the left to obtain

$$x = \text{the unknown quantity}$$

Since $x = 1 \cdot x$, we can isolate x by multiplying both sides of the equation by the reciprocal of 3, which is $\frac{1}{3}$.

$$\frac{1}{3} \cdot 3 \cdot x = 7 \cdot 9 \cdot \frac{1}{3}$$

$$1 \cdot x = \frac{63}{3}$$

$$x = 21$$

To check whether or not our solution is correct, we substitute the value obtained for x into the original equation and determine if the resulting proportion is a true statement.

$$\frac{3}{7} \stackrel{?}{=} \frac{9}{21}$$

$$(3)(21) \stackrel{?}{=} (7)(9)$$

$$63 = 63 \checkmark$$

This procedure is summarized in the box.

To Solve a Proportion for an Unknown Term

1. Cross multiply to set the product of the extremes equal to the product of the means.
2. Multiply both sides of the equation by the reciprocal of the coefficient of the unknown.
3. Check the solution by substituting the value obtained for the variable into the original equation.

example 2

Find the value of the unknown term in each of the following proportions.

(a) $\frac{5}{9} = \frac{x}{4}$

$5(4) = 9x$ — Cross multiply.

$9x = 5(4)$ — Transpose the left side and right side of the equation so the variable will be on the left.

$9x = 20$ — The coefficient of x is 9.

The reciprocal of 9 is $\frac{1}{9}$.

$\frac{1}{9}(9x) = (20)\frac{1}{9}$ — Multiply both sides of the equation by $\frac{1}{9}$.

$1 \cdot x = \frac{20}{9}$

$x = \frac{20}{9}$

CHECK

$\frac{5}{9} \; ? \; \frac{\frac{20}{9}}{4}$

$5(4) \; ? \; 9\left(\frac{20}{9}\right)$

$20 = 20 \checkmark$

(b) 8 is to y as 10 is to 3.

$\frac{8}{y} = \frac{10}{3}$

$8(3) = 10y$ — Cross multiply.

$10y = 24$ — The coefficient of y is 10.

The reciprocal of 10 is $\frac{1}{10}$.

$\left(\frac{1}{10}\right)10y = 24\left(\frac{1}{10}\right)$ — Multiply both sides of the equation by $\frac{1}{10}$.

$1 \cdot y = \frac{24}{10}$

$y = \frac{12}{5}$ — Reduce to lowest terms.

CHECK

$\frac{8}{\frac{12}{5}} \; ? \; \frac{10}{3}$

$8(3) \; ? \; \frac{12}{5}(10)$

$24 = 24 \checkmark$

(c)

$$\frac{\frac{z}{3/4}}{} \quad \frac{\frac{z}{3}}{\frac{}{4}} = \frac{\frac{7}{5}}{\frac{}{6}}$$

$\frac{5}{6}z = \frac{3}{4}(7)$	Cross multiply.
$\frac{5}{6}z = \frac{21}{4}$	The coefficient of z is $\frac{5}{6}$.
$\left(\frac{6}{5}\right)\left(\frac{5}{6}\right)z = \left(\frac{21}{4}\right)\left(\frac{6}{5}\right)$	The reciprocal of $\frac{5}{6}$ is $\frac{6}{5}$. Multiply both sides by $\frac{6}{5}$.
$1 \cdot z = \left(\frac{21}{\cancel{4}_2}\right)\left(\frac{\cancel{6}^3}{5}\right)$	Simplify.
$z = \frac{63}{10}$	

CHECK

$$\frac{\frac{63}{10}}{\frac{3}{4}} \overset{?}{=} \frac{7}{\frac{5}{6}}$$

$$\left(\frac{\cancel{63}^{21}}{\cancel{10}_2}\right)\left(\frac{\cancel{5}}{\cancel{6}_2}\right) \overset{?}{=} \left(\frac{3}{4}\right)(7)$$

$$\frac{21}{4} = \frac{21}{4} \checkmark$$

(d)

$$\frac{2\frac{3}{8}}{x} = \frac{19}{8}$$

$\left(2\frac{3}{8}\right)(8) = 19x$	Cross multiply.
$\left(\frac{19}{8}\right)(8) = 19x$	Change $2\frac{3}{8}$ to an improper fraction.
$19 = 19x$	
$19x = 19$	Transpose left side and right side. The coefficient of x is 19.
$\left(\frac{1}{19}\right)19x = 19\left(\frac{1}{19}\right)$	The reciprocal of 19 is $\frac{1}{19}$. Multiply both sides by $\frac{1}{19}$.
$1 \cdot x = 1$	
$x = 1$	

CHECK

$$\frac{2\frac{3}{8}}{1} \overset{?}{=} \frac{19}{8}$$

$$\left(2\frac{3}{8}\right)(8) \ ? \ (1)(19)$$

$$\left(\frac{19}{8}\right)(8) \ ? \ 19$$

$$19 \ ? \ 19\checkmark$$

(e)

$$\frac{1.5}{50} = \frac{0.012}{t}$$

$$1.5t = 50(0.012)$$ Cross multiply.

$$1.5t = 0.600$$

$$\left(\frac{1}{1.5}\right)(1.5)t = (0.600)\left(\frac{1}{1.5}\right)$$ Multiply both sides by $\frac{1}{1.5}$.

$$t = \frac{0.600}{1.5}$$

$$1.5_\wedge\overline{)0.6_\wedge00}\quad = 0.4$$

$$t = 0.4$$

Since the other terms are decimals, express the answer as a decimal.

CHECK

$$\frac{1.5}{50} \ ? \ \frac{0.012}{0.4}$$

$$(1.5)(0.4) \ ? \ (50)(0.012)$$

$$0.60 = 0.60\checkmark$$

●

QUICK QUIZ

Find the value of the unknown term in each of the following proportions.

1. $\dfrac{x}{7} = \dfrac{6}{21}$
2. $\dfrac{4\frac{1}{8}}{y} = \dfrac{2\frac{5}{8}}{\frac{6}{11}}$
3. $\dfrac{0.24}{0.014} = \dfrac{3.6}{t}$

ANSWERS

1. $x = 2$
2. $y = \frac{6}{7}$
3. $t = 0.21$

4.3 Exercises

Indicate which of the following statements are true and which are false. For those that are false, change the italic expression to make the statement true.

1. A *variable* is a letter that represents an unknown quantity.
2. The coefficient of a variable is the number that is *added to* the variable.

3. The first step in solving a proportion for an unknown is to *reverse multiply.*
4. To solve for the unknown term in a proportion, we *divide* both sides by the reciprocal of the coefficient of the unknown.

Identify the variables and their coefficients in each of the following mathematical expressions.

5. $5x$ **6.** $2y$ **7.** $17t$ **8.** w **9.** $\frac{5}{8}u$

10. $\frac{7}{2}v$ **11.** $1.4a$ **12.** $2.7b$ **13.** $9(2)x$ **14.** $4(5)y$

Find the value of the unknown term in each of the following proportions.

15. $\frac{3}{5}=\frac{9}{x}$ **16.** $\frac{x}{2}=\frac{15}{10}$ **17.** $\frac{4}{y}=\frac{2}{7}$ **18.** $\frac{5}{8}=\frac{w}{9}$

19. $\frac{13}{4}=\frac{10}{x}$ **20.** $\frac{6}{x}=\frac{15}{8}$ **21.** $\frac{12}{18}=\frac{y}{21}$ **22.** $\frac{81}{n}=\frac{9}{8}$

23. $\frac{6}{11}=\frac{m}{33}$ **24.** $\frac{21}{12}=\frac{28}{p}$ **25.** $\frac{54}{q}=\frac{9}{13}$ **26.** $\frac{w}{17}=\frac{81}{51}$

27. $\frac{18}{23}=\frac{3}{y}$ **28.** $\frac{4}{19}=\frac{z}{10}$ **29.** $\frac{5}{21}=\frac{7}{x}$ **30.** $\frac{x}{69}=\frac{4}{3}$

31. $\frac{12}{35}=\frac{y}{10}$ **32.** $\frac{9}{x}=\frac{25}{11}$ **33.** $\frac{18}{54}=\frac{81}{z}$ **34.** $\frac{w}{63}=\frac{42}{28}$

35. $\frac{96}{y}=\frac{60}{24}$ **36.** $\frac{36}{21}=\frac{t}{39}$ **37.** $\frac{60}{28}=\frac{q}{42}$ **38.** $\frac{18}{120}=\frac{24}{r}$

39. $\frac{\frac{4}{5}}{\frac{3}{8}}=\frac{x}{\frac{15}{16}}$ **40.** $\frac{\frac{7}{8}}{w}=\frac{\frac{2}{9}}{\frac{4}{21}}$ **41.** $\frac{\frac{5}{11}}{s}=\frac{\frac{10}{27}}{\frac{4}{15}}$ **42.** $\frac{u}{\frac{7}{16}}=\frac{\frac{12}{25}}{\frac{3}{5}}$

43. $\frac{\frac{4}{21}}{x}=\frac{2\frac{1}{7}}{4\frac{7}{8}}$ **44.** $\frac{1\frac{5}{6}}{\frac{5}{12}}=\frac{y}{3\frac{1}{3}}$ **45.** $\frac{6\frac{2}{3}}{1\frac{1}{5}}=\frac{4\frac{1}{6}}{w}$ **46.** $\frac{z}{3\frac{1}{8}}=\frac{7\frac{3}{5}}{2\frac{3}{4}}$

47. $\frac{9\frac{4}{9}}{7\frac{1}{2}}=\frac{v}{2\frac{1}{4}}$ **48.** $\frac{4\frac{4}{5}}{6\frac{1}{4}}=\frac{r}{5\frac{5}{8}}$ **49.** $\frac{3.2}{0.08}=\frac{x}{0.002}$ **50.** $\frac{x}{0.56}=\frac{40}{0.07}$

51. $\frac{0.6}{y}=\frac{12}{0.05}$ **52.** $\frac{0.81}{0.27}=\frac{0.09}{z}$ **53.** $\frac{5.4}{w}=\frac{0.21}{0.007}$ **54.** $\frac{0.036}{2.4}=\frac{u}{50}$

55. $\frac{7.2}{30}=\frac{0.006}{t}$ **56.** $\frac{400}{12}=\frac{r}{0.09}$ **57.** $\frac{0.056}{4.2}=\frac{0.08}{y}$ **58.** $\frac{x}{0.0048}=\frac{200}{3.2}$

59. $\frac{1.2}{z}=\frac{7{,}200}{0.003}$ **60.** $\frac{s}{4{,}200}=\frac{120}{0.0036}$ **61.** $\frac{0.0008}{0.024}=\frac{0.002}{t}$ **62.** $\frac{0.03}{0.00016}=\frac{u}{0.64}$

63. $\frac{168}{x}=\frac{252}{216}$ **64.** $\frac{y}{288}=\frac{192}{576}$ **65.** $\frac{1{,}344}{4{,}256}=\frac{w}{2{,}128}$ **66.** $\frac{3{,}402}{1{,}989}=\frac{4{,}536}{z}$

67. $\frac{4.71}{x}=\frac{0.94}{0.038}$ Round off x to the nearest thousandth.

68. $\dfrac{0.079}{6.21} = \dfrac{y}{84.5}$ Round off y to the nearest thousandth.

69. $\dfrac{3.54}{0.091} = \dfrac{862.7}{w}$ Round off w to the nearest whole number.

70. $\dfrac{z}{0.326} = \dfrac{53.97}{0.00248}$ Round off z to the nearest tenth.

4.4 Applications of Proportions

Proportions provide us with a powerful tool for solving a wide variety of word problems that express a relationship between two quantities. The procedure outlined in the box can be used to solve many word problems using proportions. Example 1 illustrates how to apply this procedure.

To Solve Word Problems Using Proportions

1. Determine the unknown quantity you are asked to find and assign it a letter, such as x. State what quantity x represents using the format,

   Let $x =$ unknown quantity

   This step is called *defining the variable.*
2. Set up a ratio that expresses the relationship between two known quantities given in the problem.
3. Set this ratio equal to a corresponding equivalent ratio that contains the unknown.
4. Solve the resulting proportion for the unknown.
5. Check your answer by substituting your answer for the variable in the original equation. Also check that your answer makes sense as an answer to the question asked in the problem, and label it with the appropriate units.

example 1

In a chemistry class, the ratio of men to women is 3 to 2. If there are 21 men in the class, how many women are there?

solution

Let $w =$ the number of women in the class. — Define the variable w.

Ratio of men to women is $3:2$ or $\dfrac{3}{2}$. — Set up a ratio between men and women.

Ratio of men to women is $21:w$ or $\dfrac{21}{w}$

$$\begin{matrix}\text{men}\longrightarrow \\ \text{women}\longrightarrow\end{matrix}\ \frac{3}{2} = \frac{21}{w}\ \begin{matrix}\longleftarrow\text{men} \\ \longleftarrow\text{women}\end{matrix}$$

Set this equal to a corresponding ratio containing w.

$$3w = 2(21)$$ Solve the proportion for w.

$$3w = 42$$

$$w = \frac{42}{3}$$

$$w = 14 \text{ women}$$ Label the answer.

CHECK $\frac{3}{2} \overset{?}{=} \frac{21}{14}$ Check by substituting back into the original equation.

$3(14) \overset{?}{=} 2(21)$

$42 = 42 \checkmark$

There are 14 women in the class.

In Example 1, the result of 14 is a reasonable answer, because if the ratio of men to women is 3 to 2, there must be fewer women than men in the class. An answer that is not a whole number would not make sense, since we do not refer to fractions of a person. Therefore an answer such as $14\frac{1}{2}$ would be a clue that there was probably a mistake made in performing the calculation.

example 2

If 8 pencils sell for 60¢, how much would 12 pencils cost?

solution

Let x = cost of 12 pencils. Define the variable x.

Ratio of pencils to cents is $\frac{8}{60}$. Set up a ratio between pencils and cents.

Ratio of pencils to cents is $\frac{12}{x}$.

$$\frac{8}{60} = \frac{12}{x} \quad \begin{matrix}\text{pencils} \\ \text{cents}\end{matrix}$$ Set this equal to a corresponding ratio containing x.

$$8x = 60(12)$$ Solve the proportion for x.

$$\left(\frac{1}{8}\right)8x = 720\left(\frac{1}{8}\right)$$

$$x = \frac{720}{8}$$

$$x = 90¢$$ Label the answer.

CHECK $\frac{8}{60} \overset{?}{=} \frac{12}{90}$ Check your answer.

$8(90) \overset{?}{=} 60(12)$

$720 = 720 \checkmark$

The cost of 12 pencils is 90¢.

Notice that in Example 2 we could have used a different proportion to solve the problem. We can set up a ratio between any two known quantities, as long as we also can set up a corresponding equivalent ratio that contains the unknown. For example,

$$\begin{matrix}\text{pencils} \longrightarrow \\ \text{pencils} \longrightarrow\end{matrix} \frac{8}{12} = \frac{60}{x} \begin{matrix}\longleftarrow \text{cents} \\ \longleftarrow \text{cents}\end{matrix}$$

$$8x = 12(60)$$

$$\left(\frac{1}{8}\right)8x = 720\left(\frac{1}{8}\right)$$

$$x = 90¢$$

In order to clearly indicate all of the correct proportions that could be used to solve the problem in Example 2, we need to distinguish the first quantities compared from the second quantities compared. In this case, the first quantities compared are 8 pencils and 60¢, and the second quantities compared are 12 pencils and x¢; that is,

$$\begin{matrix}\text{first} \\ \text{quantities}\end{matrix} \quad \frac{8 \text{ pencils}}{60¢} = \frac{12 \text{ pencils}}{x¢} \quad \begin{matrix}\text{second} \\ \text{quantities}\end{matrix}$$

which is the proportion we used to solve Example 2. All of the following proportions could be used also to correctly solve the problem. In each case, the first quantities compared appear in color.

1. $\dfrac{\text{pencils}}{\text{cents}} = \dfrac{\text{pencils}}{\text{cents}}$ Pencils are in both numerators. First quantities appear in the same ratio.

2. $\dfrac{\text{cents}}{\text{pencils}} = \dfrac{\text{cents}}{\text{pencils}}$ Cents are in both numerators. First quantities appear in the same ratio.

3. $\dfrac{\text{pencils}}{\text{pencils}} = \dfrac{\text{cents}}{\text{cents}}$ Pencils appear in the same ratio. First quantities are in both numerators.

4. $\dfrac{\text{pencils}}{\text{pencils}} = \dfrac{\text{cents}}{\text{cents}}$ Pencils appear in the same ratio. Second quantities are in both numerators.

Thus, a proportion is set up correctly to solve a word problem if it is set up in either of the following ways.

The same units appear in both numerators, and both first quantities appear in the same ratio (Cases 1 and 2).

OR

The same units appear in the same ratio, and both first or both second quantities appear in the numerators (Cases 3 and 4).

example 3

Set up four correct proportions that could be used to solve the following word problem. Solve one of these proportions to find the answer to the problem.
In a bag of marbles, the ratio of green marbles to yellow ones is 7 to 4. If the bag contains 12 yellow marbles, how many marbles are green?

solution

Let x = the number of green marbles.
The first quantities compared are 7 green to 4 yellow, and the second quantities compared are 12 yellow to x green. The following four proportions

could be used to correctly solve this problem. In each case, the first quantities compared appear in color.

1. $\frac{\text{green}}{\text{yellow}} = \frac{\text{green}}{\text{yellow}}$ $\qquad \frac{7}{4} = \frac{x}{12}$

2. $\frac{\text{yellow}}{\text{green}} = \frac{\text{yellow}}{\text{green}}$ $\qquad \frac{4}{7} = \frac{12}{x}$

3. $\frac{\text{green}}{\text{green}} = \frac{\text{yellow}}{\text{yellow}}$ $\qquad \frac{7}{x} = \frac{4}{12}$

4. $\frac{\text{green}}{\text{green}} = \frac{\text{yellow}}{\text{yellow}}$ $\qquad \frac{x}{7} = \frac{12}{4}$

Let us solve the fourth proportion for x.

$$\begin{matrix}\text{green} \longrightarrow \\ \text{green} \longrightarrow\end{matrix} \frac{x}{7} = \frac{12}{4} \begin{matrix}\longleftarrow \text{yellow} \\ \longleftarrow \text{yellow}\end{matrix}$$

$\frac{x}{7} = \frac{3}{1}$ — Simplify terms.

$x = 7 \cdot 3$ — Cross multiply.

$x = 21$ green marbles — Label the answer.

CHECK $\quad \frac{21}{7} \overset{?}{=} \frac{12}{4}$ — Check the answer.

$(21)4 \ ? \ 7(12)$

$84 = 84 \checkmark$

example 4

In a city in southern Arizona, all of the 6,300 registered voters are either Republicans or Democrats. If 2 out of 7 registered voters are Democrats, what is the number of registered Republican voters?

solution

Since we know the number of registered voters, and we are asked to find the number of Republicans, we must first determine the ratio of Republicans to all voters. If all the voters are either Republicans or Democrats, and 2 out of 7 voters are Democrats, then

$$7 \text{ voters} - \underbrace{2 \text{ voters}}_{\text{Democrats}} = \underbrace{5 \text{ voters}}_{\text{Republicans}}$$

Thus, for every 7 voters, 2 are Democrats and 5 are Republicans.

The ratio of Republicans to voters is 5 to 7.

Let $r =$ the number of Republicans. — Define the variable.

$$\begin{matrix}\text{Republicans} \longrightarrow \\ \text{voters} \longrightarrow\end{matrix} \frac{5}{7} = \frac{r}{6{,}300} \begin{matrix}\longleftarrow \text{Republicans} \\ \longleftarrow \text{voters}\end{matrix}$$

Set up a proportion with one known ratio and an equivalent ratio that contains the unknown.

$5(6{,}300) = 7r$ — Cross multiply.

$\frac{1}{7}(5)(6{,}300) = \frac{1}{7}(7r)$ — Multiply both sides by the reciprocal of 7.

$$\frac{5(6{,}300)}{7} = r$$

$$r = \frac{5(\overset{900}{\cancel{6{,}300}})}{\cancel{7}}$$ Transpose left side and right side. Simplify.

$r = 4{,}500$ Republicans — Label the answer.

CHECK $\frac{5}{7} \overset{?}{} \frac{4{,}500}{6{,}300}$ — Check the answer.

$5(6{,}300) \; ? \; 7(4{,}500)$

$31{,}500 = 31{,}500\checkmark$

There are 4,500 registered Republicans.

example 5

A basket of fruit contains only apples, oranges, and bananas. The ratio of oranges to apples is 3 to 2, and the ratio of bananas to apples is 4 to 3. There are 18 oranges in the basket.

(a) How many apples are in the basket?

We know that there are 18 oranges in the basket and that the ratio of oranges to apples is 3 to 2. Let us set up a proportion using oranges and apples.

Let a = the number of apples in the basket. — Define the variable.

oranges ⟶ 3 / apples ⟶ 2 = 18 ⟵ oranges / a ⟵ apples

$$\frac{3}{2} = \frac{18}{a}$$ Set up a proportion with one known ratio and an equivalent ratio that contains the unknown.

$$3a = 2(18)$$ Cross multiply.

$$\left(\frac{1}{3}\right)3a = 36\left(\frac{1}{3}\right)$$ Multiply both sides by the reciprocal of 3.

$$1a = \frac{36}{3}$$ Simplify.

$a = 12$ apples — Label the answer.

CHECK $\frac{3}{2} \; ? \; \frac{18}{12}$ — Check.

$3(12) \; ? \; 2(18)$

$36 = 36\checkmark$

There are 12 apples in the basket.

(b) How many bananas are in the basket?

We know that there are 12 apples in the basket and that the ratio of bananas to apples is 4 to 3. Let us set up a proportion using bananas and apples.

Let b = the number of bananas in the basket. — Define the variable.

bananas ⟶ 4 / apples ⟶ 3 = b ⟵ bananas / 12 ⟵ apples

$$\frac{4}{3} = \frac{b}{12}$$ Set up the proportion.

$$4(12) = 3b$$ Solve for b.

$$3b = 48$$

$$\left(\frac{1}{3}\right)3b = 48\left(\frac{1}{3}\right)$$

$$1b = \frac{48}{3}$$

$$b = 16 \text{ bananas}$$ Label the answer.

CHECK

$$\frac{4}{3} \; ? \; \frac{16}{12}$$ Check.

$$4(12) \; ? \; 3(16)$$

$$48 = 48 \checkmark$$

There are 16 bananas in the basket.

(c) How many pieces of fruit are in the basket?

Since there are only oranges, apples, and bananas in the basket, add the number of pieces of each kind of fruit to find the total amount of fruit in the basket.

$18 + 12 + 16 = 46$ fruits — There are 18 oranges, 12 apples, and 16 bananas.

There are 46 pieces of fruit in the basket. ●

The technique we have established for solving word problems using proportions requires that we use only two different units. If three different units are mentioned in a problem, we must convert one of the units to one of the other two units.

example 6

A car traveling at a constant speed takes 25 minutes to travel 20 miles. How far does the car travel in 2 hours?

solution

In this problem, distance is given in miles, and time is given in both minutes and hours. Therefore, let us convert 2 hours to its equivalent in minutes, so that the problem contains only two different units.

$$2 \text{ hr} = 2 \text{ hr} \cdot \frac{60 \text{ min}}{1 \text{ hr}} = 120 \text{ min}$$ Since 1 hr = 60 min.

Let x = distance traveled in 120 min. — Define the variable.

$$\begin{matrix} \text{minutes} \longrightarrow \\ \text{miles} \longrightarrow \end{matrix} \frac{25}{20} = \frac{120}{x} \begin{matrix} \longleftarrow \text{minutes} \\ \longleftarrow \text{miles} \end{matrix}$$ Set up the proportion.

$$25x = 120(20)$$ Solve for x.

$$\frac{1}{25}(25x) = \frac{1}{25}(2{,}400)$$

$$x = \frac{2{,}400}{25}$$

$$x = 96 \text{ mi}$$

CHECK $\frac{25}{20} \stackrel{?}{} \frac{120}{96}$

$25(96) \ ? \ 20(120)$

$2{,}400 = 2{,}400 \checkmark$

The car can travel 96 miles in 2 hours.

4.4 Exercises

Indicate which of the following statements are true and which are false. For those that are false, change the italic expression to make the statement true.

1. Proportions can be used to solve problems that express a relationship between *two* quantities.
2. Assigning a letter to an unknown quantity is called *depicting the variance*.
3. There *is* more than one correct way to set up a proportion to solve a given problem.
4. If a proportion is set up correctly and both first quantities appear in the same ratio, *different* units appear in the numerators of the ratios.

Set up four different proportions that could be used to solve each of the following word problems. Solve one of the proportions in each problem to answer the question asked in the problem.

5. The ratio of a quarterback's attempted passes to completed passes is 7 to 2. If he completed 12 passes, how many did he attempt?
6. Two men have the same ratio of shirts to ties. The first man has 4 ties and 18 shirts. If the second man has 27 shirts, how many ties does he have?
7. If it takes a secretary 20 minutes to type 4 pages, how many pages can he type in 45 minutes?
8. A motorboat travels 8 miles in 15 minutes. How long does it take for the boat to travel 24 miles while maintaining this same speed?
9. If a 10-inch candle burns for 4 hours, how long would a 6-inch candle burn?
10. Facial tissues are on sale for a price of 3 boxes for $2.00. How much would 5 boxes of tissues cost?

Solve each of the following word problems.

11. If a 27-foot tree casts an 18-foot shadow, how long is the shadow cast by a 12-foot tree?
12. Seven bags of sand weigh 18 pounds. How much do 21 bags of sand weigh?
13. A softball player makes 3 hits in every 8 times at bat. If she is up to bat 40 times in one week, how many hits would you expect her to make?
14. If it takes a painter 2 hours to paint 36 square feet of wall space, how long would you expect it to take her to paint 63 square feet of wall space?
15. A survey reveals that 4 out of 5 dentists recommend sugarless gum to their patients who chew gum. If 300 dentists were surveyed, how many recommended sugarless gum?
16. Notebooks are on sale for a price of 3 for $3.57. How much do 5 notebooks cost?
17. For every 35 people who walk into a car dealership, 2 buy a car there. If 245 people visit a car dealership during the month of March, how many car sales were made that month at that dealership?
18. Yarn sells for a price of 6 skeins for $4.98. How much do 20 skeins of yarn cost?
19. If 3 bottles of shampoo sell for $2.55, how much would 2 bottles cost?
20. How many calories are in 12 fluid ounces of beer if 8 fluid ounces contain 116 calories?

21. One out of 3 seniors from a state university plan to go directly to graduate or professional school upon graduation. If 821 students plan to go directly to graduate or professional school, how many seniors are in the class?

22. A car can travel 82 miles on 4 gallons of gas. How far can the car travel on 16.4 gallons of gas?

23. At a tennis ball factory, for every 50 balls produced, 3 are defective. If 7,250 tennis balls are produced, how many would you expect to be defective?

24. A midwestern law school accepted 2 out of every 5 applicants to its entering class. If the school received 435 applications, how many students were accepted?

25. Eleven out of every 20 people living in an apartment building are single. If 140 people live in that building, how many are single?

26. Two out of 5 books on a student's bookshelf are mathematics books. If she has 25 books on her bookshelf, how many are math books?

27. A 6-foot tall man casts a shadow $3\frac{1}{2}$ feet long. His daughter casts a shadow 14 inches long. How tall is his daughter?

28. A survey conducted at a liberal arts college indicates that 3 out of 8 students smoke cigarettes. If 4,000 students attend the college, how many smoke cigarettes?

29. In a class of 54 students, 2 out of 9 wear glasses. How many students in that class do not wear glasses?

30. Four out of 7 people ice skating at a local rink cannot skate backwards. If 42 skaters are on the ice, how many can skate backwards?

31. A sugar solution is made by dissolving 1.5 grams of sugar in 18 grams of water. How many grams of sugar must be dissolved in 12 grams of water to make a sugar solution of the same concentration?

32. If you paid $12 for 6.8 gallons of gas, how much would 17 gallons of gas cost?

33. Three out of 5 people who participated in a taste test preferred regular cola over diet cola. If 125 people took the taste test, how many preferred diet cola?

34. Five out of 24 calls that a local fire station receives are false alarms. If the station receives 120 calls during a given month, how many are real emergencies?

35. On a map of California, $\frac{1}{2}$ inch represents 30 miles. If two cities on this map are $5\frac{1}{4}$ inches apart, what is the actual distance between them?

36. Freshly ground coffee sells for $2.28 per pound. How much does $\frac{1}{4}$ pound cost?

37. If there are 220 calories in a pint of orange juice, how many calories are in a 12-fluid ounce serving of orange juice?

38. If ribbon costs $1.86 per yard, how much do $2\frac{1}{2}$ feet of ribbon cost?

39. Ground beef sells for $1.68 per pound. How much do 10 ounces of-ground beef cost?

40. The scale on a campus map states that 1 inch represents 500 feet. If the distance from your dorm to the location of your first class is $10\frac{1}{2}$ inches on this map, do you have to walk more than a mile or less than a mile to get to your first class? By how many feet?

41. A man wants to bake a cake for his wife's birthday and decides to use a recipe that calls for 5 eggs and $1\frac{1}{4}$ cups of sugar. If he discovers that there are only 4 eggs in the refrigerator, how much sugar should he use so that these ingredients are mixed in the right proportions?

42. A carpenter earns $117 for a 6-hour job. How much would she receive for a job that takes $3\frac{1}{3}$ hours to complete?

43. Three yards of cotton fabric cost $3.88. What is the cost of $6\frac{3}{4}$ yards of the same fabric?

44. A recipe that yields 4 dozen two-inch cookies calls for $\frac{2}{3}$ cup of sugar and $1\frac{1}{2}$ cups of flour. How much sugar and flour would you use to make 3 dozen two-inch cookies?

45. On prime time television, there are $3\frac{1}{2}$ minutes of commercials for every 15 minutes of broadcast time. How many minutes of commercials would one see while watching an hour of television during prime time?

46. If $\frac{3}{4}$ gallon of paint covers 250 square feet of wall space, how much paint would be needed for a room with 510 square feet of wall space?

47. A jar contains only nickels, dimes, and quarters. The ratio of nickels to dimes is 5 to 3, and the ratio of nickels to quarters is 2 to 3. There are 12 dimes in the jar.
   a. How many nickels are in the jar?
   b. How many quarters are in the jar?
   c. What is the total number of coins in the jar?

48. A drawer contains only knives, forks, and spoons. The ratio of knives to forks is 7 to 9, and the ratio of forks to spoons is 6 to 5. There are 15 spoons in the drawer.
   a. How many forks are in the drawer?
   b. How many knives are in the drawer?
   c. How many pieces of silverware are in the drawer?

49. In a class of 84 students composed of only sophomores, juniors, and seniors, 2 out of 7 students are juniors, and the ratio of sophomores to juniors is 3 to 4.
   a. How many students are juniors?
   b. How many students are sophomores?
   c. How many students are seniors?

50. In a bowl that contains only walnuts, brazil nuts, and pecans, the ratio of walnuts to pecans is 2 to 3. There are a total of 27 nuts in the bowl, and 4 out of 9 are pecans.
   a. How many pecans are in the bowl?
   b. How many walnuts are in the bowl?
   c. How many brazil nuts are in the bowl?

51. Four out of 25 women in the labor force in 1976 were professional or technical workers. If there were 38,414 women in the labor force in 1976, how many were professional or technical workers? Round your answer to the nearest whole number.

52. If 13.7 gallons of gas cost $25.89, how much gas can you buy for $5.00? Round your answer to the nearest tenth of a gallon.

53. The ratio of the area of Japan to that of Peru is 9 to 31. If Peru covers an area of 496,222 square miles, how large is Japan? Round your answer to the nearest thousand miles.

54. In the United States, 8 out of 25 babies born in 1950 had mothers between the ages of 20 and 24. If 1,155,167 mothers in that age group had babies in 1950, what was the total number of babies born in the United States that year? Round your answer to the nearest thousand.

Summary and Review

Key Terms

[4.1] A **ratio** is a fraction that expresses a relationship between two quantities.

A **unit fraction** is a fraction equivalent to one.

A **rate** is a ratio that compares two different quantities and has a denominator equal to one.

[4.2] A **proportion** is a statement of equality between two ratios. In the proportion $\frac{a}{b} = \frac{c}{d}$, the numbers a, b, c, and d are called the **terms** of the proportion. The terms a and d are called the **extremes**. The terms b and c are called the **means.**

In any proportion, **the product of the extremes equals the product of the means.** If $\frac{a}{b} = \frac{c}{d}$, then $ad = bc$. The products ad and bc are called **cross products**.

[4.3] A **variable** is a letter that represents an unknown quantity.

The **coefficient** of a variable is the number that is multiplied by the variable.

Calculations

[4.1] **To compare two quantities using a ratio**, put the first quantity in the numerator, and put the second quantity in the denominator.

[4.1] **To convert from one unit of measurement to another,** multiply by the unit fraction that has the old unit in the denominator and the new unit in the numerator.

[4.2] **To show that two ratios are equal**, show that the product of the extremes is equal to the product of the means.

To show that the first ratio is greater than the second ratio, show that the product of the extremes is greater than the product of the means.

To show that the first ratio is less than the second ratio, show that the product of the extremes is less than the product of the means.

[4.3] **To solve a proportion for an unknown term** cross multiply to set the product of the extremes equal to the product of the means. Then multiply both sides of the equation by the reciprocal of the coefficient of the unknown.

[4.4] **To solve a word problem using a proportion**, first define the variable. Then set up a ratio that expresses the relationship between two known quantities given in the problem. Set this ratio equal to a corresponding equivalent ratio that contains the unknown. Solve the resulting proportion for the unknown variable, and answer the question asked in the problem.

To determine if a proportion is set up correctly to solve a word problem, check that either the same units appear in both numerators and both first quantities appear in the same ratio, or the same units appear in the same ratio, and both first or both second quantities appear in the numerators.

Chapter 4 Review Exercises

Indicate which of the following statements are true and which are false. For those that are false, change the italic expression to make the statement true.

1. A *rate* is a ratio that compares two different quantities and has a denominator of one.
2. When comparing two ratios, if the product of the extremes exceeds the product of the means, the second ratio is *greater* than the first.
3. The coefficient of a variable is the number that is *added to* the variable.
4. The price of a single item is called a *unit price*.
5. A proportion is set up *correctly*, if the same units appear in both denominators, and both first quantities appear in the same ratio.

6. In a proportion, the product of the means and the product of the extremes are sometimes called *reverse products.*

[4.1] Express each of the following comparisons as a ratio with the same units in the numerator and denominator. Reduce to lowest terms.

7. 4 feet to 2 yards
8. 18 inches to 3 feet
9. 6 hours to 45 minutes
10. 30 days to 5 weeks
11. 10 quarts to 3 gallons
12. 2 pints to 24 fluid ounces
13. 12 nickels to 2 quarters
14. 74 dimes to 5 dollars
15. $3\frac{1}{2}$ pounds to 18 ounces
16. 54 items to $2\frac{1}{3}$ dozen
17. $\frac{3}{4}$ mile to 900 feet
18. $5\frac{1}{2}$ years to 30 months

[4.2] Determine which of the following ratios is greater, and place the appropriate sign (> or <) between them.

19. $\frac{3}{8} ? \frac{4}{11}$
20. $\frac{2}{7} ? \frac{3}{10}$
21. $\frac{4}{5} ? \frac{5}{6}$
22. $\frac{5}{9} ? \frac{4}{7}$
23. $\frac{1}{3} ? \frac{8}{25}$
24. $\frac{4}{21} ? \frac{1}{5}$
25. $\frac{2\frac{1}{4}}{6} ? \frac{\frac{2}{3}}{4}$
26. $\frac{7}{2\frac{3}{4}} ? \frac{8}{3\frac{2}{7}}$
27. $\frac{\frac{5}{3}}{\frac{3}{4}} ? \frac{2\frac{6}{7}}{\frac{3}{8}}$
28. $\frac{\frac{2}{8}}{\frac{3}{4}} ? \frac{\frac{5}{6}}{2\frac{6}{7}}$
29. $\frac{0.03}{1.1} ? \frac{0.002}{0.07}$
30. $\frac{0.13}{0.0009} ? \frac{0.4}{0.003}$
31. $\frac{1.1}{0.23} ? \frac{17}{3.47}$
32. $\frac{8{,}021}{3.19} ? \frac{16.3}{0.0065}$

[4.3] Identify the variables and their coefficients in each of the following mathematical expressions.

33. $8x$
34. $20y$
35. z
36. $4.9w$
37. $\frac{3}{5}t$
38. $\frac{2}{9}a$
39. $(3)(5)b$
40. $(4)(7)u$

[4.3] Find the value of the unknown term in each of the following proportions.

41. $\frac{3}{5} = \frac{x}{40}$
42. $\frac{w}{60} = \frac{4}{3}$
43. $\frac{4}{x} = \frac{2}{7}$
44. $\frac{6}{8} = \frac{9}{y}$
45. $\frac{9}{12} = \frac{t}{6}$
46. $\frac{6}{u} = \frac{9}{3}$
47. $\frac{24}{36} = \frac{16}{n}$
48. $\frac{28}{p} = \frac{21}{33}$
49. $\frac{\frac{2}{3}}{8} = \frac{x}{\frac{6}{7}}$
50. $\frac{y}{\frac{3}{4}} = \frac{\frac{2}{5}}{9}$
51. $\frac{2\frac{5}{8}}{5\frac{5}{6}} = \frac{y}{8\frac{1}{3}}$
52. $\frac{3\frac{3}{4}}{2\frac{1}{7}} = \frac{5\frac{4}{9}}{x}$
53. $\frac{2.7}{t} = \frac{0.09}{0.2}$
54. $\frac{w}{0.004} = \frac{54}{0.8}$
55. $\frac{0.48}{0.064} = \frac{n}{3.2}$
56. $\frac{56}{p} = \frac{4.9}{0.042}$
57. $\frac{2.73}{0.512} = \frac{x}{0.0081}$ Round off x to the nearest ten-thousandth.
58. $\frac{q}{4.37} = \frac{29}{0.0685}$ Round off q to the nearest whole number.

[4.4] Solve each of the following word problems.

59. A regular deck of 52 cards contains 4 queens. What ratio represents the fraction of the cards that are queens?

60. A calculus class has 225 students, of whom 175 are engineering majors. What ratio represents the fraction of the students who are engineering majors?

61. A 5-pound bag of flour sells for \$1.29. What is its unit price per pound?

62. A designer bought $3\frac{1}{4}$ yards of plaid fabric and 10 feet of striped fabric. Write a fraction that has the same units in the numerator and denominator that compares these two amounts. If both pieces of material have the same width, did she buy more of the plaid material or striped material?

63. A student spent $\frac{7}{8}$ hour on his math homework and 50 minutes on his chemistry homework. Write a fraction with the same units in the numerator and denominator that compares these two times. On which assignment did he spend more time?

64. A $1\frac{1}{2}$-cup serving of green beans contains 13 calories. A $\frac{2}{3}$-cup serving of broccoli contains 29 calories. For the same size serving, which contains more calories, broccoli or green beans?

65. The pitcher on a baseball team made 27 hits in 62 times at bat. The shortstop made 19 hits in 43 times at bat. Who has the higher batting average?

66. Four out of 7 sophomores at a state university have not yet decided their major. If there are 2,149 sophomores at that school, how many have decided their major?

67. If 4 dining room chairs cost \$164, how much would 6 chairs cost?

68. If 3 tires weigh 57 pounds, how much do 4 tires weigh?

69. Three out of 8 working mothers in a major U.S. city have at least one child under the age of 5. If there are 1,816 mothers employed in that city, how many have at least one child under the age of 5?

70. The scale on a map states that $1\frac{3}{4}$ inches represent 280 miles. If two cities are $2\frac{1}{2}$ inches apart on the map, what is the distance between them?

71. To make a glass of chocolate milk, you dissolve 2 teaspoons of chocolate mix in 1 cup of milk. How much chocolate mix would you need to make 1 quart of chocolate milk?

72. Two out of every 15 items sold at a local pharmacy are prescription drugs. If 8,265 items were sold last week, how many were prescription medicines?

73. For every dollar earned, a computer programmer must pay 35¢ in taxes. If her yearly salary is \$37,000, how much does she pay in taxes for one year?

74. Two out of 9 employees of a plastics company make over \$35,000 a year. If 3,483 people are employed at that company, how many make over \$35,000 a year?

75. In 1970, the state of Montana had a population of 694,409. If Montana covers 145,587 square miles, what was its average population density per square mile in 1970? Round your answer to the nearest hundredth.

76. At an international competition, a Belgian swam 500 yards in 6 minutes 35.24 seconds, and a Canadian swam 200 yards in 2 minutes 37.82 seconds. Who swam faster, the Belgian or the Canadian?

77. In the United States in 1975, the ratio of dollars spent on transportation to those spent on food was 2 to 3. If \$184.7 billion was spent on food, how much was spent on transportation? Round your answer to the nearest tenth of a billion dollars.

78. The ratio of the number of people who watch prime time television on Friday night to the number who watch on Sunday night is 17 to 21. If 84,690,000 watch prime time television on Friday night, how many people watch prime time television on Sunday night? Round your answer to the nearest ten thousand.

5 PERCENT

5.1 Meaning of Percent

The word *percent* is common to our everyday vocabulary. It is used to refer to such things as taxes, discounts, interest, commissions, and tips. In this chapter, we will discuss how the notion of a percent relates to two mathematical quantities we are already familiar with—fractions and decimals. We will also establish techniques for using percents to solve a variety of problems applicable to our daily life.

definition

The word *percent* means per one hundred, or the ratio of a number to one hundred.

Since a percent refers to the number of parts out of 100, any percent can be rewritten as a fraction with a denominator of 100. For example, 25% means 25 parts out of one hundred, or $\frac{25}{100}$, which reduces to $\frac{1}{4}$. We can rewrite any percent in this manner, as outlined in the box.

To Convert a Percent to a Fraction

1. Remove the percent sign (%).
2. Put the number over a denominator of 100.
3. Reduce to lowest terms.

example 1

Convert each of the following percents to fractions.

(a) 80%

$$80\% = \frac{80}{100} = \frac{4}{5}$$

Since 80% means 80 parts out of 100, remove % sign, put 80 over denominator of 100, and reduce.

(b) 3%

$$3\% = \frac{3}{100}$$

Remove % sign and put 3 over denominator of 100.

(c) 7.5%

$$7.5\% = \frac{7.5}{100}$$

Remove % sign and put 7.5 over denominator of 100.

$$= \frac{7.5 \times 10}{100 \times 10} = \frac{75}{1{,}000} = \frac{3}{40}$$

Simplify to remove decimal and reduce.

(d) 100%

$$100\% = \frac{100}{100} = 1$$

Notice that 100% is equal to 1.

(e) 225%

$$225\% = \frac{225}{100} = \frac{9}{4} = 2\frac{1}{4}$$

Notice that any percent greater than 100% is greater than 1.

(f) $\frac{2}{3}\%$

$$\frac{2}{3}\% = \frac{\frac{2}{3}}{100} = \frac{\frac{2}{3} \times 3}{100 \times 3} = \frac{2}{300}$$

(g) $33\frac{1}{3}\%$

$$33\frac{1}{3}\% = \frac{33\frac{1}{3}}{100} = \frac{\frac{100}{3}}{100} = \frac{\frac{100}{3} \times 3}{100 \times 3}$$

$$= \frac{100}{300} = \frac{1}{3}$$

Any fraction that has a denominator of 100 can be rewritten as a percent by removing the denominator and placing a percent sign after the numerator. For example,

$$\frac{47}{100} = 47\%$$

If a fraction does not have a denominator of 100, we first rewrite it as an equivalent fraction that does have a denominator of 100, and then convert it to a percent. For example, to convert $\frac{3}{20}$ to a percent, first multiply numerator and denominator by 5 to obtain an equivalent fraction with a denominator of 100.

$$\frac{3}{20} = \frac{3 \times 5}{20 \times 5} = \frac{15}{100} = 15\%$$

example 2

Rewrite each of the following fractions as a percent.

(a) $\frac{7}{100}$

$\frac{7}{100} = 7\%$

The denominator is 100, so remove it and place a % sign after the numerator 7.

(b) $\frac{66\frac{2}{3}}{100}$

$\frac{66\frac{2}{3}}{100} = 66\frac{2}{3}\%$

The denominator is 100, so remove it and place a % sign after the numerator $66\frac{2}{3}$.

(c) $\frac{8.1}{10}$

$\frac{8.1}{10} = \frac{8.1 \times 10}{10 \times 10} = \frac{81}{100} = 81\%$

Multiply numerator and denominator by 10 to get an equivalent fraction with a denominator of 100.

(d) $\frac{7}{25}$

$\frac{7}{25} = \frac{7 \times 4}{25 \times 4} = \frac{28}{100} = 28\%$

Multiply numerator and denominator by 4 to get an equivalent fraction with a denominator of 100.

(e) $\frac{129}{1{,}000}$

$$\frac{129}{1{,}000} = \frac{129 \times \frac{1}{10}}{1{,}000 \times \frac{1}{10}} = \frac{\frac{129}{10}}{100} = \frac{12.9}{100} = 12.9\%$$

Multiply numerator and denominator by $\frac{1}{10}$ to get an equivalent fraction with a denominator of 100.

We can also use a proportion to find an equivalent fraction with a denominator of 100 by writing a variable in place of the unknown numerator.

example 3

Convert $\frac{3}{8}$ to a percent.

solution

$\frac{3}{8} = \frac{x}{100}$

First set up a proportion to find a fraction equivalent to $\frac{3}{8}$ with a denominator of 100.

$$3 \cdot 100 = 8 \cdot x$$

Cross multiply and solve for x.

$$8x = 300$$

$$\left(\frac{1}{8}\right)8x = 300\left(\frac{1}{8}\right)$$

$$x = \frac{300}{8}$$

$$\begin{array}{r} 37.5 \\ 8\overline{)300.0} \\ \underline{24} \\ 60 \\ \underline{56} \\ 4\,0 \\ \underline{4\,0} \\ 0 \end{array}$$

$$x = 37.5$$

$$\frac{3}{8} = \frac{37.5}{100} = 37.5\%$$

Substitute 37.5 for x in the proportion and change $\frac{37.5}{100}$ to a percent. ●

The procedures that were established for converting fractions to percents are summarized in the box.

To Convert a Fraction to a Percent

1. If the fraction does not have a denominator of 100, find an equivalent fraction with a denominator of 100, using a proportion, if necessary.
2. If the fraction already has a denominator of 100, remove the denominator of 100 and write a percent sign after the numerator.

example 4

Convert each of the following fractions to percents.

(a) $\frac{5}{16}$

First set up a proportion to find an equivalent fraction with a denominator of 100.

$$\frac{5}{16} = \frac{x}{100}$$

$$5 \cdot 100 = 16 \cdot x$$

$$16x = 500$$

$$\left(\frac{1}{16}\right)16x = 500\left(\frac{1}{16}\right)$$

$$\begin{array}{r} 31.25 \\ 16\overline{)500.00} \\ \underline{48} \\ 20 \\ \underline{16} \\ 4\,0 \\ \underline{3\,2} \\ 80 \\ \underline{80} \\ 0 \end{array}$$

$$x = \frac{500}{16}$$

$$x = 31.25$$

$$\frac{5}{16} = \frac{31.25}{100} = 31.25\%$$

Substitute 31.25 for x in the proportion and change $\frac{31.25}{100}$ to a percent.

(b) $\frac{9}{8}$

$$\frac{9}{8} = \frac{x}{100}$$

First set up a proportion to find an equivalent fraction with a denominator of 100.

$$9 \cdot 100 = 8 \cdot x$$
$$8x = 900$$
$$\left(\frac{1}{8}\right)8x = 900\left(\frac{1}{8}\right)$$
$$x = \frac{900}{8}$$
$$x = 112.5$$

$$\begin{array}{r} 112.5 \\ 8\overline{)900.0} \\ \underline{8} \\ 10 \\ \underline{8} \\ 20 \\ \underline{16} \\ 4\,0 \\ \underline{4\,0} \\ 0 \end{array}$$

$$\frac{9}{8} = \frac{112.5}{100} = 112.5\%$$

Now change $\frac{112.5}{100}$ to a percent.

(c) $\frac{1}{3}$

$$\frac{1}{3} = \frac{x}{100}$$
$$1 \cdot 100 = 3 \cdot x$$
$$3x = 100$$
$$\left(\frac{1}{3}\right)3x = 100\left(\frac{1}{3}\right)$$
$$x = \frac{100}{3}$$
$$x = 33\frac{1}{3}$$

First set up a proportion to find an equivalent fraction with a denominator of 100.

$$\begin{array}{r} 33 \text{ R1} \\ 3\overline{)100} \\ \underline{9} \\ 10 \\ \underline{9} \\ 1 \end{array} = 33\frac{1}{3}$$

Notice that when the long division process continues to yield a remainder, we express the answer as a mixed number.

$$\frac{1}{3} = \frac{33\frac{1}{3}}{100} = 33\frac{1}{3}\%$$

Now change $\frac{33\frac{1}{3}}{100}$ to a percent. ●

QUICK QUIZ

Change each of the following percents to fractions reduced to lowest terms.

1. 60%

2. 5%

Change each of the following fractions to percents.

3. $\frac{73}{100}$

4. $\frac{9}{25}$

ANSWERS

1. $\frac{3}{5}$

2. $\frac{1}{20}$

3. 73%

4. 36%

5.1 Exercises

Indicate which of the following statements are true and which are false. For those that are false, change the italic expression to make the statement true.

1. Percent means *per one hundred.*
2. Any percent can be rewritten as a fraction with a *numerator* of one hundred.
3. 20% is *greater* than 2.
4. Percents greater than 100% are always equivalent to numbers greater than *10.*

Convert each of the following percents to a fraction reduced to lowest terms.

5. 75%
6. 30%
7. 2%
8. 6%
9. 57%
10. 93%
11. 64%
12. 88%
13. 120%
14. 250%
15. 500%
16. 1,800%
17. 4.4%
18. 7.6%
19. 0.45%
20. 0.28%
21. 3.07%
22. 60.1%
23. 17.3%
24. 54.9%
25. 18.75%
26. 56.25%
27. 93.75%
28. 62.50%
29. $\frac{1}{2}\%$
30. $\frac{3}{5}\%$
31. $31\frac{1}{4}\%$
32. $87\frac{1}{2}\%$
33. $70\frac{1}{5}\%$
34. $2\frac{3}{4}\%$
35. $66\frac{2}{3}\%$
36. $8\frac{1}{3}\%$

Convert each of the following fractions to a percent.

37. $\frac{13}{100}$
38. $\frac{88}{100}$
39. $\frac{1}{100}$
40. $\frac{49}{100}$
41. $\frac{3.2}{100}$
42. $\frac{0.73}{100}$
43. $\frac{5\frac{1}{3}}{100}$
44. $\frac{72\frac{2}{9}}{100}$
45. $\frac{3}{4}$
46. $\frac{8}{5}$
47. $\frac{2}{5}$
48. $\frac{3}{2}$
49. $\frac{7}{25}$
50. $\frac{9}{50}$
51. $\frac{13}{20}$
52. $\frac{11}{25}$
53. $\frac{8}{10}$
54. $\frac{15}{10}$
55. $\frac{27}{1{,}000}$
56. $\frac{318}{1{,}000}$
57. $\frac{318}{10{,}000}$
58. $\frac{64}{10{,}000}$
59. $\frac{7}{40}$
60. $\frac{1}{16}$
61. $\frac{5}{8}$
62. $\frac{3}{16}$
63. $\frac{1}{9}$
64. $\frac{2}{3}$
65. $\frac{7}{15}$
66. $\frac{11}{30}$

Answer each of the following questions.

67. At a state university, $\frac{3}{5}$ of all students work part-time while attending school. What percent of the student body has part-time jobs?
68. At a large computer firm, $\frac{1}{4}$ of all employees commute more than 10 miles to work. What percent of the company's employees commute more than 10 miles to work?
69. A 2-ounce serving of spaghetti contains 35% of the U.S. recommended daily allowance of thiamine. Express this percent as a fraction reduced to lowest terms.
70. Of all persons surveyed, 72% said they would prefer to be compensated for overtime hours by receiving extra vacation days instead of being paid for the additional hours worked. What fraction of those surveyed does this percent represent?

5.2 Fractions, Decimals, and Percents

Since any percent can be rewritten as a fraction with a denominator of 100, we can easily convert any percent to a decimal. Consider the following examples.

$$32\% = \frac{32}{100} = 0.32$$

$$6.7\% = \frac{6.7}{100} = 0.067$$

$$125\% = \frac{125}{100} = 1.25$$

These examples illustrate the procedure stated in the box.

> To convert a percent to a decimal, remove the percent sign and move the decimal point two places to the left.

example 1

Convert each of the following percents to a decimal.

(a) 19.2%

$19.2\% = .19.2 = 0.192$ — Remove % sign and move decimal point 2 places to left.

(b) 3%

$3\% = .03. = 0.03$ — Remove % sign and move decimal point 2 places to left.

(c) 0.54%

$0.54\% = .00.54 = 0.0054$ — Remove % sign and move decimal point 2 places to left.

(d) $15\frac{3}{4}\%$

$15\frac{3}{4}\% = 15.75\% = .15.75\% = 0.1575$ — First rewrite mixed number as a decimal.

We can use a similar technique to rewrite a decimal as a percent. Consider the following examples.

$$0.95 = \frac{95}{100} = 95\%$$

$$6.02 = \frac{602}{100} = 602\%$$

$$0.007 = \frac{7}{1{,}000} = \frac{0.7}{100} = 0.7\%$$

These examples illustrate the procedure stated in the box.

To convert a decimal to a percent, move the decimal point two places to the right and attach a percent sign.

example 2

Convert each of the following decimals to a percent.

(a) 0.88

0.88 = 0.88. = 88% — Move decimal point 2 places to right and add % sign.

(b) 0.0503

0.0503 = 0.05.03 = 5.03% — Move decimal point 2 places to right and add % sign.

(c) 2

2 = 2.00. = 200% — Move decimal point 2 places to right and add % sign.

In Section 5.1, we learned a technique for converting a fraction to a percent by using a proportion. Now that we know how to change a decimal to a percent, we can also use the additional technique stated in the box.

To Convert a Fraction to a Percent

1. Change the fraction to a decimal by using long division.
2. Rewrite the decimal as a percent.

example 3

Convert each of the following fractions to a percent.

(a) $\frac{5}{8}$

First change $\frac{5}{8}$ to a decimal using long division.

$\frac{5}{8} = 5 \div 8 = 0.625$

```
    .625
8 )5.000
   4 8
    20
    16
     40
     40
      0
```

0.625 = 0.62.5 = 62.5% — Now convert 0.625 to a percent.

(b) $3\frac{1}{8}$

Rewrite $3\frac{1}{8}$ as an improper fraction, and then convert it to a decimal.

$$3\frac{1}{8} = \frac{25}{8} = 25 \div 8 = 3.125$$

$$\begin{array}{r} 3.125 \\ 8\overline{)25.000} \\ \underline{24} \\ 1\,0 \\ \underline{8} \\ 20 \\ \underline{16} \\ 40 \\ \underline{40} \\ 0 \end{array}$$

$$3.125 = 3.12{\curvearrowright}5 = 312.5\%$$

Now convert 3.125 to a percent.

(c) $\frac{1}{6}$

$$\frac{1}{6} = 1 \div 6 = 0.166\ldots$$

First change $\frac{1}{6}$ to a decimal using long division.

$$\begin{array}{r} .166\ldots \\ 6\overline{)1.000} \\ \underline{6} \\ 40 \\ \underline{36} \\ 40 \\ \underline{36} \\ 4 \end{array} \qquad \begin{array}{r} .16 \\ 6\overline{)1.00} \\ \underline{6} \\ 40 \\ \underline{36} \\ 4 \end{array} \leftarrow \text{hundredths place}$$

It is apparent that the decimal 0.166... will not terminate. In a case like this, we generally carry out the division to the hundredths place, and note the place value of the remainder. We then express the quotient as the sum of a decimal and a fraction whose denominator is the divisor. That is,

$$\frac{1}{6} = 0.16 + \frac{0.04}{6}$$

$$0.16 = 0.16{\curvearrowright} = 16\%$$

We can now convert each addend to a percent separately.

$$\frac{0.04}{6} = \frac{4}{600} = \frac{\frac{4}{6}}{100} = \frac{\frac{2}{3}}{100} = \frac{2}{3}\%$$

$$\frac{1}{6} = 16\% + \frac{2}{3}\% = \left(16 + \frac{2}{3}\right)\% = 16\frac{2}{3}\%$$

Add the separate percents to obtain the answer. ●

Notice that $16\frac{2}{3}\%$ is the exact answer to the problem in Example 3(c). To express $\frac{1}{6}$ as an approximate percent correct to the nearest tenth of a percent, we would carry out the long division to the nearest ten-thousandth and then round off to the nearest thousandth before converting the decimal to a percent. This is,

$$\frac{1}{6} = 0.166\overline{6} \approx 0.167 = 0.16{\curvearrowright}7 = 16.7\%$$

QUICK QUIZ

1. Convert 7.8% to a decimal.
2. Convert 0.153 to a percent.
3. Convert $\frac{3}{5}$ to a percent.

ANSWERS

1. 0.078
2. 15.3%
3. 60%

5.2 Exercises

Indicate which of the following statements are true and which are false. For those that are false, change the italic expression to make the statement true.

1. To convert a percent to a decimal, remove the percent sign, and move the decimal point two places to the *right*.
2. To convert a decimal to a percent, move the decimal point two places to the *left*, and attach a percent sign.
3. To express a fraction as a percent correct to the nearest tenth of a percent, we first convert it to a decimal and round off to the nearest *thousandth*.
4. A repeating decimal can be expressed as an *exact* percent that contains a fraction.

Convert each of the following percents to a decimal.

5. 71%	**6.** 12%	**7.** 5%	**8.** 1%	**9.** 24.8%	**10.** 99.9%
11. 4.09%	**12.** 7.63%	**13.** 0.28%	**14.** 0.06%	**15.** 225%	**16.** 170%
17. 3,684%	**18.** 1,000%	**19.** 5.529%	**20.** 3.3%	**21.** 18.03%	**22.** 181.7%
23. $8\frac{1}{2}\%$	**24.** $2\frac{3}{4}\%$	**25.** $45\frac{3}{5}\%$	**26.** $72\frac{1}{10}\%$	**27.** $130\frac{1}{8}\%$	**28.** $205\frac{5}{16}\%$

29. $2\frac{1}{7}\%$ Round off to the nearest ten-thousandth.

30. $77\frac{2}{3}\%$ Round off to the nearest ten-thousandth.

Convert each of the following decimals to a percent.

31. 0.37	**32.** 0.18	**33.** 0.04	**34.** 0.02	**35.** 0.66	**36.** 0.95
37. 0.613	**38.** 0.809	**39.** 0.009	**40.** 0.045	**41.** 0.6	**42.** 0.1
43. 0.0675	**44.** 0.0023	**45.** 4.5	**46.** 3.6	**47.** 3	**48.** 10
49. 13.2	**50.** 65	**51.** 0.813	**52.** 0.040	**53.** 5.07	**54.** 0.301
55. 0.0605	**56.** 0.0012				

Convert each of the following fractions to a percent.

57. $\frac{1}{2}$	**58.** $\frac{3}{5}$	**59.** $\frac{7}{8}$	**60.** $\frac{9}{20}$	**61.** $\frac{1}{9}$	**62.** $\frac{2}{3}$
63. $\frac{3}{16}$	**64.** $\frac{8}{25}$	**65.** $\frac{27}{40}$	**66.** $\frac{51}{75}$	**67.** $\frac{5}{12}$	**68.** $\frac{5}{6}$
69. $\frac{2}{15}$	**70.** $\frac{11}{20}$	**71.** $\frac{2}{7}$	**72.** $\frac{5}{9}$	**73.** $\frac{5}{13}$	**74.** $\frac{4}{11}$
75. $\frac{18}{5}$	**76.** $\frac{7}{4}$	**77.** $\frac{11}{2}$	**78.** $\frac{9}{8}$	**79.** $2\frac{3}{4}$	**80.** $5\frac{1}{8}$
81. $1\frac{1}{2}$	**82.** $3\frac{1}{3}$				

83. $\frac{7}{54}$ Round off to the nearest tenth of a percent.

84. $\frac{4}{77}$ Round off to the nearest tenth of a percent.

85. $\frac{87}{144}$ Round off to the nearest tenth of a percent.

86. $\frac{62}{111}$ Round off to the nearest tenth of a percent.

87. $\frac{485}{588}$ Round off to the nearest tenth of a percent.

88. $\frac{307}{922}$ Round off to the nearest tenth of a percent.

Answer each of the following questions.

89. Three out of 8 students in a general chemistry class are also taking calculus. What percent of the students in this chemistry class are also taking calculus?

90. Of the people living in a midwestern town who voted in the last election, 36% voted Republican. What fraction of the voters in this town voted Republican?

91. One pound of a certain alloy contains 0.48 pounds of pure zinc. What percent of the alloy is pure zinc?

92. A runner completed 72% of a 1-mile race 3 minutes after the race began. Express as a decimal the distance he ran during the first 3 minutes of the race.

5.3 Equations Involving Percents

Many applications using percents will require us to solve an equation to find an unknown quantity. In this section, we will learn how to translate an expression that contains a percent into a mathematical equation that can be solved for an unknown quantity.

example 1

What number is 20% of 18?

solution

Let n = unknown number. — Define the variable.

What number is 20% of 18

$$n = 20\% \cdot 18$$ — Translate into an equation.

$$n = (0.20)(18)$$ — Convert the percent to a decimal.

$$n = 3.6$$ — Solve for the unknown.

3.6 is 20% of 18.

Look again at the answer to Example 1: 3.6 is 20% of 18. The three numbers in this statement are often referred to as the percentage, the rate, and the base.

3.6 is 20% of 18.

↑ ↑ ↑

Percentage = Rate × Base

Note that the **rate** is a percent and the **percentage** is a number. The **base** is that number following the word *of*. The percentage, rate, and base can be identified in any problem that contains a percent.

For *all* problems involving percents, the following formula holds true.

Percentage = Rate × Base

example 2

12 is 75% of what number?

solution

Let x = unknown number (base). — Define the variable.

12 is 75% of what number

$12 = 75\% \cdot x$ — Translate into an equation.

$$12 = \frac{75}{100}x$$ — Convert the percent to a fraction.

$$12 = \frac{3}{4}x$$

$$\left(\frac{4}{3}\right)\frac{3}{4}x = 12\left(\frac{4}{3}\right)$$ — Solve for the unknown.

$$x = 16$$

12 is 75% of 16.

(12 = Percentage, 75% = Rate, 16 = Base)

example 3

8 is what percent of 25?

solution

Let r = unknown rate. — Define the variable.

8 is what percent of 25

$8 = r \cdot 25$ — Translate into an equation.

$$8 = 25r$$

$$\left(\frac{1}{25}\right)25r = 8\left(\frac{1}{25}\right)$$ — Solve for the unknown.

$$r = \frac{8}{25}$$

$$r = 0.32$$ — Convert the answer to a percent.

$$r = 32\%$$

8 is 32% of 25.

(8 = Percentage, 32% = Rate, 25 = Base)

Using Proportions to Solve Percent Problems

Each of the previous examples also could have been solved by setting up a proportion that expresses a relationship involving a percent. The first ratio in the proportion is obtained by rewriting the percent (rate) as a fraction. This ratio is set equal to the ratio of the percentage to the base. The base, which is the number following the word "of," appears in the denominator of the second ratio. The procedure for using a proportion to solve a percent problem is summarized in the box.

To Use a Proportion to Solve a Percent Problem

1. Rewrite the percent (rate) as a fraction. This is the first ratio in the proportion.
2. Set this first ratio equal to the ratio of the percentage to the base.
3. Solve for the unknown.

example 4

What is 85% of 260?

solution

Let n = unknown number (percentage). — Define the variable.

n is 85% of 260 — Identify the percentage, rate, and base.

(n ↑ Percentage, 85% ↑ Rate, 260 ↑ Base)

$$\frac{85}{100} = \frac{n}{260} \begin{array}{l}\leftarrow \text{Percentage} \\ \leftarrow \text{Base}\end{array}$$

Express the rate as a fraction. Set this equal to the ratio of the percentage to the base.

$$85(260) = 100n$$

$$\left(\frac{1}{100}\right)100n = 85(260)\left(\frac{1}{100}\right)$$

Solve for the unknown.

$$n = \frac{22{,}100}{100} \qquad n = 221$$

221 is 85% of 260.

example 5

17 is $33\frac{1}{3}\%$ of what number?

solution

Let y = unknown number (base). — Define the variable.

17 is $33\frac{1}{3}\%$ of y — Identify the percentage, rate, and base.

(17 ↑ Percentage, $33\frac{1}{3}\%$ ↑ Rate, y ↑ Base)

$$\frac{33\frac{1}{3}}{100} = \frac{17}{y}$$

Express the rate as a fraction. Set this equal to the ratio of the percentage to the base.

$$33\frac{1}{3}y = 100 \cdot 17$$

$$\frac{100}{3}y = 100(17)$$

Solve for the unknown.

$$\left(\frac{3}{100}\right)\frac{100}{3}y = 100(17)\left(\frac{3}{100}\right)$$

$$y = (17)(3) \qquad y = 51$$

17 is $33\frac{1}{3}\%$ of 51.

example 6

4.5 is what percent of 30?

Let p = unknown percent. — Define the variable.

$$\underset{\text{Percentage}}{\underset{\uparrow}{4.5}} \quad \text{is} \quad \underset{\text{Rate}}{\underset{\uparrow}{p\%}} \quad \text{of} \quad \underset{\text{Base}}{\underset{\uparrow}{30}}$$

Identify the percentage, rate, and base.

$$\frac{p}{100} = \frac{4.5}{30}$$

Express the rate as a fraction. Set this equal to the ratio of the percentage to the base.

$$30p = (100)(4.5)$$

$$\left(\frac{1}{30}\right)30p = 450\left(\frac{1}{30}\right)$$

Solve for the unknown.

$$p = \frac{450}{30} \qquad p = 15$$

4.5 is 15% of 30.

Word Problems Involving Percents

We will now examine some word problems that can be solved by using equations that contain percents.

example 7

In a class of 44 students, 25% received a final grade of A. How many students got A's?

solution

We can reword the problem in the following form:

What is 25% of 44?

We will solve this problem using the first technique we discussed, in which we translate the problem directly into a mathematical equation.

Let a = number of students getting A's. — Define the variable.

$$\begin{array}{ccccc} a & \text{is} & 25\% & \text{of} & 44 \\ \downarrow & \downarrow & \downarrow & \downarrow & \downarrow \\ a & = & 25\% & \cdot & 44 \end{array}$$

Translate into an equation.

$$a = \left(\frac{25}{100}\right)(44)$$

Change the rate to a fraction.

$$a = \left(\frac{1}{4}\right)(44)$$

Solve.

a = 11 students

Thus, 11 students got A's.

example 8

At a major university 19.3% of all seniors are undecided as to whether they should look for a job or apply to graduate school. If 386 seniors are undecided, how many students are in the senior class?

solution We can reword the problem in the following form:

386 is 19.3% of what number?

We will solve this problem by setting up a proportion.

Let s = number of seniors.

Define the variable.

$$\underset{\text{Percentage}}{\underset{\uparrow}{386}} \text{ is } \underset{\text{Rate}}{\underset{\uparrow}{19.3\%}} \text{ of } \underset{\text{Base}}{\underset{\uparrow}{s}}$$

Identify percentage, rate, and base.

$$\frac{19.3}{100} = \frac{386}{s}$$

Express the rate as a fraction. Set it equal to the ratio of the percentage to the base.

$$(19.3)s = (100)(386)$$

Solve for the unknown.

$$\left(\frac{1}{19.3}\right)(19.3)s = (38{,}600)\left(\frac{1}{19.3}\right)$$

$$s = \frac{38{,}600}{19.3} \qquad s = 2{,}000 \text{ seniors}$$

Thus, there are 2,000 seniors in the class. ●

example 9 In a recent survey, 62 out of 300 drivers responded that they do not regularly use safety belts. What percent of the drivers surveyed do not regularly use safety belts?

solution We can reword the problem in the following form:

62 is what percent of 300?

Let r = unknown rate.

62 is what percent of 300

$$\underset{\downarrow}{62}\ \underset{\downarrow}{=}\ \ \underset{\downarrow}{r}\ \underset{\downarrow}{\cdot}\ \underset{\downarrow}{300}$$

We will solve this problem by translating it directly into an equation.

$$62 = 300r$$

$$\left(\frac{1}{300}\right)300r = 62\left(\frac{1}{300}\right)$$

$$r = \frac{62}{300}$$

$$300\overline{)62.00} = 0.20 + \frac{2.00}{300}$$

60
2.00

$$r = 0.20 + \frac{2}{300}$$

Express the quotient as the sum of a decimal and a fraction whose denominator is the divisor.

$$r = 0.20 + \frac{\frac{2}{3}}{100}$$

$$r = 20\% + \frac{2}{3}\%$$

Convert each addend to a percent separately.

$$r = 20\frac{2}{3}\%$$

Add the separate percents to obtain the answer.

Thus, $20\frac{2}{3}\%$ of the people surveyed do not use safety belts. ●

QUICK QUIZ

Find the unknown in each of the following word problems.

1. What number is 12% of 700?
2. 0.08 is what percent of 32?
3. 66 is 1.5% of what number?

ANSWERS

1. 84
2. 0.25%
3. 4,400

5.3 Exercises

Indicate which of the following statements are true and which are false. For those that are false, change the italic expression to make the statement true.

1. The product of the rate and the base equals the *percentage.*
2. In setting up an equation to solve a percent problem, the word "of" translates to the operation *division.*
3. The number following the word "of" is called the *percentage.*
4. In a percent problem, the *rate* is usually expressed as a percent.

Find the unknown in each of the following problems.

5. What number is 80% of 65?
6. What number is 45% of 30?
7. 18 is 24% of what number?
8. 9 is 72% of what number?
9. 14 is what percent of 56?
10. 98 is what percent of 200?
11. 3.8 is 16% of what number?
12. What number is 3.5% of 78?
13. 0.017 is what percent of 5?
14. 0.008 is 0.025% of what number?
15. 5.6 is what percent of 64?
16. What number is 1.6% of 85?
17. 19 is 19% of what number?
18. 67 is 100% of what number?
19. 31 is 200% of what number?
20. What number is 150% of 21?
21. 85 is what percent of 25?
22. 24 is what percent of 8?
23. 13 is 16% of what number?
24. What number is 84% of 73?
25. 1.8 is 0.06% of what number?
26. What number is 0.053% of 26?
27. 7.5 is what percent of 400?
28. 2.2 is what percent of 50?
29. What number is $33\frac{1}{3}\%$ of 63?
30. What number is $9\frac{1}{11}\%$ of 88?
31. 72 is what percent of 28.8?
32. 37 is what percent of 5.92?
33. 45 is 72% of what number?
34. What number is 47% of 29?
35. $5\frac{5}{8}$ is 30% of what number?
36. What number is 60% of $9\frac{1}{4}$?
37. $8\frac{5}{9}$ is what percent of 11?
38. 6 is what percent of $7\frac{1}{5}$?
39. 12 is $66\frac{2}{3}\%$ of what number?
40. What number is $8\frac{1}{7}\%$ of 49?
41. What number is 46.92% of 1.837?
42. 6.3841 is what percent of 51.0728?

43. 36.84 is 9.276% of what number? Round to the nearest hundredth.

44. 2.843 is what percent of 52.89? Round to the nearest hundredth of a percent.

Solve each of the following word problems.

45. It was found that 12% of all eighth graders who took a standardized reading test scored below a sixth grade reading level. If 150 students took the test, how many scored below a sixth grade reading level?

46. Last summer, a softball team won 5 out of the 25 games that were played. What percent of the games played did they win?

47. Fifteen percent of the teenagers attending a suburban high school smoke cigarettes. If 72 students smoke cigarettes, what is the school's total enrollment?

48. Of all dentists responding to a recent survey, 84% said that they offer nitrous oxide (laughing gas) to patients who need to have cavities filled. If 225 dentists were surveyed, how many use nitrous oxide?

49. Of the people employed at an automobile plant, 14% filed their income tax returns late. If 84 people filed late tax returns, how many are employed at that plant?

50. Twenty-eight out of 32 members of a football team weigh over 200 pounds. What percent of the team weighs over 200 pounds?

51. Of the students enrolled in a 9 A.M. calculus class, 72% eat breakfast before coming to class. If there are 200 students in the class, how many eat breakfast before class?

52. Nine families in a suburban neighborhood moved out within the last two years. If 60 families lived in that neighborhood, what percent have moved elsewhere?

53. Twenty-six out of 160 students failed general chemistry. What percent of the class failed?

54. Of the people working for a downtown bank, 88% take public transportation to work. If 550 people are employed at the bank, how many take public transportation?

55. Eighty-two percent of the people attending a banquet prefer coffee to tea. If 287 people prefer coffee, how many people are at the banquet?

56. In a second grade class, 12 out of 32 students come from families in which the parents are divorced or separated. What percent of the class does this represent?

57. On a test worth a total of 60 points, a student obtained a score of 45. What is his grade expressed as a percent?

58. A physician reported that 52% of a company's employees are overweight. If 450 people work at that company, how many are overweight?

59. A 250-milliliter acid solution contains 5 milliliters of pure acid. What percent of the solution is pure acid?

60. Of the students in a university, 14% participate in intercollegiate sports. If 1,250 students are enrolled at the school, how many participate in intercollegiate sports?

61. A well-known business school accepted 185 students out of an applicant pool of 592. What percent of all applicants were accepted?

62. An engineer earning a salary of $38,500 received a 9% raise. What is the dollar value of the raise?

63. In a town of 13,000 eligible workers, the unemployment rate is 8.5%. How many eligible workers are without jobs?

64. A social worker's monthly gross salary is $1,200, of which 6.8% is deducted every month for social security? What is the dollar value of this deduction?

65. One biscuit of shredded wheat combined with $\frac{1}{2}$ cup of whole milk contains 10% of the U.S. recommended daily allowance (RDA) of protein. If each biscuit contains 2 grams of protein, and $\frac{1}{2}$ cup of whole milk contains 4 grams of protein, what is the recommended daily allowance of protein?

66. At a private college, 60 students out of a junior class of 1,250 studied abroad during their junior year. What percent of the junior class does this represent?

67. An airline estimates that 5.6% of the people making flight reservations will not keep them. If 125 people book flights, how many people does the airline expect will cancel or not show up?

68. A four-year study done by the justice department reported that 21% of all violent crimes committed in the United States happen between friends or relatives. If 17.9 million cases were studied, how many involved friends or relatives?

69. The Internal Revenue Service examined 1,844,986 individual federal income tax returns for 1978. If 87,338,611 returns were filed that year, what percent were examined by the IRS? Round off your answer to the nearest tenth of a percent.

70. In the fall of 1979, 28.8% of all households with televisions in the United States watched the top-rated prime time program. If there were a total of 76,299,500 households with television in the United States that year, how many watched the number one prime time show?

71. The hydroelectric plant at the Grand Coulee Dam currently is producing 6,263 megawatts of power. Its ultimate capacity is 10,080 megawatts of power. What percent of its ultimate capacity is its current level of power production? Round off your answer to the nearest tenth of a percent.

72. Fresh water covers 68,490 square miles of the Canadian province of Ontario. If this represents 16.6% of the total area of Ontario, what is the total area of this province? Round off your answer to the nearest square mile.

5.4 Applications of Percent

We will now examine problems that illustrate some of the most common applications of percent. These include figuring out taxes, calculating amounts earned in commission, determining interest on money invested or borrowed, and calculating how much to tip. In these problems that involve percents, as in all others, the formula percentage = rate × base holds true.

Taxes

Taxes are one of the most frequently occurring examples of percents in our lives. They appear as a percent of money earned (income tax), a percent of the value of goods purchased (sales tax), a percent of assessed property value (property tax), a percent of the value of imported products (tariff), and a percent of an inheritance (inheritance tax). In sales tax problems, the percentage is the amount of the tax, the rate is a percent, and the base is a price (see box).

> To calculate a sales tax, multiply the tax rate by the price.
>
> Sales tax = Rate × Price

example 1

A couple wants to buy a sofa that sells for $399. What is the total amount they must pay if the sales tax is 5%?

solution

First determine the sales tax.

Let t = sales tax. *(Define the variable.)*

Sales tax = % of Price

$t = 5\%$ of $399

$t = 0.05 \times 399$ *(Translate into an equation. Sales tax = rate × price.)*

$t = 19.95$ *(Solve for t.)*

Now find the total amount they must pay.

$$\begin{array}{rl} \$399.00 & \text{sofa} \\ +\quad 19.95 & \text{tax} \\ \hline \$418.95 & \text{total} \end{array}$$

Add the sales tax of \$19.95 to \$399, the cost of the sofa.

Thus, the total cost is \$418.95 including tax. ●

Figure 5.1 is a copy of the tax rate schedules published by the Internal Revenue Service for determining federal income tax. Schedule X is for single taxpayers, Schedule Y is for married persons and qualified widows and widowers, and Schedule Z is for unmarried heads of households. Refer to Figure 5.1 as we solve the problem in Example 2.

example 2

A single man is filling out his federal income tax Form 1040. He shows an amount of \$52,000 on line 36. What is his income tax?

solution

Since the man is single, we look on Schedule X and find the line that represents an income of \$52,000.

His tax is:
\$5,304 + 35% of the amount over \$27,000.

$$\begin{array}{rl} \$52{,}000 & \\ 27{,}000 & \\ \hline \$25{,}000 & \text{excess} \end{array}$$

First determine the amount over \$27,000.

$$35\% \text{ of } \$25{,}000 = .35 \times 25{,}000$$
$$= 8750$$

Then calculate 35% of the excess.

$$\begin{array}{r} \$\ 5{,}304 \\ +\quad 8{,}750 \\ \hline \$14{,}054 \end{array}$$

Then add \$5,304 to \$8,750 to find the tax he must pay.

His income tax is \$14,054. ●

Commission

A **commission** is a bonus paid to a salesperson as an incentive to maximize individual sales productivity. Commission is expressed as a percent of sales and can be calculated by multiplying the percent by the sales (see box). In commission problems, the percentage is the amount of the commission, the rate is a percent, and the base is the value of sales.

> To calculate a commission, multiply the commission rate by the sales.
>
> Commission = Rate × Sales

example 3

A salesman earns a weekly commission of 11% on all sales exceeding \$1,500. Last week he sold \$3,200 worth of goods. What was his commission for the week?

Tax Rate Schedules

Schedule X—Single Taxpayers

Use this schedule if you checked **Filing Status Box 1** on Form 1040—

If the amount on Form 1040, line 36 is: Over—	But not over—	Enter on Form 1040, line 37	*of the amount over—*
$0	$1,800	.......**11%**	**$0**
1,800	16,800	**$198 + 15%**	**1,800**
16,800	27,000	**2,448 + 28%**	**16,800**
27,000	54,000	**5,304 + 35%**	**27,000**
54,000	....	**14,754 + 38.5%**	**54,000**

Schedule Z—Heads of Household

(including certain married persons who live apart—see page 7 of the Instructions)

Use this schedule if you checked **Filing Status Box 4** on Form 1040—

If the amount on Form 1040, line 36 is: Over—	But not over—	Enter on Form 1040, line 37	*of the amount over—*
$0	$2,500	........**11%**	**$0**
2,500	23,000	**$275 + 15%**	**2,500**
23,000	38,000	**3,350 + 28%**	**23,000**
38,000	80,000	**7,550 + 35%**	**38,000**
80,000	....	**22,250 + 38.5%**	**80,000**

Schedule Y—Married Taxpayers and Qualifying Widows and Widowers

Married Filing Joint Returns and Qualifying Widows and Widowers

Use this schedule if you checked **Filing Status Box 2 or 5** on Form 1040—

If the amount on Form 1040, line 36 is: Over—	But not over—	Enter on Form 1040, line 37	*of the amount over—*
$0	$3,000	.......**11%**	**$0**
3,000	28,000	**$330 + 15%**	**3,000**
28,000	45,000	**4,080 + 28%**	**28,000**
45,000	90,000	**8,840 + 35%**	**45,000**
90,000	....	**24,590 + 38.5%**	**90,000**

Married Filing Separate Returns

Use this schedule if you checked **Filing Status Box 3** on Form 1040—

If the amount on Form 1040, line 36 is: Over—	But not over—	Enter on Form 1040, line 37	*of the amount over—*
$0	$1,500	.......**11%**	**$0**
1,500	14,000	**$165 + 15%**	**1,500**
14,000	22,500	**2,040 + 28%**	**14,000**
22,500	45,000	**4,420 + 35%**	**22,500**
45,000	....	**12,295 + 38.5%**	**45,000**

figure 5.1

solution First determine the value of the sales that will earn a commission, that is, the amount exceeding $1,500.

$3,200	value of total sales
− 1,500	value of sales not subject to commission
$1,700	value of sales subject to commission

To determine the commission, find 11% of $1,700.

Let c = commission.

$$\text{Commission} = \% \text{ of Sales}$$
$$c = 11\% \text{ of } \$1{,}700$$
$$c = 0.11 \times 1{,}700$$
$$c = 187$$

$$\begin{array}{r} 1700 \\ \times \quad 0.11 \\ \hline 1700 \\ 1700 \\ \hline 187.00 \end{array}$$

Thus, his commission is $187. ●

example 4

A shoe saleswoman is given the option of either working for a straight monthly salary of $1,150 or working for a monthly salary of $850 plus a 6% commission on all shoes she sells. If she chooses to work for a commission, what is the minimum amount of merchandise she must sell per month in order to make at least as much as she would if working for a straight salary?

solution First determine the value of her minimum monthly commission in order for both salaries to be equal.

$1,150	straight monthly salary
− 850	base salary before commission
$ 300	minimum monthly commission

Now determine the sales necessary to generate a $300 commission.

Let s = sales.

$$\text{Commission} = \% \text{ of Sales}$$
$$\$300 = 6\% \text{ of } s$$
$$300 = \frac{6}{100} \cdot s$$
$$\frac{6}{100}s = 300$$
$$\left(\frac{100}{6}\right)\frac{6}{100}s = \overset{50}{\cancel{300}}\left(\frac{100}{\cancel{6}}\right)$$
$$s = 5{,}000$$

She therefore must sell at least $5,000 worth of shoes every month. ●

Interest

Interest is an amount you pay to borrow money, or an amount you earn for allowing someone else to use your money, for example, when you deposit your money in a bank or purchase investments. The amount of money you borrow or deposit on which interest is calculated is called the **principal**. As the amount of time the money is being used increases, so does the interest. To calculate the amount of interest, we multiply the rate by the principal by the time (see box). The percentage is the interest, the rate is a percent, the base is the principal, and there is a new factor—time.

$$\text{Interest} = \text{Rate} \times \text{Principal} \times \text{Time}$$

We will use the letter i to represent interest, r to represent rate, p to represent principal, and t to represent the length of time for which the money is borrowed or invested.

To calculate an amount of interest, use the formula

$$i = r \cdot p \cdot t$$

where i = interest, r = rate, p = principal, and t = time.

example 5

An insurance agent deposited \$3,500 in a savings account that pays an annual interest rate of $5\frac{1}{2}\%$. How much interest will he earn in 2 years?

solution

$r = 5\frac{1}{2}\%$

$p = \$3{,}500$

$t = 2$ years

$i = r \cdot p \cdot t$

Identify the values for the rate, principal, and time, and substitute them into the formula $i = r \cdot p \cdot t$.

$i = 5\frac{1}{2}\% \cdot 3{,}500 \cdot 2$

Solve for i.

$i = \dfrac{5\frac{1}{2}}{\cancel{100}} \cdot \cancel{3{,}500} \cdot 2$

$i = \left(5\frac{1}{2}\right)(70)$

$i = \left(\frac{11}{\cancel{2}}\right)(\cancel{70}^{35})$

$i = 385$

$$\begin{array}{r} 35 \\ \times\ 11 \\ \hline 35 \\ 35 \\ \hline 385 \end{array}$$

Thus, he would earn \$385 in interest in 2 years.

Example 5 illustrates how much interest would be earned if the interest was calculated, or *compounded*, just once, at the end of 2 years. Banks used to advertise that their interest was compounded semi-annually (twice a year), quarterly (four times a year), monthly, or even daily. You earn more interest the more frequently it is compounded, because the interest earned in the first period (the length of time over which the interest is calculated) becomes part of the principal for the next interest period, and you begin earning interest on interest.

For example, suppose you deposit $10,000 at an annual interest rate of 5% for 1 year. If interest is compounded only at the end of the year (simple interest), you compute the interest as follows:

$$i = r \cdot p \cdot t$$
$$i = (5\%)(10{,}000)(1)$$
$$i = (0.05)(10{,}000)(1)$$
$$i = 500$$

You earn $500 interest for the entire year.

If interest is compounded semi-annually, then the interest is computed twice. Let us calculate the interest for the first half of the year $\left(\frac{1}{2}\text{ yr}\right)$.

$$i = r \cdot p \cdot t$$
$$i = (0.05)(\overset{5{,}000}{\cancel{10{,}000}})\left(\frac{1}{\cancel{2}}\right)$$
$$i = 250$$

$$\begin{array}{r} 5{,}000 \\ \times \quad 0.05 \\ \hline 250.00 \end{array}$$

Now we add the $250 interest to the principal of $10,000 to obtain a new principal of $10,250. The interest for the second half of the year is computed on this new principal.

$$i = r \cdot p \cdot t$$
$$i = (0.05)(10{,}250)\left(\frac{1}{2}\right)$$
$$i = (0.05)(5{,}125)$$
$$i = 256.25$$

$$\begin{array}{r} {\scriptstyle 1\,2} \\ 5{,}125 \\ \times \quad .05 \\ \hline 256.25 \end{array}$$

Thus, the total interest for the year is calculated as follows:

$$\begin{array}{rl} \$250.00 & \text{first half} \\ +\ 256.25 & \text{second half} \\ \hline \$506.25 & \text{interest for the entire year} \end{array}$$

Notice that you would earn $6.25 more in interest if it was compounded twice a year instead of once a year ($256.25 − $250.00 = $6.25).

Today most banks use high-speed computers to compound interest continuously to give you the highest yield possible. This is why banks now advertise both the annual interest rate and the effective annual yield to tell you the actual amount of money earned for a given time period.

example 6

An electrician wishes to take out a home improvement loan for $5,000 at an annual interest rate of 18%. If he pays back the loan 9 months later, what is the total amount he will have to pay back?

solution

First determine the interest on the loan.

Identify the values for rate, principal, and time, and substitute them into the formula $i = r \cdot p \cdot t$.

$r = 18\% = 0.18$

$p = \$5{,}000$

$$t = 9 \text{ mo} = 9 \not{\text{mo}} \times \frac{1 \text{ yr}}{12 \not{\text{mo}}}$$

$$= \frac{9}{12} \text{ yr} = \frac{3}{4} \text{ yr}$$

Since 18% is an annual interest rate, meaning for 1 year, we must express the time period for the loan in the same units, so convert 9 months to $\frac{3}{4}$ year.

$i = r \cdot p \cdot t$

$$i = (0.18)(5{,}000)\left(\frac{3}{4}\right)$$

Solve for i.

$$i = \frac{(0.18)(\overset{1{,}250}{\not{5{,}000}})(3)}{\not{4}}$$

$$\begin{array}{r} 1{,}250 \\ \times \quad 3 \\ \hline 3750 \\ \times \quad 0.18 \\ \hline 300\ 00 \\ 375\ 0 \\ \hline 675.00 \end{array}$$

$i = 675$

The interest is therefore $675.

Now determine the total amount he will have to pay back.

$$\begin{array}{r} \$5{,}000 \\ + \quad 675 \\ \hline \$5{,}675 \end{array}$$

Add the interest to the amount borrowed.

The electrician must pay back $5,675. ●

Tipping

A *tip* is an extra amount of money paid to someone in return for a service. People we usually tip include waiters and waitresses, cab drivers, hairdressers, and porters. The amount we tip is often considered to be a standard percent of the cost of the service. For example, in a restaurant, the accepted practice is to leave a tip equivalent to 15% of the bill.

example 7

Suppose your bill in a restaurant amounts to $6.83. How much money should you leave for a tip of 15%?

solution

Find 15% of $6.83.

Tip = 15% × 6.83

Tip = (0.15)(6.83)

$$\begin{array}{r} 6.83 \\ \times \quad 0.15 \\ \hline 3415 \\ 683 \\ \hline 1.0245 \end{array}$$

Tip = 1.0245 ≅ 1.02

Round off to the nearest cent.

You should leave a tip of $1.02. ●

It is unnecessary, and often impractical, to figure out an exact tip. To estimate a 15% tip, we can use the following procedure.

1. Round off the bill to the nearest dollar.
2. Calculate 10% of the amount in step 1 by moving the decimal point one place to the left.
3. Calculate 5% of the bill by finding one-half of the amount in step 2.
4. Add the results of steps 2 and 3 to obtain the amount of a 15% tip.

Let us now use this procedure to estimate the tip for a bill of \$6.83.

$\$6.83 \cong \$7.00$	Round off to the nearest dollar.
$10\% \times \$7.00 = \$0.70$	Calculate 10% of \$7.00.
$\frac{1}{2}(\$0.70) = \$0.35$	Calculate 5% of \$7.00.
$\$0.70 + \$0.35 = \$1.05$	Add to find the 15% tip.

The tip estimate is \$1.05. Notice that this is quite close to the exact answer of \$1.02 that we obtained in Example 7.

example 8

Estimate the tip for a bill of \$17.59.

solution

$\$17.59 \cong \$18$	Round off to the nearest dollar.
$10\% \times \$18 = \$1.80$	Calculate 10% of \$18.
$\frac{1}{2}(\$1.80) = \$0.90$	Calculate 5% of \$18.
$\$1.80 + \$0.90 = \$2.70$	Add to find the 15% tip.

The tip is \$2.70. ●

example 9

At a restaurant, a table of 5 people receive a bill for \$81.07, not including tax or tip. If they agree to split the cost evenly, how much should each person contribute? Assume a 7% tax and a 15% tip.

solution

First determine the tax and tip.

$7\% + 15\% = 22\%$	Add the rates for tax and tip to determine the total rate that the bill does not include.
$\$81.07 \cong \$81$	Round off bill.
Find 22% of \$81.	
20% of $\$81 = 0.20(81) = \$16.20$	Find 20% of \$81.
$+\ 2\%$ of $\$81 = 0.02(81) = \$\ 1.62$	Find 2% of \$81.
22% of $\$81 = \$17.82$	Add to find 22% of \$81.
$\$17.82 \cong \$18.00$	Round off answer.
Find the total cost of the meal.	
$\$81 + \$18 = \$99$	Add \$18 to bill for tax and tip.
Determine how much each person should pay.	
$\$99 \cong \$100$	To split the bill 5 ways, round off and divide by 5.
$\$100 \div 5 = \$20$	

Each person should contribute \$20. ●

It is worthwhile to practice doing an exercise like this in your head.

5.4 Exercises

Indicate which of the following statements are true and which are false. For those that are false, change the italic expression to make the statement true.

1. Interest is calculated by finding a percentage of the *principal*, which is the amount of money borrowed or invested.
2. A bonus that is calculated to be a percentage of a salesperson's sales is called a *commissary*.
3. The standard tip in a restaurant is *9%*.
4. To calculate the tax on an item sold in a city where the sales tax is 5%, you multiply the price of the item by *0.005*.

Use Figure 5.1 to determine how much income tax each of the following people must pay given that the following amounts appear on line 36 of Form 1040.

5. \$50,500 single
6. \$68,000 married, joint return
7. \$75,600 married, separate return
8. \$60,800 unmarried head of household
9. \$98,400 married, joint return
10. \$57,200 single
11. \$109,855 unmarried head of household
12. \$112,623 married, separate return

Estimate a 15% tip for each of the following restaurant bills.

13. \$15.13
14. \$20.78
15. \$3.92
16. \$8.04
17. \$59.62
18. \$80.19
19. \$174.71
20. \$206.21

Solve each of the following word problems.

21. A comedian bought a new car that cost \$7,550. If sales tax is 7%, what is the total amount that the comedian must pay?
22. A teenager bought a stereo system on sale for \$649.00 and paid a total of \$681.45. How much sales tax did the teenager pay? What is the tax rate for that city?
23. A New England state has an automobile excise tax that is 13% of the market value of any car registered in that state. How much excise tax would a person pay on a car with a market value of \$2,350?
24. All automobiles imported into the United States from Japan are subject to an 11% tariff. If a car dealer paid a \$37,840 tariff on a shipment of new cars, what is the value of the cars that the dealer received?
25. A doctor who earns a salary of \$86,000 claims to be in a 43% tax bracket. How much income tax does the doctor pay?
26. An artist bought art supplies totaling \$87.25 and had to pay a sales tax of \$3.49. What is the tax rate?
27. A man bought a new shirt for \$17.99 in a state with a 5% sales tax on clothes. How much change will he get from a \$20 bill?
28. An architect owns a house that has a market value of \$120,000. The county has set the assessed value of a house to be 25% of the market value. If the property tax is 4% of the assessed value, compute the property tax on this house.
29. A used car salesman makes a 9% commission on all cars he sells. If he sells a car for \$4,500, what is his commission?
30. A real estate agent earns a 6% commission. If she sells a house for \$125,700, what is her commission?
31. A salesclerk for a men's clothing store earned \$437.78 in commission for \$6,254 worth of sales. What is the salesclerk's commission rate?
32. A cosmetics saleswoman earned a commission of \$75 for selling \$1,250 worth of merchandise. What is her commission rate?
33. A housewife earns a $12\frac{1}{2}\%$ commission on all housewares that she sells by organizing and conducting home sales parties. If in one year she earned \$150 in commission, what was the value of the housewares that she sold?
34. A textbook sales representative who earns a 4% commission sold a text to a major university that will order 2,000 copies. If each book will sell for \$29.95, what is the sales representative's commission?
35. A salesman for a machine tool company has a choice of working for a salary of \$11,000 plus 12% commission or for a

salary of \$17,000 plus 8% commission. He anticipates that he can sell \$72,000 worth of heavy machinery that year. Which option should he choose?

36. An encyclopedia salesman decides to work solely on a commission basis at a rate of 25% instead of working for a straight salary of \$14,000 a year. If encyclopedias sell for \$350 a set, how many sets must he sell in order for him to make at least as much money as he would if working for a straight salary?

37. An engineer wishes to take out an \$8,900 loan at an annual interest rate of $13\frac{1}{2}\%$. How much will she have to repay after 3 years?

38. Suppose you deposit \$6,400 in a 2-year account that has an effective annual yield of 13.45%. How much money will be in the account at the end of 2 years?

39. A woman deposited \$200 in a savings account exactly 1 year ago, and now has \$211 in her account. What is the annual interest rate?

40. After 1 year, a pianist had to pay \$325 in interest on a loan of \$2,500. What is the interest rate for the loan?

41. An electrician wishes to buy a \$6,800 car and make a down payment of \$1,500. If he borrows the remainder at an annual interest rate of 12%, how much interest will he owe after 1 year?

42. Interest on a \$5,000 investment is compounded semi-annually. The annual interest rate is 18%. How much interest is earned in 1 year?

43. After 2 years, a businesswoman earned \$6,800 in dividends, which is equivalent to an annual rate of return of 17%. What is the value of her original investment?

44. A couple buys a \$135,000 house with a down payment of \$18,000. They take out a 14% mortgage for the remainder. After 1 month, in addition to the principal, how much interest will they owe?

45. Cab fare to the airport is \$16.50. How much is a 20% tip on the fare?

46. An actress gets her hair done at a beauty salon for \$35. If she wants to leave a 20% tip, how much will it cost her?

47. A businessman left an \$8 tip for the desk clerk at a hotel where he stayed. The tip amounted to 5% of his bill. What was his hotel bill?

48. A lawyer takes a cab home from work and the fare amounts to \$4.50. If she wants to give the driver a 20% tip, how much change should she ask for from a \$10 bill?

49. A party of 8 people at a restaurant gets a bill of \$119.56, not including tax and tip. The meal tax is 7%, and they agree to tip 15%. Estimate how much each person should pay if the bill is split evenly.

50. Three people at a restaurant receive a bill of \$23.72, not including tax and tip. They agree to tip 20%, and the meal tax is 5%. If they split the bill evenly, estimate how much each person owes.

51. A student went to the bookstore and purchased books and supplies for the following amounts: \$1.89, \$0.49, \$3.57, \$35.99, \$21.99, and \$0.63. If the sales tax is 6%, what was his total bill? Round off your answer to the nearest cent.

52. A stockbroker sold investments totaling \$44,178 and received \$7,687 in commission. What was his rate of commission? Round off your answer to the nearest tenth of a percent.

53. An advertising executive invests \$15,000 in a money market fund that accumulates dividends at an annual rate of 16% compounded quarterly. How much will she earn in dividends after 1 year? Round your answer to the nearest dollar.

54. A couple left a \$3.00 tip on a restaurant bill totaling \$17.82. What percent did they tip? Round off your answer to the nearest tenth of a percent.

5.5 Percent Increase and Decrease

Some of the most common applications of percent involve expressing an increase or decrease in a particular quantity as a percent. In order to find the amount of increase or decrease, we multiply the rate by the original amount (see box).

To calculate an amount of increase or decrease, multiply the rate by the original amount.

Rate × Original amount = Amount of Increase or Decrease

example 1

A college admissions officer reported an 8% decrease in freshman applications from the previous year. If 3,850 applications were received last year, how many fewer were received this year?

solution

The rate of decrease is 8%.

3,850 is the original amount.

Rate × Original amount = Amount of Decrease

8% × 3,850 = Amount of Decrease

(0.08)(3,850) = 308 fewer

The college received 308 fewer applications. ●

example 2

The U.S. Department of Labor reported that unemployment during the month of May rose one-tenth of one percent. If 14,500,000 were unemployed at the end of April, how many were unemployed at the end of May?

solution

One tenth of one percent = 0.1%.

This is the rate of increase.

The original amount is 14,500,000.

Rate × Original amount = Amount of Increase

0.1% × 14,500,000 = Amount of Increase

(0.001)(14,500,000) = 14,500 more

14,500 more people were unemployed at the end of May.

Find the total number of unemployed at the end of May.

14,500,000	unemployed in April	Add the increase in May to the number of unemployed in April.
+ 14,500	increase	
14,514,500	unemployed in May	

Thus, 14,514,500 persons were unemployed at the end of May. ●

A common example of a percent decrease is a discount, or markdown. A **discount (markdown)** is a percentage deducted from the original price when an item goes on sale. A discount is calculated by multiplying the discount rate by the original price.

Rate × Original price = Discount

The sale price is calculated by subtracting the discount from the original price (see box).

> To determine a sale price, subtract the discount from the original price.
>
> Original price − Discount = Sale price

example 3

Sweaters that normally sell for \$35 each are marked down 15%. What is the sale price?

solution

First calculate the discount.

Rate × Original price = Discount

$15\% \times 35 = \text{discount}$

$(0.15)(35) = \$5.25 \text{ discount}$

Now find the sale price.

Original price − Discount = Sale price

\$35.00	original price
− 5.25	discount
\$29.75	sale price

The sale price is therefore \$29.75.

example 4

A pair of jeans marked down 25% is on sale for \$12. What was the original price?

solution

In this problem, we know the discount rate and the sale price.

Original price − Discount = Sale price

100% − 25% = 75%

Since the discount is 25% of the original price, the sale price is 75% of the original price.

Therefore, 75% of original price = sale price

75% of original price = 12

We know the sale price is \$12.

Let x = original price.

$75\% \text{ of } x = 12$

$$\frac{75}{100} \cdot x = 12$$

$$\frac{3}{4}x = 12$$

$$\left(\frac{4}{3}\right)\frac{3}{4}x = \overset{4}{\cancel{12}}\left(\frac{4}{\cancel{3}}\right)$$

$$x = 16$$

Therefore, the original price was \$16.

A common example of a percent increase is a profit, or markup. A **profit (markup)** is a percentage added to the cost of a product before it is sold to the consumer. A profit is calculated by multiplying the profit rate by the cost.

Rate × Cost = Profit

The selling price is determined by adding the profit to the cost (see box).

> To determine a selling price, add the profit to the cost.
>
> Cost + Profit = Selling price

example 5

A camera that sells for \$228 costs the dealer \$120. What percent profit does the dealer make when he sells this camera?

solution

First determine the profit in dollars.

Cost + Profit = Selling price

Selling price − Cost = Profit

Rearrange the formula to find the difference between the selling price and the cost.

\$228	selling price
− 120	cost
\$108	profit

Now determine the profit rate.

Let r = profit rate.

$$\text{Rate} \times \text{Cost} = \text{Profit}$$
$$r \times 120 = 108$$
$$\left(\frac{1}{120}\right)120r = 108\left(\frac{1}{120}\right)$$
$$r = \frac{108}{120}$$
$$r = 0.90$$
$$r = 90\% \text{ profit}$$

Therefore, the dealer will make a 90% profit. ●

5.5 Exercises

Indicate which of the following statements are true and which are false. For those that are false, change the italic expression to make the statement true.

1. The selling price is determined by finding the *difference between* the cost and the profit.
2. When an item goes on sale, its price is *decreased.*
3. Profit is a percentage of the *cost.*
4. A discount is *subtracted from* the original price to obtain the sale price.

Solve each of the following word problems.

5. If bus fares increase from 50¢ to 65¢, what is the percent increase?
6. At a small college, the number of students majoring in English dropped from 25 to 18. What was the percent decrease in English majors?

7. During the first year, a new car that cost $8,995 depreciates in value 30%. What is the market value of this car when it is one year old?
8. The number of people buying new cars at a local dealership dropped 40% from last year. If 325 new cars were sold last year, how many were sold this year?
9. A new car with an original sticker price of $7,599 is marked down 10%. What is its sale price?
10. A department store is advertising a 25% sale on all carpeting. If the carpeting you want originally costs $349, how much would you pay for it during the sale?
11. A bicycle store manager wants to sell a ten-speed bike at a 21% profit. If the cost of the bike is $254, what should its selling price be?
12. The cost of manufacturing an electric typewriter is $116. What should its selling price be to ensure a 55% profit?
13. The phone company has been granted permission to increase monthly service charges 2.5%. If your monthly service charge is now $18, what will it be after the increase?
14. A mechanic making $16.50 per hour received an 8% raise. What is his new hourly wage?
15. A furniture store is advertising a sale of 15% off on all merchandise. What is the sale price of a sofa that originally sells for $499?
16. A coat originally selling for $250 is on sale for $197.50. What is the discount rate?
17. It costs a donut shop $1.60 to make a dozen donuts that sell for $2.60. What percent profit is made on one dozen donuts?
18. A jeweler is selling for $39 a quartz watch that costs him $24. What is his rate of profit?
19. During one year, doctors at a large city hospital reported a 12% increase in heart attack patients. If 125 heart attack victims were admitted during the previous year, how many were treated at that hospital this year?
20. During the month of December, the cost of home heating oil rose 3.5%. If you pay $96 a month for heating oil, what will your monthly bill be after the price increase takes effect?
21. The cost of beef rose 2.5% last month. If you paid $4.80 for a piece of beef last month, how much would it cost now?
22. The number of applicants to a well-known law school dropped 4% from last year. If 575 people applied last year, how many applied this year?
23. A car dealer is advertising a sports car for $8,995, which is 20% off the sticker price. What is the sticker price?
24. A new refrigerator originally priced at $600 is on sale for $459. What is the discount rate?
25. An appliance store makes a 38% profit on all washing machines it sells. If the cost of one machine is $175, what is the selling price?
26. A boat dealer makes a 22% profit on all sailboats. If the cost of a sailboat is $3,200, what is its selling price?
27. This past year, the market value of a house that was worth $142,000 increased 17%. What is the present market value of the house?
28. Tuition at a private college is scheduled to increase 14% next year. If tuition is now $6,200, what will it be next year?
29. Last year, the cost of gas rose from $1.80 per gallon to $1.89 per gallon. What was the percent increase?
30. If the cost of a head of lettuce rose from 64¢ to $1.12, what was the percent increase?
31. A dress marked down 20% is on sale for $24. What was the original price?
32. A sports jacket originally selling for $89.95 is marked down 20%. If the sales tax is 5%, how much will it cost to buy this sports jacket?
33. This year, there are 16% more students enrolled in a local public high school than in a private high school. If there are 550 students in the private high school, how many attend the public high school?
34. Next year's defense budget is scheduled to be 18% more than this year's budget. If this year we are spending 220 billion dollars on defense, how much will we be spending next year?
35. Last week the value of a utility stock rose $62\frac{1}{3}\%$. If the stock sold for $33 a share at the beginning of the week, what was its price at the end of the week?

36. A real estate agency reported a $6\frac{1}{4}\%$ decrease in the number of people buying homes this year through its agency. If 112 homes were sold through the agency last year, how many were sold this year?

37. In 1978, the expenditures of U.S. travelers to foreign countries increased 10% over expenditures in 1977. If U.S. travelers spent $11,930,000 in 1977, how much did they spend in 1978?

38. During 1977, dentist fees increased an average of 7.5%. If your dentist charged $20 for a checkup at the beginning of 1977, how much would you expect her to charge at the end of the year?

39. The median annual income for women in 1978 was 40.5% less than that for men. If the median salary for men was $15,730 in 1978, what was it for women?

40. The number of patents issued for inventions increased 22% from 1920 to 1930. If 37,050 patents were issued in 1920, how many were issued in 1930?

41. A boat that retails for $4,299 is marked down 25%. If the sales tax is $6\frac{1}{2}\%$, what is the cost of the boat? Round off to the nearest cent.

42. In 1960, the population of the United States was 180,671,000, and in 1970, it was 204,879,000. By what percent did the population increase from 1960 to 1970? Round off to the nearest tenth of a percent.

43. In 1970, there were 2,954,000 farms in the United States, and in 1975, there were 2,808,000 farms. By what percent did the number of farms in the United States decrease from 1970 to 1975? Round off your answer to the nearest hundredth of a percent.

44. The average number of daily telephone conversations increased 72% from 1940 to 1950. If the average number of daily telephone conversations was 98,775 in 1940, what was it in 1950?

5.6 Reading Graphs

A **graph** is a diagram used to represent numerical data so that it can be easily interpreted. It is easier to compare different quantities if the information is presented in a graph rather than in a table containing long columns of numbers. Usually, the numbers in a graph cannot be represented to a high degree of accuracy, but they are sufficient to give us an overall picture of the relationships between these quantities.

Graphs appear frequently in newspapers and magazines and are widely used by businesses, governmental agencies, and educational institutions to pictorially display large amounts of numerical data. Different kinds of graphs are used to emphasize different characteristics of the information being presented. The most common types are bar graphs, pictograms, line graphs, and circle graphs. These are all illustrated in the examples that follow.

A **bar graph** is used to show comparisons between different quantities. In a bar graph, numerical quantities are represented by thick bars which run either horizontally or vertically. The length or height of the bar is proportional to the size of the quantity represented. The information contained in the graph can be determined by reading the title of the graph and the labels for the horizontal and vertical axes, which are the lines that run along the bottom and the left-hand side, respectively, of the graph. Each axis is marked off in segments which represent a range of numerical values. Example 1 illustrates a bar graph representing the rainfall in inches for a one-year period. The questions that follow refer to that graph.

example 1

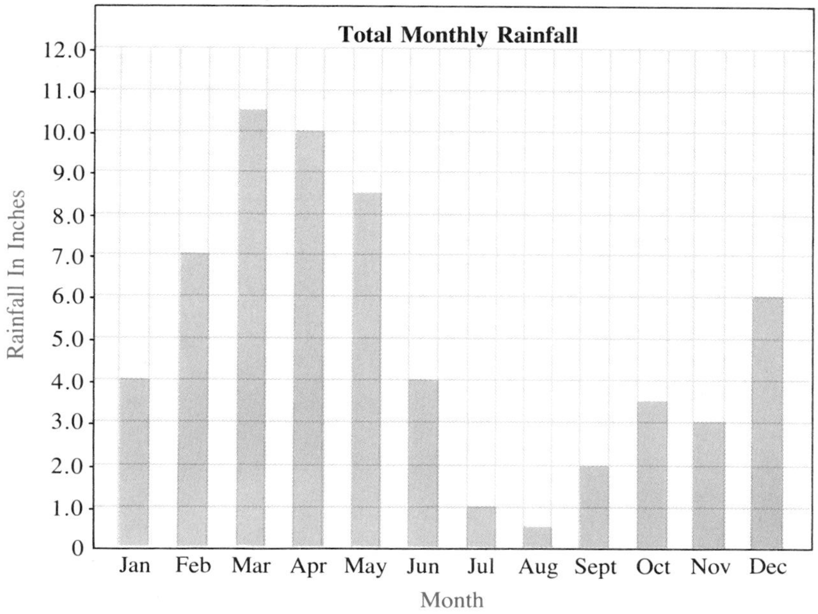

(a) During which month did the most amount of rain fall? How much rain was recorded that month?

The most amount of rain fell in March and amounted to 10.5 inches.

(b) During which month did the least amount of rain fall? How much rain was recorded that month?

The least amount of rain fell in August and amounted to 0.5 inches.

(c) How much more rain fell in May than in October?

May:	8.5 inches
Oct.:	3.5 inches
Difference:	5.0 inches

5.0 more inches of rain fell in May than in October.

(d) What was the average monthly rainfall for that year? Add the total rainfall for each month and divide the sum by 12:

$$4.0 + 7.0 + 10.5 + 10.0 + 8.5 + 4.0 + 1.0 + 0.5 +$$
$$2.0 + 3.5 + 3.0 + 6.0 = 60.0 \text{ total inches of rain}$$
$$60.0 \div 12 = 5.0 \text{ inches per month}$$

The average monthly rainfall was 5.0 inches. ●

Another type of graph, which is similar to a bar graph, is a **pictogram**. Instead of using a bar to represent a numerical quantity, pictograms use a series of pictures or symbols which usually run horizontally across the graph. The number of symbols is proportional to the size of the quantity. A **key** is used to indicate the number represented by a single symbol. A part of a symbol is often used to stand for a fractional part of a given quantity. Even though the numbers portrayed in pictograms are generally not as precise as those in other graphs, pictograms convey a lot of information at a quick glance. Example 2 illustrates a pictogram showing population estimates for a few major cities.

example 2

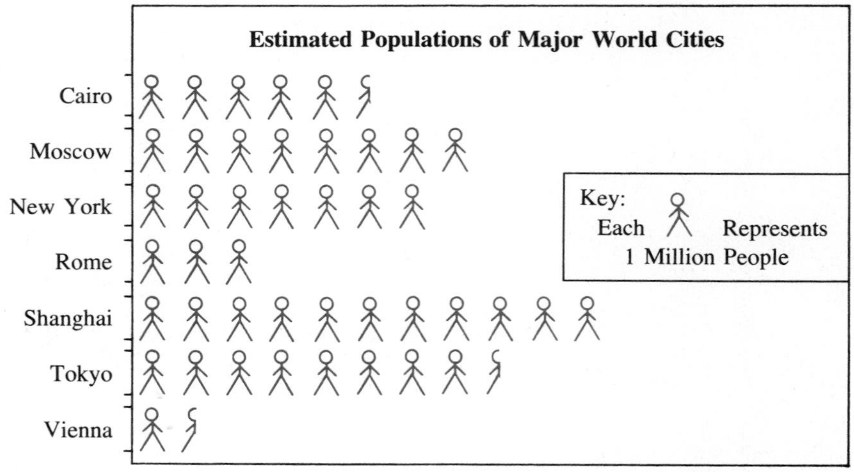

Source: United Nations Demographic Yearbook, 1978

(a) What is the approximate population of Cairo?

Cairo is represented by five and one-half . Therefore, its approximate population is

$5.5 \times 1{,}000{,}000 = 5{,}500{,}000.$

(b) Which city has a population of 2,914,600?

Since each represents 1,000,000 people, we are looking for a city represented by three . This city is Rome.

(c) Which two cities are closest in population and by approximately how much do they differ in size?

Moscow is represented by eight indicating a population of 8,000,000. Tokyo is represented by eight and one-half indicating a population of 8,500,000. Therefore, the cities Moscow and Tokyo are closest in population and differ in size by approximately 500,000 people. ●

The next type of graph we will consider is a **line graph**. A line graph is used to show how a given quantity changes over time. Each numerical value in a line graph is represented by a dot on the graph, and the dots plotted are connected by straight lines. When information is displayed using a line graph, it is easy to compare rates at which quantities increase or decrease during specified time intervals.

Line graphs often use more than one line to compare trends for different items. The meaning of each line is given in the key. Example 3 illustrates a line graph comparing the quiz grades for two students over a one-week period. Notice the break in the vertical axis indicating that there are no quiz grades between 0 and 10.

example 3

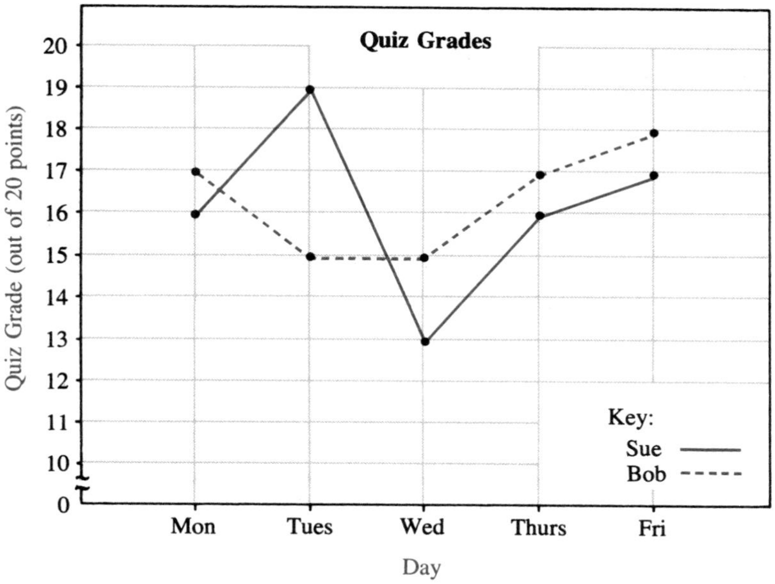

(a) Which student showed the sharpest decline in quiz scores and when did it occur?

Sue showed the sharpest decline from Tuesday to Wednesday.

(b) What was the percent increase in Bob's score from Wednesday to Thursday?

To find the percent increase in score, we find the amount of increase from Wednesday to Thursday and then divide by the original score, which is Wednesday's:

Thursday:	17
Wednesday:	−15
Amount of increase:	2 points

$$\frac{\text{Amount of increase}}{\text{Wednesday's score}} = \frac{2}{15} = 0.13333\ldots = 13\frac{1}{3}\%$$

Bob's score increased $13\frac{1}{3}\%$ from Wednesday to Thursday.

(c) Which student has the highest quiz average for the week? By how much do their averages differ?

Find the sum of the quiz grades for Bob and Sue and divide each sum by 5:

Bob:

$17 + 15 + 15 + 17 + 18 = 82$

$82 \div 5 = 16.4$

Sue:

$16 + 19 + 13 + 16 + 17 = 81$

$81 \div 5 = 16.2$

Bob has the highest quiz average. The averages differ by 0.2 points. ●

The final type of graph we will consider is a **circle graph**. Circle graphs are used to show the relationship between parts of a quantity and the whole. Circle graphs are also known as **pie charts**. A circle graph is divided into **sectors**, which resemble pieces of a pie. The size of each sector is directly proportional to the size of the quantity represented. Each piece of the circle has a percent associated with it, and the sum of all the pieces is 100%. Example 4 illustrates a circle graph that depicts the budget for federal expenditures for 1981.

example 4

Budget for Federal Expenditures 1981

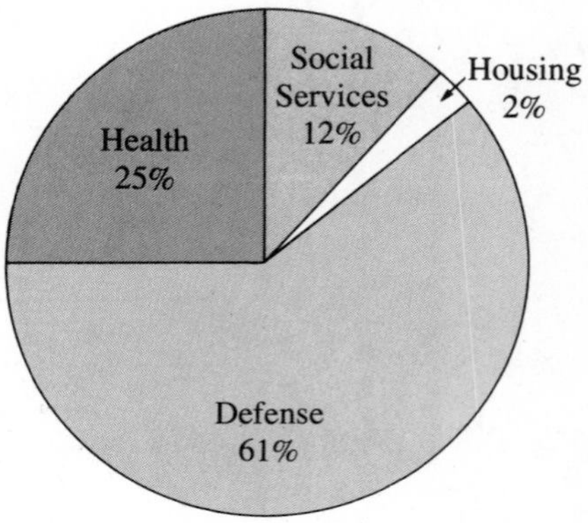

Source: U.S. Office of Management and Budget

(a) If the total budget for federal expenditures for 1981 was \$261 billion, what amount was spent in each category?

Multiply each of the percents by 261 billion:

Defense: $61\% \times 261$ billion $= (0.61)(261) = \$159.21$ billion
Health: $25\% \times 261$ billion $= (0.25)(261) = \$65.25$ billion
Social Services: $12\% \times 261$ billion $= (0.12)(261) = \$31.32$ billion
Housing: $2\% \times 261$ billion $= (0.02)(261) = \$5.22$ billion

(b) How much more money was spent on defense than on health in 1981?

Defense	\$159.21	billion
Health	− 65.25	billion
	\$ 93.96	billion more was spent on defense than on health care.

(c) What percent of the budget was spent on items other than defense? How much did this amount to?

Find the difference between 100% and 61% and multiply the result by 261 billion:

$100\% - 61\% = 39\%$
$39\% \times 261$ billion $= (0.39)(261) = \$101.79$ billion

39% of the budget was spent on items other than defense and this amounted to \$101.79 billion.

5.6 Exercises

Answer each of the following questions by reading the corresponding graph for each exercise.

1.

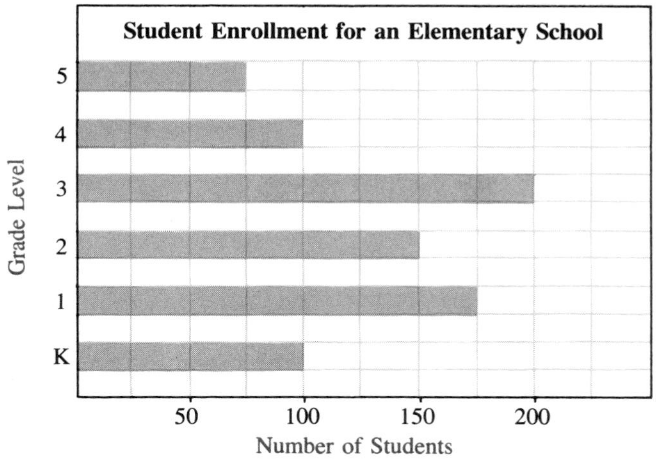

a. What grade has the largest number of students? How many students are in that grade?
b. What is the total number of students represented by this graph?
c. What percent of the students are enrolled in the fourth grade?

2.

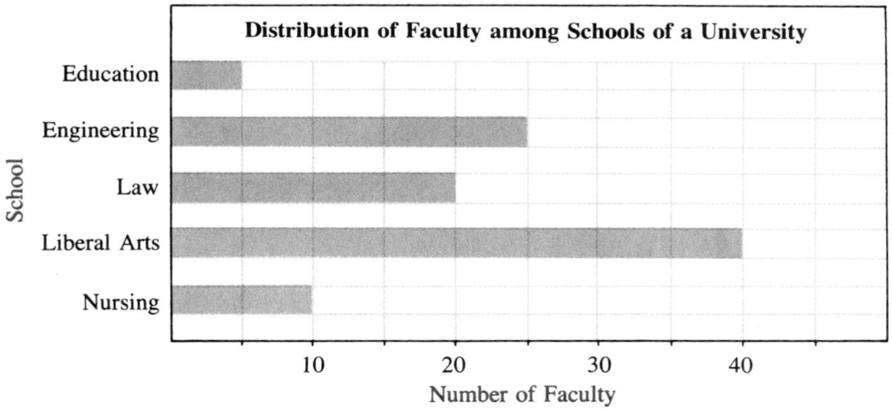

a. What school within the university has the smallest number of faculty? How many faculty members are employed by that school?
b. How many more faculty members are in the school of liberal arts than are in the law school?
c. What is the total number of faculty in the university?
d. What percent of the faculty is employed by the school of engineering?

3.

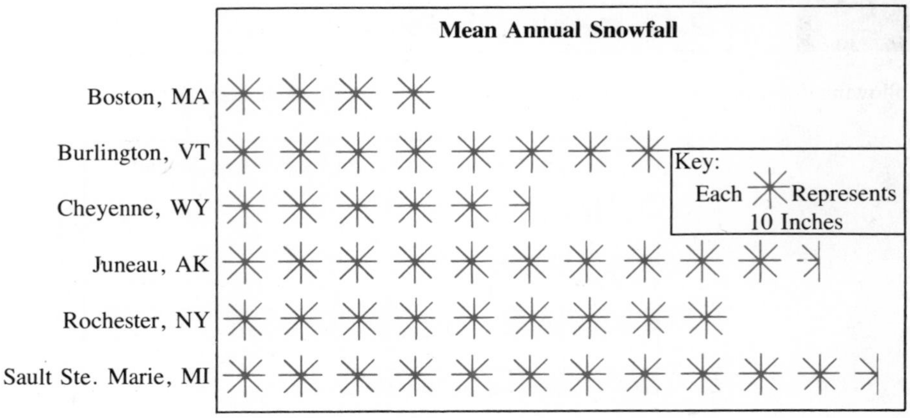

Source: National Oceanic and Atmospheric Administration, U.S. Commerce Department

a. Which city has the highest mean annual snowfall? What does it amount to?
b. Which city has a mean annual snowfall of 55 inches?
c. On the average, how much more snow falls per year in Juneau than in Boston?

4.

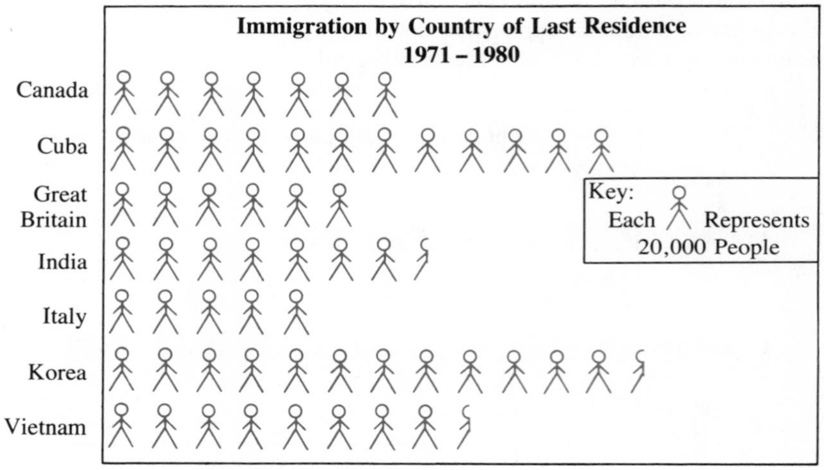

Source: U.S. Immigration and Naturalization Service

a. From which country did the greatest number of immigrants come? How many persons came from that country?
b. From which country did 150,000 immigrants come?
c. How many more immigrants came from Cuba than from Vietnam?

5.

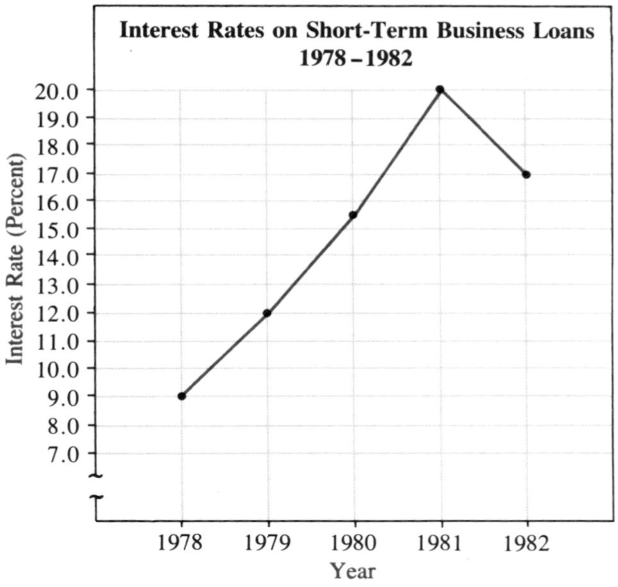

Source: Federal Reserve System

a. During what year were interest rates on short-term business loans the highest?
b. By how much did the interest rate increase from 1978 to 1981?
c. What is the average interest rate for the period 1978–1982?

6.

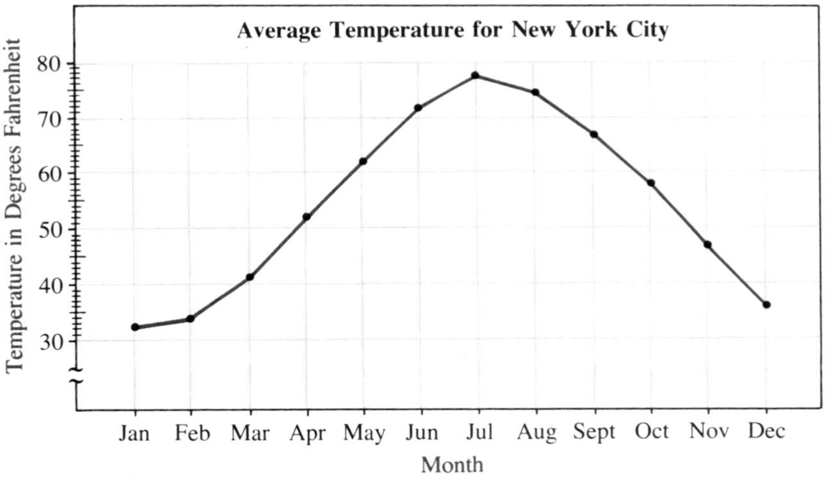

Source: National Oceanic and Atmospheric Administration, U.S. Commerce Department

a. What is the lowest average monthly temperature for New York City? During which month does it occur?
b. What is the difference between the highest average monthly temperature and the lowest average monthly temperature?
c. Which two months show the smallest change in temperature?
d. By how many degrees does the average monthly temperature increase from May to June?

7.

Persons Living Alone by Age in 1970

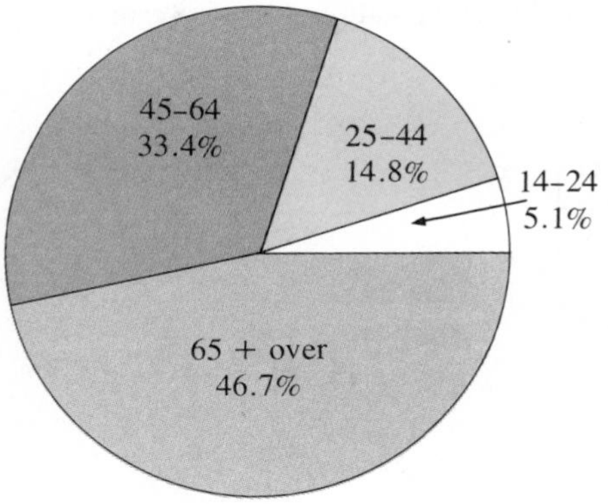

Source: Department of Commerce, Bureau of the Census

a. If a total of 10,851,000 people were living alone in 1970, how many in each age group were living alone?
b. How many more people living alone were 65 and over than were aged 45–64?
c. What percent of the population living alone was not in the 25–44 age bracket? How many people does this amount to?

8.

Deaths Due to Heart Disease in 1981

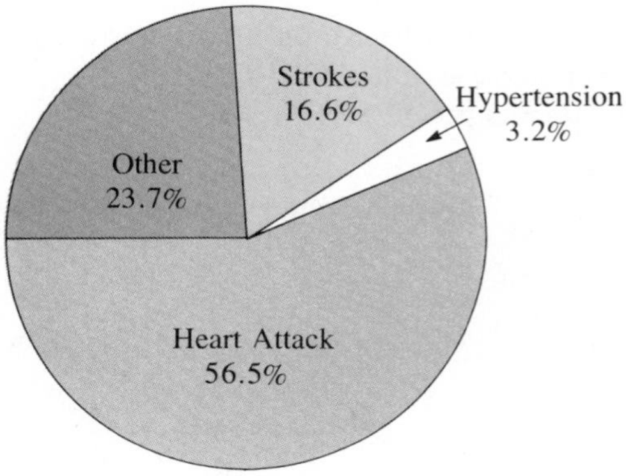

Source: National Center for Health Statistics, U.S. Department of Health and Human Services

a. If 986,610 people died of heart disease in 1981, how many of these deaths were due to heart attacks?
b. How many more deaths were due to strokes than were due to hypertension?

9.

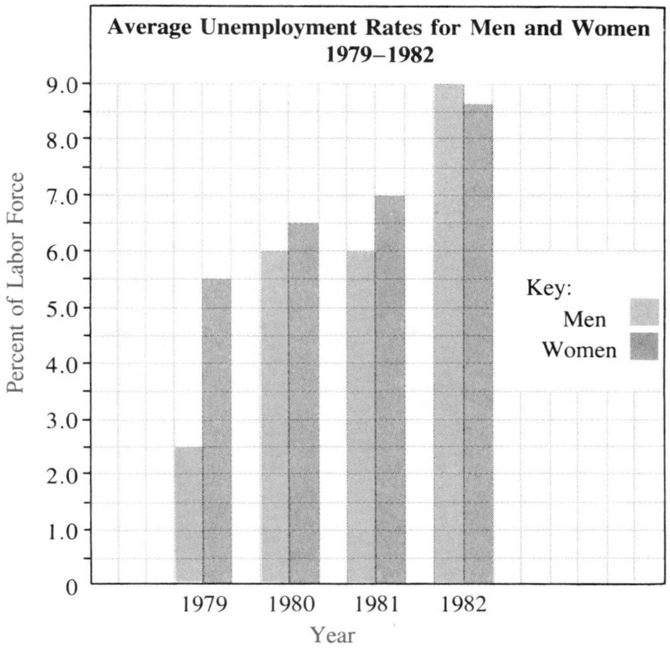

Source: Bureau of Labor Statistics, U.S. Labor Department

a. For how many years did the average unemployment rate for women exceed that for men?
b. What was the average unemployment rate for women in the 4-year period 1979–1982?
c. If the male labor force was 58 million in 1982, how many men were unemployed that year? Round off your answer to the nearest million.
d. How many times greater was the unemployment rate for men in 1981 than in 1979?

10.

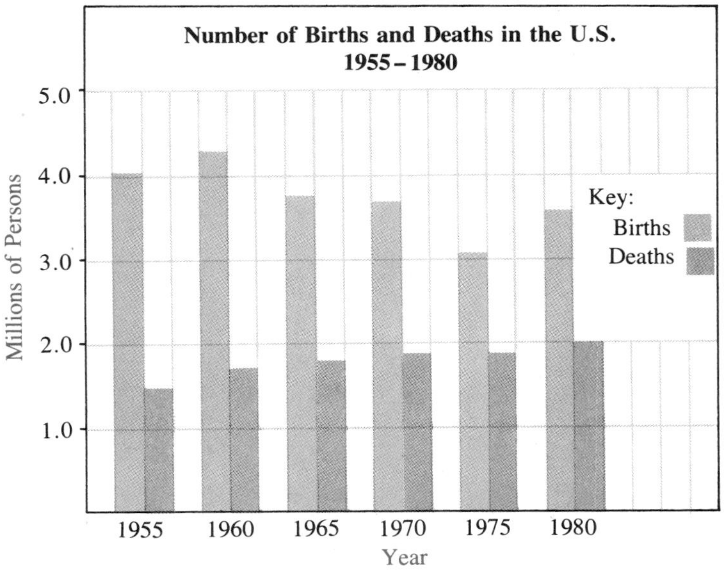

Source: National Center for Health Statistics, U.S. Department of Health and Human Services

a. In what year did the smallest number of deaths occur? How many deaths occurred that year?
b. During what 5-year interval did the number of births decline most severely? Find the percent decrease in births over this 5-year interval. Round off your answer to the nearest percent.

c. How many more people were born than died in 1965?
d. What was the average number of births per year for the period 1955–1980? Round off your answer to the nearest tenth of a million.

11.

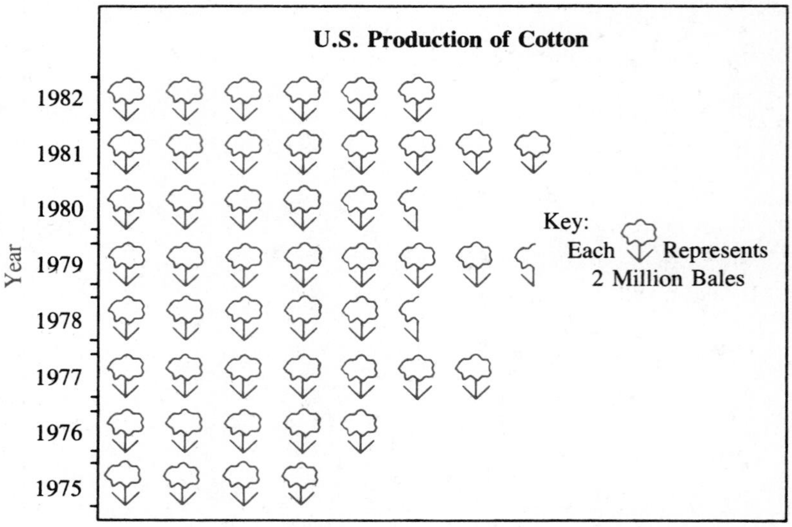

Source: Economics, Statistics and Cooperatives Service, U.S. Agriculture Department

a. During what years did the United States produce the same amount of cotton? How much cotton was produced these years?
b. How much more cotton was produced in 1981 than in 1980?
c. During what year was U.S. cotton production approximately 15 million bales?
d. What was the average yearly production of cotton for the years 1975–1982?

12.

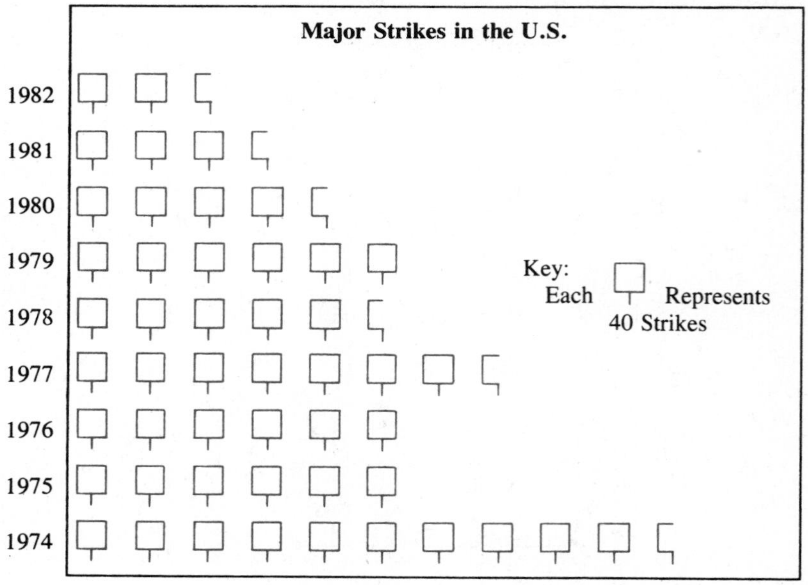

Source: Bureau of Labor Statistics, U.S. Labor Department

a. Approximately how many strikes occurred in 1978?
b. In what year did the greatest number of strikes occur? How many strikes occurred that year?
c. In what year did approximately 140 strikes occur?
d. How many times more strikes occurred in 1974 than in 1981?

13.

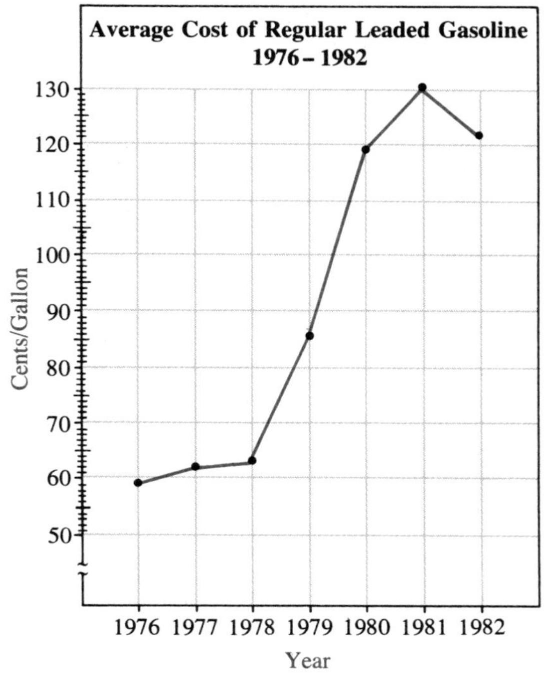

Source: Energy Information Administration, U.S. Energy Department

a. In what year was the cost of gas the highest?
b. Between what 2 years did the greatest increase in the cost of gas occur? How much was this increase?
c. What is the percent increase in the cost of gas from 1978 to 1979? Round off your answer to the nearest percent.

14.

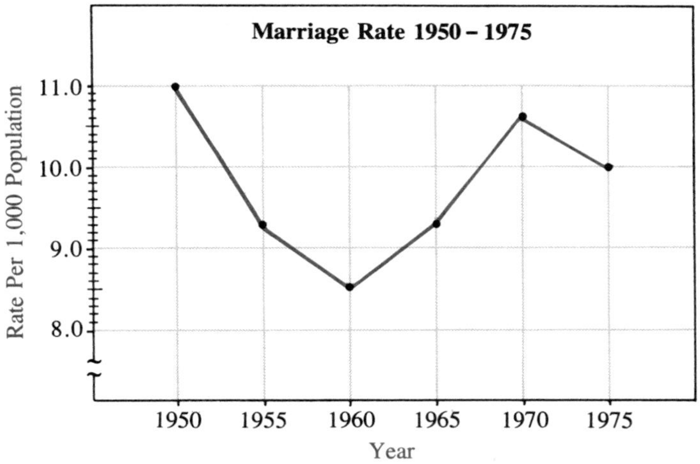

Source: National Center for Health Statistics, Public Health Service

a. In what year was the marriage rate the highest? What is this rate?
b. During what 5-year period did the sharpest decrease in the marriage rate occur?
c. What is the average marriage rate for the years 1950–1975?
d. What is the percent increase in the marriage rate from 1965 to 1970?

15.

Budget for a College Student

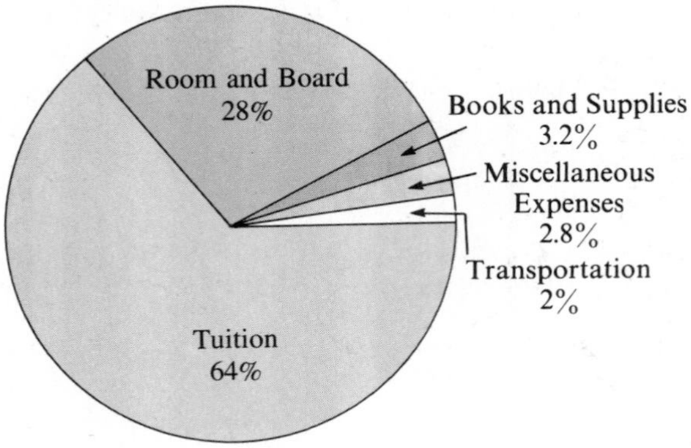

a. If the student's total budget was $12,500, how much was allocated to each item?
b. How much more money was allocated to tuition than to room and board?
c. What percent of the student's budget was not allocated to tuition or room and board? How much does this amount to?

16.

Summary of U.S. Receipts in 1981

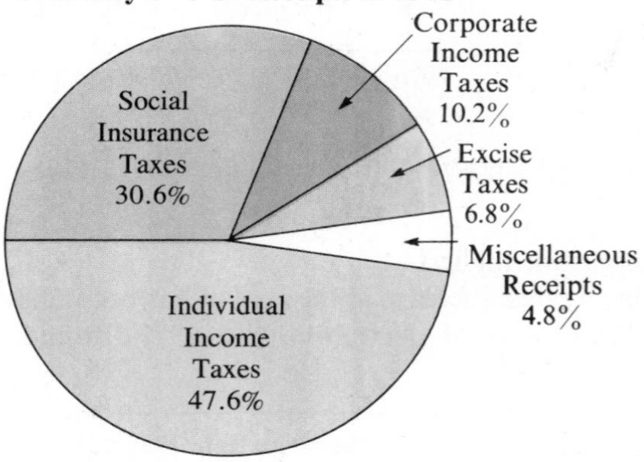

Source: U.S. Treasury Department, Bureau of Govt. Financial Operations

a. If the total income received by the federal government in 1981 was $599,272 million, how much came from individual income taxes? How much came from social insurance taxes? Round off your answers to the nearest million.
b. How much more money did the federal government receive from individual income taxes than from corporate income taxes?
c. What percent of the money received by the federal government came from sources other than corporate income taxes? How much does this amount to?

Summary and Review

Key Terms

[5.1] The word **percent** means per one hundred, or the ratio of a number to one hundred.

[5.4] A **commission** is a bonus paid to a salesperson as an incentive to maximize individual sales productivity.

Interest is an amount you pay to borrow money, or an amount you earn for allowing someone else to use your money when you deposit it in a bank or purchase investments.

The **principal** is the amount of money you borrow or deposit upon which interest is calculated.

[5.5] A **discount**, or **markdown**, is a percentage deducted from the original price when an item goes on sale.

A **profit**, or **markup**, is a percentage added to the cost of a product before it is sold to the consumer.

[5.6] A **graph** is a diagram used to represent numerical data so that it can be easily interpreted.

Calculations

[5.1] **To convert a percent to a fraction**, remove the percent sign (%), put the number over a denominator of 100, and reduce to lowest terms.

To convert a fraction to a percent, find an equivalent fraction with a denominator of 100, remove the denominator, and write a percent sign after the numerator.

OR

Change the fraction to a decimal and rewrite the decimal as a percent.

[5.2] **To convert a percent to a decimal**, remove the percent sign and move the decimal point two places to the left.

To convert a decimal to a percent, move the decimal point two places to the right and attach a percent sign.

[5.3] **To calculate the percentage or value of a percent**, multiply the rate by the base.

$$\text{Percentage} = \text{Rate} \times \text{Base}$$

To use a proportion to solve a percent problem, rewrite the percent as a fraction (the first ratio) and set this equal to the ratio of the percentage to the base. Solve for the unknown.

[5.4] **To calculate a sales tax**, multiply the tax rate by the price.

$$\text{Sales tax} = \text{Rate} \times \text{Price}$$

To calculate a commission, multiply the commission rate by the sales.

$$\text{Commission} = \text{Rate} \times \text{Sales}$$

To calculate an amount of interest, multiply the rate by the principal by the time.

$$i = r \cdot p \cdot t$$

[5.5] **To calculate an amount of increase or decrease**, multiply the rate by the original amount.

$$\text{Rate} \times \text{Original amount} = \text{Amount of Increase or Decrease}$$

To determine a sale price, subtract the discount from the original price.

$$\text{Original price} - \text{Discount} = \text{Sale price}$$

To determine a selling price, add the profit to the cost.

$$\text{Cost} + \text{Profit} = \text{Selling price}$$

Chapter 5 Review Exercises

Indicate which of the following statements are true and which are false. For those that are false, change the italic expression to make the statement true.

1. One hundred percent is equivalent to the number *1*.

2. To determine the profit, find the *sum of* the selling price and the cost.

3. A commission is paid to a *consumer*.

4. A discount always *reduces* the original price.

5. When using a proportion to solve a percent problem, the fractional equivalent of the percent is set equal to a ratio with the base in the *numerator*.

6. To convert a fraction to a percent, change the fraction to a decimal, move the decimal point two places to the *left*, and attach a % sign.

[5.1] Convert each of the following percents to a fraction reduced to lowest terms.

7. 57% **8.** 4% **9.** 8.2% **10.** 0.9% **11.** 180%

12. 72.25% **13.** $\frac{3}{4}\%$ **14.** $7\frac{1}{2}\%$ **15.** $83\frac{1}{3}\%$ **16.** $12\frac{1}{7}\%$

[5.1, 5.2] Convert each of the following fractions to a percent.

17. $\frac{9}{100}$ **18.** $\frac{3}{10}$ **19.** $\frac{4}{5}$ **20.** $\frac{7}{25}$ **21.** $2\frac{1}{2}$

22. $1\frac{3}{4}$ **23.** $\frac{7}{16}$ **24.** $\frac{1}{8}$ **25.** $\frac{5}{13}$ **26.** $\frac{7}{11}$

27. $\frac{107}{576}$ Round off to the nearest tenth of a percent.

28. $\frac{348}{795}$ Round off to the nearest tenth of a percent.

[5.2] Convert each of the following percents to a decimal.

29. 18% **30.** 80% **31.** 2% **32.** 4.4% **33.** 123%

34. 2,000% **35.** 0.09% **36.** 6.831% **37.** $18\frac{1}{5}\%$ **38.** $92\frac{3}{8}\%$

[5.2] Convert each of the following decimals to a percent.

39. 0.28 **40.** 0.07 **41.** 0.319 **42.** 0.602 **43.** 0.8

44. 5.0 **45.** 0.0043 **46.** 0.0001 **47.** 7.2 **48.** 26.8

[5.3] Find the unknown in each of the following questions.

49. What number is 60% of 154?

50. What number is 25% of 76?

51. 9 is 8% of what number?

52. 0.065 is 65% of what number?

53. 7.2 is what percent of 45?

54. 0.054 is what percent of 36?

55. What number is 250% of 6.2?

56. 0.98 is 0.32% of what number?

57. 670 is what percent of $22\frac{1}{3}$?

58. What number is $33\frac{1}{3}\%$ of 267?

59. 3.6289 is what percent of 267.9? Round off to the nearest hundredth of a percent.

60. 0.7634 is 82.719% of what number? Round off to the nearest ten-thousandth.

[5.3–5.5] Solve each of the following word problems.

61. Three out of 5 people working for a small computer company chose a four-day workweek over the traditional five-day workweek. What percent of the company's employees opted for a four-day workweek?

62. One out of every 3 students at a local community college is a part-time student. What percent of the student body is part-time?

63. Of the people who flew the 9 A.M. shuttle from Boston to New York, 88% were traveling on business. If 125 people were on the flight, how many were flying for business purposes?

64. Sixteen percent of the people in a part-time M.B.A. program were working on their second masters degree. If 28 people were working on their second degree, how many people were in the program?

65. A photographer bought some darkroom supplies totaling $37.80 before tax. If sales tax is 5%, how much will the supplies cost her?

66. A tax rate schedule states that if a single person's taxable income is over $23,500 but not over $28,800, his tax is $5,356 + 39% of the amount over $23,500. What would be the income tax for an actuary who is unmarried and has a taxable income of $26,700?

67. In September the number of students enrolled in a local elementary school was 12% less than last year. If 450 pupils attended the school last year, how many are enrolled this year?

68. Last week, the price of broccoli rose from 75¢ a bunch to 99¢ a bunch. What was the percent increase in price?

69. Sixteen out of 36 people who belong to a photography club said that they became interested in photography as teenagers. What percent of the members does this represent?

70. Out of 432 people staying at a hotel, 72 registered to stay more than two nights. What percent of the hotel's occupants are booked for more than two nights?

71. If you deposit $600 in a bank account at an annual interest rate of $5\frac{3}{4}\%$, how much money will be in your account 4 months later?

72. A party of 6 accumulates a bill of $59.75 at a restaurant, not including tax and tip. Tax is 5%, and they decide to leave a 15% tip. Estimate how much each person should contribute if the bill is split evenly.

73. A pair of shoes originally selling for $39.95 is on sale for 20% off. What is the sale price?

74. A television set costs a wholesale appliance store $80. If the television is sold for $122, what percent profit does the store make?

75. Of all the people surveyed at a local supermarket, $66\frac{2}{3}\%$ said that they prefer brand-name products instead of generic brands. If 822 people were surveyed, how many prefer brand-name products?

76. Of all the people in a small town in New Mexico, $14\frac{2}{7}\%$ who voted in the last gubernatorial election voted for the Democratic candidate. If the Democratic candidate received 216 votes from that town, how many people from that town voted in the election?

77. Between 1960 and 1970, the number of families in the United States increased from 45,111,000 to 51,586,000. What was the percent increase? Round off your answer to the nearest hundredth of a percent.

78. In 1975, the United States spent $286,547,000,000 on social welfare programs. If this figure represented 18.9% of the gross national product (GNP) for that year, what was the GNP in 1975? Round off your answer to the nearest billion dollars.

79.

Total Family Income 1978

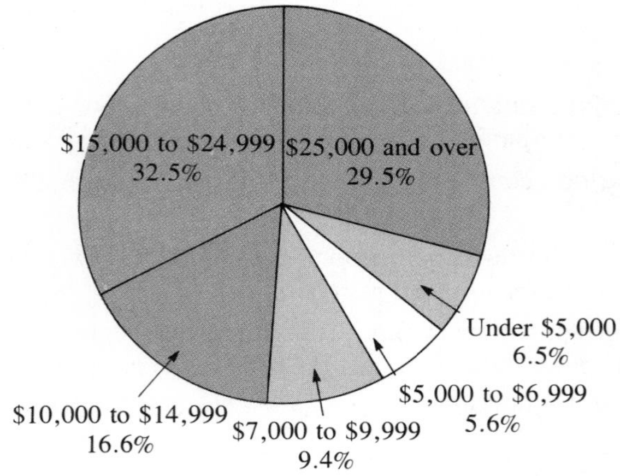

Source: Department of Commerce, Bureau of the Census

a. If there were a total of 50,910,000 families in 1978, how many had an income of $25,000 or over?
b. How many more families had an income under $5,000 than had an income of $5,000 to $6,999?

80.

Source: Department of Commerce, Bureau of the Census, and Department of Labor, Bureau of Labor Statistics

a. During what year was 22% of the female population in the labor force?
b. During what 10-year period did the percentage of women in the labor force increase the most?
c. By how much did the percent of female population in the working force increase from 1960 to 1970?

6 POSITIVE AND NEGATIVE NUMBERS

6.1 The Number Line

Thus far, our calculations have involved only **positive numbers**, which are numbers greater than zero, located to the right of zero on the number line. In this section, we will also be working with **negative numbers**, which are numbers less than zero, located to the left of zero on the number line. Figure 6.1 shows the relative location of the positive and negative numbers on the number line. The number 0 is considered neither positive nor negative.

figure *6.1*

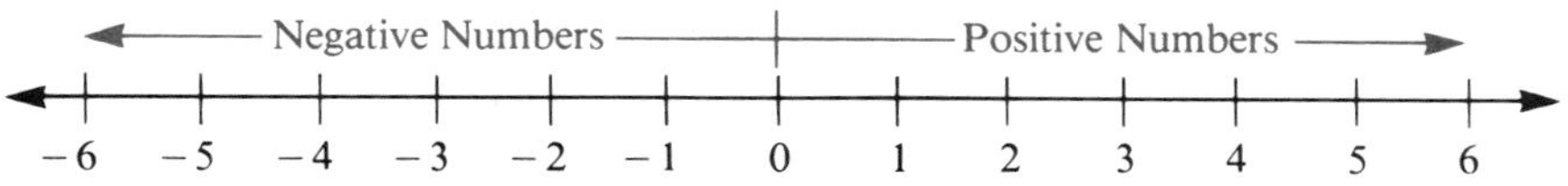

Positive numbers are sometimes preceded by a positive sign (+). If no sign precedes a number, it is assumed to be positive. Thus, +2 (read as "positive 2") and 2 represent the same number, which is located two units to the *right* of zero on a number line.

Negative numbers are always preceded by a negative sign (−). Thus, −7 (read as "negative 7") represents the number located seven units to the *left* of zero on the number line.

The set of **integers** consists of $\{\ldots, -5, -4, -3, -2, -1, 0, 1, 2, 3, 4, 5, \ldots\}$. The symbol ... indicates that there are an infinite number of positive and negative integers.

example *1*

Compare each of the following pairs of numbers on this number line. Use the appropriate sign (=, >, or <) to indicate the relationship between them.

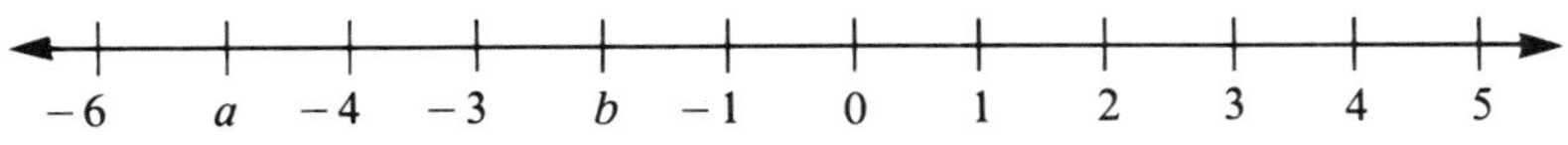

(a) $-3 \; ? \; 1$

$-3 < 1$

Since −3 is to the left of 1 on the number line, −3 is less than 1.

(b) $-1 \; ? \; -6$

$-1 > -6$

Since −1 is to the right of −6 on the number line, −1 is greater than −6.

(c) $b \; ? \; -2$

$b = -2$

Since b is at the same location as -2 on the number line, b is equal to -2.

(d) $-4 \; ? \; -1$

$-4 < -1$

Since -4 is to the left of -1 on the number line, -4 is less than -1.

(e) $-3 \; ? \; 3$

$-3 < 3$

Since -3 is to the left of 3 on the number line, -3 is less than 3.

(f) $0 \; ? \; a$

$0 > a$

Since 0 is to the right of a on the number line, 0 is greater than a.

(g) $b \; ? \; a$

$b > a$

Since b is to the right of a on the number line, b is greater than a. ●

Every negative number is the **opposite** of a corresponding positive number that is the same distance away from zero on the number line as is the negative number. For example,

-2 is the opposite of 2.

$-\frac{3}{5}$ is the opposite of $\frac{3}{5}$.

-6.3 is the opposite of 6.3.

Likewise, every positive number is the opposite of a negative number. For example,

8 is the opposite of -8.

$\frac{2}{3}$ is the opposite of $-\frac{2}{3}$.

7.9 is the opposite of -7.9.

The opposite of a number is also called its *additive inverse*. More will be said about additive inverses in Section 6.2.

To indicate the opposite of a number, we write a $-$ sign in front of the number. Thus, the mathematical expression that states that the opposite of negative 5 is positive 5 is written as

$$-(-5) = 5$$

(first $-$: the opposite of; second $-$: negative)

Notice that each $-$ sign in our example has a distinct meaning. You now know three different uses of the $-$ sign. The $-$ sign is used

1. To indicate the operation subtraction.
2. To indicate a negative number.
3. To indicate the opposite of a number.

Whenever a $-$ sign precedes a variable, we interpret it to mean the opposite of that variable. For example, $-a$ is read as "the opposite of a" and *not* as "negative a." The reason for this is that since the value of a is unknown, its opposite could be a negative number or a positive number. For example, if

$$a = -3$$

then

$$-a = -(-3) = 3$$

the opposite of — the opposite of — negative

The statement in the box summarizes our discussion of opposites. For example, the opposite of 20 is indicated by -20, and the opposite of -8 is indicated by $-(-8) = 8$.

> For any number a, the **opposite** of a is indicated by $-a$.

example 2

Find the opposite of each of the following numbers and state whether the opposite is a positive or negative number.

(a) 11.2

The opposite of 11.2 is -11.2.
-11.2 is a negative number.

(b) $-\frac{4}{5}$

The opposite of $-\frac{4}{5}$ is $-\left(-\frac{4}{5}\right) = \frac{4}{5}$.

$\frac{4}{5}$ is a positive number.

(c) $-(-1)$

$-(-1) = 1$ — The given expression simplifies to 1.
The opposite of 1 is -1.
-1 is a negative number.

(d) x if $x > 0$

The opposite of x is $-x$.
x is positive. — Since $x > 0$.
Therefore, $-x$ is negative.

(e) y if $y < 0$

The opposite of y is $-y$.
y is negative. — Since $y < 0$.
Therefore, $-y$ is positive.

(f) $-z$ if $z > 0$

The opposite of $-z$ is $-(-z) = z$.
z is positive

Since any number and its opposite are both located equal distances away from zero on a number line, they are said to have the same **absolute value**.

definition

The *absolute value* of a number is the distance between that number and zero on a number line.

We indicate the absolute value of a number by placing the symbol | | around the number. For example, the absolute value of 3 is written as $|3|$.

Since we measure distances with positive numbers, the *absolute value* of a number *is always positive.* For example,

$$|-4| = 4 \quad \text{and} \quad |4| = 4$$

The numbers -4 and 4 both have an absolute value of 4 since both numbers are four units away from zero on the number line, as shown in Figure 6.2.

figure *6.2*

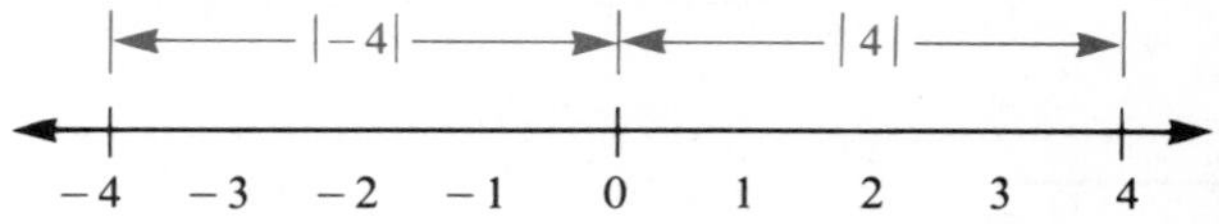

The absolute value of 0 is 0, since 0 is 0 units away from itself.

$$|0| = 0$$

The absolute value of a positive number is equal to the number itself. For example,

$$|9| = 9$$
$$|7.2| = 7.2$$
$$\left|\frac{1}{3}\right| = \frac{1}{3}$$

The absolute value of a negative number is equal to its opposite, which is a positive number. For example,

$$|-12| = 12$$
$$|-0.2| = 0.2$$
$$\left|-\frac{7}{8}\right| = \frac{7}{8}$$

The concept of absolute value is often summarized by the following definition.

definition

For any number a, the absolute value of a is defined as

$$|a| = \begin{cases} a & \text{if } a \geq 0 \\ -a & \text{if } a < 0 \end{cases}$$

The first part of the definition,

$$|a| = a \quad \text{if } a \geq 0$$

states that the absolute value of zero or a positive number is equal to itself. Thus,

$$|7| = 7 \quad \text{since } 7 > 0$$

The second part of the definition,

$$|a| = -a \quad \text{if } a < 0$$

states that the absolute value of a negative number is equal to its opposite. Thus,

$$|-5| = -(-5) = 5 \quad \text{since } -5 < 0$$

example 3

Simplify each of the following expressions.

(a) $|-6.5|$

$|-6.5| = -(-6.5) = 6.5$ — The absolute value of -6.5 is equal to its opposite, 6.5.

(b) $-|20|$

$|20| = 20$ — The absolute value of 20 is 20.

$-|20| = -20$ — The opposite of 20 is -20. Notice that we find the absolute value before we find the opposite. We begin with the innermost expression and work outwards.

(c) $-|-73|$

$|-73| = 73$ — The absolute value of -73 is 73.

$-|-73| = -73$ — The opposite of 73 is -73.

(d) $\left|-\left(-\frac{1}{2}\right)\right|$

$-\left(-\frac{1}{2}\right) = \frac{1}{2}$ — The opposite of $-\frac{1}{2}$ is $\frac{1}{2}$.

$\left|\frac{1}{2}\right| = \frac{1}{2}$ — The absolute value of $\frac{1}{2}$ is $\frac{1}{2}$.

Notice that we find the opposite before we find the absolute value. We begin with the innermost expression and work outwards. ●

example 4

Use the $>$, $<$, or $=$ sign to indicate the relationship between each of the following pairs of numbers.

(a) $|-18| \ ? \ |18|$

$18 \ ? \ |18|$ — The absolute value of -18 is 18.

$18 \ ? \ 18$ — The absolute value of 18 is 18.

$18 = 18$ — Any number is equal to itself.

$|-18| = |18|$ — Since $18 = 18$.

(b) $|-9| \ ? \ |-3|$

$9 \ ? \ |-3|$ — The absolute value of -9 is 9.

$9 \ ? \ 3$ — The absolute value of -3 is 3.

$9 > 3$ — Since 9 is to the right of 3 on the number line.

$|-9| > |-3|$ — Since $9 > 3$.

(c) $-\|66\| \; ? \; \|-66\|$	
$-(66) \; ? \; \|-66\|$	The absolute value of 66 is 66.
$-66 \; ? \; \|-66\|$	
$-66 \; ? \; 66$	The absolute value of -66 is 66.
$-66 < 66$	Since -66 is to the left of 66 on the number line.
$-\|66\| < \|-66\|$	Since $-66 < 66$.

QUICK QUIZ

Use the $>$, $<$, or $=$ sign to indicate the relationship between each of the following pairs of numbers.

1. $-5 \; ? \; 2$
2. $|-7| \; ? \; |-3|$
3. $-|-6| \; ? \; |-(-6)|$

ANSWERS

1. $-5 < 2$
2. $|-7| > |-3|$
3. $-|-6| < |-(-6)|$

6.1 Exercises

Indicate which of the following statements are true and which are false. For those that are false, change the italic expression to make the statement true.

1. The opposite of a positive number is *less than* zero.
2. The absolute value of a negative number *is* equal to the opposite of the number.
3. Numbers greater than zero have absolute values *less than* zero.
4. A negative number is located to the *right* of its opposite on a number line.

Compare each of the following pairs of numbers on the following number line. Use the $=$, $>$, or $<$ sign to indicate the relationship between them.

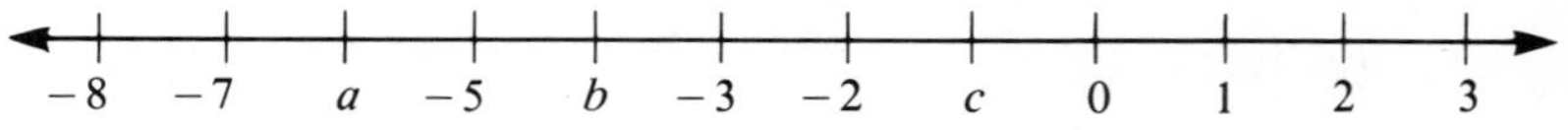

5. $-2 \; ? \; 2$ **6.** $-3 \; ? \; -7$ **7.** $b \; ? \; -4$ **8.** $c \; ? \; -1$

9. $a \; ? \; -8$ **10.** $c \; ? \; 0$ **11.** $b \; ? \; c$ **12.** $a \; ? \; b$

Find the opposite of each of the following numbers and state whether the opposite is a positive or a negative number.

13. -7 **14.** 3 **15.** -51 **16.** -82 **17.** 5.8

18. 2.1 **19.** $\frac{1}{5}$ **20.** $-\frac{2}{9}$ **21.** $-3\frac{1}{2}$ **22.** $7\frac{2}{3}$

23. 0.23 **24.** -0.097 **25.** $-(-6)$ **26.** $-(-12)$ **27.** a if $a < 0$

28. $-b$ if $b < 0$

Simplify each of the following expressions.

29. $|-3|$ **30.** $|5|$ **31.** $|51|$ **32.** $|-27|$ **33.** $\left|\frac{3}{5}\right|$

34. $\left|-\frac{7}{8}\right|$ **35.** $|3.9|$ **36.** $|-8.2|$ **37.** $-|62|$ **38.** $-|-44|$

39. $-\left|-\frac{5}{9}\right|$ **40.** $-\left|\frac{2}{3}\right|$ **41.** $|-82|$ **42.** $|-(-35)|$ **43.** $-|52|$

44. $-|-79|$ **45.** $-\left|8\frac{7}{9}\right|$ **46.** $-\left|-5\frac{1}{4}\right|$ **47.** $-\left|9\frac{3}{5}\right|$ **48.** $\left|-\left(-6\frac{1}{2}\right)\right|$

49. $-|5.4|$ **50.** $-|-6.7|$ **51.** $-|-1.85|$ **52.** $-|6.73|$ **53.** $-|5.48|$

54. $|-(-4.17)|$ **55.** $-|-(-28)|$ **56.** $-|-(-61)|$

Use the $=$, $>$, or $<$ sign to indicate the relationship between each of the following pairs of numbers.

57. $|8| \; ? \; |-8|$ **58.** $-|4| \; ? \; |-4|$ **59.** $|-9| \; ? \; -|3|$ **60.** $-|-2| \; ? \; -|5|$

61. $|-8| \; ? \; |2|$ **62.** $|7| \; ? \; -|5|$ **63.** $|-23| \; ? \; |23|$ **64.** $-|41| \; ? \; |-41|$

65. $|-58| \; ? \; -|72|$ **66.** $|-46| \; ? \; -|27|$ **67.** $-|3.8| \; ? \; |-5.6|$ **68.** $-|7.1| \; ? \; -|-9.2|$

69. $-|-0.052| \; ? \; -|0.075|$ **70.** $-|0.37| \; ? \; -|0.62|$ **71.** $-\left|-\frac{7}{9}\right| \; ? \; -\left|\frac{7}{9}\right|$

72. $\left|-\frac{3}{5}\right| \; ? \; -\left|\frac{3}{5}\right|$ **73.** $\left|-\left(-\frac{5}{8}\right)\right| \; ? \; \left|-\frac{1}{2}\right|$ **74.** $-\left|\frac{3}{4}\right| \; ? \; \left|-\frac{2}{3}\right|$

6.2 Addition of Signed Numbers

The operation addition can be illustrated by using a number line. Refer to the number line in Figure 6.3 as we work Example 1.

example 1

Add $2 + 5$.

solution

We will use the number line shown in Figure 6.3 to solve this problem. We begin at 0 and move 2 units to the right. This brings us to the first addend, 2. To add 5, we then move 5 additional units to the right. Our final location, 7, is the sum of the two numbers added together.

figure 6.3

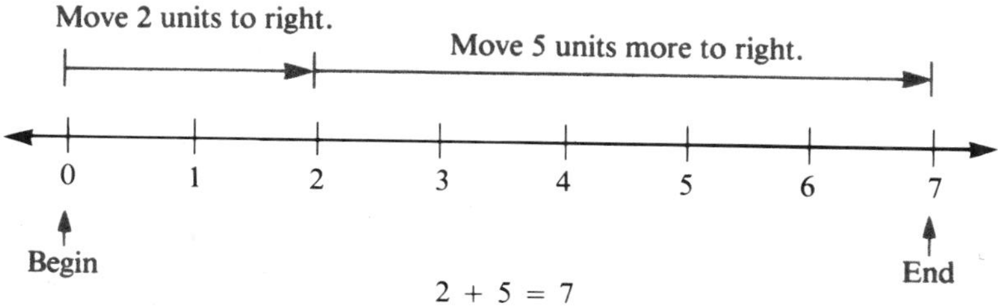

$2 + 5 = 7$

●

Thus, to add positive numbers on a number line, begin at zero, and move to the right the distance represented by each addend. In Example 1, the distance we moved along the number line can be represented as the sum of the absolute values of the addends. We can state that

$$2 + 5 = |2| + |5| = 7$$

We therefore have demonstrated the procedure summarized in the box.

> To add two positive numbers, add their absolute values.

example 2 Add $-3 + (-2)$.

solution Notice that whenever a + and − sign appear in succession, parentheses are used to indicate that the sign closest to the number is the sign of the number (-2), and that the other sign is the operation to be performed (addition). Thus, the problem is read as "negative 3 plus negative 2." Using the number line shown in Figure 6.4 to help with our addition, we begin at zero and move 3 units to the left. This brings us to the first addend, -3. To add -2, we then move 2 additional units to the left. Our final location, -5, is the sum of the two numbers.

figure 6.4

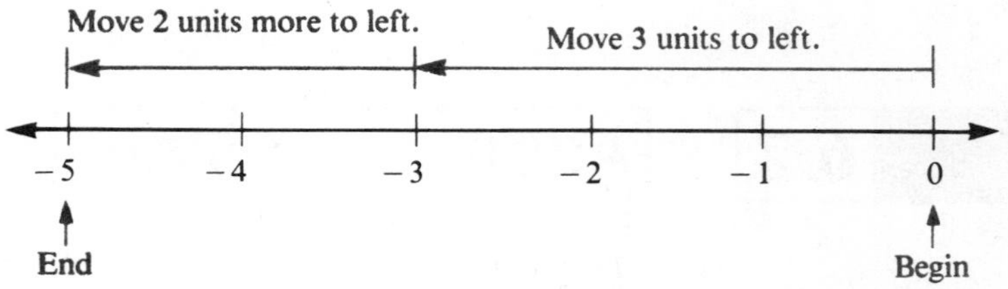

$$-3 + (-2) = -5$$

Notice that to solve the problem in Example 2, we moved a total of 5 units, which is the sum of the absolute values of -3 and -2.

$$|-3| + |-2| = 5$$

In Example 2, we found that $-3 + (-2) = -5$. The − sign in front of the 5 indicates we moved a total of 5 units to the left. We therefore have demonstrated the procedure stated in the box.

> To add two negative numbers, add their absolute values and place a negative sign in front of the result.

example 3 Add each of the following pairs of numbers.

(a) $-4 + (-9)$

$|-4| + |-9| = 4 + 9 = 13$ — Find the sum of the absolute values of the addends.

$-4 + (-9) = -13$ — Put a − sign in front of the result.

(b) $-8.5 + (-6.2)$

$|-8.5| + |-6.2| = 8.5 + 6.2 = 14.7$ — Find the sum of the absolute values of the addends.

$-8.5 + (-6.2) = -14.7$ — Put a − sign in front of the result. ●

We will now look at some examples of addition in which one addend is negative and the other is positive.

example 4

Add $9 + (-6)$.

solution

Using the number line shown in Figure 6.5 to help with our addition, we begin at zero and move 9 units to the right. This brings us to the first addend, 9. To add -6, we move 6 units to the left. Our final location, 3, is the sum of the two numbers.

figure 6.5

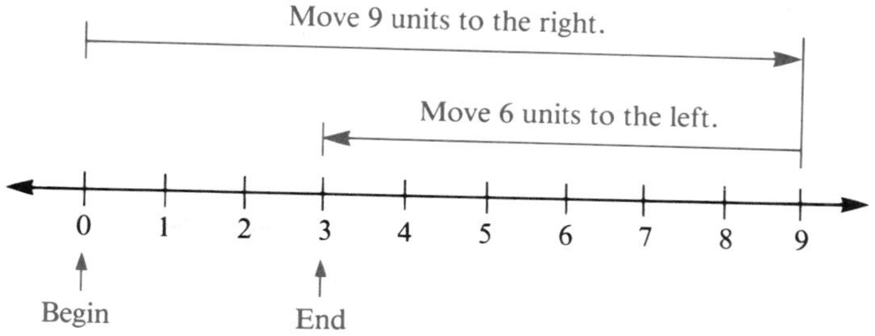

$9 + (-6) = 3$ ●

Notice that when we added a positive and a negative number together in Example 4, we moved in two different directions, first to the right, and then to the left. To determine the distance between the final location and zero, we can find the difference between the absolute values of the two addends.

$|9| - |-6| = 9 - 6 = 3$

Since we moved farther to the right (9 units), than we moved to the left (6 units), the sum is a positive number, which is located to the right of zero, our starting point. The sign of the sum is therefore the same as the addend with the larger absolute value.

$9 + (-6) = 3$

example 5

Add $-6 + 4$.

solution

Using the number line shown in Figure 6.6 to help us with our addition, we begin at zero and move 6 units to the left. This brings us to the first addend, -6. To add 4, we move 4 units to the right. Our final location, -2, is the sum of the two numbers.

figure 6.6

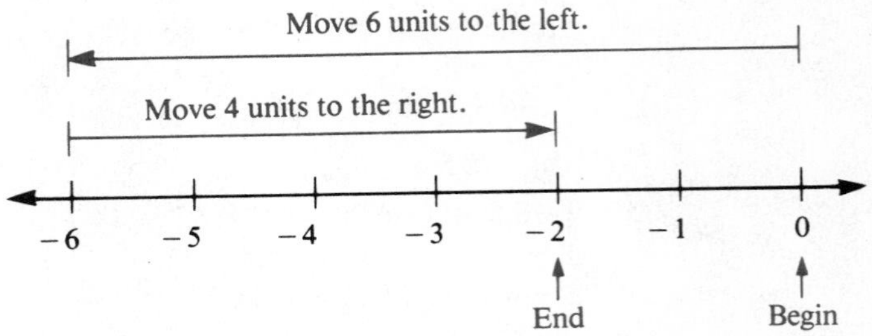

$-6 + 4 = -2$

Notice in Example 5 that by first moving a distance of 6 units to the left and then 4 units to the right, our final location is 2 units away from zero. Since we moved first to the left, and then to the right, the distance between our final location and zero, our starting point, is the same as the difference between the absolute values of the two addends.

$$|-6| - |4| = 6 - 4 = 2$$

Since we moved farther to the left (6 units), than we did to the right (4 units), the sum is a negative number, which is located to the left of zero. The sign of the sum, therefore, is the same as the addend with the larger absolute value.

$$-6 + 4 = -2$$

We therefore have established the procedure outlined in the box.

> To add a negative number and a positive number, subtract the smaller absolute value from the larger absolute value. The sum will have the same sign as the addend with the larger absolute value.

example 6

Add each of the following pairs of numbers.

(a) $4 + (-17)$

$|-17| - |4| = 17 - 4 = 13$ — Subtract the smaller absolute value from the larger.

$|-17| > |4|$ — The addend with the larger absolute value is negative.

$4 + (-17) = -13$ — The sum is also negative.

(b) $-12 + 36$

$|36| - |-12| = 36 - 12 = 24$ — Subtract the smaller absolute value from the larger.

$|36| > |-12|$ — The addend with the larger absolute value is positive.

$-12 + 36 = 24$ — The sum is also positive.

(c) $-5.7 + 1.3$

$|-5.7| - |1.3| = 5.7 - 1.3 = 4.4$ — Subtract the absolute values.

$|-5.7| > |1.3|$ — The addend with the larger absolute value is negative.

$-5.7 + 1.3 = -4.4$ — The sum is also negative. ●

Let us observe what happens when we add a number to its opposite. For example,

$$-8 + 8 = |-8| - |8| = 8 - 8 = 0$$

$$\frac{5}{6} + \left(-\frac{5}{6}\right) = \left|\frac{5}{6}\right| - \left|-\frac{5}{6}\right| = \frac{5}{6} - \frac{5}{6} = 0$$

$$-3.2 + 3.2 = |-3.2| - |3.2| = 3.2 - 3.2 = 0$$

Thus, whenever we add a number to its opposite, we obtain a sum of zero. The opposite of a number is also called its **additive inverse**.

definition

The *additive inverse*, or *opposite*, of a number is the number you must add to it to obtain a sum of zero.

example 7

Find the additive inverse of each of the following numbers.

(a) -37

37 is the additive inverse of -37. — Since $-37 + 37 = 0$.

(b) 0.02

-0.02 is the additive inverse of 0.02. — Since $0.02 + (-0.02) = 0$.

(c) $-\frac{1}{8}$

$\frac{1}{8}$ is the additive inverse of $-\frac{1}{8}$. — Since $-\frac{1}{8} + \frac{1}{8} = 0$. ●

You now know how to add numbers with the same sign and numbers with different signs. The rules for adding signed numbers are summarized in the box.

To Add Signed Numbers

1. If both numbers are positive, add their absolute values.
2. If both numbers are negative, add their absolute values and place a negative sign in front of the result.
3. If one number is positive and the other is negative, subtract the smaller absolute value from the larger absolute value. The sum will have the same sign as the addend with the larger absolute value.

example 8

Perform each of the following additions.

(a) $7 + (-9) + (-8)$

$7 + (-9) + (-8) = -2 + (-8) = -10$

(b) $-2 + (-4) + 7$

$$-2 + (-4) + 7 = -6 + 7$$
$$= 1$$

(c) $-8 + 4 + (-3) + (-2)$

$$[-8 + 4] + [(-3) + (-2)] = -4 + (-5)$$
$$= -9$$

example 9

A student received a bank statement indicating that he was \$29 overdrawn. He then deposited \$100 in his account. What was his new balance?

solution

An account that is overdrawn \$29 has a balance of $-\$29$. To find the new balance, add $-29 + 100$.

$|100| - |-29| = 100 - 29 = 71$ — Find the difference between the absolute values.

$|100| > |-29|$ — The number with the larger absolute value is positive.

$-29 + 100 = 71$ — The answer is positive.

His new balance is \$71.

QUICK QUIZ

Perform each of the following additions.

1. $-9 + 3$
2. $-8 + (-7)$
3. $3 + (-5) + (-2) + 6$
4. $-2.3 + 5.7 + (-7.8)$

ANSWERS

1. -6
2. -15
3. 2
4. -4.4

6.2 Exercises

Indicate which of the following statements are true and which are false. For those that are false, change the italic expression to make the statement true.

1. The sum of two negative numbers is *positive*.
2. To add a negative and a positive number, find the *difference between* their absolute values. values.
3. When adding a negative number on a number line, you move to the *left*.
4. The additive inverse of a number is the number you add to it to obtain a sum of *one*.

Find each of the following sums.

5. $7 + (-5)$
6. $6 + (-2)$
7. $-8 + (-3)$
8. $-5 + (-9)$

9. $-3+4$
10. $-6+9$
11. $-5+(-9)$
12. $-3+(-8)$
13. $17+(-6)$
14. $-21+(-5)$
15. $-36+(-5)$
16. $4+(-23)$
17. $41+(-20)$
18. $-18+(-55)$
19. $-62+37$
20. $63+(-25)$
21. $89+(-45)$
22. $-27+31$
23. $-23+(-59)$
24. $-14+85$
25. $-5.3+2.8$
26. $-2.9+(-5.7)$
27. $7.7+(-4.8)$
28. $-1.9+8.5$
29. $7.9+(-1.3)$
30. $-5.8+2.2$
31. $-0.48+0.36$
32. $-0.92+(-0.09)$
33. $367+(-145)$
34. $-185+647$
35. $\frac{5}{9}+\left(-\frac{1}{3}\right)$
36. $-\frac{3}{5}+\left(-\frac{1}{2}\right)$
37. $-\frac{3}{4}+\frac{7}{8}$
38. $-\frac{1}{6}+\frac{2}{3}$
39. $-\frac{5}{8}+\frac{3}{8}$
40. $-\frac{2}{7}+\left(-\frac{5}{7}\right)$
41. $2\frac{1}{2}+\left(-1\frac{1}{4}\right)$
42. $-3\frac{5}{6}+\left(-2\frac{1}{3}\right)$
43. $5+(-7)+(-3)$
44. $-8+9+(-5)$
45. $-4+14+(-9)$
46. $11+(-7)+18$
47. $24+(-37)+19$
48. $-84+69+13$
49. $-402+(-287)+154$
50. $-114+567+(-382)$
51. $\frac{1}{3}+\left(-\frac{5}{12}\right)+\frac{3}{4}$
52. $-\frac{3}{5}+\left(-\frac{7}{15}\right)+\frac{3}{4}$
53. $81{,}065{,}050+(-46{,}729{,}486)$
54. $-27{,}364{,}873+(-48{,}316{,}092)$
55. $-5.00218+3.28947$
56. $7.21843+(-8.01241)$

Find the additive inverse of each of the following numbers.

57. 28
58. -91
59. $-\frac{3}{5}$
60. $-8\frac{3}{4}$
61. 0.675
62. 0.003
63. $-534{,}987$
64. 283,005

Solve each of the following word problems using addition of signed numbers.

65. At 9 A.M. the temperature in Anchorage, Alaska, was $-12°$F. During the next 3 hours, the temperature rose 5°. What was the temperature 3 hours later?

66. An elevator in an observation tower at Niagara Falls began at ground level, went down 8 floors to pick up passengers, and then went up 14 floors to an observation deck where they got out. How many floors above ground level is this observation deck?

67. The highest point in California, Mount Whitney, is 14,776 feet higher than the lowest point in California. The lowest point in the state is Death Valley, which is 282 feet below sea level. What is the elevation of Mount Whitney?

68. A tourist on vacation in Las Vegas won $25 at the blackjack table. That same evening, she lost $51 and then won an additional $33. What were her net winnings for that evening?

69. A football team was penalized 15 yards and then gained 8 yards on the next play. If the ball was on the 22-yard line before the penalty, where was the ball after the next play?

70. A stock opened at $30\frac{1}{2}$ points. During that day's trading session, it gained $1\frac{1}{4}$ points, lost $\frac{3}{8}$ point, and then lost an additional $\frac{3}{4}$ point. What was the closing price of the stock?

6.3 Subtraction of Signed Numbers

Since addition and subtraction are inverse operations, any subtraction problem can be expressed as an addition problem. Let us consider the following examples.

$$7 - 6 = 1 \quad \text{since} \quad 1 + 6 = 7$$
$$3 - 9 = -6 \quad \text{since} \quad -6 + 9 = 3$$
$$-4 - 5 = -9 \quad \text{since} \quad -9 + 5 = -4$$
$$5 - (-2) = 7 \quad \text{since} \quad 7 + (-2) = 5$$

Notice that we can obtain the same results by adding the opposite of the quantity being subtracted.

$$7 - 6 = 7 + (-6) = 1$$
$$3 - 9 = 3 + (-9) = -6$$
$$-4 - 5 = -4 + (-5) = -9$$
$$5 - (-2) = 5 + (+2) = 7$$

Thus, to subtract a number from a given quantity, add its opposite to that quantity.

For any signed numbers a and b,

$$a - b = a + (-b)$$

example 1

Perform each of the following subtractions.

(a) $2 - 8$

$2 - 8 = 2 + (-8)$
$= -6$

To subtract 8, add −8. Two minus eight equals two plus negative eight.

(b) $-3 - 9$

$-3 - 9 = -3 + (-9)$
$= -12$

To subtract 9, add −9. Negative three minus nine equals negative three plus negative nine.

(c) $5 - (-6)$

$5 - (-6) = 5 + (+6)$
$= 11$

To subtract −6, add +6. Five minus negative six equals five plus positive six.

(d) $-78-(-21)$

$-78-(-21) = -78+(+21)$ To subtract -21, add $+21$.

$= -57$

(e) $-8.7-9.5$

$-8.7-9.5 = -8.7+(-9.5)$ To subtract 9.5, add -9.5.

$= -18.2$

(f) $\frac{1}{4}-\left(-\frac{7}{8}\right)$

$\frac{1}{4}-\left(-\frac{7}{8}\right)=\frac{1}{4}+\left(+\frac{7}{8}\right)$ To subtract $-\frac{7}{8}$, add $+\frac{7}{8}$.

$=\frac{2}{8}+\frac{7}{8}$

$=\frac{9}{8}$ ●

To solve problems that involve both addition and subtraction, change each subtraction sign to an addition sign and change the number following that sign to its opposite.

example 2

Perform each of the operations indicated.

(a) $5-(-7)+(-4)$

$5-(-7)+(-4) = 5+(+7)+(-4)$ To subtract -7, add $+7$.

$= 12+(-4)$

$= 8$

(b) $-6-20-(-47)$

$-6-20-(-47)$

$= -6+(-20)-(-47)$ To subtract 20, add -20.

$= -6+(-20)+(+47)$ To subtract -47, add $+47$.

$= -26+47$

$= 21$

(c) $4.2+(-9.8)-(-6.5)$

$4.2+(-9.8)-(-6.5)$

$= 4.2+(-9.8)+(+6.5)$ To subtract -6.5, add $+6.5$.

$= -5.6+6.5$

$= 0.9$ ●

example 3

Mercury has a melting point of $-38.87°C$. Helium has a melting point of $-272.20°C$. What is the difference between the melting point of mercury and the melting point of helium.

solution Solve the problem using subtraction of signed numbers.

$$-38.37 - (-272.20) = -38.87 + (+272.20)$$

To subtract -272.20, add $+272.20$.

$$|272.20| - |38.87| = 272.20 - 38.87 = 233.33$$

Subtract the smaller absolute value from the larger.

$$\begin{array}{r} 272.20 \\ -\ \ 38.87 \\ \hline 233.33 \end{array}$$

Mercury $\quad$ *Helium*

$$-38.87 - (-272.20) = 233.33$$

Thus, the melting point of helium is 233.33°C less than that of mercury. ●

QUICK QUIZ

Perform each of the following subtractions.

1. $-8 - 9$

2. $4 - (-7)$

Perform each of the operations indicated.

3. $-2 + (-5) - (-3)$

4. $3 - 7 - (-9) + (-2)$

ANSWERS

1. -17

2. 11

3. -4

4. 3

6.3 Exercises

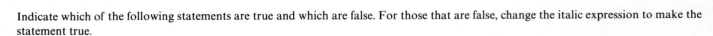

Indicate which of the following statements are true and which are false. For those that are false, change the italic expression to make the statement true.

1. Subtracting a negative number from a given quantity is the same as *adding* its opposite to that quantity.

2. The operation subtraction is the inverse of *division.*

3. When we subtract a negative number from its opposite, the result is *less than* zero.

4. Adding the opposite of a positive number to a given quantity is the same as subtracting a *negative* number from that quantity.

Perform each of the following subtractions.

5. $5 - 7$

6. $3 - 8$

7. $-4 - 2$

8. $-9 - 1$

9. $0 - 7$

10. $8 - (-6)$

11. $5 - (-9)$

12. $0 - (-4)$

13. $-8 - (-7)$

14. $-4 - (-4)$

15. $12 - 35$

16. $-25 - 41$

17. $-17 - (-53)$

18. $78 - (-39)$

19. $3.8 - 5.6$

20. $-2.9 - 7.1$

21. $5.4 - (-8.7)$

22. $-6.9 - (-8.3)$

23. $-453 - 298$

24. $-307 - (-154)$

25. $-0.036 - 0.298$

26. $-0.263 - (-0.457)$

27. $0.059 - 0.584$

28. $-0.037 - (-0.069)$

29. $1{,}658 - 3{,}924$

30. $-4{,}781 - 3{,}742$

31. $\frac{3}{8} - \frac{5}{8}$

32. $-\frac{7}{9} - \frac{8}{9}$

33. $-\frac{4}{5} - \frac{7}{10}$

34. $-\frac{5}{12} - \frac{3}{4}$

35. $-3\frac{1}{2} - 5\frac{1}{8}$

36. $6\frac{2}{3} - 8\frac{5}{6}$

37. $7{,}362{,}051 - 10{,}000{,}000$

38. $-29{,}684{,}317 - (-28{,}671{,}953)$

39. $-4.80671 - 7.35826$

40. $5.09543 - (-6.21985)$

Perform each of the operations indicated.

41. $3 + 5 - 6$

42. $2 - 7 + 4$

43. $1 + 8 - 9$

44. $6 - 9 + 3$

45. $2 - 0 - 7$

46. $0 - 5 - 3$

47. $3 - (-8) + (-4)$

48. $-2 + (-9) - 7$

49. $2.2 - (-1.4) - 7.5$

50. $-3.9 - (-5.4) + (-6.1)$

51. $45 - 63 + (-32)$

52. $-81 + (-97) + 25$

53. $-\frac{1}{2} - \left(-\frac{3}{8}\right) + \frac{1}{4}$

54. $\frac{1}{3} - \frac{2}{9} + \left(-\frac{5}{6}\right)$

55. $-2\frac{3}{4} + 1\frac{2}{3} - \left(-5\frac{7}{12}\right)$

56. $-4\frac{7}{18} + \left(-3\frac{2}{9}\right) - 4\frac{5}{6}$

57. $6{,}285{,}736 - 8{,}943{,}628 + 4{,}821{,}946$

58. $7{,}843{,}621 + (-5{,}287{,}453) - (-2{,}893{,}186)$

Solve each of the following word problems using subtraction of signed numbers.

59. On Monday, the average temperature was $-8°$F. On Tuesday, the average temperature was 3°F. How much warmer was the average temperature on Tuesday than on Monday?

60. If a bank account is overdrawn \$37, how much money must be deposited to bring the balance up to \$179?

61. In 1975, a small business showed a loss of \$35,540. In 1976, it showed a profit of \$9,380. By how much did the earning power of the business increase from 1975 to 1976?

62. The highest point in the Sahara Desert has an elevation of 11,000 feet. The lowest point is 440 feet below sea level. What is the difference between the highest and lowest points in the Sahara Desert?

6.4 Multiplication of Signed Numbers

In order to determine how to multiply signed numbers, we will use the definition of multiplication as repeated addition. Consider the following examples.

$(3)(4)$ means the sum of three 4's: $4 + 4 + 4 = 12$

$(3)(-4)$ means the sum of three (-4)'s: $(-4) + (-4) + (-4) = -12$

$(-3)(4) = (4)(-3)$ means the sum of four (-3)'s:

$(-3) + (-3) + (-3) + (-3) = -12$

The procedure we have demonstrated is summarized in the box.

> To multiply a positive number and a negative number, multiply their absolute values and place a negative sign in front of the result.

example 1

Multiply each of the following numbers.

(a) $(2)(-5)$

$(2)(-5) = -10$

(b) $\left(-\frac{1}{3}\right)\left(\frac{5}{8}\right)$

$\left(-\frac{1}{3}\right)\left(\frac{5}{8}\right) = -\frac{5}{24}$ ●

In order to determine the product obtained when we multiply two negative numbers together, we will use the following example that illustrates the distributive property.

$$(-7)[3 + (-3)] = (-7)(3) + (-7)(-3)$$

The left-hand side simplifies to zero.

$$(-7)[3 + (-3)] = (-7)(0) = 0$$

Therefore, the right-hand side must also equal zero.

$$(-7)(3) + (-7)(-3) = 0$$

or $\quad -21 + \quad ? \quad = 0$

We can now see that the product $(-7)(-3)$ is equal to some number that when added to -21 produces a sum of zero. Since

$$-21 + 21 = 0,$$

this number is 21. Therefore,

$$(-7)(-3) = 21$$

We have now shown that the product of two negative numbers is a positive number. When multiplying two negative numbers together, therefore, we can ignore the signs of the factors (see box).

> To multiply two negative numbers, find the product of their absolute values.

example 2

Perform each of the following multiplications.

(a) $(-8)(-7)$

$(-8)(-7) = 56$

(b) $(-0.03)(-0.6)$

$(-0.03)(-0.6) = 0.018$

(c) $\left(-\frac{3}{5}\right)\left(-\frac{2}{7}\right)$

$\left(-\frac{3}{5}\right)\left(-\frac{2}{7}\right) = \frac{6}{35}$

Let us now investigate what happens when we multiply more than two signed numbers together.

example 3

Perform each of the following multiplications.

(a) $(-3)(7)(-2)(5)$

$$\begin{aligned}(-3)(7)(-2)(5) &= [(-3)(7)][(-2)(5)] \\ &= (-21)(-10) \\ &= 210\end{aligned}$$

Multiply two numbers at a time.

Notice that there were two negative factors, and the product was positive.

(b) $(2)(-6)(-3)(-4)$

$$\begin{aligned}(2)(-6)(-3)(-4) &= [(2)(-6)][(-3)(-4)] \\ &= (-12)(12) \\ &= -144\end{aligned}$$

Notice that there were three negative factors, and the product was negative.

(c) $(-2)(-4)(-10)(-6)$

$$\begin{aligned}(-2)(-4)(-10)(-6) &= [(-2)(-4)][(-10)(-6)] \\ &= (8)(60) \\ &= 480\end{aligned}$$

Notice that there were four negative factors, and the product was positive.

(d) $(-1)(-7)(-2)(-3)(-4)$

$$\begin{aligned}(-1)(-7)(-2)(-3)(-4) &= [(-1)(-7)][(-2)(-3)](-4) \\ &= (7)(6)(-4) \\ &= (42)(-4) \\ &= -168\end{aligned}$$

Notice that there were five negative factors, and the product was negative.

The problems in Example 3 illustrate the methods outlined in the box.

To Multiply Signed Numbers

1. Multiply their absolute values.
2. If there is an even number of negative factors, the product is positive.
3. If there is an odd number of negative factors, the product is negative.

example 4

Perform each of the following multiplications.

(a) $(-4)(5)(2)(-10)$

$(-4)(5)(2)(-10) = +[4 \cdot 5 \cdot 2 \cdot 10]$
$= 400$

First determine the sign of the answer. Since there are 2 negative factors, the product is positive. Multiply, ignoring the signs of the factors.

(b) $\left(-\frac{3}{5}\right)\left(\frac{7}{9}\right)\left(-\frac{5}{14}\right)\left(-\frac{3}{4}\right)$

$\left(-\frac{3}{5}\right)\left(\frac{7}{9}\right)\left(-\frac{5}{14}\right)\left(-\frac{3}{4}\right)$

$= -\left(\frac{3}{5} \cdot \frac{7}{9} \cdot \frac{5}{14} \cdot \frac{3}{4}\right)$

Since there are 3 negative factors, the product is negative.

$= -\frac{3 \cdot 7 \cdot 5 \cdot 3}{5 \cdot 9 \cdot 14 \cdot 4}$

Multiply, ignoring the signs of the factors.

$= -\frac{1}{8}$

(c) $(-0.03)(-20)(-0.1)(-7)$

$(-0.03)(-20)(-0.1)(-7)$
$= +[(0.03)(20)(0.1)(7)]$
$= (0.6)(0.7) = 0.42$

Since there are 4 negative factors, the product is positive. ●

By using the rules for multiplying signed numbers, we can determine what happens when a negative number is raised to a power.

$(-2)^2 = (-2)(-2) = 4$ Two negative factors. Product is positive.

$(-2)^3 = (-2)(-2)(-2) = -8$ Three negative factors. Product is negative.

$(-2)^4 = (-2)(-2)(-2)(-2) = 16$ Four negative factors. Product is positive.

$(-2)^5 = (-2)(-2)(-2)(-2)(-2) = -32$ Five negative factors. Product is negative.

Our observations are stated in the box.

> A negative number raised to an even power simplifies to a positive number. A negative number raised to an odd power simplifies to a negative number.

example 5

Simplify each of the following expressions.

(a) $(-3)^3$

$(-3)^3 = -(3 \cdot 3 \cdot 3)$
$= -27$

Since the power is odd, the result is negative.

(b) $(-10)^6$

$(-10)^6 = 1{,}000{,}000$

Since the power is even, the result is positive.

(c) $(-1)^{57}$

$(-1)^{57} = -1$ — Since the power is odd, the result is negative.

Let us now look at some examples that involve simplifying arithmetic expressions using the order of operations rules.

example 6

Simplify each of the following expressions.

(a) $-8(-7+3)$

$(-8)(-7+3) = (-8)(-4)$ — Simplify what is in ( ) first.

$= 32$ — Then do multiplication.

(b) $7(-3)^2 - (-2)$

$7(-3)^2 - (-2) = 7(9) - (-2)$ — Raise to powers first.

$= 63 - (-2)$ — Then do multiplication.

$= 63 + (+2)$ — Finally, do subtraction.

$= 65$

(c) $-2\{[-5-(-3)]^3 + (-4-6)^2\}$

$-2\{[-5-(-3)]^3 + (-4-6)^2\}$

$= -2\{[-5+(+3)]^3 + [-4+(-6)]^2\}$ — Rewrite subtraction as addition of opposite.

$= -2[(-2)^3 + (-10)^2]$ — Simplify within [ ].

$= -2(-8+100)$ — Raise to powers.

$= -2(92)$ — Simplify within ( ).

$= -184$ — Multiply.

QUICK QUIZ	ANSWERS
Perform each of the following multiplications.	
1. $(-3)(-9)$	**1.** 27
2. $(-5)(-2)(-7)$	**2.** -70
Perform each of the operations indicated.	
3. $(-3)^2 + (-2)^3$	**3.** 1
4. $-5[8+(-5)]$	**4.** -15

6.4 Exercises

Indicate which of the following statements are true and which are false. For those that are false, change the italic expression to make the statement true.

1. The product of two negative numbers is *always* positive.

2. When multiplying signed numbers, the sign of the product is determined by the number of *positive* factors.
3. A negative number raised to an odd power simplifies to a *negative* number.
4. The product of an even number of negative factors and an odd number of positive factors is *positive*.

Perform each of the following multiplications.

5. $(-3)(7)$
6. $(4)(-9)$
7. $(-6)(-5)$
8. $(-8)(-2)$
9. $(0.7)(-0.3)$
10. $(-0.5)(-0.9)$
11. $(8.2)(-3)$
12. $(-2)(6.1)$
13. $(-4.2)(0.4)$
14. $(-0.3)(-5.1)$
15. $(1.1)(-0.6)$
16. $(0.8)(-0.12)$
17. $\left(-\frac{1}{2}\right)\left(-\frac{3}{5}\right)$
18. $\left(\frac{2}{3}\right)\left(-\frac{2}{7}\right)$
19. $\left(-\frac{3}{8}\right)\left(-\frac{4}{15}\right)$
20. $\left(\frac{7}{9}\right)\left(-\frac{3}{14}\right)$
21. $\left(-1\frac{3}{8}\right)\left(-2\frac{2}{11}\right)$
22. $\left(5\frac{1}{2}\right)\left(-2\frac{2}{3}\right)$
23. $\left(-2\frac{2}{5}\right)\left(3\frac{3}{4}\right)$
24. $\left(-4\frac{4}{7}\right)\left(-2\frac{5}{8}\right)$
25. $(-5)(-1)(9)$
26. $(-8)(4)(2)$
27. $(7)(0)(-6)$
28. $(-5)(-4)(-7)$
29. $(-3)(-9)(-2)$
30. $(-1)(8)(6)$
31. $(-3.0)(-0.2)(-40)$
32. $(60)(-0.04)(-0.3)$
33. $(-0.08)(-500)(1.2)$
34. $(0.11)(-0.4)(-20)$
35. $\left(-\frac{5}{8}\right)\left(\frac{4}{9}\right)\left(-\frac{3}{10}\right)$
36. $\left(\frac{3}{4}\right)\left(-\frac{2}{15}\right)\left(\frac{8}{9}\right)$
37. $\left(-\frac{7}{8}\right)\left(\frac{5}{14}\right)\left(-\frac{4}{15}\right)$
38. $\left(-\frac{2}{9}\right)\left(-\frac{5}{16}\right)\left(-\frac{3}{10}\right)$
39. $(-5)(-6)(2)(-7)$
40. $(-7)(8)(0)(-4)$
41. $(-3)(5)(2)(-8)$
42. $(-4)(-6)(-2)(-10)$
43. $(-4)(-6)(-1)(-2)(-3)$
44. $(-5)(4)(-7)(-3)(-2)$
45. $(-35.47)(-2.169)$
46. $(438.7)(-72.45)$
47. $(-4.8)(37.6)(-813)$
48. $(-5.4)(-81.4)(-293)$

Simplify each of the following expressions.

49. $(-3)^4$
50. $(-6)^2$
51. $(-1)^{32}$
52. $(-5)^3$
53. $4[2+(-7)]$
54. $6(-3+8)$
55. $5(-9+2)$
56. $8[-4+(-5)]$
57. $(-3)(-10)^2$
58. $(-2)^5(6)$
59. $(-4)^2-(-3)(7)$
60. $(-8)(-5)+(-6)^2$
61. $(-7-3)(-2+5)$
62. $[5+(-2)][-4-(-3)]$
63. $4(-3)^2-(-7)^2(2)$
64. $6(-10)^3+(-7)(-10)^2$
65. $[7-(-2)]^2-[5+(-3)]^3$
66. $(-4-6)^3+[-3+(-4)]^2$
67. $2\{6-(-3)+[7+(-5)]^3\}$
68. $-5\{(-2+5)^3-[6-(-1)]\}$
69. $(-15)^3-(-23)(42)$
70. $(-51)(39)+(-27)^2$
71. $-843[-341-(-706)]$
72. $-675[947+(-253)]$

6.5 Division of Signed Numbers

Since multiplication and division are inverse operations, we can rewrite any division problem as a multiplication problem. For example,

$$30 \div 6 = (30)\left(\frac{1}{6}\right) = 5$$

$$-72 \div (-9) = (-72)\left(-\frac{1}{9}\right) = 8$$

$$48 \div (-8) = (48)\left(-\frac{1}{8}\right) = -6$$

$$-6 \div \frac{1}{4} = (-6)(4) = -24$$

Therefore, to determine the sign of a quotient, we can use the same rules we established for determining the sign of a product (see box).

To Divide Signed Numbers

1. Find the quotient of their absolute values.
2. If both numbers have the same sign, the quotient is positive.
3. If the numbers have opposite signs, the quotient is negative.

example 1

Find each of the following quotients.

(a) $-144 \div 12$

$-144 \div 12 = -12$

Since one number is negative and the other is positive, the quotient is negative.

(b) $-3.5 \div (-0.7)$

$-3.5 \div (-0.7) = 5$

Since both numbers are negative, the quotient is positive.

(c) $\frac{1}{3} \div \left(-\frac{5}{9}\right)$

$$\frac{1}{3} \div \left(-\frac{5}{9}\right) = -\left(\frac{1}{3} \cdot \frac{9}{5}\right)$$

$$= -\left(\frac{1 \cdot \overset{3}{\cancel{9}}}{\cancel{3} \cdot 5}\right) = -\frac{3}{5}$$

Since one number is positive and the other is negative, the quotient is negative.

A division problem can also be expressed as a fraction. For example,

$$\frac{-56}{8} = -56 \div 8 = -7$$

$$\frac{56}{-8} = 56 \div (-8) = -7$$

$$-\frac{56}{8} = -(56 \div 8) = -7$$

Notice that regardless of whether the negative sign appears in front of the numerator, the denominator, or the entire fraction, the fraction simplifies to the same negative number. If the numerator and denominator contain the product of signed numbers, we can use the procedure outlined to find the quotient.

To Simplify a Fraction That Expresses the Quotient of Products of Signed Numbers

1. First determine the sign of the answer. If there is an even number of negative factors in the numerator and denominator, the result is positive. If there is an odd number of negative factors, the result is negative.
2. Reduce to lowest terms, ignoring the signs of the individual factors.

example 2

Simplify $\dfrac{(-2)(8)(27)}{(6)(-9)(-32)}$.

solution

$$\frac{(-2)(8)(27)}{(6)(-9)(-32)} = -\left(\frac{\cancel{2}\cdot\cancel{8}^{\,\cancel{3}}\cdot\cancel{27}}{\cancel{6}_{\,\cancel{2}}\cdot\cancel{9}\cdot\cancel{32}_{\,4}}\right)$$

Since there are 3 negative factors, the result is negative.

$$= -\frac{1}{4}$$

example 3

Simplify $\dfrac{(-64)(-36)(-63)(12)}{(-16)(-21)(48)(-72)}$.

solution

$$\frac{(-64)(-36)(-63)(12)}{(-16)(-21)(48)(-72)}$$

$$= +\frac{\cancel{64}^{\,\cancel{4}}\cdot\cancel{36}\cdot\cancel{63}^{\,3}\cdot\cancel{12}}{\cancel{16}\cdot\cancel{21}\cdot\cancel{48}_{\,\cancel{4}}\cdot\cancel{72}_{\,2}}$$

Since there are 6 negative factors, the result is positive.

$$= \frac{3}{2}$$

Let us now look at some problems that require us to use the order of operations rules.

example 4

Simplify each of the following expressions.

(a) $12 \div (-4) + (-2)(-7)$

$$12 \div (-4) + (-2)(-7) = -3 + 14$$
$$= 11$$

Do multiplication and division first. Then do addition.

(b) $[8 + (-2)]^2 \div (-3)^2 - 5$

$[8 + (-2)]^2 \div (-3)^2 - 5$

$= 6^2 \div (-3)^2 - 5$ — Simplify what is in [ ] first.

$= 36 \div 9 - 5$ — Then raise to powers.

$= 4 - 5$ — Then do division.

$= 4 + (-5)$ — Finally, do subtraction.

$= -1$

(c) $[5 - (-7)] \div [6(-3) + (-2)(-7)]$

$[5 - (-7)] \div [6(-3) + (-2)(-7)]$ — Rewrite subtraction in [ ].

$= [5 + (+7)] \div [-18 + 14]$ — Do multiplication in [ ].

$= 12 \div (-4)$ — Do addition in [ ].

$= -3$ — Finally, do division. ●

Problems in which division is the last operation to be performed can also be expressed in fractional form. Thus, Example 4(c) can be written as follows:

$$[5 - (-7)] \div [6(-3) + (-2)(-7)] = \frac{5 - (-7)}{6(-3) + (-2)(-7)}$$

> To simplify fractions that involve more than one operation, first simplify the numerator and denominator separately using the order of operations rules, and then do the division.

example 5

Simplify $\dfrac{(-6)(7) + (-3)}{-1 - (2)(-5)}$.

solution

$$\frac{(-6)(7) + (-3)}{-1 - (2)(-5)} = \frac{-42 + (-3)}{-1 - (-10)}$$ First do multiplication.

$$= \frac{-45}{-1 + (+10)}$$ Then do addition and subtraction.

$$= \frac{-45}{9}$$

$$= -5$$ Finally, do division. ●

example 6

Simplify $\dfrac{(3 - 8) - (7)(-2)}{9 + (-2) + (-6 + 2)}$.

solution

$$\frac{(3-8)-(7)(-2)}{9+(-2)+(-6+2)}$$

$$=\frac{-5-(7)(-2)}{9+(-2)+(-4)}$$ Do operations in ( ) first.

$$=\frac{-5-(-14)}{9+(-2)+(-4)}$$ Do multiplication.

$$=\frac{-5+(+14)}{7+(-4)}$$ Do addition and subtraction.

$$=\frac{9}{3}$$

$$=3$$ Finally, do division.

QUICK QUIZ

Find each of the following quotients.

1. $-18 \div (-6)$
2. $-\frac{3}{5} \div \frac{3}{4}$
3. $\frac{(-6)(8)}{(12)(-2)}$
4. $\frac{4(-9+2)}{(-3)(-14)}$

ANSWERS

1. 3
2. $-\frac{4}{5}$
3. 2
4. $-\frac{2}{3}$

6.5 Exercises

Indicate which of the following statements are true and which are false. For those that are false, change the italic expression to make the statement true.

1. If the divisor and quotient are negative, the dividend is *negative.*
2. When dividing two numbers of opposite signs, the quotient is *positive.*
3. If the numerator and denominator of a fraction contain an *even* number of negative factors, the fraction simplifies to a positive number.
4. To simplify a fraction that involves more than one operation, simplify the numerator and denominator separately *before* doing the division.

Find each of the following quotients.

5. $-32 \div 4$ **6.** $27 \div (-3)$ **7.** $-72 \div (-9)$

8. $63 \div (-7)$ **9.** $400 \div (-80)$ **10.** $-490 \div 70$

11. $-144 \div (-12)$ **12.** $121 \div (-11)$ **13.** $-2.4 \div (-0.8)$

14. $-0.42 \div (-70)$ **15.** $-8.1 \div 90$ **16.** $6.3 \div (-90)$

17. $-0.056 \div (0.07)$

18. $-0.048 \div (-0.06)$

19. $-\frac{3}{8} \div \frac{9}{4}$

20. $\left(-\frac{5}{7}\right) \div \left(-\frac{10}{21}\right)$

21. $\frac{1}{12} \div \left(-\frac{5}{6}\right)$

22. $-\frac{8}{9} \div \frac{2}{3}$

23. $2\frac{2}{5} \div \left(-1\frac{3}{8}\right)$

24. $\left(-5\frac{1}{3}\right) \div \left(-2\frac{2}{9}\right)$

25. $\frac{(3)(-8)}{(-2)}$

26. $\frac{(-7)(-6)}{(-3)}$

27. $\frac{(-9)}{(-6)(4)}$

28. $\frac{(-2)}{(8)(3)}$

29. $\frac{(-4)(9)}{(-6)(-8)}$

30. $\frac{(5)(-7)}{(14)(-10)}$

31. $\frac{(-21)(-5)}{(45)(-7)}$

32. $\frac{(49)(-9)}{(7)(-72)}$

33. $\frac{(-25)(-64)}{(-30)(-16)}$

34. $\frac{(-42)(24)}{(36)(-56)}$

35. $\frac{(-9)(16)(-35)}{(20(-7)(-36)}$

36. $\frac{(54)(-21)(-8)}{(32)(-24)(14)}$

37. $\frac{(81)(-45)(8)}{(-18)(24)(-90)}$

38. $\frac{(-64)(4)(-27)}{(12)(-72)(-8)}$

39. $\frac{(-5)(-3)(-12)(-6)}{(2)(-4)(-15)(-7)}$

40. $\frac{(-10)(3)(-21)(9)}{(-18)(5)(-4)(7)}$

41 $\frac{(-5)(-18)(3)(-12)}{(36)(-15)(4)(-6)}$

42. $\frac{(81)(-7)(5)(-56)}{(-8)(45)(63)(-3)}$

43. $\frac{(-145)(31)(102)}{(17)(-29)(-186)}$

44. $\frac{(-91)(83)(-228)}{(-332)(-57)(13)}$

Simplify each of the following expressions.

45. $(3 - 7) \div (-4)$

46. $[4 + (-9)] \div 5$

47. $-72 \div (-5 - 3)$

48. $-42 \div [3 - (-4)]$

49. $(-6)^2 \div (-3)^2$

50. $(-4)^3 \div (-2)^4$

51. $-18 \div (-3) + (5)(-7)$

52. $-36 \div (-9) - (4)(-2)$

53. $[(6)(-2) - 3] \div 5$

54. $[(-7)(3) + (-6)] \div (-9)$

55. $[7 + (-2) - (-9)] \div (-2)$

56. $[8 - (-6) - (-2)] \div 4$

57. $\frac{(3)(-3) + (-5)}{9 - (-2)(6)}$

58. $\frac{(8)(-3) - (-4)}{(-3)(-4) + (-2)}$

59. $\frac{(-5 - 3)^2}{-13 - (-5)}$

60. $\frac{-9 - (-1)}{(5 - 7)^4}$

61. $\frac{[8 + (-2)]^2}{(-7)(-6)}$

62. $\frac{(-2)(9)}{(-9 + 6)^3}$

63. $\frac{(3 - 7) + (4)(-2)}{(-7)(2) + (-9 + 5)}$

64. $\frac{(-5)(-3) + (-5 - 1)}{(-2 + 6) - (-8)(3)}$

Summary and Review

Key Terms

[6.1] **Positive numbers** are numbers greater than zero, located to the right of zero on a number line.

Negative numbers are numbers less than zero, located to the left of zero on a number line.

The set of **integers** includes $\{\ldots, -3, -2, -1, 0, 1, 2, 3, \ldots\}$.

The **opposite** of a number is a number located on the other side of zero on a number line, the same distance away from zero as the original number. For any number a, the opposite of a is indicated by $-a$.

The **absolute value** of a number is the distance between that number and zero on a number line. For any number a, the absolute value of a is defined as follows:

$$|a| = \begin{cases} a & \text{if } a \geq 0 \\ -a & \text{if } a < 0 \end{cases}$$

The **additive inverse**, or **opposite**, of a number is the number you must add to it to obtain a sum of zero.

Calculations

[6.2] **To add two positive numbers**, add their absolute values.

To add two negative numbers, add their absolute values and place a negative sign in front of the result.

To add a positive number and a negative number, subtract the smaller absolute value from the larger absolute value. The sum will have the same sign as the addend with the larger absolute value.

[6.3] **To subtract a number from a given quantity**, add its opposite to the quantity.

[6.4] **To multiply signed numbers**, multiply their absolute values. If there is an even number of negative factors, the product is positive. If there is an odd number of negative factors, the product is negative.

A negative number raised to an even power simplifies to a positive result.

A negative number raised to an odd power simplifies to a negative result.

[6.5] **To divide signed numbers**, find the quotient of their absolute values. If both numbers have the same sign, the quotient is positive. If the numbers have opposite signs, the quotient is negative.

To simplify a fraction that expresses the quotient of products of signed numbers, first determine the sign of the answer. If there is an even number of negative factors in the numerator and denominator, the result is positive. If there is an odd number of negative factors, the result is negative. Then reduce to lowest terms, ignoring the signs of the factors.

To simplify fractions that involve more than one operation, first simplify the numerator and denominator separately using the order of operations rules. Then do the division.

Chapter 6 Review Exercises

[6.2–6.5] State which of the following problems are correct and which are incorrect. For those that are incorrect, replace the underlined answer with the correct one.

1. $9 + (-7) = \underline{-2}$

2. $-4 - (-5) = \underline{-1}$

3. $(-7)(-9) = \underline{-63}$

4. $54 \div (-6) = \underline{-9}$

5. $9(-2) - (-5) = \underline{-13}$

6. $-25 - (3)(-4) = \underline{-37}$

7. $(-5)^2 + (-3)^2 = \underline{34}$

8. $(-9)^2 - (-2)^4 = \underline{65}$

9. $-48 \div (-4)(-3) = \underline{4}$

10. $(-6)(4) \div (-8) = \underline{-3}$

[6.1] Use the $=$, $>$, or $<$ sign to indicate the relationship between each of the following pairs of numbers.

11. $-7 \ ? \ 1$

12. $3 \ ? \ -5$

13. $-3 \ ? \ -|3|$

14. $-7 \ ? \ |-7|$

15. $|4| \ ? \ |-9|$

16. $|-2| \ ? \ |-8|$

17. $4 \ ? \ -8$

18. $0 \ ? \ -2$

19. $|-2| \ ? \ 2$

20. $|-23| \; ? \; -|23|$
21. $-|15| \; ? \; |-5|$
22. $-|-6| \; ? \; |-(-6)|$
23. $-|9| \; ? \; -(-9)$
24. $-(-6) \; ? \; -6$

[6.2–6.5] Perform each of the operations indicated.

25. $8 + (-6)$
26. $-9 + (-8)$
27. $2 - 7$
28. $4 - (-5)$
29. $(-6)(-7)$
30. $(8)(-5)$
31. $24 \div (-8)$
32. $-63 \div (-7)$
33. $48 + (-53)$
34. $-61 + (-25)$
35. $17 - 86$
36. $32 - (-74)$
37. $(0.8)(-0.7)$
38. $(-0.05)(-0.6)$
39. $(-2.8) \div (0.07)$
40. $(-0.72) \div (-0.008)$
41. $\frac{1}{2} + \left(-\frac{1}{3}\right)$
42. $-\frac{5}{8} + \left(-\frac{1}{4}\right)$
43. $\frac{3}{5} - \frac{1}{6}$
44. $\frac{2}{9} - \left(-\frac{5}{12}\right)$
45. $\left(\frac{5}{8}\right)\left(-\frac{4}{15}\right)$
46. $\left(-\frac{7}{9}\right)\left(-\frac{6}{7}\right)$
47. $\left(-\frac{7}{8}\right) \div \left(-\frac{21}{4}\right)$
48. $\left(-\frac{5}{9}\right) \div \left(\frac{20}{27}\right)$
49. $-7 + (-3) - (-9)$
50. $-11 + 8 - (-5)$
51. $4 - 9 + (-5)$
52. $9 - (-4) + (-3)$
53. $(-3)(7)(-2)$
54. $(-8)(-3)(-4)$
55. $(-4)(-5)(-9)$
56. $(-2)(-6)(7)$
57. $\frac{(-8)(14)}{(-6)(7)}$
58. $\frac{(-9)(2)}{(3)(-4)}$
59. $\frac{(15)(-21)(6)}{(7)(-5)(-42)}$
60. $\frac{(-15)(49)(-6)}{(-56)(-5)(36)}$
61. $-3(-2 + 9)$
62. $5[3 + (-8)]$
63. $(-4)^2 - (2)(-5)$
64. $(7)(-6) + (-3)^3$
65. $-5(-3 - 1)^2$
66. $4[7 + (-2)]^2$
67. $(5)(-10)^3 + (3)(-10)^2$
68. $(-2)(-3)^3 - (8)(-1)^9$
69. $[5 - (-2)](3 - 7)$
70. $[8 + (-2)](-4 - 5)$
71. $63 \div (-5 - 2)$
72. $-48 \div [8 + (-2)]$
73. $\frac{(-4) + (6)(-2)}{(5)(-3) - (-3)}$
74. $\frac{(4)(-6) - (-3)}{7 - (-2 - 5)}$
75. $-356 + (437)(-192)$
76. $-853 - (-781) + (-349)$
77. $(-38)^3 - (-57)^4$
78. $[-493 + (-918)] \div (-17)$

7 ROOTS OF NUMBERS

7.1 Square Roots

In Chapter 1, we defined the square of a number as the product of that number multiplied by itself. When we square a number, we raise it to the second power. The exponent 2 indicates that the number being squared, or base, appears as a factor twice. For example,

$$7^2 = 7 \cdot 7 = 49$$
$$(-5)^2 = (-5)(-5) = 25$$
$$\left(\frac{1}{3}\right)^2 = \frac{1}{3} \cdot \frac{1}{3} = \frac{1}{9}$$

In this section, we will discuss the inverse of squaring a number, which is taking the **square root** of a number. We indicate the square root of a number by a radical sign, $\sqrt{}$

definition

The *square root* of a given number is a number whose square is equal to the given number. For any number *a*, the square root of *a* is indicated by $\sqrt{a}$.

For example,

$\sqrt{4}$ indicates the square root of 4.

$\sqrt{65}$ indicates the square root of 65.

The easiest square roots to determine are square roots of **perfect squares**.

definition

A *perfect square* is a number whose square root is a whole number.

Here are some examples of square roots of perfect squares.

$\sqrt{25} = 5$ since $5^2 = 5 \cdot 5 = 25$

$\sqrt{64} = 8$ since $8^2 = 8 \cdot 8 = 64$

$\sqrt{121} = 11$ since $11^2 = 11 \cdot 11 = 121$

If we reexamine closely our definition of square root, we realize that finding the square root of a number produces two results—a positive number and its opposite. For example, the square root of 16 is 4 since $4^2 = 16$. The square root of 16 is also -4 since $(-4)^2 = 16$. To indicate the positive square root, or **principal square root**, we use only the radical sign. To indicate the negative square root, we place a negative sign in front of the radical. To indicate both roots, we place a $\pm$ sign in front of the radical. Generally, we are only concerned with the principal square root of a number. If no negative sign appears in front of the radical, find only the positive square root. For example,

$$\sqrt{49} = 7$$
$$-\sqrt{49} = -7$$
$$\pm\sqrt{49} = \pm 7$$

example 1

Simplify each of the following square roots.

(a) $\sqrt{81}$

Find the positive square root of 81. — The radical is unsigned.

$\sqrt{81} = 9$ — Since $9^2 = 9 \cdot 9 = 81$.

(b) $-\sqrt{144}$

Find the negative square root of 144. — There is a $-$ sign in front of the radical.

$-\sqrt{144} = -12$ — Since $(-12)^2 = (-12)(-12) = 144$.

(c) $\pm\sqrt{36}$

Find both square roots of 36. — There is a $\pm$ sign in front of the radical.

$\pm\sqrt{36} = \pm 6$ — Since $6^2 = 36$ and $(-6)^2 = 36$. ●

The squares of all number we have worked with thus far are always positive, so we can find square roots of positive numbers. We cannot take square roots of negative numbers unless we introduce a new system of numbers called imaginary numbers. An **imaginary number** is the square root of a negative number. A detailed discussion of imaginary numbers is beyond the scope of this book, but you will learn more about them in more advanced mathematics courses.

To determine square roots of numbers that lie between two perfect squares, we can use a table that lists square roots accurate to a certain number of decimal places. Table 7.1 lists squares and square roots of the numbers 0 through 99. You should be able to recognize the squares of all numbers between 1 and 20. (In Section 7.3 we will learn the procedure for determining square roots of numbers.)

example 2

Use Table 7.1 to find each of the following square roots.

(a) $\sqrt{68}$

Find 68 in the column labeled "number." Look to the right to the column labeled "square root."

$\sqrt{68} = 8.246$ correct to the nearest thousandth.

(b) $\sqrt{1{,}849}$

Since the largest number in the column labeled "number" is 99, find 1,849 in the column labeled "square." Since the square root of 1,849 is the number whose square is 1,849, look to the left to the column labeled "number."

$\sqrt{1{,}849} = 43$ — Since $(43)^2 = 43 \cdot 43 = 1{,}849$.

table 7.1
Square Root Table

NUMBER n	SQUARE n^2	SQUARE ROOT $\sqrt{n}$	NUMBER n	SQUARE n^2	SQUARE ROOT $\sqrt{n}$
0	0	0.000	50	2,500	7.071
1	1	1.000	51	2,601	7.141
2	4	1.414	52	2,704	7.211
3	9	1.732	53	2,809	7.280
4	16	2.000	54	2,916	7.348
5	25	2.236	55	3,025	7.416
6	36	2.449	56	3,136	7.483
7	49	2.646	57	3,249	7.550
8	64	2.828	58	3,364	7.616
9	81	3.000	59	3,481	7.681
10	100	3.162	60	3,600	7.746
11	121	3.317	61	3,721	7.810
12	144	3.464	62	3,844	7.874
13	169	3.606	63	3,969	7.937
14	196	3.742	64	4,096	8.000
15	225	3.873	65	4,225	8.062
16	256	4.000	66	4,356	8.124
17	289	4.123	67	4,489	8.185
18	324	4.243	68	4,624	8.246
19	361	4.359	69	4,761	8.307
20	400	4.472	70	4,900	8.367
21	441	4.583	71	5,041	8.426
22	484	4.690	72	5,184	8.485
23	529	4.796	73	5,329	8.544
24	576	4.899	74	5,476	8.602
25	625	5.000	75	5,625	8.660
26	676	5.099	76	5,776	8.718
27	729	5.196	77	5,929	8.775
28	784	5.292	78	6,084	8.832
29	841	5.385	79	6,241	8.888
30	900	5.477	80	6,400	8.944
31	961	5.568	81	6,561	9.000
32	1,024	5.657	82	6,724	9.055
33	1,089	5.745	83	6,889	9.110
34	1,156	5.831	84	7,056	9.165
35	1,225	5.916	85	7,225	9.220
36	1,296	6.000	86	7,396	9.274
37	1,369	6.083	87	7,569	9.327
38	1,444	6.164	88	7,744	9.381
39	1,521	6.245	89	7,921	9.434
40	1,600	6.325	90	8,100	9.487
41	1,681	6.403	91	8,281	9.539
42	1,764	6.481	92	8,464	9.592
43	1,849	6.557	93	8,649	9.644
44	1,936	6.633	94	8,836	9.695
45	2,025	6.708	95	9,025	9.747
46	2,116	6.782	96	9,216	9.798
47	2,209	6.856	97	9,409	9.849
48	2,304	6.928	98	9,604	9.899
49	2,401	7.000	99	9,801	9.950

(c) $\sqrt{5{,}041}$

Find 5,041 in the column labeled "square." Look to the left to the column labeled "number."

$\sqrt{5{,}041} = 71$ Since $(71)^2 = 71 \cdot 71 = 5{,}041$.

Today, square roots are frequently computed with a pocket calculator. If your calculator has a $\sqrt{x}$ button, simply enter the number whose square root you wish to determine, and then press the $\sqrt{x}$ button.

QUICK QUIZ

Use Table 7.1 to find each of the following square roots.

1. $\sqrt{9}$
2. $\pm\sqrt{100}$
3. $-\sqrt{169}$
4. $\sqrt{7{,}921}$

ANSWERS

1. 3
2. ± 10
3. -13
4. 89

7.1 Exercises

Indicate which of the following statements are true and which are false. For those that are false, change the italic expression to make the statement true.

1. A number whose square root is a whole number is called a *complete square.*
2. If no sign appears in front of the radical, find only the *positive* square root of the number under the radical.
3. The principal square root of a number is its *negative* square root.
4. The square root of a negative number is *a negative* number.

Find each of the following square roots.

5. $\sqrt{1}$ 6. $\sqrt{0}$ 7. $\sqrt{49}$ 8. $\sqrt{16}$ 9. $-\sqrt{81}$
10. $-\sqrt{36}$ 11. $\pm\sqrt{64}$ 12. $\pm\sqrt{25}$ 13. $\sqrt{144}$ 14. $\sqrt{100}$
15. $-\sqrt{9}$ 16. $-\sqrt{121}$ 17. $\sqrt{324}$ 18. $\sqrt{256}$ 19. $\sqrt{225}$
20. $\sqrt{196}$ 21. $\sqrt{289}$ 22. $-\sqrt{289}$ 23. $\sqrt{169}$ 24. $\sqrt{400}$
25. $\sqrt{900}$ 26. $\sqrt{625}$

Use Table 7.1 to find each of the following square roots. Round off square roots of numbers that are not perfect squares to the nearest hundredth.

27. $\sqrt{2}$ 28. $\sqrt{3}$ 29. $\sqrt{5}$ 30. $\sqrt{7}$ 31. $\sqrt{15}$
32. $\sqrt{37}$ 33. $\sqrt{85}$ 34. $\sqrt{28}$ 35. $\sqrt{62}$ 36. $\sqrt{77}$
37. $\sqrt{63}$ 38. $\sqrt{88}$ 39. $\sqrt{576}$ 40. $\sqrt{961}$ 41. $\sqrt{729}$
42. $\sqrt{676}$ 43. $\sqrt{1{,}024}$ 44. $\sqrt{3{,}025}$ 45. $\sqrt{4{,}096}$ 46. $\sqrt{6{,}084}$

47. $\sqrt{5{,}041}$ **48.** $\sqrt{1{,}936}$ **49.** $\sqrt{3{,}844}$ **50.** $\sqrt{1{,}296}$ **51.** $\sqrt{9{,}025}$
52. $\sqrt{6{,}889}$

Use a pocket calculator to determine each of the following square roots. Round off each answer to the nearest hundredth.

53. $\sqrt{5.4}$ **54.** $\sqrt{7.7}$ **55.** $\sqrt{61.5}$ **56.** $\sqrt{86.3}$ **57.** $\sqrt{188}$
58. $\sqrt{524}$ **59.** $\sqrt{762}$ **60.** $\sqrt{393}$ **61.** $\sqrt{554}$ **62.** $\sqrt{821}$
63. $\sqrt{638.4}$ **64.** $\sqrt{997.2}$ **65.** $\sqrt{5{,}348}$ **66.** $\sqrt{6{,}792}$ **67.** $\sqrt{28{,}427}$
68. $\sqrt{39{,}615}$ **69.** $\sqrt{0.43}$ **70.** $\sqrt{0.97}$ **71.** $\sqrt{0.0318}$ **72.** $\sqrt{0.0689}$
73. $\sqrt{0.00272}$ **74.** $\sqrt{0.00546}$ **75.** $\sqrt{7{,}365{,}217}$ **76.** $\sqrt{4{,}028{,}641}$

7.2 Properties of Square Roots

Square roots possess two important properties that will enable us to determine the square roots of a wide variety of numbers that are multiples of perfect squares. Consider the following example.

$$\sqrt{4{,}900} = 70 \quad \text{since } 70 \cdot 70 = 4{,}900.$$

Notice that 4,900 can be expressed as the product of two perfect squares.

$$4{,}900 = 49 \cdot 100$$

The square root of 4,900, 70, is the product of the square roots of each of the perfect square factors.

$$\begin{aligned}\sqrt{4{,}900} &= \sqrt{49 \cdot 100}\\ &= \sqrt{49} \cdot \sqrt{100}\\ &= 7 \cdot 10\\ &= 70\end{aligned}$$

We have thus illustrated the property stated in the box.

For any positive numbers a and b,

$$\sqrt{a \cdot b} = \sqrt{a} \cdot \sqrt{b}$$

In other words, the square root of a product is the product of the square roots of the factors.

For example, $\sqrt{4 \cdot 9} = \sqrt{4} \cdot \sqrt{9}$

Let us now look at some examples of how this property can be used to calculate square roots of numbers.

example 1

Find each of the following square roots.

(a) $\sqrt{900}$

$$\begin{aligned}\sqrt{900} &= \sqrt{9 \cdot 100} \\ &= \sqrt{9} \cdot \sqrt{100} \\ &= 3 \cdot 10 \\ &= 30\end{aligned}$$

Express 900 as the product of two perfect squares.

Find the square root of each factor.

Find the product of the square roots.

(b) $\sqrt{14{,}400}$

$$\begin{aligned}\sqrt{14{,}400} &= \sqrt{144 \cdot 100} \\ &= \sqrt{144} \cdot \sqrt{100} \\ &= 12 \cdot 10 \\ &= 120\end{aligned}$$

Express 14,400 as the product of two perfect squares.

Find the square root of each factor.

Find the product of the square roots.

(c) $\sqrt{360{,}000}$

$$\begin{aligned}\sqrt{360{,}000} &= \sqrt{36 \cdot 10{,}000} \\ &= \sqrt{36}\sqrt{10{,}000} \\ &= 6 \cdot 100 = 600\end{aligned}$$

If no operation sign appears between two radicals, multiplication is implied.

We can apply this same property to express the square root of a number in simplest form.

> To express a square root in simplest form, we determine the square root of the largest perfect square that is a factor of the number under the radical sign. The other factor, which is not a perfect square, remains under the radical sign.

For example,

$$\begin{aligned}\sqrt{12} &= \sqrt{4 \cdot 3} \\ &= \sqrt{4}\sqrt{3} \\ &= 2\sqrt{3}\end{aligned}$$

Notice that even though the multiplication sign does not appear, it is understood to be there.

Thus, $2\sqrt{3}$ is $\sqrt{12}$ expressed in simplest form. We read $2\sqrt{3}$ as "two times the square root of three."

example 2

Write each of the following square roots in simplest form.

(a) $\sqrt{18}$

$$\begin{aligned}\sqrt{18} &= \sqrt{9 \cdot 2} \\ &= \sqrt{9}\sqrt{2} = 3\sqrt{2}\end{aligned}$$

9 is the largest perfect square that is a factor of 18.

(b) $\sqrt{32}$

$$\begin{aligned}\sqrt{32} &= \sqrt{16 \cdot 2} \\ &= \sqrt{16}\sqrt{2} = 4\sqrt{2}\end{aligned}$$

16 is the largest perfect square that is a factor of 32.

(c) $\sqrt{250}$

$$\sqrt{250} = \sqrt{25 \cdot 10}$$
$$= \sqrt{25}\sqrt{10}$$
$$= 5\sqrt{10}$$

25 is the largest perfect square that is a factor of 250.

We will now examine a second property of square roots that will enable us to determine square roots of fractions. Consider the following example.

$$\sqrt{\frac{4}{9}} = \frac{2}{3} \quad \text{since } \frac{2}{3} \cdot \frac{2}{3} = \frac{4}{9}$$

Notice that $\frac{4}{9}$ is the quotient of two perfect squares, 4 and 9. The square root of $\frac{4}{9}$, $\frac{2}{3}$, is the quotient of the square root of the numerator 4, and the square root of the denominator 9.

$$\sqrt{\frac{4}{9}} = \frac{\sqrt{4}}{\sqrt{9}} = \frac{2}{3}$$

We have thus illustrated the following property.

For any positive number a and b,

$$\sqrt{\frac{a}{b}} = \frac{\sqrt{a}}{\sqrt{b}}$$

In other words, the square root of a quotient is the quotient of the square roots of the numerator and the denominator.

For example,

$$\sqrt{\frac{9}{25}} = \frac{\sqrt{9}}{\sqrt{25}} = \frac{3}{5}$$

To determine the square roots of certain decimals, we can rewrite the decimal as a fraction, and then use this property to simplify the fraction. For example,

$$\sqrt{0.64} = \sqrt{\frac{64}{100}} = \frac{\sqrt{64}}{\sqrt{100}} = \frac{8}{10} = 0.8$$

example 3

Find each of the following square roots.

(a) $\sqrt{\frac{1}{16}}$

$$\sqrt{\frac{1}{16}} = \frac{\sqrt{1}}{\sqrt{16}} = \frac{1}{4}$$

Find the square root of the numerator and denominator and then find the quotient.

(b) $\sqrt{\dfrac{4}{49}}$

$$\sqrt{\frac{4}{49}} = \frac{\sqrt{4}}{\sqrt{49}} = \frac{2}{7}$$

(c) $\sqrt{\dfrac{36}{169}}$

$$\sqrt{\frac{36}{169}} = \frac{\sqrt{36}}{\sqrt{169}} = \frac{6}{13}$$

(d) $\sqrt{0.04}$

$$\sqrt{0.04} = \sqrt{\frac{4}{100}} = \frac{\sqrt{4}}{\sqrt{100}} = \frac{2}{10} = 0.2$$

First, express the decimal as a fraction.
Express the answer as a decimal.

(e) $\sqrt{0.0081}$

$$\sqrt{0.0081} = \sqrt{\frac{81}{10{,}000}} = \frac{\sqrt{81}}{\sqrt{10{,}000}} = \frac{9}{100} = 0.09$$

(f) $\sqrt{1.44}$

$$\sqrt{1.44} = \sqrt{\frac{144}{100}} = \frac{\sqrt{144}}{\sqrt{100}} = \frac{12}{10} = 1.2$$

QUICK QUIZ

Simplify each of the following square roots.

1. $\sqrt{6{,}400}$
2. $\sqrt{75}$
3. $\sqrt{\dfrac{25}{81}}$
4. $\sqrt{0.36}$

ANSWERS

1. 80
2. $5\sqrt{3}$
3. $\dfrac{5}{9}$
4. 0.6

7.2 Exercises

Indicate which of the following statements are true and which are false. For those that are false, change the italic word to make the statement true.

1. The square root of a product is the *sum* of the square roots of each factor.
2. The square root of a fraction is equivalent to the square root of the numerator *multiplied* by the square root of the denominator.
3. When a square root is expressed in *simplest* form, the number under the radical does not have any factors that are perfect squares.
4. The number $5\sqrt{3}$ is read as "five *plus* the square root of three."

Use the properties of square roots to find each of the following square roots.

5. $\sqrt{400}$ 6. $\sqrt{1{,}600}$ 7. $\sqrt{2{,}500}$ 8. $\sqrt{3{,}600}$ 9. $\sqrt{8{,}100}$

10. $\sqrt{10{,}000}$ 11. $\sqrt{12{,}100}$ 12. $\sqrt{16{,}900}$ 13. $\sqrt{640{,}000}$ 14. $\sqrt{490{,}000}$
15. $\sqrt{\frac{1}{25}}$ 16. $\sqrt{\frac{1}{49}}$ 17. $\sqrt{\frac{4}{81}}$ 18. $\sqrt{\frac{9}{64}}$ 19. $\sqrt{\frac{36}{121}}$
20. $\sqrt{\frac{49}{144}}$ 21. $\sqrt{\frac{64}{169}}$ 22. $\sqrt{\frac{16}{225}}$ 23. $\sqrt{0.25}$ 24. $\sqrt{0.81}$
25. $\sqrt{0.49}$ 26. $\sqrt{0.16}$ 27. $\sqrt{1.21}$ 28. $\sqrt{2.25}$ 29. $\sqrt{0.09}$
30. $\sqrt{0.01}$ 31. $\sqrt{0.0064}$ 32. $\sqrt{0.0036}$ 33. $\sqrt{0.0009}$ 34. $\sqrt{0.0001}$

Write each of the following square roots in simplest form.

35. $\sqrt{48}$ 36. $\sqrt{27}$ 37. $\sqrt{20}$ 38. $\sqrt{45}$ 39. $\sqrt{50}$ 40. $\sqrt{72}$
41. $\sqrt{80}$ 42. $\sqrt{63}$ 43. $\sqrt{98}$ 44. $\sqrt{28}$ 45. $\sqrt{128}$ 46. $\sqrt{108}$
47. $\sqrt{360}$ 48. $\sqrt{640}$ 49. $\sqrt{288}$ 50. $\sqrt{242}$

7.3 Procedure for Calculating Square Roots (Optional)

Thus far, we have learned how to calculate square roots of numbers that express a product or quotient of perfect squares. In this section, we will learn an arithmetic procedure that can be used to calculate the square root of a number that is not a perfect square to a desired degree of accuracy. This procedure is outlined in the box. The numbered steps in the box correspond to the numbered steps in Examples 1 and 2.

To Calculate a Square Root

1. Block off every group of two digits to the left and to the right of the decimal point. If there is an odd number of digits to the left of the decimal point, the leftmost block will contain only one digit.
2. Since each block will correspond to one digit in the calculated square root, add an appropriate number of digits to the right of the decimal point so that the number of blocks to the right of the decimal point is one more than the number of decimal places needed for the approximation for the square root. Put a decimal point directly above the decimal point under the radical sign.
3. Find the largest number whose square is less than or equal to the number in the leftmost block. Write it above the block and subtract it from the number in the block.
4. Bring down the next block of two digits.
5. Double the number above the radical, write it to the left of the difference, and attach a blank box to the right of it.
6. Find the largest number to put in the box so that the product of the number with the attached box and the number in the box is less than or equal to the difference on the right.
7. Subtract that product from the difference on the right.
8. Repeat steps 4–7 until a number appears above every block of two digits.
9. Round off the result to obtain the desired square root.

example 1 Calculate $\sqrt{51.7}$ correct to the nearest tenth.

solution

$\sqrt{51.70\,00}$

1. Block off groups of two digits.
2. Attach 3 zeros to the right of the decimal point. Put a decimal point above the decimal point under the radical sign.

$$\begin{array}{r} 7. \\ \sqrt{51.70\,00} \\ 49 \\ \hline 2 \end{array}$$

3. 49 is the largest perfect square ≤ 51. Put 7 above the first block. Subtract $51 - 49 = 2$.

$$\begin{array}{rr} & 7. \\ & \sqrt{51.70\,00} \\ & 49 \\ 14\square & 2\,70 \end{array}$$

4. Bring down the next block: 70.
5. Double the number above the radical: $7 \times 2 = 14$. Write 14 to the left of 270 and attach a box to its right.

$$\begin{array}{rr} & 7.1 \\ & \sqrt{51.70\,00} \\ & 49 \\ 14\boxed{1} & 2\,70 \\ & 1\,41 \\ \hline & 1\,29 \end{array}$$

6. $\begin{array}{r} 14\boxed{2} \\ \times\ \boxed{2} \\ \hline 28\,4 > 270 \end{array}$ $\qquad$ $\begin{array}{r} 14\boxed{1} \\ \times\ \boxed{1} \\ \hline 14\,1 < 270 \end{array}$

   Write 1 in the box and above the radical.
7. Subtract $270 - 141 = 129$.

$$\begin{array}{rr} & 7.\,1\ \ 9 \\ & \sqrt{51.70\,00} \\ & 49 \\ 14\boxed{1} & 2\,70 \\ & 1\,41 \\ 142\boxed{9} & 1\,29\,00 \\ & 1\,28\,61 \\ \hline & 39 \end{array}$$

8. Bring down the next block: 00. Double the number above the radical: $71 \times 2 = 142$. Write 142 to the left of 12900 and attach a box to its right.

   $\begin{array}{r} 142\boxed{9} \\ \times\ \boxed{9} \\ \hline 1286\,1 < 12900 \end{array}$

   Write 9 in box and above the radical.
9. Round off the result.

$7.19 \approx 7.2$

Thus $\sqrt{5.17} \approx 7.2$.

CHECK

$$\begin{array}{r} 7.2 \\ \times\ \ 7.2 \\ \hline 1\,4\,4 \\ 50\,4 \\ \hline 51.8\,4 \checkmark \end{array}$$

To check the accuracy of the answer we can square it.

Since $51.84 \approx 51.7$, then 7.2 is a reasonable approximation for $\sqrt{51.7}$. ●

example 2 Calculate $\sqrt{250}$ correct to the nearest tenth.

solution

$\sqrt{2\,50.00\,00}$

1. Block off groups of two digits.
2. Attach 2 blocks of zeros to the right of the decimal point. Put a decimal point above the decimal point under the radical sign.

$$\begin{array}{l} \quad 1\ . \\ \sqrt{\overset{\frown}{2}\overset{\frown}{50}.\overset{\frown}{00}\,\overset{\frown}{00}} \\ \quad 1 \\ \hline \quad 1 \end{array}$$

3. 1 is the largest perfect square ≤ 2. Write 1 above the number 2 under the radical. Subtract $2 - 1 = 1$.

$$\begin{array}{r|l} & \quad 1\ . \\ & \sqrt{\overset{\frown}{2}\overset{\frown}{50}.\overset{\frown}{00}\,\overset{\frown}{00}} \\ & \quad 1 \\ 2\,\square & \quad 150 \end{array}$$

4. Bring down the next block: 50.

5. Double the number above the radical: $1 \times 2 = 2$. Write 2 to the left of 1 and attach a box to its right.

$$\begin{array}{r|l} & \quad 1\,5\ . \\ & \sqrt{\overset{\frown}{2}\overset{\frown}{50}.\overset{\frown}{00}\,\overset{\frown}{00}} \\ & \quad 1 \\ 2\,\boxed{5} & \quad 150 \\ & \quad 125 \\ \hline & \quad\ \ 25 \end{array}$$

6. $\begin{array}{r} 2\boxed{5} \\ \times \quad \boxed{5} \\ \hline 12\,5 < 150 \end{array}$ $\qquad$ $\begin{array}{r} 2\boxed{6} \\ \times \quad \boxed{6} \\ \hline 15\,6 > 150 \end{array}$

Write 5 in the box and above the radical.

7. Subtract $150 - 125 = 25$.

$$\begin{array}{r|l} & \quad 1\,5\,.\,8 \\ & \sqrt{\overset{\frown}{2}\overset{\frown}{50}.\overset{\frown}{00}\,\overset{\frown}{00}} \\ & \quad 1 \\ 2\,\boxed{5} & \quad 150 \\ & \quad 125 \\ 30\,\boxed{8} & \quad\ \ 25\,00 \\ & \quad\ \ 24\,64 \\ \hline & \qquad\quad 36 \end{array}$$

8. Bring down the next block: 00. Double the number above the radical: $15 \times 2 = 30$. Write 30 to the left of 2500 and attach a box to its right.

$$\begin{array}{r} 30\boxed{8} \\ \times \quad \boxed{8} \\ \hline 246\,4 < 2500 \end{array}$$

Write 8 in the box and above the radical. Subtract $2500 - 2464 = 36$.

$$\begin{array}{r|l} & \quad 1\,5\,.\,8\ \ 1 \\ & \sqrt{\overset{\frown}{2}\overset{\frown}{50}.\overset{\frown}{00}\,\overset{\frown}{00}} \\ & \quad 1 \\ 2\,\boxed{5} & \quad 150 \\ & \quad 125 \\ 30\,\boxed{8} & \quad\ \ 25\,00 \\ & \quad\ \ 24\,64 \\ 316\,\boxed{1} & \qquad\quad 36\,00 \\ & \qquad\quad 31\,61 \\ \hline & \qquad\quad\ \ 4\,39 \end{array}$$

Bring down next block: 00. Double the number above the radical: $158 \times 2 = 316$. Write 316 to the left of 3600 and attach a box to its right.

$$\begin{array}{r} 316\boxed{1} \\ \times \quad \boxed{1} \\ \hline 316\,1 < 3600 \end{array}$$

Write 1 in the box and above the radical.

$15.81 \approx 15.8$

9. Round off the result.

Thus, $\sqrt{250} \approx 15.8$.

CHECK Square 15.8.

$$\begin{array}{r} 15.8 \\ \times \quad 15.8 \\ \hline 12\,6\,4 \\ 79\,0 \\ 158 \\ \hline 249.6\,4 \end{array} \checkmark$$

Since $249.64 \approx 250$, then 15.8 is a reasonable approximation for $\sqrt{250}$. ●

This procedure is very easy to use for determining the square root of perfect squares, as Example 3 illustrates.

example 3

solution

Calculate $\sqrt{3{,}240{,}000}$.

$$\begin{array}{rr|l} & & \;\;1\;\;8\;\;0\;\;0. \\ & & \sqrt{3\;24\;00\;00.} \\ & & \;\;1 \\ 2 & \boxed{8} & \;\;2\;24 \\ & & \;\;2\;24 \\ & & \;\;\;\;0 \end{array}$$

$$\begin{array}{r} 2\boxed{8} \\ \times\;\;\boxed{8} \\ \hline 22\,4 = 224 \end{array}$$

Thus, $\sqrt{3{,}240{,}000} = 1{,}800$.

Notice that the 2 zeros are necessary placeholders in the answer.

CHECK

$$\begin{array}{r} 1800 \\ \times\;\;1800 \\ \hline 1440000 \\ 18\;\;\;\;\;\;\; \\ \hline 3240000\;\checkmark \end{array}$$

example 4

solution

Calculate $\sqrt{116.2084}$.

$$\begin{array}{rr|l} & & \;\;1\;0.\;\;7\;\;8 \\ & & \sqrt{1\;16.20\;84} \\ & & \;\;1 \\ 2 & \boxed{0} & \;\;16 \\ & & \;\;\;0 \\ 20 & \boxed{7} & \;\;16\;20 \\ & & \;\;14\;49 \\ 214 & \boxed{8} & \;\;1\;71\;84 \\ & & \;\;1\;71\;84 \\ & & \;\;\;\;\;\;\;\;0 \end{array}$$

$$\begin{array}{r} 20\boxed{8} \\ \times\;\;\boxed{8} \\ \hline 166\,4 > 1620 \end{array} \qquad \begin{array}{r} 20\boxed{7} \\ \times\;\;\boxed{7} \\ \hline 144\,9 < 1620 \end{array}$$

$$\begin{array}{r} 214\boxed{8} \\ \times\;\;\boxed{8} \\ \hline 1718\,4 = 17184 \end{array}$$

Thus, $\sqrt{116.2084} = 10.78$.

example 5

solution

Calculate $\sqrt{0.734}$ correct to the nearest thousandth.

$$\begin{array}{rr|l} & & \;\;0.\;\;8\;\;5\;\;6\;\;7 \\ & & \sqrt{0.73\;40\;00\;00} \\ & & \;\;64 \\ 16 & \boxed{5} & \;\;9\;40 \\ & & \;\;8\;25 \\ 170 & \boxed{6} & \;\;1\;15\;00 \\ & & \;\;1\;02\;36 \\ 1712 & \boxed{7} & \;\;12\;64\;00 \\ & & \;\;11\;98\;89 \\ & & \;\;\;\;\;\;65\;11 \end{array}$$

$$\begin{array}{r} 16\boxed{6} \\ \times\;\;\boxed{6} \\ \hline 99\,6 > 940 \end{array} \qquad \begin{array}{r} 16\boxed{5} \\ \times\;\;\boxed{5} \\ \hline 82\,5 < 940 \end{array}$$

$$\begin{array}{r} 170\boxed{7} \\ \times\;\;\boxed{7} \\ \hline 1194\,9 > 11500 \end{array} \qquad \begin{array}{r} 170\boxed{6} \\ \times\;\;\boxed{6} \\ \hline 1023\,6 < 11500 \end{array}$$

$$\begin{array}{r} 1712\boxed{7} \\ \times\;\;\boxed{7} \\ \hline 11988\,9 < 126400 \end{array}$$

$0.8567 \approx 0.857$

Thus, $\sqrt{0.734} \approx 0.857$.

QUICK QUIZ

1. Calculate $\sqrt{34.81}$.
2. Calculate $\sqrt{752}$ correct to the nearest tenth.

ANSWERS

1. 5.9
2. 27.4

7.3 Exercises

Calculate each of the following square roots. Check your answers by squaring.

1. $\sqrt{27.04}$	**2.** $\sqrt{90.25}$	**3.** $\sqrt{44.89}$	**4.** $\sqrt{75.69}$	**5.** $\sqrt{985.96}$
6. $\sqrt{561.69}$	**7.** $\sqrt{1{,}303.21}$	**8.** $\sqrt{4{,}678.56}$	**9.** $\sqrt{127{,}449}$	**10.** $\sqrt{170{,}569}$
11. $\sqrt{630{,}436}$	**12.** $\sqrt{833{,}569}$	**13.** $\sqrt{270{,}400}$	**14.** $\sqrt{136{,}900}$	**15.** $\sqrt{0.3844}$
16. $\sqrt{0.961}$	**17.** $\sqrt{0.0841}$	**18.** $\sqrt{0.9216}$	**19.** $\sqrt{0.481636}$	**20.** $\sqrt{0.714025}$
21. $\sqrt{0.004624}$	**22.** $\sqrt{0.008281}$	**23.** $\sqrt{1{,}440{,}000}$	**24.** $\sqrt{7{,}290{,}000}$	

Calculate each of the following square roots. Round off as indicated.

25. $\sqrt{5.7}$ nearest tenth
26. $\sqrt{3.2}$ nearest hundredth
27. $\sqrt{17.4}$ nearest tenth
28. $\sqrt{82.5}$ nearest tenth
29. $\sqrt{67.9}$ nearest hundredth
30. $\sqrt{43.6}$ nearest hundredth
31. $\sqrt{923}$ nearest tenth
32. $\sqrt{546}$ nearest tenth
33. $\sqrt{372.4}$ nearest hundredth
34. $\sqrt{789.1}$ nearest hundredth
35. $\sqrt{4{,}317}$ nearest tenth
36. $\sqrt{6{,}749}$ nearest tenth
37. $\sqrt{58{,}402}$ nearest whole number
38. $\sqrt{90{,}357}$ nearest whole number
39. $\sqrt{0.73}$ nearest hundredth
40. $\sqrt{0.45}$ nearest thousandth
41. $\sqrt{0.273}$ nearest hundredth
42. $\sqrt{0.458}$ nearest thousandth
43. $\sqrt{0.013}$ nearest thousandth
44. $\sqrt{0.059}$ nearest thousandth
45. $\sqrt{0.0462}$ nearest hundredth
46. $\sqrt{0.0898}$ nearest hundredth
47. $\sqrt{0.005274}$ nearest hundredth
48. $\sqrt{0.007329}$ nearest thousandth
49. $\sqrt{6{,}741{,}308}$ nearest whole number
50. $\sqrt{2{,}316{,}452}$ nearest whole number

7.4 Higher Roots of Numbers

Just as we can raise any whole number to an exponent that is a positive integer, we can also take the nth root of a number, where n is a positive integer.

definition

The nth root of a number is some number whose nth power is equal to the given number. For any number a, the nth root of a is indicated by $\sqrt[n]{a}$.

For example, $\sqrt[3]{8}$ indicates the cube root of 8. It is equivalent to some number that yields a product of 8 when cubed. Thus,

$$\sqrt[3]{8} = 2 \quad \text{since } 2^3 = 2 \cdot 2 \cdot 2 = 8.$$

In this section, we will deal only with higher roots of numbers that can be calculated on sight or by using a table. Table 7.2, which lists the values obtained when the numbers 1 through 10 are raised to the 2nd, 3rd, 4th, and 5th powers, can be used to calculate the square, cube, 4th, and 5th roots of numbers in the corresponding columns.

table 7.2
Powers of Numbers

n	n^2	n^3	n^4	n^5
1	1	1	1	1
2	4	8	16	32
3	9	27	81	243
4	16	64	256	1,024
5	25	125	625	3,125
6	36	216	1,296	7,776
7	49	343	2,401	16,807
8	64	512	4,096	32,768
9	81	729	6,561	59,049
10	100	1,000	10,000	100,000

example 1

Use Table 7.2 to find each of the following roots.

(a) $\sqrt[4]{81}$

Look down the column labeled n^4 until you find 81.

$\sqrt[4]{81} = 3$ — Since $3^4 = 81$.

(b) $\sqrt[3]{216}$

Look down the column labeled n^3 until you find 216.

$\sqrt[3]{216} = 6$ — Since $6^3 = 216$.

(c) $\sqrt[5]{100{,}000}$

$\sqrt[5]{100{,}000} = 10$ — Since $10^5 = 100{,}000$. ●

The same properties that apply to square roots of numbers also apply to higher roots of numbers. We restate these basic properties in the boxes.

> For any positive numbers a and b and positive integer n,
>
> $$\sqrt[n]{a \cdot b} = \sqrt[n]{a}\,\sqrt[n]{b}$$

For example,

$$\sqrt[3]{8 \cdot 5} = \sqrt[3]{8}\,\sqrt[3]{5} = 2\sqrt[3]{5}$$

For any positive numbers a and b and positive integer n,

$$\sqrt[n]{\frac{a}{b}} = \frac{\sqrt[n]{a}}{\sqrt[n]{b}}$$

For example,

$$\sqrt[4]{\frac{16}{81}} = \frac{\sqrt[4]{16}}{\sqrt[4]{81}} = \frac{2}{3}$$

example 2

Simplify each of the following radical expressions. (Use Table 7.2.)

(a) $\sqrt[3]{\frac{1}{27}}$

$$\sqrt[3]{\frac{1}{27}} = \frac{\sqrt[3]{1}}{\sqrt[3]{27}} = \frac{1}{3}$$

Find the cube roots of the numerator and denominator.

(b) $\sqrt[5]{96}$

$$\sqrt[5]{96} = \sqrt[5]{32 \cdot 3} = \sqrt[5]{32}\sqrt[5]{3} = 2\sqrt[5]{3}$$

Find the 5th root of each factor. Simplify.

(c) $\sqrt[4]{0.0016}$

$$\sqrt[4]{0.0016} = \sqrt[4]{\frac{16}{10{,}000}} = \frac{\sqrt[4]{16}}{\sqrt[4]{10{,}000}} = \frac{2}{10} = 0.2$$

Rewrite decimal as a fraction. Find 4th roots of numerator and denominator. Express answer as decimal.

(d) $\sqrt[3]{54}$

$$\sqrt[3]{54} = \sqrt[3]{27 \cdot 2} = \sqrt[3]{27}\sqrt[3]{2} = 3\sqrt[3]{2}$$

Find the cube root of each factor. Simplify. ●

Recall that a negative number raised to an even power produces a positive result. In this book, we have calculated even roots of only positive numbers, because the calculation of even roots of negative numbers involves working with imaginary numbers. However, a negative number raised to an odd power produces a negative result. Therefore, a negative number has an odd root that is also a negative number. For example,

$$\sqrt[3]{-8} = -2 \quad \text{since } (-2)^3 = -8$$

example 3

Find each of the following roots.

(a) $\sqrt[5]{-1}$

$$\sqrt[5]{-1} = -1$$

Since $(-1)^5 = -1$.

(b) $\sqrt[3]{-125}$

$$\sqrt[3]{-125} = -5$$

Since $(-5)^3 = -125$. ●

QUICK QUIZ

Simplify each of the following radical expressions.

1. $\sqrt[4]{16}$
2. $\sqrt[3]{\dfrac{1}{64}}$
3. $\sqrt[5]{-10{,}000}$

ANSWERS

1. 2
2. $\dfrac{1}{4}$
3. -10

7.4 Exercises

Use Table 7.2 to find each of the following roots.

1. $\sqrt[3]{27}$ **2.** $\sqrt[5]{32}$ **3.** $\sqrt[4]{16}$ **4.** $\sqrt[3]{125}$ **5.** $\sqrt[4]{256}$

6. $\sqrt[3]{343}$ **7.** $\sqrt[4]{625}$ **8.** $\sqrt[5]{243}$ **9.** $\sqrt[4]{10{,}000}$ **10.** $\sqrt[5]{3{,}125}$

11. $\sqrt[4]{\dfrac{1}{16}}$ **12.** $\sqrt[5]{\dfrac{1}{32}}$ **13.** $\sqrt[3]{\dfrac{1}{1{,}000}}$ **14.** $\sqrt[3]{\dfrac{27}{125}}$ **15.** $\sqrt[3]{0.008}$

16. $\sqrt[4]{0.0081}$ **17.** $\sqrt[5]{0.01024}$ **18.** $\sqrt[3]{0.125}$ **19.** $\sqrt[5]{-32}$ **20.** $\sqrt[3]{-8}$

21. $\sqrt[3]{-512}$ **22.** $\sqrt[5]{-1{,}024}$ **23.** $\sqrt[3]{-\dfrac{1}{64}}$ **24.** $\sqrt[3]{-\dfrac{8}{27}}$

Write each of the following in simplest form.

25. $\sqrt[3]{16}$ **26.** $\sqrt[4]{80}$ **27.** $\sqrt[3]{24}$ **28.** $\sqrt[5]{64}$ **29.** $\sqrt[3]{250}$

30. $\sqrt[4]{162}$ **31.** $\sqrt[3]{270}$ **32.** $\sqrt[3]{128}$ **33.** $\sqrt[4]{20{,}000}$ **34.** $\sqrt[3]{3{,}000}$

35. $\sqrt[5]{500{,}000}$ **36.** $\sqrt[5]{486}$

7.5 Number Systems Used in Mathematics

Throughout our discussions of various arithmetic operations, we have encountered many different kinds of numbers that are all included in the set of **real numbers**. Each of the number systems we have used thus far are identified in the box on the following page.

Since all of the number systems we described are part of the real number system, they are sometimes called subsets of the real numbers. Set A is a *subset* of set B if every element of set A is also an element of set B. For example, since every whole number is included in the set of rational numbers, the whole numbers are a subset of the rational numbers. However, the rational numbers are not a subset of the whole numbers, because there are many rational numbers that are not whole numbers. Figure 7.1 illustrates which numbers are included in each of the subsets of the real number system.

Number Systems

1. The set of **natural numbers**, or **counting numbers**, consists of $\{1, 2, 3, 4, 5, 6, \ldots\}$. (Remember that the symbol ... indicates that there is an infinite number of natural numbers.)
2. The set of **whole numbers** consists of $\{0, 1, 2, 3, 4, \ldots\}$. The whole numbers include all the counting numbers in addition to zero.
3. The set of **integers** consists of $\{\ldots, -4, -3, -2, -1, 0, 1, 2, 3, 4, \ldots\}$. The integers include all the whole numbers and their opposites. Note that there is an infinite number of negative and positive integers.
4. The set of **rational numbers** consists of all numbers that can be written in the form $\frac{a}{b}$, where a and b are integers and $b \neq 0$. For example, $\frac{3}{5}$, $-\frac{9}{8}$, 7, and $-4\frac{1}{2}$ are all rational numbers. All terminating and repeating decimals are included in the set of rational numbers. For example, 0.06, -3.72, and $0.3\bar{3}$ are rational numbers since they can be rewritten as $\frac{6}{100}$, $-\frac{3{,}720}{1{,}000}$, and $\frac{1}{3}$, respectively.
5. The set of **irrational numbers** consists of all numbers that can be expressed as a nonterminating, nonrepeating decimal, but cannot be expressed in the form of a rational number. For example $\sqrt{2} = 1.414213562\ldots$ is an irrational number since it cannot be expressed in the form $\frac{a}{b}$, where a and b are integers and $b \neq 0$.
6. The set of **real numbers** consists of all rational and irrational numbers.

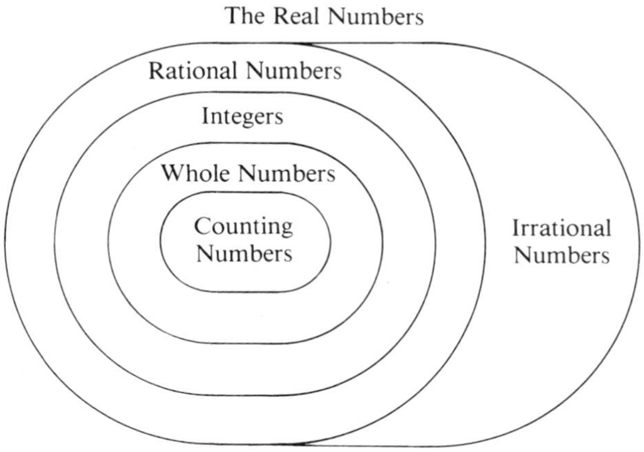

figure 7.1

example 1

List all of the number systems in which each of the following numbers is included.

(a) -7

$$-7 = -\frac{7}{1}$$

-7 is an integer, a rational number, and a real number.

(b) 0.28

$$0.28 = \frac{28}{100}$$

0.28 is a rational number and a real number.

(c) $\sqrt{36}$

$$\sqrt{36} = 6 = \frac{6}{1}$$

$\sqrt{36}$ is a counting number, a whole number, an integer, a rational number, and a real number.

(d) 0.454545...

0.454545... is a repeating decimal.

$0.4\overline{545}$ is a rational number and a real number.

(e) $\sqrt{3}$

$\sqrt{3} = 1.73205\ldots$ is a nonrepeating, nonterminating decimal.

$\sqrt{3}$ is an irrational number and a real number.

All real numbers can be represented on a number line. Since there is an infinite number of positive and negative real numbers, the number line extends infinitely to the right and to the left. There is also an infinite number of real numbers located between any two real numbers. Figure 7.2 shows the location of a few real numbers on a number line.

figure 7.2

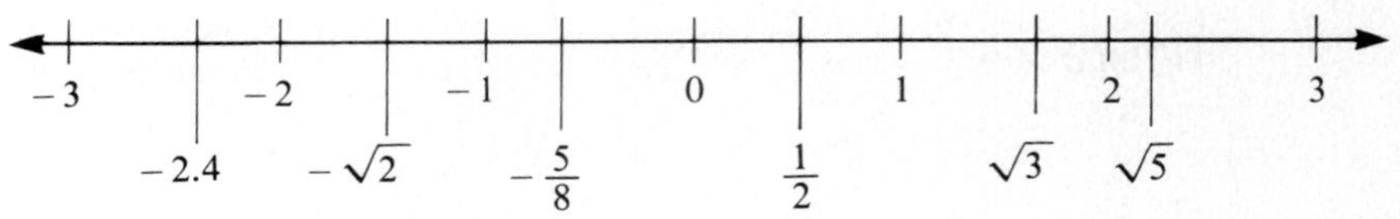

example 2

Indicate the relationship between each of the following pairs of numbers, using the $>$, $<$, or $=$ sign.

(a) $2 \ ? \ \sqrt{3}$

$2 > \sqrt{3}$ — Since $\sqrt{3} \approx 1.732$.

(b) $-2.4 \ ? \ -\sqrt{9}$

$-2.4 > -\sqrt{9}$ — Since $-\sqrt{9} = -3$.

(c) $\sqrt{\frac{1}{4}} \;?\; \frac{1}{2}$

$\sqrt{\frac{1}{4}} = \frac{1}{2}$ — Since $\frac{1}{2} \cdot \frac{1}{2} = \frac{1}{4}$.

(d) $\sqrt{5} \;?\; 3$

$\sqrt{5} < 3$ — Since $\sqrt{5} \approx 2.236$.

(e) $\sqrt{0.26} \;?\; 0.5$

$\sqrt{0.26} > 0.5$ — Since $0.5 = \sqrt{0.25}$.

QUICK QUIZ

List all the number systems in which each of the following is included.

1. $\sqrt{7}$
2. $-\sqrt{25}$

Use the >, <, or = sign to indicate the relationship between each of the following pairs of numbers.

3. $\sqrt{0.16} \;?\; 0.4$
4. $-\sqrt{49} \;?\; -7.1$

ANSWERS

1. irrational number, real number
2. integer, rational number, real number
3. $\sqrt{0.16} = 0.4$
4. $-\sqrt{49} > -7.1$

7.5 Exercises

Indicate which of the following statements are true and which are false. For those that are false, change the italic expression to make the statement true.

1. The set of counting numbers *does not* include zero.
2. Every integer is also a *whole* number.
3. Every whole number is also a *rational* number.
4. A number cannot belong to both the rational numbers and the *real* numbers.
5. The quotient of any two nonzero integers is always *an integer.*
6. The quotient of any two nonzero rational numbers is always a *rational* number.
7. The irrational numbers are a subset of the *rational* numbers.
8. The counting numbers are a subset of the *integers.*

List all of the number systems in which each of the following numbers is included.

9. -9	**10.** 4	**11.** $\sqrt{5}$	**12.** $-\frac{1}{2}$	**13.** 3.7
14. -6.0	**15.** $-\sqrt{23}$	**16.** $11\frac{3}{5}$	**17.** $0.6\overline{6}$	**18.** $\frac{7}{8}$

19. -3.15 **20.** $5\sqrt{2}$ **21.** $\frac{32}{4}$ **22.** $-\sqrt{49}$ **23.** -0.4397

24. $-\frac{72}{8}$ **25.** $\sqrt{50}$ **26.** $-9.48\overline{48}$ **27.** 0 **28.** $-\sqrt{1.44}$

Use the >, <, or = sign to indicate the relationship between each of the following pairs of numbers.

29. $\sqrt{3}$? 2 **30.** $-\sqrt{2}$? -1 **31.** -3 ? $-\sqrt{9}$ **32.** $\sqrt{4}$? 2

33. $\sqrt{0.09}$? 0.03 **34.** 0.7 ? $\sqrt{0.49}$ **35.** $\sqrt{\frac{1}{25}}$? $\frac{1}{5}$ **36.** $\frac{1}{7}$? $\sqrt{\frac{1}{36}}$

37. $\sqrt{8}$? 3 **38.** -4 ? $-\sqrt{16}$ **39.** $\sqrt{0.48}$? 0.7 **40.** -0.9 ? $-\sqrt{0.82}$

41. $\sqrt{\frac{9}{64}}$? $\frac{3}{7}$ **42.** $\frac{1}{4}$? $\sqrt{\frac{3}{16}}$ **43.** $\sqrt{0.01}$? 0.01 **44.** $\sqrt{\frac{1}{9}}$? $\frac{1}{9}$

45. $-\sqrt{4}$? -2.1 **46.** -5.9 ? $-\sqrt{36}$ **47.** $\sqrt{1.44}$? 1.2 **48.** 0.08 ? $\sqrt{0.064}$

Summary and Review

Key Terms

[7.1] The **square root** of a given number is a number whose square is equal to the given number. For any number a, the square root of a is indicated by $\sqrt{a}$.

A **perfect square** is a number whose square root is a whole number.

The **principal square root** of a number is its positive square root.

An **imaginary number** is the square root of a negative number.

[7.4] The ***n*th root** of a given number is some number whose nth power is equal to the given number. For any number a, the nth root of a is indicated by $\sqrt[n]{a}$.

Calculations

[7.2] **To find the square root of a product**, find the product of the square roots of the factors.

To express a square root in simplest form, determine the square root of the largest perfect square that is a factor of the number under the radical sign. The other factor, which is not a perfect square, remains under the radical sign.

To find the square root of a quotient, find the quotient of the square roots of the numerator and the denominator.

Properties of Roots

[7.4] For any positive numbers a and b and positive integer n,

$$\sqrt[n]{a \cdot b} = \sqrt[n]{a}\sqrt[n]{b}$$

In other words, the nth root of a product is the product of the nth roots of the factors.

For any positive numbers a and b and positive integer n,

$$\sqrt[n]{\frac{a}{b}} = \frac{\sqrt[n]{a}}{\sqrt[n]{b}}$$

In other words, the nth root of a quotient is the quotient of the nth roots of the numerator and the denominator.

Number Systems

[7.5] **Whole Numbers**: $\{0, 1, 2, 3, 4, 5, \ldots\}$

Natural Numbers, or **Counting Numbers**: $\{1, 2, 3, 4, 5, \ldots\}$

Integers: $\{\ldots, -4, -3, -2, -1, 0, 1, 2, 3, 4, \ldots\}$

Rational Numbers: Numbers that can be written in the form $\frac{a}{b}$, where a and b are integers and $b \neq 0$.

Irrational Numbers: Nonterminating, nonrepeating decimals that cannot be expressed in the form of a rational number.

Real Numbers: All rational and irrational numbers.

Chapter 7 Review Exercises

[7.1–7.3] Determine each of the following square roots. Round off where indicated.

1. $\sqrt{49}$
2. $-\sqrt{81}$
3. $\pm\sqrt{\frac{9}{25}}$
4. $\sqrt{\frac{1}{16}}$
5. $-\sqrt{0.36}$
6. $\sqrt{0.64}$
7. $-\sqrt{4{,}900}$
8. $\pm\sqrt{6{,}400}$
9. $-\sqrt{0.0009}$
10. $\sqrt{0.0004}$
11. $\sqrt{10{,}000}$
12. $\sqrt{810{,}000}$
13. $\sqrt{13.69}$
14. $\sqrt{51.84}$
15. $\sqrt{324}$
16. $\sqrt{4{,}489}$
17. $\sqrt{16{,}129}$
18. $\sqrt{383{,}161}$
19. $\sqrt{0.0784}$
20. $\sqrt{0.5776}$
21. $\sqrt{43{,}296{,}400}$
22. $\sqrt{6{,}604{,}900}$
23. $\sqrt{0.133225}$
24. $\sqrt{0.471969}$
25. $\sqrt{5.9}$ nearest hundredth
26. $\sqrt{37.1}$ nearest hundredth
27. $\sqrt{637}$ nearest tenth
28. $\sqrt{4{,}712}$ nearest tenth
29. $\sqrt{38{,}895}$ nearest whole number
30. $\sqrt{617{,}204}$ nearest whole number
31. $\sqrt{0.281}$ nearest thousandth
32. $\sqrt{0.0174}$ nearest thousandth
33. $\sqrt{53{,}742{,}109}$ nearest whole number
34. $\sqrt{1{,}347{,}269}$ nearest whole number
35. $\sqrt{0.634812}$ nearest thousandth
36. $\sqrt{0.004517}$ nearest ten-thousandth

[7.2, 7.4] Write each of the following radical expressions in simplest form.

37. $\sqrt{200}$
38. $\sqrt{450}$
39. $\sqrt{8}$
40. $\sqrt{54}$
41. $\sqrt{24}$
42. $\sqrt{125}$
43. $\sqrt[4]{16}$
44. $\sqrt[5]{32}$
45. $\sqrt[4]{81}$
46. $\sqrt[3]{125}$
47. $\sqrt[5]{-1}$
48. $\sqrt[3]{-27}$

[7.5] List all of the number systems in which each of the following numbers is included.

49. $-\sqrt{2}$
50. $\sqrt{25}$
51. 3.78
52. -917
53. $5\frac{7}{8}$
54. $\frac{21}{3}$
55. $\sqrt{0.09}$
56. $0.87\overline{87}$
57. $\sqrt{100}$
58. -3.0
59. $-5\sqrt{15}$
60. $\sqrt{3{,}600}$

[7.5] Use the >, <, or = sign to indicate the relationship between each of the following pairs of numbers.

61. $-\sqrt{3}\ ?\ -2$

62. $2\ ?\ \sqrt{5}$

63. $\sqrt{49}\ ?\ 7$

64. $-\sqrt{37}\ ?\ -6$

65. $\sqrt{0.16}\ ?\ 0.4$

66. $\sqrt{0.9}\ ?\ 0.3$

67. $-\sqrt{\frac{4}{49}}\ ?\ -\frac{3}{7}$

68. $\frac{5}{8}\ ?\ \sqrt{\frac{25}{81}}$

69. $\sqrt{0.049}\ ?\ 0.07$

70. $\sqrt{0.121}\ ?\ 0.11$

71. $\sqrt{65}\ ?\ 8$

72. $-4\ ?\ -\sqrt{17}$

8 MEASUREMENT IN THE ENGLISH AND METRIC SYSTEMS

To **measure** something is to find a number that characterizes one of its properties. In order to establish a uniform convention for assigning numbers to things we need to measure, it is necessary to define a **unit of measurement** that indicates the size of a known quantity. Units of measurement have been used to indicate the magnitude of various properties such as length, area, weight, volume, and temperature. A *system of measurement* defines the relationship between the standard units that are used to describe the various properties and all other units that are derived from them. In this chapter, we will discuss two systems of measurement—the **English system**, which is used primarily in the United States, and the **metric system**, which is used throughout the rest of the world and will eventually become the standard system in the United States as well.

8.1 Length

The standard units of length in the English system are the inch, foot, yard, and mile. Table 8.1 summarizes equivalent measures of length in the English system.

table *8.1*
English Units of Length

1 foot (ft)	= 12 inches (in.)
1 yard (yd)	= 3 feet (ft)
1 mile (mi)	= 5,280 feet (ft)

In Section 4.1, we showed how to convert from one unit to another by multiplying the quantity to be converted by a unit fraction, that is, a fraction equivalent to one. Whenever we multiply a unit of measurement by a series of unit fractions to convert it to another unit of measurement, we are using a technique called *dimensional analysis.* This technique is illustrated in Example 1.

example *1*

Convert 7 yd to inches.

$$7 \text{ yd} = 7 \cancel{\text{yd}} \times \frac{3 \cancel{\text{ft}}}{1 \cancel{\text{yd}}} \times \frac{12 \text{ in.}}{1 \cancel{\text{ft}}} = 252 \text{ in.}$$

Since 1 yd = 3 ft and 1 ft = 12 in., multiply 7 yd by the unit fractions $\frac{3 \text{ ft}}{1 \text{ yd}}$ and $\frac{12 \text{ in.}}{1 \text{ ft}}$ so that yd and ft will cancel. ●

The standard unit of length in the metric system is the **meter** (m), which is a little longer than a yard. Other units of length in the metric system are derived by multiplying or dividing a meter by powers of 10. These other units are named by attaching the appropriate prefix to the word *meter*. The metric prefixes and their meanings are given in Table 8.2. The equivalent units of length in the metric system are given in Table 8.3.

To facilitate our work with the metric system, it is helpful to be able to visualize the lengths of everyday objects in terms of metric units.

table 8.2
Metric Prefixes

PREFIX	MEANING
milli	thousandth
centi	hundredth
deci	tenth
deka	ten
hecto	hundred
kilo	thousand

For example,

The length of this book is about 28 cm.
The height of the Empire State Building is 381 m.
The diameter of a penny is about 19 mm.

table 8.3
Metric Units of Length

1 millimeter (mm)	$= \frac{1}{1,000}$ meter (m)	or	1,000 mm = 1 m
1 centimeter (cm)	$= \frac{1}{100}$ meter (m)	or	100 cm = 1 m
1 decimeter (dm)	$= \frac{1}{10}$ meter (m)	or	10 dm = 1 m
1 dekameter (dam)	= 10 meters (m)		
1 hectometer (hm)	= 100 meters (m)		
1 kilometer (km)	= 1,000 meters (m)		

Figure 8.1 shows how a ruler that is 1 decimeter or 10 centimeters in length compares to another ruler that is 4 inches long.

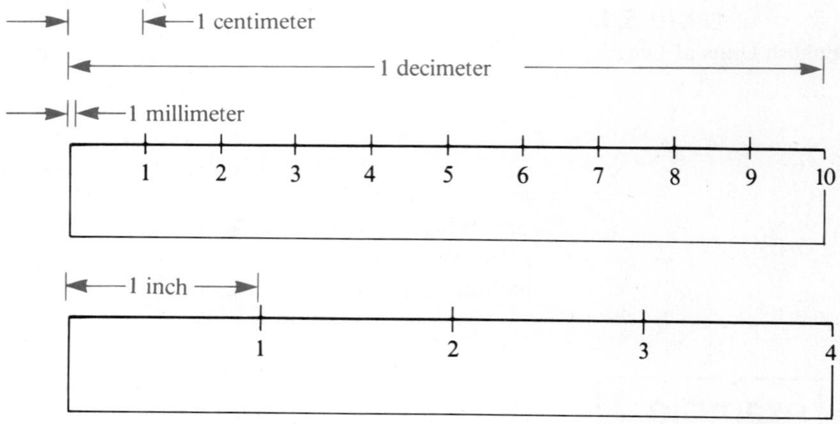

figure 8.1

To convert from one unit of length to another in the metric system, we simply multiply and/or divide the unit to be converted by powers of 10. Therefore, conversion from one unit to another is much simpler in the metric system than in the English system.

example 2

Convert each of the following metric units as instructed.

(a) Convert 8 m to centimeters.

$$8 \text{ m} = 8 \not{\text{m}} \times \frac{100 \text{ cm}}{1 \not{\text{m}}} = 800 \text{ cm}$$

Since 1 m = 100 cm, multiply 8 m by the unit fraction $\frac{100 \text{ cm}}{1 \text{ m}}$ so that m cancel.

(b) Convert 1.5 hm to millimeters.

$$1.5 \text{ hm} = 1.5 \not{\text{hm}} \times \frac{100 \not{\text{m}}}{1 \not{\text{hm}}} \times \frac{1{,}000 \text{ mm}}{1 \not{\text{m}}}$$

$$= 150{,}000 \text{ mm}$$

Since 1 hm = 100 m and 1 m = 1,000 mm, multiply 1.5 hm by the unit fractions $\frac{100 \text{ m}}{1 \text{ hm}}$ and $\frac{1{,}000 \text{ mm}}{1 \text{ m}}$ so that hm and m cancel.

To multiply by 100,000 move the decimal point 5 places to the right.

(c) Convert 82 dm to kilometers.

$$82 \text{ dm} = 82 \not{\text{dm}} \times \frac{1 \not{\text{m}}}{10 \not{\text{dm}}} \times \frac{1 \text{ km}}{1{,}000 \not{\text{m}}}$$

$$= 0.0082 \text{ km}$$

Since 1 m = 10 dm and 1 km = 1,000 m, multiply 82 dm by the unit fractions $\frac{1 \text{ m}}{10 \text{ dm}}$ and $\frac{1 \text{ km}}{1{,}000 \text{ m}}$ so that dm and m cancel.

To divide by 10,000 move the decimal point 4 places to the left. ●

The relationship between the units of length in the English and metric systems is given in Table 8.4. Each of the values in Table 8.4 is correct to four places of accuracy, or four significant digits. Therefore, when using Table 8.4 for a calculation, round off your answer so that it contains no more than four significant digits.

table 8.4
Equivalent Units of Length

ENGLISH TO METRIC	METRIC TO ENGLISH
1 in. = 2.540 cm	1 cm = 0.3937 in.
1 ft = 0.3048 m	1 m = 39.37 in.
1 yd = 0.9144 m	1 m = 1.094 yd
1 mi = 1.609 km	1 km = 0.6214 mi

To convert from one system to another, we can also multiply the unit to be converted by an appropriate unit fraction, as shown in Example 3.

example 3

Convert 357 mi to kilometers.

solution

$$357 \text{ mi} = 357 \not{\text{mi}} \times \frac{1.609 \text{ km}}{1 \not{\text{mi}}}$$

$$= 574.413 \text{ km} \approx 574.4 \text{ km}$$

Since 1 mi = 1.609 km, multiply 357 mi by the unit fraction $\frac{1.609 \text{ km}}{1 \text{ mi}}$ so that mi cancel.

Since the conversion factor has only four significant digits, our final answer can have no more than four significant digits. Therefore, we rounded off to the nearest tenth. ●

It is very important to be able to mentally estimate conversions between the English and metric systems based on knowledge of some approximate conversion factors. Since 1 m = 39.37 in. and 1 yd = 36 in., a meter is a little longer than a yard.

$$1 \text{ m} \approx 1 \text{ yd}$$

Since 1 in. = 2.54 cm, we can think of 1 in. as approximately equal to $2\frac{1}{2}$ cm. Therefore,

$$2 \text{ in.} \approx 5 \text{ cm}$$

and

$$1 \text{ ft} \approx 30 \text{ cm}$$

Since 1 km = 0.6214 mi, we can think of 1 km as approximately equal to $\frac{5}{8}$ mi $\left(\frac{5}{8} = 0.625\right)$ or

$$5 \text{ mi} \approx 8 \text{ km}$$

Some approximations that are less exact but easier to remember are

$$1 \text{ mi} \approx 2 \text{ km}$$

$$1 \text{ km} \approx \frac{1}{2} \text{ mi}$$

example 4

Approximate each of the following conversions between English and metric units. Then compare your answers to the values obtained by using Table 8.4.

(a) Convert 15 cm to inches.

Since 5 cm ≈ 2 in.

then 15 cm ≈ 6 in.

$\left(\text{Think, } \frac{5 \text{ cm}}{2 \text{ in.}} = \frac{15 \text{ cm}}{?}\right)$

Use Table 8.4 to find the exact answer.

$$15 \text{ cm} = 15 \cancel{\text{cm}} \times \frac{0.3937 \text{ in.}}{1 \cancel{\text{cm}}}$$

$$= 5.9055 \text{ in.}$$

Multiply 15 cm by the unit fraction $\frac{0.3937 \text{ in.}}{1 \text{ cm}}$ so that cm cancel.

15 cm = 5.9055 in. ≈ 5.906 in. correct to four significant digits. Our approximation is off by about 0.1 in.

6 in. − 5.906 in. ≅ 0.1 in.

(b) Convert 5 ft to centimeters.

Since 1 ft ≈ 30 cm,

then 5 ft ≈ 150 cm

$\left(\text{Think, } \frac{1 \text{ ft}}{30 \text{ cm}} = \frac{5 \text{ ft}}{?}\right)$

Use Table 8.4 to find the exact answer.

$$5 \text{ ft} = 5 \cancel{\text{ft}} \times \frac{0.3048 \cancel{\text{m}}}{1 \cancel{\text{ft}}} \times \frac{100 \text{ cm}}{1 \cancel{\text{m}}}$$

$$= 152.4 \text{ cm}$$

Multiply 5 ft by the unit fractions $\frac{0.3048 \text{ m}}{1 \text{ ft}}$ and $\frac{100 \text{ cm}}{1 \text{ m}}$ so that ft and m cancel.

5 ft = 152.4 cm correct to four significant digits. Our approximation is off by 2.4 cm.

152.4 − 150 cm = 2.4 cm

(c) Convert 2,500 mi to kilometers.

Since 5 mi ≈ 8 km,

2,500 mi ≈ 4,000 km

$\left(\text{Think, } \frac{5 \text{ mi}}{8 \text{ km}} = \frac{2{,}500 \text{ mi}}{?}\right)$

Use Table 8.4 to find the exact answer.

$$2{,}500 \text{ mi} = 2{,}500 \cancel{\text{mi}} \times \frac{1.609 \text{ km}}{1 \cancel{\text{mi}}}$$

$$= 4{,}022.5 \text{ km}$$

Multiply 2,500 mi by the unit fraction $\frac{1.609 \text{ km}}{1 \text{ mi}}$ so that mi cancel.

2,500 mi = 4,022.5 km ≈ 4,023 km correct to four significant digits. Our approximation is off by about 23 km.

4,023 km − 4,000 km = 23 km ●

example 5

The captain of a swim team is 6 ft 2 in. tall, and his coach is 171 cm tall. Who is taller?

solution

Approximate the height of the swim team captain in cm.

Since 1 ft ≈ 30 cm,

6 ft ≈ 180 cm

Since 2 in. ≈ 5 cm,

6 ft 32 in. ≈ 180 cm + 5 cm = 185 cm

The swim team captain is taller than his coach. Since 185 cm > 171 cm. ●

QUICK QUIZ

Perform each of the following conversions.

1. 7 ft = ____ yd

2. 6,000 mm = ____ m

3. 5.2 km = ____ cm

4. 8 in. ≈ ____ cm

ANSWERS

1. $2\frac{1}{3}$

2. 6

3. 520,000

4. 20

8.1 Exercises

Indicate which of the following statements are true and which are false. For those that are false, change the italic word to make the statement true.

1. The standard unit of length in the metric system is the *meter*.

2. One yard is about 3 inches *longer* than a meter.

3. One meter is equivalent to 100 *millimeters.*

4. Thirty centimeters is roughly equivalent to 1 *inch.*

Convert each of the following English units as indicated.

5. 8 ft = ____ in.
6. 60 in. = ____ ft
7. 7 ft = ____ yd
8. 528 ft = ____ mi
9. $5\frac{2}{3}$ yd = ____ ft
10. $1\frac{3}{4}$ mi = ____ ft
11. 7 ft 2 in. = ____ in.
12. 3 yd 2 ft = ____ ft
13. 7 yd 1 ft = ____ yd
14. 9 ft 9 in. = ____ ft
15. $\frac{3}{5}$ mi = ____ ft
16. 1,584 ft = ____ mi
17. 23.5 ft = ____ in.
18. 8.5 mi = ____ ft
19. $\frac{3}{4}$ ft = ____ in.
20. $\frac{7}{9}$ yd = ____ in.
21. $\frac{1}{2}$ mi = ____ yd
22. $5\frac{2}{3}$ yd = ____ in.
23. 114 in. = ____ ft
24. 2,256 in. = ____ yd
25. 540 in. = ____ yd
26. 0.05 yd = ____ in.

Convert each of the following metric units as indicated.

27. 5 m = ____ cm
28. 870 cm = ____ m
29. 19 m = ____ mm
30. 6,400 mm = ____ m
31. $3\frac{1}{2}$ km = ____ m
32. $7\frac{3}{4}$ cm = ____ mm
33. 3.8 km = ____ hm
34. 9.5 cm = ____ dm
35. 53 mm = ____ cm
36. 98 dam = ____ hm
37. $\frac{7}{10}$ km = ____ m
38. $\frac{3}{5}$ m = ____ cm
39. 590 cm = ____ m
40. 4.5 hm = ____ m
41. 0.2 km = ____ cm
42. 8,800 mm = ____ m
43. 857 dm = ____ m
44. 631 dam = ____ m
45. 720 cm = ____ dm
46. 3,500 mm = ____ dm
47. 8.1 hm = ____ dam
48. 2.7 cm = ____ hm
49. 0.7 km = ____ mm
50. 0.3 dm = ____ dkm

Indicate which measurement most closely approximates each of the following lengths.

51. The length of a playing card: 20 cm 9 cm 9 mm
52. The diameter of a quarter: 3.8 mm 2.5 cm 1.4 cm
53. The length of a football field: 91 m 80 m 107 m
54. The average height of an adult woman: 300 cm 163 cm 250 cm
55. The length of a tennis racquet: 68.5 cm 6.85 m 68.5 mm
56. The width of this book: 19.5 cm 19.5 mm 1.95 m
57. The height of a classroom: 320 mm 3.2 m 0.32 km
58. The height of a 100-story building: 343 km 3.43 km 343 m
59. The length of a pen: 15 cm 150 cm 15 mm
60. The distance between Boston and Philadelphia: 43.6 km 80.5 km 436 km
61. The diameter of the earth: 12,755 m 127,550 m 12,755 km
62. The height of a desk: 77 cm 7 m 77 dm

Approximate each of the following conversions between English and metric units. Then compare your results to the values obtained by using Table 8.4.

63. 35 cm = ____ in.
64. 8 in. = ____ cm
65. 5 ft = ____ cm
66. 210 cm = ____ ft
67. 2 yd = ____ m
68. 8 m = ____ yd
69. 450 mi = ____ km
70. 64 km = ____ mi
71. 5 yd = ____ cm
72. 20 mi = ____ m

Solve each of the following word problems.

73. The distance from San Francisco to Minneapolis is 1,997 mi, and the distance from San Francisco to Chicago is 2,989 km. Which city is farther from San Francisco, Minneapolis or Chicago?

74. A Canadian athlete ran a 100-m race in 10.83 min. An American athlete ran a 100-yd race in the same time. Who ran faster?

75. A room is 5.98 m long. How many tiles must be laid end to end to reach from one end of the room to the other if each tile is 23 cm in length?

76. A stack of 17 identical books is 76.5 cm high. What is the thickness of each book?

77. A roll of transparent tape is 20.3 m long. How many feet of tape are in this roll? Round off your answer to the nearest tenth of a foot.

78. A speed limit is posted as 55 mi/hr. What is the speed limit in km/hr? Round off your answer to the nearest kilometer.

79. If fabric sells for \$4.98 a meter, what is the cost of a piece that measures $3\frac{1}{2}$ yd in length? Round off your answer to the nearest cent.

80. Fencing material sells for \$12 a foot. How much would it cost to construct a fence that is 4 m long?

8.2 Weight

The **weight** of an object is the force of the earth's gravitational pull acting upon it. The weight of an object decreases as its distance from the center of the earth increases. Weight is directly proportional to *mass*, which is the quantity of matter in an object. The mass of an object remains constant, regardless of the forces acting upon it.

The standard units of weight in the English system are the ounce, pound, and ton. Table 8.5 summarizes equivalent units of weight in the English system.

table 8.5
English Units of Weight

1 pound (lb)	= 16 ounces (oz)
1 ton (T)	= 2,000 pounds (lb)

example 1

Convert $7\frac{5}{8}$ lb to ounces.

solution

$$7\frac{5}{8}\text{ lb} = 7\frac{5}{8}\text{ lb} \times \frac{16\text{ oz}}{1\text{ lb}}$$

Multiply $7\frac{5}{8}$ lb by the unit fraction $\frac{16\text{ oz}}{1\text{ lb}}$ so that lb cancel.

$$= \frac{61}{8} \times 16\text{ oz} = 122\text{ oz}$$

●

The standard unit of weight in the metric system is the **gram** (g). Other metric units of weight are named by attaching the appropriate prefix to the word *gram*. The equivalent units of weight in the metric system are listed in Table 8.6. The units decigram $\left(\frac{1}{10}\text{ g}\right)$, dekagram (10 g) and hectogram (100 g) have been omitted from Table 8.6 since they are rarely used.

table 8.6
Metric Units of Weight

1 milligram (mg)	$= \frac{1}{1{,}000}$ gram (g)	or	1,000 mg = 1 g
1 centigram (cg)	$= \frac{1}{100}$ gram (g)	or	100 cg = 1 g
1 kilogram (kg)	= 1,000 grams (g)		
1 metric ton (t)	= 1,000,000 grams (g)	or	1 t = 1,000 kg

The weights of a few well-known objects in metric units of measurement are as follows:

An aspirin weighs about 500 mg.

An egg weighs about 50 g.

A football player weighs about 130 kg.

example 2

Convert each of the following metric units as instructed.

(a) Convert 17 mg to grams.

$$17 \text{ mg} = 17 \cancel{\text{mg}} \times \frac{1 \text{ g}}{1{,}000 \cancel{\text{mg}}} = 0.017 \text{ g}$$

Since 1 g = 1,000 mg, multiply 17 mg by the unit fraction $\frac{1 \text{ g}}{1{,}000 \text{ mg}}$ so that mg cancel.

To divide by 1,000, move the decimal point 3 places to the left.

(b) Convert 3.5 t to kilograms.

$$3.5 \text{ t} = 3.5 \cancel{\text{t}} \times \frac{1{,}000 \text{ kg}}{1 \cancel{\text{t}}} = 3{,}500 \text{ kg}$$

Since 1 t = 1,000 kg, multiply 3.5 t by the unit fraction $\frac{1{,}000 \text{ kg}}{1 \text{ t}}$ so that t cancel.

To multiply by 1,000, move the decimal point 3 places to the right. ●

The relationship between the units of weight in the English and metric system is given in Table 8.7. Each value is accurate to four significant digits.

table 8.7
Equivalent Units of Weight

ENGLISH TO METRIC	METRIC TO ENGLISH
1 oz = 28.35 g	1 g = 0.03527 oz
1 lb = 0.4536 kg	1 kg = 2.205 lb
1 T = 0.9072 t	1 t = 1.102 T

From this table, we can derive the following approximations that will enable us to mentally estimate conversions from one system to the other.

$1 \text{ oz} \approx 30 \text{ g}$

$1 \text{ lb} \approx 0.45 \text{ kg}$

$$1 \text{ kg} \approx 2.2 \text{ lb}$$
$$1 \text{ T} \approx 1 \text{ t}$$

Some approximations that are less exact but easier to remember are

$$1 \text{ lb} \approx \frac{1}{2} \text{ kg}$$

$$1 \text{ kg} \approx 2 \text{ lb}$$

example 3

Approximate the following conversions. Then compare your results to the values obtained by using Table 8.7.

(a) Convert 10 lb to kilograms.

Since 1 lb ≈ 0.45 kg, $\left(\text{Think, } \frac{1 \text{ lb}}{0.45 \text{ kg}} = \frac{10 \text{ lb}}{?}\right)$

then 10 lb ≈ 4.5 kg

Use Table A.7 to find the exact answer.

$$10 \text{ lb} = 10 \cancel{\text{lb}} \times \frac{0.4536 \text{ kg}}{1 \cancel{\text{lb}}}$$

Multiply 10 lb by the unit fraction $\frac{0.4536 \text{ kg}}{1 \text{ lb}}$ so that lb cancel.

$$= 4.536 \text{ kg}$$

10 lb = 4.536 kg correct to four significant digits. Our approximation is off by 0.036 kg.

4.536 kg − 4.5 kg = 0.036 kg

(b) Convert 8 oz to grams.

Since 1 oz ≈ 30 g, $\left(\text{Think, } \frac{1 \text{ oz}}{30 \text{ g}} = \frac{8 \text{ oz}}{?}\right)$

8 oz ≈ 240 g

Use Table A.7 to find the exact answer.

$$8 \text{ oz} = 8 \cancel{\text{oz}} \times \frac{28.35 \text{ g}}{1 \cancel{\text{oz}}} = 226.8 \text{ g}$$

Multiply 8 oz by the unit fraction $\frac{28.35 \text{ g}}{1 \text{ oz}}$ so that oz cancel.

8 oz = 226.8 g correct to four significant digits. Our approximation is off by 13.2 g.

240 g − 226.8 g = 13.2 g ●

example 4

According to an international agreement, an airline is required to compensate a passenger for lost baggage at a rate not exceeding $20 per kilogram. If a 22-lb bag was lost by an airline, what is the maximum amount its owner can expect to receive in compensation?

solution

$$22 \text{ lb} = 22 \cancel{\text{lb}} \times \frac{1 \text{ kg}}{2.2 \cancel{\text{lb}}} = 10 \text{ kg}$$

Convert 22 lb to kg using the conversion factor 1 kg ≈ 2.2 lb.

$$10 \text{ kg} \times \frac{\$20}{\text{kg}} = \$200$$

Then multiply the result by 20.

The maximum amount the owner can expect is $200. ●

QUICK QUIZ

Perform each of the following conversions.

1. 8 oz = ____ lb
2. 20 mg = ____ g
3. 9,000 g = ____ kg
4. 2 kg ≈ ____ lb

ANSWERS

1. $\frac{1}{2}$
2. 0.02
3. 9
4. 4.4

8.2 Exercises

Indicate which of the following statements are true and which are false. For those that are false, change the italic expression to make the statement true.

1. As the distance between an object and the center of the earth increases, the weight of the object *increases.*
2. The standard unit of weight in the *English* system is the gram.
3. One kilogram is roughly equivalent to 2.2 *ounces.*
4. One kilogram is equivalent to *1,000,000* milligrams.

Convert each of the following English units as instructed.

5. 32 oz = ____ lb
6. 5 lb = ____ oz
7. $\frac{1}{2}$ T = ____ lb
8. 3,000 lb = ____ T
9. $7\frac{3}{4}$ lb = ____ oz
10. 4.5 T = ____ lb
11. 8 lb 9 oz = ____ oz
12. 20 lb 2 oz = ____ lb
13. 0.053 T = ____ lb
14. 0.0069 T ≅ ____ lb
15. $5\frac{3}{8}$ lb = ____ oz
16. 108 oz = ____ lb
17. 250 lb = ____ T
18. 50,000 lb = ____ T
19. 80 oz = ____ lb
20. 0.72 lb = ____ oz
21. $3\frac{2}{5}$ T = ____ lb
22. $2\frac{7}{8}$ lb = ____ oz
23. 4,800 lb = ____ T
24. 7,200 oz = ____ lb

Convert each of the following metric units as instructed.

25. 5 kg = ____ g
26. 28 g = ____ mg
27. 350 cg = ____ g
28. 9.3 g = ____ cg
29. 5 t = ____ kg
30. 3,800 kg = ____ t
31. 6.7 mg = ____ g
32. 7.2 cg = ____ g
33. 750 mg = ____ g
34. 0.4 g = ____ cg
35. 347 g = ____ mg
36. 218 cg = ____ mg
37. 4.18 kg = ____ g
38. 0.257 g = ____ mg
39. 3 kg = ____ mg
40. 5,000 mg = ____ kg
41. 88,000 g = ____ t
42. 7,000 kg = ____ t
43. $9\frac{1}{2}$ mg = ____ g
44. $8\frac{2}{5}$ g = ____ cg
45. $\frac{5}{8}$ kg = ____ g
46. $3\frac{3}{4}$ g = ____ mg
47. 0.07 t = ____ g
48. 0.02 kg = ____ mg

Indicate which measurement most closely approximates each of the following weights.

49. The weight of a box of crackers: 454 g 4.54 kg 4,540 mg

50. The weight of a cold capsule: 650 mg 6.5 g 0.065 kg
51. The weight of one-half dozen eggs: 1 kg 100 g 300 g
52. The weight of a newborn baby: 4 kg 40 kg 40 g
53. The weight of a nickel: 50 mg 5 g 0.5 kg
54. The average weight of an adult man: 860 g 86 kg 8.6 kg
55. The weight of an automobile: 500 kg 1 t 50,000 g
56. The weight of a paper clip: 5 g 50 mg 0.5 g

Approximate each of the following conversions. Then compare your results to the values obtained by using Table 8.7.

57. 30 lb = ____ kg **58.** 90 g = ____ oz **59.** 4 kg = ____ lb **60.** 66 lb = ____ kg
61. 7.2 T = ____ t **62.** 23 t = ____ T **63.** 700 oz = ____ g **64.** 200 kg = ____ lb
65. 3 kg = ____ oz **66.** 240 g = ____ lb

Solve each of the following word problems.

67. If one aspirin weighs 500 mg, how many grams would 100 aspirins weigh?

68. Which weighs more, 10 lb of potatoes or 5 kg of potatoes?

69. A chocolate bar that contains nuts weighs 45 g. Another chocolate bar that does not contain nuts weighs 2 oz. Which chocolate bar weighs more?

70. A 10-year-old boy weighs 60 lb, and a 10-year-old girl weighs 30 kg. Who weighs more?

71. If a dozen bricks weigh 54 kg, how much does a single brick weigh?

72. A package containing 16 slices of cheese weighs 340 g. What is the weight of a single slice?

73. A 12-oz package of chocolate chip cookies costs $1.69. A 420-g package of chocolate chip cookies sells for the same price. Which is the better buy?

74. A 2-lb box of sugar sells for $1.29. How much would a box that weighs 2 kg cost? Round off your answer to the nearest cent.

8.3 Capacity

The standard units of capacity (volume) in the English system are the cup, pint, quart, gallon, fluid ounce, tablespoon, and teaspoon. Table 8.8 summarizes equivalent units of capacity in the English system.

table 8.8
English Units of Capacity

1 pint (pt)	= 2 cups (c)
1 quart (qt)	= 2 pints (pt)
1 gallon (gal)	= 4 quarts (qt)
1 pint (pt)	= 16 fluid ounces (fl oz)
1 tablespoon (tbsp)	= 3 teaspoons (tsp)

example 1

Convert 22 fl oz to pints.

solution

$$22 \text{ fl oz} = 22 \not{\text{fl oz}} \times \frac{1 \text{ pt}}{16 \not{\text{fl oz}}} = \frac{22}{16} \text{ pt}$$

$$= \frac{11}{8} \text{ pt} = 1\frac{3}{8} \text{ pt}$$

Multiply 22 fl oz by the unit fraction $\frac{1 \text{ pt}}{16 \text{ fl oz}}$ so that fl oz cancel.

The standard unit of capacity in the metric system is the *liter* (L), which is a little more than a quart. Other metric units of capacity are named by attaching the appropriate prefix to the word *liter*. The equivalent units of capacity in the metric system are listed in Table 8.9.

table 8.9
Metric Units of Capacity

$$1 \text{ milliliter (mL)} = \frac{1}{1{,}000} \text{ liter (L)} \quad \text{or} \quad 1{,}000 \text{ mL} = 1 \text{ L}$$

$$1 \text{ hectoliter (hL)} = 100 \text{ liters (L)}$$

$$1 \text{ kiloliter (kL)} = 1{,}000 \text{ liters (L)}$$

The units centiliter $\left(\frac{1}{100}\text{ L}\right)$, deciliter $\left(\frac{1}{10}\text{ L}\right)$, and dekaliter (10 L) have been omitted from Table 8.9 since they are rarely used.

The volumes of a few well-known items in metric units of measurement are as follows:

A bottle of eye drops contains 15 mL.
A can of soda contains 355 mL.
A bathtub can hold 150 L.

example 2

Convert each of the following metric units as instructed.

(a) Convert 3.5 L to milliliters.

$$3.5 \text{ L} = 3.5 \cancel{\text{L}} \times \frac{1{,}000 \text{ mL}}{1 \cancel{\text{L}}} = 3{,}500 \text{ mL}$$

Multiply 3.5 L by the unit fraction $\frac{1{,}000 \text{ mL}}{1 \text{ L}}$ so that L cancel.

(b) Convert 80,000 mL to hectoliters.

80,000 mL

$$= 80{,}000 \cancel{\text{mL}} \times \frac{1 \cancel{\text{L}}}{1{,}000 \cancel{\text{mL}}} \times \frac{1 \text{ hL}}{100 \cancel{\text{L}}}$$

$$= 0.8 \text{ hL}$$

Multiply 80,000 mL by the unit fractions $\frac{1 \text{ L}}{1{,}000 \text{ mL}}$ and $\frac{1 \text{ hL}}{100 \text{ L}}$ so that mL and L cancel.

To divide by 100,000 move the decimal point 5 places to the left. ●

The relationship between units of capacity in the English and metric systems is given in Table 8.10. Each value is accurate to four significant digits.

table 8.10
Equivalent Units of Capacity

ENGLISH TO METRIC	METRIC TO ENGLISH
1 qt = 0.9463 L	1 L = 1.057 qt
1 gal = 3.785 L	1 L = 0.2642 gal
1 fl oz = 29.57 mL	1 mL = 0.03381 fl oz
1 tsp = 4.929 mL	1 mL = 0.2029 tsp
1 tbsp = 14.79 mL	1 mL = 0.06763 tbsp

From this table, we can derive the following approximations that will enable us to mentally estimate conversions from one system to the other.

$$1 \text{ qt} \approx 1 \text{ L}$$
$$1 \text{ fl oz} \approx 30 \text{ mL}$$
$$1 \text{ tsp} \approx 5 \text{ mL}$$
$$1 \text{ tbsp} \approx 15 \text{ mL}$$

Since teaspoons and tablespoons are not measured exactly, the values in Table 8.10 for converting teaspoon and tablespoon to milliliter are given only for the purpose of deriving these approximations.

example 3

Approximate each of the following conversions. Then compare your results to the values obtained by using Table 8.10.

(a) Convert 60 mL to fluid ounces.

Since 30 mL ≈ 1 fl oz,

then 60 mL ≈ 2 fl oz

$\left(\text{Think, } \dfrac{30 \text{ mL}}{1 \text{ fl oz}} = \dfrac{60 \text{ mL}}{?}\right)$

Use Table A.10 to find the exact answer.

$$60 \text{ mL} = 60 \text{ mL} \times \frac{0.03381 \text{ fl oz}}{1 \text{ mL}}$$

$$= 2.0286 \text{ fl oz}$$

Multiply 60 mL by the unit fraction $\dfrac{0.03381 \text{ fl oz}}{1 \text{ mL}}$ so that mL cancel.

2.0286 fl oz ≈ 2.029 fl oz correct to four significant digits. Our approximation is off by about 0.03 fl oz.

$2.0286 \text{ fl oz} - 2 \text{ fl oz} \approx 0.03 \text{ fl oz}$

(b) Convert $2\frac{1}{2}$ gal to litres.

$$2\frac{1}{2} \text{ gal} = \frac{5}{2} \text{ gal} \times \frac{4 \text{ qt}}{1 \text{ gal}} = \frac{5}{2} \times 4 \text{ qt} = 10 \text{ qt}$$

Since 1 gal = 4 qt and 1 qt ≈ 1 L, we will first convert $2\frac{1}{2}$ gal to qt.

Thus, $2\frac{1}{2}$ gal = 10 qt ≈ 10 L.

Use Table 8.10 to find the exact answer.

$$2\frac{1}{2} \text{ gal} = \frac{5}{2} \text{ gal} \times \frac{3.785 \text{ L}}{1 \text{ gal}}$$

$$= 9.4625 \text{ L}$$

Multiply $2\frac{1}{2}$ gal by the unit fraction $\dfrac{3.785 \text{ L}}{1 \text{ gal}}$ so that gal cancel.

9.4625 L ≈ 9.463 L correct to four significant digits. Our approximation was off by about 0.5 L.

$10 \text{ L} - 9.463 \text{ L} \approx 0.5 \text{ L}$

example 4

A pitcher of iced tea that holds $3\frac{1}{2}$ L is divided evenly among 14 people. How much iced tea does each person receive?

$$3\frac{1}{2}\text{ L} = 3.5\text{ L} \times \frac{1{,}000\text{ mL}}{1\text{ L}} = 3{,}500\text{ mL}$$

Since a serving of iced tea is measured in mL, first convert $3\frac{1}{2}$ L to mL and then divide by 14.

3,500 mL ÷ 14 people = 250 mL/person.
Thus, each person receives 250 mL of iced tea. ●

QUICK QUIZ

Peform each of the following conversions.

1. 14 pt = ____ gal
2. 50 mL = ____ L
3. 8,536 L = ____ kL
4. 2 tsp ≈ ____ mL

ANSWER

1. $1\frac{3}{4}$
2. 0.05
3. 8.536
4. 10

8.3 Exercises

Indicate which of the following statements are true and which are false. For those that are false, change the italic expression to make the statement true.

1. A liter is a little *less than* a quart.
2. A teaspoon is a unit of capacity in the *metric* system.
3. A hectoliter is equivalent to $\frac{1}{100}$ liter.
4. One hundred milliliters is *equivalent to* $\frac{1}{10}$ liter.

Convert each of the following English units as instructed.

5. 36 gal = ____ qt
6. 20 pt = ____ c
7. 5 pt = ____ qt
8. 18 qt = ____ gal
9. 28 fl oz = ____ pt
10. $3\frac{1}{2}$ pt = ____ fl oz
11. 5.25 gal = ____ qt
12. 9.5 qt = ____ pt
13. 4 tbsp = ____ tsp
14. 12 tsp = ____ tbsp
15. $5\frac{1}{2}$ qt = ____ pt
16. $8\frac{3}{4}$ gal = ____ qt
17. $4\frac{5}{8}$ pt = ____ fl oz
18. $\frac{3}{4}$ qt = ____ fl oz
19. 1 pt 10 fl oz = ____ pt

20. 9 gal 2 qt = ____ qt
21. 9 pt = ____ gal
22. 20 c = ____ qt
23. 48 c = ____ gal
24. 80 fl oz = ____ gal

Convert each of the following metric units as instructed.

25. 85 L = ____ hL
26. 250 L = ____ kL
27. 62 mL = ____ L
28. 0.9 L = ____ mL
29. 57.6 kL = ____ L
30. 325.8 mL = ____ L
31. 8,800 mL = ____ L
32. 4.37 hL = ____ L
33. 0.897 L = ____ mL
34. 74.8 kL = ____ L
35. $7\frac{1}{2}$ kL = ____ L
36. $\frac{3}{8}$ L = ____ mL
37. $5\frac{3}{4}$ L = ____ mL
38. $\frac{3}{5}$ kL = ____ L
39. 362 L = ____ hL
40. 2.81 L = ____ kL
41. 0.069 L = ____ mL
42. 0.052 hL = ____ mL
43. 92,000 mL = ____ hL
44. 0.008 kL = ____ mL

Indicate which measurement most closely approximates each of the following capacities.

45. A six-pack of beer: 2,100 mL 2,100 L 21 L
46. Capacity of a swimming pool: 75 kL 750 L 7,500 L
47. One gallon of paint: 3.8 L 38 L 380 mL
48. A dosage of cough syrup: 50 mL 5 mL 500 mL
49. Capacity of a gas tank: 20 L 70 L 15 hL
50. A cup of coffee: 2 L 600 mL 250 mL
51. Refrigerator capacity: 40 L 0.4 kL 400 mL
52. A bottle of wine: 1.5 L 15 mL 15 L

Approximate each of the following conversions. Then compare your results to the values obtained by using Table 8.10.

53. 9 qt = ____ L
54. 12 fl oz = ____ mL
55. 25 mL = ____ tsp
56. 2 tbsp = ____ mL
57. 900 mL = ____ fl oz
58. 8 tsp = ____ mL
59. 300 mL = ____ tbsp
60. 7 gal = ____ L
61. 3 L = ____ fl oz
62. 150 fl oz = ____ L

Solve each of the following word problems.

63. A carton of cream contains $\frac{1}{2}$ pt. A pitcher of milk contains 250 mL. Is there more milk or cream?
64. A 32-fl oz bottle of dishwashing detergent sells for \$0.69. Would a 1-L bottle cost more or less?
65. A bottle of cough syrup contains 118 mL. If the recommended dosage is 1 tsp, approximately how many doses are contained in this bottle?
66. A carafe containing 1 L of wine was shared equally among five people. How many milliliters of wine did each person drink?
67. A half gallon of ice cream is equivalent to how many liters?
68. A cookie recipe calls for $\frac{3}{4}$ tsp of ginger. This amount is equivalent to how many milliliters?
69. If gas sells for \$1.89 per gallon, what is the cost of 1 L of gas? Round off your answer to the nearest cent.
70. A can of spray paint that contains 335 mL costs \$5.95. Another can that contains 250 mL costs \$4.35. Which size is the better buy?

8.4 Temperature

Temperature indicates the degree to which something is either hot or cold. Scales that use the freezing and boiling points of water as reference points have been devised to measure differences in temperature.

The temperature scale used in the English system is the **Fahrenheit scale**, named after Gabriel Fahrenheit, the inventor of the mercury thermometer. On this scale, the freezing point of water is 32°F (read as "32 degrees Fahrenheit"), and the boiling point of water is 212°F.

The temperature scale used in the metric system is the **Celsius** (or **centigrade**) **scale**, which was devised by Anders Celsius. On this scale, the freezing point of water is 0°C (zero degrees Celsius), and the boiling point of water is 100°C.

Other useful temperatures to know in both Fahrenheit and Celsius are as follows:

Room temperature 68°F = 20°C

Normal body temperature 98.6°F = 37°C

Oven temperature for roasting 350°F = 175°C

A comparison of the Celsius and Fahrenheit temperature scales is shown in Figure 8.2.

By looking at Figure 8.2, we can see that a change in temperature of 1°C is roughly equivalent to a change of 2°F. For example, 10°C (which is 10°C above the freezing point of water) is roughly equivalent to 32°F (the freezing point of water) plus 20°F (a change of 10°C is approximately equal to 20°F since 1°C $\approx$ 2°F), which equals 52°F.

$$10°C \approx 32°F + 20°F$$
$$10°C \approx 52°F$$

We also can see that 200°F (which is 12°F less than the boiling point of water) is roughly

figure 8.2

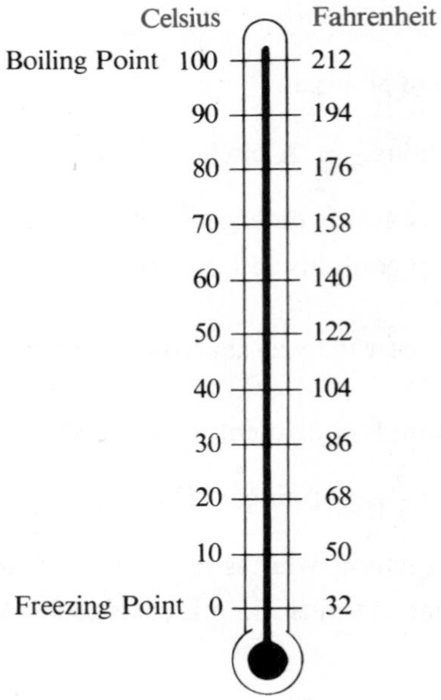

equivalent to 100°C (the boiling point of water) minus 6°C (12°F ≈ 6°C since 2°F ≈ 1°C), which equals 94°C.

$$200°F \approx 100°C - 6°C$$
$$200°F \approx 94°C$$

To determine more exact values when converting from one temperature scale to another, we can use the procedures outlined in the boxes.

To Convert from Celsius to Fahrenheit

1. Multiply degrees Celsius by $\frac{9}{5}$.
2. Add 32.

This can be summarized by the formula

$$F = \frac{9}{5}C + 32$$

where F = °F and C = °C.

example 1 Convert 25°C to °F.

solution

$F = \frac{9}{5}C + 32$ — Use the formula for converting from Celsius to Fahrenheit.

$F = \frac{9}{5}(25) + 32$ — Substitute 25 for C.

$F = \frac{9}{\cancel{5}}(\overset{5}{\cancel{25}}) + 32$ — Perform multiplication before addition. Multiply 25 by $\frac{9}{5}$.

$F = 45 + 32$ — Add 32 and 45.

$F = 77$

Thus, 25°C = 77°F. ●

To Convert from Fahrenheit to Celsius

1. Subtract 32 from degrees Fahrenheit.
2. Multiply the result by $\frac{5}{9}$.

This can be summarized by the formula.

$$C = \frac{5}{9}(F - 32)$$

where C = °C and F = °F.

example 2

Convert 59°F to °C.

solution

$C = \frac{5}{9}(F - 32)$ — Use the formula for converting from Fahrenheit to Celsius.

$C = \frac{5}{9}(59 - 32)$ — Substitute 59 for F.

$C = \frac{5}{9}(27)$ — Do subtraction in ( ) first.

$C = \frac{5}{\cancel{9}}(\overset{3}{\cancel{27}})$ — Multiply 27 by $\frac{5}{9}$.

$C = 15$

Thus, 59°F = 15°C.

example 3

Approximate each of the following conversions. Then determine the exact answers.

(a) Convert 0°F to °C.

$0°F \approx 0°C - 16°C$

$0°F \approx -16°C$

0°F is 32°F below the freezing point of water. Since 1°C ≈ 2°F, subtract 16°C from 0°C (the freezing point of water).

Determine the exact answer.

$C = \frac{5}{9}(F - 32)$ — Use the formula.

$C = \frac{5}{9}(0 - 32)$

$C = \frac{5}{9}(-32)$

$C = -\frac{160}{9} = -17\frac{7}{9}$

Thus, $0°F = -17\frac{7}{9}°C$

Our approximation is off by $1\frac{7}{9}°C$. — $-17\frac{7}{9}°C - (-16°C) = -1\frac{7}{9}°C$

(b) Convert 150°C to °F.

$150°C \approx 212°F + 100°F$

$150°C \approx 312°F$

150°C is 50°C above the boiling point of water. Since 1°C ≈ 2°F, add 100°F to 212°F (the boiling point of water).

Determine the exact answer.

$F = \frac{9}{5}C + 32$ — Use the formula.

$F = \frac{9}{5}(150) + 32$

$F = 270 + 32$

$F = 302$

Thus, 150°C = 302°F.

Our approximation is off by 10°F. 312°F − 302°F = 10°F

QUICK QUIZ

Perform each of the following conversions.

1. 85°C = ____ °F

2. 5°F = ____ °C

3. −10°C = ____ °F

ANSWERS

1. 185

2. −15

3. 14

8.4 Exercises

Indicate which of the following statements are true and which are false. For those that are false, change the italic expression to make the statement true.

1. The freezing point of water is 0°*F*.

2. The *metric* system uses the Fahrenheit scale.

3. A change of temperature of 2°F is roughly equivalent to a change of *1°C*.

4. 120°F is *warmer than* 120°C.

Indicate which measurement most closely approximates each of the following temperatures.

5. Low fever: 99°C 38°C 41°C

6. Ice skating rink: −1°C 32°C 20°C

7. Baking temperature for cookies: 120°C 190°C 375°C

8. Broiling temperature: 190°C 550°C 290°C

9. Hot summer day: 30°C 85°C 20°C

10. Hot coffee: 200°C 40°C 90°C

Convert each of the following temperatures as instructed. Round off to the nearest degree.

11. 68°F = ____°C
12. 95°F = ____°C
13. 23°F = ____°C
14. 41°F = ____°C
15. 35°C = ____°F
16. 45°C = ____°F
17. 115°C = ____°F
18. 95°C = ____°F
19. −4°F = ____°C
20. −22°F = ____°C
21. −5°C = ____°F
22. −20°C = ____°F
23. 67°C = ____°F
24. 82°C = ____°F
25. 24°C = ____°F
26. 13°C = ____°F
27. 52°F = ____°C
28. 89°F = ____°C
29. 130°F = ____°C
30. 106°C = ____°F
31. −2°C = ____°F
32. −15°F = ____°C

Approximate each of the following conversions. Then determine the exact answers.

33. 77°F = ____°C
34. 113°F = ____°C
35. 65°C = ____°F
36. 30°C = ____°F
37. 59°F = ____°C
38. 239°F = ____°C
39. 90°C = ____°F
40. 75°C = ____°F
41. −40°F = ____°C
42. −15°C = ____°F
43. 28°F = ____°C
44. 91°F = ____°C
45. 141°C = ____°F
46. −8°C = ____°F

8.5 Addition and Subtraction of Units of Measurement

Many applications of addition and subtraction involve working with units of measurement. The list of equivalent units of measurement printed on the inside cover of this book will be useful in solving these kinds of problems. Abbreviations for each unit are given in parentheses. You should memorize these relationships, if you do not already know them.

Addition of Units of Measurement

All that you have learned about addition can be applied directly to the operation of adding units of measurement. The following examples illustrate how the notion of carrying can be extended to adding units of measurement.

example 1

Add 1 hr 45 min + 2 hr 50 min.

solution

$$\begin{array}{r} 1 \text{ hr } 45 \text{ min} \\ +2 \text{ hr } 50 \text{ min} \\ \hline 3 \text{ hr } 95 \text{ min} \end{array}$$

Write the problem vertically and line up like units of measurement.

Add the numbers in the minutes and hours columns separately.

$= 3 \text{ hr} + \overbrace{60 \text{ min} + 35 \text{ min}}$

Since 60 min = 1 hr, rewrite 95 min as 1 hr + 35 min.

$= \underbrace{3 \text{ hr} + 1 \text{ hr}} + 35 \text{ min}$

$= 4 \text{ hr} + 35 \text{ min}$

$= 4 \text{ hr } 35 \text{ min}$

Carry 1 to the hours column to obtain a final answer of 4 hr 35 min. ●

example 2

Add 4 yd 1 ft 7 in. + 3 yd 2 ft 9 in.

solution

$$\begin{array}{r} 4 \text{ yd } 1 \text{ ft } \; 7 \text{ in.} \\ +3 \text{ yd } 2 \text{ ft } \; 9 \text{ in.} \\ \hline 7 \text{ yd } 3 \text{ ft } 16 \text{ in.} \end{array}$$

Write the problem vertically and line up like units of measurement.

Add the numbers in the inches, feet, and yards columns separately.

$= 7 \text{ yd} + 3 \text{ ft} + \overbrace{12 \text{ in.} + 4 \text{ in.}}$

Since 12 in. = 1 ft, rewrite 16 in. as 1 ft 4 in.

$= 7 \text{ yd} + \underbrace{3 \text{ ft} + 1 \text{ ft}} + 4 \text{ in.}$

$= 7 \text{ yd} + 4 \text{ ft} + 4 \text{ in.}$

Carry 1 to the feet column to get 7 yd + 4 ft + 4 in.

$= 7 \text{ yd} + \overbrace{1 \text{ yd} + 1 \text{ ft}} + 4 \text{ in.}$

Since 3 ft = 1 yd, rewrite 4 ft as 1 yd + 1 ft.

$= 8 \text{ yd} + 1 \text{ ft} + 4 \text{ in.}$

$= 8 \text{ yd } 1 \text{ ft } 4 \text{ in.}$

Carry 1 to the yards column to obtain a final answer of 8 yd 1 ft 4 in. ●

example 3

While on vacation, a tourist bought 2 pounds 7 ounces of chocolate fudge, 10 ounces of vanilla fudge, 1 pound 4 ounces of butterscotch fudge, and 1 pound 5 ounces of strawberry fudge. What was the total amount of fudge purchased?

solution Add 2 lb 7 oz + 10 oz + 1 lb 4 oz + 1 lb 5 oz.

$$\begin{array}{r} 2\text{ lb}\ \ 7\text{ oz} \\ 10\text{ oz} \\ 1\text{ lb}\ \ 4\text{ oz} \\ +1\text{ lb}\ \ 5\text{ oz} \\ \hline 4\text{ lb } 26\text{ oz} \end{array}$$

$= 4\text{ lb} + \overbrace{16\text{ oz} + 10\text{ oz}}$

$= \underbrace{4\text{ lb} + 1\text{ lb}} + 10\text{ oz}$

Since 16 oz = 1 lb, rewrite 26 oz as 1 lb 10 oz.

$= \quad 5\text{ lb} + 10\text{ oz}$

$= \quad 5\text{ lb } 10\text{ oz}$

The tourist purchased 5 pounds 10 ounces of fudge. ●

Subtraction of Units of Measurement

All that you have learned about subtraction can be applied directly to the operation of subtracting units of measurement. The following examples illustrate how the notion of borrowing can be extended to subtracting units of measurement.

example 4

Subtract 5 ft 3 in. − 2 ft 9 in.

solution

$$\begin{array}{rr} 5\text{ ft } 3\text{ in.} = & 4\text{ ft}\ \ 15\text{ in.} \\ -\ 2\text{ ft } 9\text{ in.} = & 2\text{ ft}\ \ \ 9\text{ in.} \\ \hline & 2\text{ ft}\ \ \ 6\text{ in.} \end{array}$$

Write the problem vertically and line up like units of measurement.

1. Since 3 in. < 9 in., we must borrow. Since 12 in. = 1 ft, we borrow 1 ft from the feet column to get 4 ft, and add 12 in. to the inches column to get 15 in.
2. Subtract the numbers in the inches column: 15 in. − 9 in. = 6 in.
3. Subtract the numbers in the feet column: 4 ft − 2 ft = 2 ft. The result is 2 ft 6 in. ●

example 5

A swimmer began a race at 3:55:44 P.M. and finished at 4:03:18 P.M. How long did it take her to finish the race?

solution

3:55:44 means 3 hr 55 min 44 sec.
4:03:18 means 4 hr 3 min 18 sec.
Subtract 4 hr 3 min 18 sec − 3 hr 55 min 44 sec.

$$\begin{array}{rrr} 4\text{ hr}\ \ 3\text{ min } 18\text{ sec} = & 4\text{ hr}\ \ 2\text{ min } 78\text{ sec} = & 3\text{ hr } 62\text{ min } 78\text{ sec} \\ -\ 3\text{ hr } 55\text{ min } 44\text{ sec} = & -\ 3\text{ hr } 55\text{ min } 44\text{ sec} = & -\ 3\text{ hr } 55\text{ min } 44\text{ sec} \\ \hline & 34\text{ sec} & 7\text{ min } 34\text{ sec} \end{array}$$

1. Since 18 sec < 44 sec, we must borrow. Since 1 min = 60 sec, borrow 1 min from the minutes column to get 2 min, and add 60 sec to the seconds column to get 78 sec.
2. Subtract the numbers in the seconds column: 78 sec − 44 sec = 34 sec.
3. Since 2 min < 55 min, we must borrow. Since 1 hr = 60 min, borrow 1 hr from the hours column to get 3 hr, and add 60 min to the minutes columns to get 62 min.
4. Subtract the numbers in the minutes column: 62 min − 55 min = 7 min.
5. Subtract the numbers in the hours column: 3 hr − 3 hr = 0 hr.
6. The result is 7 min 34 sec. ●

QUICK QUIZ

Perform the operations indicated.

1. 18 min 40 sec + 25 min 28 sec
2. 5 lb 2 oz − 3 lb 9 oz

ANSWERS

1. 44 min 8 sec
2. 1 lb 9 oz

8.5 Exercises

Solve each of the following addition problems.

1. 7 ft 7 in.
+8 ft 6 in.
2. 3 hr 20 min
+5 hr 45 min
3. 4 lb 12 oz
+1 lb 8 oz
4. 6 gal 1 qt
+9 gal 3 qt
5. 7 wk 5 da
+4 wk 3 da
6. 8 qt 1 pt
+1 qt 1 pt
7. 5 hr 20 min 45 sec
+1 hr 40 min 15 sec
8. 2 yd 1 ft 8 in.
+3 yd 2 ft 6 in.

Solve each of the following subtraction problems.

9. 8 ft 3 in.
−4 ft 7 in.
10. 6 hr 15 min
−2 hr 45 min
11. 9 lb 5 oz
−3 lb 10 oz
12. 5 gal 1 qt
−1 gal 3 qt
13. 4 wk 2 da
−3 wk 5 da
14. 5 qt
−2 qt 1 pt
15. 9 hr 18 min 21 sec
−3 hr 25 min 48 sec
16. 5 yd 1 ft 2 in.
−1 yd 1 ft 10 in.

Solve each of the following word problems.

17. If a student worked on his math homework 50 minutes on Saturday and 1 hour 35 minutes on Sunday, what is the total amount of time he spent on his math homework that weekend?
18. If the electric cord on a lamp is 2 feet 8 inches long and is attached to an extension cord 3 feet 6 inches in length, how far from the nearest electrical outlet can the lamp be placed?
19. A woman buys the following amounts of nuts: 1 pound 3 ounces of pistachios, 2 pounds of peanuts, 10 ounces of cashews, and 1 pound 5 ounces of walnuts. What is the total weight of the nuts she has purchased?
20. At a company picnic, employees brought the following amounts of beverages: 2 gallons 1 quart of lemonade, 3 quarts of cola, and 1 gallon 2 quarts of iced tea. What is the total quantity of beverages brought to the picnic?
21. If a man is 6 feet 2 inches tall, and his son is 5 feet 9 inches tall, how much taller is the father than his son?
22. If a student began studying at 2:45 P.M. and finished at 5:10 P.M., how long did she study?
23. A bag of apples weighs 8 pounds 5 ounces, and a bag of oranges weighs 5 pounds 12 ounces. How much more do the apples weigh than the oranges?
24. An athlete made 4 gallons of cranberry-apple juice by mixing apple juice with 2 gallons 3 quarts of cranberry juice. How much apple juice did he use?

Summary and Review

Key Terms

[8.1] To **measure** something is to find a number that characterizes one of its properties.
A **unit of measurement** indicates the size of a known quantity.

The **English system** is the system of measurement used in the United States.

The **metric system** is the system of measurement used throughout most of the world.

The **meter** is the standard unit of length in the metric system.

[8.2] The **weight** of an object is the force of the earth's gravitational pull acting upon it.

The **gram** is the standard unit of weight in the metric system.

[8.3] The **liter** is the standard unit of capacity (volume) in the metric system.

[8.4] **Temperature** indicates the degree to which something is either hot or cold.

The **Fahrenheit** scale is used to measure temperature in the English system.

The **Celsius** (or **centigrade**) scale is used to measure temperature in the metric system.

[8.1] **TABLE** Metric Prefixes

milli—thousandth	deka—ten
centi—hundredth	hecto—hundred
deci—tenth	kilo—thousand

Calculations

[8.1] **To convert from one unit of measurement to another**, multiply by the appropriate unit fraction(s).

[8.4] **To convert from Celsius to Fahrenheit**, multiply degrees Celsius by $\frac{9}{5}$ and then add 32.

$$F = \frac{9}{5}C + 32$$

To convert from Fahrenheit to Celsius, subtract 32 from degrees Fahrenheit, and then multiply the result by $\frac{5}{9}$.

$$C = \frac{5}{9}(F - 32).$$

Chapter 8 Review Exercises

Indicate which of the following statements are true and which are false. For those that are false, change the italic expression to make the statement true.

1. The prefix centi means *hundred*.
2. One kilogram is a little *less than* 2 pounds.
3. Four liters is a little *more than* one gallon.
4. The boiling point of water is *100°*F.

[8.1–8.4] Indicate which measurement most closely approximates each of the following lengths, weights, capacities, or temperatures.

5. Length of a tennis court: 2.38 m 23.8 m 238 m
6. Length of skis: 175 cm 17.5 m 175 mm
7. Weight of a tube of toothpaste: 2.32 kg 232 g 2.32 g
8. Weight of a vitamin tablet: 250 mg 25 g 2.5 mg
9. Capacity of a tea kettle: 5 L 2 L 200 mL
10. Capacity of a trash can: 14 L 40 L 114 L
11. Length of a birthday candle: 0.6 m 600 mm 6 cm
12. Small can of frozen orange juice: 177 mL 17 mL 1.7 L

13. Normal body temperature: 37°C 40°C 98.6°C
14. Cold winter day: −2°C 20°C 32°C

[8.1–8.4] Convert each of the following units of measurement as instructed.

15. 8 in. = ____ ft
16. $2\frac{1}{2}$ mi = ____ ft
17. 36 oz = ____ lb
18. 0.4 lb = ____ oz
19. 6 pt = ____ qt
20. 2 gal = ____ pt
21. 300 m = ____ km
22. 72 mm = ____ cm
23. 8,200 g = ____ kg
24. 6,400 cg = ____ g
25. 53 L = ____ mL
26. 2.7 hL = ____ L
27. 22 yd = ____ ft
28. 40 in. = ____ yd
29. $5\frac{1}{4}$ T = ____ lb
30. 7,500 lb = ____ T
31. 5 tsp = ____ tbsp
32. 60 fl oz = ____ pt
33. 9.2 dam = ____ dm
34. 0.59 hm = ____ mm
35. 0.23 g = ____ mg
36. 75 mg = ____ cg
37. 0.86 kL = ____ L
38. 450 L = ____ kL
39. 50°F = ____ °C
40. 176°F = ____ °C
41. 85°C = ____ °F
42. −10°C = ____ °F
43. $7\frac{1}{2}$ m = ____ dm
44. $\frac{3}{4}$ km = ____ hm
45. 0.018 kg = ____ mg
46. 5,400 cg = ____ kg
47. 63,000 mL = ____ kL
48. 0.08 hL = ____ mL
49. 21°F = ____ °C
50. 102°C = ____ °F

[8.1–8.4] Approximate each of the following conversions. Then use tables to determine more precise answers.

51. 150 cm = ____ in.
52. 500 mi = ____ km
53. 44 lb = ____ kg
54. 7 oz = ____ g
55. 3 gal = ____ L
56. 75 mL = ____ tsp
57. 72°F = ____ °C
58. 5°C = ____ °F

[8.1–8.5] Solve each of the following word problems.

59. A delicatessen uses 250 g of roast beef for every roast beef sandwich they make. About how many sandwiches can be made from an 8.8-lb piece of roast beef?

60. A box that contains 24 tea bags weighs 37 g. If the box weighs 7 g, what is the weight of a single tea bag?

61. The distance from Tokyo to Berlin is 5,540 mi. What is this distance in kilometers?

62. If an extension cord is 2.4 m long, will it reach from an outlet to a lamp cord that is $8\frac{1}{2}$ ft away?

63. A 237-mL bottle of suntan lotion costs $2.28. A 120-mL bottle costs $1.19. Which size bottle is the better buy?

64. A 2-lb box of sugar costs $1.89. What is the cost of 1 kg of sugar? Round off your answer to the nearest cent.

65. A nurse found a patient's temperature to be 39°C. What is the equivalent temperature in °F?

66. How much will it cost to fill a 70-L gas tank if gas costs $1.79 per gallon? Round off your answer to the nearest cent.

67. If a long-distance runner began a race at 1:47 P.M. and finished at 4:02 P.M., how long did it take him to run the race?

68. If one bag of groceries weighs 5 pounds 7 ounces and a second bag weighs 4 pounds 10 ounces, what is the combined weight of both bags of groceries?

9 INTRODUCTION TO GEOMETRY

9.1 Geometric Figures

In this section, we will learn to identify geometric figures known as **polygons**.

definition

A *polygon* is a closed figure that has many sides. A *regular polygon* is a polygon in which all sides and all angles are equal.

The names of certain polygons indicate the number of sides that form the figure. For example,

*Tri*angle: *three*-sided figure

*Quad*rilateral: *four*-sided figure

*Pent*agon: *five*-sided figure

*Hex*agon: *six*-sided figure

*Oct*agon: *eight*-sided figure

Figure 9.1 shows an example of each of these polygons.

figure 9.1

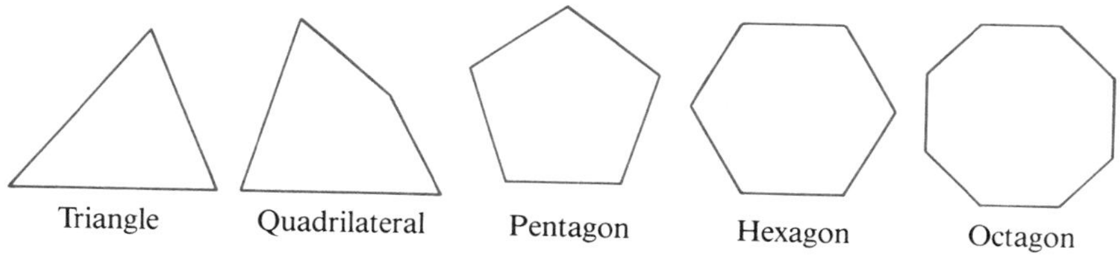

The shape of any polygon is determined by the length of each side and the size of each **angle**.

definition

An *angle* is formed whenever two lines meet. The two lines are called the *sides* of the angle, and the point at which they meet is called the *vertex* of the angle.

A typical angle is illustrated in Figure 9.2. Notice that each corner of a polygon is the vertex of an angle.

figure 9.2

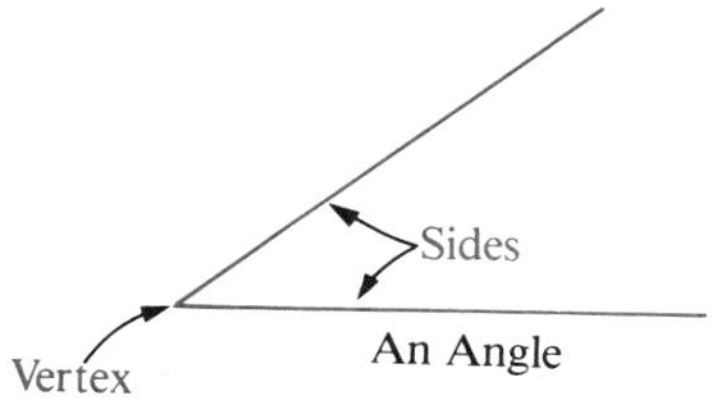

The size of an angle is measured in **degrees** (°). If you were to stand in one place, but pivot on your feet to turn one complete revolution, you would have turned 360°. One-half of a revolution is equivalent to 180°, and one-quarter of a revolution is equivalent to 90°. Notice that a straight line represents a 180° angle, and the intersection of a horizontal and a vertical line represents a 90° angle, as shown in Figure 9.3.

figure 9.3

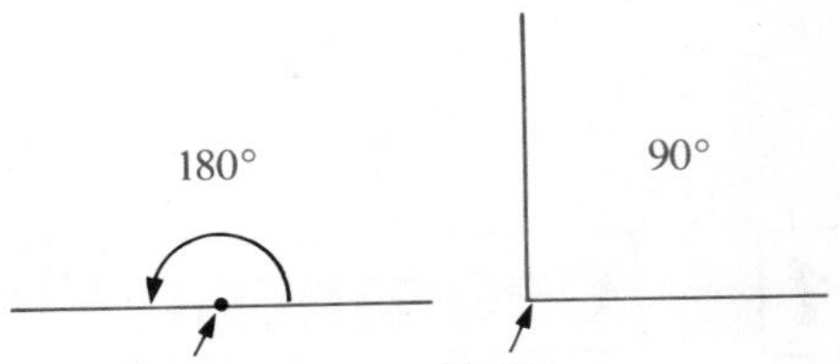

A 90° angle is sometimes called a **right angle**. Angles that are less than 90° are called **acute**, and angles that are greater than 90° and less than 180° are called **obtuse**. These are illustrated in Figure 9.4. Notice that a right angle is indicated by forming a small box at the vertex. Lines that meet in a right angle are said to be **perpendicular**.

figure 9.4

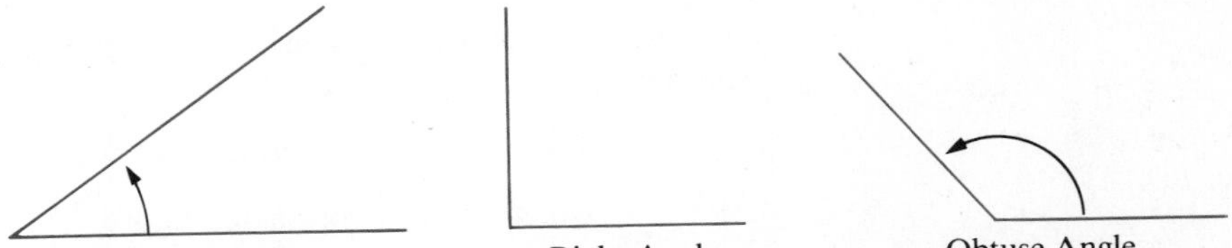

Acute Angle Right Angle Obtuse Angle

Now that we know the different types of angles, we can classify the different kinds of triangles as follows:

An **equilateral** triangle has three equal sides and three equal angles.

An **isosceles** triangle has two equal sides and two equal angles.

A **right** triangle has a right angle.

These triangles are illustrated in Figure 9.5.

figure 9.5

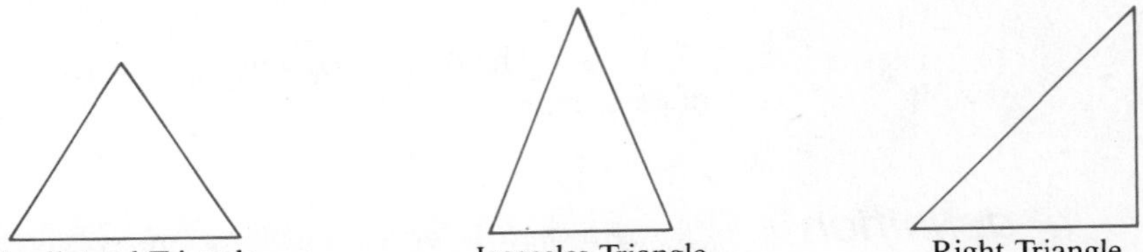

Equilateral Triangle Isosceles Triangle Right Triangle

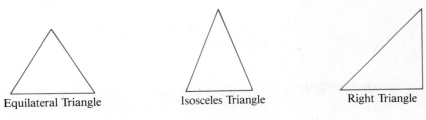

Many quadrilaterals contain lines that are **parallel**.

definition

***Parallel lines* are lines in the same plane that never meet.**

Figure 9.6 shows a pair of parallel lines. Notice that parallel lines are always the same distance apart, regardless of how far they are extended.

figure 9.6

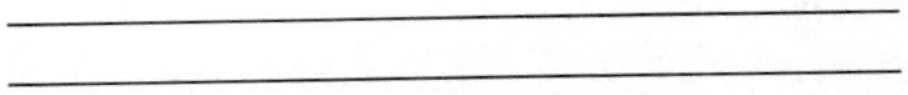

Parallel Lines

Now that we understand parallel lines, we can classify the different kinds of quadrilaterals as follows:

A **trapezoid** is a quadrilateral that has two parallel sides.

A **parallelogram** is a quadrilateral whose opposite sides are equal and parallel.

A **rectangle** is a parallelogram that has four right angles.

A **square** is a rectangle that has four equal sides.

These quadrilaterals are illustrated in Figure 9.7.

figure 9.7

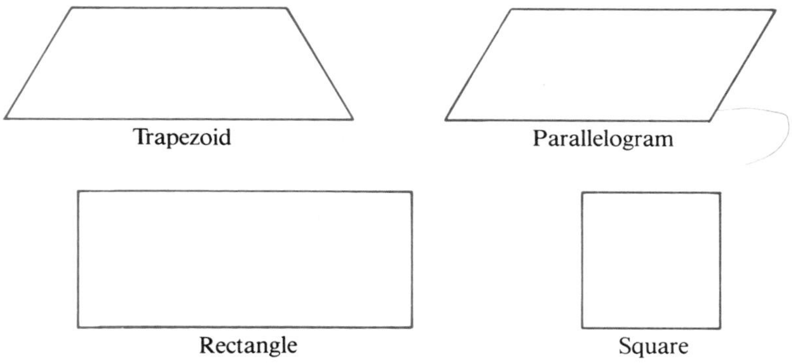

9.1 Exercises

Indicate whether each of the following statements are true or false. For those that are false, change the italic expression to make the statement true.

1. Lines that meet in a right angle are said to be *parallel.*
2. The sides of a *regular* polygon are all equal.
3. An isosceles triangle has *three* equal sides.
4. A rectangle has four equal *sides.*
5. Perpendicular lines meet in a *45°* angle.
6. A straight line represents a *180°* angle.
7. A pentagon has *six* sides.
8. A *trapezoid* is a quadrilateral whose opposite sides are parallel.
9. A right triangle has a *45°* angle.
10. The point where the two sides of an angle meet is called the *apex.*
11. An obtuse angle is greater than *90°.*
12. One complete revolution is a *180°* turn.

Choose the measurement that correctly indicates the size of each of the following angles.

13.

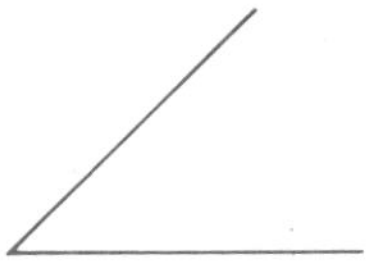

45° 90° 120°

14.

60° 90° 100°

15.

100° 120° 180°

16.

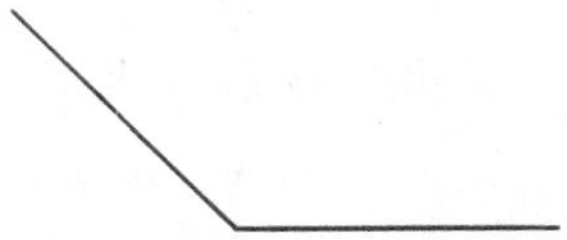

90° 135° 180°

17.

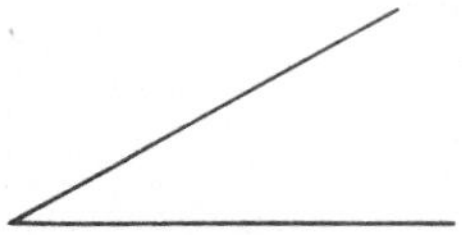

30° 75° 90°

18.

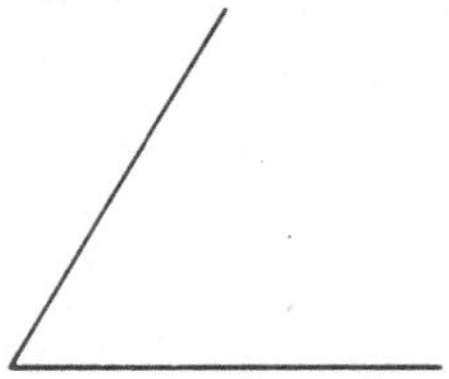

15° 60° 90°

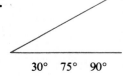

Choose the word from the following list that most completely describes each of the figures in Exercises 19–26.

triangle	octagon	parallelogram
quadrilateral	equilateral triangle	trapezoid
pentagon	isosceles triangle	rectangle
hexagon	right triangle	square

19.

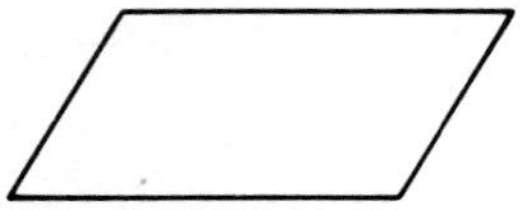

20.

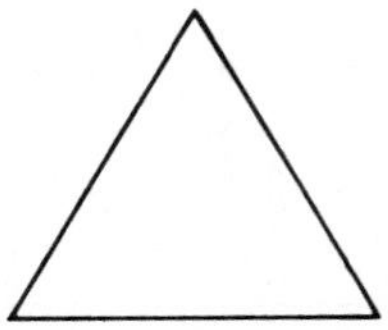

21.

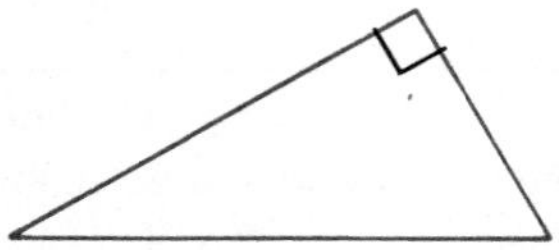

22.

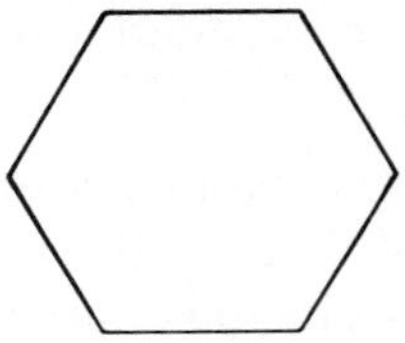

23.

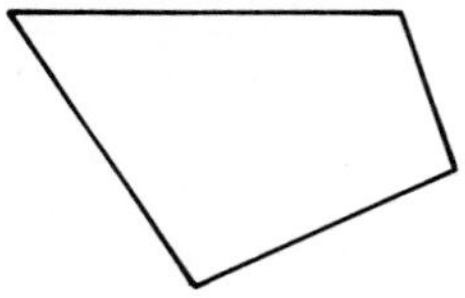

24.

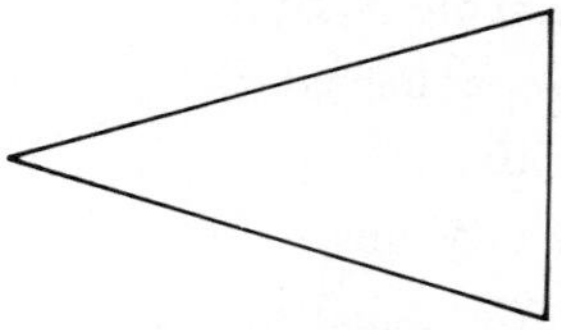

25.

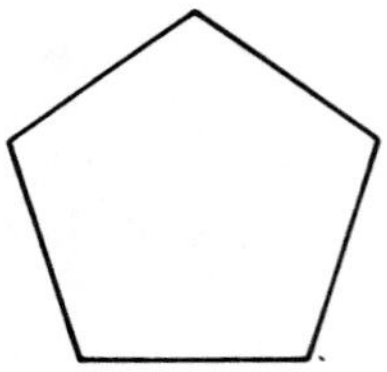

26.

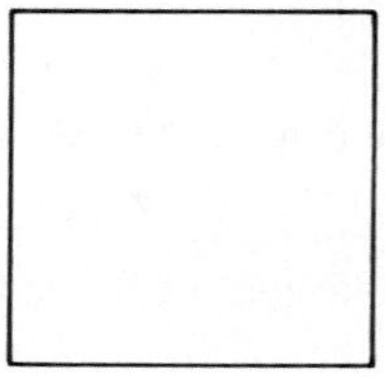

9.2 Perimeter

We will now discuss some characteristics of geometric figures that have widespread applications. The first of these is **perimeter**.

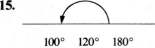

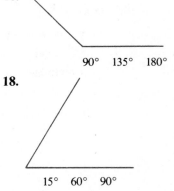

definition

> ***Perimeter* is the distance around a geometric figure.**

To find the perimeter of any figure, we find the sum of the lengths of the sides. Since length is measured in linear, or one-dimensional, units, perimeter is also a one-dimensional quantity.

example 1

Find the perimeter of the accompanying figure.

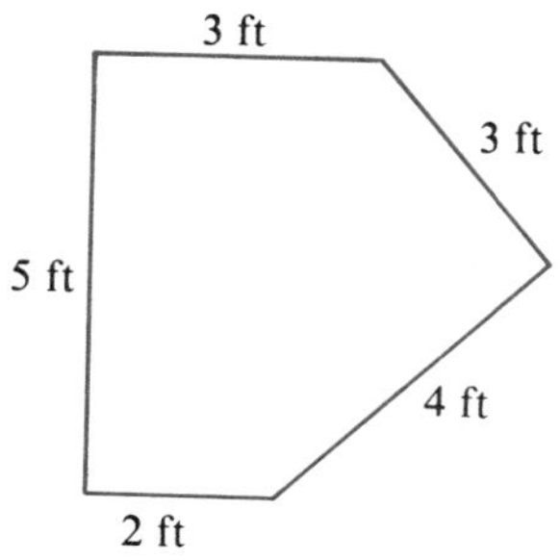

solution

$P = 3 + 3 + 4 + 2 + 5$ — Find the sum of the sides.

$P = 17$ ft — Since all dimensions are given in feet, the perimeter is also in feet. ●

We will now establish some formulas for finding the perimeters of certain geometric figures. To find the perimeter of a rectangle whose length is indicated by the letter l, and whose width is indicated by the letter w, as shown in Figure 9.8, we can find the sum of the sides.

figure 9.8

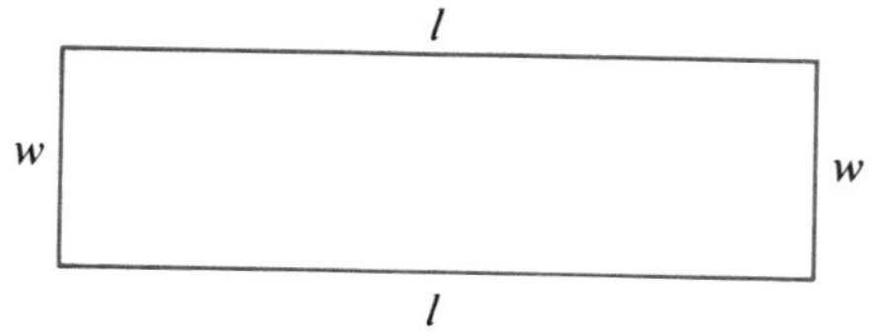

$$P = l + w + l + w$$
$$P = l + l + w + w$$
$$P = 2l + 2w$$

We have thus established the formula in the box.

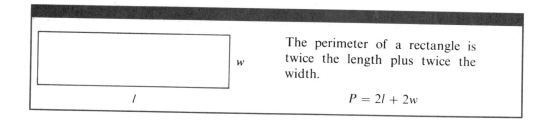

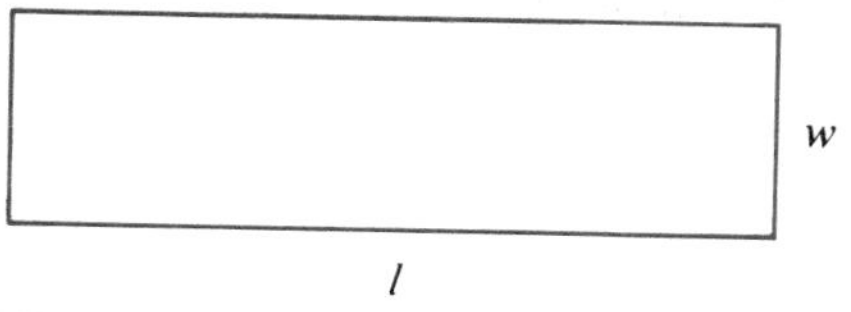

The perimeter of a rectangle is twice the length plus twice the width.

$$P = 2l + 2w$$

example 2

Find the perimeter of a rectangle whose length is 7 m and whose width is 3 m.

solution

$P = 2l + 2w$ — Use the formula for finding the perimeter of a rectangle.

$P = 2(7) + 2(3)$ — Substitute 7 for l and 3 for w.

$P = 14 + 6$ — Do multiplication before addition.

$P = 20$ m — Since l and w are given in meters, so is the perimeter.

example 3

Find the perimeter of a regular pentagon whose sides measure 7 cm each.

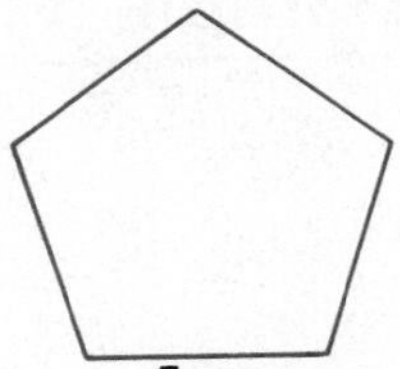

solution

$P = 7 + 7 + 7 + 7 + 7$

$P = 5 \cdot 7$

$P = 35$ cm

Since a regular pentagon has five equal sides, each 7 cm long, the perimeter of this figure is equal to the sum of five 7s, which simplifies to the product $5 \cdot 7$.

The formula for finding the perimeter of *any regular polygon* with n sides, each of length s, is presented in the box.

> The perimeter of any regular polygon is the product of the number of sides, n, times the length of each side, s.
>
> $$P = ns$$

For example,

$P = 3s$ gives the perimeter of an equilateral triangle.

$P = 4s$ gives the perimeter of a square.

$P = 5s$ gives the perimeter of a regular pentagon.

$P = 6s$ gives the perimeter of a regular hexagon.

$P = 8s$ gives the perimeter of a regular octagon.

example 4

How many yards of trimming are needed to decorate the border of a 54-inch square tablecloth?

solution

Find the perimeter of the tablecloth.

$P = 4s$ — Use the formula for finding the perimeter of a square.

$P = 4(54)$ — Substitute 54 for s.

$P = 216$ in.

Find the number of yards of trimming needed.

$$216 \text{ in.} = 216 \text{ in.} \times \frac{1 \text{ yd}}{36 \text{ in.}}$$

We must convert 216 in. to yds.

$$= \frac{216}{36} \text{ yd} = 6 \text{ yd}$$

example 5

The backyard of a suburban home is 40 ft long and 12 yd wide. How much fencing is needed to enclose the yard?

solution

Find the perimeter.

$$12 \text{ yd} = 12 \text{ yd} \times \frac{3 \text{ ft}}{1 \text{ yd}} = 36 \text{ ft}$$

First convert 12 yd to ft.

$$P = 2l + 2w$$

Use the formula for finding the perimeter of a rectangle.

$$P = 2(40) + 2(36)$$

Substitute $l = 40$ and $w = 36$.

$$P = 80 + 72$$

Do multiplication before addition.

$$P = 152 \text{ ft}$$

Since both length and width are in ft, so is the perimeter.

QUICK QUIZ

Find the perimeters of each of the following figures.

1. Rectangle of length 5 in. and width 8 in.
2. Equilateral triangle whose side measures 9 m.
3. Regular hexagon whose side measures 30 cm.

ANSWERS

1. 26 in.
2. 27 m
3. 180 cm

9.2 Exercises

State which of the following statements are true and which are false. For those that are false, change the italic expression to make the statement true.

1. The distance around a figure is called its *perimeter*.
2. The perimeter of a rectangle is always *less than* its length.
3. The perimeter of any object is the *product* of the lengths of all of the sides.
4. The perimeter of a regular *octagon* is six times the length of a side.

Find the perimeters of each of the following polygons. All figures are drawn to scale.

5.

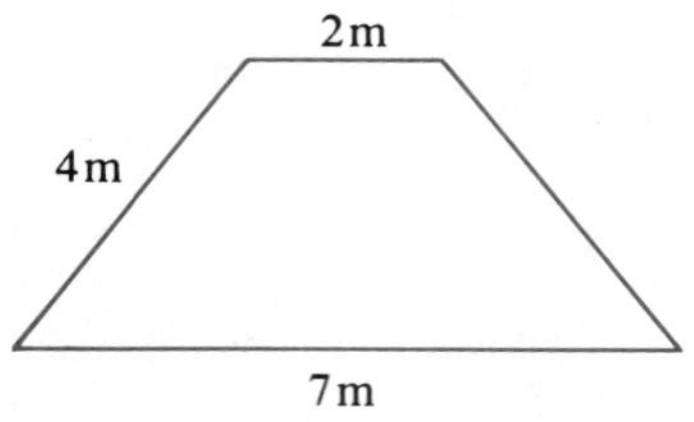

6.

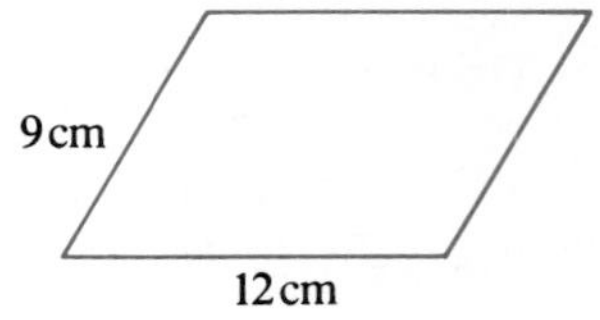

7.

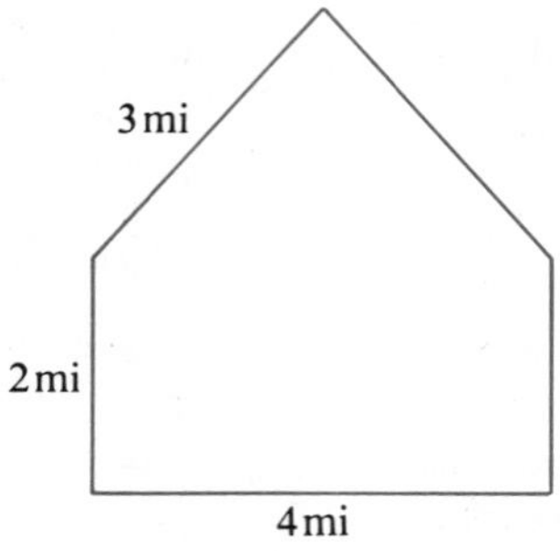

8.

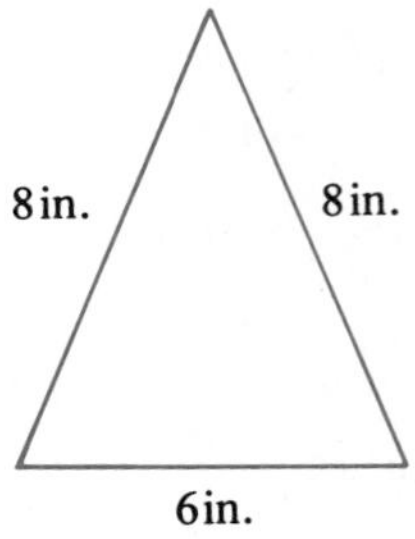

9.

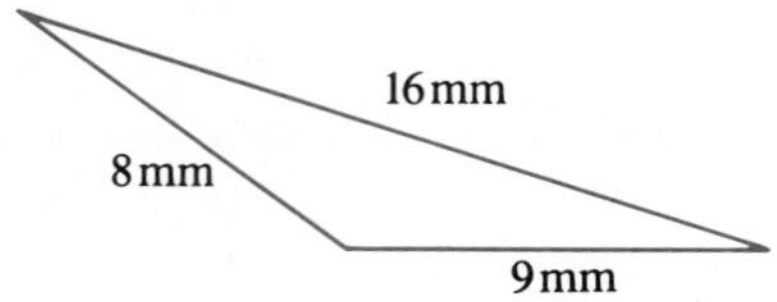

10.

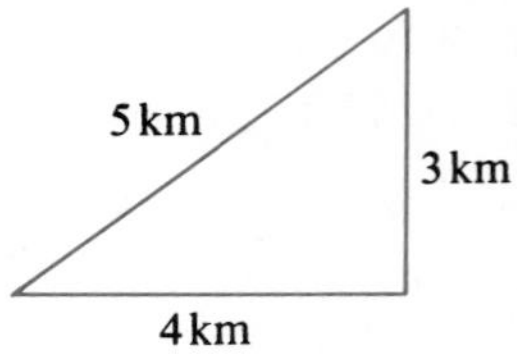

11.

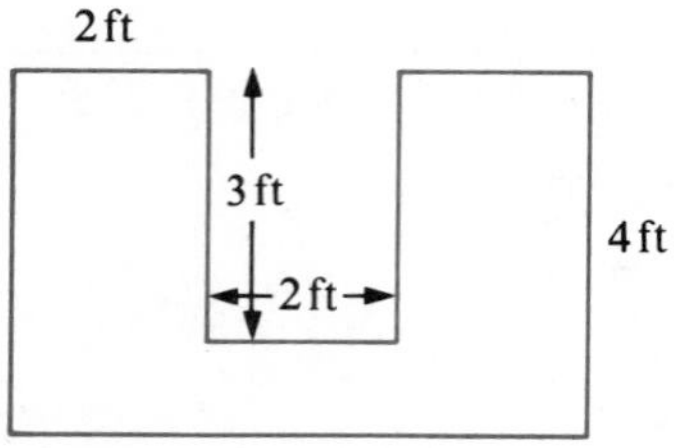

12.

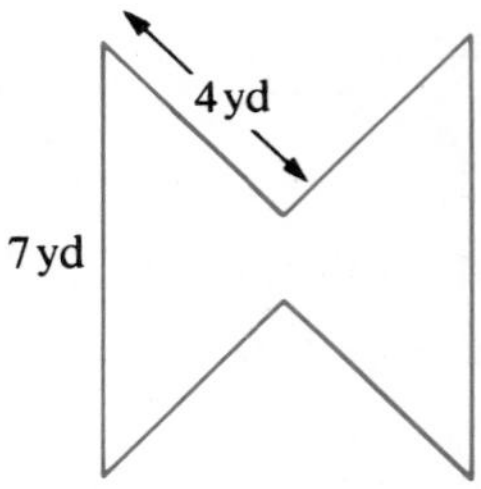

Find the perimeter of each of the following figures.

13. Rectangle of length 9 in. and width 5 in.

14. Rectangle of length 5 yd and width 2 yd.

15. Equilateral triangle whose side measures 8 km.

16. Equilateral triangle whose side measures 15 hm.

17. Square whose side measures 1.2 m.

18. Square whose side measures 3.1 cm.

19. Rectangle of length $7\frac{1}{2}$ yd and width $1\frac{1}{4}$ yd.

20. Rectangle of length $3\frac{5}{8}$ ft and width $2\frac{1}{8}$ ft.

21. Regular hexagon whose side measures 3 cm.

22. Regular hexagon whose side measures 20 ft.

23. Square whose side measures 15 mm.

24. Square whose side measures 42 m.

25. Regular pentagon whose side measures 9 dm.

26. Regular octagon whose side measures 20 dam.

27. Rectangle of length 2 ft and width 9 in.

28. Rectangle of length 1 yd and width 2 ft.

Solve each of the following word problems.

29. A garden measures 10 ft long and 8 ft wide. How many feet of fencing are needed to enclose the garden?

30. A farmer wishes to fence in a corral that is 80 ft long and 60 ft wide. How much fencing does he need?

31. A television screen that is 42 cm long and 36 cm wide is surrounded by a chrome border. How many centimeters of chrome were used to make the border?
32. A tablecloth that is 72 in. long and 60 in. wide has a decorated border. How many feet of trimming were used to make it?
33. How many feet of caulking are needed to insulate a rectangular window of length 4 ft and width $2\frac{1}{2}$ ft?
34. A shopkeeper wishes to put a string of colored lights along the border of a sign that measures $1\frac{1}{2}$ m long and $\frac{3}{4}$ m wide. How long must the light string be?
35. The front of a house is decorated with a wooden trimming in the shape of an equilateral triangle. If one side is 7.28 m long, how many meters of wood were needed to construct the triangle?
36. How many meters of baseboard are needed for a room that is 4.15 m long and 2.85 m wide?

9.3 Area

Another characteristic of geometric figures with practical applications is **area**.

definition

***Area* is the measurement of the surface of a figure.**

To measure area we use square units of measurement, because area is a two-dimensional quantity. For example, a square inch (abbreviated as sq in. or in.^{2}) represents the area of a square whose sides are each 1 inch in length. A square centimeter (abbreviated as sq cm or cm^2) represents the area of a square whose sides are each 1 centimeter in length. These are both illustrated in Figure 9.9.

figure 9.9

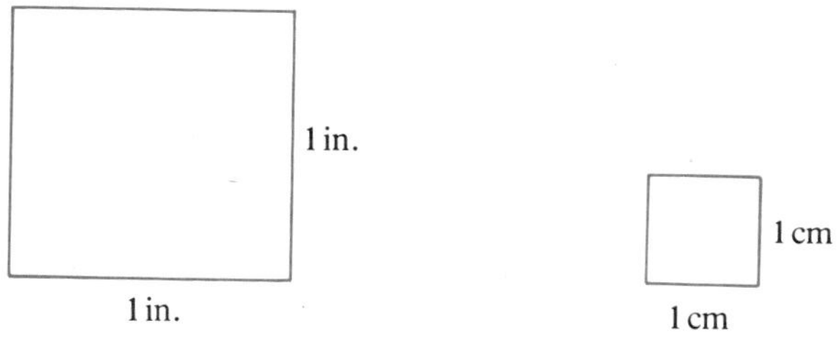

1 square inch (1 sq in. or 1 in.^{2})

1 square centimeter (1 sq cm or 1 cm^2)

Let us now consider how to find the area of the rectangle shown in Figure 9.10. Here we have a rectangle 7 inches long and 3 inches wide. The problem of finding the area of this rectangle is the same as determining how many square blocks,

figure 9.10

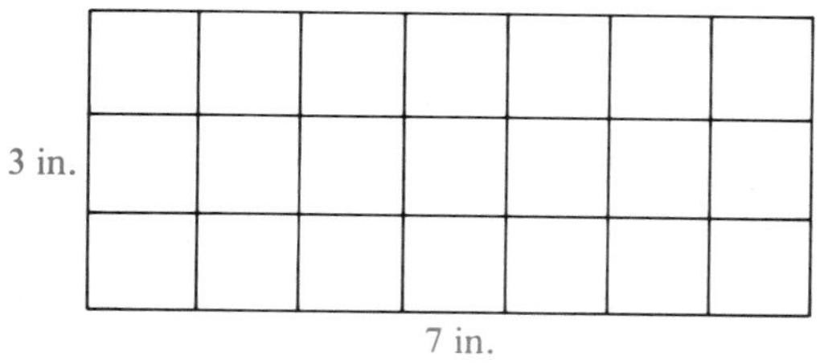

each measuring 1 inch along an edge, must be used to completely cover the region enclosed by the rectangle. The area of each of these blocks is 1 square inch. Looking at Figure 9.10, we see that 21 of these blocks are needed. Thus, the area of this figure is 21 square inches. Notice that the same result is obtained by multiplying the length of the rectangle by its width.

$$A = (7 \text{ in.})(3 \text{ in.})$$
$$A = 21 \text{ in.}^2$$

We therefore have established the formula in the box for finding the area of a rectangle.

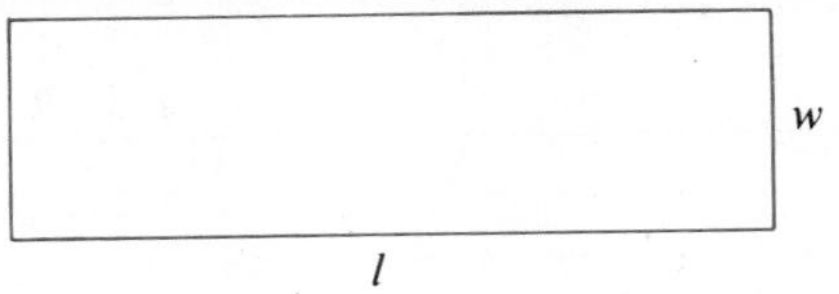

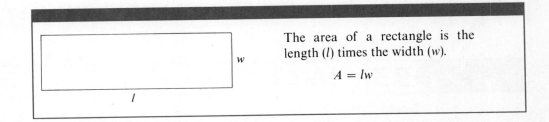

The area of a rectangle is the length (l) times the width (w).

$$A = lw$$

example 1 Find the area of a rectangle that is $1\frac{1}{2}$ ft long and 10 in. wide.

solution To find the area, we must express the length and width in terms of the same unit of measurement.

$1\frac{1}{2}\text{ ft} = \frac{3}{2}\cancel{\text{ft}} \times \frac{12 \text{ in.}}{1\ \cancel{\text{ft}}} = 18 \text{ in.}$ — Convert $1\frac{1}{2}$ ft to in.

$A = lw$ — Use the formula for finding the area of a rectangle.

$A = (18)(10)$ — Substitute $l = 18$ and $w = 10$.

$A = 180$ sq in. — Since length and width are given in inches, the resulting area is in square inches.

To illustrate how equivalent units of area are derived, let us consider a square whose sides are each 1 yard long, as shown in Figure 9.11. By definition, this square has an area of 1 square yard. Since 1 yard = 3 feet, we can also determine this area in feet by multiplying the length, which is 3 feet, by the width, which is also 3 feet.

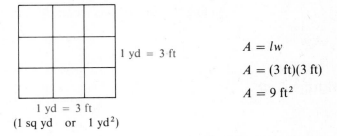

figure 9.11

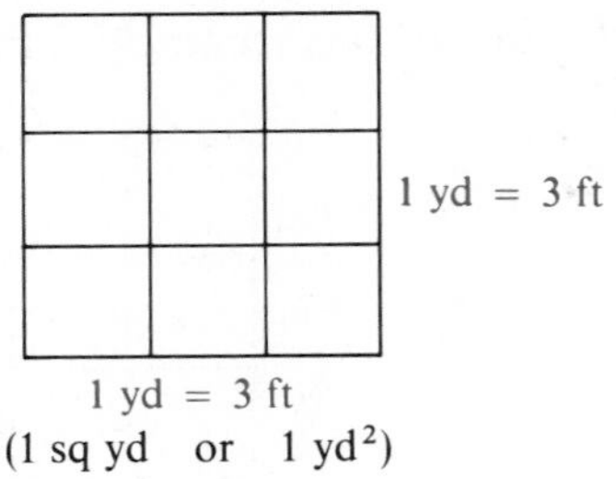

$$A = lw$$
$$A = (3 \text{ ft})(3 \text{ ft})$$
$$A = 9 \text{ ft}^2$$

Therefore, an area of 1 square yard is equivalent to 9 square feet.

Notice that since the length and width of a square are the same, to determine its area, we can simply square the measurement of one of its sides (see box). Using this procedure, we can also show that 1 square foot = 144 square inches.

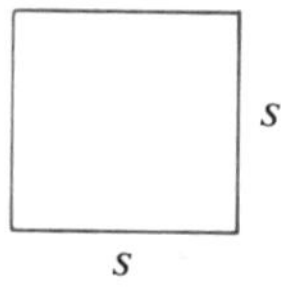

The area of a square is the square of one of its sides (s).

$$A = s^2$$

table 9.1
English Units of Area

1 ft^2	= 144 $in.^2$
1 yd^2	= 9 ft^2
1 acre	= 4,840 yd^2
1 acre	= 43,560 ft^2
1 mi^2	= 640 acres

Table 9.1 summarizes equivalent units of area in the English system. Square inches, square feet, and square yards are used to measure relatively small areas. Acres and square miles are used to measure large areas of land.

example 2

A biologist owns a piece of land that measures 330 ft × 330 ft (the × is read as "by"). How many acres does she own?

solution

Find this area first in terms of square feet.

$A = s^2$ — 330 ft × 330 ft gives the dimensions of a square. Use the formula for finding the area of a square.

$A = (330)^2$ — Substitute $s = 330$.

$A = 108{,}900$ sq ft — Since s is given in ft, the resulting area is in sq ft.

Now convert 108,900 sq ft to acres.

$$108{,}900 \text{ sq ft} = 108{,}900 \not{\text{sq ft}} \times \frac{1 \text{ acre}}{43{,}560 \not{\text{sq ft}}} = 2.5 \text{ acres}$$

The biologist owns 2.5 acres of land. ●

In the metric system, each unit of area differs from the next larger unit by a factor of 100. For example, 1 square centimeter is equivalent to 100 square millimeters, as shown in Figure 9.12. The metric units that are used to measure large areas of land are the

figure 9.12

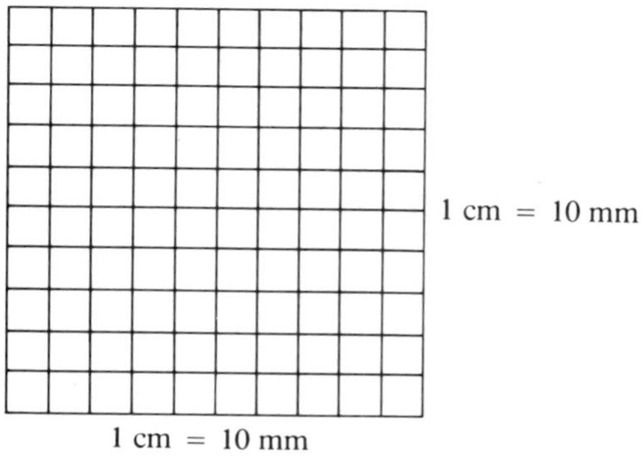

$1 \text{ cm}^2 = 100 \text{ mm}^2$
1 square centimeter

table 9.2
Metric Units of Area

1 cm^2	=	100 mm^2
1 dm^2	=	100 cm^2
1 m^2	=	100 dm^2
1 a	=	100 m^2
1 ha	=	100 a

are (a), which is pronounced "air," and the hectare (ha). Table 9.2 summarizes equivalent metric units of area.

We will now determine formulas for finding areas of parallelograms, triangles, and trapezoids. To find the area of a parallelogram, we need to know two dimensions—the base (b) which is the measurement of one of its sides, and the height (h), which is the distance between the base and the side parallel to it.

Figure 9.13 shows a parallelogram with a base of 6 cm and a height of 3 cm. We can determine the area of this figure by cutting off the shaded piece on the left and attaching it to the right of this figure to form a rectangle, as shown in Figure 9.14.

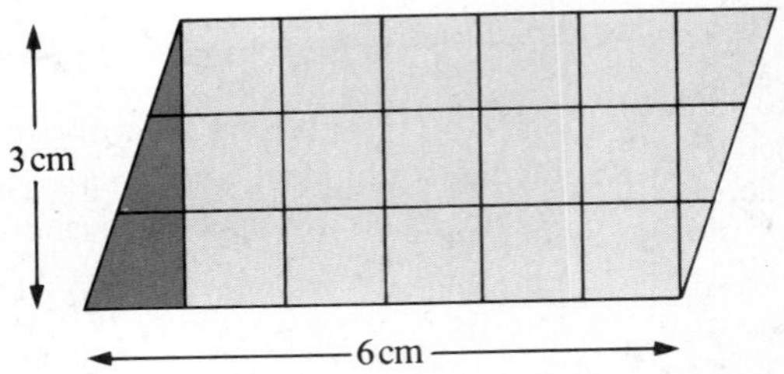

figure 9.13

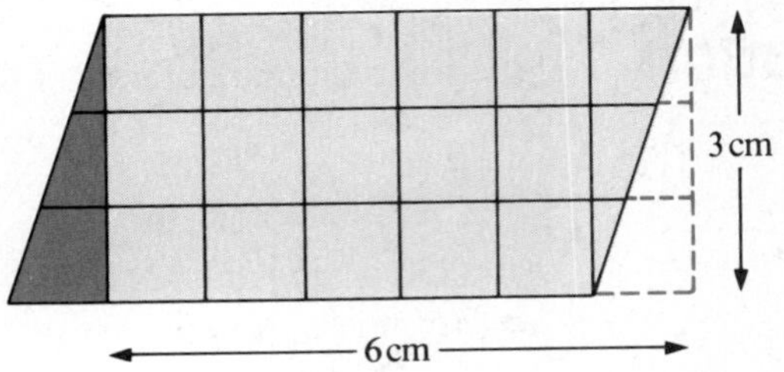

figure 9.14

The area of the resulting parallelogram is simply the product of the length (6 cm) and the width (3 cm), which is 18 cm^2:

$$A = (6 \text{ cm})(3 \text{ cm})$$
$$A = 18 \text{ cm}^2$$

Notice that the same result is obtained by finding the product of the base and the height of the original parallelogram. The formula for determining the area of a parallelogram is given in the box.

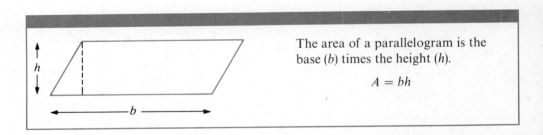

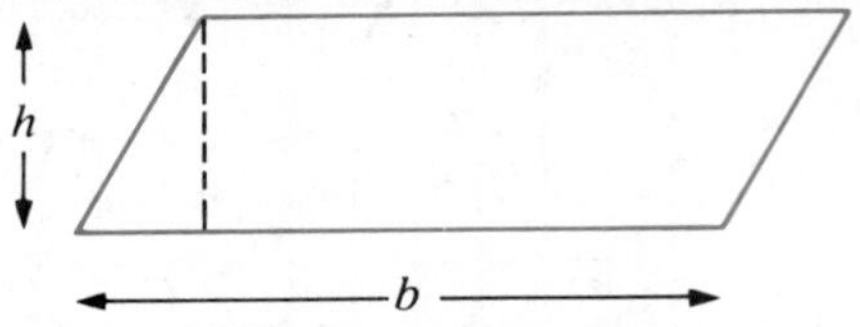

The area of a parallelogram is the base (b) times the height (h).

$$A = bh$$

example 3

Find the area of a parallelogram with a base of 70 m and a height of 60 m. Express your answer in areas.

solution

First find the area in terms of square meters.

$A = bh$ — Use the formula for finding the area of a parallelogram.

$A = (70)(60)$ — Substitute $b = 70$ m and $h = 60$ m.

$A = 4{,}200 \text{ m}^2$

Now convert 4,200 m^2 to ares.

$$4{,}200 \text{ m}^2 = 4{,}200 \cancel{\text{m}^2} \times \frac{1 \text{ a}}{100 \cancel{\text{m}^2}} = 42 \text{ a}$$

The area of the parallelogram is 42 a.

example 4

The area of a parallelogram with an 80-in. base is 4,800 sq in. What is its height?

solution

$A = bh$ — Use the formula for finding the area of a parallelogram.

$4{,}800 = 80h$ — Substitute $A = 4{,}800$ and $b = 80$, and solve for h.

$80h = 4{,}800$

$$\frac{1}{80}(80h) = \frac{1}{80}(4{,}800)$$

$h = 60$ in. — Since area is given in sq in. and the base is in in., the height must also be in in.

The height of the parallelogram is 60 in.

To find the area of a triangle we also need to know the base (b), which is one of its sides, and the height (h), which is the perpendicular distance from the base to the vertex opposite the base. Figure 9.15 shows the base and height of three different triangles. Figure 9.16 shows that the area of each of these triangles is exactly equal to half of the area of a parallelogram with the same base b and height h. We have thus established the formula in the box for calculating the area of a triangle.

figure 9.15

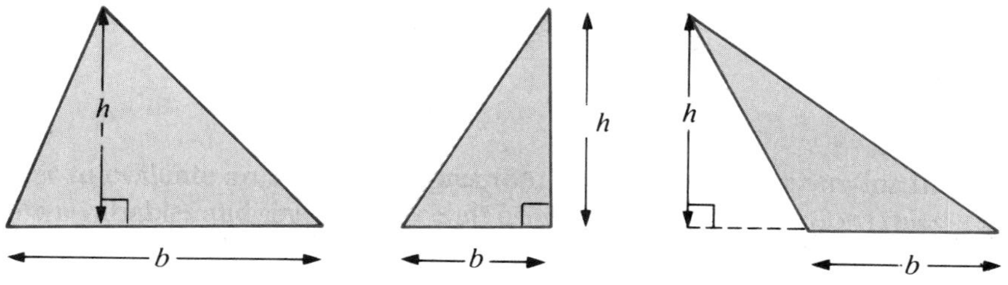

figure 9.16

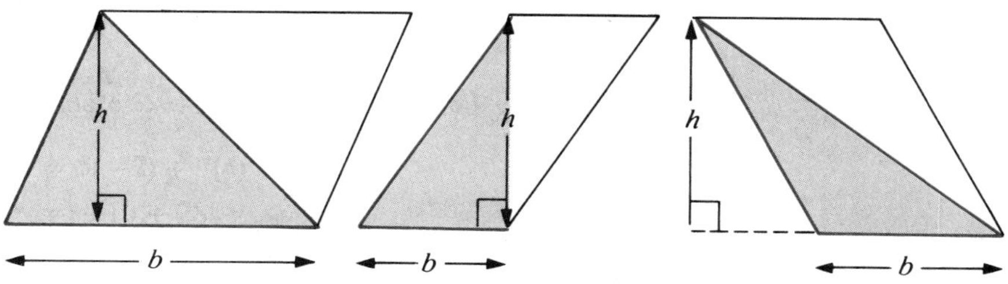

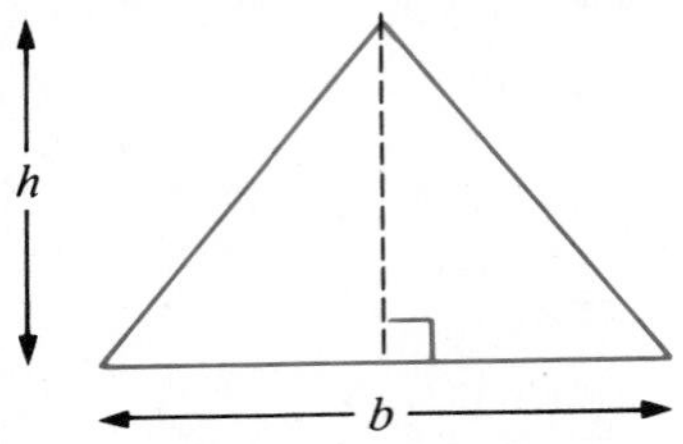

The area of a triangle is one-half the base (b) times the height (h).

$$A = \frac{1}{2}bh$$

example 5

Find the area of the accompanying triangle.

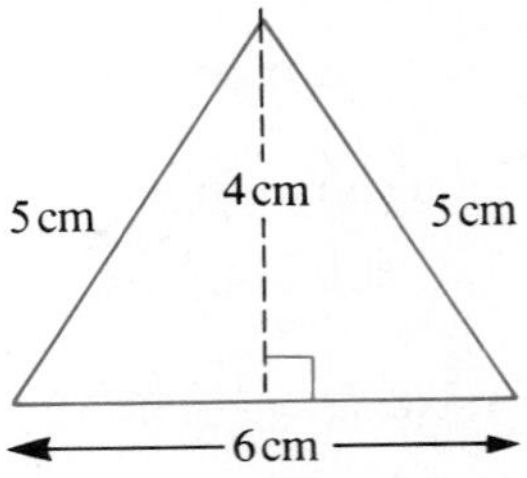

solution

The dimensions needed are the base, which is 6 cm, and the height, which is 4 cm.

$A = \frac{1}{2}bh$ — Use the formula for finding the area of a triangle.

$A = \frac{1}{2}(6)(4)$ — Substitute $b = 6$ and $h = 4$.

$A = \frac{1}{2}(24)$

$A = 12 \text{ cm}^2$ — Since the base and height are given in cm, the area is in cm^2.

To find the area of a trapezoid, we need to know three dimensions—the two parallel sides, which are sometimes referred to as the lower base (b_1) and the upper base (b_2), and the height (h), which is the distance between the bases. The area of a trapezoid can be calculated by dividing it into two triangles, as shown in Figure 9.17, and calculating the area of each triangle. The area of a trapezoid is the sum of these two areas.

$$A = A_1 + A_2$$

$$A = \frac{1}{2}b_1h + \frac{1}{2}b_2h$$

$$A = \frac{1}{2}h(b_1 + b_2)$$

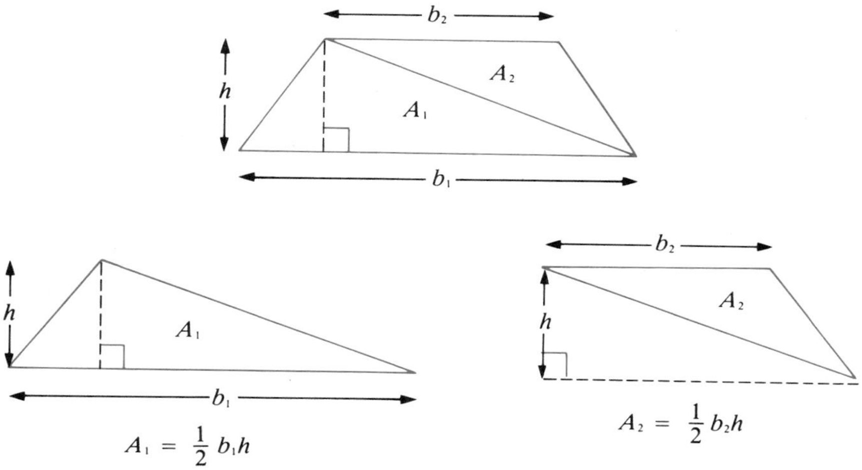

figure 9.17

The last line in our calculation is the formula for finding the area of a trapezoid. To show that it is equivalent to the previous line, multiply it out, using the distributive property. The formula is summarized in the box.

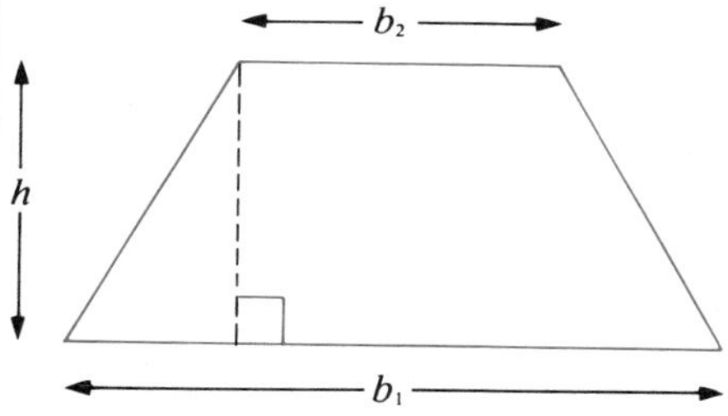

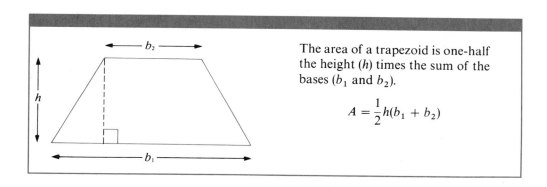

The area of a trapezoid is one-half the height (h) times the sum of the bases (b_1 and b_2).

$$A = \frac{1}{2}h(b_1 + b_2)$$

example 6

Find the area of a trapezoid that has bases of 8 yd and 2 yd and a height of 9 ft.

solution

To find the area, all dimensions must be expressed in the same units.

$$9 \text{ ft} = 9 \cancel{\text{ft}} \times \frac{1 \text{ yd}}{3 \cancel{\text{ft}}} = 3 \text{ yd}$$ First convert 9 ft to yd.

$$A = \frac{1}{2}h(b_1 + b_2)$$ Use the formula for finding the area of a trapezoid.

$$A = \frac{1}{2}(3)(8 + 2)$$ Substitute $h = 3$, $b_1 = 8$, and $b_2 = 2$.

$$A = \frac{1}{2}(3)(10)$$

$A = \frac{1}{2}(30)$

$A = 15 \text{ yd}^2$ — Since the dimensions were given in yd, the resulting area is in yd^2.

example 7

A remnant of fabric is shaped as illustrated. What is its area?

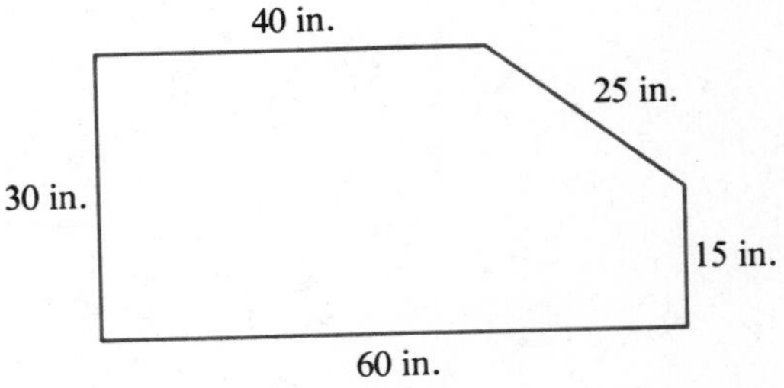

solution

There are a number of different ways to solve this problem. One way is to view the figure as a trapezoid sitting on top of a rectangle. The dimensions needed to find the area of each are shown in the illustration below. Compare this to the original above to see that the height and lower base of the trapezoid were obtained by knowing the dimensions of the rectangle. Now find the area of each piece.

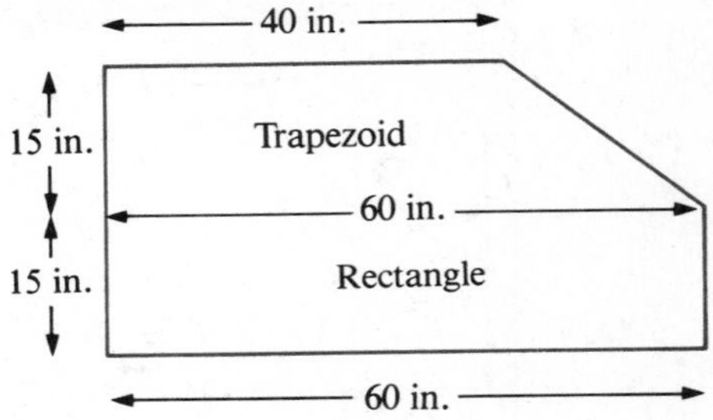

Find the area of the rectangle.

$A_R = lw$ — Use A_R to indicate the area of the rectangle.

$A_R = (60)(15)$ — Substitute $l = 60$ and $w = 15$.

$A_R = 900 \text{ in.}^2$

Find the area of the trapezoid.

$A_T = \frac{1}{2}h(b_1 + b_2)$ — Use A_T to indicate the area of the trapezoid.

$A_T = \frac{1}{2}(15)(60 + 40)$ — Substitute $h = 15$, $b_1 = 60$, and $b_2 = 40$.

$A_T = \frac{1}{2}(15)(100)$

$A_T = \frac{1}{2}(1{,}500)$

$A_T = 750 \text{ in.}^2$

To find the total area, add the two pieces together.

$A = A_R + A_T$

$A = 900 + 750$

$A = 1{,}650 \text{ in.}^2$

The area of the remnant is 1,650 square inches.

QUICK QUIZ

Find the area of each of the following polygons.

1. Rectangle of length 8 m and width 7 m.
2. Square whose sides are 1.3 mi long.
3. Triangle with base $2\frac{1}{2}$ in. and height 3 in.
4. Trapezoid with bases 5 cm and 7 cm long and a height of 4 cm.

ANSWERS

1. 56 m^2
2. 1.69 mi^2
3. $3\frac{3}{4} \text{ in.}^2$
4. 24 cm^2

9.3 Exercises

Indicate which of the following statements are true and which are false. For those that are false, change the italic expression to make the statement true.

1. Area is measured in *linear* units of measurement.
2. The area of a triangle is *one-third* of the area of a parallelogram with the same base and height.
3. An *are* is equivalent to 100 square meters.
4. One square yard is equivalent to *3* square feet.

Find the area of each of the following polygons.

5. Rectangle of length 12 ft and width 9 ft.
6. Rectangle of length 21 in. and width 16 in.
7. Rectangle of length 8.1 cm and width 7 cm.
8. Rectangle of length 5.6 m and width 3 m.
9. Rectangle of length $4\frac{3}{4}$ mi and width $2\frac{1}{2}$ mi.
10. Rectangle of length $8\frac{1}{2}$ km and width $5\frac{1}{4}$ km.
11. Rectangle of length 4 yd and width 10 ft.
12. Rectangle of length 9 cm and width 30 mm.
13. Square whose sides are 1.3 ft long.
14. Square whose sides are 25 in. long.
15. Square whose sides are 44 cm long.
16. Square whose sides are 1.5 km long.
17. Square whose sides are $4\frac{2}{3}$ yd long.
18. Square whose sides are $7\frac{1}{2}$ m long.
19. Parallelogram with base 3.6 dm and height 2.1 dm.
20. Parallelogram with base 0.4 mi and height 0.3 mi.
21. Parallelogram with base $2\frac{1}{2}$ yd and height 12 ft.
22. Parallelogram with base 70 m and height 9 dam.
23. Triangle with base 2 yd and height 3 yd.
24. Triangle with base 9 dm and height 7 dm.

25. Triangle with base $3\frac{1}{2}$ m and height $2\frac{1}{4}$ m.

26. Triangle with base $4\frac{3}{8}$ ft and height $4\frac{4}{5}$ ft.

27. Triangle with base 2 ft and height 1 yd.

28. Triangle with base 6 m and height 80 dm.

29. Trapezoid with bases 4 ft and 8 ft long and a height of $7\frac{1}{2}$ ft.

30. Trapezoid with bases $4\frac{1}{2}$ in. and $9\frac{1}{2}$ in. long and a height of 4 in.

31. Trapezoid with bases 6.2 cm and 4.4 cm long and a height of 3.1 cm.

32. Trapezoid with bases 51 m and 39 m long and a height of 20 m.

33. Trapezoid with bases 8 in. and 1 ft long and a height of 5 in.

34. Trapezoid with bases 60 mm and 4 cm long and a height of 8 cm.

Convert each of the following units of measurement as indicated.

35. 18 ft^2 = ____ yd^2

36. 3 acres = ____ yd^2

37. 2 sq ft = ____ sq in.

38. 320 acres = ____ sq mi

39. 300 mm^2 = ____ cm^2

40. 8 m^2 = ____ dm^2

41. 75 dm^2 = ____ m^2

42. 42 cm^2 = ____ mm^2

43. 7 a = ____ m^2

44. 4 ha = ____ a

45. 36 in.^{2} = ____ ft^2

46. 1,210 yd^2 = ____ acres

47. 0.75 acres = ____ ft^2

48. $5\frac{1}{4}$ mi^2 = ____ acres

49. 800 mm^2 = ____ m^2

50. 3,500 cm^2 = ____ a

Solve each of the following word problems.

51. How many square feet of fabric are needed to make a tablecloth that measures 70 in. long and 54 in. wide?

52. How many square feet of shelf paper are needed to cover four shelves, each $3\frac{1}{2}$ ft long and $1\frac{1}{4}$ ft wide?

53. A shawl is made out of a triangular piece of fabric that has a base of 54 in. and a height of 18 in. What is the area of the shawl?

54. The mainsail of a boat is shaped like a right triangle that has a 5-ft base and a 12-ft height. What is the area of the sail?

55. How many square tiles measuring 20 cm along an edge are needed to cover a floor 5 m long and 7 m wide?

56. How large must a piece of plywood be to build a door that is 6 ft 6 in. long and 2 ft 8 in. wide?

57. A rectangular piece of land measures 3 km × 5 km. What is its area in hectares?

58. A square piece of land measures 8 hm on a side. What is its area in ares?

59. How much would it cost to carpet a room that measures 18 ft × 12 ft if carpeting sells for $21.95 per square yard?

60. One gallon of paint covers 300 sq ft. How much paint is needed to cover the ceiling of a room that measures 10 ft × 9 ft?

61. A rectangle with a width of 9 in. has an area of 108 sq in. What is its length?

62. A square has an area of 225 sq ft. What is the length of one of its sides?

63. A parallelogram with a base of $7\frac{1}{2}$ cm has an area of 12 cm^2. What is its height?

64. A rectangle with an area of $9\frac{3}{4}$ m^2 is $6\frac{1}{4}$ m long. What is its width?

65. A triangle whose base measures 2.4 in. has an area of 3.6 in.^{2} What is its height?

66. A triangle whose area is 84 mm^2 has a height of 16 mm. How long is its base?

67. A trapezoid with bases that measure 3 dm and 7 dm has an area of 45 dm^2. What is its height?

68. A trapezoid with an area of 480 yd^2 has two bases that measure 17 yd and 13 yd. What is its height?

69. How many square feet of wallpaper are needed to cover a wall that measures $9\frac{1}{2}$ ft × $7\frac{3}{4}$ ft and has a window that measures $4\frac{1}{2}$ ft × 3 ft? (See the accompanying figure.)

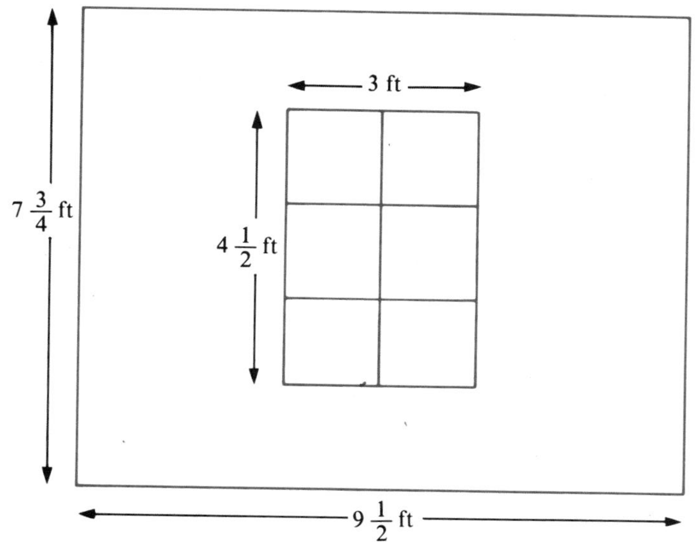

70. How many square feet of Formica(TM) are needed to cover the counter shown in **(a)**.

71. How much paneling would it take to panel a wall shaped like the accompanying drawing **(b)**?

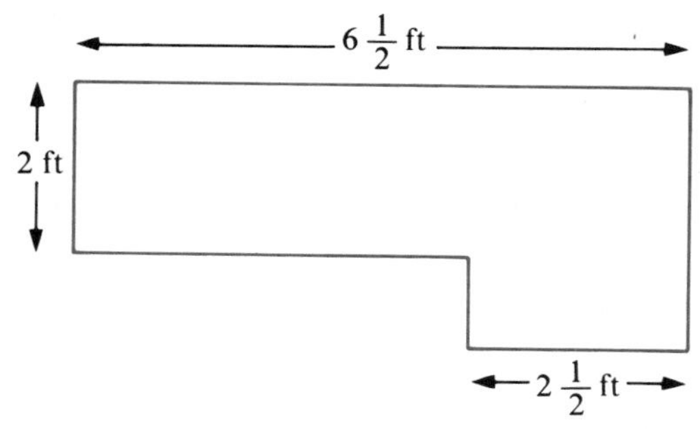

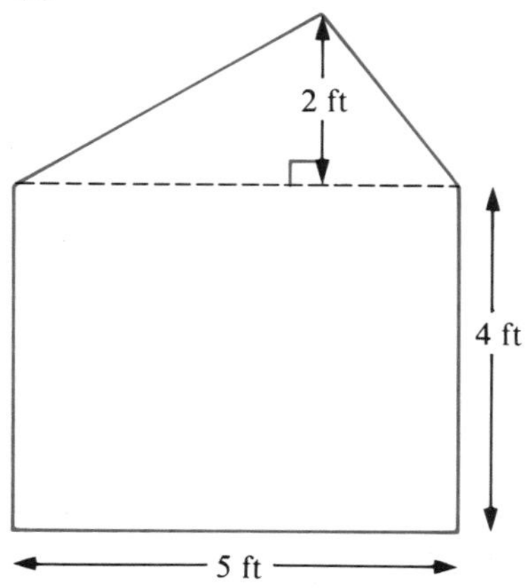

72. How many square feet of carpeting are needed to cover four steps that measure 42 in. long, 11 in. wide, and 9 in. high, as shown in this picture?

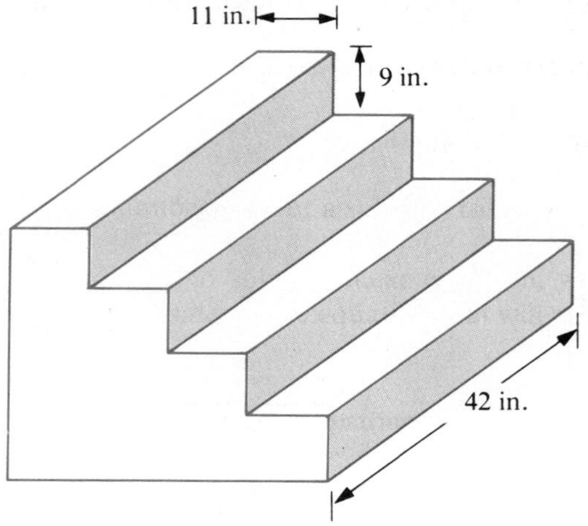

9.4 The Pythagorean Theorem

One of the most famous theorems in mathematics was discovered by the Greek mathematician Pythagoras around 500 B.C. This theorem states a formula that describes the relationship between the sides of any right triangle. The two sides of a right triangle that meet in a right angle are called the **legs**, and the longest side, which joins the two legs, is called the **hypotenuse**.

The Pythagorean Theorem is illustrated in Figure 9.18. Here we have a right triangle with legs 3 units and 4 units in length and a hypotenuse of 5 units. A square has been constructed on each side of the triangle. Notice that the sum of the areas of the two squares drawn adjacent to the two legs is equal to the area of the square drawn adjacent to the hypotenuse.

$$9 + 16 = 25$$

figure 9.18

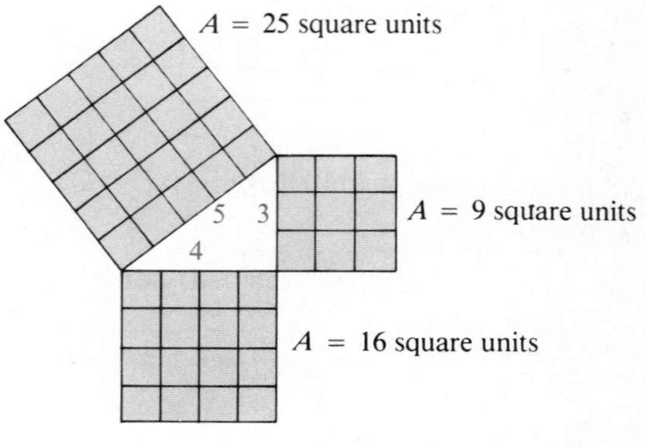

Therefore, the sum of the squares of the two legs is equal to the square of the hypotenuse.

$$(3)^2 + (4)^2 = (5)^2$$

This same relationship holds true for any right triangle. To generalize this result, we let the letters a and b represent the legs of a right triangle and the letter c represent the hypotenuse. Then we write the formula

$$a^2 + b^2 = c^2$$

or

$$c^2 = a^2 + b^2$$

The Pythagorean Theorem is summarized in the box.

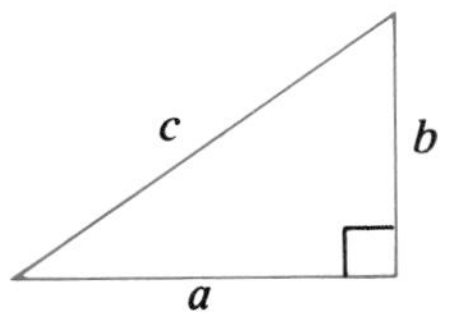

The Pythagorean Theorem states that for any right triangle with legs a and b and hypotenuse c, the square of the hypotenuse is equal to the sum of the squares of the two legs.

$$c^2 = a^2 + b^2$$

example 1

Find the hypotenuse of a right triangle whose legs measure 9 m and 12 m.

solution

$c^2 = a^2 + b^2$	Use the Pythagorean Theorem.
$c^2 = 9^2 + 12^2$	Substitute $a = 9$ and $b = 12$.
$c^2 = 81 + 144$	Raise to powers first.
$c^2 = 225$	Then do addition.
$c = \sqrt{225}$	Solve for c by taking the square root of both sides.
$c = 15$ m	

The hypotenuse is 15 m. ●

Generally, when we take the square root of both sides of an equation, we obtain both a positive and a negative result. However, since we are dealing with dimensions of a geometric figure in these examples, only the positive square root makes sense as a solution for c.

example 2

A triangle has sides that measure 7 cm, 8 cm, and 11 cm in length. Is it a right triangle?

solution

In order for the figure to be a right triangle, it must satisfy the Pythagorean Theorem.

$a^2 + b^2 \ ? \ c^2$	
$7^2 + 8^2 \ ? \ 11^2$	Substitute $a = 7$, $b = 8$, and $c = 11$.
$49 + 64 \ ? \ 121$	See if both sides of the resulting statement are equal.
$113 \neq 121$	
The figure is not a right triangle.	Since the two sides are unequal, the Pythagorean Theorem is not satisfied. ●

The formula $c^2 = a^2 + b^2$ enables us to find the hypotenuse c, given the measurements of the two legs, a and b. If we are given the measurement of the hypotenuse and one leg, we must rearrange the formula to find the other leg as follows:

$$a^2 + b^2 = c^2$$

$$a^2 + b^2 - b^2 = c^2 - b^2$$ Subtract b^2 from both sides.

$$a^2 + 0 = c^2 - b^2$$

$$a^2 = c^2 - b^2$$

$$a = \sqrt{c^2 - b^2}$$ Take the square root of both sides.

Remember that since dimensions of geometric figures are always positive, we are only concerned with the positive square root. Therefore, to find the measure of leg a given hypotenuse c and leg b, use the formula $a = \sqrt{c^2 - b^2}$. Similarly, it can be shown that to find the measure of leg b given the hypotenuse c and leg a, we use the formula $b = \sqrt{c^2 - a^2}$.

example 3

The hypotenuse of a right triangle is 100 ft long, and one leg is 80 ft long. How long is the other leg?

solution

$a = \sqrt{c^2 - b^2}$ Use the alternate form for the Pythagorean Theorem.

$a = \sqrt{100^2 - 80^2}$ Substitute $c = 100$ and $b = 40$.

$a = \sqrt{10{,}000 - 6{,}400}$ Simplify expression under the radical. First raise to powers.

$a = \sqrt{3{,}600}$ Then do subtraction.

$a = 60$ ft Finally, take the square root.

The other leg is 60 ft long.

example 4

A kite is flying 15 ft above the person holding the string. How long is the string if the horizontal distance between the kite flyer and the kite is 20 ft?

solution

Draw a diagram to illustrate the problem. The length of the string is the hypotenuse of a right triangle. Since it is unknown, assign it the variable c.

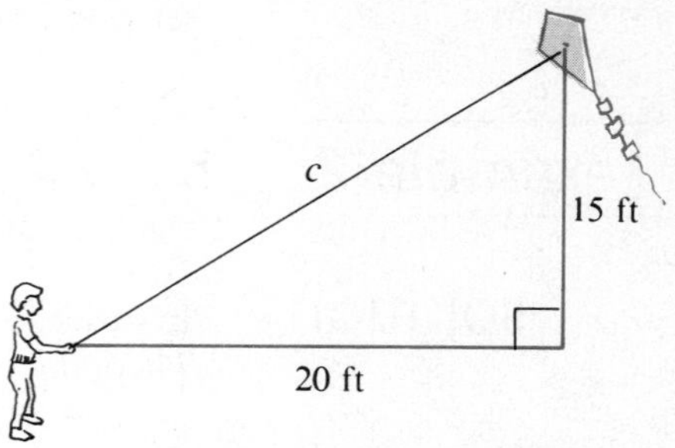

Let c = length of the string.

$c^2 = a^2 + b^2$ Use the Pythagorean Theorem.

$c^2 = 20^2 + 15^2$ Substitute $a = 20$ and $b = 15$.

$c^2 = 400 + 225$ First raise to powers. Then do addition.

$c^2 = 625$

$c = \sqrt{625}$ Take the square root of both sides.

$c = 25$ ft

The string is 25 ft long.

QUICK QUIZ

Given a right triangle with legs a and b and hypotenuse c, find the following dimensions.

1. Find c if $a = 5$ cm and $b = 12$ cm.
2. Find a if $c = 17$ in. and $b = 15$ in.

ANSWERS

1. $c = 13$ cm
2. $a = 8$ in.

9.4 Exercises

For each of the following right triangles with legs a and b and hypotenuse c, use the Pythagorean Theorem to find the unknown side.

1. Find c if $a = 6$ in. and $b = 8$ in.
2. Find c if $a = 8$ cm and $b = 15$ cm.
3. Find c if $a = 12$ m and $b = 16$ m.
4. Find c if $a = 30$ ft and $b = 40$ ft.
5. Find a if $c = 13$ yd and $b = 5$ yd.
6. Find b if $c = 15$ mm and $a = 12$ mm.
7. Find b if $c = 2.5$ cm and $a = 2.0$ cm.
8. Find a if $c = 3.0$ mi and $b = 1.8$ mi.
9. Find c if $a = 900$ mm and $b = 1{,}200$ mm.
10. Find b if $c = 1{,}000$ in. and $b = 800$ in.
11. Find c if $a = 12.5$ cm and $b = 17.7$ cm. Round off to nearest tenth.
12. Find c if $a = 35.8$ ft and $b = 28.4$ ft. Round off to nearest tenth.
13. Find a if $c = 62.4$ yd and $b = 47.6$ yd. Round off to nearest tenth.
14. Find b if $c = 8.94$ m and $a = 3.61$ m. Round off to nearest hundredth.

State whether or not each of the following triangles with sides a, b, and c are right triangles. Assume that the sides of each triangle are given in terms of the same unit of measurement.

15. $a = 6$, $b = 8$, $c = 12$
16. $a = 5$, $b = 6$, $c = 8$
17. $a = 9$, $b = 40$, $c = 41$
18. $a = 5$, $b = 13$, $c = 14$
19. $a = 21$, $b = 28$, $c = 35$
20. $a = 18$, $b = 22$, $c = 27$
21. $a = 90$, $b = 120$, $c = 150$
22. $a = 50$, $b = 120$, $c = 130$

Solve each of the following word problems.

23. The base of a ladder is 5 ft away from a wall, and its top edge, which rests against the wall, is 12 ft above the ground. How long is the ladder?
24. To straighten a tree, a rope is tied around its trunk at a point 6 ft above the ground and pulled taut. The other end of the rope is tied to a stake in the ground that is 8 ft away from the base of the trunk. What is the length of the rope measured from the tree to the stake?
25. Two airplanes pass each other in the sky, one headed due north at a speed of 600 mph and the other headed due east at a speed of 800 mph. What is the distance between them 1 hr later?
26. A baseball diamond is in the shape of a square whose vertices are the bases and home plate. If the distance between first and second base is 90 ft, what is the distance between second base and home plate? Round off your answer to the nearest foot.
27. A 20-ft long guy wire that supports a telephone pole is attached to a stake in the ground 12 ft away from the base of the pole. How high above the ground is the other end of the guy wire attached to the pole?
28. The diagonal of a rectangle is 300 cm, and its width is 180 cm. What is the length of the rectangle? (The diagonal of a rectangle is a straight line that connects opposite vertices.)
29. A right triangle has a hypotenuse measuring 7.5 m and a leg measuring 4.5 m. What is its area?
30. A rectangle has a 12.5-in. diagonal and a length of 10 in. What is its area?

31. Find a and b for the triangle shown here. Then find its perimeter and area.

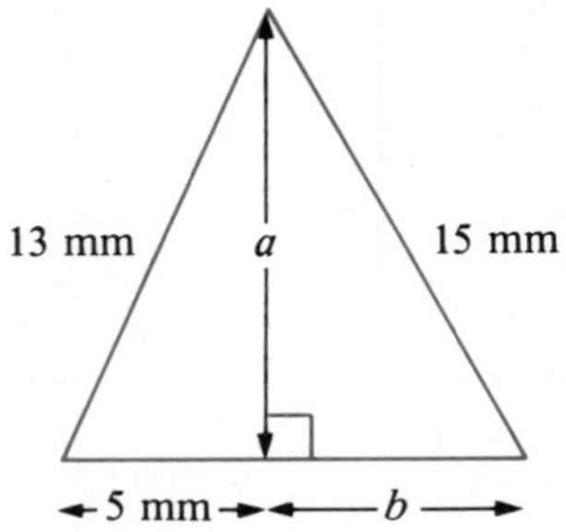

32. Find the value of h for the parallelogram shown here. Then find its perimeter and area.

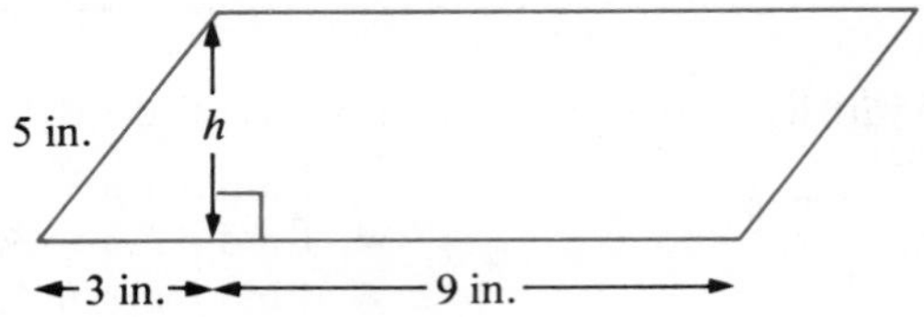

9.5 Circles

In this section, we will learn how to find the perimeter and area of figures called **circles**.

definition

A *circle* is a geometric figure that includes all points that are the same distance away from a fixed point called the *center*. The distance from the center to any point on the circle is called the *radius*. A line that passes through the center and connects two points on a circle is called the *diameter*.

Figure 9.19 illustrates the radius (r) and diameter (d) of a circle. Notice that the diameter is exactly equal to twice the radius. For example, a circle with a diameter of 18 in. has a radius of 9 in. A circle with a radius of 21 cm has a diameter of 42 cm.

The perimeter of a circle, or distance around it, is called the **circumference**. Early mathematicians discovered that the circumference of any circle is a little more than three times its diameter. In fact, they found that for any circle, the ratio of the circumference to the diameter is equal to the same fixed number. This number is

figure 9.19

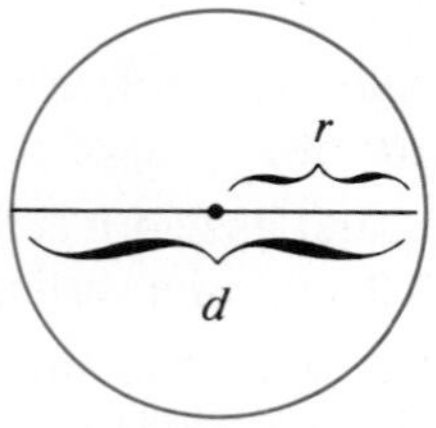

A Circle

$d = 2r$

represented by the symbol π, which is the Greek letter pi. Pi is an irrational number that can be expressed as the following nonrepeating decimal.

$$\pi = 3.14159265\ldots$$

Thus, the ratio of the circumference (C) to the diameter (d) of any circle is equal to π.

$$\frac{C}{d} = \pi$$

To find a formula for the circumference of a circle, we can multiply both sides of this equation by d.

$$d \cdot \frac{C}{d} = \pi \cdot d$$

$$\cancel{d} \cdot \frac{C}{\cancel{d}} = \pi d$$

$$C = \pi d$$

Since the diameter is equal to twice the radius, another formula for the circumference can be obtained by substituting $2r$ for d.

$$C = \pi(2r) \qquad \text{or} \qquad C = 2\pi r$$

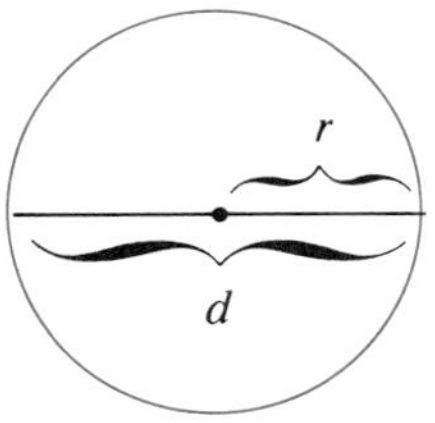

The circumference (perimeter) of a circle is π times the diameter (d) or 2π times the radius (r).

$$C = \pi d$$

$$C = 2\pi r$$

example 1 Find the circumference of a circle whose radius is 8 cm.

solution

$C = 2\pi r$	Use the formula for finding circumference in terms of r.
$C = 2\pi(8)$	Substitute $r = 8$.
$C = 16\pi$ cm	Since the radius is given in cm, so is the circumference.
	Notice that 16π represents the product $16 \cdot \pi$.

In Example 1, we expressed the circumference in terms of π, which is the exact answer. This is not a convenient way to express a measurement for practical applications. Therefore, we frequently use the decimal 3.14 or the fraction $\frac{22}{7}$ to approximate π.

example 2

A circular tablecloth has a diameter of 70 in. How many feet of trimming are needed to decorate its border?

solution

First find the circumference of the table in inches.

$C = \pi d$ — Use the formula for finding circumference in terms of d.

$C = \pi(70)$ — Substitute $d = 70$.

$C = \frac{22}{7}(70)$ — Use $\frac{22}{7}$ to approximate π.

$C = 220$ in.

Now convert 220 in. to feet.

$$220 \text{ in.} = 220 \not{\text{in.}} \times \frac{1 \text{ ft}}{12 \not{\text{in.}}} = \frac{220}{12} \text{ ft} = \frac{55}{3} \text{ ft} = 18\frac{1}{3} \text{ ft}$$

Thus, $18\frac{1}{3}$ ft of trimming are needed. ●

example 3

A record with a diameter of 12 in. is moving at a speed of $33\frac{1}{3}$ revolutions per minute (rpm). After 1 min, how far will a bug that is sitting on the edge of the record have traveled?

solution

After one revolution, the bug will have traveled a distance that is equal to the circumference of the circle.

$C = \pi d$

$C = \pi(12)$ — Substitute $d = 12$.

$C = 12\pi$ in.

In 1 min, the record will make $33\frac{1}{3}$ revolutions. Therefore, the distance traveled in 1 min will equal the circumference times $33\frac{1}{3}$.

$$\text{distance} = 12\pi \cdot 33\frac{1}{3}$$

$$= 12\pi \cdot \frac{100}{3}$$

$$= \frac{\overset{4}{\not{12}}\pi(100)}{\not{3}}$$

$$= 400\pi$$

$$= 400(3.14)$$ — Use $\pi \approx 3.14$.

$$\text{distance} = 1{,}256 \text{ in.}$$

After 1 min, the bug will have traveled 1,256 in. ●

To find the area of a circle, we can divide it up into a number of wedge-shaped pieces called *sectors* and rearrange them as shown in Figure 9.20. Notice that the resulting figure approximates a parallelogram with a base equal to one-half the circumference, and a height equal to the radius.

figure *9.20*

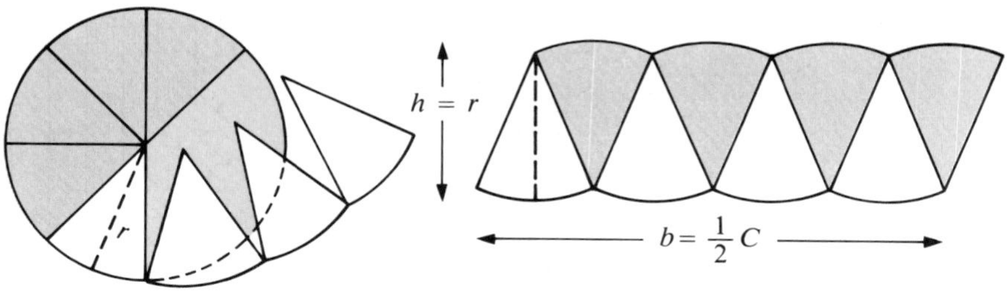

If we divide the circle up into a larger number of sectors, the figure formed in this way more closely resembles a parallelogram, since the curved edges of the sectors that form the bases will approach straight lines. We can find the area of this parallelogram as follows:

$$A = bh$$

$$A = \frac{1}{2}Cr \qquad \text{Substitute } b = \frac{1}{2}C \text{ and } h = r.$$

$$A = \frac{1}{2}(2\pi r)r \qquad \text{Substitute } C = 2\pi r.$$

$$A = \pi r^2$$

Our result is the formula for finding the area of a circle, which is restated in the box.

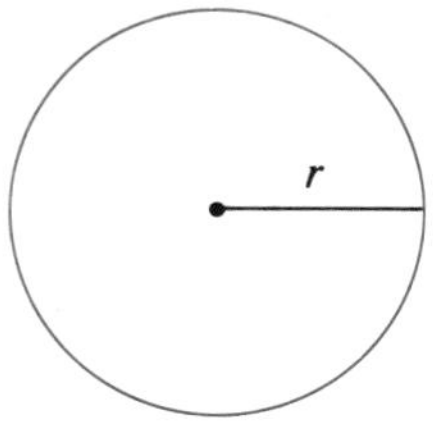

The area of a circle is π times the square of the radius.

$$A = \pi r^2$$

example *4*

Find the area of a circle whose radius is 15 ft. Use 3.14 to approximate π.

solution

$A = \pi r^2$ — Use the formula for finding the area of a circle.

$A = \pi(15)^2$ — Substitute $r = 15$.

$A = (3.14)(225)$ — Use $\pi \approx 3.14$.

$A = 706.5$ sq ft — Since the radius is given in ft, the area is in sq ft.

The area is 706.5 sq ft. ●

example 5

The area of a circle is 900π sq yd. What is its diameter?

solution

Find the radius.

$A = \pi r^2$ — Use the formula for finding the area of a circle.

Substitute $A = 900\pi$.

$900\pi = \pi r^2$

$\pi r^2 = 900\pi$

$\frac{1}{\pi}\pi r^2 = 900\pi\frac{1}{\pi}$ — Multiply both sides by $\frac{1}{\pi}$ so that the π's cancel.

$\frac{1}{\pi}\pi r^2 = 900\pi\frac{1}{\pi}$

$r^2 = 900$

$r = \sqrt{900}$ — Take the square root of both sides.

$r = 30$ yd — Since area is given in sq yd, the radius is in yd.

Find the diameter.

$d = 2r$ — Multiply the radius by 2.

$d = 2(30)$

$d = 60$ yd

Therefore the diameter is 60 yd.

example 6

Find the area of the smallest circle that circumscribes a square with a diagonal of 6 cm.

solution

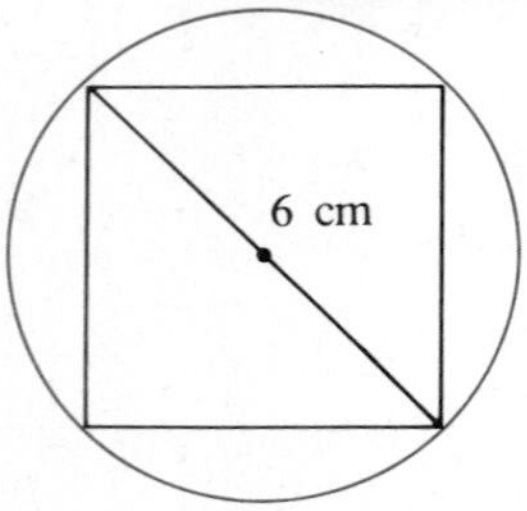

First draw a sketch. Notice that the vertices of the square touch the circle that circumscribes it. Therefore, the diagonal of the square is also the diameter of the circle. Thus, the radius of the circle is 3 cm.

$A = \pi r^2$

$A = \pi 3^2$ — Substitute $r = 3$.

$A = 9\pi$ cm^2

The area of the circle is 9π cm^2.

QUICK QUIZ

1. Find the circumference of a circle with a radius of 56 mm. Use $\pi \approx \frac{22}{7}$.
2. Find the area of a circle with a radius of 13 ft.
3. Find the area of a circle with a diameter of 20 cm. Use $\pi \approx 3.14$.

ANSWERS

1. 352 mm
2. 169π sq ft
3. 314 cm^2

9.5 Exercises

Indicate which of the following statements are true and which are false. For those that are false, change the italic word to make the statement true.

1. The area of a circle is π times the square of its *diameter*.

2. The diameter of a circle is equal to *twice* its radius.

3. For any circle, the ratio of the circumference to the *radius* is equal to π.

4. If the radius of a circle is doubled, its area is *quadrupled*.

Find the circumference of each of the following circles given its diameter d or radius r. Express your answers in terms of π unless otherwise indicated.

5. $d = 31$ in.

6. $d = 8.4$ m

7. $r = 7.2$ cm

8. $r = 18$ ft.

9. $d = 63$ in. Use $\pi \approx \frac{22}{7}$.

10. $d = 2.1$ dm Use $\pi \approx \frac{22}{7}$.

11. $d = 3\frac{1}{2}$ km Use $\pi \approx \frac{22}{7}$.

12. $d = 6\frac{1}{8}$ mi Use $\pi \approx \frac{22}{7}$.

13. $r = 3.5$ ft Use $\pi \approx 3.14$.

14. $d = 9$ yd Use $\pi \approx 3.14$.

15. $d = 8\frac{3}{4}$ m Use $\pi \approx 3.14$.

16. $r = 4\frac{1}{2}$ cm Use $\pi \approx 3.14$.

17. $r = 0.84$ hm Use $\pi \approx \frac{22}{7}$.

18. $d = 98$ in. Use $\pi \approx \frac{22}{7}$.

19. $d = 542$ ft Use $\pi \approx 3.14$.

20. $r = 265$ mm Use $\pi \approx 3.14$.

Find the area of each of the following circles given its diameter d or radius r. Express your answers in terms of π unless otherwise indicated.

21. $r = 19$ dm

22. $r = 150$ in.

23. $d = 5.2$ km

24. $d = 87$ yd

25. $r = 70$ mm Use $\pi \approx \frac{22}{7}$.

26. $r = 28$ ft Use $\pi \approx \frac{22}{7}$.

27. $r = 2\frac{5}{8}$ yd Use $\pi \approx \frac{22}{7}$.

28. $r = 6\frac{2}{9}$ dm Use $\pi \approx \frac{22}{7}$.

29. $r = 300$ cm Use $\pi \approx 3.14$.

30. $r = 0.7$ mi Use $\pi \approx 3.14$.

31. $d = 8\frac{2}{5}$ hm Use $\pi \approx \frac{22}{7}$.

32. $d = 11\frac{1}{5}$ yd Use $\pi \approx \frac{22}{7}$.

33. $d = 6$ ft Use $\pi \approx 3.14$.

34. $d = 40$ cm Use $\pi \approx 3.14$.

35. $r = 45.3$ mi Use $\pi \approx 3.14$.

36. $d = 6.2$ m Use $\pi \approx 3.14$.

Solve each of the following word problems. Leave your answers in terms of π unless otherwise indicated.

37. How much gold braid is needed to decorate the top and bottom edges of a lampshade that has an 11-in. diameter? Use $\pi \approx 3.14$.

38. How much fencing is needed to enclose a circular flower bed if the distance from the center to the edge is 1.4 m? Use $\pi \approx \frac{22}{7}$.

39. What is the area of the top of the stump of a tree with a diameter of $3\frac{1}{2}$ ft? Use $\pi \approx \frac{22}{7}$.

40. How much wood is needed to construct the top of a circular table that has a 70-in. diameter? Use $\pi \approx \frac{22}{7}$.

41. A bicycle wheel has a diameter of 26 in. How far does the air nozzle on the edge of the wheel move when the wheel makes one revolution? Use $\pi \approx 3.14$.

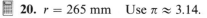

42. A fan is shaped like a quarter of a circle with a radius of 20 cm. What is the area of the fan? Use $\pi \approx 3.14$.

43. What is the radius of a circle that has a circumference of 74π cm?

44. What is the radius of a circle that has an area of 196π mm^2?

45. A skating rink is shaped like a rectangle bounded by two semicircles, as shown in the accompanying drawing. Find its area. Use $\pi \approx 3.14$.

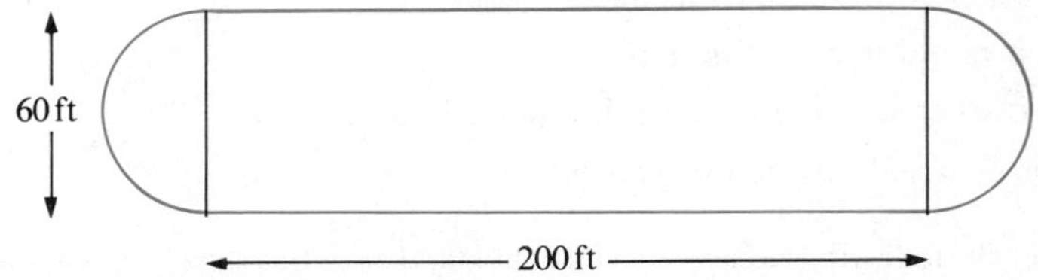

46. Find the area of the largest circle that can be inscribed in a square whose side measures 5 in., as shown in the accompanying drawing.

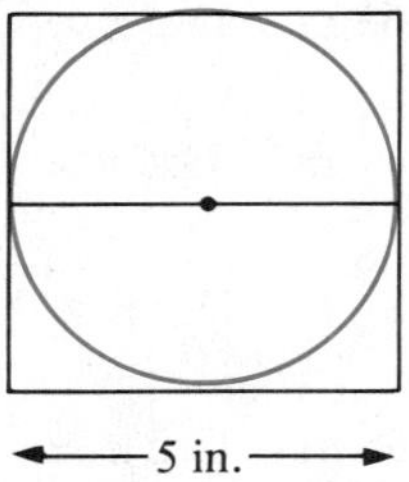

47. Find the area of a sidewalk that surrounds a circular flower bed, as shown in the accompanying drawing. Use $\pi \approx 3.14$.

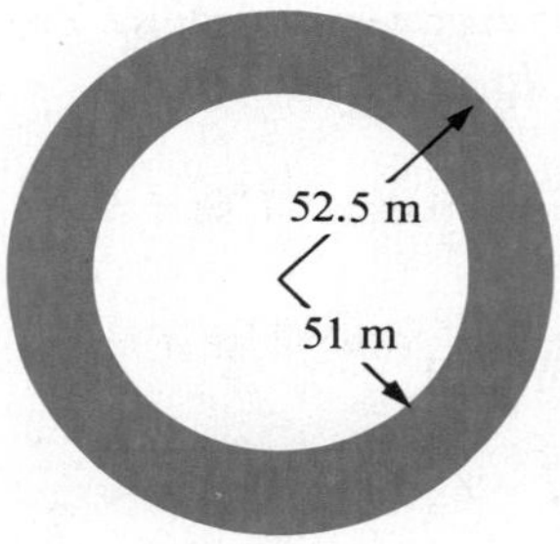

48. The planet Mercury completes one revolution about the sun in 88 days, moving at a speed of 30 mi/sec. What is the distance between Mercury and the sun? Use $\pi \approx 3.14$. Round off your answer to the nearest million miles.

9.6 Volume

Our discussion of capacity in Section 8.3 involved a characteristic of three-dimensional figures known as **volume**.

definition

Volume **is the measurement of the space enclosed by a solid, that is, the space enclosed by a three dimensional figure.**

To measure volume, we use cubic units of measurement, because volume is a three-dimensional quantity. For example, a cubic inch (abbreviated as cu in. or in.^{3}) represents the volume of a cube whose edges are each 1 inch in length. A cubic centimeter (abbreviated as cu cm or cm^3) represents the volume of a cube whose edges are each 1 centimeter in length. These are both illustrated in Figure 9.21.

figure *9.21*

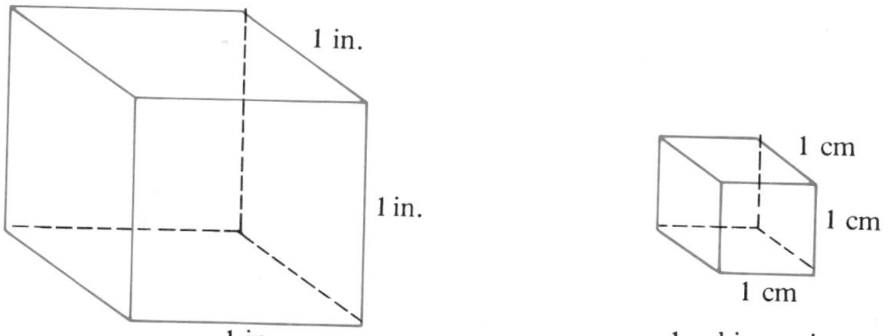

1 cubic inch
(1 cu in. or 1 $in.^3$)

1 cubic centimeter
(1 cu cm or 1 cm^3)

Let us now consider how to find the volume of the rectangular box shown in Figure 9.22. Here we have a box that has a length of 5 inches, a width of 3 inches, and a height of 4 inches. The problem of finding the volume of the box is the same as determining how many cubes, each measuring 1 inch along every edge, are needed to completely fill the box. The volume of each of these cubes is one cubic inch. The bottom layer of cubes that forms the base of the box has five cubes along its length and three cubes along its width, making a total of $5 \times 3 = 15$ cubes. The entire box consists of four of these layers of 15 cubes each, making a total of $15 \times 4 = 60$ cubes. Thus, the volume of this box is 60 cubic inches.

figure *9.22*

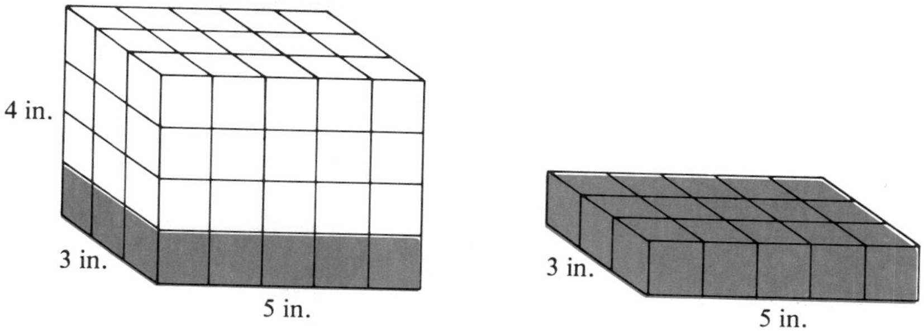

Notice that this same result can be obtained simply by finding the product of the length, width, and height.

$$V = (5 \text{ in.})(3 \text{ in.})(4 \text{ in.})$$

$$V = 60 \text{ in.}^3$$

We have thus established the formula for calculating the volume of a rectangular solid (see box).

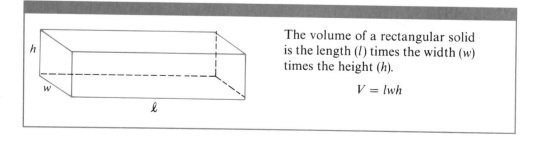

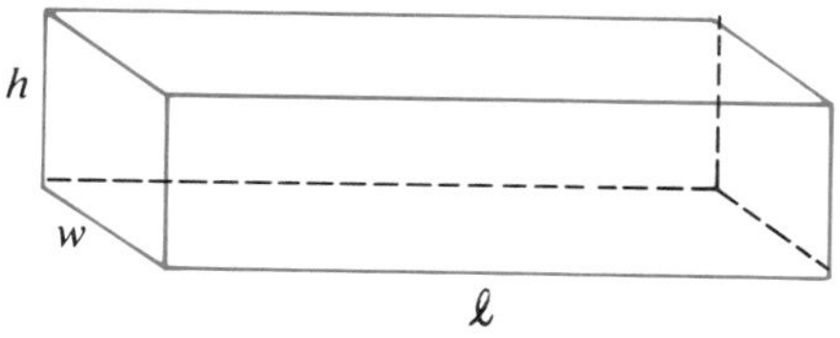

The volume of a rectangular solid is the length (l) times the width (w) times the height (h).

$$V = lwh$$

To illustrate how equivalent units of volume are derived, let us consider a cube whose edges are each 1 yard long, as shown in Figure 9.23. By definition, this cube has a volume of 1 cubic yard. Since 1 yard = 3 feet, we can also determine this volume by multiplying the length, which is 3 feet, times the width, which is 3 feet, times the height, which is also 3 feet.

figure 9.23

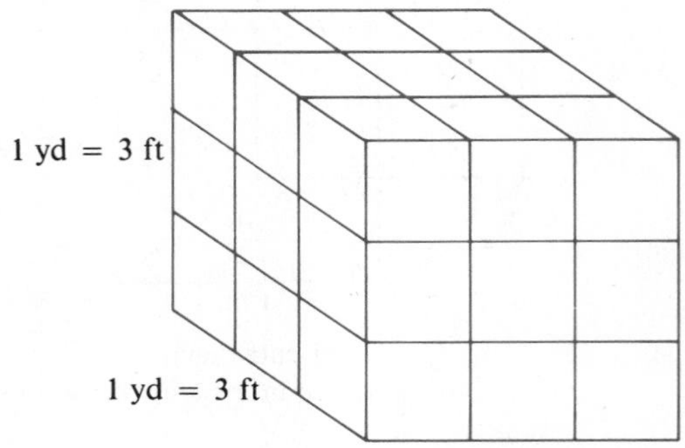

$V = lwh$

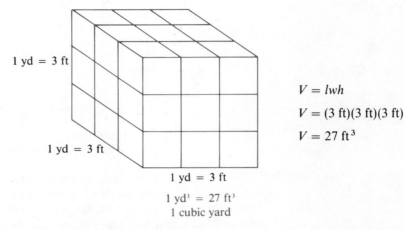

$V = (3\text{ ft})(3\text{ ft})(3\text{ ft})$

$V = 27\text{ ft}^3$

$1\text{ yd}^3 = 27\text{ ft}^3$
1 cubic yard

Therefore, the volume of 1 cubic yard is equivalent to 27 cubic feet.

Notice that since the length, width, and height of a cube are all the same, to determine its volume, we can simply cube the measurement of one of its edges (see box).

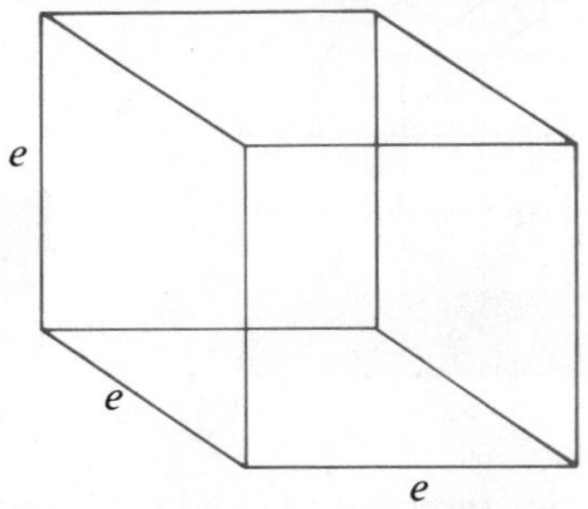

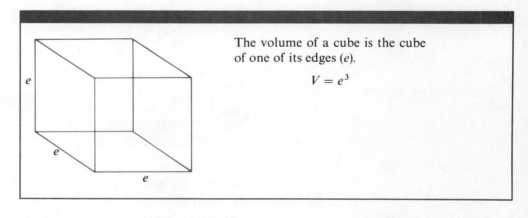

The volume of a cube is the cube of one of its edges (e).

$$V = e^3$$

table 9.3
English Units of Volume

1 ft^3	$= 1{,}728\text{ in.}^3$
1 yd^3	$= 27\text{ ft}^3$
1 ft^3	$= 7.48\text{ gal}$
1 gal	$= 231\text{ in.}^3$

Using this procedure, we can also show that $1\text{ ft}^3 = 1{,}728\text{ in.}^3$ Table 9.3 summarizes equivalent units of volume in the English system. Notice that the last two relationships can be used for conversion to gallons, which were introduced as a unit of capacity in Section 8.3.

example 1

An aquarium has a length of 30 in., a width of 15.4 in., and a height of 20 in. How many gallons of water does it hold when full?

solution

First determine the volume of the aquarium in cubic inches.

$V = lwh$ — Use the formula for finding the volume of a rectangular solid.

$V = (30)(15.4)(20)$ Substitute $l = 30$, $w = 15.4$, and $h = 20$.

$V = 9{,}240 \text{ in.}^3$

Now convert $9{,}240 \text{ in.}^3$ to gallons.

$$9{,}240 \text{ in.}^3 = 9{,}240 \cancel{\text{in.}^3} \times \frac{1 \text{ gal}}{231 \cancel{\text{in.}^3}} = \frac{9{,}240}{231} \text{ gal} = 40 \text{ gal}$$

Thus, the aquarium holds 40 gal when full. ●

In the metric system, each unit of volume differs from the next larger unit by a factor of 1,000. For example, 1 cubic centimeter is equivalent to 1,000 cubic millimeters, as shown in Figure 9.24.

figure 9.24

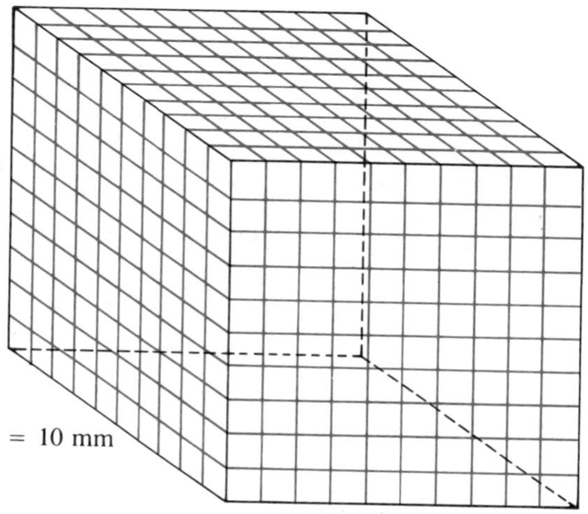

$1 \text{ cm}^3 = 1{,}000 \text{ mm}^3$
1 cubic centimeter

One liter (L) is sometimes defined to be the volume contained in a cube that measures one decimeter along every edge. Thus,

$$1 \text{ L} = 1 \text{ dm}^3$$

Since $1 \text{ dm}^3 = 1{,}000 \text{ cm}^3$

$$1 \text{ L} = 1{,}000 \text{ cm}^3$$

and $\frac{1}{1{,}000} \text{ L} = 1 \text{ mL} = 1 \text{ cm}^3$

These relationships are illustrated in Figure 9.25.

Similarly, since $1 \text{ kL} = 1{,}000 \text{ L}$

$$1 \text{ kL} = 1{,}000 \text{ dm}^3$$

and $1 \text{ kL} = 1 \text{ m}^3$

Table 9.4 summarizes equivalent metric units of volume.

table 9.4
Metric Units of Volume

1 cm^3	$= 1{,}000 \text{ mm}^3$
1 dm^3	$= 1{,}000 \text{ cm}^3$
1 m^3	$= 1{,}000 \text{ dm}^3$
1 L	$= 1 \text{ dm}^3 = 1{,}000 \text{ cm}^3$
1 mL	$= 1 \text{ cm}^3$
1 kL	$= 1{,}000 \text{ dm}^3 = 1 \text{ m}^3$

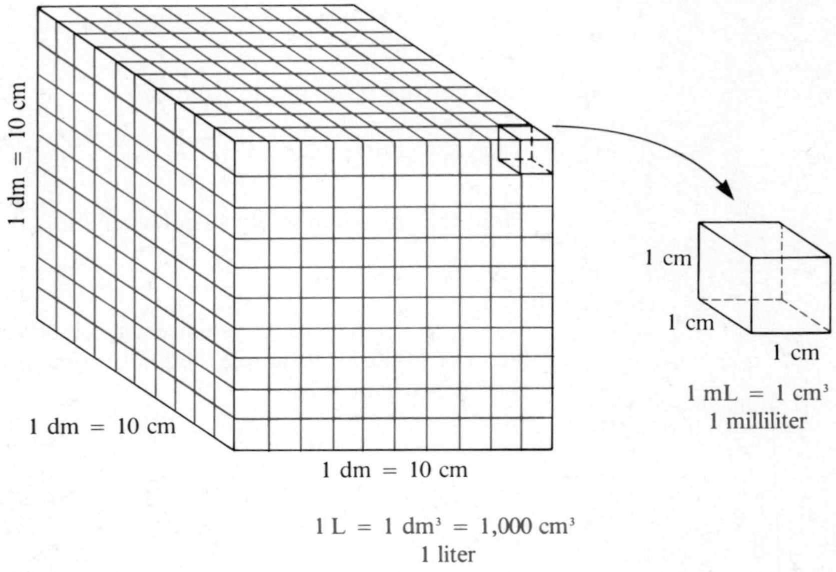

figure 9.25

example 2

Convert each of the following metric units of measurement as instructed.

(a) Convert 500 mL to cubic centimeters.

$$500 \text{ mL} = 500 \cancel{\text{mL}} \times \frac{1 \text{ cm}^3}{1 \cancel{\text{mL}}}$$

$$= 500 \text{ cm}^3$$

(b) Convert 30 cm^3 to cubic millimeters.

$$30 \text{ cm}^3 = 30 \cancel{\text{cm}^3} \times \frac{1{,}000 \text{ mm}^3}{1 \cancel{\text{cm}^3}}$$

$$= 30{,}000 \text{ mm}^3$$

(c) Convert 7 m^3 to cubic centimeters.

$$7 \text{ m}^3 = 7 \cancel{\text{m}^3} \times \frac{1{,}000 \text{ dm}^3}{1 \cancel{\text{m}^3}}$$

$$= \frac{7{,}000 \cancel{\text{dm}^3}}{1} \times \frac{1{,}000 \text{ cm}^3}{1 \cancel{\text{dm}^3}}$$

$$= 7{,}000{,}000 \text{ cm}^3$$

We will now present the formulas for finding the volume of a cylinder, a sphere, and a cone. A *cylinder* is a figure that is shaped like a can. It has two circular bases joined by sides that are perpendicular to them. The radius of a cylinder (r) is the radius of the circular base. The height of a cylinder (h) is the distance between the two bases. To find the volume of a cylinder, we can multiply the area of the base, which is πr^2, by the height h (see the box at the top of the following page).

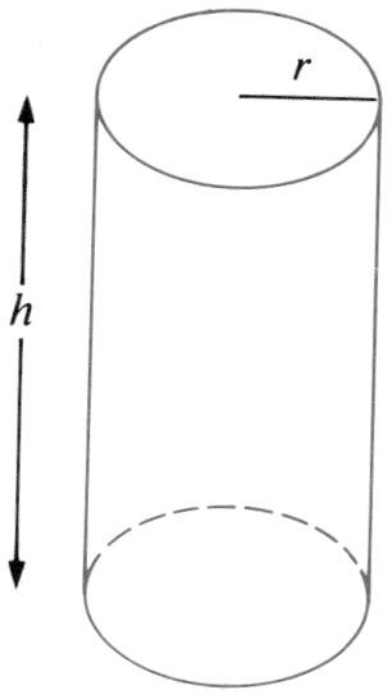

The volume of a cylinder is the area of the area of the base (πr^2) times the height (h).

$$V = \pi r^2 h$$

example 3

Find the volume of a cylinder with a radius of 3 cm and a height of 80 mm.

solution

Both radius and height must be in terms of the same units.

$$80 \text{ mm} = 80 \, \cancel{\text{mm}} \times \frac{1 \text{ cm}}{10 \, \cancel{\text{mm}}} = 8 \text{ cm}$$ Convert 80 mm to cm.

$V = \pi r^2 h$

$V = \pi(3)^2(8)$ Substitute $r = 3$ and $h = 8$.

$V = \pi(9)(8)$ Raise to powers before multiplying.

$V = 72\pi \text{ cm}^3$ Since the dimensions are in cm, the volume is in cm^3.

A *sphere* is a figure shaped like a ball. The radius (r) of a sphere is the distance from the center to any point on the sphere. Since the technique needed to derive the formula for the volume of a sphere requires a knowledge of calculus, we simply state the result in the box below.

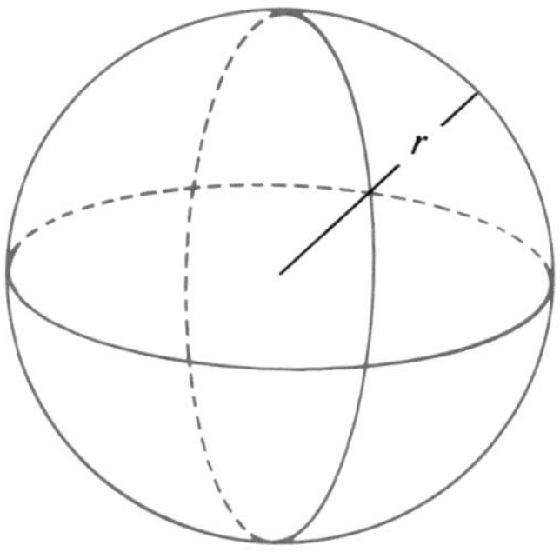

The volume of a sphere is $\frac{4}{3}\pi$ times the cube of the radius (r).

$$V = \frac{4}{3}\pi r^3$$

example 4

Find the volume of a hemisphere with a 3-ft radius.

solution

Find the volume of the sphere and take one half of the result. — Since a hemisphere is one-half of a sphere.

$V = \frac{4}{3}\pi r^3$

$V = \frac{4}{3}\pi(3)^3$ — Substitute $r = 3$.

$V = \frac{4}{3}\pi(27)$

$V = 36\pi \text{ ft}^3$ — 36π ft^3 is the volume of a sphere of radius 3 ft.

$\frac{1}{2}(36\pi) = 18\pi$ — Find one-half of the volume of the sphere.

The volume of the hemisphere is 18π ft^3. ●

A *cone* is a figure that is shaped like a funnel. It has a circular base and sides that come to a point directly above or below the center of the base. The radius (r) of a cone is the radius of the circle that forms the base. The height (h) of a cone is the distance from the center of the base to the point where the sides meet. The radius and height are perpendicular to each other. In order to derive the formula for the volume of a cone, one must know calculus, so we simply state the result in the box.

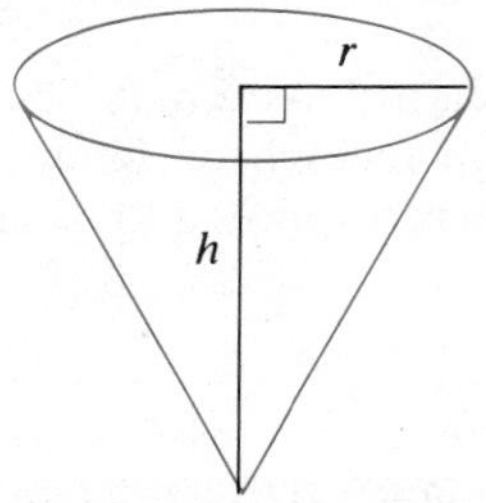

The volume of a cone is $\frac{1}{3}\pi$ times the square of the radius (r) times the height (h).

$$V = \frac{1}{3}\pi r^2 h$$

example 5

Find the volume of a cone that has a height of 6 in. and a base diameter of 4 in.

solution

If the base has a diameter of 4 in., its radius is 2 in.

$V = \frac{1}{3}\pi r^2 h$

$V = \frac{1}{3}\pi(2)^2(6)$ — Substitute $r = 2$ and $h = 6$.

$$V = \frac{1}{3}\pi(4)(6)$$

$$V = 8\pi \text{ in.}^3$$

example 6

Find the volume of the accompanying figure that is composed of a cylinder of height 10 cm and radius 2 cm, topped by a cone with the same radius and a height of 9 cm. Use $\pi \approx 3.14$.

solution

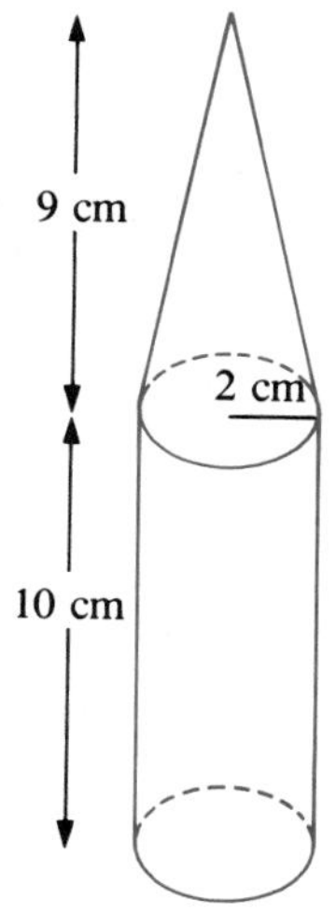

First find the volume of the cylinder and cone separately.

$V_{cyl} = \pi r^2 h$ — Formula for finding the volume of a cylinder.

$V_{cyl} = \pi(2)^2(10)$ — Substitute $r = 2$ and $h = 10$.

$V_{cyl} = (3.14)(4)(10)$ — Use $\pi \approx 3.14$.

$V_{cyl} = 125.6 \text{ cm}^3$

$V_{cone} = \frac{1}{3}\pi r^2 h$ — Formula for finding the volume of a cone.

$V_{cone} = \frac{1}{3}\pi(2)^2(9)$ — Substitute $r = 2$ and $h = 9$.

$V_{cone} = \frac{1}{3}(3.14)(4)(9)$ — Use $\pi \approx 3.14$.

$V_{cone} = 37.68 \text{ cm}^3$

The total volume is equal to the volume of the cylinder plus the volume of the cone.

$V_{total} = V_{cyl} + V_{cone}$

$V_{total} = 125.6 + 37.68$

$V_{total} = 163.28 \text{ cm}^3$

QUICK QUIZ

Find the volume of each of the following figures.

1. Rectangular solid with length 9 cm, width 5 cm, and height 6 cm.
2. Cylinder with radius 3 in. and height 7 in.
3. Sphere with radius of 6 ft.
4. Cone with radius 7 m and height 12 m. Use $\pi \approx \frac{22}{7}$.

ANSWERS

1. 270 cm^3
2. $63\pi \text{ in.}^3$
3. $288\pi \text{ ft}^3$
4. 616 m^3

9.6 Exercises

Indicate which of the following statements are true and which are false. For those that are false, change the italic expression to make the statement true.

1. Volume is measured in *square* units of measurement.
2. A liter is equivalent to one cubic *meter*.
3. A cubic yard is equivalent to 27 cubic *inches*.
4. A cubic centimeter is equivalent to *100* cubic millimeters.

Find the volume of each of the following geometric solids. Leave your answers in terms of π unless otherwise indicated.

5. Rectangular solid with length 8 in., width 4 in., and height 3 in.
6. Rectangular solid with length 11 dm, width 2 dm, and height 4 dm.
7. Rectangular solid with length $8\frac{1}{3}$ yd, width $4\frac{1}{5}$ yd, and height $3\frac{1}{3}$ yd.
8. Rectangular solid with length $5\frac{1}{2}$ m, width $2\frac{3}{4}$ m, and height 6 m.
9. Rectangular solid with length 6 ft, width 3 ft, and height 9 in.
10. Rectangular solid with length 9 cm, width 20 mm, and height 7 cm.
11. Cube that measures 4 cm along every edge.
12. Cube that measures 5 ft along every edge.
13. Cube that measures 1.2 in. along every edge.
14. Cube that measures 80 mm along every edge.
15. Cube that measures $2\frac{1}{4}$ dm along every edge.
16. Cube that measures $1\frac{1}{2}$ yd along every edge.
17. Cylinder with radius 2 cm and height 9 cm.
18. Cylinder with radius 3 in. and height 7 in. Use $\pi \approx \frac{22}{7}$.
19. Cylinder with radius 1 ft and height 5 ft. Use $\pi \approx 3.14$.
20. Cylinder with radius $2\frac{1}{2}$ m and height 40 dm.
21. Cylinder with diameter 8 dm and height 120 cm.
22. Cylinder with diameter 6 ft and height $1\frac{1}{3}$ ft. Use $\pi \approx 3.14$.
23. Sphere with radius 7 ft.
24. Sphere with radius 1.2 m.
25. Sphere with radius 3 cm. Use $\pi \approx 3.14$.
26. Sphere with radius 6 in. Use $\pi \approx 3.14$.
27. Sphere with diameter 15 in.
28. Sphere with diameter 18 dm.
29. Hemisphere with radius 9 ft. Use $\pi \approx 3.14$.
30. Hemisphere with diameter 6 yd.
31. Cone with radius 10 ft and height 30 ft. Use $\pi \approx 3.14$.
32. Cone with radius 1.5 dm and height 2.5 dm.
33. Cone with radius 4 in. and height 12 in. Use $\pi \approx 3.14$.
34. Cone with diameter 14 m and height 9 m. Use $\pi \approx \frac{22}{7}$.
35. Cone with diameter 8 in. and height 1 ft.
36. Cone with radius 3 cm and height 80 mm.
37. Cone with radius 5.7 cm and height 9.2 cm. Use $\pi \approx 3.14$. Round off to the nearest tenth.
38. Sphere with diameter 8.5 in. Use $\pi \approx 3.14$. Round off to the nearest tenth.

Convert each of the following units of measurement as instructed.

39. $3\text{ ft}^3 = \_\_\_\_ \text{ in.}^3$
40. $54\text{ ft}^3 = \_\_\_\_ \text{ yd}^3$
41. $\frac{1}{2}\text{ gal} = \_\_\_\_ \text{ in.}^3$
42. $7\text{ ft}^3 = \_\_\_\_ \text{ gal}$
43. $9\text{ cm}^3 = \_\_\_\_ \text{ mm}^3$
44. $200\text{ cm}^3 = \_\_\_\_ \text{ dm}^3$

45. $7{,}000 \text{ mm}^3 = \_\_\_\_ \text{ cm}^3$

46. $8 \text{ dm}^3 = \_\_\_\_ \text{ cm}^3$

47. $19 \text{ cm}^3 = \_\_\_\_ \text{ mL}$

48. $80 \text{ dm}^3 = \_\_\_\_ \text{ L}$

49. $60 \text{ m}^3 = \_\_\_\_ \text{ kL}$

50. $75 \text{ cm}^3 = \_\_\_\_ \text{ L}$

51. $800 \text{ mm}^3 = \_\_\_\_ \text{ mL}$

52. $7 \text{ dm}^3 = \_\_\_\_ \text{ mL}$

53. $600 \text{ m}^3 = \_\_\_\_ \text{ cm}^3$

54. $70{,}000 \text{ mm}^3 = \_\_\_\_ \text{ dm}^3$

Solve each of the following word problems.

55. A swimming pool is 200 ft long, 100 ft wide, and $3\frac{1}{2}$ ft deep. How many gallons of water are needed to fill the pool?

56. A refrigerator is 2.5 ft long, 2 ft wide, and 5.5 ft high. What is its capacity in cubic feet?

57. A beach ball has a 21-in. diameter. What is its volume? Use $\pi \approx \frac{22}{7}$.

58. A dome is shaped like a hemisphere that has a 20-ft diameter. What is the volume it encloses? Use $\pi \approx 3.14$. Round off to the nearest cubic foot.

59. A can of soda is 12 cm high and 7 cm in diameter. How many milliliters can it hold? Use $\pi \approx \frac{22}{7}$.

60. A cylindrical gas tank has a radius of 2 m and a height of 5 m. How many liters of gas are in the tank when it is full? Use $\pi \approx 3.14$.

61. A sand pile is shaped like a cone that is 10 ft high and has a radius of 6 ft. How many cubic feet of sand are in the pile? Use $\pi \approx 3.14$.

62. A water tank in the shape of a cone is 7 m high and has a radius of 3 m. How many kiloliters of water can it hold? Use $\pi \approx \frac{22}{7}$.

63. Find the volume of the accompanying figure that is composed of a cylinder of height 10 in. and radius 6 in., topped by a hemisphere with the same radius. Use $\pi \approx 3.14$.

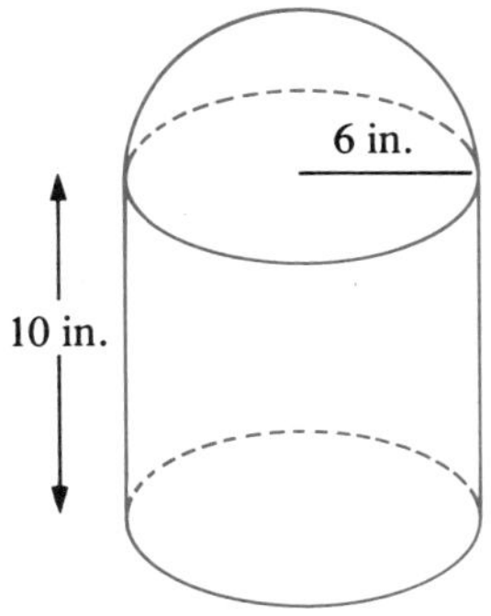

64. Find the volume of the accompanying figure that is a cone of height 8 m and radius 3 m, topped by a hemisphere with the same radius. Use $\pi \approx 3.14$.

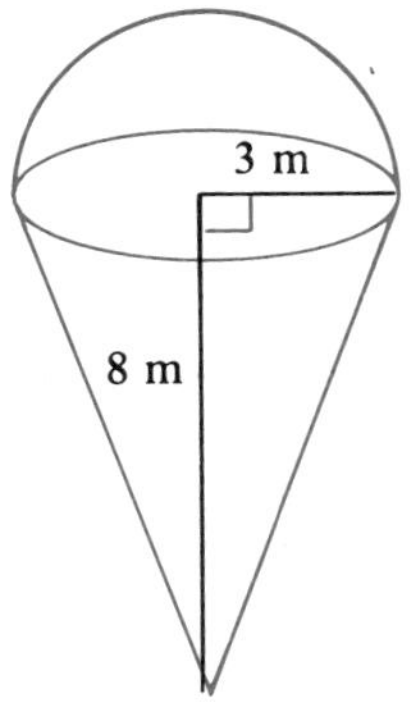

Summary and Review

Key Terms

[9.1] A **polygon** is a closed figure that has many sides. A **regular polygon** is a polygon in which all sides and all angles are equal.

A **triangle** is a three-sided figure.

A **quadrilateral** is a four-sided figure.

A **pentagon** is a five-sided figure.

A **hexagon** is a six-sided figure.

An **octagon** is an eight-sided figure.

An **angle** is formed whenever two lines meet. The two lines are called the sides of the angle, and the point at which they meet is called the **vertex** of the angle.

A **degree** is a unit of measurement that is used to indicate the size of an angle.

A **right angle** is a 90° angle.

An **acute angle** is an angle less than 90°.

An **obtuse angle** is an angle greater than 90° and less than 180°.

Perpendicular lines meet in a right angle.

An **equilateral triangle** has three equal sides and three equal angles.

An **isosceles triangle** has two equal sides and two equal angles.

A **right triangle** has a right angle.

Parallel lines are lines in the same plane that never meet.

A **trapezoid** is a quadrilateral that has two parallel sides.

A **parallelogram** is a quadrilateral whose opposite sides are equal and parallel.

A **rectangle** is a parallelogram that has four right angles.

A **square** is a rectangle that has four equal sides.

[9.2] **Perimeter** is the distance around a geometric figure.

[9.3] **Area** is the measurement of the surface of a figure.

[9.4] The **legs** of a right triangle are the two sides that meet in a right angle.

The **hypotenuse** of a right triangle is the longest side, which joins the two legs.

[9.5] A **circle** is a geometric figure that includes all points that are the same distance away from a fixed point called the *center.*

The **radius** of a circle is the distance from the center to any point on the circle.

The **diameter** of a circle is a line that connects two points on a circle and passes through its center.

The **circumference** of a circle is the perimeter of, or distance around, a circle.

Pi (π) is the ratio of the circumference of a circle to its diameter. π is an irrational number approximately equal to 3.14.

[9.6] **Volume** is the measurement of the space enclosed by a solid, that is, the space enclosed by a three-dimensional figure.

Calculations

[9.2] **To find the perimeter of a rectangle** of length l and width w,

$$P = 2l + 2w$$

To find the perimeter of a regular polygon with n sides, each of length s,

$$P = ns$$

[9.3] **To find the area of a rectangle** of length l and width w,

$$A = lw$$

To find the area of a square with sides of length s,

$$A = s^2$$

To find the area of a parallelogram with base b and height h,

$$A = bh$$

To find the area of a triangle with base b and height h,

$$A = \frac{1}{2}bh$$

To find the area of a trapezoid with bases b_1 and b_2 and height h,

$$A = \frac{1}{2}h(b_1 + b_2)$$

[9.4] The **Pythagorean Theorem** states that for any right triangle with legs a and b and hypotenuse c,

$$c^2 = a^2 + b^2$$

[9.5] **To find the circumference of a circle** of diameter d and radius r,

$$C = \pi d \quad \text{or} \quad C = 2\pi r$$

[9.6] **To find the area of a circle** of radius r,

$$A = \pi r^2$$

To find the volume of a rectangular solid of length l, width w, and height h,

$$V = lwh$$

To find the volume of a cube with edges of length e,

$$V = e^3$$

To find the volume of a cylinder of radius r and height h,

$$V = \pi r^2 h$$

To find the volume of a sphere of radius r,

$$V = \frac{4}{3}\pi r^3$$

To find the volume of a rectangular solid of length l, width w, and height h,

$$V = \frac{1}{3}\pi r^2 h$$

Chapter 9 Review Exercises

Indicate which of the following statements are true and which are false. For those that are false, change the italic word to make the statement true.

1. Parallel lines *sometimes* intersect.
2. Perimeter is measured in *linear* units of measurement.
3. The region enclosed by a two-dimensional figure is called *volume.*

4. The Pythagorean Theorem applies to *equilateral* triangles.
5. The circumference of a circle is another name for its *area.*
6. If the radius of a sphere is doubled, its volume is increased by a factor of *eight*.

[9.3, 9.6] Convert each of the following units of measurement as instructed.

7. $2\ mi^2 =$ ____ acres
8. $72\ in.^2 =$ ____ ft^2
9. $400\ m^2 =$ ____ a
10. $3.5\ dm^2 =$ ____ cm^2
11. $36\ ft^2 =$ ____ yd^2
12. $36\ in.^2 =$ ____ ft^2
13. $250\ cm^2 =$ ____ m^2
14. $0.8\ ha =$ ____ m^2
15. $0.2\ ft^3 =$ ____ $in.^3$
16. $693\ in.^3 =$ ____ gal
17. $2.7\ ft^3 =$ ____ yd^2
18. $748\ gal =$ ____ ft^3
19. $14\ mL =$ ____ cm^3
20. $9{,}000\ cm^3 =$ ____ L
21. $5.4\ kL =$ ____ m^3
22. $400\ mm^3 =$ ____ cm^3

[9.2, 9.5] Find the perimeter (or circumference) of each of the following figures. Leave answers in terms of π where appropriate.

23. Rectangle of length 8 cm and width 3 cm.
24. Square whose sides are each 4.3 m long.
25. Triangle with sides 7 in., $4\frac{1}{2}$ in., and $6\frac{1}{4}$ in. long.
26. Regular octagon whose sides are each 10.5 ft long.
27. Circle of radius 7 mm.
28. Circle of diameter 5 in.
29. Right triangle whose legs measure 3 ft and 4 ft.
30. Right triangle whose legs measure 5 cm and 12 cm.

[9.3, 9.5] Find the area of each of the followng figures. Leave answers in terms of π where appropriate.

31. Rectangle of length 7 ft and width 9 ft.
32. Square whose sides are 12 cm long.
33. Parallelogram with base 9.5 m and height 6 m.
34. Triangle with base 10 in. and height 3.2 in.
35. Trapezoid with bases 2 ft and 8 ft long and height 6 ft.
36. Circle of radius 9 mm.
37. Right triangle with hypotenuse 10 m and leg 8 m.
38. Right triangle with hypotenuse 13 ft and leg 12 ft.
39. Circle of diameter 9 in.
40. Right triangle whose legs measure 5 in. and 7 in.
41. Rectangle of length 90 cm and width 3 dm.
42. Parallelogram with base 2 ft and height 8 in.

[9.6] Find the volume of each of the following solids. Leave answers in terms of π where appropriate.

43. Cube that measures 8 in. on every edge.
44. Cylinder with radius 4 ft and height 7 ft.
45. Sphere with radius 3 in.
46. Hemisphere with radius 9 cm.
47. Cone with radius 2 m and height 6 m.
48. Sphere with diameter 12 ft.
49. Cylinder with radius 9 cm and height 5 dm.
50. Cone with radius 8 in. and height $1\frac{1}{2}$ ft.

[9.2–9.6] Solve each of the following word problems.

51. A door that measures 2.15 m long and 90 cm wide is to be decorated with a string of lights. How long must the light string be in order to reach around the door?
52. A rectangular piece of land that measures 160 yd long × 120 yd wide is bounded on one side by a river that runs parallel to the longest side. How much fencing is needed to enclose the three remaining sides?
53. The top of a 13-ft ladder rests against a wall at a point that is 12 ft above the ground. How far away from the wall is the base of the ladder?

54. Two boats leave from the same starting point, one headed due south at a speed of 40 mph, and the other headed due east at a speed of 30 mph. How far apart are they $\frac{1}{2}$ hr later?

55. How many square feet of carpeting are needed to cover the floor of a room shaped like the accompanying figure?

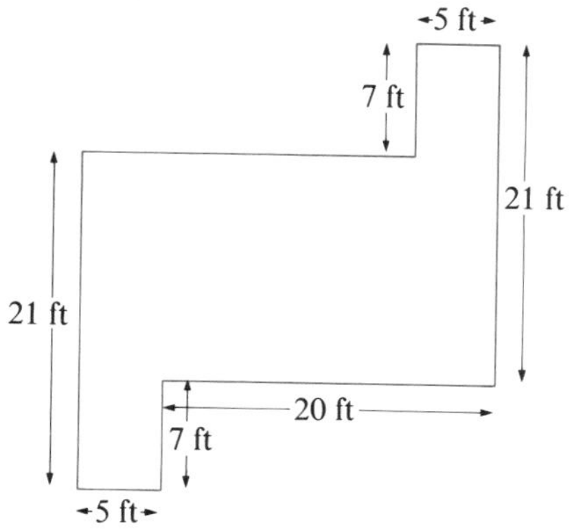

56. How much would it cost to paint the four walls of a room that measures 20 ft long, 18 ft wide, and $8\frac{1}{2}$ ft high if paint sells for $12 per gallon, and 1 gal covers 200 sq ft? Assume that there are no windows in the room and that the door is to be painted.

57. A square has a perimeter of 150 cm. What is the length of one of its sides?

58. An equilateral triangle has a perimeter of 54 in. What is the length of one of its sides?

59. A triangle with a 72-ft base has an area of 2,160 ft^2. What is its height?

60. A rectangle with an area of 69 cm^2 has a length of 10 cm. What is its width?

61. If the side of a square measures 5 in., what is the length of its diagonal? Round off to the nearest tenth of an inch.

62. The diagonal of a rectangle measures 20 cm. If the width of the rectangle is 12 cm, what is its length?

63. What is the diameter of a circle whose circumference measures 62π m?

64. What is the radius of a circle whose area is 81π ft^2?

65. The volume of a cube is 64 cm^3. What is the length of one of its edges?

66. The volume of a cylinder with a 3-in. radius is 72π $in.^3$ What is its height?

67. The earth has a diameter of 7,927 mi at the equator. What is the circumference of the earth measured along the equator? Use $\pi \approx 3.14$. Round off your answer to the nearest mile.

68. The moon makes one revolution around the earth in about 27 days while traveling at a speed of 0.63 miles per second. What is the approximate distance between the earth and the moon? Use $\pi \approx 3.14$. Round off your answer to the nearest thousand miles.

69. A punch bowl is shaped like a hemisphere with a radius of 20 cm. How many liters of punch can it hold when full? Use $\pi \approx 3.14$. Round off to the nearest tenth of a liter.

70. A barrel is shaped like a cylinder that has a diameter of 85 cm and a height of 120 cm. How many liters can it hold when full? Use $\pi \approx 3.14$. Round off your answer to the nearest liter.

10 INTRODUCTION TO ALGEBRA

10.1 The Language of Algebra

Algebra is considered to be a generalization of arithmetic. It is a branch of mathematics that provides us with an extremely powerful tool for solving a wide variety of problems that involve finding an unknown quantity. Before we begin our discussion of algebra, it is important that we understand the vocabulary that is used to explain the ideas that pertain to it. The notation used in algebraic expressions consists of **constants**, which are numbers, and **variables**, which are letters that represent unknown quantities. The **coefficient** of a variable is the number that is multiplied by the variable. For example, the expression $7x$ means 7 multiplied by x. The variable is x and the constant is 7. The coefficient of x is 7.

The **terms** of an algebraic expression are the quantities that are added. For example, the expression

$$\underbrace{4x + 9y}_{\text{terms}}$$

has two terms, $4x$ and $9y$. The expression

$$\underbrace{8a - 9b + 5c}_{\text{terms}}$$

has three terms, $8a$, $-9b$, and $5c$. The term $8a$ is composed of two factors, the constant 8 and the variable a. Eight is also the coefficient of a. In the term $-9b$, -9 is the coefficient of the variable b. In the term $5c$, 5 is the coefficient of the variable c.

The procedure for evaluating an algebraic expression is given in the box.

> In order to evaluate an algebraic expression, we substitute numbers for the unknown variables and simplify the result using the order of operations rules.

example 1 Evaluate $5x + 7y$ when $x = -2$ and $y = 4$.

solution

$5x + 7y = 5(-2) + 7(4)$	Substitute -2 for x and 4 for y.
$= -10 + 28$	First do multiplication.
$= 18$	Then do addition.

example 2

Evaluate $5(2a - b) + 4c$ when $a = 4$, $b = -2$, and $c = -5$.

solution

$5(2a - b) + 4c = 5[2(4) - (-2)] + 4(-5)$	Substitute $a = 4$, $b = -2$, $c = -5$.
$= 5[8 - (-2)] + 4(-5)$	Simplify expression in [ ] first by multiplying, then subtracting.
$= 5(10) + 4(-5)$	
$= 50 + (-20)$	Do multiplication.
$= 30$	Finally, do addition.

Since algebra is an important tool for solving a wide variety of word problems, we will now illustrate how words can be translated into algebraic expressions.

example 3

Translate each of the following word expressions into algebraic expressions.

(a) Three more than x.

$x + 3$

(b) The product of eight and n.

$8n$

Notice that the multiplication sign is omitted when expressing the product of a number and a variable.

(c) Five times the sum of seven and y.

$5(7 + y)$

Notice that the parentheses indicate that addition is to be performed before multiplication.

(d) Two-fifths the quantity eight minus t.

$\frac{2}{5}(8 - t)$

Notice that the expression that follows the words *the quantity* is enclosed in parentheses.

Let us now look at some examples of algebraic expressions that can be translated into words.

example 4

Translate each of the following algebraic expressions into words.

solution

(a) $5 + 3w$

Five plus the product of three and w.

(b) $6(a + b)$

Six times the sum of a and b, or six times the quantity a plus b.

Remember that the words *the quantity* indicate that the expression that follows appears in parentheses.

QUICK QUIZ	ANSWERS
1. Evaluate $5x + 7y - 3z$ when $x = 2$, $y = -1$, and $z = -2$.	1. 9
2. Write an algebraic expression that represents eight subtracted from t.	2. $t - 8$
3. Translate the expression $(x + 5) \div 2$ into words.	3. The quantity x plus five, divided by two.

10.1 Exercises

Indicate which of the following statements are true and which are false. For those that are false, change the italic word to make the statement true.

1. A number that appears in an algebraic expression is called a *constant*.
2. Quantities that are added or *multiplied* are called terms.
3. The *coefficient* of a variable is a number that is multiplied by a variable.
4. Algebra is a generalization of *geometry*.

Indicate how many terms are in each of the following algebraic expressions. For each term, identify the variable and its coefficient.

5. $3w$ **6.** t **7.** $2x + 5y$ **8.** $8a - 7b$

9. $3p + 9q + r$ **10.** $x + 6y - 4z$ **11.** $\frac{1}{2}c - d$ **12.** $x + \frac{3}{4}y + \frac{1}{8}z - 12w$

Evaluate each of the following algebraic expressions for the values given.

13. $a + 4b$ when $a = -8, b = 3$

14. $5x - y$ when $x = 2, y = 6$

15. $4s - 7t$ when $s = 5, t = -1$

16. $3w + 2z$ when $w = -6, z = -5$

17. $\frac{1}{2}x - \frac{2}{3}y$ when $x = 4, y = 9$

18. $\frac{3}{4}a + \frac{7}{8}b$ when $a = 12, b = -8$

19. $4p - 3q - 5r$ when $p = 3, q = 7, r = -1$

20. $-7a - 9b - 11c$ when $a = -2, b = 3, c = 1$

21. $-5x + y - 2z$ when $x = \frac{4}{5}, y = 9, z = -\frac{1}{2}$

22. $s - 3t - 7u$ when $s = -4, t = \frac{2}{3}, u = \frac{5}{7}$

23. $-2(a + 5b)$ when $a = -15, b = 12$

24. $4(3x - y)$ when $x = 21, y = 30$

25. $5(2s - 5t + 3u)$ when $s = 3, t = -2, u = 6$

26. $-7(3x - 4y - 2z)$ when $x = -5, y = -4, z = -1$

27. $3(w - 2t) + 5(3w + t)$ when $w = -1, t = 6$

28. $9(x - y) - 2(3x - 4y)$ when $x = 4, y = -7$

Translate each of the following word expressions into algebraic expressions.

29. x decreased by five.

30. Nine more than y.

31. The product of t and negative two.

32. Thirteen divided by w.

33. Two-thirds of z.

34. One-half of q.

35. Thirty-five less than x.

36. Twenty-four increased by y.

37. Five more than the product of seven and a.

38. Two less than the product of negative four and n.

39. Negative eight times the sum of a and b.

40. Seven times the quantity x minus y.

41. The quantity fourteen minus x, divided by two.
42. The quantity y plus nine, divided by four.
43. Twice the sum of a and five.
44. The quantity nine minus t, tripled.
45. Five-eights the quantity u plus seven.
46. Three-fifths the quantity one minus w.

Translate each of the following algebraic expressions into words.

47. $x - 4$
48. $y + 9$
49. $6a$
50. $14 \div a$
51. $9 + t$
52. $3 - n$
53. $w \div 11$
54. $-5z$
55. $2y + 8$
56. $7x - 13$
57. $3n - 2$
58. $4p + 19$
59. $5(z - 7)$
60. $6(t + 4)$
61. $(4 - u) \div 3$
62. $(y + 9) \div 2$
63. $\frac{3}{8}(x + 5)$
64. $\frac{1}{3}(w - 8)$

10.2 Simplifying Algebraic Expressions

In this section, we will show how to simplify algebraic expressions using a technique called **combining like terms.**

definition

***Like terms*, or *similar terms*, are terms whose variable factors are the same.**

For example, the terms $8x$ and $\frac{3}{5}x$ are like terms since they have the same variable factor, x. The terms $7xy$ and $-2xz$ are not like terms, because the variable factors of $7xy$ are xy, and the variable factors of $-2xz$ are xz. The procedure for combining like terms is summarized in the box and illustrated in the following examples.

> To simplify an algebraic expression by combining like terms, we first rewrite the expression using the distributive property, and then add or subtract the coefficients of the variable.

example 1

Simplify each of the following algebraic expressions by combining like terms.

(a) $8x + 3x$

$8x + 3x = (8 + 3)x$ — Rewrite using the distributive property.

$= 11x$ — Add the coefficients of x.

(b) $9y - 4y$

$9y - 4y = (9 - 4)y$ — Rewrite using the distributive property.

$= 5y$ — Subtract the coefficients of y.

(c) $5x + x + 8x$

$5x + x + 8x = (5 + 1 + 8)x$ — Rewrite using the distributive property.

$= 14x$ — Add the coefficients of x. Keep in mind that the second term, x, has a coefficient of 1.

(d) $3x - 7x - 9x$

$3x - 7x - 9x = (3 - 7 - 9)x$ — Think, $3 - 7 - 9 = 3 + (-7) + (-9)$.

$= -13x$ ●

Notice that when we simplify algebraic expressions, we usually do not write the operation subtraction as adding the opposite, even though mentally we may be using that technique to combine the coefficients. (See Example 1(d).)

Sometimes, before we can combine like terms, we need to use the commutative property of addition to change the order of the terms when rewriting the problem as shown in Example 2.

example 2

Simplify each of the following algebraic expressions.

(a) $5x - 8 + 7x + 3$

$5x - 8 + 7x + 3 = 5x + 7x - 8 + 3$ — By the commutative property of addition.

$= (5 + 7)x - 8 + 3$ — By the distributive property.

$= 12x - 8 + 3$ — Add the coefficients of x.

$= 12x - 5$ — Combine the constant terms.

(b) $2a - 5 - 8a + 9 - 4a - 3$

$2a - 5 - 8a + 9 - 4a - 3$

$= 2a - 8a - 4a - 5 + 9 - 3$ — By the commutative property of addition.

$= (2 - 8 - 4)a - 5 + 9 - 3$ — By the distributive property.

$= -10a - 5 + 9 - 3$ — Combine the coefficients of a.

$= -10a + 1$ — Combine the constant terms. ●

If an algebraic expression that appears in parentheses cannot be simplified, we multiply each term in parentheses by the factor that precedes the parentheses, using the distributive property. We then simplify the resulting expression by combining like terms as shown in Example 3.

example 3

Simplify each of the following algebraic expressions.

(a) $3(2y - 5) + 8$

$3(2y - 5) + 8 = 3(2y) - 3(5) + 8$ — Use the distributive property to remove the parentheses by multiplying each term in parentheses by 3.

$= (3 \cdot 2)y - 3(5) + 8$ — By the associative property.

$= 6y - 15 + 8$ — Do multiplication first.

$= 6y - 7$ — Add constant terms.

(b) $2(7x - 3) + 5(-2x + 5)$

$2(7x - 3) + 5(-2x + 5)$

$= 2(7x) - 2(3) + 5(-2x) + 5(5)$ — Use the distributive property to remove parentheses.

$= 14x - 6 + (-10x) + 25$

$= 14x + (-10x) - 6 + 25$ — By the commutative property.

$= (14 - 10)x - 6 + 25$ — To add $-10x$, we subtract $10x$.

$= 4x + 19$

(c) $8 - (4w + 3)$

$8 - (4w + 3) = 8 + (-1)(4w + 3)$ — To subtract a quantity in parentheses, we add the opposite of each term in parentheses. This can be shown using the distributive property. Multiply each term in parentheses by -1.

$= 8 + (-1)(4w) + (-1)(3)$

$= 8 + (-4w) + (-3)$

$= -4w + 8 - 3$ — By commutative property.

$= -4w + 5$

(d) $-8(x - 9) - 3(x + 5)$

$-8(x - 9) - 3(x + 5)$

$= -8(x - 9) + (-3)(x + 5)$

$= -8x - 8(-9) + (-3)x + (-3)(5)$ — Use the distributive property to remove parentheses.

$= -8x + 72 - 3x - 15$

$= -8x - 3x + 72 - 15$ — By the commutative property.

$= -11x + 57$ — Combine like terms.

QUICK QUIZ

Simplify each of the following algebraic expressions by combining like terms.

1. $-2x + 9x - 4x$
2. $5z - 3 - 7z + 8$
3. $4(5t + 2) - (t - 3)$

ANSWERS

1. $3x$
2. $-2z + 5$
3. $19t + 11$

10.2 Exercises

Indicate which of the following statements are true and which are false. For those that are false, change the italic word to make the statement true.

1. Like terms are terms in which the *constant* factors are the same.
2. When combining like terms, we add or *subtract* the coefficients of the variable.
3. To multiply a quantity in parentheses, we use the *distributive* property.
4. To change the order of terms when simplifying an algebraic expression, we use the *associative* property.

Simplify each of the following algebraic expressions by combining like terms.

5. $8x + 9x$
6. $7y - 2y$
7. $-3a + 5a$
8. $-4b - 6b$
9. $t - 9t$
10. $5w + w$
11. $\frac{3}{4}a - \frac{1}{4}a$
12. $\frac{1}{5}z + \frac{3}{5}z$
13. $-35u - 18u$
14. $-61t + 23t$
15. $7.2p - 4.9p$
16. $5.1z - 3.4z$
17. $8y + 4y + 3y$
18. $7x + 5x + 9x$
19. $-2z + 4z + 12z$
20. $8s - 3s - 9s$
21. $23c + c + 19c$
22. $69r - 47r - 51r$
23. $-8.5u + u - 6.7u$
24. $w - 4.8w + 2.5w$
25. $5(y + 2) - 3$
26. $7(q - 6) + 4$
27. $8(5 - n) + 3$
28. $-3(7 + y) - 5$
29. $5 - (9w + 4)$
30. $2 - (5 + 8t)$
31. $-3 - (5p - 8)$
32. $-7 - (4 - d)$
33. $7x - 5(2x - 4)$
34. $-3z + 7(-4z - 3)$
35. $4s - 3(5s - 2)$
36. $-2u - 4(-7u + 5)$
37. $-3(6x + 7) + 9(2x + 1)$
38. $8(3a + 4) - 4(6a + 2)$
39. $6(2z - 4) + 8(7z - 5)$
40. $2(7m - 6) - 4(3m - 2)$
41. $4(-3t + 6) - 2(5t - 8)$
42. $-6(3b - 5) + 7(5b - 3)$
43. $-7(6k + 2) + 5(1 - 4k)$
44. $9(2a - 3) - 8(-a - 1)$

10.3 The Addition Principle of Equality

The term **equation** was defined in Section 1.2 as a mathematical expression that states that two quantities are equal. We have already worked with some simple equations in Section 4.4, which dealt with finding the unknown term in a proportion, and in Section 5.3, which dealt with equations involving percent. In the next three sections, we will learn how to solve a variety of similar equations that are called **linear equations**.

definition

A *linear equation* in one variable is an equation that can be written in the form $ax + b = c$, where a, b, and c are constants and $a \neq 0$.

Some examples of linear equations in one variable are:

$$3x + 9 = 0 \qquad 6x + 5 = 2x - 8$$
$$x - 7 = 2 \qquad 3(x - 2) = 4$$

Notice that each of these equations has only a single variable, x, and that the exponent of x is 1 since $x = x^1$. For this reason, equations of this form are often called first-degree equations of a single variable.

definition

The *solution* to a linear equation in one variable is the value that can be substituted for the unknown variable so that the resulting statement is true.

For example, the number 4 is a solution to the linear equation $3x - 5 = 7$, since, the resulting statement is true when we substitute 4 for x.

$$3x - 5 = 7$$
$$3(4) - 5 = 7$$
$$12 - 5 = 7$$
$$7 = 7 \qquad \text{This is a true statement.}$$

example 1

Determine if the number 2 is a solution to the linear equation $7x - 11 = 8 - 3x$.

solution

$$7x - 11 = 8 - 3x$$

$$7(2) - 11 = 8 - 3(2) \qquad \text{Substitute 2 for } x.$$

$$14 - 11 = 8 - 6$$

$$3 = 2 \qquad \text{This is a false statement.}$$

Therefore, the number 2 is not a solution to the linear equation. ●

In order to solve a linear equation, we will perform a series of algebraic operations on both sides of the equation that will result in a series of **equivalent equations**.

definition

***Equivalent equations* are equations that have the same solution.**

For example, the equations

$$-5x + 3 = 18 \quad \text{and} \quad x = -3$$

are equivalent equations since they have the same solution, -3. That is, when -3 is substituted for x in both equations, the resulting statements are true.

example 2

Determine if $x = -\frac{1}{2}$ and $2(2 - 3x) = 3 - 8x$ are equivalent equations.

solution

$$2(2 - 3x) = 3 - 8x$$

$$2\left[2 - 3\left(-\frac{1}{2}\right)\right] = 3 - 8\left(-\frac{1}{2}\right)$$

Substitute $-\frac{1}{2}$ for x in the second equation and determine if the resulting statement is true.

$$2\left[2 - \left(-\frac{3}{2}\right)\right] = 3 - (-4)$$

$$2\left(2 + \frac{3}{2}\right) = 3 + 4$$

$$2\left(\frac{7}{2}\right) = 7$$

$$7 = 7 \qquad \text{This is a true statement.}$$

Therefore, $x = -\frac{1}{2}$ and $2(2 - 3x) = 3 - 8x$ are equivalent equations. ●

The first type of linear equation that we will consider can be solved using the **addition principle of equality**.

According to the **addition principle of equality**, if the same quantity is added to both sides of an equation, the resulting equation is equivalent to the original one.

Recall from Section 4.3 that the goal in solving an equation is to isolate the variable on one side of the equation so that the number that is a solution to the equation appears on the other side. If an equation is of the form $x + b = c$ where b and c are constants, we can isolate the variable x on the left by adding the opposite of b, or $-b$, to both sides of the equation as follows:

$$x + b + (-b) = c + (-b)$$ By addition property of equality.

$$x + 0 = c + (-b)$$ Since $b + (-b) = 0$.

$$x = c - b$$ Combine constants on the right.

This procedure is illustrated by Example 3.

example 3

Solve $x + 5 = 3$.

solution

$$x + 5 = 3$$

$$x + 5 + (-5) = 3 + (-5)$$ Add the opposite of 5, or -5, to both sides.

$$x + 0 = -2$$

$$x = -2$$

example 4

Solve $-6 = a - 2$.

solution

$$-6 + 2 = a - 2 + 2$$ Add the opposite of -2, or 2, to both sides.

$$-4 = a + 0$$

$$-4 = a$$

$$a = -4$$ Notice that the statements $-4 = a$ and $a = -4$ represent the same solution.

Though it is correct to express the solution with the variable on either the left- or right-hand side of the equal sign, the variable usually appears on the left.

Often, we need to combine like terms on both sides of the equation before applying the addition principle of equality, as shown in Example 5.

example 5

Solve $-8y - 3 + 9y - 7 = 4$.

solution

$$-8y - 3 + 9y - 7 = 4$$

$$-8y + 9y - 3 - 7 = 4$$ By the commutative property.

$$y - 10 = 4$$ Combine like terms.

$$y - 10 + 10 = 4 + 10$$ Add the opposite of -10, or 10, to both sides.

$$y + 0 = 14$$

$$y = 14$$

Whenever variable terms appear on both sides of the equation, we use the addition principle of equality to move all variable terms to one side and all constant terms to the other side, as shown in the following examples.

example 6

solution

Solve $-5x = 8 - 6x$.

$$-5x = 8 - 6x$$

$$-5x + 6x = 8 - 6x + 6x$$ Add the opposite of $-6x$, or $6x$, to both sides.

$$x = 8 + 0$$

$$x = 8$$

example 7

solution

Solve $5x - 2 = 9 + 4x$.

$$5x - 2 = 9 + 4x$$

$$5x - 2 + (-4x) = 9 + 4x + (-4x)$$ Add the opposite of $4x$, or $(-4x)$, to both sides.

$$5x - 2 + (-4x) = 9 + 0$$

$$5x + (-4x) - 2 = 9$$ By the commutative property.

$$x - 2 = 9$$ Combine like terms.

$$x - 2 + 2 = 9 + 2$$ Add 2 to both sides.

$$x + 0 = 11$$

$$x = 11$$

Whenever quantities appear in parentheses on either side of the equation, we must first multiply them using the distributive property, and then combine like terms. This procedure is illustrated in the following examples.

example 8

solution

Solve $7(x - 3) = 6x$.

$$7(x - 3) = 6x$$

$$7x - 21 = 6x$$ By the distributive property.

$$(-6x) + 7x - 21 = 6x + (-6x)$$ Add the opposite of $6x$, or $-6x$, to both sides.

$$x - 21 = 0$$ Combine like terms.

$$x - 21 + 21 = 0 + 21$$ Add the opposite of -21, or 21, to both sides.

$$x + 0 = 21$$

$$x = 21$$

example 9

solution

Solve $-3(t + 2) + 2(2t - 1) = -5$.

$$-3(t + 2) + 2(2t - 1) = -5$$

$$-3t - 6 + 4t - 2 = -5$$ By the distributive property.

$$-3t + 4t - 6 - 2 = -5$$ By the commutative property.

$$t - 8 = -5$$ Combine like terms.

$$t - 8 + 8 = -5 + 8$$ Add 8 to both sides.

$$t + 0 = 3$$

$$t = 3$$

QUICK QUIZ

Solve each of the following linear equations.

1. $x - 9 = 6$

2. $11 - 7y + 8y = 5$

3. $-4t + 3 = 7 - 5t$

4. $5z = 4(z + 6)$

ANSWERS

1. $x = 15$

2. $y = -6$

3. $t = 4$

4. $z = 24$

10.3 Exercises

Indicate which of the following statements are true and which are false. For those that are false, change the italic word to make the statement true.

1. Equivalent equations have the same *variables.*
2. Linear equations are sometimes called *second* degree equations.
3. To move a term to the other side of a linear equation, we add its *opposite* to both sides.
4. The *solution* to a linear equation in one variable is the value that can be substituted for the variable so that the resulting statement is true.

Determine whether or not each of the following numbers is a solution to the linear equation that follows it.

5. 2: $3x + 8 = 14$

6. $\frac{1}{2}$: $2y + 5 = 4$

7. -5: $2z + 1 = z - 3$

8. 3: $4x + 3 = 18 - x$

9. -11: $7(t - 3) = 9t$

10. $-\frac{2}{3}$: $-2(3x + 5) + 7 = 1$

Determine whether or not each of the following pairs of equations is equivalent.

11. $x = -4$ and $3x + 7 = 5$

12. $x = \frac{13}{5}$ and $5x - 9 = 3$

13. $z = \frac{8}{3}$ and $-2z + 7 = 4z - 9$

14. $y = -2$ and $9y = 2y - 14$

15. $a = 1$ and $3(2a - 5) + 2 = -8$

16. $b = -3$ and $2(b - 1) = 9(2b + 5)$

Use the addition principle of equality to solve each of the following linear equations.

17. $x - 2 = 6$

18. $x - 9 = 1$

19. $y + 5 = 2$

20. $y + 3 = 11$

21. $w - 6 = -8$

22. $t + 4 = -7$

23. $5 = x + 7$

24. $8 = x - 3$

25. $-3 = y - 11$

26. $-6 = y + 5$

27. $a + 19 = 64$

28. $b + 61 = 24$

29. $z - 52 = 29$

30. $w - 33 = 47$

31. $x + 3.6 = 8.4$

32. $y - 7.2 = 2.3$

33. $5x + 8 - 4x = 3$

34. $9x + 3 - 8x = 5$

35. $-8y - 6 + 9y = -2$

36. $-3y - 7 + 4y = -6$

37. $7z + 2 - 6z - 1 = 4$

38. $5z - 8 - 4z + 3 = 7$

39. $8 - 2t + 5 + 3t = 9$

40. $-3 + 6t + 8 - 5t = 2$

41. $5 = 4t - 6 - 3t + 7$
42. $7 = 9t + 5 - 8t + 1$
43. $3p - 9p + 7p + 5 = -8$
44. $5p - 8p - 3 + 4p = -5$
45. $-7 = 4w + 9 + 3w - 4 - 6w$
46. $-4 = 8w + 5 - 5w - 7 - 2w$
47. $8 - 5x + 6 - 2x + 8x = 3$
48. $7 - 3x - 4 - 5x + 9x = -6$
49. $9x + 2 = 8x$
50. $5x - 7 = 4x$
51. $5x = 4x - 9$
52. $7x = 6x - 5$
53. $7c + 3 = 6c - 5$
54. $3c - 8 = 2c + 5$
55. $8 - 4d = -5d - 3$
56. $11 - 2d = -3d + 4$
57. $-3x = -4x + 7$
58. $-4x = -5x + 9$
59. $9 - 5w = 11 - 6w$
60. $7 - 2w = 1 - 3w$
61. $8a - 3 - a + 11 = 9 - 3a + 5 + 9a$
62. $5a - 9 + a - 7 = 3 + 9a - 6 - 4a$
63. $9 - 5b + 3 + 8b = 7b + 3 - 5b$
64. $3 - 7b + 6 + 5b = 2b + 8 - 5b$

Solve each of the following linear equations by first using the distributive property to remove parentheses.

65. $5(x - 7) - 4x = 5$
66. $3(x - 8) - 2x = 1$
67. $3(x + 5) - 2x = -9$
68. $9(x + 3) - 8x = -5$
69. $-6(x - 8) + 7x = 4$
70. $-2(k + 6) + 3k = -1$
71. $3(4x - 7) = 11x - 8$
72. $6(3x + 8) = 17x + 9$
73. $19x = 6(3x + 5)$
74. $25x = 8(3x - 1)$
75. $5(2t + 3) - 9t = 7$
76. $7u - 2(3u + 4) = 4$
77. $4(5x - 7) + 6 = 6(2x - 1) + 7x$
78. $3(4x + 5) - 7 = 5(3x + 1) - 4x$
79. $2(8x - 3) - 5x = 5(2x + 1)$
80. $-3(2x + 7) + 11x = 2(2x - 5)$

10.4 The Multiplication Principle of Equality

Notice that in each of the examples presented in Section 10.3, once we simplified the example by combining like terms, the coefficient of the variable was 1. We were then able to solve the equation using the addition principle of equality. To solve linear equations in which the coefficient of the variable is not 1, we must use the **multiplication principle of equality** to isolate the variable.

> According to the **multiplication principle of equality**, if both sides of an equation are multiplied by the same nonzero quantity, the resulting equation is equivalent to the original one.

Recall that we used this principle in Sections 4.4 and 5.3 to solve equations of the form $ax = c$ where a and c are constants. To isolate the variable x, we multiplied both sides by the reciprocal of the coefficient of x, which is $\frac{1}{a}$, as shown.

$$\frac{1}{a} \cdot ax = c \cdot \frac{1}{a}$$ By multiplication principle of equality.

$$1 \cdot x = c \cdot \frac{1}{a}$$ Since $\frac{1}{a} \cdot a = 1$.

$$x = \frac{c}{a}$$

We will use the following examples to review this technique.

example 1

Solve $6x = 108$.

solution

$6x = 108$

$\left(\frac{1}{6}\right)(6x) = (108)\left(\frac{1}{6}\right)$ Multiply both sides by the reciprocal of 6, which is $\frac{1}{6}$.

$1 \cdot x = \frac{108}{6}$

$x = 18$ Reduce to lowest terms. ●

example 2

Solve $-\frac{x}{4} = \frac{3}{8}$.

solution

$-\frac{1}{4}x = \frac{3}{8}$ Recall that $-\frac{x}{4} = -\frac{1}{4}x$.

$(-4)\left(-\frac{1}{4}x\right) = \left(\frac{3}{8}\right)(-4)$ Multiply both sides by the reciprocal of $-\frac{1}{4}$, which is -4.

$1 \cdot x = -\frac{(3)(4)}{8}$

$x = -\frac{3}{2}$ Reduce to lowest terms. ●

example 3

Solve $\frac{2}{3}z = \frac{5}{9}$.

solution

$\frac{2}{3}z = \frac{5}{9}$

$\left(\frac{3}{2}\right)\left(\frac{2}{3}z\right) = \left(\frac{5}{9}\right)\left(\frac{3}{2}\right)$ Multiply both sides by the reciprocal of $\frac{2}{3}$, which is $\frac{3}{2}$.

$1 \cdot z = \frac{15}{18}$ Since $\left(\frac{3}{2}\right)\left(\frac{2}{3}\right) = 1$.

$z = \frac{5}{6}$ Reduce to lowest terms. ●

example 4

Solve $-5\frac{1}{4}x = 63$.

solution

$-5\frac{1}{4}x = 63$

$-\frac{21}{4}x = 63$ Rewrite $-5\frac{1}{4}$ as an improper fraction.

$$\left(-\frac{4}{21}\right)\left(-\frac{21}{4}x\right) = (63)\left(-\frac{4}{21}\right)$$ Multiply both sides by the reciprocal of $-\frac{21}{4}$, which is $-\frac{4}{21}$.

$$x = \frac{-(63)(4)}{21}$$

$$x = -12$$ Reduce to lowest terms. ●

example 5

Solve $3.7x = 18.5$.

solution

$$3.7x = 18.5$$

$$\left(\frac{1}{3.7}\right)(3.7x) = (18.5)\left(\frac{1}{3.7}\right)$$ Multiply both sides by the reciprocal of 3.7, which is $\frac{1}{3.7}$.

$$1 \cdot x = \frac{18.5}{3.7}$$

$$x = 5$$ Reduce to lowest terms. ●

Sometimes it is necessary to combine like terms before applying the multiplication principle, as shown in the following examples.

example 6

Solve $2x - 9x = 63$.

solution

$$2x - 9x = 63$$

$$(2 - 9)x = 63$$ By the distributive property.

$$-7x = 63$$ Combine like terms.

$$\left(-\frac{1}{7}\right)(-7x) = (63)\left(-\frac{1}{7}\right)$$ Multiply both sides by the reciprocal of -7, which is $-\frac{1}{7}$.

$$1 \cdot x = \frac{-63}{7}$$

$$x = -9$$ Reduce to lowest terms. ●

example 7

Solve $2 = 8x - 4x - 7x$.

solution

$$2 = 8x - 4x - 7x$$

$$2 = (8 - 4 - 7)x$$ By the distributive property.

$$2 = -3x$$ Combine like terms.

$$-3x = 2$$ Rewrite equation so the variable term is on the left.

$$\left(-\frac{1}{3}\right)(-3x) = (2)\left(-\frac{1}{3}\right)$$ Multiply both sides by the reciprocal of -3, which is $-\frac{1}{3}$.

$$1 \cdot x = -\frac{2}{3}$$ Since $\left(-\frac{1}{3}\right)(-3) = 1$.

$$x = -\frac{2}{3}$$ ●

Notice that in Example 7, the statements $2 = -3x$ and $-3x = 2$ represent the same equation. Therefore, to get the variable term on the left, we performed no algebraic operation, but merely switched the terms to the other side of the equal sign.

QUICK QUIZ

Solve each of the following linear equations.

1. $-9x = 72$
2. $\frac{z}{3} = \frac{2}{9}$
3. $-\frac{5}{8}w = 35$
4. $-3x + 7x - 2x = 12$

ANSWERS

1. $x = -8$
2. $z = \frac{2}{3}$
3. $w = -56$
4. $x = 6$

10.4 Exercises

Use the multiplication principle of equality to solve each of the following linear equations.

1. $5x = 35$
2. $8x = 24$
3. $-7x = 63$
4. $-5x = 25$
5. $72 = 9x$
6. $56 = 7x$
7. $45 = -5x$
8. $42 = -6x$
9. $-24z = -120$
10. $-18w = -72$
11. $7p = 19$
12. $11q = 4$
13. $-96 = 16a$
14. $-132 = 33b$
15. $8x = 0$
16. $-3x = 0$
17. $\frac{1}{3}w = 5$
18. $\frac{1}{7}y = 2$
19. $\frac{3}{5}x = 9$
20. $\frac{4}{9}x = 16$
21. $-\frac{5}{8}x = 45$
22. $-\frac{3}{7}x = 15$
23. $\frac{x}{7} = 4$
24. $-\frac{x}{5} = 4$
25. $\frac{t}{6} = -\frac{2}{3}$
26. $\frac{w}{8} = \frac{3}{4}$
27. $\frac{2}{7}y = \frac{5}{9}$
28. $\frac{3}{4}y = \frac{2}{5}$
29. $\frac{7}{9}z = \frac{3}{4}$
30. $\frac{3}{8}w = \frac{5}{6}$
31. $6.7m = 53.6$
32. $2.8m = 16.8$
33. $-3.8y = 19$
34. $6.2z = -43.4$

Solve each of the following linear equations by combining like terms before applying the multiplication principle of equality.

35. $5x + 2x = 42$
36. $8x - 3x = 45$
37. $5x - 8x = 27$
38. $9w - 2w + 5w = 48$
39. $4.5t - 3.8t = 6.3$
40. $2.9t + 5.6t = 34$
41. $3w - 6w + 7w = 16$
42. $5w - 8w + 6w = 33$
43. $-8z + 3z - 6z = 55$
44. $-9z - 2z + 5z = -36$
45. $-80 = -5y - 8y + 3y$
46. $-64 = 3y - 9y - 2y$
47. $17r + 9r - 35r = 27$
48. $52s - 6s - 76s = -6$

10.5 Steps for Solving Linear Equations

We are now prepared to tackle some linear equations that require us to use both the addition principle of equality and the multiplication principle of equality. To solve these equations, we can use the following procedure.

Steps for Solving Linear Equations

1. Simplify both sides of the equation as much as possible by using the distributive property to multiply all quantities in parentheses, and then combining like terms.
2. Use the addition principle of equality to move all variable terms to one side of the equation and all constant terms to the other side. This is done by adding to both sides of the equation the opposite of each term you wish to move, and then combining like terms.
3. Use the multiplication principle of equality to isolate the variable on one side of the equation and the solution on the other side. This is done by multiplying both sides of the equation by the reciprocal of the coefficient of the variable.
4. Check that your solution is correct by substituting it back into the original equation to see if the resulting statement is true.

We will illustrate this procedure in the following examples.

example 1 Solve $9y - 7 = 11$.

solution

$9y - 7 = 11$

$9y - 7 + 7 = 11 + 7$ — Add 7 to both sides to move constant term to the right.

$9y = 18$

$\left(\frac{1}{9}\right)(9y) = (18)\left(\frac{1}{9}\right)$ — Multiply both sides by $\frac{1}{9}$.

$y = 2$

CHECK

$9y - 7 = 11$

$9(2) - 7 \ ? \ 11$ — Substitute $y = 2$.

$18 - 7 \ ? \ 11$

$11 = 11 \ \checkmark$ — This is a true statement.

Therefore, $y = 2$ is the correct solution. ●

Before we begin moving terms using the addition principle when solving more complicated linear equations, we must decide on which side to put the variable terms and on which side to put the constant terms. Otherwise, we might end up moving the same terms back and forth without simplifying the equation.

example 2 Solve $-7z = 3z - 40$.

solution Let's collect the variable terms on the left and the constant terms on the right.

$$-7z = 3z - 40$$

$$-3z + (-7z) = -3z + 3z - 40 \qquad \text{Add } -3z \text{ to both sides to move the variable term to the left.}$$

$$-10z = -40$$

$$\left(-\frac{1}{10}\right)(-10z) = (-40)\left(-\frac{1}{10}\right) \qquad \text{Multiply both sides by } -\frac{1}{10}.$$

$$z = 4$$

CHECK

$$-7z = 3z - 40$$

$$-7(4) \ ? \ 3(4) - 40 \qquad \text{Substitute } z = 4.$$

$$-28 \ ? \ 12 - 40$$

$$-28 = -28 \checkmark \qquad \text{This is a true statement.}$$

Therefore, $z = 4$ is the correct solution. ●

example 3 Solve $2x - 3 = -5x + 9$.

solution Let's collect variable terms on the left and constant terms on the right.

$$2x - 3 = -5x + 9$$

$$5x + 2x - 3 = 5x - 5x + 9 \qquad \text{Add } 5x \text{ to both sides to move variable term to the left.}$$

$$7x - 3 = 9$$

$$7x - 3 + 3 = 9 + 3 \qquad \text{Add 3 to both sides to move constant term to the right.}$$

$$7x = 12$$

$$\left(\frac{1}{7}\right)(7x) = (12)\left(\frac{1}{7}\right) \qquad \text{Multiply both sides by } \frac{1}{7}.$$

$$x = \frac{12}{7}$$

CHECK

$$2x - 3 = -5x + 9$$

$$2\left(\frac{12}{7}\right) - 3 \ ? \ -5\left(\frac{12}{7}\right) + 9 \qquad \text{Substitute } x = \frac{12}{7}.$$

$$\frac{24}{7} - \frac{21}{7} \ ? \ -\frac{60}{7} + \frac{63}{7}$$

$$\frac{3}{7} = \frac{3}{7} \checkmark \qquad \text{This is a true statement.}$$

Therefore, $x = \frac{12}{7}$ is the correct solution. ●

Let us now look at some examples that first require us to multiply expressions in parentheses using the distributive property.

example 4

Solve $4(x - 2) = 5x + 3$.

solution

$4(x - 2) = 5x + 3$	Use the distributive property to multiply the quantity in parentheses.
$4x - 8 = 5x + 3$	
$-5x + 4x - 8 = -5x + 5x + 3$	Add $-5x$ to both sides.
$-x - 8 = 3$	Combine like terms.
$-x - 8 + 8 = 3 + 8$	Add 8 to both sides.
$-x = 11$	
$(-1)(-x) = (11)(-1)$	Multiply both sides by -1.
$x = -11$	

CHECK

$4(x - 2) = 5x + 3$	
$4(-11 - 2) \ ? \ 5(11) + 3$	Substitute $x = -11$.
$4(-13) \ ? \ -55 + 3$	
$-52 = -52 \checkmark$	This is a true statement.

Therefore, $x = -11$ is the correct solution.

example 5

Solve $5(w - 3) + 7 = 7w - 2$.

solution

$5(w - 3) + 7 = 7w - 2$	Use the distributive property to multiply quantity in parentheses.
$5w - 15 + 7 = 7w - 2$	
$-7w + 5w - 15 + 7 = 7w + (-7w) - 2$	Add $-7w$ to both sides.
$-2w - 8 = -2$	Combine like terms.
$-2w - 8 + 8 = -2 + 8$	Add 8 to both sides.
$-2w = 6$	
$\left(-\frac{1}{2}\right)(-2w) = (6)\left(-\frac{1}{2}\right)$	Multiply both sides by $-\frac{1}{2}$.
$w = -3$	

CHECK

$5(w - 3) + 7 = 7w - 2$	
$5(-3 - 3) + 7 \ ? \ 7(-3) - 2$	Substitute $w = -3$.
$5(-6) + 7 \ ? \ -21 - 2$	
$-30 + 7 \ ? \ -23$	
$-23 = -23 \checkmark$	This is a true statement.

Therefore, $w = -3$ is the correct answer.

example 6

Solve $3(z - 1) = 4(8 - z)$.

solution

$3(z - 1) = 4(8 - z)$	Use the distributive property to multiply the quantity in parentheses.
$3z - 3 = 32 - 4z$	
$4z + 3z - 3 = 32 - 4z + 4z$	Add $4z$ to both sides.

$$7z - 3 = 32$$ Combine like terms.

$$7z - 3 + 3 = 32 + 3$$ Add 3 to both sides.

$$7z = 35$$

$$\left(\frac{1}{7}\right)(7z) = (35)\left(\frac{1}{7}\right)$$ Multiply both sides by $\frac{1}{7}$.

$$z = 5$$

CHECK

$$3(z - 1) = 4(8 - z)$$

$$3(5 - 1) \ ? \ 4(8 - 5)$$ Substitute $z = 5$.

$$3(4) \ ? \ 4(3)$$

$$12 = 12 \ \checkmark$$ This is a true statement.

Therefore, $z = 5$ is the correct solution.

example 7

Solve $5(x + 5) = 9 - 4(x - 3)$.

solution

$$5(x + 5) = 9 - 4(x - 3)$$ Use the distributive property to multiply quantity in parentheses.

$$5x + 25 = 9 - 4x + 12$$

$$5x + 25 = 21 - 4x$$ Combine constants on right.

$$4x + 5x + 25 = 21 - 4x + 4x$$ Add $4x$ to both sides.

$$9x + 25 = 21$$ Combine like terms.

$$9x + 25 + (-25) = 21 + (-25)$$ Add -25 to both sides.

$$9x = -4$$ Combine constants.

$$\left(\frac{1}{9}\right)(9x) = (-4)\left(\frac{1}{9}\right)$$ Multiply both sides by $\frac{1}{9}$.

$$x = -\frac{4}{9}$$

CHECK

$$5(x + 5) = 9 - 4(x - 3)$$

$$5\left(-\frac{4}{9} + 5\right) \ ? \ 9 - 4\left(-\frac{4}{9} - 3\right)$$ Substitute $x = -\frac{4}{9}$.

$$5\left(-\frac{4}{9} + \frac{45}{9}\right) \ ? \ 9 - 4\left(-\frac{4}{9} - \frac{27}{9}\right)$$

$$5\left(\frac{41}{9}\right) \ ? \ 9 - 4\left(-\frac{31}{9}\right)$$

$$\frac{205}{9} \ ? \ \frac{81}{9} + \frac{124}{9}$$

$$\frac{205}{9} = \frac{205}{9} \ \checkmark$$ This is a true statement.

Therefore, $x = -\frac{4}{9}$ is the correct solution.

QUICK QUIZ

Solve each of the following linear equations.

1. $3y + 2 = 12$
2. $-7x = 2x + 27$
3. $-2(z + 8) = 10$
4. $4(2w - 5) + 7 = 3$

ANSWERS

1. $y = \frac{10}{3}$
2. $x = -3$
3. $z = -13$
4. $w = 2$

10.5 Exercises

Indicate which of the following statements are true and which are false. For those that are false, change the italic word to make the statement true.

1. When solving linear equations, we use the addition principle of equality *before* the multiplication principle of equality.
2. To isolate the variable when solving a linear equation, we *divide* both sides by the reciprocal of its coefficient.
3. To remove parentheses in a linear equation, we use the *commutative* property.
4. When solving a linear equation, we combine like terms *after* using the multiplication principle of equality.

Use both the addition principle of equality and the multiplication principle of equality to solve each of the following linear equations.

5. $7y - 8 = 13$
6. $8k + 2 = 18$
7. $3x + 5 = -22$
8. $4a - 9 = 11$
9. $4 - 5z = 54$
10. $-2w + 9 = 15$
11. $5 + 3t = 6$
12. $8 - 7p = 3$
13. $2 = 5b - 7$
14. $12 = 4 - 3r$
15. $43n + 11 = 97$
16. $27q - 32 = 76$
17. $1.7x + 4.1 = 9.2$
18. $-2.1x + 3.7 = -8.9$
19. $8x + 3 = 11x$
20. $2x - 7 = -12x$
21. $-5x = 2x - 9$
22. $3x = 8x + 5$
23. $3a + 7 = 5a + 4$
24. $-6b - 5 = 8b - 9$
25. $8 + 7u = 12u + 23$
26. $29 - 3m = 4m - 6$
27. $-2t + 9 = 6 + 5t$
28. $11 - 3t = -7t - 4$
29. $7s + 8 - 3s = 5s$
30. $5 + 9c = 2 - 6c + 8$
31. $21 - 22p = 71 - 47p$
32. $28 - 54q = 35q - 61$

Solve each of the following linear equations by first using the distributive property to remove parentheses.

33. $5(x + 7) = 15$
34. $7(x - 3) = 42$
35. $-3(x - 8) = 36$
36. $-2(x + 4) = 12$
37. $4(3y - 8) = 16$
38. $5(2y - 9) = 45$
39. $-8 = 2(5y + 1)$
40. $-18 = 3(5y + 4)$
41. $3(4t - 8) = 15$
42. $8(4t + 7) = 83$
43. $4(3y - 5) = 8y$
44. $2(7y + 3) = 8y$
45. $-6(6 - y) = 4y$
46. $-4(5 - 2y) = -2y$
47. $9(3p - 4) = 15(6 - p)$
48. $3(6p - 1) = 9(2 + p)$
49. $-4(2 - 5z) = 2(3z + 10)$
50. $-6(2z + 1) = 2(9 - 2z)$
51. $6(2z - 3) + 3z = 22$
52. $4(5z + 2) - 7z = -18$
53. $5z - 2(3z + 1) = 18$
54. $3z - 5(2z - 3) = -6$
55. $5(2t - 7) - 1 = 3(t + 2)$
56. $7(t + 4) - 10 = 4(3t + 2)$

57. $6 + 8(2 - s) = 2(4s - 5)$

58. $5(3 - s) - 1 = 3(2s + 1)$

59. $7(2x - 1) = 9(2x - 3) - 6x$

60. $8x - 3(x - 7) = 4(3x + 2)$

61. $3(x + 2) - 2(x - 1) = 5(x + 2) + 6$

62. $-4(x - 2) + 3(x - 1) = 2(x - 7) - 5$

10.6 Formulas

Many word problems that involve finding an unknown quantity require the use of formulas. A formula is an equation that expresses a shorthand notation for performing a calculation. We use formulas to find unknown quantities in applications involving fields such as geometry, physics, and business.

example 1

The formula for finding the area of a triangle is $A = \frac{1}{2}bh$ where b is the base of the triangle and h is its height. Find the area of a triangle having a base of 8 in. and a height of 5 in.

solution

$A = \frac{1}{2}bh$ — Write the required formula.

$A = \frac{1}{2}(8)(5)$ — Substitute the values $b = 8$ and $h = 5$.

$A = 20 \text{ in.}^2$ — Simplify and label the answer with the appropriate units.

example 2

The formula for finding the simple interest earned on an amount of money invested is $i = prt$ where p is the principle or amount invested, r is the rate of interest and t is the time or number of years the money is invested. Find the number of years \$5000 would have to be invested at a rate of 6% to earn \$1200 in interest.

solution

$i = prt$ — Write the required formula.

$1200 = 5000(.06)t$ — Substitute $p = 5000$, $r = .06$, $i = 1200$

$1200 = 300.00t$ — Simplify.

$300t = 1200$

$\frac{1}{300}(300t) = \frac{1}{300}(1200)$ — Solve for t.

$t = 4$ years — Label answer with the appropriate units.

Sometimes, it is useful to rewrite a formula by expressing one variable in terms of the other variables. For example, it would have been simpler to solve the problem in Example 2 if the interest formula were expressed in terms of the variable t. To solve a formula for a given variable, we isolate that variable on one side of the equation using the same steps for solving equations discussed in Section 10.5.

example 3

Rewrite the interest formula $i = prt$ in terms of the variable t.

solution

$i = prt$	Write the required formula.
$prt = i$	Rewrite the equation so that t appears on the left.
$\frac{1}{pr}(prt) = \frac{1}{pr}(i)$	Multiply both sides of the equation by $\frac{1}{pr}$ to isolate t on the left.
$\frac{\cancel{pr}t}{\cancel{pr}} = \frac{i}{pr}$	
$t = \frac{i}{pr}$	

example 4

The formula for finding the perimeter of a rectangle is

$P = 2l + 2w$. Solve this formula for l.

solution

$2l + 2w = P$	Rewrite the equation so l appears on the left.
$2l + 2w + (-2w) = P + (-2w)$	Add $-2w$ to both sides to isolate the term containing l on the left.
$2l = P - 2w$	
$\frac{1}{2}(2l) = \frac{1}{2}(P - 2w)$	Multiply both sides by $\frac{1}{2}$.
$l = \frac{P - 2w}{2}$	

example 5

The formula for finding the volume of a cone is $V = \frac{1}{3}\pi r^2 h$ where π is a constant approximately equal to 3.14, r is the radius of the cone, and h is the height of the cone. Solve this formula for h.

solution

$V = \frac{1}{3}\pi r^2 h$	
$\frac{1}{3}\pi r^2 h = V$	Rewrite the formula so h appears on the left.
$3\left(\frac{1}{3}\pi r^2 h\right) = 3(V)$	Multiply both sides of the equation by 3.
$\pi r^2 h = 3V$	
$\frac{1}{\pi r^2}(\pi r^2 h) = \frac{1}{\pi r^2}(3V)$	Multiply both sides of the equation by $\frac{1}{\pi r^2}$ to isolate h on the left.
$h = \frac{3V}{\pi r^2}$	

QUICK QUIZ

Solve each of the following formulas for the variable indicated.

1. $A = lw$ Solve for l
2. $PV = nrt$ Solve for t
3. $s = \frac{1}{2}gt^2$ Solve for g
4. $V = \frac{4}{3}\pi r^3$ Solve for r^3

ANSWERS

1. $l = \frac{A}{w}$
2. $t = \frac{PV}{nr}$
3. $g = \frac{2s}{t^2}$
4. $r^3 = \frac{3V}{4\pi}$

10.6 Exercises

Solve each of the following formulas for the unknown variable given the values for the other variables.

1. $A = lw$ Find A if $l = 9$, $w = 8$
2. $A = bh$ Find A if $b = 5$, $h = 7$
3. $E = IR$ Find I if $E = 48$ and $R = 8$
4. $F = ma$ Find m if $F = 42$, $a = 6$
5. $d = rt$ Find t if $d = 72$, $r = 8$
6. $P = IV$ Find V if $P = 35$, $I = 5$
7. $E = mgh$ Find m if $E = 64$, $g = 2$, $h = 4$
8. $i = prt$ Find p if $i = 36$, $r = .06$, $t = 3$
9. $V = lwh$ Find V if $l = 5$, $w = 4$, $h = 7$
10. $A = \frac{1}{2}bh$ Find A if $b = 9$, $h = 6$
11. $P = 2l + 2w$ Find l if $P = 10$, $w = 2$
12. $P = 2l + 2w$ Find w if $P = 48$, $l = 16$
13. $A = \pi r^2$ Find A if $r = 8$
14. $C = 2\pi r$ Find C if $r = 7$
15. $V = \frac{1}{3}\pi r^2 h$ Find V if $r = 3$, $h = 5$
16. $V = \pi r^2 h$ Find V if $r = 2$, $h = 7$
17. $C = \frac{5}{9}(F - 32)$ Find C if $F = 86$
18. $F = \frac{9}{5}C + 32$ Find F is $C = 45$
19. $E = \frac{1}{2}mv^2$ Find m if $E = 32$, $v = 2$
20. $s = \frac{1}{2}at^2$ Find a if $s = 45$, $t = 3$

Solve each of the following formulas for the variable indicated.

21. $F = ma$ Solve for m
22. $A = lw$ Solve for l
23. $d = rt$ Solve for r
24. $w = mg$ Solve for m
25. $i = prt$ Solve for t
26. $PV = nrt$ Solve for r
27. $E = mc^2$ Solve for m
28. $A = \frac{1}{2}bh$ Solve for b
29. $V = lwh$ Solve for w
30. $P = 2l + 2w$ Solve for w
31. $E = mgh$ Solve for h
32. $s = \frac{1}{2}gt^2$ Solve for g
33. $A = P(1 + r)$ Solve for r
34. $C = 2\pi r$ Solve for r
35. $V = \frac{4}{3}\pi r^3$ Solve for r^3

36. $c^2 = a^2 + b^2$ Solve for b^2

37. $A = \pi r^2$ Solve for r^2

38. $F = \dfrac{GMm}{r^2}$ Solve for r^2

39. $S = \dfrac{a}{1-r}$ Solve for r

40. $C = \dfrac{5}{9}(F - 32)$ Solve for F

41. $V = \pi r^2 h$ Solve for r^2

42. $V = \dfrac{1}{3}\pi r^2 h$ Solve for h

43. $A = \dfrac{1}{2}(B + b)h$ Solve for h

44. $A = \dfrac{1}{2}(B + b)h$ Solve for b

Solve each of the following word problems using the formulas given.

45. Find the area of a rectangle of length 7 in. and width 4 in. $A = lw$

46. Find the circumference of a circle whose radius is 3 *cm*. Use $\pi = 3.14$. $A = 2\pi r$

47. Find the area of a trapezoid having bases 2 ft and 5 ft and height 4 ft. $A = \dfrac{1}{2}(B + b)h$

48. Find the volume of a cylinder of radius 2 in. and height 6 in. Use $\pi = 3.14$. $V = \pi r^2 h$

49. Find the base of a triangle whose area is 12 sq ft and whose height is 3 ft. $A = \dfrac{1}{2}bh$

50. Find the width of a rectangle whose perimeter is 18 *m* and whose length is 6 *m*. $P = 2l + 2w$

10.7 Solving Word Problems Using Linear Equations

In this section, we will discuss an important application of linear equations, which is solving word problems. Since it will often be impossible for us to absorb all the information given in a word problem in a single reading, it is important that we approach each problem in a systematic fashion, keeping in mind the particular details we wish to extract from the problem in each reading. The following procedure will help us to organize our thinking when solving the word problems covered in this section.

Steps for Solving Word Problems

1. Determine what you are asked to find. It may help to write this down.
2. Define all variables that you will need to solve the problem. Represent one unknown quantity with a letter and any other unknown quantities in terms of that variable.
3. Write an equation that relates the variables. Note the key words in the problem that indicate relationships between the quantities.
4. Solve the equation for the unknown and label your answer.
5. Answer the question asked in the problem. Keep in mind that this may or may not be the same as the solution to your equation. Check to make sure that your answer satisfies the conditions of the problem.

example 1

A mother is five years older than twice the age of her daughter. If the mother is 37 years old, how old is her daughter?

solution

Find the daughter's age.

Let x = daughter's age

$2x + 5$ = mother's age

mother's age = 37

1. Determine what are you asked to find.
2. Define all variables. The mother's age is 5 more than twice her daughter's.
3. Since the mother's age is 37, we can write this equation to find x.
4. Solve this equation for x.

$$2x + 5 = 37$$
$$2x + 5 - 5 = 37 - 5$$
$$2x = 32$$
$$\frac{1}{2}(2x) = \frac{1}{2}(32)$$
$$x = 16 \text{ yr}$$

$x = 16$ yr represents the daughter's age.

The daughter is 16 years old.

5. Answer the question.

CHECK $5 + 2(16) = 37$ ✓

This is the mother's age, so the answer checks.

example 2

Two numbers have a sum of 24. If the second number is three times as large as the first, what are these two numbers?

solution

Find two numbers.

Let x = the first number

$3x$ = the second number

first number + second number = 24

1. Determine what you are asked to find.
2. Define all variables. The second number is 3 times as large as the first.
3. Since the two numbers have a sum of 24, we can write this equation to find them.

Now solve this equation for x.

$$x + 3x = 24$$
$$4x = 24$$
$$\frac{1}{4}(4x) = \frac{1}{4}(24)$$
$$x = 6 \text{ is the first number}$$

$3x = 3(6) = 18$ is the second number

Since $x = 6$ is the first number.

The two numbers are 6 and 18.

5. Answer the question.

CHECK $6 + 18 = 24$ ✓

These answers check.

example 3

The length of a rectangle is four times the width. What are the dimensions of the rectangle if its perimeter is 55 ft?

solution

Find the length and width of the rectangle.

1. Determine what you are asked to find.

Let w = width

$4w$ = length

2. Define all variables. The length is 4 times the width.

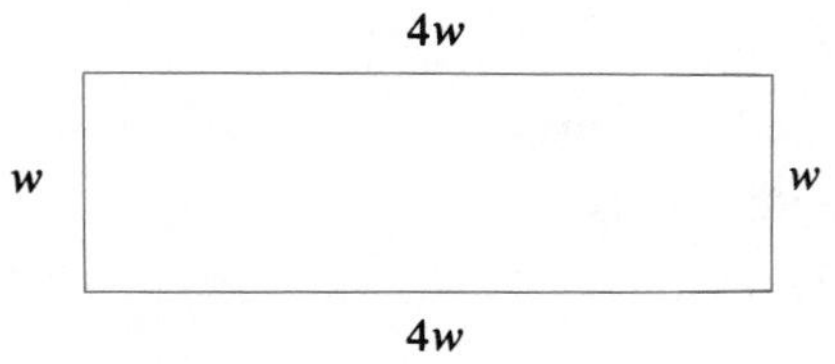

3. Draw a rectangle and label the four sides as shown. Since perimeter is the distance around a figure, we can obtain an equation just by setting the sum of the sides equal to 55.

4. Now solve this equation for w.

$$w + 4w + w + 4w = 55$$

$$10w = 55$$

$$\frac{1}{10}(10w) = \frac{1}{10}(55)$$

$$w = \frac{55}{10} = 5\frac{1}{2} \text{ ft is the width}$$

$$4w = 4\left(5\frac{1}{2}\right) = 4\left(\frac{11}{2}\right)$$

$$= 22 \text{ ft is the length}$$

Since $w = 5\frac{1}{2}$ ft is the width.

The rectangle has a length of 22 ft and a width of $5\frac{1}{2}$ ft.

5. Answer the question.

CHECK $2(22) + 2\left(5\frac{1}{2}\right) = 44 + 11 = 55$ ✓

This is the perimeter, so the answers check.

example 4

A student bought three textbooks for his history, chemistry, and mathematics courses. The chemistry text cost \$5 more than the history text. The mathematics text cost twice as much as the chemistry text. If the total bill for these three books was \$45, what is the price of each book?

solution

Find the price of each of the three books.

1. Determine what you are asked to find.

Let p = price of history text

$p + 5$ = price of the chemistry text

Let $2(p + 5)$ = price of math text

2. Define all variables. The chemistry text cost \$5 more than the history text. The math text cost twice as mucl as the chemistry text.

history text + chemistry text + math text = \$45

3. Since the total cost of the three books is \$45, we can write this equation to find the three prices.

4. Now solve this equation for p.

$$p + p + 5 + 2(p + 5) = 45$$

$$p + p + 5 + 2p + 10 = 45$$

$$4p + 15 = 45$$

$$4p = 30$$

$$p = 7.5 \text{ dollars for the history text.}$$

$$p + 5 = 7.5 + 5$$

$$= 12.5 \text{ dollars for chemistry text.}$$

Since $p = 7.5$ dollars for history text.

$$2(p + 5) = 2(7.5 + 5) = 25 \text{ dollars for math text.}$$

Therefore, the history text cost \$7.50, the chemistry text cost \$12.50, and the math text cost \$25.00.

5. Answer the question.

CHECK $7.5 + 12.5 + 25 = 45$ √

This is the total cost, so the answers check.

10.7 Exercises

Solve each of the following word problems.

1. A woman's age is seven years less than twice the age of her brother. If the woman is 29 years old, how old is her brother?
2. A doctor is two years older than three times the age of her patient. If the doctor is 35 years old, how old is her patient?
3. Two numbers have a sum of 32. If the second is eight more than seven times the first, what are the two numbers?
4. Two numbers have a sum of 20. If the second is five less than four times the first, what are the two numbers?
5. Find two consecutive integers whose sum is 55. (*Hint:* Let x be the first integer and $x + 1$ be the second.)
6. Find two consecutive even integers whose sum is 78. (*Hint:* Let x be the first integer and $x + 2$ be the second.)
7. In a statistics class, there are five more men than there are women. If there are a total of 31 students in the class, how many are men?
8. A youngster practiced his piano 15 minutes longer on Sunday than on Saturday. If he practiced a total of 41 minutes that weekend, how long did he practice on Sunday?
9. The distance between New York and Dallas is 54 miles more than twice the distance between Memphis and Pittsburgh. If New York and Dallas are 1,374 miles apart, what is the distance between Memphis and Pittsburgh?
10. A father is 7 inches shorter than his son. If the sum of their heights is 155 inches, how tall is the father?
11. The length of a rectangle is 5 feet longer than its width. Find the dimensions of the rectangle if its perimeter is 22 feet.
12. The width of a rectangle is 8 centimeters shorter than its length. If the perimeter of the rectangle is 84 centimeters, what are its dimensions?
13. A house is now selling for \$5,000 more than four times what the owners paid for it in 1950. If the selling price today is \$95,000, what did the owners pay for it in 1950?
14. A used car that is three years old is selling for \$150 more than half its original price. If the car now sells for \$2,750, what did it cost three years ago when it was brand new?
15. A student bought a notebook, a pen, and a candy bar. The pen cost 14¢ more than the candy bar, and the notebook cost twice as much as the pen. If the student was charged \$1.82 for all three items, what was the cost of each one?
16. An intern worked two more hours on Monday than on Tuesday, and twice as many hours on Wednesday as on Monday. If he worked a total of 42 hours on those three days, how many hours did he work each day?
17. An isosceles triangle has a perimeter of 14 inches. If the two equal sides are each three times as long as the third side, what is the length of each side?
18. An isoceles triangle has a perimeter 370 millimeters. The base of the triangle is 40 millimeters longer than the two equal sides. What is the length of each side?
19. A bag of apples weighs twice as much as a bag of oranges. If the apples are 3 pounds heavier than the oranges, what is the weight of the oranges?
20. A child has eight more blue marbles than red ones. If there are three times as many blue marbles as red ones, how many red marbles does he have?

Summary and Review

Key Terms

[10.1] **Algebra** is a generalization of arithmetic.

A **constant** is a number that appears in an algebraic expression.

A **variable** is a letter that represents an unknown quantity.

The **coefficient** of a variable is the number that is multiplied by the variable.

The **terms** of an algebraic expression are the quantities that are added or subtracted.

[10.2] **Like terms**, or **similar terms**, are terms whose variable factors are the same.

[10.3] An **equation** is a mathematical expression that states that two quantities are equal.

A **linear equation** is an equation in one variable that can be written in the form $ax + b = c$, where a, b, and c are constants and $a \neq 0$.

The **solution** to a linear equation in one variable is the value that can be substituted for the variable so that the resulting statement is true.

Equivalent equations are equations that have the same solution.

The **addition principle of equality** states that if the same quantity is added to both sides of an equation, the resulting equation is equivalent to the original one.

[10.4] The **multiplication principle of equality** states that if both sides of an equation are multiplied by the same nonzero quantity, the resulting equation is equivalent to the original one.

[10.6] A **formula** is an equation that expresses a shorthand notation for performing a calculation.

Calculations

[10.1] **To evaluate an algebraic expression**, we substitute numbers for the variables and simplify the result using the order of operations rules.

[10.2] **To combine like terms**, we rewrite the expression using the distributive property and then add or subtract the coefficients of the variable.

[10.5] **Steps for Solving Linear Equations**

1. Simplify both sides of the equation as much as possible by using the distributive property to multiply all quantities in parentheses, and then combining like terms.
2. Use the addition principle of equality to move all variable terms to one side of the equation and all constant terms to the other side. This is done by adding to both sides of the equation the opposite of each term you wish to move, and then combining like terms.
3. Use the multiplication principle of equality to isolate the variable on one side of the equation and the solution on the other side. This is done by multiplying both sides of the equation by the reciprocal of the coefficient of the variable.
4. Check that your solution is correct by substituting it back into the original equation to see if the resulting statement is true.

[10.6] **To solve a formula for a given variable**, isolate that variable on one side of the equation using the steps for solving linear equations.

[10.7] Steps for Solving Word Problems

1. *Determine what you are asked to find.* It may help to write this down.
2. *Define all variables* that you will need to solve the problem. Represent one unknown quantity with a letter and any other unknown quantities in terms of that variable.
3. *Write an equation* that relates the variables. Note the key words in the problem that indicate relationships between the quantities.
4. *Solve the equation* for the unknown and label your answer.
5. *Answer the question asked in the problem.* Keep in mind that this may or may not be the same as the solution to your equation. Check to make sure your solution satisfies the conditions of the problem.

Chapter 10 Review Exercises

[10.1] Evaluate each of the following algebraic expressions.

1. $5x - 2y + z$ when $x = 2, y = -3, z = -5$
2. $-7a + 4b - 9c$ when $a = -1, b = 4, c = 3$
3. $8(x - 6y)$ when $x = 5, y = 2$
4. $3(4t + 5u)$ when $t = -2, u = 4$
5. $\frac{3}{4}p - \frac{5}{8}q$ when $p = 12, q = 16$
6. $-\frac{1}{9}r + \frac{2}{3}s$ when $r = 18, s = -6$

[10.1] Translate each of the following word expressions into algebraic expressions.

7. Seven less than x.
8. The product of three and a.
9. Five times the sum of z and six.
10. Two less than the product of nine and y.
11. The quantity seven plus t, divided by three.
12. Six times the quantity x minus four.

[10.1] Translate each of the following algebraic expressions into words.

13. $-5y$
14. $t + 7$
15. $3(x - 2)$
16. $(4 + z) \div 8$
17. $6w - 2$
18. $\frac{3}{5}(n + 4)$

[10.2] Simplify each of the following algebraic expressions by combining like terms.

19. $4x - 9x$
20. $a - 7a$
21. $5y - 2 - 8y - 6$
22. $-3t + 4 - 6t - 5$
23. $4(2x - 7) + 8$
24. $3(6y + 2) - 13$
25. $2(5a - 7) - 3(4a + 5)$
26. $8(4 - t) - 7(3t - 1)$

[10.5] Solve each of the following linear equations.

27. $4x - 7 = 25$
28. $-5y + 2 = 47$
29. $9u + 1 = 28$
30. $5t - 8 = -43$
31. $7 - 3z = 31$
32. $-6 - 5w = 54$
33. $3p + 2 = 5p - 8$
34. $9q - 4 = 12 - 7q$
35. $8 - 5a = 7a + 2$
36. $6b + 2 = 9b - 4$
37. $3(2m + 5) = 5m$
38. $4(3 - n) = 11n$
39. $8(k - 7) = 2(5k + 2)$
40. $6(2s - 5) = 3(s + 4)$
41. $4(8x + 2) + 3x = 17$
42. $9 - 2(4x - 7) = 2x$

[10.6] Solve each of the following formulas for the unknown variable given the values for the other variables.

43. $A = bh$ Find b if $A = 42, h = 6$
44. $P = 2l + 2w$ Find l if $P = 28, w = 4$

45. $C = \frac{5}{9}(F - 32)$ Find C if $F = 95$

46. $i = prt$ Find p if $i = 45$, $r = .05$, $t = 4$

47. $s = \frac{1}{2}at^2$ Find a if $s = 18$, $t = 3$

48. $V = \frac{1}{3}\pi r^2 h$ Find V if $r = 3$, $h = 7$

Solve each of the following formulas for the variable indicated.

49. $d = rt$ Solve for t

50. $A = \frac{1}{2}bh$ Solve for h

51. $C = \pi d$ Solve for d

52. $c^2 = a^2 + b^2$ Solve for a^2

53. $V = \frac{1}{3}\pi r^2 h$ Solve for h

54. $F = \frac{GMm}{r^2}$ Solve for M

[**10.7**] Solve each of the following word problems.

55. Find two consecutive odd integers whose sum is 68.

56. Three numbers have a sum of 39. If the second number is four less than the first, and the third number is equivalent to three times the second, what are the three numbers?

It took a student $1\frac{1}{2}$ hours longer to complete her math homework on Tuesday night than it took her on Monday night. If both nights she spent a total of $5\frac{1}{2}$ hours doing math homework, how much time did she devote to math homework on Tuesday night?

58. Five-eighths of the people employed by a community hospital smoke cigarettes. If 138 hospital employees do not smoke cigarettes, how many people are employed by the hospital?

59. The width of a rectangle is one-third as long as its length. What are the dimensions of the rectangle if its perimeter is 32 centimeters?

60. An isoceles triangle has a perimeter of 43 inches. What is the length of each side if the base is 5 inches shorter than the sum of the two equal sides?

61. The height of the World Trade Center in New York is 123 feet less than twice the height of the Prudential Tower in Boston. If the World Trade Center is 1,377 feet high, how high is the Prudential Tower?

62. The diameter of the planet Jupiter is 1,503 miles larger than 11 times the diameter of the earth. If the diameter of Jupiter is 88,700 miles, what is the diameter of the earth?

11 EXPONENTS

11.1 Multiplication with Exponents

In Chapter 1 we discussed some properties of exponential expressions in which both the base and the exponents were constants. We will now extend our knowledge by considering expressions in which the base is a variable and the exponent is a constant.

Recall that an *exponent* indicates the number of times the base appears as a factor. For example, the expression x^3 indicates that the base x appears as a factor 3 times. The number 3 is called the exponent or power:

$$\underset{\text{base}}{x}^{\overset{\text{exponent}}{3}} = \underbrace{x \cdot x \cdot x}_{\text{three } x\text{'s}}$$

In this section, we will be working with exponential expressions in which the variable base can represent any real number and the exponent is a whole number. In Section 9.2 we will discuss exponents that are negative integers, and in Section 9.4 we will cover exponents that are rational numbers. Keep in mind the following definitions, which pertain to the exponents zero and one:

definition

For any real number $x \neq 0$, $x^0 = 1$.

definition

For any real number x, $x^1 = x$.

Different mathematical notations indicate whether or not a negative sign that appears in front of an exponential expression is considered part of the base. For example, to evaluate $(-2)^4$, the base -2 is used as a factor 4 times:

$$(-2)^4 = (-2)(-2)(-2)(-2) = 16$$

$(-2)^4$ means the fourth power of negative two. To evaluate -2^4, we raise the base 2 to the fourth power, and then take the opposite of the result:

$$-2^4 = -(2 \cdot 2 \cdot 2 \cdot 2) = -(16) = -16$$

-2^4 means the opposite of the fourth power of two. In general,

$(-x)^n$ means first take the opposite of x, and then raise the result to the nth power.

$-x^n$ means first raise x to the nth power, and then take the opposite of the result.

In order to evaluate an exponential expression, we substitute in the values given for the variables, and then simplify the result.

example 1

Evaluate each of the following if $x = -3$ and $y = 2$.

(a) $(-y)^4$

$(-y)^4 = (-2)^4$ — Substitute in $y = 2$.

$= 16$ — Raise -2 to the fourth power.

(b) $-x^2$

$-x^2 = -(-3)^2$ — Substitute in $x = -3$.

$= -(9)$ — Square -3.

$= -9$ — Find the opposite of 9.

(c) x^3y^2

$x^3y^2 = (-3)^3(2)^2$ — Substitute in $x = -3$, $y = 2$.

$= (-27)(4)$ — Raise to powers.

$= -108$ — Do multiplication.

(d) $-xy^3$

$-xy^3 = -(-3)(2)^3$ — Substitute in $x = -3$, $y = 2$.

$= -(-3)(8)$ — Cube 2.

$= -(-24)$ — Do multiplication.

$= 24$ — Find opposite of -24. ●

To illustrate the procedure for multiplying exponential expressions having the same base, let us consider the following example:

$$y^2 \cdot y^3 = (y \cdot y) \cdot (y \cdot y \cdot y) = y^5$$

Notice that the same result may be obtained by adding the exponents on the two expressions being multiplied:

$$y^2 \cdot y^3 = y^{2+3} = y^5$$

In general,

$$x^a \cdot x^b = x^{a+b}$$

To multiply exponential expressions of the same base, keep the base the same and add the exponents.

example 2

Simplify each of the following.

(a) $z^5 \cdot z^9$

$z^5 \cdot z^9 = z^{5+9} = z^{14}$

(b) $w^3 \cdot w^7 \cdot w^5$

$w^3 \cdot w^7 \cdot w^5 = w^{3+7+5} = w^{15}$

(c) $(a+b)^2 \cdot (a+b)^5$

$(a+b)^2 \cdot (a+b)^5 = (a+b)^{2+5} = (a+b)^7$

(d) $x^{2a} \cdot x^{3a}$

$x^{2a} \cdot x^{3a} = x^{2a+3a} = x^{5a}$

This next example illustrates what happens when we raise an exponential expression to a power:

$$(y^2)^3 = y^2 \cdot y^2 \cdot y^2 = y^6$$

Notice that this same result may be obtained by multiplying the exponents:

$$(y^2)^3 = y^{2\cdot 3} = y^6$$

In general,

$$(x^a)^b = x^{a\cdot b}$$

To raise an exponential expression to a power, keep the base the same and multiply the exponents.

example 3

Simplify each of the following.

(a) $(p^3)^7$

$(p^3)^7 = p^{3\cdot 7} = p^{21}$

(b) $(x^2)^5(y^3)^4$

$(x^2)^5(y^3)^4 = x^{2\cdot 5}y^{3\cdot 4} = x^{10}y^{12}$

(c) $(t^3)^5 \cdot (t^{10})^2$

$(t^3)^5 \cdot (t^{10})^2 = t^{3\cdot 5} \cdot t^{10\cdot 2}$

$= t^{15} \cdot t^{20}$

$= t^{15+20}$

$= t^{35}$

First raise exponential expressions to powers. Multiply the exponents.

Then multiply exponential expressions of the same base. Add the exponents.

(d) $(y^n)^5$

$(y^n)^5 = y^{n\cdot 5} = y^{5n}$

We will now consider how to simplify expressions in which the product of two or more factors is raised to a power. For example,

$$\begin{aligned}(x^2y^3)^3 &= (x^2y^3)(x^2y^3)(x^2y^3)\\ &= x^2 \cdot x^2 \cdot x^2 \cdot y^3 \cdot y^3 \cdot y^3\\ &= x^6y^9\end{aligned}$$

Notice that this same result may be obtained by raising each factor in parentheses to the third power:

$$(x^2y^3)^3 = (x^2)^3(y^3)^3 = x^6y^9$$

In general,

$$(x^ay^b)^c = x^{ac}y^{bc}$$

To raise a product to a power, raise each factor to that power.

example 4

Simplify each of the following.

(a) $(ab^2)^5$

$(ab^2)^5 = a^5(b^2)^5 = a^5b^{10}$

(b) $(2t^3)^4$

$(2t^3)^4 = 2^4(t^3)^4 = 16t^{12}$

(c) $(-x^4y^3)^2$

$$\begin{aligned}(-x^4y^3)^2 &= (-1)^2(x^4)^2(y^3)^2\\ &= 1 \cdot x^8y^6\\ &= x^8y^6\end{aligned}$$

Since the negative sign is in parentheses, square -1.

(d) $-(pq^3)^2$

$$\begin{aligned}-(pq^3)^2 &= -[p^2(q^3)^2]\\ &= -p^2q^6\end{aligned}$$

Since the negative sign is outside parentheses, we square pq^3 and then take its opposite.

QUICK QUIZ

Simplify each of the following.

1. $p^8 \cdot p^{11}$
2. $(z^6)^3$
3. $(7w^3)^2$
4. $\left(-\frac{1}{2}ab^2\right)^4$

ANSWERS

1. p^{19}
2. z^{18}
3. $49w^6$
4. $\frac{1}{16}a^4b^8$

11.1 Exercises

Indicate which of the following statements are true and which are false. For those that are false, change the italic or underlined expression to make the statement true.

1. Any nonzero number raised to the zero power is equal to *zero*.
2. In the expression $(-t)^6$, the base is $\underline{-t}$.
3. The expression $-y^2$ is read as *negative y squared*.
4. To raise a product to a power, we raise *only the last* factor to that power.

Evaluate each of the following expressions if $a = 4$ and $b = -1$.

5. $(-a)^2$ 6. $-b^3$ 7. $-ab^4$ 8. a^2b^5 9. $(-ab)^3$ 10. $-ab^3$

Simplify each of the following.

11. $x^5 \cdot x^4$	12. $y^3 \cdot y^8$	13. $t \cdot t^2$	14. $w^4 \cdot w^0$
15. $n^{13} \cdot n^9$	16. $p^5 \cdot p^{20}$	17. $(y^2)^8$	18. $(t^6)^2$
19. $(u^4)^{10}$	20. $(m^{11})^3$	21. $z^2 \cdot z^4 \cdot z^7$	22. $a^4 \cdot a^9 \cdot a^2$
23. $y^5 \cdot y^0 \cdot y^9$	24. $x^7 \cdot x^6 \cdot x$	25. $(a)^7$	26. $(b^0)^4$
27. $(q^4)^9$	28. $(r^7)^5$	29. $(xy^2)^3$	30. $(x^3y)^4$
31. $(3x^3)^2$	32. $(2y^2)^3$	33. $(5t)^3$	34. $(9n)^2$
35. $(x^2y^7)^3$	36. $(a^4b^5)^2$	37. $(m^5n^2)^6$	38. $(u^4v^3)^7$
39. $(-a^2b)^5$	40. $(-p^3q)^4$	41. $(-t^3)^2$	42. $-(w^5)^4$
43. $(3v)^3$	44. $(9u)^2$	45. $(-2y^5)^4$	46. $(-3z^8)^3$
47. $-(4x^2)^3$	48. $-(6x^5)^2$	49. $\left(\frac{3}{4}t\right)^3$	50. $\left(\frac{2}{5}n\right)^2$
51. $(5x^4y)^2$	52. $(2xy^6)^4$	53. $-(a^3b^4)^3$	54. $(-s^5t^7)^2$
55. $\left(\frac{1}{3}ab^2\right)^2$	56. $\left(\frac{2}{3}x^2y\right)^3$	57. $(x^3y^5z)^4$	58. $(a^7bc^3)^5$
59. $(2a^2bc^4)^5$	60. $(3r^4s^2t)^3$	61. $(x+2)^3(x+2)^5$	62. $(a-b)^7(a-b)^2$
63. $z^3 \cdot z^n$	64. $p^k \cdot p^7$	65. $(a^3)^c$	66. $(b^6)^n$
67. $(w^a)^5$	68. $(v^r)^8$	69. $t^{2k} \cdot t^{7k}$	70. $u^{5a} \cdot u^{3a}$

11.2 Division and Negative Exponents

In order to develop a procedure for dividing exponential expressions, let us consider the following example:

$$\frac{p^5}{p^2} = \frac{p \cdot p \cdot p \cdot p \cdot p}{p \cdot p} = \frac{\cancel{p} \cdot \cancel{p} \cdot p \cdot p \cdot p}{\cancel{p} \cdot \cancel{p}} = p \cdot p \cdot p = p^3$$

Notice that this same result may be obtained by subtracting the exponent in the denominator from the exponent in the numerator:

$$\frac{p^5}{p^2} = p^{5-2} = p^3$$

In general,

For any real number $x \neq 0$, $\dfrac{x^a}{x^b} = x^{a-b}$.

To divide exponential expressions of the same base, keep the base the same and subtract the exponents.

example 1

Simplify each of the following.

(a) $\dfrac{y^{11}}{y^4}$

$$\frac{y^{11}}{y^4} = y^{11-4} = y^7$$

(b) $n^6 \div n^6$

$$n^6 \div n^6 = n^{6-6} = n^0 = 1$$

(c) $\dfrac{w^5 \cdot w^4}{w^8}$

$$\frac{w^5 \cdot w^4}{w^8} = \frac{w^{5+4}}{w^8} = \frac{w^9}{w^8} = w^{9-8} = w^1 = w$$

(d) $t^{7n} \div t^{3n}$

$$t^{7n} \div t^{3n} = t^{7n-3n} = t^{4n}$$

(e) $\dfrac{(x+5)^9}{(x+5)^2}$

$$\frac{(x+5)^9}{(x+5)^2} = (x+5)^{9-2} = (x+5)^7$$

●

Thus far we have not considered how to divide exponential expressions in which the exponent in the denominator is greater than the exponent in the numerator. Such problems will require the use of negative exponents.

definition

For any real number $x \neq 0$, $x^{-n} = \dfrac{1}{x^n}$.

For example,

$$t^{-5} = \frac{1}{t^5}$$

$$7^{-2} = \frac{1}{7^2} = \frac{1}{49}$$

Notice that 7^{-2} indicates the reciprocal of 7^2 and *not* the opposite of 7^2. Negative exponents indicate reciprocals, not negative numbers.

We can illustrate the meaning of this definition by using the procedure for dividing powers of the same base to rewrite the expression $\frac{1}{x^8}$:

$$\frac{1}{x^8} = \frac{x^0}{x^8} = x^{0-8} = x^{-8}$$

Notice that the exponent on x is positive when x appears in the denominator, and negative when x appears in the numerator.

The following example illustrates how to simplify an expression having a negative exponent in the denominator:

$$\frac{1}{y^{-3}} = \frac{1}{\frac{1}{y^3}} = \frac{1 \cdot y^3}{\frac{1}{y^3} \cdot y^3} = \frac{y^3}{1} = y^3$$

Notice that the exponent on y is negative when y appears in the denominator, and positive when y appears in the numerator. We can summarize these results as follows:

> Factors having negative exponents in the numerator can be written with positive exponents in the denominator. Factors having negative exponents in the denominator can be written with positive exponents in the numerator.

When we are asked to simplify an exponential expression that contains negative exponents, this means to rewrite it using positive exponents.

example 2

Simplify each of the following.

(a) 3^{-4}

$$3^{-4} = \frac{1}{3^4} = \frac{1}{81}$$

(b) $8y^{-5}$

$$8y^{-5} = 8 \cdot \frac{1}{y^5} = \frac{8}{y^5}$$

(c) $-x^{-7}y^3$

$$-x^{-7}y^3 = -\frac{1}{x^7} \cdot y^3 = -\frac{y^3}{x^7}$$

(d) $\frac{p}{q^{-6}}$

$$\frac{p}{q^{-6}} = pq^6$$

(e) $\frac{a^3b^{-5}}{c^{-1}d^2}$

$$\frac{a^3b^{-5}}{c^{-1}d^2} = \frac{a^3c^1}{b^5d^2} = \frac{a^3c}{b^5d^2}$$

To evaluate exponential expressions containing negative exponents, first rewrite the expression using positive exponents, and then substitute in the given values for the variables.

example 3

Evaluate each of the following if $r = 4$ and $s = -2$.

(a) s^{-3}

$$s^{-3} = \frac{1}{s^3} = \frac{1}{(-2)^3} = \frac{1}{-8} = -\frac{1}{8}$$

(b) $r^{-1}s^2$

$$r^{-1}s^2 = \frac{s^2}{r} = \frac{(-2)^2}{4} = \frac{4}{4} = 1$$

(c) $-rs^{-5}$

$$-rs^{-5} = \frac{-r}{s^5} = \frac{-4}{(-2)^5} = \frac{-4}{-32} = \frac{1}{8}$$

●

Let us now look at some examples that require us to apply the procedures we have learned thus far to problems that have negative exponents.

example 4

Simplify each of the following.

(a) $w^{-5} \div w^2$

$$w^{-5} \div w^2 = w^{-5-2} = w^{-7} = \frac{1}{w^7}$$

(b) $t^7 \cdot t^{-4}$

$$t^7 \cdot t^{-4} = t^{7+(-4)} = t^3$$

(c) $(a^2)^{-3}$

$$(a^2)^{-3} = a^{2(-3)} = a^{-6} = \frac{1}{a^6}$$

(d) $(x^{-2}y)^7$

$$(x^{-2}y)^7 = x^{-2(7)}y^7 = x^{-14}y^7 = \frac{y^7}{x^{14}}$$

(e) $(2w^{-4})^{-5}$

$$(2w^{-4})^{-5} = 2^{-5}w^{-4(-5)} = \frac{1}{2^5}w^{20} = \frac{1}{32}w^{20}$$

(f) $\dfrac{(u^{-2})^3u^4}{u^8}$

$$\frac{(u^{-2})^3u^4}{u^8} = \frac{u^{-6}u^4}{u^8} = \frac{u^{-2}}{u^8} = u^{-10} = \frac{1}{u^{10}}$$

●

We will now show how to simplify expressions in which a quotient is raised to a power. Consider this example:

$$\left(\frac{a^4}{b^7}\right)^3 = \left(\frac{a^4}{b^7}\right)\left(\frac{a^4}{b^7}\right)\left(\frac{a^4}{b^7}\right) = \frac{a^4 \cdot a^4 \cdot a^4}{b^7 \cdot b^7 \cdot b^7} = \frac{a^{12}}{b^{21}}$$

Notice that this same result may be obtained by raising both numerator and denominator to the third power:

$$\left(\frac{a^4}{b^7}\right)^3 = \frac{a^{4\cdot 3}}{b^{7\cdot 3}} = \frac{a^{12}}{b^{21}}$$

In general,

For any real number $y \neq 0$,

$$\left(\frac{x^a}{y^b}\right)^c = \frac{x^{ac}}{y^{bc}}.$$

To raise a quotient to a power, raise the numerator and denominator to that power.

example 5

Simplify each of the following.

(a) $\left(\frac{2}{3}\right)^{-3}$

$$\left(\frac{2}{3}\right)^{-3} = \frac{2^{-3}}{3^{-3}} = \frac{3^3}{2^3} = \frac{27}{8}$$

(b) $\left(\frac{x^2}{4}\right)^3$

$$\left(\frac{x^2}{4}\right)^3 = \frac{x^{2\cdot 3}}{4^3} = \frac{x^6}{64}$$

(c) $\left(\frac{-2a}{b}\right)^4$

$$\left(\frac{-2a}{b}\right)^4 = \frac{(-2a)^4}{b^4} = \frac{(-2)^4a^4}{b^4} = \frac{16a^4}{b^4}$$

(d) $\left(\frac{w^3}{w^5}\right)^{-2}$

$$\left(\frac{w^3}{w^5}\right)^{-2} = \frac{w^{3(-2)}}{w^{5(-2)}} = \frac{w^{-6}}{w^{-10}} = w^{-6-(-10)} = w^4$$

Another solution may be obtained by simplifying the expression in parentheses first:

$$\left(\frac{w^3}{w^5}\right)^{-2} = (w^{3-5})^{-2} = (w^{-2})^{-2} = w^{-2(-2)} = w^4$$

(e) $\left(\frac{3x^{-2}y}{z^{-5}}\right)^{-2}$

$$\left(\frac{3x^{-2}y}{z^{-5}}\right)^{-2} = \frac{3^{-2}x^4y^{-2}}{z^{10}} = \frac{x^4}{3^2y^2z^{10}} = \frac{x^4}{9y^2z^{10}}$$

QUICK QUIZ

Simplify each of the following.

1. $p^4 \div p^{-2}$
2. $\frac{y^{-1} \cdot y^3}{y^5}$
3. $(6w^4)^{-2}$
4. $\left(\frac{a^6}{b^{-3}}\right)^2$

ANSWERS

1. p^6
2. $\frac{1}{y^3}$
3. $\frac{1}{36w^8}$
4. $a^{12}b^6$

11.2 Exercises

Indicate which of the following statements are true and which are false. For those that are false, change the italic word to make the statement true.

1. To divide exponential expressions of the same base, keep the base the same and *divide* the exponents.
2. A factor having a negative exponent in the denominator can be written with a *positive* exponent in the numerator.
3. When an exponential expression is in simplest form, all the exponents are *positive.*
4. For any nonzero value of x, x^{-2} is *negative.*

Simplify each of the following.

5. $\frac{x^9}{x^2}$
6. $\frac{y^8}{y^3}$
7. $n^{12} \div n^4$
8. $p^{17} \div p$
9. $t^9 \div t^0$
10. $k^7 \div k^7$
11. $\frac{a^4 \cdot a^7}{a^5}$
12. $\frac{b^8 \cdot b^2}{b^9}$
13. $\frac{w \cdot w^5}{w^2 \cdot w^4}$
14. $\frac{s^3 \cdot s^{11}}{s^5 \cdot s^2}$
15. $\frac{(c^3)^4}{c^5}$
16. $\frac{(k^5)^2}{k^7}$
17. $x^{6n} \div x^{2n}$
18. $y^{9k} \div y^k$
19. $a^7 \div a^t$
20. $b^n \div b^5$
21. $\frac{(x-3)^8}{(x-3)^3}$
22. $\frac{(y+7)^4}{(y+7)}$

Simplify each of the following. (Make sure your answers contain only positive exponents.)

23. 2^{-6}
24. 5^{-3}
25. 9^{-2}
26. 3^{-4}
27. $\frac{1}{3^{-2}}$
28. $\frac{1}{4^{-3}}$
29. $2x^{-7}$
30. $5y^{-1}$
31. $a^{-3}b^5$
32. r^4s^{-8}

33. $-r^5s^{-1}t^{-8}$ 34. $4x^{-9}yz^{-6}$ 35. $\frac{1}{x^{-5}}$ 36. $\frac{1}{y^{-9}}$ 37. $\frac{p^{-5}}{q^2}$

38. $\frac{a^4}{b^{-7}}$ 39. $\frac{x^{-3}y}{z^{-2}}$ 40. $\frac{m^5n^{-1}}{p^4}$ 41. $\frac{5a^{-2}b}{c^3d^{-7}}$ 42. $\frac{-9x^4y^{-3}}{z^{-2}w^{-3}}$

Evaluate each of the following if $x = -3$, $y = 2$, and $z = -1$.

43. x^{-2} 44. $-y^{-3}$ 45. $-xy^{-4}$ 46. $x^{-1}y^2$ 47. $-5x^{-2}y^2$

48. $-7x^{-1}y^{-1}$ 49. $\frac{x^{-3}y}{z^{-2}}$ 50. $\frac{x^2}{y^{-1}z^5}$ 51. $\frac{-2x}{y^3z^{-3}}$ 52. $\frac{3z^{-4}}{xy^{-2}}$

Simplify each of the following. (Make sure your answers contain only positive exponents.)

53. $a^4 \div a^9$ 54. $b \div b^7$ 55. $t^{-3} \div t^8$ 56. $w^4 \div w^{-2}$

57. $\frac{3^4}{3^7}$ 58. $\frac{2^3}{2^8}$ 59. $\frac{5^{-1}}{5^{-4}}$ 60. $\frac{6}{6^{-1}}$

61. $\frac{-y^3}{y^{-5}}$ 62. $\frac{-x^{-2}}{x^9}$ 63. $\frac{n}{n^{-7}}$ 64. $\frac{p^{-5}}{p^{-5}}$

65. $k^3 \cdot k^{-9}$ 66. $m^{-8} \cdot m^7$ 67. $x^{-8} \cdot x^{11}$ 68. $y \cdot y^{-4}$

69. $3^{-2} \cdot 3^5$ 70. $4^7 \cdot 4^{-8}$ 71. $2^7 \cdot 2^{-11}$ 72. $5^{-1} \cdot 5^{-2}$

73. $\frac{m^4 \cdot m^{-3}}{m^8}$ 74. $\frac{k^{-3} \cdot k^9}{k^{-2}}$ 75. $(r^{-3})^6$ 76. $(s^4)^{-7}$

77. $x^3 \cdot (x^{-2})^2$ 78. $(y^{-4})^{-2} \cdot y^{-5}$ 79. $(3a^5)^{-2}$ 80. $(2b^{-3})^3$

81. $(-7t^{-2})^{-2}$ 82. $-(5u^5)^{-3}$ 83. $(x^2y^{-7})^3$ 84. $(x^{-6}y)^{-2}$

85. $(rs^{-3}t^4)^{-5}$ 86. $(a^2b^{-5}c^{-1})^3$ 87. $\frac{(p^{-3})^2 \cdot p^7}{p^{-5}}$ 88. $\frac{q^{-4} \cdot (q^2)^{-3}}{q^8}$

89. $(8x^{-1}y^{-3})^{-2}$ 90. $(4x^{-3}y^{-7})^{-3}$ 91. $\frac{(w^{-5})^2}{(w^{-3})^{-4}}$ 92. $\frac{(t^7)^{-3}}{(t^{-4})^2}$

93. $\left(\frac{3}{4}\right)^{-2}$ 94. $\left(\frac{7}{8}\right)^{-1}$ 95. $\left(\frac{5}{2}\right)^{-3}$ 96. $\left(\frac{1}{3}\right)^{-4}$

97. $\left(\frac{a^5}{8}\right)^2$ 98. $\left(\frac{b^3}{2}\right)^5$ 99. $\left(\frac{r^{-5}}{s^2}\right)^{-1}$ 100. $\left(\frac{u^5}{v^{-3}}\right)^3$

101. $\left(\frac{-7x^3}{y^{-2}}\right)^2$ 102. $\left(\frac{5a^{-3}}{b^{-1}}\right)^{-3}$ 103. $\left(\frac{t^4}{t^{-3}}\right)^{-2}$ 104. $\left(\frac{n^{-5}}{n}\right)^4$

105. $\left(\frac{8k^{-3}}{k^3}\right)^{-1}$ 106. $\left(\frac{-3s^4}{s^{-2}}\right)^{-3}$ 107. $\left(\frac{4x^{-2}y}{z^{-1}}\right)^2$ 108. $\left(\frac{2xy^{-5}}{z^7}\right)^{-5}$

11.3 Operations and Monomials

In this section we will illustrate how to perform the four basic arithmetic operations on **monomials**.

definition

A monomial is an algebraic expression that expresses the product of a constant and variable(s) raised to whole number exponents.

Some examples of monomials are $5x$, $-9t^2$, x^2y^3z, and $(-2a^2b)^3$.

The **numerical coefficient** of a monomial is the number that is multiplied by the variable factors. For example, the numerical coefficient of the monomial $7xy^3$ is 7.

We will now discuss a procedure for multiplying monomials.

To Multiply Monomials

1. Rearrange the factors using the commutative and associative properties. This may be done mentally.
2. Multiply the numerical coefficients.
3. Multiply the variable factors that are exponential expressions of the same base by adding their exponents.
4. Multiply the results obtained in steps 2 and 3.

example 1

Multiply each of the following.

(a) $(-5a^3)(3a^7)$

$= (-5 \cdot 3)(a^3 \cdot a^7)$ — By commutative and associative properties.

$= -15a^{10}$ — Multiply numerical coefficients and add exponents of a.

(b) $\left(\frac{2}{3}x^2y^3\right)(27x^3y)$

$= \left(\frac{2}{3} \cdot 27\right)(x^2 \cdot x^3)(y^3 \cdot y)$ — By commutative and associative properties.

$= 18x^5y^4$ — Multiply numerical coefficients, add exponents of x and y.

(c) $(-3ab^2)(2a^3b^3)(5ab)$

$= (-3 \cdot 2 \cdot 5)(a \cdot a^3 \cdot a)(b^2 \cdot b^3 \cdot b)$ — This step is done mentally whenever possible.

$= -30a^5b^6$

(d) $(3x^2y)^2(2x^4y^3)^3$

$= 9x^4y^28x^{12}y^9$ — First raise each monomial to the power indicated.

$= 72x^{16}y^{11}$ — Then multiply the results. ●

Let us now consider a procedure to divide monomials:

To Divide Monomials

1. Divide the numerical coefficients.
2. Divide the variable factors that are powers of the same base by subtracting the exponents.
3. Multiply the results obtained in steps 1 and 2.

example 2

Divide each of the following.

(a) $\dfrac{21t^9}{7t}$

$= \dfrac{21}{7} \cdot \dfrac{t^9}{t}$ — Divide numerical coefficients and powers of t.

$= 3t^8$ — Multiply the results.

(b) $\dfrac{12x^7y^3}{3x^4y^5}$

$= \dfrac{12}{3} \cdot \dfrac{x^7}{x^4} \cdot \dfrac{y^3}{y^5}$ — Divide numerical coefficients, powers of x, and powers of y.

$= 4x^3y^{-2}$ — Multiply the results.

$= \dfrac{4x^3}{y^2}$ — Rewrite using positive exponents.

(c) $\dfrac{-5a^2b^3}{20b^5}$

$= \dfrac{-5}{20} \cdot \dfrac{a^2}{1} \cdot \dfrac{b^3}{b^5}$ — This step is done mentally whenever possible.

$= -\dfrac{1}{4}a^2b^{-2}$

$= -\dfrac{a^2}{4b^2}$

(d) $\dfrac{(6x^2y^6)^2}{(3xy^3)^3}$

$= \dfrac{36x^4y^{12}}{27x^3y^9}$ — First raise each factor to the power indicated.

$= \dfrac{36}{27}\dfrac{x^4}{x^3}\dfrac{y^{12}}{y^9}$ — Then divide the results.

$= \dfrac{4}{3}x^1y^3$

$= \dfrac{4xy^3}{3}$

(e) $\dfrac{(5a^2b^3)(2ab^4)}{6a^7b^5}$

$= \dfrac{10a^3b^7}{6a^7b^5}$ — First multiply monomials in numerator.

$= \dfrac{5}{3}a^{-4}b^2$ — Then divide.

$= \dfrac{5b^2}{3a^4}$ — Rewrite using positive exponents. ●

In order to add or subtract monomials, we use the technique of *combining like terms* which was introduced in Section 10.2. Remember that like (or similar) terms have the same variable factors where the same variables are raised to the same exponents. For example, the monomials $7x^2y^3$ and $-5x^2y^3$ are like terms since the variables x and y are raised to the same exponents in both monomials. The monomials $3a^4b$ and $3ab^4$ are not like terms since the variables a and b are raised to different exponents in each monomial. Recall that to combine like terms, we first rewrite the expression using the distributive property, and then add or subtract the numerical coefficients.

example 3

For each of the following, combine like terms.

(a) $-7x^2 + 9x^2$

$= (-7 + 9)x^2$ — Rewrite using the distributive property.

$= 2x^2$ — Add the coefficients of x^2.

(b) $5a^2b - 9a^2b$

$= (5 - 9)a^2b$

$= -4a^2b$

(c) $3xy + 5xz - xy$

$= 3xy - xy + 5xz$ — By commutative property.

$= (3 - 1)xy + 5xz$ — By distributive property.

$= 2xy + 5xz$ — These are not like terms, so they cannot be combined.

(d) $5a^3b^2 + a^2b^2 - 3a^3b^2 + 6a^2b^2$

$\underline{5a^3b^2} + \underline{\underline{a^2b^2}} - \underline{3a^3b^2} + \underline{\underline{6a^2b^2}}$ — There are two pairs of like terms.

$= (5a^3b^2 - 3a^3b^2) + (a^2b^2 + 6a^2b^2)$ — By commutative and associative properties.

$= 2a^3b^2 + 7a^2b^2$

(e) $-3x^2 + 5y^2 - 8xy - 2x^2 - 7y^2 + 6x^2$

$\underline{-3x^2} + \underline{\underline{5y^2}} - 8xy - \underline{2x^2} - \underline{\underline{7y^2}} + \underline{6x^2}$

$= (-3x^2 - 2x^2 + 6x^2) + (5y^2 - 7y^2) - 8xy$

$= x^2 - 2y^2 - 8xy$ ●

QUICK QUIZ

Perform each of the operations indicated.

1. $(2x^2y)(-3xy^2)$
2. $\dfrac{20a^2b^3}{5a}$
3. $-5x^3 + 9x^3 - 3x^3$

ANSWERS

1. $-6x^3y^3$
2. $4ab^3$
3. x^3

11.3 Exercises

Multiply each of the following.

1. $(8x^2)(-2x^6)$
2. $(-5y^3)(-4y^8)$
3. $(3a^3)(9a^5)$
4. $(2b^7)(4b^2)$
5. $(35t^8)\left(\frac{1}{7}t^3\right)$
6. $\left(\frac{1}{3}n^9\right)(12n^4)$
7. $(8u^2)(5u^7)(-2u^3)$
8. $(3w^5)(-7w^2)(2w^8)$
9. $(-5x^2y)(3x^3y^4)$
10. $(2xy^7)(-9x^4y^2)$
11. $(9a^3b^2)\left(\frac{2}{3}a^5b^7\right)$
12. $\left(\frac{3}{5}a^4b^7\right)(15ab^6)$
13. $(3x^2y^4z)(7x^3y^3z)$
14. $(6xy^3z^2)(4x^4y^5z^3)$
15. $(-5a^2b^3c^5)(2a^7b^4c^8)$
16. $(-3ab^1c^6)(-9a^4bc^2)$
17. $(2x^2y)(-5xy^2)(3x^3y^2)$
18. $(-4x^3y^2)(3xy^5)(-7x^5y^3)$
19. $(2ab^2)^3(5a^4b^3)$
20. $(7a^6b^4)(3a^2b^3)^2$
21. $(x^2y^4)^5(2x^3y^2)^3$
22. $(4x^3y^5)^2(x^2y^3)^4$
23. $(-3xy^2)^4(2x^3y^2)^2$
24. $(-2xy^2)^3(5x^4y)^2$

Divide each of the following.

25. $\frac{32w^5}{8w^2}$
26. $\frac{36n^9}{4n}$
27. $\frac{-15x^3}{30x^2}$
28. $\frac{16y^2}{12y^5}$
29. $\frac{72t^3}{27t^{11}}$
30. $\frac{-10u}{35u^8}$
31. $\frac{16x^2y^5}{-4x^3y^3}$
32. $\frac{21x^7y^3}{7x^2y^6}$
33. $\frac{9x^3y}{27x^5y^7}$
34. $\frac{42x^5y^7}{6xy^3}$
35. $\frac{56a^4b^9}{8a^7}$
36. $\frac{15a^3}{60a^4b^2}$
37. $\frac{5xy^3}{20x^4}$
38. $\frac{30x^5}{6x^3y^9}$
39. $\frac{-7x^2yz^3}{28x^5y^2z^8}$
40. $\frac{48x^5y^3z^4}{-6x^7y^2z^3}$
41. $\frac{(5a^2b)^2}{(3a^4b^7)^3}$
42. $\frac{(-2x^4y^9)^3}{(7x^2y^5)^2}$
43. $\frac{(x^3y^2z^5)^3}{(x^4y^3z^2)^2}$
44. $\frac{(a^3b^7c^2)^2}{(a^5b^2c^5)^4}$
45. $\frac{(2a^2b)(8a^3b^4)}{4a^7b}$
46. $\frac{(7x^7y^9)(4x^5y^2)}{14x^9y}$
47. $\frac{(5s^3t^4)(9s^2t^7)}{(3s^4t^2)(10st^5)}$
48. $\frac{(6p^4q^3)(3p^7q^5)}{(4pq)(9p^8q^2)}$

For each of the following, combine like terms.

49. $-8a^3 + 4a^3$
50. $7t^5 - 3t^5$
51. $9xy - 5xy$
52. $6ab + 11ab$
53. $4m^4 - 9m^4 + 2m^4$
54. $5k^7 + 3k^7 - k^7$
55. $3r^2s^2 + 9r^2s^2$
56. $10p^3q - 2p^3q$
57. $9abc - 7abc + 5abc$
58. $-3xyz + 2xyz - 8xyz$
59. $4x^2 + 8y^2 - 7x^2$
60. $-5w^3 + 4z^3 - 9w^3$
61. $6b^5 - 3b^4 - 2b^5$
62. $8c^3 + 3c^2 - 5c^3$
63. $2xy - 5yz + 9xy$
64. $-5pq + 4ps - 3pq$
65. $-5ab^3 + 7a^3b^3 + 2ab^3$
66. $4x^2y + 7x^2y - 8x^2y$
67. $4x^3y^2 + 2x^2y^3 - 8x^3y^2 + 5x^2y^3$
68. $9a^2b^4 - 3a^3b^5 - 7a^2b^4 - 5a^3b^5$
69. $5t^2 - 7st + 4s^2 - 9t^2 + 2st$
70. $-4p^2 + 3pq - 7q^2 - 5p^2 - 8pq$
71. $5w^5z^3 - 9w^3z^5 + w^5z^3$
72. $u^4v^2 - 3u^2v^4 - 9u^4v^2$

11.4 Fractional Exponents

Fractional exponents are used to indicate roots of numbers. For example, $\sqrt{x}$ may be represented by $x^{1/2}$ which is read as "x to the one-half power." By the definition of square roots introduced in Section 7.1, we know that

$$\sqrt{x} \cdot \sqrt{x} = x, x \geqslant 0$$

Similarly, if we express the square root of x using the fractional exponent 1/2, and apply the procedure for multiplying powers of the same base, we obtain the same result:

$$x^{1/2} \cdot x^{1/2} = x^{1/2+1/2} = x^1 = x$$

Thus, two notations may be used to express the root of a number: a radical sign and a fractional exponent.

definition

For any real number x and positive integer n,

$$\mathbf{x^{1/n} = \sqrt[n]{x}}$$

provided x is positive when n is an even number.

For example, $4^{1/2} = \sqrt{4}$, $27^{1/3} = \sqrt[3]{27}$, $t^{1/5} = \sqrt[5]{t}$. Remember that x must be positive if n is even.

example 1

Simplify each of the following.

(a) $64^{1/2}$

$$64^{1/2} = \sqrt{64} = 8$$

Notice that: $8^2 = 64$
$(64^{1/2})^2 = 64.$

(b) $(-32)^{1/5}$

$$(-32)^{1/5} = \sqrt[5]{-32} = -2$$

Notice that: $(-2)^5 = -32$
$[(-32)^{1/5}]^5 = -32$

(c) $\left(\frac{1}{27}\right)^{1/3}$

$$\left(\frac{1}{27}\right)^{1/3} = \sqrt[3]{\frac{1}{27}} = \frac{\sqrt[3]{1}}{\sqrt[3]{27}} = \frac{1}{3}$$

Notice that: $\left(\frac{1}{3}\right)^3 = \frac{1}{27}$
$\left[\left(\frac{1}{27}\right)^{1/3}\right]^3 = \frac{1}{27}$

We will now consider some problems in which the numerator of the fractional exponent is not equal to one. For example, to determine the value of $8^{2/3}$ we can rewrite it using the properties of exponents as follows:

$$8^{2/3} = (8^{1/3})^2 = (\sqrt[3]{8})^2 = 2^2 = 4$$

or

$$8^{2/3} = (8^2)^{1/3} = \sqrt[3]{8^2} = \sqrt[3]{64} = 4$$

Notice that the numerator of a fractional exponent indicates the power, and the denominator indicates the root. Also notice that we obtain the same answer regardless of whether we take the root first or raise to the power first.

definition

For any real number x and integers a and $b \neq 0$

$$x^{a/b} = \sqrt[b]{x^a} = (\sqrt[b]{x})^a$$

provided x is positive when b is an even number.

For example,

$$16^{3/4} = \sqrt[4]{16^3} = (\sqrt[4]{16})^3$$

To simplify this expression, we see that it is easier to take the root first and then raise to the power:

$$16^{3/4} = (\sqrt[4]{16})^3 = 2^3 = 8$$

example 2

Simplify each of the following.

(a) $-25^{3/2}$

$$-25^{3/2} = -(\sqrt{25})^3 = -(5)^3 = -125$$

This is read as "the opposite of twenty-five to the three-halves power." Notice that the base is 25.

(b) $(-8)^{4/3}$

$$(-8)^{4/3} = (\sqrt[3]{-8})^4 = (-2)^4 = 16$$

This is read as "negative eight to the four-thirds power." Notice that the base is -8.

(c) $\left(\frac{1}{32}\right)^{3/5}$

$$\left(\frac{1}{32}\right)^{3/5} = \left(\sqrt[5]{\frac{1}{32}}\right)^3 = \left(\frac{1}{2}\right)^3 = \frac{1}{8}$$

(d) $27^{2/3} + 100^{3/2}$

$= (\sqrt[3]{27})^2 + (\sqrt{100})^3$ — First take roots and raise to powers.

$= 3^2 + 10^3$

$= 9 + 1{,}000$

$= 1{,}009$ — Then do addition. ●

Negative fractional exponents are used to indicate reciprocals, just as negative integer exponents. For example,

$$8^{-2/3} = \frac{1}{8^{2/3}} = \frac{1}{(\sqrt[3]{8})^2} = \frac{1}{2^2} = \frac{1}{4}$$

example 3

Simplify each of the following.

(a) $49^{-1/2}$

$$49^{-1/2} = \frac{1}{49^{1/2}} = \frac{1}{\sqrt{49}} = \frac{1}{7}$$

(b) $-(4)^{-5/2}$

$$-(4)^{-5/2} = -\frac{1}{4^{5/2}} = -\frac{1}{(\sqrt{4})^5}$$

$$= -\frac{1}{2^5} = -\frac{1}{32}$$

This is read as the "opposite of four to the negative five-halves power." Notice that the base is 4.

(c) $\left(\frac{9}{64}\right)^{-1/2}$

$$\left(\frac{9}{64}\right)^{-1/2} = \frac{9^{-1/2}}{64^{-1/2}} = \frac{64^{1/2}}{9^{1/2}} = \frac{\sqrt{64}}{\sqrt{9}} = \frac{8}{3}$$

(d) $27^{-1/3} \cdot 8^{-2/3}$

$$= \frac{1}{27^{1/3}} \cdot \frac{1}{8^{2/3}}$$

$$= \frac{1}{\sqrt[3]{27}} \cdot \frac{1}{(\sqrt[3]{8})^2}$$

First take roots and raise to powers.

$$= \frac{1}{3} \cdot \frac{1}{2^2}$$

$$= \frac{1}{3} \cdot \frac{1}{4}$$

$$= \frac{1}{12}$$

Then do multiplication. ●

The same procedures we used to simplify exponential expressions having integer exponents apply to problems having fractional exponents. Consider these examples.

example 4

Perform each of the operations indicated.

(a) $x^{1/2} \cdot x^{1/3}$

$$x^{1/2} \cdot x^{1/3} = x^{1/2+1/3}$$

$$= x^{3/6+2/6}$$

$$= x^{5/6}$$

To multiply powers of the same base, add the exponents.

(b) $(y^{2/5})^5$

$$(y^{2/5})^5 = y^{2/5 \cdot 5}$$

$$= y^2$$

To raise an exponential expression to a power, multiply the exponents.

(c) $\frac{a^{3/4}}{a^{1/4}}$

$$\frac{a^{3/4}}{a^{1/4}} = a^{3/4-1/4}$$

$$= a^{2/4}$$

$$= a^{1/2}$$

To divide powers of the same base, subtract the exponents.

(d) $(81x^6y^8z)^{1/4}$

$$(81x^6y^8z)^{1/4} = 81^{1/4}x^{6/4}y^{8/4}z^{1/4} = 3x^{3/2}y^2z^{1/4}$$

Raise each expression in parentheses to the $\frac{1}{4}$ power.

QUICK QUIZ

Simplify each of the following.

1. $9^{3/2}$

2. $36^{-1/2}$

3. $a \cdot a^{1/2}$

4. $(4x^2y^4)^{1/2}$

ANSWERS

1. 27

2. $\frac{1}{6}$

3. $a^{3/2}$

4. $2xy^2$

11.4 Exercises

Simplify each of the following.

1. $9^{1/2}$ **2.** $25^{1/2}$ **3.** $144^{1/2}$ **4.** $100^{1/2}$ **5.** $16^{1/4}$

6. $27^{1/3}$ **7.** $125^{1/3}$ **8.** $169^{1/2}$ **9.** $-81^{1/2}$ **10.** $-32^{1/5}$

11. $\left(\frac{4}{9}\right)^{1/2}$ **12.** $\left(\frac{1}{81}\right)^{1/4}$ **13.** $(-27)^{1/3}$ **14.** $(-8)^{1/3}$ **15.** $4^{3/2}$

16. $8^{5/3}$ **17.** $32^{3/5}$ **18.** $81^{3/4}$ **19.** $100^{5/2}$ **20.** $64^{2/3}$

21. $-16^{3/4}$ **22.** $-9^{3/2}$ **23.** $(-8)^{5/3}$ **24.** $(-125)^{2/3}$ **25.** $(-64)^{2/3}$

26. $(-32)^{3/5}$ **27.** $\left(\frac{1}{4}\right)^{5/2}$ **28.** $\left(\frac{1}{125}\right)^{2/3}$ **29.** $\left(\frac{9}{16}\right)^{3/2}$ **30.** $\left(\frac{8}{27}\right)^{4/3}$

31. $-\left(\frac{1}{49}\right)^{1/2}$ **32.** $\left(-\frac{1}{32}\right)^{1/5}$ **33.** $8^{1/3} + 81^{1/2}$ **34.** $36^{1/2} + 27^{2/3}$ **35.** $121^{1/2} - 4^{3/2}$

36. $125^{1/3} - 9^{1/2}$ **37.** $16^{3/4} \cdot 100^{1/2}$ **38.** $8^{2/3} \cdot 25^{1/2}$

Simplify each of the following.

39. $64^{-1/2}$ **40.** $8^{-1/3}$ **41.** $27^{-2/3}$ **42.** $16^{-3/4}$ **43.** $121^{-1/2}$

44. $100^{-3/2}$ **45.** $(-32)^{-2/5}$ **46.** $-25^{-3/2}$ **47.** $\left(\frac{1}{49}\right)^{-1/2}$ **48.** $\left(\frac{1}{27}\right)^{-1/3}$

49. $\left(\frac{4}{9}\right)^{-3/2}$ **50.** $\left(\frac{27}{8}\right)^{-2/3}$ **51.** $25^{-1/2} \cdot 16^{-1/4}$ **52.** $32^{-2/5} \cdot 81^{-1/2}$ **53.** $36^{-1/2} + 9^{-1/2}$

54. $16^{-1/4} - 4^{-1/2}$

Perform each of the operations indicated.

55. $x^{1/5} \cdot x^{2/5}$ **56.** $y^{1/3} \cdot y^{2/3}$ **57.** $a \cdot a^{1/3}$ **58.** $b^2 \cdot b^{1/2}$ **59.** $w^{-1/2} \cdot w^{3/2}$

60. $z^{5/8} \cdot z^{-1/8}$ **61.** $\frac{t^{5/8}}{t^{1/8}}$ **62.** $\frac{n^{3/4}}{n^{1/4}}$ **63.** $\frac{k}{k^{1/2}}$ **64.** $\frac{u^{3/2}}{u}$

65. $\dfrac{p^{1/3}}{p^{2/3}}$ **66.** $\dfrac{q^{2/5}}{q^{4/5}}$ **67.** $\dfrac{x^{1/3} \cdot x^{5/3}}{x^{2/3}}$ **68.** $\dfrac{y^{3/5} \cdot y^{1/5}}{y^{2/5}}$ **69.** $(m^{3/4})^4$

70. $(c^{5/7})^7$ **71.** $(a^{-2/3})^3$ **72.** $(b^{-3/5})^5$ **73.** $(8x^2)^{1/3}$ **74.** $(25y^3)^{1/2}$

75. $(x^2y^8)^{1/2}$ **76.** $(x^6y^9)^{1/3}$ **77.** $(27a^9b^{12})^{2/3}$ **78.** $(4a^8b^6)^{3/2}$ **79.** $(-8xy^3)^{1/3}$

80. $-(25x^4y)^{1/2}$ **81.** $(49x^6yz^4)^{1/2}$ **82.** $(64xy^9z^6)^{1/3}$ **83.** $(2a^{1/3}b^{2/3})^3$ **84.** $(3a^{3/4}b^{1/2})^4$

85. $(5x^{-1/2}yz^{-2})^2$ **86.** $(3x^3y^{-2/3}z^{-1})^3$ **87.** $(4x^{1/3}y^{-1/2})^{-2}$ **88.** $(7x^{1/2}y^{3/5})^{-1}$

11.5 Scientific Notation

People who frequently work with very large or very small numbers find it convenient to express them using powers of 10, since it saves them from having to write out a large number of zeros. In Section 1.6, we learned how to express very large numbers in terms of powers of 10. For example,

$$70{,}000 = 7 \times 10{,}000 = 7 \times 10^4$$

Notice that the number that is equivalent to 7×10^4 can be obtained by moving the decimal point in the number 7 a total of 4 places to the *right*.

$$7 \times 10^4 = 7.\underbrace{0000}_{\text{4 places}}. = 70{,}000$$

Similarly, we can also express very small numbers in terms of powers of 10 using negative exponents as shown in this example:

$$0.00009 = \frac{9}{100{,}000} = \frac{9}{10^5} = 9 \times 10^{-5}$$

Notice that the number that is equivalent to 9×10^{-5} can be obtained by moving the decimal point in the number 9 a total of 5 places to the *left*.

$$9 \times 10^{-5} = .\underbrace{0000}_{\text{5 places}}9. = 0.00009$$

When we express numbers in this fashion, using powers of 10, we are using **scientific notation**.

To write a number in scientific notation we express it in the form

$$a \times 10^n$$

where s is a number such that $1 \leq a < 10$ and n is an integer.

When numbers greater than 10 are converted to scientific notation, n is a positive integer that tells you the number of places the decimal point must be moved to the *left* to obtain the value of a. For example,

$$520 = 5.20. = 5.2 \times 10^2$$

$$287{,}000 = 2.87000. = 2.87 \times 10^5$$

$$6{,}000{,}000{,}000 = 6.000000000. = 6 \times 10^9$$

When positive numbers less than 1 are converted to scientific notation, n is a negative number that tells you the number of places the decimal point must be moved to the *right* to obtain the value of a. For example,

$$0.004 = 0.004_{\curvearrowright} = 4 \times 10^{-3}$$

$$0.00000793 = 0.000007_{\curvearrowright}93 = 7.93 \times 10^{-6}$$

$$0.000000056 = 0.00000005_{\curvearrowright}6 = 5.6 \times 10^{-8}$$

Scientific notation is especially useful when we need to multiply or divide numbers that contain a large number of zeros.

example 1

Multiply each of the following.

(a) $12{,}000 \times 4{,}000{,}000$

$= (1.2 \times 10^4) \times (4 \times 10^6)$ — Rewrite each factor in scientific notation.

$= (1.2 \times 4) \times (10^4 \times 10^6)$ — By commutative and associative properties.

$= 4.8 \times 10^{10}$ — To multiply powers of 10, add the exponents.

(b) $(0.000008)(0.007)$

$= (8 \times 10^{-6})(7 \times 10^{-3})$

$= (8 \times 7)(10^{-6} \times 10^{-3})$

$= 56 \times 10^{-9}$ — To express answer in scientific notation:

$= 5.6 \times 10^1 \times 10^{-9}$ — • Rewrite 56 in scientific notation.

$= 5.6 \times 10^{-8}$ — • Multiply powers of 10.

(c) $(20{,}000)(0.00004)(0.009)$

$= (2 \times 10^4)(4 \times 10^{-5})(9 \times 10^{-3})$

$= (2 \times 4 \times 9)(10^4 \times 10^{-5} \times 10^{-3})$

$= 72 \times 10^{-4}$

$= 7.2 \times 10^1 \times 10^{-4}$

$= 7.2 \times 10^{-3}$ •

example 2

Divide each of the following.

(a) $\dfrac{6000}{0.002}$

$= \dfrac{6 \times 10^3}{2 \times 10^{-3}}$ — Rewrite numerator and denominator in scientific notation.

$= \dfrac{6}{2} \times \dfrac{10^3}{10^{-3}}$ — Group exponential factors together.

$= 3 \times 10^6$ — To divide powers of 10, subtract the exponents.

(b) $\dfrac{25{,}000}{5{,}000{,}000}$

$= \dfrac{2.5 \times 10^4}{5 \times 10^6}$

$= \dfrac{2.5}{5} \times \dfrac{10^4}{10^6}$

$= 0.5 \times 10^{-2}$

$= 5 \times 10^{-1} \times 10^{-2}$

$= 5 \times 10^{-3}$

To express answer in scientific notation:

- Rewrite 0.5 in scientific notation.
- Multiply powers of 10.

(c) $\dfrac{(600)(0.00008)}{24{,}000}$

$= \dfrac{(6 \times 10^2)(8 \times 10^{-5})}{2.4 \times 10^4}$ Rewrite each number in scientific notation.

$= \dfrac{(6)(8)}{2.4} \times \dfrac{(10^2)(10^{-5})}{10^4}$ Separate powers of 10.

$= \dfrac{\overset{20}{(\not{6})(\not{8})}}{\underset{0.3}{\not{2.4}}} \times \dfrac{10^{-3}}{10^4}$ Multiply powers of 10.

$= 20 \times 10^{-7}$ Divide powers of 10.

$= 2 \times 10^1 \times 10^{-7}$

$= 2 \times 10^{-6}$

QUICK QUIZ

1. Convert 57,200,000 to scientific notation.
2. Find the numerical equivalent of 8.5×10^{-4}.
3. Multiply $(1.1 \times 10^{-7})(4 \times 10^5)$.
4. Divide $\dfrac{6.3 \times 10^3}{7 \times 10^{-6}}$.

ANSWERS

1. 5.72×10^7
2. 0.00085
3. 4.4×10^{-2}
4. 9×10^8

11.5 Exercises

Write each of the following in scientific notation.

1. 68,000	2. 2,300,000	3. 0.00005	4. 0.0004	5. 356
6. 817	7. 4,800,000,000	8. 9,200,000,000	9. 0.0000014	10. 0.0000000039
11. 507,000	12. 0.00201	13. 0.032564	14. 7,625,900	

Write the number expressed by each of the following.

15. 5.4×10^3 **16.** 7.2×10^6 **17.** 3.8×10^{-5} **18.** 9.1×10^{-2} **19.** 3.27×10^1
20. 3.27×10^{-1} **21.** 8×10^{-6} **22.** 4.563×10^{-4} **23.** 1.275×10^7 **24.** 2×10^5
25. 6.04×10^{-3} **26.** 1.08×10^4 **27.** 5.362814×10^5 **28.** 2.213857×10^{-3}

Multiply each of the following and express all answers in scientific notation.

29. $(7 \times 10^4)(3 \times 10^5)$ **30.** $(5 \times 10^2)(4 \times 10^6)$ **31.** $(1.8 \times 10^{-3})(2 \times 10^{-7})$
32. $(6 \times 10^{-5})(1.2 \times 10^{-2})$ **33.** $(5 \times 10^7)(9 \times 10^{-4})$ **34.** $(4 \times 10^{-9})(7 \times 10^3)$
35. (600,000)(7,000,000) **36.** (3,000)(90,000,000) **37.** (0.0007)(0.000002)
38. (0.005)(0.00008) **39.** (21,000)(0.00003) **40.** (0.0045)(2,000,000)
41. $(5 \times 10^{-3})(8 \times 10^6)(3 \times 10^{-5})$ **42.** $(4 \times 10^7)(3 \times 10^{-4})(7 \times 10^4)$ **43.** (5,000,000)(700)(2,000)
44. (60,000)(500)(80,000,000) **45.** (0.004)(0.02)(0.00008) **46.** (0.03)(0.0003)(0.007)
47. (200,000)(0.003)(0.00009) **48.** (0.00004)(7,000)(0.002)

Divide each of the following and express all answers in scientific notation.

49. $\dfrac{4.2 \times 10^5}{6 \times 10^2}$ **50.** $\dfrac{3.5 \times 10^9}{7 \times 10^4}$ **51.** $\dfrac{8.1 \times 10^{-3}}{9 \times 10^{-5}}$ **52.** $\dfrac{4.8 \times 10^{-7}}{8 \times 10^{-2}}$

53. $\dfrac{4.5 \times 10^{-4}}{9 \times 10^7}$ **54.** $\dfrac{6.3 \times 10^3}{7 \times 10^{-6}}$ **55.** $\dfrac{72{,}000{,}000}{800}$ **56.** $\dfrac{3{,}600{,}000}{4{,}000}$

57. $\dfrac{0.000144}{0.012}$ **58.** $\dfrac{0.077}{0.00007}$ **59.** $\dfrac{49{,}000}{0.007}$ **60.** $\dfrac{0.0000056}{8{,}000}$

61. $\dfrac{(6 \times 10^{-3})(4 \times 10^8)}{3 \times 10^{-5}}$ **62.** $\dfrac{(8 \times 10^4)(7 \times 10^{-9})}{4 \times 10^3}$ **63.** $\dfrac{(0.00005)(12{,}000)}{1{,}500{,}000}$ **64.** $\dfrac{(600{,}000)(0.011)}{0.0000033}$

65. $\dfrac{(4 \times 10^9)(9 \times 10^{-2})}{(2 \times 10^{-5})(6 \times 10^4)}$ **66.** $\dfrac{(6 \times 10^5)(8 \times 10^7)}{(4 \times 10^{-8})(3 \times 10^{-3})}$ **67.** $\dfrac{(9 \times 10^3)(8 \times 10^5)}{(6 \times 10^{-7})(1.2 \times 10^{-4})}$ **68.** $\dfrac{(2.1 \times 10^{-2})(4 \times 10^8)}{(2 \times 10^6)(7 \times 10^{-9})}$

Answer the question asked in each of the following.

69. The mechanical equivalent of heat is 0.000239 kilocalorie/joule. How is this number written in scientific notation?

70. A light year is equivalent to 5,880,000,000,000 miles. How is this number written in scientific notation?

71. The speed of light is equivalent to 3×10^8 meters/second. What number is represented by this expression?

72. The universal gravitational constant, G, is equivalent to 6.67×10^{-11} newton-meter^2/kilogram^2. What number is represented by this expression?

73. An astronomical unit is equivalent to 93,000,000 miles. How is this number expressed in scientific notation? How many feet are in an astronomical unit?

74. An angstrom is equivalent to 0.0000001 millimeters. How is this number expressed in scientific notation? What is an angstrom equivalent to in meters?

Summary and Review

Key Terms

[11.1] An **exponent** indicates the number of times the base appears as a factor.

[11.3] A **monomial** is an algebraic expression that expresses the product of a constant and variable(s) raised to whole number exponents.

Properties of Exponents

[11.1] $x^0 = 1, x \neq 0$ — Any nonzero number raised to the zero power equals one.

$x^a \cdot x^b = x^{a+b}$ — To multiply exponential expressions of the same base, add the exponents.

$(x^a)^b = x^{a \cdot b}$ — To raise an exponential expression to a power, multiply the exponents.

$(x^a y^b)^c = x^{ac} x^{bc}$ — To raise a product to a power, raise each factor to that power.

[11.2] $\frac{x^a}{x^b} = x^{a-b}, x \neq 0$ — To divide exponential expressions of the same base, subtract the exponents.

$x^{-n} = \frac{1}{x^n}, x \neq 0$ — A negative exponent indicates a reciprocal.

$\left(\frac{x^a}{y^b}\right)^c = \frac{x^{ac}}{y^{bc}}, y \neq 0$ — To raise a quotient to a power, raise numerator and denominator to that power.

[11.4] $x^{1/n} = \sqrt[n]{x}, n \neq 0,$ $x > 0$ if n is even — The $1/n$ power indicates the nth root.

$x^{a/b} = \sqrt[b]{x^a} = (\sqrt[b]{x})^a, b \neq 0,$ $x > 0$ if b is even — The numerator of a fractional exponent indicates the power and the denominator indicates the root.

Calculations

[11.3] **To multiply monomials**, multiply the numerical coefficients, and add the exponents of the variable factors having the same base.

To divide monomials, divide the numerical coefficients, and subtract the exponents of the variable factors having the same base.

To add or subtract monomials, combine like terms using the distributive property.

[11.5] **To write a number in scientific notation**, express it in the form $a \times 10^n$ where a is a decimal such that $1 \leq a < 10$ and n is an integer.

Chapter 11 Review Exercises

[11.1, 11.2, 11.4] Evaluate each of the following if $x = 8$, $y = -4$, and $z = -1$.

1. $-x^2$ **2.** $(-y)^3$ **3.** $-xz^3$ **4.** y^3z^5 **5.** $(yz)^{-1}$

6. $-x^{-1}y^2$ **7.** $\frac{x^{-2}y^2}{z^{-5}}$ **8.** $\frac{y^{-3}z^7}{x^{-1}}$ **9.** $x^{-2/3}$ **10.** $-(8y)^{1/5}$

11. $(-2xz)^{1/4}$ **12.** $(yz)^{-3/2}$

[11.1, 11.2, 11.4] Simplify each of the following.

13. $p^8 \cdot p^5$ **14.** $n^6 \div n$ **15.** $(t^3)^7$ **16.** $u^4 \cdot u^{-6}$

17. $(a^2b^5)^3$ **18.** $-(xy^2)^5$ **19.** $(-8x^3y)^2$ **20.** $\left(\frac{3}{4}a^4b^7\right)^3$

21. $(5b^3)^{-2}$ **22.** $(4c^9)^{-3}$ **23.** $(x^{-5}y^6)^3$ **24.** $(x^4y^{-7})^{-5}$

25. $\frac{q^3 \cdot q^{-3}}{q^{-9}}$ **26.** $k^4 \cdot k^{-7} \cdot k^{-2}$ **27.** $-(z^{-3})^4 \cdot (z^9)^{-1}$ **28.** $(-w^5)^3 \cdot (w^7)^{-2}$

29. $\left(\frac{6x^{-3}y^{-5}}{z^4}\right)^2$

30. $\left(\frac{3a^2}{b^7c^{-4}}\right)^{-3}$

31. $8^{5/3}$

32. $32^{-1/5}$

33. $a^{-2/3} \div a^{1/3}$

34. $b^{3/7} \cdot b^{-1}$

35. $(49x^3y^{-6}z^8)^{-1/2}$

36. $(27x^{-9}y^{12}z^7)^{1/3}$

[11.3] Perform each of the operations indicated.

37. $(5a^4)(-3a^7)$

38. $(9x^5)(8x^2)$

39. $\frac{9t^4}{12t^9}$

40. $\frac{45n^7}{30n^5}$

41. $(2x^2y^4)(7x^6y^3)$

42. $(-6a^9b^3)(-4ab^8)$

43. $(-3x^3y)^2(2x^5y^4)^3$

44. $(10xy^3)^2(-2x^6y)^5$

45. $\frac{-56x^5y^3}{7x^2y^8}$

46. $\frac{12x^7y^5}{72x^3y}$

47. $\frac{(8a^5b^2)(5ab^6)}{4a^3b^3}$

48. $\frac{18a^7b^5}{(2ab^3)(3a^4b)}$

49. $8wz + 4wz$

50. $2rs - 3rs$

51. $5x^5 - 9x^5 + 6x^5$

52. $-3y^8 + 9y^8 - y^8$

53. $4pq^4 + p^4q - 7pq^4$

54. $-2a^3b^7 + 8a^7b^3 - 9a^3b^7$

[11.5] Perform each of the operations indicated. Express all answers using scientific notation.

55. $(5 \times 10^7)(7 \times 10^2)$

56. $(9 \times 10^{-3})(2 \times 10^{-8})$

57. $\frac{5.6 \times 10^8}{8 \times 10^2}$

58. $\frac{4.8 \times 10^5}{6 \times 10^3}$

59. $(0.00004)(300)$

60. $(6{,}000)(0.008)$

61. $\frac{0.0021}{3{,}000}$

62. $\frac{360{,}000}{0.009}$

63. $(60{,}000{,}000)(0.0002)(7{,}000)$

64. $(0.000004)(200)(30{,}000)$

65. $\frac{72{,}000}{(0.006)(4{,}000)}$

66. $\frac{(0.0008)(6{,}000{,}000)}{0.0012}$

12 POLYNOMIALS

12.1 Addition and Subtraction of Polynomials

In this chapter, we will learn how to perform various algebraic operations on expressions called *polynomials.*

definition

A *polynomial* is an algebraic expression that consists of the sum of one or more terms such that no variables appear in the denominator and all exponents of variables are whole numbers.

Different kinds of polynomials are named according to the number of terms they contain. A monomial is a polynomial that contains one term, a *binomial* is a polynomial that contains two terms, and a *trinomial* is a polynomial that contains three terms. Examples of each are shown here:

MONOMIALS	BINOMIALS	TRINOMIALS
$8x$	$x^2 - 9$	$x^2 + 6x + 5$
$-6x^2y^3$	$4a + 8b$	$-3xy + 9yz + 2zw$
$\frac{2}{3}abc^2$	$7x^2y^2 - 3x^4y^9$	$8t^5 - 2t^3 + t$

If a polynomial contains only one variable, the *degree* of that polynomial is the highest power of the variable in any of the terms. For example,

$$9y^4 - 5y^3 + 4y^2 + 8y - 2$$

is a polynomial of degree 4 since 4 is the highest power of y. Notice that this polynomial is written in *descending powers of* y. That is, the highest power of y appears first, the next highest power appears second, and so on. A polynomial is said to be in standard form when its powers appear in descending order moving from left to right.

Any constant is said to be a monomial of degree zero since it can be expressed as the product of the constant and a variable raised to the zero power. For example, 5 is a monomial of degree 0 since $5x^0 = 5$.

example 1

Write each of the following polynomials in standard form and state its degree.

(a) $4 - 5z - 2z^2$

$= -2z^2 - 5z + 4$

This is a second degree polynomial since the highest power of z is 2.

(b) $8x - 3x^3 + x^2 + 7x^5$

$= 7x^5 - 3x^3 + x^2 + 8x$

This is a fifth degree polynomial since the highest power of x is 5.

We are now ready to discuss procedures for performing the various arithmetic operations on polynomials.

To Add or Subtract Polynomials

1. Rewrite each polynomial in descending powers of the variable. Sometimes it is more convenient to write the problem vertically lining up like terms.
2. Combine like terms by adding their numerical coefficients.

example 2

Add each of the following.

(a) $(5 - 9x + 2x^2) + (4x - 7 + 3x^2)$

$(2x^2 - 9x + 5) + (3x^2 + 4x - 7)$

Write each polynomial in descending powers of x.

$$\begin{array}{r} 2x^2 - 9x + 5 \\ 3x^2 + 4x - 7 \\ \hline 5x^2 - 5x - 2 \end{array}$$

Arrange by vertically lining up like terms.

Combine like terms by adding their numerical coefficients.

Since subtracting a term is the same as adding its opposite, we can visualize this procedure as follows:

$$\begin{array}{l} 2x^2 + (-9x) + 5 \\ 3x^2 + \quad 4x + (-7) \\ \hline 5x^2 + (-5x) + (-2) \\ = 5x^2 - 5x - 2 \end{array}$$

$2 + 3 = 5 \qquad -9 + 4 = -5$

$5 + (-7) = -2$

(b) $(7y - 3y^2 + 5y^3) + (8 - 2y^2 + y) + (-4y^2 + y^3 - 6)$

$$\begin{array}{r} 5y^3 - 3y^2 + 7y \\ - 2y^2 + \ \ y + 8 \\ y^3 - 4y^2 - 6 \\ \hline 6y^3 - 9y^2 + 8y + 2 \end{array}$$

Write the problem vertically, leaving spaces for missing powers of y in order to line up like terms.

Combine like terms by adding their numerical coefficients.

(c) $(5x + 7x^2 - 2x^4) + (-8x - x^3 + 9x^2)$

We will do this problem using the horizontal format:

$= (-2x^4 + 7x^2 + 5x) + (-x^3 + 9x^2 - 8x)$ Rewrite each polynomial in descending powers of x.

$= -2x^4 + (-x^3) + (7x^2 + 9x^2) + [5x + (-8x)]$ By commutative and associative properties.

$= -2x^4 - x^3 + 16x^2 - 3x$ Combine like terms.

Now let us look at some examples of subtracting polynomials using both the vertical and horizontal formats.

example 3

Subtract $2a + 3b$ from $7a - 5b$.

solution

$$\begin{array}{r} 7a - 5b \\ \underline{2a + 3b} \\ 5a - 8b \end{array}$$

Write the problem vertically lining up like terms.

Notice that $7a - 5b$ appears on top.

Combine like terms by subtracting their numerical coefficients:
$7 - 2 = 7 + (-2) = 5$
$-5 - 3 = -5 + (-3) = -8$

Notice that subtracting a polynomial is the same as adding the opposite of each of its terms. We can therefore change this to an addition problem by multiplying each term in the polynomial being subtracted by -1 using the distributive property:

$(7a - 5b) - (2a + 3b)$

$= 7a + (-5b) + (-1)(2a + 3b)$ — By definition of subtraction.

$= 7a + (-5b) + (-2a) - 3b$ — By distributive property.

$= [7a + (-2a)] + (-5b) - 3b$ — By commutative and associative properties of addition.

$= 5a + (-8b) = 5a - 8b$ — Combine like terms. By definition of subtraction.

We can summarize these results as follows:

> To subtract one polynomial from another, add the opposite of each term in the polynomial being subtracted.

example 4

Subtract $1 - 5x^2 - 2x$ from $4x - 7x^2 - 8$.

solution

Let us first solve the problem using the vertical format.

$$\begin{array}{r} -7x^2 + 4x - 8 \\ \underline{-5x^2 - 2x + 1} \end{array}$$

Write each polynomial in descending powers of x.

Line up like terms

$$\begin{array}{r} -7x^2 + 4x - 8 \\ \underline{\overset{\oplus}{-}5x^2 \overset{\oplus}{-} 2x \overset{\ominus}{+} 1} \\ -2x^2 + 6x - 9 \end{array}$$

To indicate that we add the opposite of each term in the second polynomial, we write the new signs in circles above the original ones:
$-7 - (-5) = -7 + 5 = -2$
$4 - (-2) = 4 + 2 = 6$
$-8 - 1 = -8 + (-1) = -9$

Writing the problem horizontally, we obtain:

$(-7x^2 + 4x - 8) - (-5x^2 - 2x + 1)$	The second polynomial must appear in parentheses
$= -7x^2 + 4x - 8 + 5x^2 + 2x - 1$	Add the opposite of each term in the second polynomial
$= (-7x^2 + 5x^2) + (4x + 2x) + (-8 - 1)$	By commutative and associative properties
$= -2x^2 + 6x - 9$	Combine like terms ●

example 5

Subtract $2y^2 + 5y - 7$ from $y^3 - 9y + 2$.

solution

$$\begin{array}{l} y^3 \qquad - 9y + 2 \\ \quad 2y^2 + 5y - 7 \\ \hline \end{array}$$

Write the problem vertically, leaving space for missing terms.

$$\begin{array}{l} y^3 + 0y^2 - \ \ 9y + 2 \\ \ominus 2y^2 \overset{\ominus}{+} \ 5y \overset{\oplus}{-} 7 \\ \hline y^3 - 2y^2 - 14y + 9 \end{array}$$

Notice that the coefficient of the missing term on top is 0.

Add the opposite of each term in the second polynomial

Writing the problem horizontally we obtain:

$y^3 - 9y + 2 - (2y^2 + 5y - 7)$	Only the second polynomial must appear in parentheses.
$= y^3 - 9y + 2 - 2y^2 - 5y + 7$	Add the opposite of each term in parentheses.
$= y^3 - 2y^2 - 14y + 9$	Combine like terms. ●

To simplify algebraic expressions involving both addition and subtraction of polynomials, we can follow these guidelines for removing parentheses:

1. For expressions in parentheses preceded by a + sign or no sign, simply remove the parentheses.
2. For expressions in parentheses preceded by a − sign, change the sign of each term to its opposite, and remove the parentheses.
3. For nested sets of parentheses, remove one set at a time, beginning with the innermost and working outwards.

example 6

Simplify each of the following.

(a) $7x - (3y + 5x) + (x - y)$

$= 7x - 3y - 5x + x - y = 3x - 4y$	Remove parentheses. Combine like terms.

(b) $3a^2 - [5a + (7 - 2a^2) - (a + 4)]$

$= 3a^2 - [5a + 7 - 2a^2 - a - 4]$	Remove innermost ( ).
$= 3a^2 - [4a + 3 - 2a^2]$	Combine like terms in [ ].
$= 3a^2 - 4a - 3 + 2a^2$	Remove [ ].
$= 5a^2 - 4a - 3$	Combine like terms.

(c) $-\{3z - [5 - (9z^2 + 3) - 4z + z^2]\}$

$= -\{3z - [5 - 9z^2 - 3 - 4z + z^2]\}$	Remove innermost ( ).
$= -\{3z - [-8z^2 - 4z + 2]\}$	Combine like terms in [ ].
$= -\{3z + 8z^2 + 4z - 2\}$	Remove [ ].
$= -\{8z^2 + 7z - 2\} = -8z^2 - 7z + 2$	Combine like terms in { }. Remove { }. ●

QUICK QUIZ

Perform each of the operations indicated.

1. $(5a - b) + (-7a + 4b)$
2. $(3x - x^3 + 4x^2) + (7 - 8x^3 - 2x)$
3. $(5y + 2y^2 - 7) - (1 + y^2 - 6y)$
4. $8w - [7z + (5w - 9z)]$

ANSWERS

1. $-2a + 3b$
2. $-9x^3 + 4x^2 + x + 7$
3. $y^2 + 11y - 8$
4. $3w + 2z$

12.1 Exercises

Indicate which of the following statements are true and which are false. For those that are false, change the italic word to make the statement true.

1. A *trinomial* is a polynomial that contains two terms.
2. The degree of a polynomial containing a single variable is the *lowest* power of that variable.
3. When a polynomial appears in standard form, it is written in *ascending* powers of the variable.
4. To subtract one polynomial from another, we *add* the opposite of each term in the polynomial being subtracted.

Write each of the following polynomials in standard form and state its degree.

5. $7 - 4x^2$
6. $5 + 3y$
7. $2a + 4a^3 - 9$
8. $2b - 7b^4 + 5b^2$
9. $-3t^2 + 6t^5 - 4t + 2t^3$
10. $8z^3 - 1 + 7z^9 - 3z + 4z^5$

Add each of the following.

11. $\begin{array}{r} 5a - 7b \\ -2a + 9b \\ \hline \end{array}$

12. $\begin{array}{r} -3x + 8y \\ -9x - y \\ \hline \end{array}$

13. $\begin{array}{r} -3x^2 - 2x + 1 \\ 4x^2 + x - 8 \\ \hline \end{array}$

14. $\begin{array}{r} 7z^2 + 9z - 4 \\ -5z^2 + 3z - 6 \\ \hline \end{array}$

15. $\begin{array}{r} -8y^3 + 3y^2 - 5y + 2 \\ 4y^3 - y^2 - 8y - 9 \\ \hline \end{array}$

16. $\begin{array}{r} w^3 - 8w^2 + 5w - 7 \\ -3w^3 - 2w^2 + 6w - 4 \\ \hline \end{array}$

17. $(7x - 8y) + (-9x + 4y)$
18. $(-3a + 9b) + (2a - 7b)$
19. $(4t^2 - 5t - 1) + (-7t^2 + 8t + 4)$
20. $(-2p^2 + p - 9) + (-6p^2 + 5p - 3)$
21. $(5c - 7a + 3b) + (-5b - 6c - 2a)$
22. $(3x - 5z + 2y) + (-6y - 4z + 7x)$
23. $(8z - z^2 + 5) + (3 - 5z^2 + 2z)$
24. $(9 - 5w + 6w^2) + (w^2 - 4 - 7w)$

25. $\begin{array}{r} -6r + 4s + 3t \\ 9r - 7s - 2t \\ 5r + 9s - 8t \\ \hline \end{array}$

26. $\begin{array}{r} 5a - 7b + 4c \\ -8a + 2b - 3c \\ -6a - 5b + 9c \\ \hline \end{array}$

27. $(7x - 2x^3 - 4) + (3 + 5x^3 - 8x^2)$
28. $(3y^2 - 5y + y^3) + (-2y + 4y^3 - 6y^2)$
29. $(-5w + 7w^2) + (3 - 5w^2) + (-2w + 6w^2)$
30. $(9 - 4m^2) + (-2m - 7m^2) + (5m - 8)$

Subtract each of the following.

31. $\begin{array}{r} 5x - 3y \\ 7x + 4y \\ \hline \end{array}$

32. $\begin{array}{r} 9a + 2b \\ -3a - 4b \\ \hline \end{array}$

33. $\begin{array}{r} 4t^2 - 7t + 3 \\ -2t^2 + 9t - 8 \\ \hline \end{array}$

34. $\begin{array}{r} -w^2 - 5w + 6 \\ -4w^2 + 2w + 3 \\ \hline \end{array}$

35. $\begin{array}{r} -6z^3 - z^2 + 8z - 5 \\ 2z^3 + 4z^2 - 3z - 1 \\ \hline \end{array}$

36. $\begin{array}{r} 3p^3 - 2p^2 + 7p + 4 \\ 7p^3 + 4p^2 - 2p + 8 \\ \hline \end{array}$

37. $(7r + 3s) - (2r - 5s)$

38. $(-2w - 6z) - (-5w - 7z)$

39. $(9a^2 - 6a + 3) - (2a^2 + 5a - 8)$

40. $(-5b^2 - 7b - 2) - (3b^2 + 4b - 6)$

41. $(7x^2 + 5xy - 2y^2) - (-3x^2 + xy - 6y^2)$

42. $(-3a^2 - 2ab + b^2) - (6a^2 - 5ab + 4b^2)$

43. $(3y - 7x + 4z) - (2x - 8z - 5y)$

44. $(-4r + 5p + 7q) - (-3p - 8r + 4q)$

45. $(-9y - y^2 + 5) - (3 - 2y + 6y^2)$

46. $(5x^2 + 2 - 2x) - (3x - 7x^2 + 4)$

47. $(3z - 2z^3 + 5) - (7z^2 - 9z^3 + 4z)$

48. $(-9n^2 + 4n^3 - 6) - (-5n - 8n^3 + 3)$

49. $(8p - 6p^2 + 5 - 3p^3) - (4p^2 - 8 + p^3 - 7p)$

50. $(3q^2 + 7q^3 - 9 - 5q) - (-5q + 7q^2 - 3q^3 + 6)$

Simplify each of the following.

51. $4a - (2a + b) + (a - 7b)$

52. $(3r + 2s) - 7s - (6r - 5s)$

53. $(5x - 7) - (-2x + 9) + (4x - 3)$

54. $(-2y + 5) + (-6y - 7) - (3y + 8)$

55. $(6x^2 - 1) - (8x + 4) + (9 - 3x^2)$

56. $-(2m - 5m^2) + (7 - 3m^2) - (8m + 9m^2)$

57. $4x - [3y + (2x + 7y)]$

58. $9w + [5z - (3w - 8z)]$

59. $-(8t - 7) + 9t - [5 + (6t + 3)]$

60. $-3p - [7p + (p - 8)] - (4p - 1)$

61. $8w - [3w^2 + (6 - 9w) + 8]$

62. $-6z^2 + [5 - (7z - 3) - 2z^2]$

63. $9a - \{7b + [3a - (6b - 2a) - 9b]\}$

64. $-7y - \{4 - [2y - 5 - (8y + 3)]\}$

12.2 Multiplication of Polynomials

In this section we will discuss procedures for multiplying a polynomial by a monomial, and a polynomial by another polynomial, using the distributive property.

Recall from Chapter 1 that, according to the distributive property,

$$a(b + c) = ab + ac$$

For example,

$$\begin{aligned} 5x(2y + 7) &= 5x(2y) + 5x(7) \\ &= 10xy + 35x \end{aligned}$$

We have therefore illustrated the following:

> To multiply a polynomial by a monomial, multiply each term in the polynomial by the monomial and add the results.

example 1

Multiply each of the following.

(a) $3a(5a^2 - 7a + 2)$

$$3a(5a^2 - 7a + 2) = 3a(5a^2) + 3a(-7a) + 3a(2)$$ By distributive property.

$$= 15a^3 - 21a^2 + 6a$$

(b) $(8t + 5u)4$

$$(8t + 5u)4 = (8t)4 + (5u)4$$ By distributive property.

$$= 32t + 20u$$

(c) $2x^2y^3(3xy^2 - 7x^3y)$

$$2x^2y^3(3xy^2 - 7x^3y)$$
$$= (2x^2y^3)(3xy^2) + (2x^2y^3)(-7x^3y)$$
$$= 6x^3y^5 - 14x^5y^4$$

If a problem involves more than one operation, we perform multiplication before addition and subtraction, according to the order of operation rules. Remove nested sets of parentheses beginning with the innermost and working outwards.

example 2

Simplify each of the following.

(a) $-4t(2t - 5) + 6t(3 - 7t)$

$$-4t(2t - 5) + 6t(3 - 7t)$$ Perform multiplication first using the distributive property.

$$= -4t(2t) + (-4t)(-5) + 6t(3) + 6t(-7t)$$
$$= -8t^2 + 20t + 18t - 42t^2$$
$$= -50t^2 + 38t$$ Combine like terms.

(b) $5x[3x^2 - 2x(x - 7) - 4]$

$$5x[3x^2 + (-2x)(x - 7) - 4]$$ Remove ( ) in [ ] using the distributive property.

$$= 5x[3x^2 - 2x^2 + 14x - 4]$$
$$= 5x[x^2 + 14x - 4]$$ Combine like terms in [ ].

$$= 5x^3 + 70x^2 - 20x$$ Remove [ ] using distributive property.

If we want to multiply two binomials together, we can use the distributive property twice:

$$(a + b)(c + d) = a(c + d) + b(c + d)$$
$$= ac + ad + bc + bd$$

For example,

$$(x + 3)(y - 7) = x(y - 7) + 3(y - 7)$$
$$= xy - 7x + 3y - 21$$

Notice that each term of the first binomial is multiplied by each term of the second. This same technique can be applied when we multiply polynomials that contain more than two terms.

> To multiply two polynomials, multiply each term of the first polynomial by each term of the second polynomial, and add the results.

example 3

Multiply $(5x - 3)(2x^2 + 4x - 9)$.

solution

$5x(2x^2 + 4x - 9) + (-3)(2x^2 + 4x - 9)$

$= 5x(2x^2) + 5x(4x) + 5x(-9)$
$\quad + (-3)(2x^2) + (-3)(4x) + (-3)(-9)$

Multiply each term in the trinomial by $5x$ and then multiply each term in the trinomial by -3.

$= 10x^3 + 20x^2 - 45x - 6x^2 - 12x + 27$

$= 10x^3 + 14x^2 - 57x + 27$

Combine like terms. Notice that we express the answer in descending powers of x.

This same problem can be done using a vertical format similar to that used for multiplying two-digit numbers.

$$\begin{array}{rrrr} & 2x^2 + & 4x & - 9 \\ & & 5x & - 3 \\ \hline 10x^3 + & 20x^2 - & 45x & \\ - & 6x^2 - & 12x & + 27 \\ \hline 10x^3 + & 14x^2 - & 57x & + 27 \end{array}$$

Place the polynomial with the larger number of terms on top.

Multiply each term on the top by the first term on the bottom: $5x(2x^2 + 4x - 9)$.

Multiply each term on the top by the second term on the bottom, lining up like terms: $-3(2x^2 + 4x - 9)$.

Add the like terms in each column. ●

example 4

Multiply $(4x^2 + 5x - 1)(3x^2 - 2x + 7)$.

solution

We will use a vertical format:

$$\begin{array}{rrrrr} & & 3x^2 - & 2x & + 7 \\ & & 4x^2 + & 5x & - 1 \\ \hline 12x^4 - & 8x^3 + & 28x^2 & & \\ & 15x^3 - & 10x^2 + & 35x & \\ & - & 3x^2 + & 2x & - 7 \\ \hline 12x^4 + & 7x^3 + & 15x^2 + & 37x & - 7 \end{array}$$

$4x^2(3x^2 - 2x + 7)$

$5x(3x^2 - 2x + 7)$

$-1(3x^2 - 2x + 7)$

Add the like terms in each column. ●

If more than two polynomials are multiplied, we first multiply any two together, and then multiply that result by the remaining factors.

example 5

Multiply each of the following.

(a) $(4t^2)(7t - 9)(2t)$

It is easiest to multiply the monomial factors first.

$= [(4t^2)(2t)](7t - 9)$

By commutative and associative properties.

$= 8t^3(7t - 9)$

Find product of the two monomials.

$= 56t^4 - 72t^3$

Multiply result by the binomial using the distributive property.

(b) $(x-3)^3$

$= (x-3)(x-3)(x-3)$	To cube a binomial, use it as a factor 3 times.
$= (x-3)[(x-3)(x-3)]$	By associative property.
$= (x-3)[x(x-3) - 3(x-3)]$	Use distributive property twice.
$= (x-3)[x^2 - 3x - 3x + 9]$	
$= (x-3)(x^2 - 6x + 9)$	Combine like terms in [ ].
$= x(x^2 - 6x + 9) - 3(x^2 - 6x + 9)$	By distributive property.
$= x^3 - 6x^2 + 9x - 3x^2 + 18x - 27$	
$= x^3 - 9x^2 + 27x - 27$	Combine like terms.

(c) $2x(x+5)^2$

$2x(x+5)^2 = 2x(x+5)(x+5)$	This expression represents the product of 3 factors.
$= 2x[(x+5)(x+5)]$	By associative property.
$= 2x[x(x+5) + 5(x+5)]$	Use distributive property twice.
$= 2x(x^2 + 5x + 5x + 25)$	
$= 2x(x^2 + 10x + 25)$	Combine like terms in ( ).
$= 2x^3 + 20x^2 + 50x$	By distributive property.

QUICK QUIZ

Simplify each of the following.

1. $-2xy(5x - 7y)$
2. $7a^2 - 3a(9 - 2a + a^2)$
3. $(x^2 + 4x - 7)(x - 2)$
4. $-3x(x+2)^2$

ANSWERS

1. $-10x^2y + 14xy^2$
2. $-3a^3 + 13a^2 - 27a$
3. $x^3 + 2x^2 - 15x + 14$
4. $-3x^3 - 12x^2 - 12x$

12.2 Exercises

Multiply each of the following.

1. $3a(7 - 2b)$
2. $-5x(2y - 4)$
3. $-2w(8w - 9)$
4. $3y(6y + 5)$
5. $(7p - 4q)3$
6. $(-9x + 2y)8$
7. $-2a(a^2 - 6a - 3)$
8. $4b(2b^2 + 3b - 5)$
9. $3xy(-5x + 4y)$
10. $-6pq(8p - 7q)$
11. $ab^2(2a^2 - 8b)$
12. $s^2t(5t - 3s^2)$
13. $5y(4xy - 7xy^2)$
14. $-2x(-x^2y^2 + 3xy)$
15. $(-t^2 + 4t - 8)(5t)$
16. $(2w^2 - 5w + 1)(-3w)$
17. $3a^2b(-5ab - 7b^2 + 5a^3b)$
18. $-4pq^3(3p^2 - 4pq + p^3q^2)$
19. $2abc(a^2b - 7bc^2 - 3ac)$
20. $-5yz(3x^2y - 6xyz + 2xz^2)$

Simplify each of the following.

21. $5x^2 - 3x(7 - 4x)$
22. $4y(9y - 2) - 8y^2$
23. $-3s^2(9t - s) + 4s^2t$
24. $2p^2q - 8pq(5 - 3p)$
25. $3(2x - y) - 5(-4x - 7y)$
26. $-2(5a + 6b) + 3(8a - 2b)$

27. $5n(n-8)+3n(9-2n)$
28. $-6p(5-9p)-2p(3p+7)$
29. $-2w(5-7w+w^2)-4(7-w^3)$
30. $6z(2z^3+3)-z^2(5z-8+2z^2)$
31. $2xy(3x^2-5y)-7x^2(4xy-3y^2)$
32. $-5y^2(2x^2-6xy)+3xy(2x^2-7xy)$
33. $5z[2z-7(3z+5)]$
34. $-3w[8-2w(3-8w)]$
35. $xy^2-y[x^2y-x(y-7)]$
36. $x[y(x^2-3y)-8]+9xy^2$
37. $4a[7a^2-8a(a-9)]+3a^2(6a-7)$
38. $-9b^3(2-5b)-2b[3b^2-8(7b+5)]$

Multiply each of the following.

39. $(x+4)(x-2)$
40. $(x-7)(x+5)$
41. $(3a-4)(b-8)$
42. $(s+3)(2s+5)$
43. $(x+8)(x-8)$
44. $(y-3)(y+3)$
45. $(5y-9)(2y+3)$
46. $(8y+4)(2y-7)$
47. $(3x+2y)(-5x+7y)$
48. $(4a-6b)(-2a+5b)$
49. $(7t-3)^2$
50. $(5t+4)^2$
51. $(x+2)(x^2-3x+4)$
52. $(x-5)(x^2+4x-2)$
53. $(3z^2-7z-4)(z+5)$
54. $(3w^2+5w-1)(w-8)$
55. $(5m+4)(8m^2-3m+2)$
56. $(9p-5)(6p^2-2p+7)$
57. $(4x^2-6x+1)(x^2-5x-3)$
58. $(y^2-5y+2)(2y^2+4y-5)$
59. $(4w)(2w-5)(-9w)$
60. $(-6n)(4-2n)(3n)$
61. $(-3x)(x-7)(2x+1)$
62. $(2y)(3y-7)(y+4)$
63. $(y+2)^3$
64. $(x-5)^3$
65. $(2p+3)(p-7)(3p+5)$
66. $(4z-1)(2z+3)(z+6)$
67. $4x(x-3)^2$
68. $-2x(x+6)^2$
69. $-5x(x+2)^2$
70. $-3x(x-4)^2$

12.3 Products of Binomials

In this section, we will introduce a technique that will enable us to quickly determine the product of two binomials. Whenever we multiply two binomials using the distributive property, we obtain the sum of four terms: the product of the *first* terms in each binomial, the product of the two *outer* terms, the product of the two *inner* terms, and finally, the product of the *last* terms in each binomial. Consider this example:

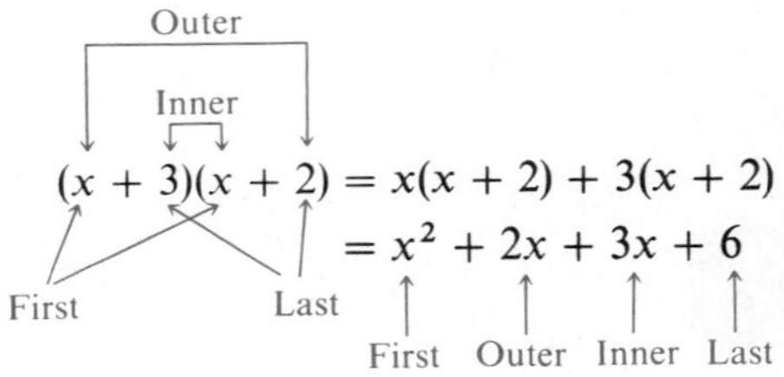

From now on, whenever we multiply two binomials, we will not write down the first step, but immediately express the product as the sum of the products of the First, Outer, Inner, and Last terms. This technique is called the FOIL method.

Notice that the two middle terms in the product we just calculated can be combined to obtain the following:

$$(x+3)(x+2) = x^2 + \underbrace{2x + 3x} + 6$$
$$= x^2 + 5x + 6$$

As we become more familiar with the FOIL method, we will be able to do this step in our head and write down only the final answer.

example 1

Multiply each of the following.

(a) $(x - 7)(x + 4)$

$$(x - 7)(x + 4) = \underset{\text{First}}{x^2} + \underset{\text{Outer}}{4x} - \underset{\text{Inner}}{7x} - \underset{\text{Last}}{28} = x^2 - 3x - 28$$

(b) $(3a - 5)(2a - 1)$

$$(3a - 5)(2a - 1) = \underset{\text{First}}{6a^2} - \underset{\text{Outer}}{3a} - \underset{\text{Inner}}{10a} + 5 = 6a^2 - 13a + 5$$

(c) $(7 + 2y)(5 - 4y)$

$$(7 + 2y)(5 - 4y) = 35 - 28y + 10y - 8y^2$$
$$= -8y^2 - 18y + 35$$

(d) $(x + 5y)(4x + 9y)$

$$(x + 5y)(4x + 9y) = 4x^2 + 9xy + 20xy + 45y^2$$
$$= 4x^2 + 29xy + 45y^2$$

Let us now look at some examples that involve the squares of binomials:

$$(x + 7)^2 = (x + 7)(x + 7) = x^2 + 7x + 7x + 49 = x^2 + 14x + 49$$
$$(y - 4)^2 = (y - 4)(y - 4) = y^2 - 4y - 4y + 16 = y^2 - 8y + 16$$
$$(5a + 2)^2 = (5a + 2)(5a + 2) = 25a^2 + 10a + 10a + 4 = 25a^2 + 20a + 4$$
$$(8 - 3b)^2 = (8 - 3b)(8 - 3b) = 64 - 24b - 24b + 9b^2 = 64 - 48b + 9b^2$$

Notice that in each case the result is a trinomial in which the first term is the square of the first term in the binomial, the second term is twice the product of the two terms in the binomial, and the third term is the square of the last term in the binomial. We can summarize these results by the following two formulas.

Squares of Binomials

$$(a + b)^2 = a^2 + 2ab + b^2$$
$$(a - b)^2 = a^2 - 2ab + b^2$$

The square of a binomial is equal to the first term squared, plus twice the product of the two terms, plus the last term squared.

Sometimes these formulas are referred to as *special products*. Though they can always be derived using the FOIL method, it is a good idea to remember them, since they will be used frequently.

example 2

Square each of the following binomials.

(a) $(z - 9)^2$

$$(z - 9)^2 = (z)^2 - 2(z)(9) + (9)^2$$
$$= z^2 - 18z + 81$$

(b) $(2x + 5)^2$

$$(2x + 5)^2 = (2x)^2 + 2(2x)(5) + (5)^2$$
$$= 4x^2 + 20x + 25$$

(c) $(3a - 7b)^2$

$$(3a - 7b)^2 = (3a)^2 - 2(3a)(7b) + (7b)^2$$
$$= 9a^2 - 42ab + 49b^2$$

Another special product is obtained whenever we multiply the sum of two numbers by the difference of these numbers. Consider these examples:

$$(x + 4)(x - 4) = x^2 - 4x + 4x - 16 = x^2 - 16$$
$$(2y - 1)(2y + 1) = 4y^2 + 2y - 2y - 1 = 4y^2 - 1$$
$$(8a + 3b)(8a - 3b) = 64a^2 - 24ab + 24ab - 9b^2 = 64a^2 - 9b^2$$

Notice that in each case, the product obtained using the FOIL method has two middle terms that add up to zero. The resulting product simplifies to the square of the first term of the binomial minus the square of the second term of the binomial. This result can be summarized by the following formula which should also be memorized:

Difference of Two Squares

$$(a + b)(a - b) = a^2 - b^2$$

This product of the sum of two terms and the difference of two terms is the square of the first term, minus the square of the second term.

The expression on the right, $a^2 - b^2$ is often called the *difference of two squares.*

example 3

Multiply each of the following.

(a) $(z - 6)(z + 6)$

$$(z - 6)(z + 6) = (z)^2 - (6)^2$$
$$= z^2 - 36$$

(b) $(5 + 2a)(5 - 2a)$

$$(5 + 2a)(5 - 2a) = (5)^2 - (2a)^2$$
$$= 25 - 4a^2$$

(c) $(t^2 - 8)(t^2 + 8)$

$$(t^2 - 8)(t^2 + 8) = (t^2)^2 - (8)^2$$
$$= t^4 - 64$$

A common mistake in algebra is to confuse the sum or difference of two squares $a^2 + b^2$ or $a^2 - b^2$, with the square of a binomial, $(a + b)^2$ or $(a - b)^2$.

$$(a + b)^2 \neq a^2 + b^2 \quad \text{since} \quad (a + b)^2 = a^2 + 2ab + b^2$$
$$(a - b)^2 \neq a^2 - b^2 \quad \text{since} \quad (a - b)^2 = a^2 - 2ab + b^2$$

For example, $(x - 5)^2 \neq x^2 - 25$ since $(x - 5)^2 = x^2 - 10x + 25$. Therefore, whenever you square a binomial, do not forget the middle term. Only the product of the sum and difference of two terms lacks the middle term.

QUICK QUIZ

Perform each of the operations indicated.

1. $(2x + 7)(5x - 3)$
2. $(3y - 1)^2$
3. $(7t - 6)(7t + 6)$

ANSWERS

1. $10x^2 + 29x - 21$
2. $9y^2 - 6y + 1$
3. $49t^2 - 36$

12.3 Exercises

Multiply each of the following using the FOIL method.

1. $(x + 5)(x + 7)$ **2.** $(x + 8)(x + 2)$ **3.** $(x - 9)(x + 3)$ **4.** $(x + 5)(x - 1)$
5. $(x - 4)(x - 2)$ **6.** $(x - 6)(x - 5)$ **7.** $(a + 6)(a - 7)$ **8.** $(y - 3)(y + 5)$
9. $(t - 1)(t - 8)$ **10.** $(w + 4)(w + 9)$ **11.** $(z + 9)(z + 2)$ **12.** $(n - 6)(n - 8)$
13. $(8 - x)(9 + x)$ **14.** $(4 + y)(7 - y)$ **15.** $(3x + 4)(x + 7)$ **16.** $(5x - 1)(x + 3)$
17. $(y - 5)(2y + 9)$ **18.** $(a - 7)(6a + 5)$ **19.** $(6x + 1)(2x - 7)$ **20.** $(4x + 5)(3x - 2)$
21. $(6 + 3y)(y + 7)$ **22.** $(x + 3)(9 - 3x)$ **23.** $(3y - 2)(7y - 1)$ **24.** $(8a + 3)(2a + 5)$
25. $(4 + 7w)(2w + 3)$ **26.** $(6z + 1)(8 - 3z)$ **27.** $(8 - 5x)(4 + 3x)$ **28.** $(4 - 9x)(2 - 7x)$
29. $(9x + y)(x - 4y)$ **30.** $(x - 8y)(3x - y)$ **31.** $(8x + 3y)(2x + 7y)$ **32.** $(5x - 2y)(4x + 3y)$
33. $(a^2 + 5)(a^2 - 3)$ **34.** $(z^2 - 1)(z^2 + 4)$ **35.** $(5x^2 - 2)(x - 8)$ **36.** $(x^2 + 2)(7x + 3)$

For each of the following, use the formulas for special products to perform the operations indicated.

37. $(x + 4)^2$ **38.** $(x + 9)^2$ **39.** $(y - 8)^2$ **40.** $(y - 3)^2$
41. $(z + 6)^2$ **42.** $(z - 6)^2$ **43.** $(3 + y)^2$ **44.** $(4 - y)^2$
45. $(9 - p)^2$ **46.** $(2 + q)^2$ **47.** $(4x + 1)^2$ **48.** $(3x + 2)^2$
49. $(8x - 3)^2$ **50.** $(6x - 1)^2$ **51.** $(2x + 7)^2$ **52.** $(5x - 4)^2$
53. $(5 - 8y)^2$ **54.** $(2 - 5y)^2$ **55.** $(x + 3y)^2$ **56.** $(x - 8y)^2$
57. $(5x - y)^2$ **58.** $(9x + y)^2$ **59.** $(3x + 2y)^2$ **60.** $(5x + 3y)^2$
61. $(9x - 4y)^2$ **62.** $(4x - 7y)^2$ **63.** $(5a + 8b)^2$ **64.** $(8a - 3b)^2$
65. $(x^2 + 3)^2$ **66.** $(x^2 - 7)^2$ **67.** $(2x^2 - 1)^2$ **68.** $(5x^2 + 2)^2$
69. $(x - 8)(x + 8)$ **70.** $(x + 5)(x - 5)$ **71.** $(x + 1)(x - 1)$ **72.** $(x - 9)(x + 9)$
73. $(3 - y)(3 + y)$ **74.** $(7 + z)(7 - z)$ **75.** $(4x - 3)(4x + 3)$ **76.** $(5x - 2)(5x + 2)$
77. $(9x + 2)(9x - 2)$ **78.** $(4x + 7)(4x - 7)$ **79.** $(3 - 5a)(3 + 5a)$ **80.** $(8 + 7z)(8 - 7z)$
81. $(x + 7y)(x - 7y)$ **82.** $(6x - y)(6x + y)$ **83.** $(4x + 3y)(4x - 3y)$ **84.** $(2x - 5y)(2x + 5y)$
85. $(x^2 - 7)(x^2 + 7)$ **86.** $(x^2 + 5)(x^2 - 5)$

12.4 Division of a Polynomial by a Monomial

To divide a polynomial by a monomial, we can multiply the polynomial by the reciprocal of the monomial using the distributive property. Consider the following example:

$$\frac{9x^2 + 12x}{3x} = \frac{1}{3x}(9x^2 + 12x)$$

$$= \frac{1}{3x}(9x^2) + \frac{1}{3x}(12x) \qquad \text{By distributive property.}$$

$$= \frac{9x^2}{3x} + \frac{12x}{3x} \qquad \text{To multiply by } \frac{1}{3x}\text{, divide by } 3x.$$

$$= 3x + 4 \qquad \text{Simplify each monomial.}$$

Notice that this procedure can be simplified by just dividing each term in the polynomial by the monomial.

> To divide a polynomial by a monomial, divide each term in the polynomial by the monomial.

example 1

Divide $\frac{36a + 16b + 8c}{4}$.

solution

$$\frac{36a + 16b + 8c}{4} = \frac{36a}{4} + \frac{16b}{4} + \frac{8c}{4}$$

$$= 9a + 4b + 2c$$

example 2

Divide $\frac{45y^3 + 30y^2 - 60y}{15y}$.

solution

$$\frac{45y^3 + 30y^2 - 60y}{15y} = \frac{45y^3}{15y} + \frac{30y^2}{15y} - \frac{60y}{15y}$$

$$= 3y^2 + 2y - 4$$

example 3

Divide $(3z^5 - 18z^3 + 27) \div 6z^2$.

solution

Rewrite the problem using the fractional format.

$$\frac{3z^5 - 18z^3 + 27}{6z^2} = \frac{3z^5}{6z^2} - \frac{18z^3}{6z^2} + \frac{27}{6z^2}$$

$$= \frac{z^3}{2} - 3z + \frac{9}{2z^2}$$

example 4

Divide $3x^2y - 36xy^3$ by $12xy$.

solution

$$\frac{3x^2y - 36xy^3}{12xy} = \frac{3x^2y}{12xy} - \frac{36xy^3}{12xy}$$

$$= \frac{x}{4} - 3y^2$$

A common algebra mistake is to forget to divide *every* term of the polynomial by the monomial. $\frac{ax + b}{a} \neq x + b$ since $\frac{ax + b}{a} = \frac{ax}{a} + \frac{b}{a} = x + \frac{b}{a}$.

For example, $\frac{4x + 5}{4} \neq x + 5$ since $\frac{4x + 5}{4} = \frac{4x}{4} + \frac{5}{4} = x + \frac{5}{4}$.

QUICK QUIZ

Divide each of the following.

1. $\frac{8x + 10y}{2}$
2. $\frac{21a^3 - 7a^2 + 35a}{14a}$
3. $\frac{64x^3y - 24x^2y^2 - 32xy^3}{8xy^2}$

ANSWERS

1. $4x + 5y$
2. $\frac{3a^2}{2} - \frac{a}{2} + \frac{5}{2}$
3. $\frac{8x^2}{y} - 3x - 4y$

12.4 Exercises

Divide each of the following.

1. $\frac{18x + 12y}{3}$
2. $\frac{25a - 15b}{5}$
3. $\frac{8a + 16b}{-4}$
4. $\frac{30w + 12z}{-6}$
5. $\frac{14p + 28q - 7r}{7}$
6. $\frac{8x - 4y + 6z}{2}$
7. $(-25x - 30y) \div 5$
8. $(18a + 33b) \div (-3)$
9. $(30t^2 - 12t) \div (-6t)$
10. $(56q^2 + 72q) \div 8q$
11. $\frac{32x^2 + 16x}{8x}$
12. $\frac{9x^2 - 27x}{3x}$
13. $\frac{25a^3 - 15a^2 - 10a}{5a}$
14. $\frac{12y^3 + 20y^2 - 32y}{4y}$
15. $\frac{4z^4 - 7z^3 + 8z^2}{-z}$
16. $\frac{6w^3 + 7w^2 - w}{-w}$
17. $\frac{3p^2 - 27p + 18}{9p}$
18. $\frac{-4s^2 + 16s - 24}{8s}$

19. $(42c^3 - 21c^2 - 63c) \div 7c$

20. $(-27p^3 + 9p^2 - 18p) \div 9p$

21. $(9t^5 - 5t^3 + 4t^2) \div t^2$

22. $(-5b^4 + 3b^3 - 6b^2) \div b^2$

23. $(-7y^6 + 3y^4 + 2y^2) \div y^3$

24. $(4z^5 - 2z^3 - 7z) \div z^3$

25. $\dfrac{5x^5 - 10x^3 - 20x}{10x^2}$

26. $\dfrac{8t^8 - 6t^4 + 12t^2}{-4t^2}$

27. $\dfrac{8a^5 - 36a^3 - 24a^2}{-12a^3}$

28. $\dfrac{-36y^4 + 6y^3 - 27y}{9y^3}$

29. $\dfrac{-x^4y^3 + x^3y^2}{-x^2y^2}$

30. $\dfrac{x^3y^4 - x^2y^5}{x^2y^3}$

31. $\dfrac{15a^3b^3 - 20a^2b}{5a^3b^2}$

32. $\dfrac{24x^3y + 16x^2y^4}{4xy^3}$

33. $\dfrac{a^5b^4 - a^3b - a^2b^3}{ab}$

34. $\dfrac{x^4y^2 + x^3y^3 - xy^2}{-xy}$

35. $\dfrac{5x^3y - 3x^2y^3 - 2x^2y^2}{-x^2y}$

36. $\dfrac{-9a^3b^2 + 5a^2b^3 - 7ab^5}{ab^2}$

37. $\dfrac{12x^4y^3 - 15x^3y^2 + 9x^2y^3}{3x^2y^2}$

38. $\dfrac{25x^3y^4 - 10x^2y^3 + 15xy^4}{5xy^2}$

39. $\dfrac{16a^5b^2 - 4a^3b + 12ab^2}{8a^3b^2}$

40. $\dfrac{18a^3b - 6a^2b^4 - 3ab^3}{9a^2b^3}$

12.5 Division of a Polynomial by a Polynomial

The procedure for dividing a polynomial by a polynomial is similar to the technique of long division for dividing whole numbers. Consider the following example:

example 1

Divide $(5x^2 - 7x - 6) \div (x - 2)$.

solution

$$
\begin{array}{r}
5x + 3 \\
x - 2\overline{)5x^2 - 7x - 6} \\
\ominus 5x^2 \oplus 10x \\
\hline
3x - 6 \\
\ominus 3x \oplus 6 \\
\hline
0
\end{array}
$$

1. Write the problem in divisor $\overline{)\text{dividend}}$ notation.
2. Find the first term of the quotient by dividing: $\dfrac{5x^2}{x} = 5x$.
3. Multiply this term by the divisor: $5x(x - 2) = 5x^2 - 10x$.
4. Subtract the product from the dividend: $(5x^2 - 7x) - (5x^2 - 10x) = 3x$. Bring down the -6.
5. Find the next term of the quotient by dividing: $\dfrac{3x}{x} = 3$. Multiply $3(x - 2) = 3x - 6$. Subtract $(3x - 6) - (3x - 6) = 0$.

The quotient is $5x + 3$ with a remainder of 0.

To divide a polynomial we can follow this stepwise technique:

Procedure for Dividing a Polynomial by a Polynomial

1. Write the problem in divisor $\overline{)\text{dividend}}$ notation arranging each polynomial in descending powers of the variable. Include missing powers of the variable in the dividend using zero as their coefficient.
2. Find the first (or next) term of the quotient by dividing the first term of the dividend by the first term of the divisor. Write it above the dividend.
3. Multiply this term by the entire divisor and write the product under the dividend lining up like terms.
4. Subtract this product from the dividend bringing down the next term to obtain the new dividend.
5. Repeat steps 2–4 until the degree of the new dividend (or remainder) is less than the degree of the divisor. Express your answer in the form:

$$\text{quotient} + \frac{\text{remainder}}{\text{divisor}}$$

To Check the Answer to a Division Problem

Multiply the quotient by the divisor and add the remainder. The result should equal the dividend.

$$\text{quotient} \times \text{divisor} + \text{remainder} = \text{dividend}$$

example 2

Divide $(3x^3 + 7x^2 + 4x - 5) \div (x + 3)$.

$$\begin{array}{rl}
 & 3x^2 - 2x + 10 \\
x + 3 & \overline{)3x^3 + 7x^2 + 4x - 5} \\
 & \ominus 3x^3 \overset{\ominus}{+} 9x^2 \\
\hline
 & -2x^2 + 4x \\
 & \overset{\oplus}{-} 2x^2 \overset{\oplus}{-} 6x \\
\hline
 & 10x - 5 \\
 & \ominus 10x \overset{\ominus}{+} 30 \\
\hline
\text{remainder} & -35
\end{array}$$

1. Write the problem in divisor $\overline{)\text{dividend}}$ notation.
2. Divide $\frac{3x^3}{x} = 3x^2$.
3. Multiply $3x^2(x + 3) = 3x^3 + 9x^2$.
4. Subtract $(3x^3 + 7x^2) - (3x^3 + 9x^2) = -2x^2$. Bring down $4x$.
5. Divide $\frac{-2x^2}{x} = -2x$.
   Multiply $-2x(x + 3) = -2x^2 - 6x$.
   Subtract $(-2x^2 + 4x) - (-2x^2 - 6x) = 10x$.
   Bring down -5.
   Divide $\frac{10x}{10} = 10$.
   Multiply $10(x + 3) = 10x + 30$.
   Subtract $(10x - 5) - (10x + 30) = -35$.

The quotient is $3x^2 - 2x + 10$ and the remainder is -35.

The answer is:

$$3x^2 - 2x + 10 - \frac{35}{x + 3}$$

CHECK

$$
\begin{array}{rrrrl}
3x^2 & -\,2x & +\,10 & & \text{quotient} \\
 & x & +\ \ 3 & & \text{divisor} \\
\hline
3x^3 - 2x^2 & +\,10x & & & \\
 & 9x^2 - & 6x & +\,30 & \\
\hline
3x^3 + 7x^2 + & 4x & & +\,30 & \text{remainder} \\
 & & & -\,35 & \text{dividend } \surd \\
\hline
3x^3 + 7x^2 + & 4x & & -\ \ 5 &
\end{array}
$$

example 3

Divide $\dfrac{4x^3 - 8x^2 + x + 3}{2x + 1}$.

solution

$$
\begin{array}{r}
2x^2 - 5x + 3 \\
2x + 1\,\overline{)\,4x^3 - 8x^2 + x + 3} \\
\ominus 4x^3 \overset{\ominus}{+} 2x^2 \qquad\qquad \\
\hline
-\,10x^2 + x \qquad \\
\overset{\oplus}{-}\,10x^2 \overset{\oplus}{-} 5x \qquad \\
\hline
6x + 3 \\
\ominus 6x \overset{\ominus}{+} 3 \\
\hline
0
\end{array}
$$

The quotient is $2x^2 - 5x + 3$.

CHECK

$$
\begin{array}{rrrrl}
2x^2 - & 5x & +\,3 & & \text{quotient} \\
 & 2x & +\,1 & & \text{divisor} \\
\hline
4x^3 - 10x^2 & +\,6x & & & \\
 & 2x^2 - & 5x & +\,3 & \\
\hline
4x^3 - \ \ 8x^2 & + & x & +\,3 & \text{dividend } \surd
\end{array}
$$

example 4

Divide $\dfrac{x^3 + 8}{x - 4}$.

solution

$$
\begin{array}{r}
x^2 + 4x + 16 \\
x - 4\,\overline{)\,x^3 + 0x^2 + 0x + 8} \\
\ominus x^3 \overset{\oplus}{-} 4x^2 \qquad\qquad \\
\hline
4x^2 + 0x \qquad \\
\ominus 4x^2 \overset{\oplus}{-} 16x \qquad \\
\hline
16x + 8 \\
\ominus 16x \overset{\oplus}{-} 64 \\
\hline
72
\end{array}
$$

Include missing powers of x using zero as their coefficient.

The answer is:

$$x^2 + 4x + 16 + \frac{72}{x - 4}$$

CHECK

$$\begin{array}{rl}
x^2 + 4x \;\; + 16 & \text{quotient} \\
x \;\; - \;\; 4 & \text{divisor} \\
\hline
x^3 + 4x^2 + 16x & \\
- 4x^2 - 16x - 64 & \\
\hline
x^3 + 0x^2 + \;\; 0x - 64 & \\
+ 72 & \text{remainder} \\
\hline
x^3 + 0x^2 + \;\; 0x + \;\; 8 & \text{dividend} \surd
\end{array}$$

example 5

Divide $\dfrac{x^3 + x^2y - 21xy^2 + 4y^3}{x - 4y}$.

solution

$$\begin{array}{r}
x^2 + 5xy - y^2 \\
x - 4y \overline{\big)\, x^3 + x^2y - 21xy^2 + 4y^3} \\
\ominus x^3 \overset{\oplus}{-} 4x^2y \\
\hline
5x^2y - 21xy^2 \\
\ominus 5x^2y \overset{\oplus}{-} 20xy^2 \\
\hline
- \;\; xy^2 + 4y^3 \\
\overset{\oplus}{-} \;\; xy^2 \overset{\ominus}{+} 4y^3 \\
\hline
0
\end{array}$$

The quotient is $x^2 + 5xy - y^2$.

CHECK

$$\begin{array}{rl}
x^2 + 5xy - \;\; y^2 & \text{quotient} \\
x \;\; - \;\; 4y^2 & \text{divisor} \\
\hline
x^3 + 5x^2y - \;\; xy^2 & \\
- 4x^2y - 20xy^2 + 4y^3 & \\
\hline
x^3 + \;\; x^2y - 21xy^2 + 4y^3 & \text{dividend} \surd
\end{array}$$

QUICK QUIZ

Divide each of the following:

1. $(9x^2 - 38x - 35) \div (x - 5)$
2. $\dfrac{4a^3 + 5a^2 + 10a - 1}{4a - 3}$
3. $\dfrac{x^3 + 27}{x + 3}$

ANSWERS

1. $9x + 7$
2. $a^2 + 2a + 4 + \dfrac{11}{4a - 3}$
3. $x^2 - 3x + 9$

12.5 Exercises

Divide each of the following and check your answers.

1. $(8x^2 - 19x - 15) \div (x - 3)$
2. $(6x^2 + 7x - 10) \div (x + 2)$
3. $(9x^2 + 58x + 24) \div (x + 6)$
4. $(3x^2 - 5x - 28) \div (x - 4)$
5. $(18y^2 - 23y - 6) \div (2y - 3)$
6. $(32z^2 + 12z - 5) \div (8z + 5)$
7. $\dfrac{4t^3 - 33t^2 + 37t - 14}{t - 7}$
8. $\dfrac{3w^3 + 4w^2 - 35w - 12}{w + 4}$
9. $\dfrac{7x^3 + 40x^2 + 29x + 17}{x + 5}$

10. $\dfrac{8x^3 - 51x^2 + 13x + 25}{x - 6}$

11. $\dfrac{15a^3 - 24a^2 - 22a - 4}{5a + 2}$

12. $\dfrac{8b^3 - 2b^2 + 11b - 56}{4b - 7}$

13. $\dfrac{9x^3 + 35x^2 - 31x + 7}{9x - 1}$

14. $\dfrac{7x^3 + 65x^2 - 17x - 11}{7x + 2}$

15. $\dfrac{2x^3 - 5x^2 + 9x + 6}{6x + 3}$

16. $\dfrac{4x^3 + 39x^2 - 34x + 6}{8x - 2}$

17. $\dfrac{x^3 + 64}{x + 4}$

18. $\dfrac{x^3 - 27}{x - 3}$

19. $\dfrac{x^2 + 36}{x + 6}$

20. $\dfrac{x^2 + 81}{x + 9}$

21. $\dfrac{3x^3 - 7x + 5}{x - 2}$

22. $\dfrac{2x^3 - 13x + 7}{x + 3}$

23. $\dfrac{z^3 - 8}{z + 2}$

24. $\dfrac{y^3 + 1}{y - 1}$

25. $\dfrac{5x^3 + 14x^2 - 9x - 18}{x^2 + 4x + 3}$

26. $\dfrac{7x^3 - 12x^2 - 39x - 10}{x^2 - 2x - 5}$

27. $\dfrac{a^5 + 9a^2 - 5}{a^2 - 3}$

28. $\dfrac{t^5 - 4t^3 + 7}{t^2 + 2}$

29. $\dfrac{3x^3 - 20x^2y - xy^2 + 2y^3}{3x + y}$

30. $\dfrac{5x^3 - 2x^2y - 15xy^2 - 2y^3}{x - 2y}$

Summary and Review

Key Terms

[12.1] A **polynomial** is an algebraic expression that consists of the sum of one or more terms such that no variables appear in the denominator and all exponents of variables are whole numbers.

A **monomial** is a polynomial that contains one term.

A **binomial** is a polynomial that contains two terms.

A **trinomial** is a polynomial that contains three terms.

The **degree** of a polynomial that contains only one variable is the highest power of that variable.

Calculations

[12.1] **To add or subtract polynomials**, rewrite each polynomial in descending powers of the variable, and combine like terms by adding or subtracting their numerical coefficients.

To subtract one polynomial from another, add the opposite of each term in the polynomial being subtracted.

[12.2] **To multiply a polynomial by a monomial**, multiply each term in the polynomial by the monomial and add the results.

To multiply two polynomials, multiply each term of the first polynomial by each term of the second polynomial, and add the results.

[12.3] **To multiply two binomials using the FOIL method**, add the products of the first terms, the outer terms, the inner terms, and the last terms.

[12.4] **To divide a polynomial by a monomial**, divide each term in the polynomial by the monomial.

[12.5] To divide a polynomial by a polynomial, first divide the first term of the dividend by the first term of the divisor. Then multiply the result by the entire divisor. Finally, subtract that product from the dividend bringing down the next term to obtain the new dividend. Repeat these steps until the degree of the remainder is less than that of the divisor.

[12.6] To check the answer to a division problem, multiply the quotient by the divisor, and add the remainder. The result should equal the dividend.

quotient × divisor + remainder = dividend

Special Products

[12.3] $(a + b)^2 = a^2 + 2ab + b^2$ Square of a binomial

$(a - b)^2 = a^2 - 2ab + b^2$ Square of a binomial

$(a + b)(a - b) = a^2 - b^2$ Difference of two squares

Common Mistakes

[12.3] $(a + b)^2 \neq a^2 + b^2$

$(a - b)^2 \neq a^2 - b^2$

[12.4] $\dfrac{ax + b}{a} \neq x + b$

Chapter 12 Review Exercises

Indicate which of the following statements are true and which are false. For those that are false, change the italic expression to make the statement true.

1. The square of a binomial is *equal* to the difference of two squares.
2. To divide a polynomial by a monomial, divide *only the first* term of the polynomial by that monomial.
3. If a division problem has been solved correctly, the quotient times the divisor plus the remainder equals the *dividend.*
4. To multiply a monomial by a polynomial, use the *distributive* property.

[12.1-12.5] Perform each of the operations indicated.

5. $(5x + 3y) + (-2x + 7y)$
6. $(4a - 3b) - (-2a - 5b)$
7. $(9a - 2b) - (5a - 3b)$
8. $(2x + y) - (8x + 3y)$
9. $(5x^2 - 6x + 2) + (3x^2 - 4x + 9)$
10. $(y^2 - 6y - 4) + (3y^2 - 8y + 1)$
11. $(z^2 - 8z - 5) - (4z^2 + 7z + 2)$
12. $(3w^2 - 6w + 5) - (5w^2 + 2w - 7)$
13. $7x(4x - 8)$
14. $-5y(2y + 5)$
15. $-3ab(5a^2 - 2b^2)$
16. $4xy(-2x^3 + 5y)$
17. $6x^2(2x^3 - 5x^2y + 7xy^2)$
18. $-2pq^3(4p^2q + 7pq^2 - 3q^3)$
19. $-8x(5x + 2) + 2x(x - 5)$
20. $3y(4 - 7y) + 6y(5y + 2)$
21. $(-3n)(2n - 7)(4n)$
22. $(5p)(8 - 3p)(-9p)$
23. $-5x - [4x + (x - 6)] - (2x + 7)$
24. $-(8 - 3x) - 7y + [4 - (3x - 1)]$
25. $(a - 3)(a + 5)$
26. $(b - 9)(b - 2)$
27. $(z + 4)(z + 8)$
28. $(w + 3)(w - 6)$
29. $(5x - 7)(2x - 5)$
30. $(8x + 1)(5x + 2)$
31. $(3 + 4y)(9 - 7y)$
32. $(5 - 3z)(6 + 7z)$
33. $(5a + b)(2a + 9b)$
34. $(2x + 3y)(x - 4y)$
35. $(t + 7)^2$
36. $(s - 4)^2$
37. $(3n - 8)^2$
38. $(5p + 2)^2$
39. $(7x + 4y)^2$
40. $(3a - 8b)^2$
41. $(a + 7)(a - 7)$
42. $(t - 3)(t + 3)$
43. $(2x - 5)(2x + 5)$
44. $(5y + 9)(5y - 9)$
45. $(4a + 5b)(4a - 5b)$
46. $(p - 8q)(p + 8q)$

47. $\dfrac{8x + 12y}{4}$

48. $\dfrac{21a - 9b}{3}$

49. $\dfrac{9x^2 - 15x}{3x}$

50. $\dfrac{25y^2 + 10y}{-5y}$

51. $\dfrac{36a^3 + 12a^2 - 42a}{-6a}$

52. $\dfrac{24b^3 + 64b^2 - 16b}{8b}$

53. $\dfrac{7t^5 - 5t^3 + 3t^2}{t^2}$

54. $\dfrac{9z^6 + 4z^5 + 2z^3}{z^3}$

55. $\dfrac{5x^3y^2 + 6x^2y + 9xy^3}{xy^2}$

56. $\dfrac{2x^3y - 7x^2y^3 - 9xy}{x^2y}$

57. $\dfrac{2x^2 + 3x - 25}{x - 3}$

58. $\dfrac{5x^2 + 33x - 10}{x + 7}$

59. $\dfrac{72x^2 + 29x - 21}{8x - 3}$

60. $\dfrac{30x^2 + 13x + 53}{5x + 8}$

61. $\dfrac{x^3 + 10x^2 + 9x - 18}{x + 2}$

62. $\dfrac{x^3 + 2x^2 - 8x - 35}{x - 5}$

63. $\dfrac{x^3 - 1}{x - 1}$

64. $\dfrac{x^3 + 8}{x + 2}$

13 FACTORING

Recall from Chapter 1 that *factors* are numbers you multiply together to produce a product. The process of *factoring* involves rewriting an expression as a product of its factors.

13.1 Monomial Factors

In Section 12.2, we learned how to multiply a monomial by a polynomial using the distributive property. For example,

$$5x(x - 3) = 5x^2 - 15x$$

To factor the expression $5x^2 - 15x$, we use the distributive property in reverse and rewrite it as the product of two factors: $5x$ and $x - 3$.

$$5x^2 - 15x = 5x(x - 3)$$

Notice that $5x$ is the largest monomial that divides both $5x^2$ and $15x$. Therefore, $5x$ is the **greatest common factor** of the polynomial $5x^2 - 15x$.

To Find the Greatest Common Factor (GCF) of a Polynomial

1. Find the largest number that divides each of the numerical coefficients in the polynomial.
2. For each variable appearing in every term of the polynomial, indicate the lowest power that occurs.
3. Multiply the results found in steps 1 and 2.

example 1

Find the GCF of each of the following polynomials.

(a) $36y^5 - 12y^4 + 18y^3$

The largest number that divides 36, 12, and 18 is 6.

The lowest power of y is y^3.

The GCF is $6y^3$.

(b) $8a^3b^2 + 28a^2b^5 + 20ab^3$

The largest number that divides 8, 28, and 20 is 4.

The lowest power of a is a.

The lowest power of b is b^2.

The GCF is $4ab^2$.

(c) $x^5y^3z^2 - 3x^3y^2$

The largest number that divides 1 and 3 is 1.

The lowest power of x is x^3.

The lowest power of y is y^2.

The GCF is x^3y^2.

Notice that since the variable z does not appear in the second term of the polynomial, it cannot appear in the GCF. ●

Sometimes polynomials have a common factor that contains more than one term. In these cases, the GCF may be a polynomial, or the product of a polynomial and a monomial.

example 2

Find the GCF of each of the following polynomials.

(a) $3x(a + b) + 2y(a + b)$

The GCF is the binomial $(a + b)$.

(b) $4x^3(y - 3) + 6x^2(y - 3)$

The common monomial factor is $2x^2$.

The common binomial factor is $(y - 3)$.

The GCF is $2x^2(y - 3)$. ●

Once we have found the greatest common factor of a polynomial, we complete the factoring process by dividing each term of the polynomial by the GCF, and writing the result in parentheses to the right of the GCF. Consider the following example.

example 3

Factor $16y^3 - 8y^2$.

solution

The GCF is $8y^2$. — Find the GCF.

$$\frac{16y^3 - 8y^2}{8y^2} = \frac{16y^3}{8y^2} - \frac{8y^2}{8y^2} = 2y - 1$$

Divide each term of the polynomial by the GCF.

$$16y^3 - 8y^2 = 8y^2(2y - 1)$$

Factored form is the product of $8y^2$ and $2y - 1$. ●

As we become more familiar with this procedure, we should be able to divide the polynomial by the GCF mentally.

We can check our result by multiplying the expression on the right using the distributive property. The result should be the same as the original problem.

$$8y^2(2y - 1) = 16y^3 - 8y^2$$

Whenever we factor, we should always check our answers mentally using this technique.

We now summarize the procedure for factoring out the greatest common factor of a polynomial.

To Factor Out the GCF of a Polynomial

1. Find the GCF.
2. Divide each term in the polynomial by the GCF to obtain the other factor of the polynomial.
3. Express the polynomial as the product of the factors found in steps 1 and 2.

example 4

Factor each of the following expressions.

(a) $18a^4 - 54a^3 + 30a^2$

The GCF is $6a^2$. — Find the GCF.

$$\frac{18a^4 - 54a^3 + 30a^2}{6a^2} = 3a^2 - 9a + 5$$

Divide each term of the polynomial by the GCF.

$18a^4 - 54a^3 + 30a^2$
$= 6a^2(3a^2 - 9a + 5)$

Factored form is the product of $6a^2$ and $3a^2 - 9a + 5$.

(b) $-4x^5 - 12x^3$

The GCF is $-4x^3$. — Find the GCF. Since -1 is a factor of each term, the GCF is $-4x^3$.

$$\frac{-4x^5 - 12x^3}{-4x^3} = x^2 + 3$$

Divide each term of the polynomial by the GCF.

$-4x^5 - 12x^3 = -4x^3(x^2 + 3)$

Factored form is the product of $-4x^3$ and $x^2 + 3$.

(c) $42x^3y^2 - 35x^2y^5$

The GCF is $7x^2y^2$. — Find the GCF.

$$\frac{42x^3y^2 - 35x^2y^5}{7x^2y^2} = 6x - 5y^3$$

Divide each term of the polynomial by the GCF.

$42x^3y^2 - 35x^2y^5 = 7x^2y^2(6x - 5y^3)$

Factored form is the product of $7x^2y^2$ and $6x - 5y^3$.

(d) $56a^5b^2c^3 + 24a^3b^3c^4 - 32a^2bc^5$

The GCF is $8a^2bc^3$. — Find the GCF.

$$\frac{56a^5b^2c^3 + 24a^3b^3c^4 - 32a^2bc^5}{8a^2bc^3}$$
$$= 7a^3b + 3ab^2c - 4c^2$$

Divide each term of the polynomial by the GCF.

$56a^5b^2c^3 + 24a^3b^3c^4 - 32a^2bc^5$
$= 8a^2bc^3(7a^3b + 3ab^2c - 4c^2)$

Factored form is the product of $8a^2bc^3$ and $7a^3b + 3ab^2c - 4c^2$. ●

Let us now look at some examples of factoring polynomials that have common factors containing more than one term.

example 5

Factor each of the following expressions.

(a) $5a(x + y) + 3b(x + y)$

The GCF is $(x + y)$ — Find the GCF.

$$\frac{5a(x + y) + 3b(x + y)}{(x + y)}$$

Divide each term of the polynomial by the GCF.

$$= \frac{5a(x + y)}{(x + y)} + \frac{3b(x + y)}{(x + y)}$$

Factored form is the product of $x + y$ and $5a + 3b$.

$$= 5a + 3b$$

$$5a(x + y) + 3b(x + y) = (x + y)(5a + 3b)$$

(b) $3t^5(t + 5) + 9t^3(t + 5)$

The GCF is $3t^3(t + 5)$ — Find the GCF.

$$\frac{3t^5(t + 5) + 9t^3(t + 5)}{3t^3(t + 5)} = t^2 + 3$$

Divide each term of the polynomial by the GCF.

$$3t^5(t + 5) + 9t^3(t + 5) = 3t^3(t + 5)(t^2 + 3)$$

Factored form is the product of $3t^3(t + 5)$ and $t^3 + 3$.

We will now use a method called *factoring by grouping* to factor certain polynomial expressions containing four terms.

To Factor a Polynomial by Grouping

1. **Rearrange the polynomial so that the first two terms have a common factor and the last two terms have a common factor.**
2. **Factor out the greatest common monomial factor from the first two terms and from the last two terms.**
3. **Factor out the greatest common binomial factor from the two remaining terms.**

Note that not all polynomials containing four terms can be factored by grouping.

example 6

Use factoring by grouping to factor each of the following expressions.

(a) $x^2 + 3x + 8x + 24$

$$x^2 + 3x + 8x + 24 = x(x + 3) + 8(x + 3)$$

Notice that there is no common factor to all four terms. However, the first two terms have a common factor of x and the last two terms have a common factor of 8.

$$= (x + 3)(x + 8)$$

Factor out $(x + 3)$ from the two remaining terms.

(b) $x^2 - 30 + 6x - 5x$

$x^2 - 30 + 6x - 5x = x^2 + 6x - 5x - 30$ — Rearrange the polynomial so that the first two terms have a common factor of x and the last two terms have a common factor of -5.

$= x(x + 6) - 5(x + 6)$ — Factor out x from the first two terms and -5 from the last two terms.

$= (x + 6)(x - 5)$ — Factor out $(x + 6)$ from each term.

(c) $3x^2 + 18x + 4x + 24$

$3x^2 + 18x + 4x + 36 = 3x(x + 6) + 4(x + 6)$ — Factor out $3x$ from the first two terms and 4 from the last two terms.

$= (x + 6)(3x + 4)$ — Factor out $(x + 6)$ from each term.

(d) $5x^3 + 21 - 15x^2 - 7x$

$5x^3 + 21 - 15x^2 - 7x = 5x^3 - 15x^2 - 7x + 21$ — Rearrange the polynomial.

$= 5x^2(x - 3) - 7(x - 3)$ — Factor out $5x^2$ from the first two terms and -7 from the last two terms.

$= (x - 3)(5x^2 - 7)$ — Factor out $(x - 3)$ from each term.

QUICK QUIZ

Factor each of the following expressions.

1. $15x + 40y$
2. $16w^4 - 36w^3 + 24w^2$
3. $8a^6b + 2a^5b^4 - 6a^3b^3$
4. $2x(a + b) - y(a + b)$
5. $x^2 + 32 - 4x - 8x$

ANSWERS

1. $5(3x + 8y)$
2. $4w^2(4w^2 - 9w + 6)$
3. $2a^3b(4a^3 + a^2b^3 - 3b^2)$
4. $(a + b)(2x - y)$
5. $(x - 8)(x - 4)$

13.1 Exercises

Find the GCF of each of the following polynomials.

1. $5a + 10b$
2. $21x - 7y$
3. $9x^2 - 3x$
4. $4y^3 + 20y$
5. $12z^4 + 8z^3 - 28z^2$
6. $15w^3 - 6w^2 + 18w$
7. $a^3b^2 + a^2b^3$
8. $x^4y^2 - x^2y$
9. $42x^4y^2 - 18x^3y^3 + 6xy^5$
10. $32a^5b^3 + 8a^4b^2 - 40a^3b^3$
11. $3x(a + b) + 5(a + b)$
12. $2(x - y) + 7x(x - y)$
13. $9p^3(p + 2) - 6p^2(p + 2)$
14. $8x^5(a - b) - 4x^3(a - b)$

Factor each of the following expressions.

15. $9x + 3y$
16. $12a - 16b$
17. $8p - 24$
18. $25t + 5$
19. $9 + 27t$
20. $35 - 5p$
21. $ax + bx$
22. $sx + sy$
23. $8x^2 + 8y^2$
24. $9p^2 + 9q^2$
25. $15a^2 - 5a$
26. $6b^3 + 36b$
27. $n^5 + n^3$
28. $w^2 + w$
29. $-3x^2 - 6x$

30. $-8y^3 - 4y$
31. $14x^3 + 21x^2$
32. $42y^5 + 36y^3$
33. $40p^4 - 8p^3$
34. $9t^3 - 27t$
35. $15y^3 + 30y^2 - 20y$
36. $28p^4 - 14p^3 + 42p^2$
37. $72t^8 + 36t^5 + 54t$
38. $64w^7 - 40w^4 - 56w^2$
39. $x^5y^2 + 2x^3y$
40. $3x^3y^4 - x^3y^2$
41. $3x^3y^4 - 6x^2y^3$
42. $12a^5b^2 + 8a^3b$
43. $63a^2b^3 + 27ab^5$
44. $24x^5y^3 - 54x^3y^2$
45. $-13x^3y^5 - 26x^2y$
46. $-9a^5b^2 - 18a^4b^3$
47. $4x^5y^3 + x^4y^2 + 3x^3y^2$
48. $ab^4 - 7a^3b^2 - 3ab$
49. $24a^2b^2 - 48ab + 18ab^3$
50. $21x^3y^2 + 14x^2y^3 - 56xy^4$
51. $63x^2y^5 - 27x^3y^2 - 18x^2y$
52. $56a^3b^4 + 48ab^3 + 64ab^2$
53. $42x^3y^2z^2 + 18x^2yz^4 - 36x^2y^3z^5$
54. $10a^5b^2c - 25a^3b^2 + 15a^3b^4c^2$
55. $3x(a - b) + y(a - b)$
56. $a(x + y) + 2b(x + y)$
57. $5x(x + y) + 10(x + y)$
58. $12a(a - b) + 4(a - b)$
59. $9a^2(a - 5) + 6a(a - 5)$
60. $5y^3(y + 2) - 15y^2(y + 2)$
61. $x^2 + 2x + 7x + 14$
62. $x^2 + 4x + 3x + 12$
63. $x^2 - 8x + 5x - 40$
64. $x^2 + 9x - 4x - 36$
65. $x^2 + 28 - 7x - 4x$
66. $x^2 + 18 - 6x - 3x$
67. $3x^2 + 15x + 4x + 20$
68. $5x^2 + 12 + 30x + 2x$
69. $4x^2 + 21 - 28x - 3x$
70. $2x^2 + 4x - 5x - 10$
71. $3x^3 + 12x^2 + 8x + 32$
72. $5x^3 - 30x^2 - 2x + 12$

Indicate which of the following statements are correct and which are incorrect. For those that are incorrect, change the underlined statement to make the statement correct.

73. $5x - 15y = \underline{5(x - 3y)}$
74. $12a + 3b = \underline{3(4a + 3b)}$
75. $21x^2 - 7x = \underline{7x(3x - 7)}$
76. $t^5 + 3t^3 = \underline{3t^3(t^2 + 1)}$
77. $25y^3 + 30y^2 + 5y = \underline{5y(5y^2 + 6y)}$
78. $36x^3 - 18x^2 - 12x = \underline{6x(6x^2 - 3x - 2)}$
79. $-3z^2 - 6z = \underline{-3z(z^2 + 2)}$
80. $-5p^3 + 20p^2 = \underline{-5p^2(p + 4)}$
81. $4x^3y^2 - 8x^2y^5 + 12xy = \underline{4xy^2(x^2y - 2xy^3 + 3xy)}$
82. $18a^2b^2 + 6a^3b^4 - 24a^2b^5 = \underline{6a^2b^2(3 + ab^2 - 4b^3)}$
83. $7a(x + y) + 14(x + y) = \underline{7(x + y)(a + 2)}$
84. $3x^2(a + b) - x(a + b) = \underline{3x(a + b)(x - 1)}$

13.2 Factoring the Special Products

In Section 12.3 we introduced three formulas for special products: the difference of two squares, the square of a binomial sum, and the square of a binomial difference. We can factor expressions that are in the form of these special products using these formulas in reverse.

Factoring an expression that is in the form of the difference of two squares, $a^2 - b^2$, is the reverse of multiplying the sum of two numbers $(a + b)$ by the difference of these two numbers $(a - b)$. This can be represented by the formula:

Difference of Two Squares

$$a^2 - b^2 = (a + b)(a - b)$$

For example,

$$x^2 - 49 = (x + 7)(x - 7)$$
$$100 - y^2 = (10 + y)(10 - y)$$
$$4a^2 - 1 = (2a + 1)(2a - 1)$$

Notice that when a binomial is in the form of the difference of two squares, both the first and second terms are perfect squares, and there is a minus sign between them. The first term of each factor is the square root of the first term in the binomial, and the second term of each factor is the square root of the second term in the binomial.

example 1

Factor each of the following expressions if possible.

(a) $z^2 - 144$

$z^2 - 144 = (z)^2 - (12)^2 = (z + 12)(z - 12)$

(b) $t^2 - 8$

This cannot be factored so that the terms of each factor are integers since 8 is not a perfect square.

(c) $9x^2 - 25$

$9x^2 - 25 = (3x)^2 - (5)^2 = (3x + 5)(3x - 5)$

(d) $p^2 + 49$

This cannot be factored since it is the *sum* of two perfect squares. ●

Trinomial expressions that are equal to the square of a binomial can be factored using one of these two formulas:

Perfect Square Trinomials

$$a^2 + 2ab + b^2 = (a + b)^2$$
$$a^2 - 2ab + b^2 = (a - b)^2$$

For example,

$$x^2 + 6x + 9 = (x + 3)^2$$
$$y^2 - 10y + 25 = (y - 5)^2$$

Notice that trinomials that can be factored using these special product formulas have these characteristics: the first and last terms are perfect squares, and the middle term is twice the product of the square roots of the first and last terms. If the middle term is positive, the trinomial factors into the square of the sum of the square roots of the first and last terms. If the middle term is negative, the trinomial factors into the square of the difference of the square roots of the first and last terms.

Whenever we factor an expression using the special product formulas, we should check our answer mentally by multiplying the factors using the FOIL method.

example 2

Factor each of the following expressions.

(a) $t^2 + 16t + 64$

$$t^2 + 16t + 64 = (t)^2 + 2(8)(t) + (8)^2$$
$$= (t + 8)^2$$

(b) $w^2 - 20w + 100$

$$w^2 - 20w + 100 = (w)^2 - 2(10)(w) + (10)^2$$
$$= (w - 10)^2$$

(c) $4x^2 - 20x + 25$

$$4x^2 - 20x + 25 = (2x)^2 - 2(2x)(5) + (5)^2$$
$$= (2x - 5)^2$$

(d) $9x^2 + 6xy + y^2$

$$9x^2 + 6xy + y^2 = (3x)^2 + 2(3x)(y) + (y)^2$$
$$= (3x + y)^2$$

To factor an expression in the form of the difference of the difference of two cubes we can use this formula.

Difference of Two Cubes

$$a^3 - b^3 = (a - b)(a^2 + ab + b^2)$$

We can verify the accuracy of this formula by multiplying the right hand side using the distributive property.

$$(a - b)(a^2 + ab + b^2) = a(a^2 + ab + b^2) - b(a^2 + ab + b^2)$$
$$= a^3 + a^2b + ab^2 - a^2b - ab^2 - b^3$$
$$= a^3 - b^3$$

Whenever we factor the difference of two cubes, we can check our answer by multiplying the factors using the distributive property.

example 3 Factor each of the following expressions.

(a) $x^3 - 27$

$$x^3 - 27 = (x)^3 - (3)^3$$
$$= (x - 3)(x^2 + 3x + 9)$$

(b) $8a^3 - 125$

$$8a^3 - 125 = (2a)^3 - (5)^3$$
$$= (2a - 5)(4a^2 + 10a + 25)$$

(c) $27x^3 - 64y^3$

$$27x^3 - 64y^3 = (3x)^3 - (4y)^3$$
$$= (3x - 4y)(9x^2 + 12xy + 16y^2)$$

To factor an expression in the form of the sum of two cubes we can use the following formula.

Sum of Two Cubes

$$a^3 + b^3 = (a + b)(a^2 - ab + b^2)$$

We can verify the accuracy of this formula by multiplying the right hand side using the distributive property.

$$\begin{aligned}(a+b)(a^2-ab+b^2) &= a(a^2-ab+b^2)+b(a^2-ab+b^2)\\ &= a^3-a^2b+ab^2+a^2b-ab^2+b^3\\ &= a^3+b^3\end{aligned}$$

Whenever we factor an expression in the form of the sum of two cubes, we can check our answer by multiplying the factors using the distributive property.

example 4

Factor each of the following expressions.

(a) t^3+64

$$\begin{aligned}t^3+64 &= (t)^3+(4)^3\\ &= (t+4)(t^2-4t+16)\end{aligned}$$

(b) $1000x^3+y^3$

$$\begin{aligned}1000x^3+y^3 &= (10x)^3+(y)^3\\ &= (10x+y)(100x^2-10xy+y^2)\end{aligned}$$

(c) $8a^3+27b^3$

$$\begin{aligned}8a^3+27b^3 &= (2a)^3+(3b)^3\\ &= (2a+3b)(4a^2-6ab+9b^2)\end{aligned}$$

Sometimes when we factor an expression, we may be able to factor one of the factors we obtained. An expression is *factored completely* when none of its factors can be factored further.

> To factor an expression completely, first factor out the greatest common monomial factor. Then factor the polynomial factor.

example 5

Factor each of the following expressions completely.

(a) $3x^2-12$

$$3x^2-12 = 3(x^2-4)$$ First factor out 3 from both terms.

$$= 3(x+2)(x-2)$$ Then factor the difference between two squares.

(b) $4y^2-24y+36$

$$4y^2-24y+36 = 4(y^2-6y+9)$$ First factor out 4.

$$= 4(y-3)^2$$ Then factor the trinomial.

(c) $2w^3+16w^2+32w$

$$2w^3+16w^2+32w$$

$$= 2w(w^2+8w+16)$$ First factor out 2*w*.

$$= 2w(w+4)^2$$ Then factor the trinomial.

(d) $a^4 - b^4$

$$a^4 - b^4 = (a^2)^2 - (b^2)^2$$

$$= (a^2 + b^2)(a^2 - b^2)$$ Factor the difference between two squares.

$$= (a^2 + b^2)(a + b)(a - b)$$ Factor the difference between two squares again.

QUICK QUIZ

Factor the following expressions.

1. $a^2 - 64$
2. $t^2 + 12t + 36$
3. $2x^2 - 28x + 98$
4. $w^3 - 216$
5. $125 + y^3$

ANSWERS

1. $(a + 8)(a - 8)$
2. $(t + 6)^2$
3. $2(x - 7)^2$
4. $(w - 6)(w^2 + 6w + 36)$
5. $(5 + y)(25 - 5y + y^2)$

13.2 Exercises

Factor each of the following expressions using the special product formulas.

1. $x^2 - 9$ **2.** $x^2 - 16$ **3.** $a^2 - 100$ **4.** $b^2 - 36$
5. $t^2 - 1$ **6.** $p^2 - 121$ **7.** $49 - n^2$ **8.** $25 - s^2$
9. $64 - p^2$ **10.** $81 - m^2$ **11.** $4x^2 - 81$ **12.** $9x^2 - 100$
13. $25z^2 - 144$ **14.** $16w^2 - 169$ **15.** $25 - 36y^2$ **16.** $1 - 49z^2$
17. $16x^2 - 121y^2$ **18.** $4a^2 - 9b^2$ **19.** $x^2 + 14x + 49$ **20.** $x^2 + 10x + 25$
21. $y^2 - 12y + 36$ **22.** $y^2 - 6y + 9$ **23.** $a^2 + 4a + 4$ **24.** $a^2 - 4a + 4$
25. $z^2 - 2z + 1$ **26.** $z^2 + 2z + 1$ **27.** $t^2 + 20t + 100$ **28.** $s^2 - 24s + 144$
29. $p^2 - 16p + 64$ **30.** $w^2 - 8w + 16$ **31.** $y^2 + 26y + 169$ **32.** $x^2 + 18x + 81$
33. $9a^2 + 30a + 25$ **34.** $4a^2 + 24a + 36$ **35.** $16x^2 - 24x + 9$ **36.** $25y^2 - 40y + 16$
37. $49t^2 - 28t + 4$ **38.** $64z^2 + 16z + 1$ **39.** $36x^2 + 12xy + y^2$ **40.** $49x^2 - 14xy + y^2$
41. $a^2 - 10ab + 25b^2$ **42.** $a^2 + 16ab + 64b^2$ **43.** $9x^2 + 24xy + 16y^2$ **44.** $25x^2 - 20xy + 4y^2$
45. $t^3 - 125$ **46.** $a^3 - 1$ **47.** $64 - w^3$ **48.** $27 - b^3$
49. $x^3 + 1000$ **50.** $y^3 + 8$ **51.** $216 + y^3$ **52.** $27 + x^3$
53. $a^3 + 8b^3$ **54.** $216x^3 - y^3$ **55.** $27a^3 - b^3$ **56.** $x^3 + 1000y^3$
57. $64x^3 + 125y^3$ **58.** $27x^3 + 64y^3$ **59.** $216a^3 - 125b^3$ **60.** $8a^3 - 27b^3$

Factor each of the following expressions completely.

61. $3x^2 - 75$ **62.** $5x^2 - 45$ **63.** $2y^2 - 32$ **64.** $6w^2 - 600$ **65.** $72 - 2t^2$
66. $27 - 3p^2$ **67.** $3x^2 + 18x + 27$ **68.** $4y^2 + 40y + 100$ **69.** $5b^2 - 40b + 80$ **70.** $3a^2 - 12a + 12$
71. $2t^2 + 24t + 72$ **72.** $2w^2 - 48w + 288$ **73.** $x^3 - 14x^2 + 49x$ **74.** $y^3 + 18y^2 + 81y$ **75.** $2t^3 + 32t^2 + 128t$
76. $5p^3 - 20p^2 + 20p$ **77.** $2x^3 - 72x$ **78.** $3z^3 - 48z$ **79.** $x^4 - 16$ **80.** $y^4 - 81$
81. $a^3b - 25ab^3$ **82.** $49xy^3 - x^3y$ **83.** $2x^2y - 98y^3$ **84.** $300x^3 - 3xy^2$

Indicate which of the following statements are correct and which are incorrect. For those that are incorrect, change the underlined expression to make the statement correct.

85. $\underline{x^2 + 9} = (x + 3)(x - 3)$ **86.** $y^2 - 25 = \underline{(y - 5)(y + 5)}$

87. $t^2 - 16t + 64 = (t - 8)(t - 8)$

88. $p^2 + 7p + 49 = (p + 7)^2$

89. $4x^2 - 9y^2 = (2x - 3y)(2x + 3y)$

90. $25a^2 - 8b^2 = (5a + 4b)(5a - 4b)$

91. $3w^2 - 30w + 75 = 3(w - 5)(w + 5)$

92. $4x^2 + 12x + 36 = 4(x + 3)^2$

93. $2p^2 - 18 = 2(p + 3)(p - 3)$

94. $3q^2 + 12 = 3(q + 2)(q - 2)$

13.3 Factoring Trinomials

We will learn how to factor trinomials that are not special products. Since factoring is the reverse of multiplication, rewriting a trinomial as the product of two binomial factors merely involves using the FOIL method in reverse. For example,

Multiplying: $(x + 8)(x + 3) = x^2 + 11x + 24$

Factoring: $x^2 + 11x + 24 = (x + 8)(x + 3)$

Notice that the product of the last terms of the binomial factors is the last term of the trinomial and that the sum of the last terms of the binomial factors is the coefficient of the middle term of the trinomial:

$$8 \cdot 3 = 24 \quad \text{and} \quad 8 + 3 = 11$$

In general,

since $\quad (x + a)(x + b) = x^2 + ax + bx + ab$
$\qquad\qquad\qquad\qquad = x^2 + (a + b)x + ab,$

then $\quad x^2 + (a + b)x + ab = (x + a)(x + b)$

Notice that trinomials that can be factored in the form $(x + a)(x + b)$ have the following characteristics: the first term is x^2, the coefficient of the middle term is the sum $(a + b)$, and the last term is the product ab.

Thus, to factor second degree trinomials whose first term has a coefficient of 1, we need to find two numbers whose product is the last term of the trinomial and whose sum is the coefficient of the middle term of the trinomial. Keep in mind that whenever we factor a trinomial, we should always check our answer mentally by multiplying the factors using the FOIL method. Also, remember that not all trinomials are factorable.

example 1

Factor each of the following expressions.

(a) $x^2 + 10x + 21$

We need two numbers whose product is 21 and whose sum is 10. Since the product and sum are positive, both numbers must be positive. Therefore, the factors are of the form $(x + \quad)(x + \quad)$. Since $3 \cdot 7 = 21$ and $3 + 7 = 10$, the numbers are 3 and 7. Thus,

$$x^2 + 10x + 21 = (x + 3)(x + 7)$$

(b) $y^2 + 5y - 36$

We need two numbers whose product is -36 and whose sum is 5. Since the product is negative, one number is positive and the other is negative. Therefore, the factors are of the form $(y + \quad)(y - \quad)$. Since the sum is positive, the positive number must have the larger absolute value.

POSSIBLE NUMBERS	PRODUCT	SUM
36 and −1	−36	35
18 and −2	−36	16
12 and −3	−36	9
6 and −6	−36	0
9 and −4	−36	5

Therefore, the numbers are 9 and −4. Thus,

$$y^2 + 5y - 36 = (y + 9)(y - 4)$$

(c) $p^2 - p - 56$

We need two numbers whose product is −56 and whose sum is −1. Since the product is negative, one number must be negative and the other must be positive. Therefore, the factors are of the form $(p - \quad)(p + \quad)$. Since the sum is negative, the negative number must have the larger absolute value. Since $(-8)(7) = -56$ and $-8 + 7 = -1$, the numbers are −8 and 7. Thus,

$$p^2 - p - 56 = (p - 8)(p + 7)$$

(d) $t^2 - 13t + 42$

We need two numbers whose product is 42 and whose sum is −13. Since the product is positive and the sum is negative, both numbers must be negative. Therefore, the factors are of the form $(t - \quad)(t - \quad)$. Since $(-6)(-7) = 42$ and $-6 + (-7) = -13$, the numbers are −6 and −7. Thus,

$$t^2 - 13t + 42 = (t - 6)(t - 7)$$

(e) $x^2 + 5xy + 6y^2$

The factors are of the form $(x + \quad y)(x + \quad y)$. To fill in the blanks, we need two numbers whose product is 6 and whose sum is 5. These numbers are 2 and 3. Thus,

$$x^2 + 5xy + 6y^2 = (x + 2y)(x + 3y)$$

Sometimes, before we can factor a trinomial into its binomial factors, we must first factor out the greatest common monomial factor.

example 2

Completely factor each of the following expressions.

(a) $3x^2 + 3x - 90$

The greatest common factor is 3.

$3x^2 + 3x - 90 = 3(x^2 + x - 30)$ First factor out 3.

$= 3(x + 6)(x - 5)$ Then factor the trinomial.

(b) $-y^2 + 11y - 28$

The greatest common factor is −1.

$-y^2 + 11y - 28$

$= -(y^2 - 11y + 28)$ First factor out −1.

$= -(y - 4)(y - 7)$ Then factor the trinomial.

NOTE: When factoring second degree trinomials whose first term has a coefficient of -1, factor out -1 before attempting to find the binomial factors.

(c) $4x^3 + 32x^2 + 60x$

The greatest common factor is $4x$.

$4x^3 + 32x^2 + 60x$

$= 4x(x^2 + 8x + 15)$ — First factor out $4x$.

$= 4x(x + 3)(x + 5)$ — Then factor the trinomial.

(d) $-2a^2b + 10ab + 48b$

The greatest common factor is $-2b$.

$-2a^2b + 10ab + 48b$

$= -2b(a^2 - 5a - 24)$ — First factor out $-2b$.

$= -2b(a + 3)(a - 8)$ — Then factor the trinomial.

It is more difficult to find the binomial factors of second degree trinomials whose first term is not equal to 1. For these problems, we use a process of trial and error as illustrated by this example.

example 3

Factor $5x^2 + 33x + 18$.

Since the product of the first terms of the binomial factors must be $5x^2$, the factors are of the form $(5x + \quad)(x + \quad)$. To fill in the blanks, we need two numbers whose product is 18. We will list all the possibilities along with the product obtained using the FOIL method.

POSSIBLE FACTORS	FIRST TERM	MIDDLE TERM	LAST TERM
$(5x + 1)(x + 18)$	$5x^2$	$+91x$	$+18$
$(5x + 18)(x + 1)$	$5x^2$	$+23x$	$+18$
$(5x + 2)(x + 9)$	$5x^2$	$+47x$	$+18$
$(5x + 9)(x + 2)$	$5x^2$	$+19x$	$+18$
$(5x + 3)(x + 6)$	$5x^2$	$+33x$	$+18$
$(5x + 6)(x + 3)$	$5x^2$	$+21x$	$+18$

Notice that the fifth line gives us the middle term that appears in the trinomial we are trying to factor. Thus,

$$5x^2 + 33x + 18 = (5x + 3)(x + 6)$$

example 4

Factor $2x^2 + 7x - 15$.

solution

The product of the first terms of the binomial factors must be $2x^2$, and the product of the last terms must be -15. We will list the possible factors and the middle term that is found when they are multiplied using the FOIL method.

POSSIBLE FACTORS	MIDDLE TERM
$(2x + 1)(x - 15)$	$-29x$
$(2x - 1)(x + 15)$	$+29x$
$(2x + 15)(x - 1)$	$+13x$
$(2x - 15)(x + 1)$	$-13x$
$(2x + 3)(x - 5)$	$-7x$
$(2x - 3)(x + 5)$	$+7x$
$(2x + 5)(x - 3)$	$-x$
$(2x - 5)(x + 3)$	$+x$

The sixth line gives us the appropriate middle term. Thus,

$$2x^2 + 7x - 15 = (2x - 3)(x + 5)$$

example 5

Factor $4t^2 - t - 5$.

solution

The product of the first terms of the binomial factors must be $4t^2$, and the product of the last terms must be -5.

POSSIBLE FACTORS	MIDDLE TERM
$(2t + 1)(2t - 5)$	$-8t$
$(2t - 1)(2t + 5)$	$+8t$
$(4t + 1)(t - 5)$	$-19t$
$(4t - 1)(t + 5)$	$+19t$
$(4t + 5)(t - 1)$	$+t$
$(4t - 5)(t + 1)$	$-t$

The last line gives us the appropriate middle term. Thus,

$$4t^2 - t - 5 = (4t - 5)(t + 1).$$

example 6

Factor $3x^2 - 17xy + 10y^2$.

solution

The product of the first terms of the binomial factors must be $3x^2$, and the product of the last terms must be $10y^2$. Since the middle term of the trinomial is negative, both binomials are differences.

POSSIBLE FACTORS	MIDDLE TERM
$(3x - y)(x - 10y)$	$-31xy$
$(3x - 10y)(x - y)$	$-13xy$
$(3x - 5y)(x - 2y)$	$-11xy$
$(3x - 2y)(x - 5y)$	$-17xy$

The fourth line gives us the appropriate middle term. Thus,

$$3x^2 - 17xy + 10y^2 = (3x - 2y)(x - 5y).$$

example 7

Factor $-4x^2y - 30xy - 56y$.

solution The greatest common factor is $-2y$.

$$-4x^2y - 30xy - 56y$$
$$= -2y(2x^2 + 15x + 28) \quad \text{First factor out } -2y$$
$$= -2y(2x + \quad)(x + \quad)$$

To factor $2x^2 + 15x + 28$, we will list some of the possible factors of a trinomial having a first term of $2x^2$ and a last term of 28, until we obtain a middle term of $+15x$.

POSSIBLE FACTORS	MIDDLE TERM
$(2x + 4)(x + 7)$	$+18x$
$(2x + 7)(x + 4)$	$+15x$

Thus,

$$2x^2 + 15x + 28 = (2x + 7)(x + 4)$$

and

$$-4x^2y - 30xy - 56y = -2y(2x^2 + 15x + 28)$$
$$= -2y(2x + 7)(x + 4)$$

QUICK QUIZ

Factor each of the following expressions.

1. $x^2 - 4x - 45$
2. $-y^3 - 5y^2 + 14y$
3. $5t^2 + 23t + 12$
4. $6x^2 - 39x + 63$

ANSWERS

1. $(x + 5)(x - 9)$
2. $-y(y - 2)(y + 7)$
3. $(5t + 3)(t + 4)$
4. $3(2x - 7)(x - 3)$

13.3 Exercises

Factor each of the following trinomials.

1. $x^2 + 9x + 20$ **2.** $y^2 + 12y + 27$ **3.** $w^2 + 5w - 14$ **4.** $z^2 - 7z - 8$
5. $a^2 + 2a - 24$ **6.** $b^2 + 4b - 21$ **7.** $t^2 - 17t + 72$ **8.** $p^2 - 11p + 28$
9. $x^2 - 7x - 18$ **10.** $x^2 + 15x + 56$ **11.** $y^2 - y - 20$ **12.** $t^2 - 8t + 7$
13. $w^2 - 6w - 16$ **14.** $s^2 - 2s - 63$ **15.** $n^2 + 11n + 30$ **16.** $q^2 - 12q + 32$
17. $u^2 - 15u + 54$ **18.** $p^2 + 9p + 18$ **19.** $r^2 - 3r - 10$ **20.** $c^2 + 2c - 63$
21. $x^2 + x - 6$ **22.** $y^2 + 4y - 12$ **23.** $z^2 + 12z + 35$ **24.** $w^2 - 6w - 27$
25. $s^2 - 3s - 18$ **26.** $t^2 - 17t + 72$ **27.** $x^2 + 8xy + 15y^2$ **28.** $a^2 + 15ab + 56b^2$
29. $a^2 + 2ab - 63b^2$ **30.** $x^2 - xy - 20y^2$ **31.** $s^2 - 11st + 24t^2$ **32.** $p^2 - 11pq + 30q^2$
33. $x^2 + 8xy - 9y^2$ **34.** $x^2 - 4xy - 5y^2$ **35.** $a^2 + 10ab + 21b^2$ **36.** $a^2 - 8ab + 12b^2$
37. $s^2 - 13st + 36t^2$ **38.** $p^2 + 8pq + 15q^2$ **39.** $x^2 - 3xy - 40y^2$ **40.** $x^2 - xy - 56y^2$
41. $2x^2 + 9x + 10$ **42.** $3x^2 + 11x + 8$ **43.** $3y^2 - 19y - 72$ **44.** $2x^2 + 5x - 42$
45. $4a^2 - 9a + 2$ **46.** $5b^2 - 17b + 6$ **47.** $7t^2 + 34t - 5$ **48.** $2w^2 - 7w - 30$
49. $3s^2 - 4s - 32$ **50.** $4p^2 + 5p - 21$ **51.** $4x^2 - 16x + 15$ **52.** $6x^2 - 13x - 28$

53. $5y^2 + 29y - 6$
54. $4z^2 + 20z + 9$
55. $6a^2 - 11a - 35$
56. $7m^2 + 26m - 8$
57. $12y^2 + 31y + 7$
58. $15z^2 - 31z + 10$
59. $9t^2 - 6t - 8$
60. $16w^2 - 8w - 15$
61. $20x^2 - 67x + 56$
62. $18x^2 + 39x + 20$
63. $2x^2 - 3xy - 5y^2$
64. $3x^2 + xy - 14y^2$
65. $5a^2 + 33ab + 18b^2$
66. $4a^2 - 9ab + 5b^2$
67. $4x^2 + 8xy - 45y^2$
68. $9x^2 + 9xy - 28y^2$
69. $6a^2 - 25ab + 24b^2$
70. $8a^2 + 38ab + 35b^2$

Completely factor each of the following expressions by first finding the greatest common monomial factor.

71. $2x^2 + 24x + 70$
72. $3y^2 + 39y + 108$
73. $5w^2 - 50w + 80$
74. $4z^2 - 44z + 120$
75. $6p^2 - 24p - 126$
76. $8t^2 + 8t - 160$
77. $-x^2 + 4x + 5$
78. $-x^2 + 7x + 18$
79. $-2z^2 + 4z + 70$
80. $-3w^2 + 9w + 12$
81. $x^3 + 15x^2 + 54x$
82. $y^3 + 4y^2 - 21y$
83. $a^4 + 11a^3 + 28a^2$
84. $b^5 - 11b^4 + 24b^3$
85. $3x^3 - 30x^2 + 48x$
86. $5y^3 + 25y^2 + 30y$
87. $3x^2y - 18xy - 48y$
88. $2xy^2 + 4xy - 30x$
89. $a^3 - 7a^2b + 10ab^2$
90. $x^3 + 5x^2y + 6xy^2$
91. $4x^2 + 14x - 30$
92. $15x^2 - 57x - 12$
93. $16a^2 - 28a + 12$
94. $6b^2 + 50b + 56$
95. $30w^2 + 5w - 5$
96. $18z^2 - 12z - 48$
97. $-4x^2 - 16x - 15$
98. $-7x^2 - 40x + 12$
99. $-15y^2 + 5y + 50$
100. $-16t^2 + 44t - 24$
101. $4x^3 - 4x^2 - 15x$
102. $6y^3 + 29y^2 + 28y$
103. $3a^5 - 14a^4 + 8a^3$
104. $4a^6 + 7a^5 - 15a^4$
105. $15t^3 + 33t^2 + 18t$
106. $18w^3 - 6w^2 - 40w$
107. $16z^3 + 44z^2 + 24z$
108. $10s^3 - 55s^2 - 105s$
109. $2x^3 + 9x^2y + 4xy^2$
110. $3a^3 - 13a^2b - 10ab^2$

13.4 A General Factoring Strategy

It is important to be able to recognize which of the factoring techniques we have already discussed is appropriate for factoring a particular polynomial. The following strategy summarizes these techniques.

Steps for Factoring a Polynomial

1. Factor out the greatest common monomial factor.
2. If the polynomial has two terms, determine whether it is the difference of two squares, the difference of two cubes, or the sum of two cubes.
3. If the polynomial has three terms, determine whether it is a perfect square trinomial. If not, use trial and error.
4. If the polynomial has four terms, attempt to factor by grouping.
5. Make sure that the polynomial is completely factored, that is, no factor containing more than one term can be factored further.

example 1

Factor each of the following polynomials completely.

(a) $8x^5 - 8x$

$8x^5 - 8x = 8x(x^4 - 1)$ — Factor out the greatest common factor, $8x$.

$= 8x(x^2 + 1)(x^2 - 1)$ — The factor $x^4 - 1$ has two terms and is the difference of two squares.

$= 8x(x^2 + 1)(x + 1)(x - 1)$ — The factor $x^2 - 1$ can be factored further.

(b) $x^4 + x^3 - 56x^2$

$x^4 + x^3 - 56x^2 = x^2(x^2 + x - 56)$ — Factor out the greatest common factor, x^2.

$= x^2(x + 8)(x - 7)$ — The trinomial $x^2 + x - 56$ has three terms and can be factored by trial and error.

(c) $-3x^2 - 9x - 15x - 45$

$-3x^2 - 9x - 15x - 45$

$= -3(x^2 + 3x + 5x + 15)$ — Factor out the greatest common factor, -3

$= -3[x(x + 3) + 5(x + 3)]$ — The polynomial $x^2 + 3x + 5x + 15$ can be factored by grouping. Note the use of the square brackets to indicate that -3 is a factor of the entire expression.

$= -3(x + 3)(x + 5)$ — Factor out $(x + 3)$ from each term in square brackets.

(d) $2x^2y - 28xy + 98y$

$2x^2y - 28xy + 98y$

$= 2y(x^2 - 14x + 49)$ — Factor out the greatest common factor, $2y$.

$= 2y(x - 7)^2$ — The polynomial $x^2 - 14x + 49$ has three terms and is a perfect square trinomial.

(e) $x^3 + 4x^2 - 9x - 36$

$x^3 + 4x^2 - 9x - 36$

$= x^2(x + 4) - 9(x + 4)$ — There is no factor common to all four terms but we can factor by grouping.

$= (x + 4)(x^2 - 9)$ — Factor out $(x + 4)$ from each term.

$= (x + 4)(x + 3)(x - 3)$ — The factor $x^2 - 9$ can be factored further.

(f) $2x^3y - 16y^4$

$2x^3y - 16y^4 = 2y(x^3 - 8y^3)$ — Factor out the greatest common factor, $2y$.

$= 2y[(x)^3 - (2y)^3]$ — Factor the difference of two cubes.

$= 2y(x - 2y)(x^2 + 2xy + 4y^2)$ ●

QUICK QUIZ

Factor each of the following expressions completely.

1. $3x^3 + 30x^2 + 75x$
2. $x^4 - 16$
3. $-7x^4 - 28x^3 + 84x^2$
4. $x^3 - 3x^2 - 9x^2 + 27x$
5. $x^3y^2 - 10x^2y^2 + 21xy^2$
6. $125a^5 + a^2b^3$

ANSWERS

1. $3x(x + 5)^2$
2. $(x^2 + 4)(x - 2)(x + 2)$
3. $-7x^2(x + 6)(x - 2)$
4. $x(x - 9)(x - 3)$
5. $xy^2(x - 3)(x - 7)$
6. $a^2(5a + b)(25a^2 - 5ab + b^2)$

13.4 Exercises

Factor each of the following polynomials completely.

1. $2x^2 + 6x - 56$
2. $x^3 + 11x^2 + 30x$
3. $4x^3 - 100x$
4. $5x^3 - 20x$
5. $x^3 - 16x^2 + 64x$
6. $4x^2 - 32x + 64$
7. $x^2 + 7x + 9x + 63$
8. $x^2 - 3x + 8x - 24$
9. $3x^4 - 48x^2$
10. $2x^3 - 128x$
11. $2x^2 + 13x + 21$
12. $3x^2 - 14x - 5$
13. $x^4 - 81$
14. $x^4 - 1$
15. $x^5 + 13x^4 + 42x^3$
16. $x^4 - 13x^3 + 40x^2$
17. $9x^3 - 36x^2 + 36x$
18. $2x^3 + 20x^2 + 50x$
19. $3x^2 + 6x - 7x - 14$
20. $2x^2 + 5x - 6x - 15$
21. $-x^3 + 2x^2 + 48x$
22. $-x^3 - 13x^2 - 36x$
23. $8x^3y + 32x^2y + 24xy$
24. $5x^3y^2 - 10x^2y^2 - 40xy^2$
25. $x^3y^2 - 36xy^2$
26. $x^4y - 4x^2y$
27. $5x^4 + 30x^3 + 45x^2$
28. $3x^5 + 24x^4 + 48x^3$
29. $4x^3 - 28x^2 - 8x^2 + 56x$
30. $x^4 + 8x^3 - 9x^3 - 72x^2$
31. $3x^2y - 11xy - 4y$
32. $2x^2y - xy - 6y$
33. $5x^3 - 40x^2 + 3x - 24$
34. $3x^3 + 15x^2 - 2x - 10$
35. $x^3 - 7x^2 - x + 7$
36. $x^3 + 3x^2 - 4x - 12$
37. $5x^3y + 5x^2y - 60xy$
38. $x^4y - 15x^3y + 56x^2y$
39. $x^3y^2 - 8x^2y^2 + 6x^2y^2 - 48xy^2$
40. $3x^2y^2 + 12xy^2 - 6xy^2 - 24y^2$
41. $5x^3y + 40y$
42. $2t^4 - 128t$
43. $a^3b^2 - 27b^5$
44. $216x^4y + xy^4$

13.5 Solving Quadratic Equations by Factoring

In this section, we will use the techniques of factoring to solve *quadratic equations*. Quadratic equations are second degree equations of a single variable, or equations in which the highest power of the variable is 2. The *standard form* for a quadratic equation is

$$ax^2 + bx + c = 0$$

where a, b, and c are constants and $a \neq 0$.

Once a quadratic equation is in standard form, if we can factor the left-hand side, we can use the following principle to determine the values of the unknown variable.

> If the product of two numbers is zero, then at least one of the numbers must be zero.

If we let a and b represent real numbers, we can summarize this principle as follows:

$$\text{If } ab = 0, \quad \text{then} \quad a = 0 \quad \text{or} \quad b = 0$$

Let us now use this principle to solve the following equation.

example 1 Solve $x^2 - 3x - 28 = 0$ for x.

solution The equation is already in standard form, so we can factor the left-hand side:

$$(x - 7)(x + 4) = 0$$

Since the two factors, which represent two real numbers, have a product of zero, at least one factor must be equal to zero. We can therefore find all possible solutions to this equation by setting each factor equal to zero and solving for x:

$$x - 7 = 0 \quad \text{or} \quad x + 4 = 0$$
$$x = 7 \quad \text{or} \quad x = -4$$

The solution set is $\{7, -4\}$.

We can check both solutions by substituting each value for x in the original equation.

CHECK

$$x^2 - 3x - 28 = 0$$
$$x = 7: (7)^2 - 3(7) - 28 \ ? \ 0$$
$$49 - 21 - 28 \ ? \ 0$$
$$0 = 0 \ \surd$$

$$x^2 - 3x - 28 = 0$$
$$x = -4: (-4)^2 - 3(-4) - 28 \ ? \ 0$$
$$16 + 12 - 28 \ ? \ 0$$
$$0 = 0 \ \surd$$

We will now summarize this procedure for solving quadratic equations.

Procedure for Solving Quadratic Equations by Factoring

1. Rewrite the equation in standard form, $ax^2 + bx + c = 0$.
2. Factor the left-hand side of the equation.
3. Set each factor that contains a variable equal to zero.
4. Solve each of the resulting equations for the variable.

Notice that sometimes the solutions to an equation are called the *roots* of the equation.

definition

A *root* is a solution to an equation having one variable.

Also keep in mind that since we cannot factor all polynomials, not all quadratic equations can be solved by factoring. We will discuss additional techniques for solving quadratic equations in Chapter 17.

example 2

Find the roots of $2x^2 + x = 15$.

$2x^2 + x - 15 = 0$ — 1. Rewrite in standard form.

$(2x - 5)(x + 3) = 0$ — 2. Factor the left-hand side.

$2x - 5 = 0 \quad x + 3 = 0$ — 3. Set each factor equal to zero.

$2x = 5 \quad x = -3$ — 4. Solve each equation for x.

$$x = \frac{5}{2}$$

The roots of the equation are $\frac{5}{2}$ and -3.

CHECK

$x = \frac{5}{2}$: $2\left(\frac{5}{2}\right)^2 + \frac{5}{2} \; ? \; 15$

$2\left(\frac{25}{4}\right) + \frac{5}{2} \; ? \; 15$

$\frac{25}{2} + \frac{5}{2} \; ? \; 15$

$15 = 15 \; \checkmark$

$x = -3$: $2(-3)^2 + (-3) \; ? \; 15$

$2(9) + (-3) \; ? \; 15$

$18 - 3 \; ? \; 15$

$15 = 15 \; \checkmark$

example 3

Find the solution set of $4y^2 = 8y$.

solution

$4y^2 - 8y = 0$ — 1. Rewrite in standard form.

$4y(y - 2) = 0$ — 2. Factor the left-hand side.

$4y = 0 \qquad y - 2 = 0$ — 3. Set each factor equal to zero.

$y = 0 \qquad y = 2$ — 4. Solve each equation for y.

The solution set is $\{0, 2\}$.

CHECK

$y = 0$: $4(0)^2 \; ? \; 8(0)$

$0 = 0 \; \checkmark$

$y = 2$: $4(2)^2 \; ? \; 8(2)$

$4(4) \; ? \; 16$

$16 = 16 \; \checkmark$

Sometimes we must use the distributive property to remove parentheses in an equation before we can put it in standard form. Consider this example:

example 4

Solve $x(x + 8) = -16$.

solution

$x(x + 8) = -16$ — Multiply the left-hand side using the distributive property.

$x^2 + 8x = -16$

$x^2 + 8x + 16 = 0$ — Rewrite in standard form.

$(x + 4)(x + 4) = 0$ — Factor the left-hand side.

$x + 4 = 0$ — Set factor equal to zero.

$x = -4$ — Solve for x.

Notice that since both factors are identical, there is only one solution, or root, to this equation.

CHECK

$-4(-4 + 8) \; ? \; -16$

$-4(4) \; ? \; -16$

$-16 = -16 \; \checkmark$

We can also solve some higher degree equations by factoring as shown in this example.

example 5 Solve $3t^3 = 24t^2 - 21t$ for t.

solution

$3t^3 - 24t^2 + 21t = 0$ Rewrite equation with zero on right.

$3t(t^2 - 8t + 7) = 0$ Factor out $3t$.

$3t(t - 1)(t - 7) = 0$ Factor trinomial.

$3t = 0 \quad t - 1 = 0 \quad t - 7 = 0$ Set each factor equal to zero.

$t = 0 \quad t = 1 \quad t = 7$ Solve each equation for t.

The solution set is $\{0, 1, 7\}$.

CHECK

$t = 0$: $3(0)^3 \ ? \ 24(0)^2 - 21(0)$
$0 \ ? \ 0 - 0$
$0 = 0$ √

$t = 1$: $3(1)^3 \ ? \ 24(1)^2 - 21(1)$
$3(1) \ ? \ 24 - 21$
$3 = 3$ √

$t = 7$: $3(7)^3 \ ? \ 24(7)^2 - 21(7)$
$3(343) \ ? \ 1{,}176 - 147$
$1{,}029 = 1{,}029$ √

QUICK QUIZ

Solve each of the following equations by factoring.

1. $x^2 + 5x = 36$
2. $3a^2 = 7a$
3. $2y^3 - 15y^2 + 18y = 0$

ANSWERS

1. $x = 4, -9$
2. $a = 0, \frac{7}{3}$
3. $y = 0, \frac{3}{2}, 6$

13.5 Exercises

Solve each of the following quadratic equations by factoring and check your solutions.

1. $x^2 + 13x + 40 = 0$
2. $y^2 + 10y + 21 = 0$
3. $z^2 - 10z + 24 = 0$
4. $w^2 - 11w + 18 = 0$
5. $t^2 + 3t = 18$
6. $p^2 - 2p = 35$
7. $m^2 - 8m = 9$
8. $n^2 + n = 42$
9. $-x^2 - 13x - 42 = 0$
10. $-x^2 - 8x - 15 = 0$
11. $y^2 = -7(2y + 7)$
12. $z^2 = 12(z - 3)$
13. $t^2 = -5t$
14. $p^2 = 2p$
15. $y^2 - 49 = 0$
16. $y^2 - 9 = 0$
17. $a^2 = 25$
18. $b^2 = 64$
19. $2x^2 + 13x + 20 = 0$
20. $3y^2 + 19y - 14 = 0$
21. $5z^2 - 16z = -12$
22. $4w^2 - 7w = -3$
23. $7a^2 = 10 - 9a$
24. $2b^2 = 27 + 3b$
25. $0 = 3x^2 - 24x$
26. $0 = 4x^2 + 16x$
27. $8y^2 = -8y$
28. $6y^2 = 18y$
29. $16x^2 = 25$
30. $9x^2 = 64$
31. $-3a^2 - 30a - 63 = 0$
32. $-5b^2 - 50b - 80 = 0$
33. $-6w^2 - 36w = 54$
34. $-4z^2 + 40z = 100$
35. $6t^2 = -23t - 21$
36. $12p^2 = -19p - 5$
37. $6x = 14x^2$
38. $3x = 15x^2$
39. $5(5x^2 - 14x) = -49$
40. $3(3x^2 + 8x) = -16$

Solve each of the following higher degree equations by factoring and check your solutions.

41. $7x^3 - 35x^2 = 0$
42. $5y^3 + 10y^2 = 0$
43. $y^3 - 5y^2 - 24y = 0$
44. $x^3 + 2x^2 - 35x = 0$
45. $5a^3 - 35a^2 = -60a$
46. $3b^3 + 27b^2 = -60b$
47. $2z^3 = -19z^2 - 42z$
48. $3w^3 = 20w^2 - 32w$
49. $10t^3 = 6t + 28t^2$
50. $8p^3 = 24p - 4p^2$
51. $15x^2 - 18x^3 = -42x$
52. $160y - 20y^2 = 15y^3$
53. $42w^3 + 4w^2 = 6w$
54. $24z^3 + 18z^2 = 15z$
55. $25a^3 = 4a$
56. $64b^3 = 25b$
57. $14x^4 + 27x^3 + 9x^2 = 0$
58. $18y^4 - 21y^3 + 5y^2 = 0$

13.6 Word Problems Involving Quadratic Equations

In this section, we will look at a variety of word problems that can be solved using quadratic equations. To solve the quadratic equations in this section, we will use the factoring technique presented in Section 13.5. Consider these examples:

example 1

Three times the product of two consecutive even integers is 144. Find the integers.

solution

The problem asks us to find two integers.

Let $x =$ first integer — Define the variables.

$x + 2 =$ second integer

$$3x(x+2) = 144$$ Write an equation relating the variables.

$$3x^2 + 6x = 144$$

$$3x^2 + 6x - 144 = 0$$ Rewrite the quadratic equation in standard form.

$$3(x^2 + 2x - 48) = 0$$

$$3(x-6)(x+8) = 0$$ Solve the equation by factoring.

$$x - 6 = 0 \quad \text{or} \quad x + 8 = 0$$

$$x = 6 \quad \text{or} \quad x = -8$$

The solution presents us with two sets of answers.

$x = 6$ is the first integer or $x = -8$ is the first integer
$x + 2 = 8$ is the second integer or $x + 2 = -6$ is the second integer

The two integers are 6 and 8 or -8 and -6.

CHECK

$3(6)(8) = 3(48) = 144$ ✓
$3(-8)(-6) = 3(48) = 144$ ✓

The next three examples involve use of formulas involving geometric figures. Students who are not familiar with these formulas should refer to Chapter 9, Introduction to Geometry.

example 2

The length of a rectangle is 5 in. longer than twice its width. If the area of the rectangle is 52 in.², what are the dimensions of the rectangle?

solution

The problem asks us to find the length and width of a rectangle.

Let $x = \text{width}$

$2x + 5 = \text{length}$

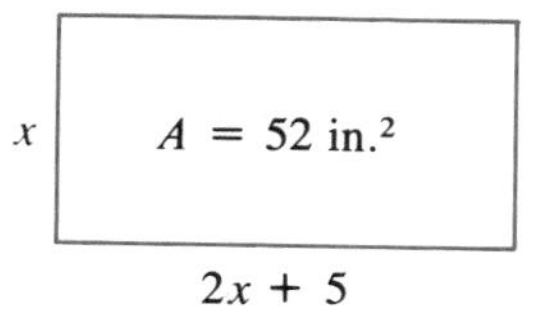

$$A = l \cdot w$$ The area of a rectangle is length times width.

$$52 = (2x + 5)x$$
$$52 = 2x^2 + 5x$$ Substitute into the formula to obtain an equation relating the variables.

$$2x^2 + 5x - 52 = 0$$ Rewrite the quadratic equation in standard form.

$$(2x + 13)(x - 4) = 0$$ Solve the quadratic equation by factoring.

$$2x + 13 = 0 \quad \text{or} \quad x - 4 = 0$$
$$x = -\frac{13}{2} \quad \text{or} \quad x = 4$$

Reject the solution $x = -\frac{13}{2}$ since the dimensions must be positive.

$x = 4$ in. is the width

$2x + 5 = 13$ in. is the length

The width of the rectangle is 4 in. and its length is 13 in.

CHECK $A = l \cdot w = 13 \cdot 4 = 52$ √

example 3

One leg of a right triangle is 3 ft longer than the other leg. If the hypotenuse is 15 ft long, find the length of the two legs.

solution

The problem asks you to find the two legs of a right triangle.

Let $x = \text{length of one leg}$

$x + 3 = \text{length of other leg}$

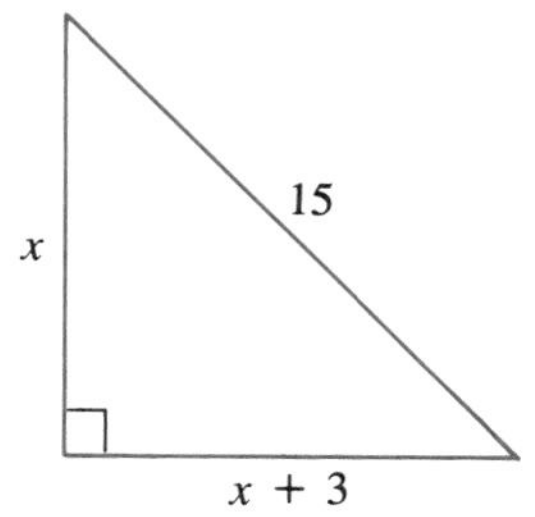

To solve this problem we will use the Pythagorean Theorem.

$$a^2 + b^2 = c^2$$ Given legs a and b and hypotenuse c.

$$x^2 + (x + 3)^2 = 15^2$$
$$x^2 + x^2 + 6x + 9 = 225$$ Substitute into the Pythagorean Theorem to obtain an equation relating the variables.

$$2x^2 + 6x - 216 = 0$$
$$2(x^2 + 3x - 108) = 0$$
$$2(x + 12)(x - 9) = 0$$
$$x + 12 = 0 \quad \text{or} \quad x - 9 = 0$$
$$x = -12 \quad \text{or} \quad x = 9$$

Rewrite the quadratic equation in standard form.

Solve the quadratic equation by factoring.

Reject the solution $x = -12$ since the dimensions must be positive.

$x = 9$ ft is the length of one leg

$x + 3 = 12$ ft is the length of the other leg

The legs measure 9 ft and 12 ft.

CHECK $9^2 + 12^2 = 81 + 144 = 225 = (15)^2$ ✓

example 4

The base of a triangle is 4 times its height. If the base is decreased by 2 cm and the height is increased by 5 cm, the area of the new triangle is 90 cm^2. Find the area of the original triangle.

solution

The problem asks us to find the area of the original triangle. First we must find its dimensions.

Let x = height of original triangle

$4x$ = base of original triangle

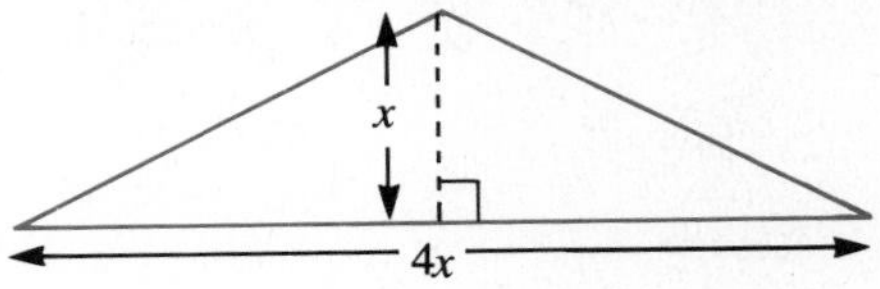

Let $x + 5$ = height of new triangle

$4x - 2$ = base of new triangle

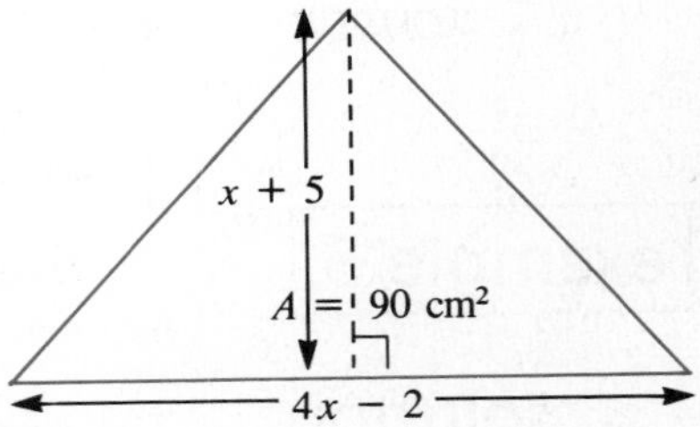

$$A = \frac{1}{2} b \cdot h$$

The area of a triangle is $\frac{1}{2}$ base times height.

$$90 = \frac{1}{2}(4x - 2)(x + 5)$$

Substitute the dimensions and area of the new triangle into the formula to obtain an equation relating the variables.

$$90 = (2x - 1)(x + 5)$$
$$90 = 2x^2 + 9x - 5$$
$$2x^2 + 9x - 95 = 0$$

Rewrite the quadratic equation in standard form.

$$(2x + 19)(x - 5) = 0$$

Solve the quadratic equation by factoring.

$$2x + 19 = 0 \quad \text{or} \quad x - 5 = 0$$
$$x = -\frac{19}{2} \quad \text{or} \quad x = 5$$

Reject the solution $x = -\frac{19}{2}$ since the dimensions must be positive.

$x = 5$ cm is the height of the original triangle.
$4x = 20$ cm is the base of the original triangle.

To find the area of the original triangle we use the formula:

$$A = \frac{1}{2}b \cdot h$$

$$A = \frac{1}{2}(20)(5)$$

$A = 50\text{ cm}^2$ is the area of the original triangle.
The area of the original triangle is 50 cm^2.

CHECK Verify that the area of the new triangle is 90 cm^2 if $x = 5$:

$$A = \frac{1}{2}b \cdot h = \frac{1}{2}(x + 5)(4x - 2) = \frac{1}{2}(10)(18) = 90\text{ cm}^2 \quad \checkmark$$

13.6 Exercises

Solve each of the following word problems.

1. The product of two consecutive integers is 42. Find the integers.
2. The product of two consecutive even integers is 80. Find the integers.
3. Four less than the product of two consecutive odd integers is 59. Find the integers.
4. Twice the product of two consecutive integers is 10 more than their sum. Find the integers.
5. One integer is 2 less than three times another integer. Their product is 65. Find the integers.
6. Nine times a certain integer is 18 less than twice the square of the integer. Find the integer.
7. The sum of a father's age and that of his daughter is 42. If the product of their ages is 320, how old is the father and how old is the daughter?
8. The sum of a mother's age and that of her son is 32. If the product of their ages is 60, how old is the mother and how old is the son?
9. The length of a rectangle is 5 more than twice the width. If the area of the rectangle is 33 in.^{2}, what are the dimensions of the rectangle?
10. The length of a rectangle is 7 less than four times its width. If the area of the rectangle is 36 ft^2, what are the dimensions of the rectangle?
11. The base of a triangle is 5 m longer than its height. If the area of the triangle is 12 m^2, find its base and height.
12. The base of a triangle is 8 cm less than its height. If the area of the triangle is 42 cm^2, find its base and height.
13. The perimeter of a rectangle is 30 yd and its area is 56 yd^2. Find the dimensions of the rectangle.
14. The perimeter of a rectangle is 30 ft and its area is 54 ft^2. Find the dimensions of the rectangle.
15. One leg of a right triangle exceeds the other leg by 4 cm. If the hypotenuse of the triangle is 20 cm, find the length of the two legs.
16. One leg of a right triangle is 7 m shorter than the other leg. If the hypotenuse of the triangle is 17 m, find the length of the two legs.

17. A rectangle is 7 ft long and 3 ft wide. If the length and width of the rectangle are both increased by the same number of feet, the area of the new rectangle is 96 ft^2. By how many feet should the length and width be increased?

18. A picture frame 9 in. long and 6 in. wide is surrounded by a frame of uniform width. If the area enclosed by the frame is 88 $in.^2$, how wide is the frame?

19. The base of a parallelogram is 3 times its height. If the base is decreased by 6 cm and the height is increased by 2 cm, the area of the resulting parallelogram is 63 cm^2. What is the area of the original parallelogram?

20. The base of a triangle is 3 m longer than its height. If the base is increased by 2 m and the height is decreased by 2 m, the resulting triangle has an area of 60 m^2. What is the area of the original triangle?

Summary and Review

Key Terms

[13.1] **Factoring** is rewriting an expression as a product of its factors.

The **greatest common factor (GCF)** of a polynomial is the largest polynomial that divides every term of the polynomial.

[13.2] An expression is **factored completely** when none of its factors can be factored further.

[13.5] A **quadratic equation** is a second degree equation that can be written in the form $ax^2 + bx + c = 0$ where a, b, and c are constants and $a \neq 0$.

A **root** is a solution to an equation of one variable.

Calculations

[13.1] **To find the greatest common factor (GCF) of a polynomial**, multiply the largest number that divides each of the numerical coefficients in the polynomial by the lowest power that occurs for each variable appearing in every term of the polynomial.

To factor out the GCF of a polynomial, express the polynomial as the product of the GCF and the factor obtained when every term of the polynomial is divided by the GCF.

[13.1] **To factor a polynomial by grouping**, rearrange the polynomial so that the first two terms have a common factor and the last two terms have a common factor. Factor out the greatest common monomial factor from the first two terms and from the last two terms. Then factor out the greatest common binomial factor from the two remaining terms.

[13.2] **To factor an expression completely**, first factor out the greatest common monomial factor, and then factor the polynomial factor.

[13.4] **Steps for Factoring a Polynomial**

1. Factor out the greatest common monomial factor.
2. If the polynomial has two terms, determine whether it is the difference of two squares, the difference of two cubes, or the sum of two cubes.
3. If the polynomial has three terms, determine whether it is a perfect square trinomial. If not, use trial and error.
4. If the polynomial has four terms, attempt to factor by grouping.
5. Make sure that the polynomial is completely factored, that is, no factor containing more than one term can be factored further.

[13.5] **To solve a quadratic equation by factoring**, rewrite the equation in standard form, factor the left-hand side of the equation, set each factor equal to zero, and solve the resulting equations for the unknown variable.

Factoring Formulas

[13.2] Difference of Two Squares $a^2 - b^2 = (a + b)(a - b)$

Perfect Square Trinomials $a^2 + 2ab + b^2 = (a + b)^2$

$a^2 - 2ab + b^2 = (a - b)^2$

Difference of Two Cubes $a^3 - b^3 = (a - b)(a^2 + ab + b^2)$

Sum of Two Cubes $a^3 + b^3 = (a + b)(a^2 - ab + b^2)$

[13.3] Factorable Trinomial $x^2 + (a + b)x + ab = (x + a)(x + b)$

Chapter 13 Review Exercises

Indicate which of the following statements are true and which are false. For those that are false, change the italic word to make the statement true.

1. To factor a polynomial completely, find the greatest common monomial factor *last*.
2. A *root* is a solution to one equation having one variable.
3. A quadratic equation is a *fourth* degree equation.
4. *All* quadratic equations can be solved by factoring.
5. When we factor a polynomial, we use the *distributive* property in the reverse direction.
6. The greatest common factor of a polynomial includes the *lowest* power that occurs for each variable appearing in every term of the polynomial.

[13.1–13.3] Factor each of the following expressions.

7. $6x - 54$
8. $2y + 8$
9. $15a^2 + 5a$
10. $4b^2 - 16b$
11. $12w^6 + 9w^3 - 24w^2$
12. $-6t^5 + 8t^4 - 18t^3$
13. $-4x^2y^3 + 20x^5y^2$
14. $27x^3y^2 - 9x^4y$
15. $2x(a + b) + 8y(a + b)$
16. $9a(x + y) + 3b(x + y)$
17. $x^2 + 8x + 4x + 32$
18. $x^2 - 3x + 7x - 21$
19. $y^2 - 64$
20. $z^2 - 144$
21. $4x^2 - 25$
22. $9x^2 - 49$
23. $t^2 + 18t + 81$
24. $w^2 + 12w + 36$
25. $p^2 - 22p + 121$
26. $q^2 - 8q + 16$
27. $a^2 + 6ab + 9b^2$
28. $x^2 - 16xy + 64y^2$
29. $w^3 + 125$
30. $216 - t^3$
31. $x^2 + 7x + 12$
32. $z^2 - 13z + 40$
33. $w^2 - 2w - 24$
34. $w^2 + 6w - 27$
35. $t^2 + 16t + 63$
36. $p^2 + 12p + 32$
37. $x^2 + 2xy - 15y^2$
38. $x^2 - 3xy - 28y^2$
39. $2x^2 + 17x + 30$
40. $3y^2 - 13y + 14$
41. $3x^2 - 2x - 15x + 10$
42. $2x^2 + 3x + 12x + 18$
43. $5w^2 + 27w - 18$
44. $4z^2 + 4z - 15$
45. $9a^2 - 18x + 8$
46. $10b^2 + 9b + 2$
47. $4x^2 - 25x - 21$
48. $7x^2 - 11x - 6$
49. $2x^2 - 5xy - 12y^2$
50. $3x^2 - 7xy - 20y^2$
51. $x^3 - 7x^2 + 8x - 56$
52. $x^3 + 3x^2 - 9x - 27$

[13.2–13.4] Factor each of the following expressions completely.

53. $2x^2 - 50$
54. $3y^2 - 48$
55. $4y^2 + 24y + 36$
56. $5x^2 + 100x + 500$
57. $7z^2 - 28z + 28$
58. $9w^2 - 18w + 9$
59. $x^3y - 49xy^3$
60. $64a^3b - ab^3$

61. $x^3 - 7x^2 - 6x^2 + 42x$
62. $3x^2 + 6x + 24x + 48$
63. $4x^2 - 8x - 60$
64. $3x^2 + 15x - 42$
65. $-y^2 - 10y - 24$
66. $-z^2 - 12z - 27$
67. $a^3 - 8a^2 + 15a$
68. $b^3 - 4b^2 - 32b$
69. $5t^3 + 15t^2 - 90t$
70. $2w^3 - 26w^2 + 72w$
71. $12x^2 + 66x + 30$
72. $12x^2 + 4x - 56$
73. $-5y^2 + 19y - 12$
74. $-4z^2 - 27z - 18$
75. $4p^3 - 10p^2 - 84p$
76. $9t^3 - 24t^2 + 15t$
77. $2x^3 + 18x^2 - 6x - 54$
78. $x^4 - 7x^3 + 5x^2 - 35x$
79. $x^3 - 2x^2y - 15xy^2$
80. $a^3 + 3a^2b - 28ab^2$

[13.5] Solve each of the following equations by factoring.

81. $x^2 - 2x = 35$
82. $y^2 + 7y = 18$
83. $a^2 = 8(2a - 8)$
84. $b^2 = -3(2b + 3)$
85. $5w^2 = 20w$
86. $8z^2 = 48z$
87. $2x^2 + 15x + 28 = 0$
88. $3x^2 + 23x + 40 = 0$
89. $5p^2 = 20p + 60$
90. $8t^2 = 32 - 24t$
91. $x^3 + 17x^2 = -72x$
92. $y^3 + 10y^2 = -21y$
93. $2a^3 = 98a$
94. $5b^3 = 20b$
95. $5t^4 - 21t^3 + 18t^2 = 0$
96. $2z^4 + 11z^3 - 21z^2 = 0$

[13.6] Solve each of the following word problems.

97. The product of two consecutive odd integers is 7 more than their sum. Find the integers.

98. Nine times a certain integer is 12 less than three times the square of the integer. Find the integer.

99. The length of a rectangle is 6 less than five times the width. If the area of the rectangle is 27 ft^2, what are its dimensions?

100. One leg of a right triangle is 7 in. less than the length of the other leg. If the hypotenuse of the triangle is 13 in., what are the lengths of the two legs?

14 GRAPHING

14.1 Inequalities and Line Graphs

In Chapter 10 we learned a technique for solving first degree equations containing a single variable, which are also known as linear equations. Recall that the solution set for a linear equation is a real number.

In this section, we will learn how to solve *linear inequalities* that contain a single variable. The solution set for a linear inequality can be represented by an *interval* on a number line, which contains an infinite number of real numbers. The following examples illustrate how to graph the solution set to a linear inequality on a number line.

example 1

Graph the solution set represented by each of the following on a number line.

(a) $x \leqslant 4$

The solution set is all real numbers less than or equal to 4. We graph the solution set by shading the portion of a number line to the left of and including 4.

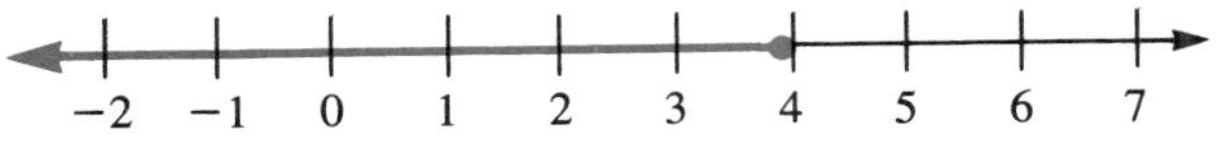

The closed circle at 4, the endpoint of the interval, indicates that 4 is included in the solution set.

(b) $y < \frac{1}{2}$

The solution set is all real numbers less than $\frac{1}{2}$. Shade the region of the number line to the left of $\frac{1}{2}$.

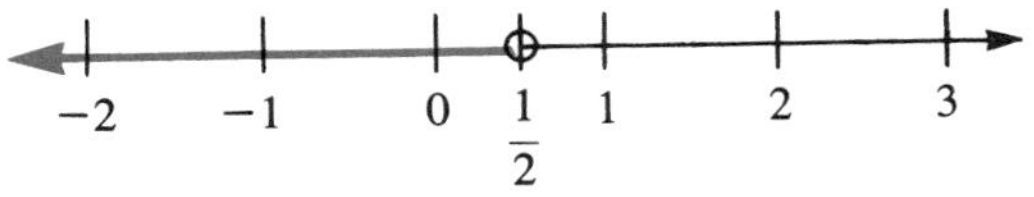

The open circle at $\frac{1}{2}$, the endpoint of the interval, indicates that $\frac{1}{2}$ is not included in the solution set.

(c) $t \geqslant -5$

The solution set is all real numbers greater than or equal to -5. Shade the region of the number line to the right of and including -5.

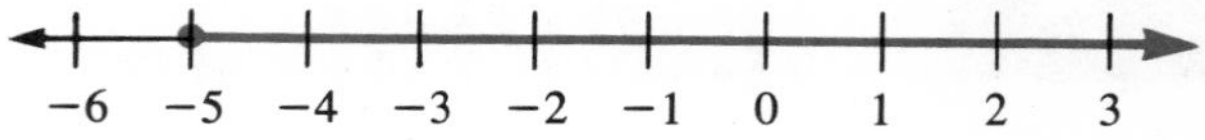

Remember that the closed circle at -5, the endpoint of the interval, indicates that -5 is included in the solution set.

(d) $2.3 > w$

Rewrite the inequality with the variable on the left. Remember that the open end of the inequality sign must face the larger quantity, which is 2.3.

$w < 2.3$

The solution set is all real numbers less than 2.3. Shade the interval of the number line to the left of 2.3.

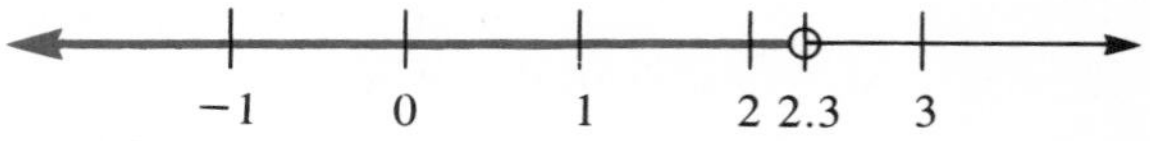

The open circle at 2.3, the endpoint of the interval, indicates that 2.3 is not included in the solution set.

The technique we used to solve a linear equation involved adding the same quantity to both sides of the equation until all variable terms were on one side and the constants were on the other side. To solve for the unknown, we then multiplied both sides by the reciprocal of the coefficient of the variable.

We will now use a similar technique to solve linear inequalities of a single variable. Before we can proceed, we must first determine what happens when a quantity is added to both sides of the inequality. For example, let us consider the expression, $2 < 8$, which states that 2 is less than 8. If we add the number 5 to both sides, we obtain the following:

$$2 < 8$$
$$2 + 5 \ ? \ 8 + 5$$
$$7 < 13$$

Thus, when a positive number is added to both sides, the left-hand side of the inequality is still less than the right. If we add the number -9 to both sides of the original inequality, we obtain the following:

$$2 < 8$$
$$2 + (-9) \ ? \ 8 + (-9)$$
$$-7 < -1$$

Thus, when a negative number ~~is~~ added to both sides, the left-hand side of the inequality is still less than the right. This property, called the **addition property of inequalities**, may be stated as follows:

definition

According to the *addition property of inequalities*, if any quantity is added to both sides of an inequality, the inequality sign remains unchanged.

Using symbols, we can let a, b, and c represent any algebraic expressions and summarize this property as follows:

Addition Property of Inequalities

If $a < b$,
then $a + c < b + c$

If $a > b$,
then $a + c > b + c$

The addition property of inequalities also holds for $\geqslant$ and $\leqslant$.

Let us now use the addition property to solve some linear inequalities.

example 2

Solve each of the following linear inequalities, and graph the solution set on a number line.

(a) $x + 5 \leqslant 3$

$x + 5 + (-5) \leqslant 3 + (-5)$ Add -5 to both sides.

$x \leqslant -2$

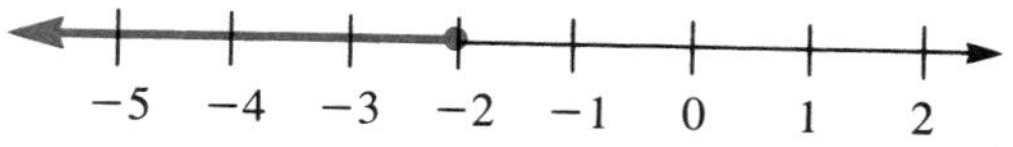

(b) $3w + 2(w - 4) > 4w$

$3w + 2w - 8 > 4w$ Remove ( ) using distributive property.

$5w - 8 > 4w$ Combine like terms on left.

$-4w + 5w - 8 > -4w + 4w$ Add $-4w$ to both sides.

$w - 8 > 0$ Combine like terms on left.

$w - 8 + 8 > 0 + 8$ Add 8 to both sides.

$w > 8$

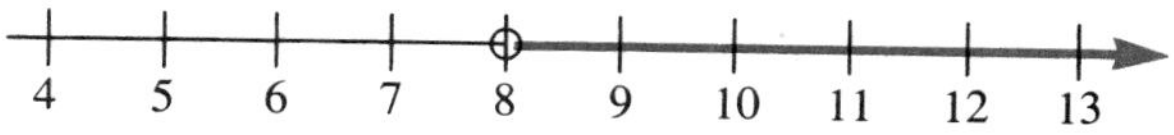

Let us now examine what happens when both sides of a linear inequality are multiplied by a positive number or a negative number. If we begin with the inequality $2 < 8$ and multiply both sides by 3, we obtain the following:

$$2 < 8$$
$$2(3) \;?\; 8(3)$$
$$6 < 24$$

Thus, when both sides of the inequality are multiplied by a positive number, the left-hand side remains less than the right-hand side. If we multiply both sides of the original inequality by -1, we obtain the following:

$$2 < 8$$
$$2(-1) \ ? \ 8(-1)$$
$$-2 > -8$$

Thus, when both sides of the inequality are multiplied by a negative number, the inequality *reverses*, and the left-hand side becomes greater than the right-hand side. This property is called the **multiplication property of inequalities** and may be stated as follows:

definition

According to the *multiplication property of inequalities*, if both sides of an inequality are multiplied by a positive number, the inequality sign remains unchanged. If both sides are multiplied by a negative number, the inequality sign reverses.

Using symbols, we can let a, b, and c represent any algebraic expressions and summarize this property as follows:

Multiplication Property of Inequalities

If	$a < b$,	
then	$ac < bc$,	when c is positive
and	$ac > bc$,	when c is negative
If	$a > b$,	
then	$ac > bc$	when c is positive
and	$ac < bc$	when c is negative

The multiplication property of inequalities also holds for $\leqslant$ and $\geqslant$.

We will now use the multiplication property to solve some linear inequalities.

example 3

Solve each of the following linear inequalities and graph the solution set on a number line.

(a) $-7z \leqslant 14$

$$\left(-\frac{1}{7}\right)(-7z) \geqslant \left(-\frac{1}{7}\right)(14)$$ Multiply both sides by $-\frac{1}{7}$ and reverse the inequality sign.

$$z \geqslant -2$$

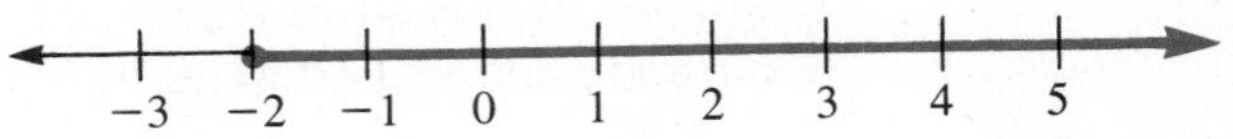

(b) $2 < \frac{1}{3}t$

$$(3)(2) < (3)\left(\frac{1}{3}t\right)$$ Multiply both sides by 3. The inequality sign remains unchanged.

$6 < t$ — Rewrite inequality with variable on left.

$t > 6$ — Notice that $6 < t$ and $t > 6$ mean the same thing. In each case, the open end of the inequality sign faces the larger quantity, t.

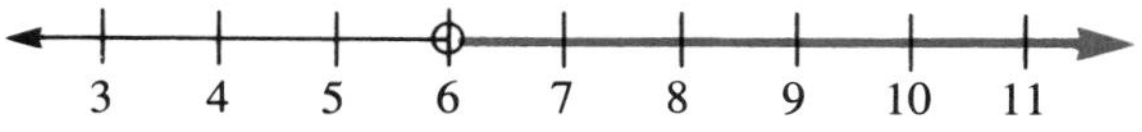

Many linear inequalities will require us to see both the addition property and the multiplication property in order to solve them. For these problems, we can use the following procedure:

Procedure for Solving Linear Inequalities

1. Simplify both sides as much as possible using the distributive property to multiply all quantities in parentheses and combine like terms.
2. Use the addition property of inequalities to move all variable terms to one side of the inequality and all constant terms to the other side.
3. Use the multiplication property of inequalities to isolate the variable on one side of the inequality. Remember to reverse the inequality sign if you multiply both sides by a negative number.
4. Graph the solution set on a number line.

Let us now look at some examples that illustrate this procedure.

example 4

Solve each of the following linear inequalities and graph the solution set on a number line.

(a) $5x - 7 \leqslant 2x + 8$

$-2x + 5x - 7 \leqslant -2x + 2x + 8$ — Add $-2x$ to both sides.

$3x - 7 \leqslant 8$

$3x - 7 + 7 \leqslant 8 + 7$ — Add 7 to both sides.

$3x \leqslant 15$

$\frac{1}{3}(3x) \leqslant \frac{1}{3}(15)$ — Multiply both sides by $\frac{1}{3}$.

$x \leqslant 5$

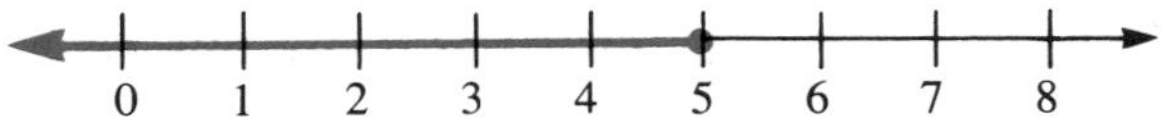

(b) $-4(x + 3) + 7 > 5x + 2$

$-4x - 12 + 7 > 5x + 2$ — Remove ( ) using distributive property.

$-4x - 5 > 5x + 2$

$-5x - 4x - 5 > -5x + 5x + 2$ — Add $-5x$ to both sides.

$-9x - 5 > 2$

$$-9x - 5 + 5 > 2 + 5$$ Add 5 to both sides.

$$-9x > 7$$

$$\left(-\frac{1}{9}\right)(-9x) < \left(-\frac{1}{9}\right)(7)$$ Multiply both sides by $-\frac{1}{9}$ and reverse inequality sign.

$$x < -\frac{7}{9}$$

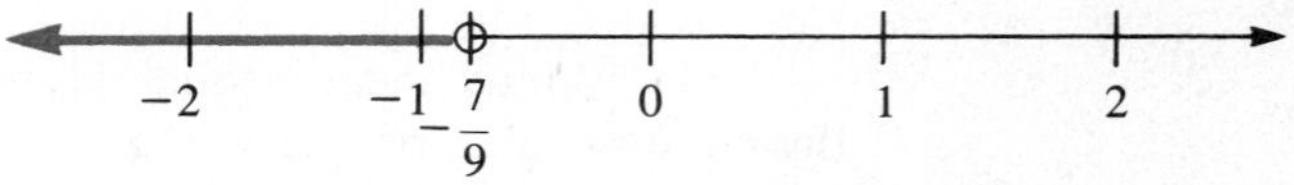

example 5 The sum of three times a number and eight is greater than five times the number. What numbers satisfy this condition?

solution Let x = the number

$$3x + 8 > 5x$$

$$-2x + 8 > 0$$ Add $-5x$ to both sides.

$$-2x > -8$$ Add -8 to both sides.

$$x < 4$$ Multiply both sides by $-\frac{1}{2}$ and reverse inequality sign.

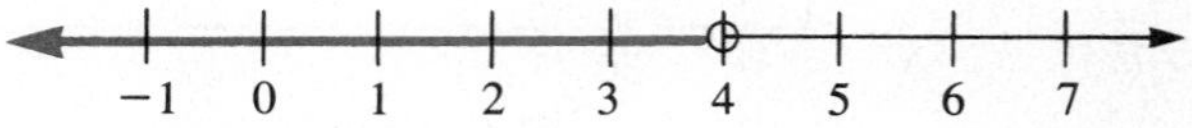

Any number less than 4 satisfies this condition.

QUICK QUIZ

Find the solution set for each of the following linear inequalities and graph the solution on a number line.

1. $x + 3 \leqslant 2$

2. $-\frac{3}{4}y < 9$

3. $5t + 2 < 4 - t$

4. $7(t - 3) \geqslant 4(1 + 2t)$

ANSWERS

1. $x \leqslant -1$

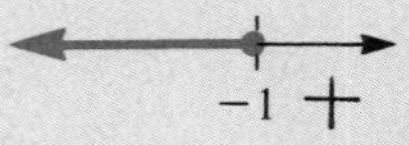

2. $y > -12$

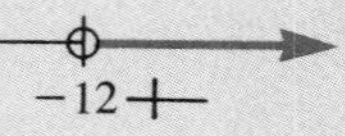

3. $t < \frac{1}{3}$

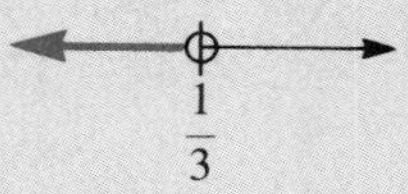

4. $t \leqslant -25$

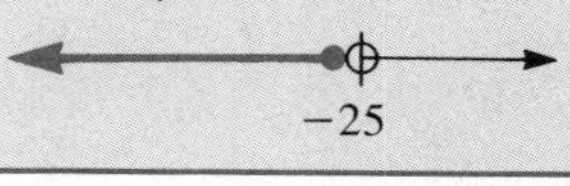

14.1 Exercises

Graph the solution set represented by each of the following on a number line.

1. $x > 7$
2. $x < 3$
3. $y \leqslant -2$
4. $y > -5$
5. $t \geqslant \frac{1}{3}$
6. $z \leqslant 1.5$
7. $p < -4.9$
8. $w \geqslant -\frac{3}{4}$
9. $9 \geqslant x$
10. $2 < x$

Use the addition property of inequalities to solve each of the following and graph the solution set on a number line.

11. $x + 3 > 9$
12. $x + 8 \leqslant 2$
13. $x - 7 \leqslant 5$
14. $x - 5 < 1$
15. $5y - 2 \geqslant 4y + 7$
16. $8y + 6 > 7y + 3$
17. $9 < t + 6$
18. $5 \geqslant t - 4$
19. $2(3t - 5) \leqslant 5t$
20. $4(2t - 1) \leqslant 7t$
21. $8t - 3(2t + 1) > 5(t - 2) - 4t$
22. $6(t - 3) + 3t < 2(4t + 5) - 7$

Use the multiplication property of inequalities to solve each of the following and graph your solution set on a number line.

23. $3x \leqslant 12$
24. $4x < 32$
25. $-7t < 21$
26. $-5w \geqslant 35$
27. $\frac{1}{4}y > 8$
28. $\frac{3}{5}p < 6$
29. $-\frac{2}{3}x \geqslant 16$
30. $\frac{1}{7}y \geqslant 4$
31. $9 < 3p$
32. $6 > -2q$
33. $-2z \leqslant -5$
34. $-4t \geqslant -7$

Solve each of the following linear inequalities and graph your solution set on a number line.

35. $5x - 3 \geqslant 7$
36. $3x + 5 \leqslant 2$
37. $2 - 7x < 16$
38. $-4x - 7 > 5$
39. $8y + 5 \geqslant 2$
40. $7y - 8 < -3$
41. $4t + 9 < 7t - 2$
42. $2t - 5 \geqslant 6t - 3$
43. $4(t - 5) \leqslant 8$
44. $5(t + 3) > 10$
45. $3(2y + 8) < 5$
46. $4(3y + 7) \geqslant -3$
47. $-2(x - 7) > 4$
48. $-3(x - 5) < 12$
49. $5(y + 3) - 2y < 8y + 9$
50. $6(y - 4) + y \geqslant 5y - 4$
51. $4(7 - t) \leqslant -3(2t + 5)$
52. $5(2t + 1) < -2(9 - t)$
53. $2(4w + 5) - 3w \geqslant -5(7 - 2w)$
54. $-3(9 - 2z) + z < 4(3z - 8)$

Solve each of the following word problems.

55. The sum of three times a number and five is less than twice the number. What numbers satisfy this condition?
56. The sum of eight times a number and nine is greater than or equal to five times the number. What numbers satisfy this condition?
57. The product of a number and seven is greater than the number plus twelve. Find all possible solutions.
58. The product of a number and four is less than or equal to the number minus six. Find all possible solutions.
59. A number decreased by two is greater than or equal to three times the number minus sixteen. What numbers satisfy this condition?
60. A number increased by five is less than seven times the number plus forty-one. What numbers satisfy this condition?

14.2 The Cartesian Coordinate System

We have already seen, in Section 7.5, that any real number corresponds to a point on a number line. For example, the points that represent the numbers $3, \frac{1}{2}, -2, 0$, and $\sqrt{3}$ are shown in the number line illustrated in Figure 14.1.

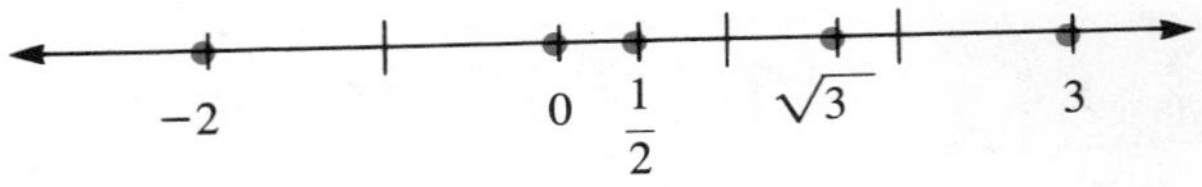

figure 14.1

In this section, we will learn how to graphically represent an *ordered pair* of real numbers. Just as a real number is associated with a point on a number line, an ordered pair is associated with a point in a two-dimensional plane called the *Cartesian Coordinate System* or *Rectangular Coordinate System.*

To draw the Cartesian Coordinate System, we join two number lines, one vertical and the other horizontal, at their zeros. The horizontal number line is called the x-axis, and the vertical number line is called the y-axis. The x- and y-axes are perpendicular, that is, they meet in a right angle. Their point of intersection is called the *origin.* The x- and y-axes divide the Cartesian plane into four quadrants numbered I, II, III, and IV moving counterclockwise as shown in Figure 14.2.

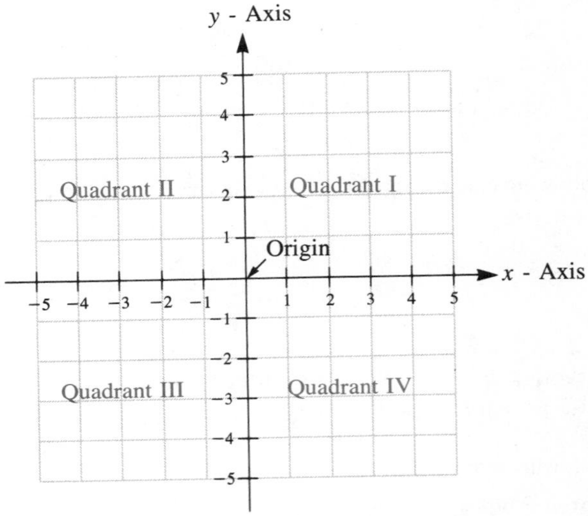

figure 14.2

To locate a point in the Cartesian plane (or x-y plane), we must know its x-coordinate, which is the horizontal distance from the y-axis, and the y-coordinate, which is the vertical distance from the x-axis. In general, any point in the Cartesian plane can be represented by the ordered pair (x, y), where x is the x-coordinate or *abscissa*, and y is the y-coordinate or *ordinate.* Note that the x-coordinate always appears before the y-coordinate in the ordered pair.

Points having positive x-coordinates are located to the right of the y-axis, those having negative x-coordinates are located to the left of the y-axis, and those for which the x-coordinate is zero are *on* the y-axis. Points having positive y-coordinates are located above the x-axis, those having negative y-coordinates are located below the x-axis, and those for which the y-coordinate is zero are *on* the x-axis. The origin has coordinates $(0, 0)$.

For example, in the ordered pair $(2, 3)$, 2 is the x-coordinate or abscissa and 3 is the y-coordinate or ordinate. To graph this point in the x-y plane we begin at the origin and move 2 units to the right of the y-axis and 3 units up from the x-axis, as shown in Figure 14.3. The points associated with the ordered pairs $(-3, 1)$, $(1, -3)$, $(-2, -4)$ and $(4, 0)$ are also shown in Figure 14.3.

figure 14.3

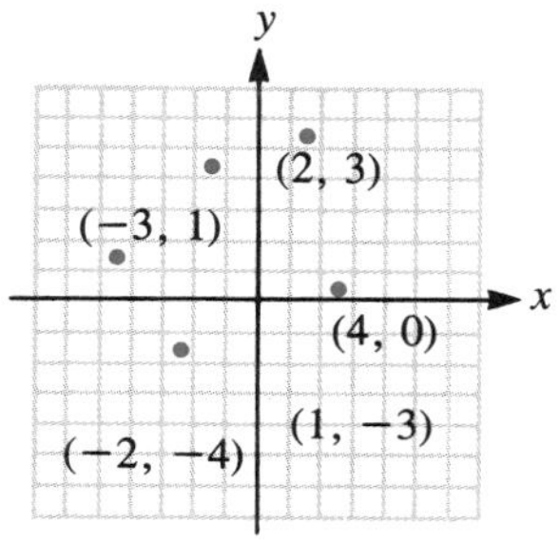

The order of the coordinates in an ordered pair specifies the location of the point it represents. If the x- and y-coordinates are reversed and $x \neq y$, a different point is represented. For example, note that the ordered pairs $(-3, 1)$ and $(1, -3)$ are associated with two different points in Figure 14.3.

This procedure of graphing a point in the x-y plane is also called **plotting** a point. Let us now consider these examples.

example 1

Plot the points associated with the ordered pairs $(5, 2)$, $(-3, 0)$, $(-1, -4)$, $(0, 6)$, and $(2, -2)$.

solution

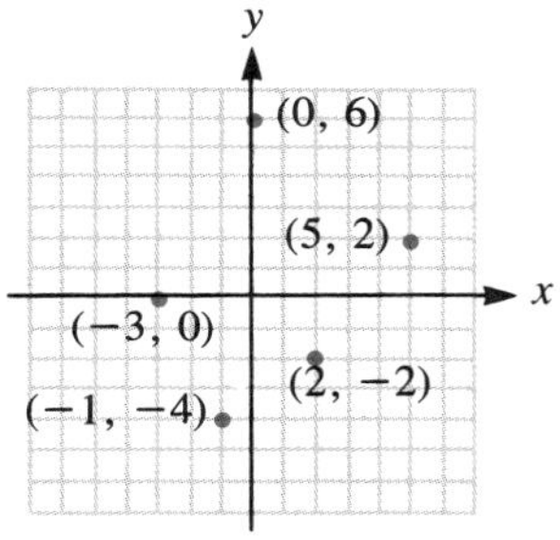

example 2

Give the ordered pairs represented by the points A, B, C, D, E, and F shown in this diagram.

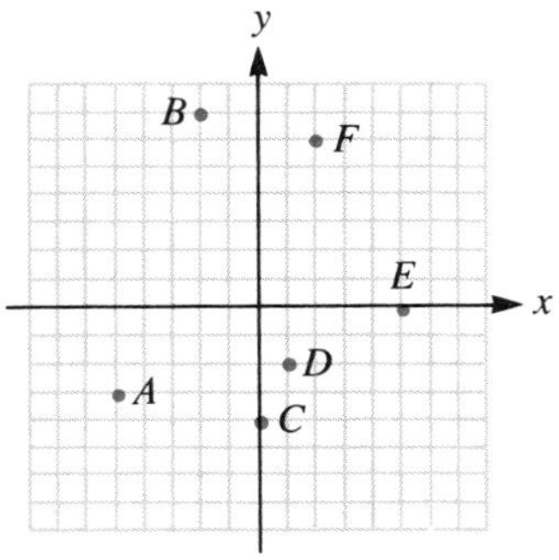

solution

$A(-5, -3)$ $B(-2, 7)$ $C(0, -4)$
$D(1, -2)$ $E(5, 0)$ $F(2, 6)$

14.2 Exercises

Indicate which of the following statements are true and which are false. For those that are false, change the italic word to make the statement true.

1. The rectangular coordinate system is also called the *Cartesian* coordinate system.
2. In an ordered pair that represents a point in the x-y plane, the y-coordinate appears *first*.
3. The x-coordinate of an ordered pair is also called the *ordinate*.
4. The Cartesian plane is divided into *four* quadrants by the x- and y-axes.
5. If a point is located in the second quadrant, its abscissa is *greater* than zero.
6. The *ordinate* of every point on the x-axis is zero.
7. The *y-coordinate* of every point on the y-axis is zero.
8. Both the x- and y-coordinates of every point in the *fourth* quadrant are less than zero.

Plot the points associated with each of the following ordered pairs. Be sure to label each point you plot.

9. (6, 1) **10.** (1, 6) **11.** (5, 0) **12.** (0, 8) **13.** (−3, 4) **14.** (7, −2)
15. (5, −9) **16.** (−4, 4) **17.** (−8, −5) **18.** (−3, −7) **19.** (0, −6) **20.** (−9, 0)
21. (−5, 7) **22.** (7, −5) **23.** (6, 9) **24.** (8, 1) **25.** (−4, 0) **26.** (5, 0)
27. (−9, −9) **28.** (−7, 7)

Give the ordered pairs represented by the points shown in this diagram.

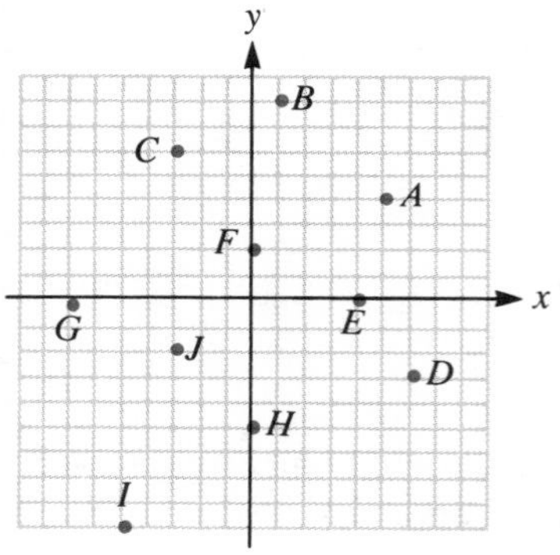

29. *A* **30.** *B* **31.** *C* **32.** *D* **33.** *E* **34.** *F*
35. *G* **36.** *H* **37.** *I* **38.** *J*

14.3 Graphing Linear Equations

In Chapter 10, we found that the solution to a linear equation in one variable is a single real number. For example, the solution to the equation

$$2x + 3 = 5$$

is the number 1. When 1 is substituted in for the variable x, the resulting statement is true:

$$2(1) + 3 \;?\; 5$$
$$2 + 3 \;?\; 5$$
$$5 = 5 \;\checkmark$$

The solution set for a linear equation in two variables is a set of ordered pairs. For example, one solution to the linear equation

$$x + 2y = 7$$

is the ordered pair (3, 2). This solution can also be expressed in the form $x = 3$, $y = 2$. Thus, when 3 is substituted in for the variable x and 2 is substituted in for the variable y, the resulting statement is true:

$$3 + 2(2) \;?\; 7$$
$$3 + 4 \;?\; 7$$
$$7 = 7 \;\checkmark$$

There are an infinite number of ordered pairs that are solutions to a linear equation in two variables

> To determine whether an ordered pair is a solution to a linear equation in two variables, substitute the first coordinate for x and the second coordinate for y. If the resulting statement is true, the ordered pair is a solution to the equation.

Other solutions to the linear equation

$$x + 2y = 7$$

are the ordered pairs (7, 0), (11, −2), (−1, 4), and $\left(0, \frac{7}{2}\right)$. Notice that for each ordered pair, the statements that result from substituting the appropriate values for x and y are true:

(7, 0): $7 + 2(0) \;?\; 7$, $7 + 0 \;?\; 7$, $7 = 7 \;\checkmark$

(11, −2): $11 + 2(-2) \;?\; 7$, $11 + -4 \;?\; 7$, $7 = 7 \;\checkmark$

(−1, 4): $-1 + 2(4) \;?\; 7$, $-1 + 8 \;?\; 7$, $7 = 7 \;\checkmark$

$\left(0, \frac{7}{2}\right)$: $0 + 2\left(\frac{7}{2}\right) \;?\; 7$, $0 + 7 \;?\; 7$, $7 = 7 \;\checkmark$

example 1

Determine whether or not each of the following ordered pairs are solutions to the linear equations given.

(a) (1, −2): $5x - 3y = 8$

$$5(1) - 3(-2) \;?\; 8$$
$$5 + 6 \;?\; 8$$
$$11 \neq 8$$

Thus, (1, −2) is not a solution to the equation $5x - 3y = 8$.

(b) $(-1, 9)$: $y = -8x + 1$

$$9 \ ? \ -8(-1) + 1$$
$$9 \ ? \ 8 + 1$$
$$9 = 9 \ \checkmark$$

Thus, $(-1, 9)$ is a solution to the equation $y = -8x + 1$. ●

To find a solution to a linear equation in two variables, we can use this procedure:

1. Choose a value for either the x- or y-coordinate of the ordered pair.
2. Substitute into the given equation the value chosen for the first variable, and then solve the equation for the other coordinate of the ordered pair.

Sometimes we are given one coordinate of an ordered pair that is a solution to a linear equation, and we are asked to find the other coordinate, as shown in the following examples:

example 2

Find the missing coordinates in the ordered pairs that are solutions to the following equations.

(a) $3x + y = 7$: $(0, \ \ )$

Substitute in $x = 0$ and solve for y.

$$3(0) + y = 7$$
$$0 + y = 7$$
$$y = 7$$

The solution is $(0, 7)$.

(b) $y = 9x + 2$: $(\ \ , -1)$

Substitute in $y = -1$ and solve for x.

$$-1 = 9x + 2$$
$$9x + 2 = -1$$
$$9x = -3$$
$$x = -\frac{1}{3}$$

The solution is $\left(-\frac{1}{3}, -1\right)$. ●

The graph of a linear equation is a straight line which includes the set of points representing the ordered pairs of the solution set. Since a straight line is determined by two points, we need to find at least two ordered pairs that are solutions to the linear equation in order to graph it. A third ordered pair may be found to check that the solution is correct.

To Graph a Linear Equation

1. Find at least two ordered pairs that are solutions to the linear equation.
2. Plot the points that correspond to those ordered pairs on the x-y plane.
3. Draw a straight line through those points.

example 3 Graph the equation $4x - 3y = 6$.

solution Find a few ordered pairs of the solution set:

Let $x = 0$: $4(0) - 3y = 6$
$-3y = 6$
$y = -2$
$(0, -2)$ is a solution.

Let $x = 3$: $4(3) - 3y = 6$
$12 - 3y = 6$
$-3y = -6$
$y = 2$
$(3, 2)$ is a solution.

Let $y = 0$: $4x - 3(0) = 6$
$4x = 6$
$x = \frac{3}{2}$

$\left(\frac{3}{2}, 0\right)$ is a solution.

When graphing linear equations, the solutions are often listed in a table as shown here. Plot these three points and draw a straight line through them to obtain the graph.

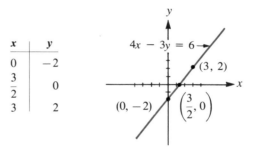

x	y
0	-2
$\frac{3}{2}$	0
3	2

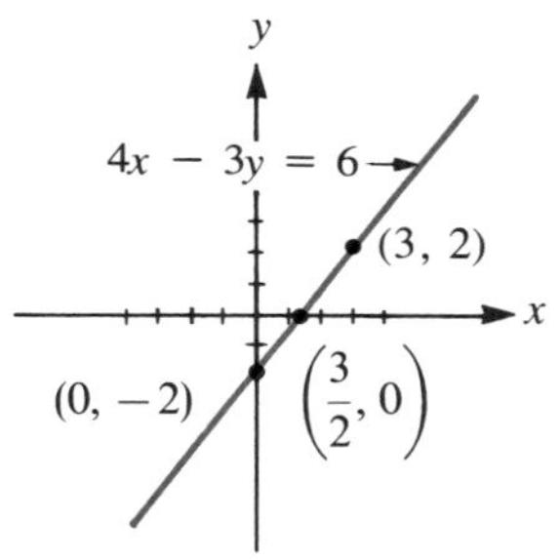

Though it is only necessary to find 2 points, the third point acts as a check. If the three points did not all lie in a straight line, we made a mistake in calculating one of the coordinates. ●

The x-coordinate of the point where the graph of an equation crosses the x-axis is called the *x-intercept*, and the y-coordinate of the point where it crosses the y-axis is called the *y-intercept*. In Example 3, the equation $4x - 3y = 6$ crosses the x-axis at the point $\left(\frac{3}{2}, 0\right)$, and it crosses the y-axis at the point $(0, -2)$. Thus, the x-intercept is $\frac{3}{2}$ and the y-intercept is -2.

In general, if a is an x-intercept of a graph, then the graph passes through the point $(a, 0)$. If b is a y-intercept of a graph, then the graph passes through the point $(0, b)$. This is illustrated in Figure 14.4.

Therefore, to find the x-intercept of an equation, set $y = 0$ and solve for x. To find the y-intercept, set $x = 0$ and solve for y. *The simplest way to graph many linear equations is to find the x- and y-intercepts and connect them with a straight line.*

figure 14.4

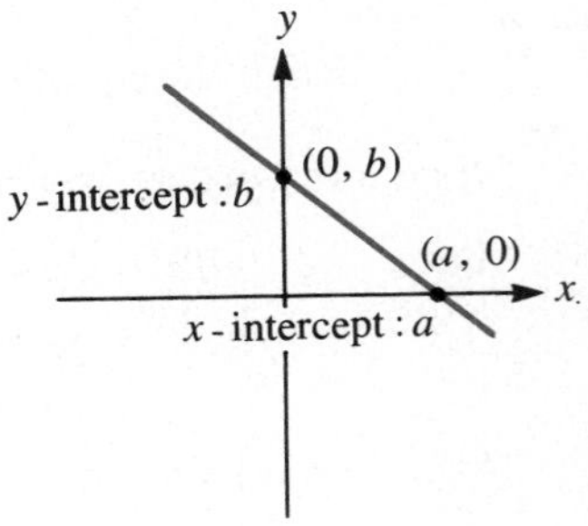

example 4

Graph the equation $3x + 5y = 15$.

solution

Find the x-intercept:

Let $y = 0$: $3x + 5(0) = 15$

$3x = 15$

$x = 5$

Find the y-intercept:

Let $x = 0$: $3(0) + 5y = 15$

$5y = 15$

$y = 3$

Plot (5, 0) and (0, 3) and draw a straight line through them.

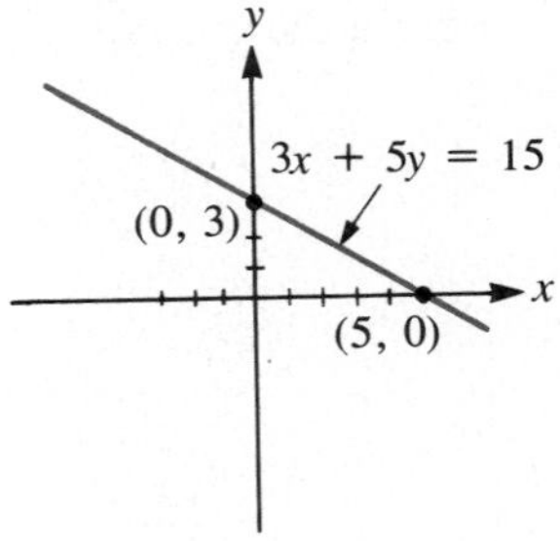

example 5

Graph the equation $y = -5x + 7$.

solution

Choose values for x and mentally calculate the corresponding y-coordinates. List the solutions using a table format. Plot these points and draw a straight line through them.

x	y
0	7
1	2
2	−3

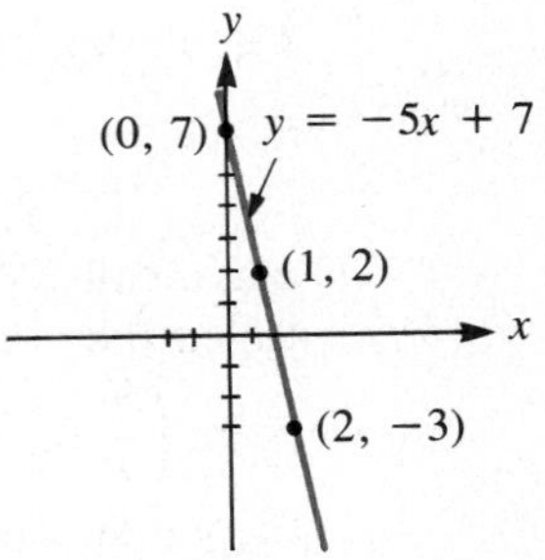

example 6

Graph the equation $x = 3$.

solution

The x-coordinate of every point on this line must be 3.
The y-coordinate can be anything.
Thus, two points on this line are (3, 0) and (3, 5).

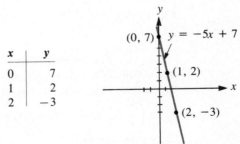

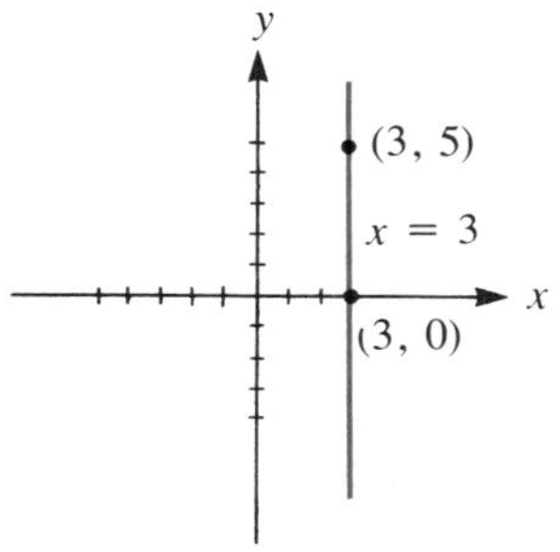

Notice that the graph of the equation $x = 3$ is a vertical line. ●

example 7

Graph the equation $y = -1$.

solution

The y-coordinate of every point on this line must be -1.
The x-coordinate can be anything.
Thus, two points on this line are $(-2, -1)$ and $(2, -1)$.

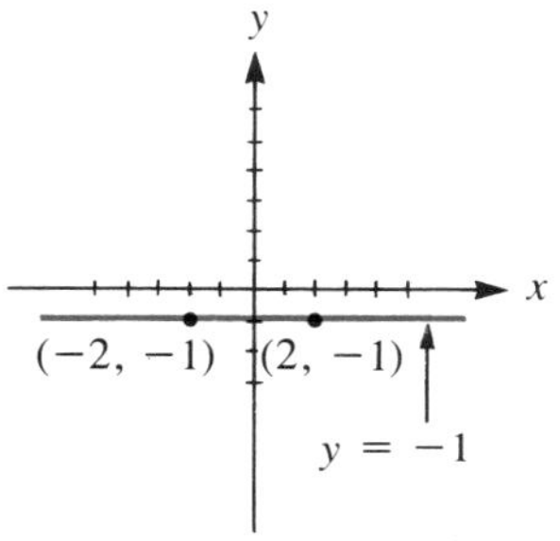

Notice that the graph of the equation $y = -1$ is a horizontal line. ●

Examples 6 and 7 illustrate the following:

> The equation of a vertical line is $x = a$, where a is the x-coordinate of every point on the line. The equation of a horizontal line is $y = b$, where b is the y-coordinate of every point on the line.

example 8

Find the equation of the vertical line that passes through the point $(-5, 2)$.

solution

The equation of a vertical line is $x = a$ where a is the x-coordinate of every point on the line. Since one point on this line has an x-coordinate of -5, the equation of this line is $x = -5$. ●

14.3 Exercises

Indicate which of the following statements are true and which are false. For those that are false, change the italic expression to make the statement true.

1. A solution to a linear equation in two variables is an *ordered pair.*
2. To find the x-intercept of an equation, set the *variable* x equal to zero.
3. A *horizontal* line has the equation $y = b$ where b is the y-coordinate of every point on the line.
4. The *x-intercept* of the line $y = b$ is b.

Determine whether or not each of the following ordered pairs are solutions to the linear equation given.

5. $(1, -8)$: $y = -8x$
6. $(3, 1)$: $y = 3x$
7. $(4, 1)$: $x + y = 4$
8. $(2, -4)$: $x - y = 6$
9. $(-2, 3)$: $x + 4y = 10$
10. $(4, 1)$: $2x + y = 8$
11. $(-3, -8)$: $y = 5x + 7$
12. $(-2, -9)$: $y = 3x - 4$
13. $(3, 4)$: $5x - 2y = 6$
14. $(2, 1)$: $-3x + 7y = 1$

Find the missing coordinates in the ordered pairs which are solutions to the following equations.

15. $y = -\frac{1}{3}x$ $(6, \ ), (\ , 0)$
16. $y = \frac{2}{5}x$ $(\ , 3), (5, \ )$
17. $x + 2y = 6$ $(\ , 2), (-4, \ )$
18. $x - 3y = 14$ $(-2, \ ), (\ , 3)$
19. $y = 5x$ $(-1, \ ), (\ , -10)$
20. $y = -8x$ $(\ , -8), (0, \ )$
21. $y = 5x + 2$ $(3, \ ), (-2, \ )$
22. $y = 8x - 4$ $(-1, \ ), (3, \ )$
23. $4x + y = 11$ $(\ , 3), (-6, \ )$
24. $-7x + y = 5$ $(\ , -9), (1, \ )$
25. $3x + 6y = 18$ $(\ , 3), (-2, \ )$
26. $2x - 4y = 8$ $(4, \ ), (\ , 3)$
27. $-3x + 2y = -8$ $(\ , -4), (\ , 5)$
28. $-5x - 4y = 3$ $(\ , 3), (\ , -2)$
29. $x = 7$ $(\ , -5), (7, \ )$
30. $y = -3$ $(\ , -3), (9, \ )$

Find the x-intercept and y-intercept of each of the following linear equations.

31. $x - y = 8$
32. $x + y = 3$
33. $y = 5x$
34. $y = -7x$
35. $y = 3x + 2$
36. $y = -5x + 1$
37. $x + 4y = 8$
38. $3x - y = 6$
39. $2x - 5y = 10$
40. $4x + 7y = 28$
41. $3x + 7y = 21$
42. $9x - 2y = 18$

Graph each of the following linear equations.

43. $x + y = 7$
44. $x - y = 4$
45. $y = 3x + 6$
46. $y = -5x + 10$
47. $y = -5x$
48. $y = 3x$
49. $3x + 4y = 12$
50. $3x + 6y = 18$
51. $5x - 3y = 15$
52. $7x - 2y = 14$
53. $y = 5x - 2$
54. $y = 4x + 1$
55. $3x + y = 9$
56. $4x - y = 8$
57. $x - 5y = 10$
58. $x + 2y = 6$
59. $x = 7$
60. $y = 3$
61. $y = -4$
62. $x = -6$

Answer the following questions.

63. What is the equation of the vertical line that passes through the point $(2, -5)$?
64. What is the equation of the horizontal line that passes through the point $(-1, -7)$?
65. What is the equation of the horizontal line that passes through the point $\left(\frac{1}{2}, \frac{2}{3}\right)$?
66. What is the equation of the vertical line that passes through the point $(-3.5, -6.8)$?
67. Plot three points on the x-axis. What is the y-coordinate of every point on the x-axis? What is the equation of the x-axis?
68. Plot three points on the y-axis. What is the x-coordinate of every point on the y-axis? What is the equation of the y-axis?

69. Graph the lines $y = 2x + 3$ and $y = 2x - 5$ on the same set of coordinate axes. What do you notice about these two lines?

70. Graph the lines $y = -3x - 1$ and $y = -3x + 4$ on the same set of coordinate axes. What do you notice about these two lines?

14.4 Slope of a Line

An important property by which we can characterize straight lines is *slope*. The slope of a line is a number that indicates the steepness of the line. Slope is defined as the ratio of the vertical change to the horizontal change as we move from one point to another along a line.

figure 14.5

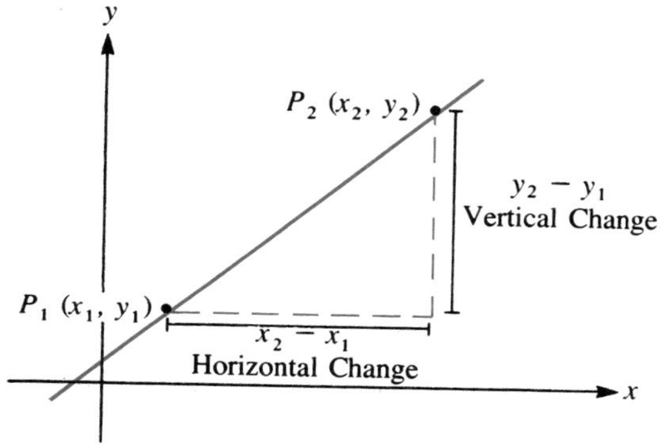

Thus, to find the slope of a line, we need to know the coordinates of two points on the line. Let us use the notation (x_1, y_1) and (x_2, y_2) to indicate the coordinates of two points P_1 and P_2 on the line shown in Figure 14.5. The horizontal change is the difference between the two x-coordinates $x_2 - x_1$. The vertical change is the difference between the two y-coordinates $y_2 - y_1$. The slope of the line may be calculated as follows:

$$\text{slope} = \frac{\text{vertical change}}{\text{horizontal change}} = \frac{y_2 - y_1}{x_2 - x_1}, \quad x_2 \neq x_1$$

Sometimes we use the shorthand notation Δy (read as "delta y") to indicate the vertical change $y_2 - y_1$ and Δx (read as "delta x") to indicate the horizontal change $x_2 - x_1$. The symbol Δ means "change in." Slope is usually indicated by the letter m. Therefore, the slope of a line that passes through the points (x_1, y_1) and (x_2, y_2) may be defined as follows:

definition

The *slope* of a line that passes through the points (x_1, y_1) and (x_2, y_2) is given by:

$$\textbf{slope} = m = \frac{y_2 - y_1}{x_2 - x_1} = \frac{\Delta y}{\Delta x}, \quad x_2 \neq x_1$$

example 1

Find the slope of a line that passes through the points (3, 1) and (5, 7).

solution

Let $(x_1, y_1) = (3, 1)$ and $(x_2, y_2) = (5, 7)$
Substitute these values into the equation for slope:

$$m = \frac{y_2 - y_1}{x_2 - x_1} = \frac{7 - 1}{5 - 3} = \frac{6}{2} = 3$$

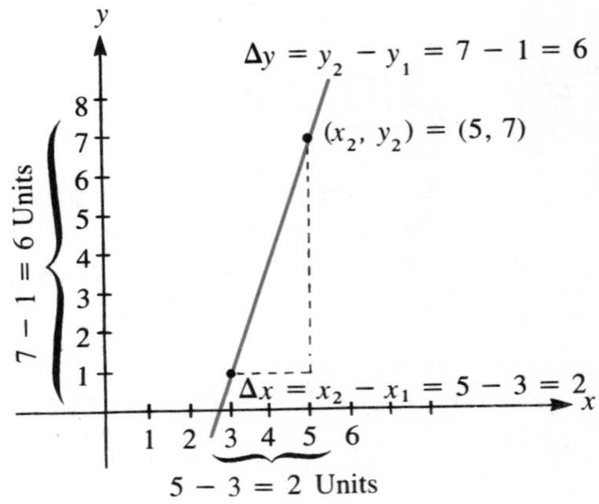

Notice that this line has a positive slope and runs up to the right. ●

example 2

Find the slope of the line $2x + 3y = 6$.

solution

First find two points on this line:

when $x = 0$, $y = 2$: $(0, 2) = (x_1, y_1)$
when $y = 0$, $x = 3$: $(3, 0) = (x_2, y_2)$

Then substitute these values into the equation for slope.

$$m = \frac{y_2 - y_1}{x_2 - x_1} = \frac{0 - 2}{3 - 0} = -\frac{2}{3}$$

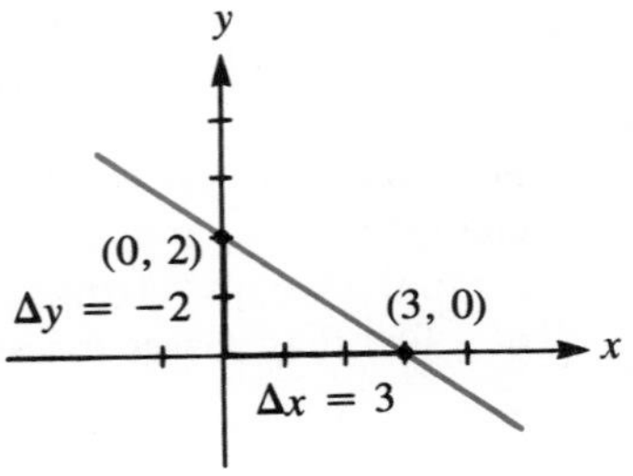

Notice that this line has a negative slope and runs down to the right. ●

We can draw the graph of a line if we are given the slope and one point on the line as shown in the following examples.

example 3

Draw the graph of a line that passes through the point $(-3, 2)$ and has a slope of $\frac{4}{5}$.

solution

First plot the point $(-3, 2)$ on the coordinate axes.

Since slope $= \frac{\Delta y}{\Delta x} = \frac{4}{5}$,

horizontal change = 5 and vertical change = 4.

To find another point on the line, begin at $(-3, 2)$ and move 5 units to the right and 4 units up.
This brings you to (2, 6).
Then draw a line through the points $(-3, 2)$ and (2, 6).

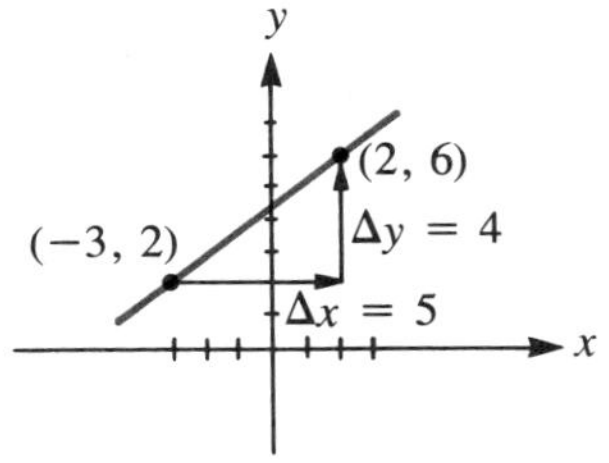

This line has a positive slope and runs up to the right. ●

example 4

Draw the graph of a line that passes through the point (1, 4) and has a slope of -2.

solution

First plot the point (1, 4) on the coordinate axes.

Since slope $= \frac{\Delta y}{\Delta x} = -2 = \frac{-2}{1}$,

horizontal change = 1 and vertical change = -2.

To find another point on the line, begin at (1, 4) and move 1 unit to the right and 2 units down.
This brings you to (2, 2).
Then draw a line through the points (1, 4) and (2, 2).

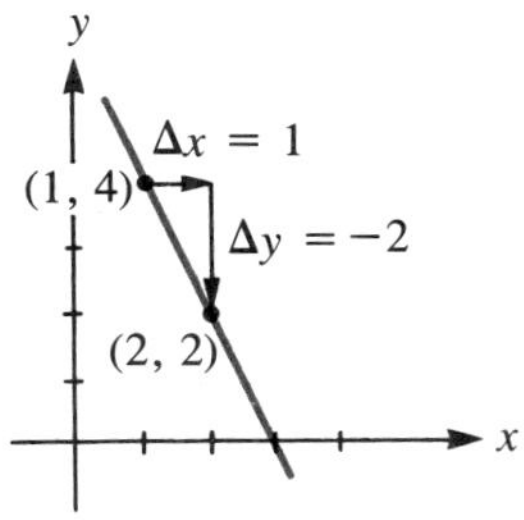

This line has a negative slope and runs down to the right. ●

The next two examples illustrate how to find the slope of a horizontal and vertical line.

example 5

Find the slope of the line $y = 5$.

solution

This is the equation of a horizontal line.
Let $(x_1, y_1) = (0, 5)$ and $(x_2, y_2) = (3, 5)$ be 2 points on this line. Substituting into the equation for slope we obtain:

$$m = \frac{\Delta y}{\Delta x} = \frac{y_2 - y_1}{x_2 - x_1} = \frac{5-5}{3-0} = \frac{0}{3} = 0$$

Notice that for any horizontal line, $\Delta y = 0$, and $\Delta x \neq 0$.
Thus, the slope of any horizontal line is 0.

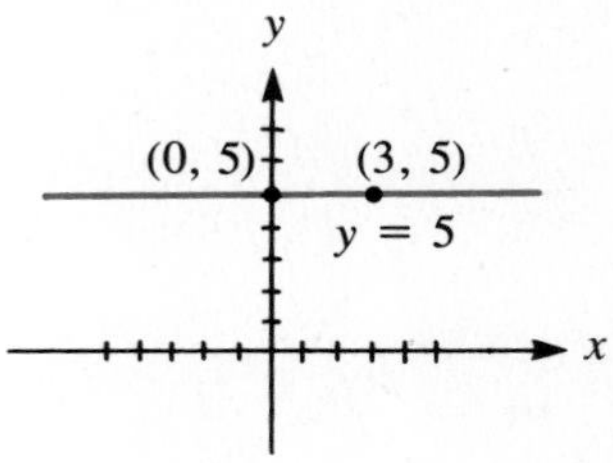

example 6

Find the slope of the line $x = -2$.

solution

Let $(x_1, y_1) = (-2, 4)$ and $(x_2, y_2) = (-2, 0)$ be two points on this line. Substituting into the equation for slope we obtain:

$$m = \frac{\Delta y}{\Delta x} = \frac{y_2 - y_1}{x_2 - x_1} = \frac{4-0}{-2-(-2)} = \frac{4}{0}, \text{ which is undefined.}$$

Notice that for any vertical line $\Delta y \neq 0$ and $\Delta x = 0$.
Thus, the slope of any vertical line is undefined.

Examples 5 and 6 illustrate the following:

> The slope of any horizontal line is zero.
> The slope of any vertical line is undefined.

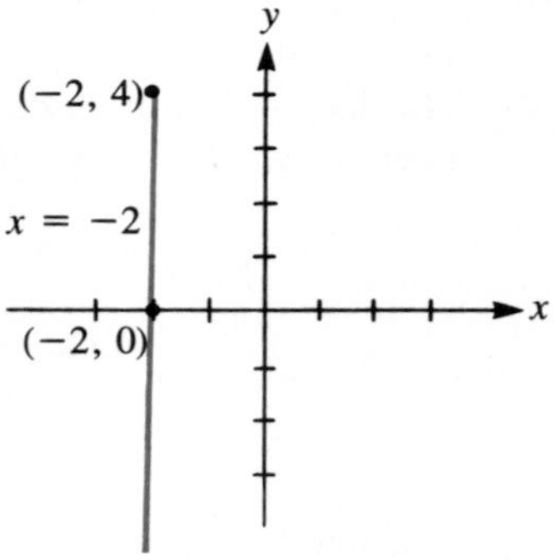

example 7

Graph the lines $y = 3x + 2$ and $y = 3x - 4$ on the same set of axes and find the slope of each.

solution Find the x- and y-intercept of each line.

$y = 3x + 2$

To find the y-intercept, let $x = 0$:

$y = 3(0) + 2$
$y = 2$

Thus, $(x_1, y_1) = (0, 2)$
To find the x-intercept, let $y = 0$:

$0 = 3x + 2$
$3x = -2$
$x = -\frac{2}{3}$

Thus, $(x_2, y_2) = \left(-\frac{2}{3}, 0\right)$

$y = 3x - 4$

To find the y-intercept, let $x = 0$:

$y = 3(0) - 4$
$y = -4$

Thus, $(x_1, y_1) = (0, -4)$.
To find the x-intercept, let $y = 0$:

$0 = 3x - 4$
$3x = 4$
$x = \frac{4}{3}$

Thus, $(x_2, y_2) = \left(\frac{4}{3}, 0\right)$

Use these points to graph each line and find its slope:

$$m = \frac{y_2 - y_1}{x_2 - x_1} = \frac{0 - 2}{-\frac{2}{3} - 0} = \frac{-2}{-\frac{2}{3}} = 3 \qquad m = \frac{y_2 - y_1}{x_2 - x_1} = \frac{0 - (-4)}{\frac{4}{3} - 0} = \frac{4}{\frac{4}{3}} = 3$$

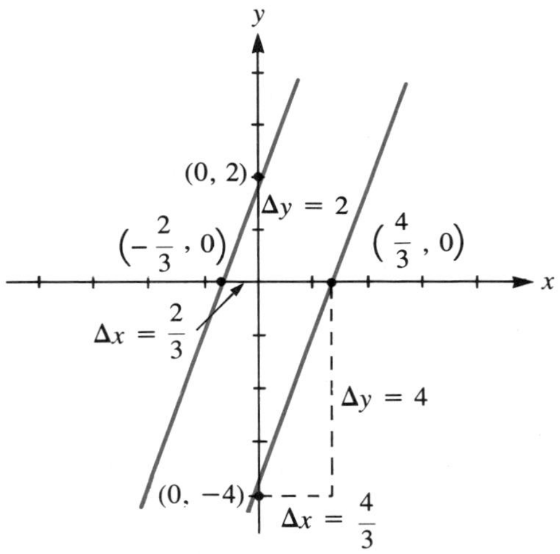

Notice that both lines have the same slope.
Looking at the graph, we can see that the lines are also parallel.
This example illustrates that parallel lines have the same slope. ●

We now summarize some properties of lines that can be determined from their orientations.

1. Lines having a positive slope run up to the right.
2. Lines having a negative slope run down to the right.
3. The slope of any horizontal line is zero.
4. The slope of any vertical line is undefined.
5. Parallel lines have the same slope.

QUICK QUIZ

1. Find the slope of the line that passes through the points $(0, -5)$ and $(-1, 3)$.
2. Find the slope of the line $x - 3y = 6$.
3. Find another point on a line that passes through the point $(3, -1)$ and has a slope of $\frac{5}{2}$.

ANSWERS

1. $m = -8$
2. $m = \frac{1}{3}$
3. $(5, 4)$

14.4 Exercises

Indicate which of the following statements are true and which are false. For those that are false, change the italic expression to make the statement true.

1. Lines that run up to the right have a *negative* slope.
2. Slope is defined as the change in y *divided by* the change in x.
3. The slope of a horizontal line is *undefined.*
4. Parallel lines have *the same* slope.

For each of the following, find the slope of the line that passes through the given points.

5. $(3, 1)$ and $(7, 5)$
6. $(2, 4)$ and $(5, 9)$
7. $(4, 5)$ and $(3, 9)$
8. $(5, 7)$ and $(4, 8)$
9. $(3, 0)$ and $(8, 1)$
10. $(0, 4)$ and $(6, 2)$
11. $(-4, 2)$ and $(6, -3)$
12. $(3, -2)$ and $(-8, 4)$
13. $(-3, -6)$ and $(1, 2)$
14. $(2, 7)$ and $(-1, -2)$
15. $(3, -5)$ and $(0, -7)$
16. $(6, -2)$ and $(-4, 0)$
17. $(7, -4)$ and $(5, 6)$
18. $(8, -6)$ and $(9, 2)$
19. $(8, 7)$ and $(8, -2)$
20. $(3, -5)$ and $(-6, -5)$
21. $(-2, -1)$ and $(-5, -2)$
22. $(-4, -3)$ and $(-1, -2)$
23. $(2, 9)$ and $(-7, 9)$
24. $(-4, 7)$ and $(-4, -3)$

Find the slope of the line defined by the given equation.

25. $x + y = 7$
26. $x - y = 4$
27. $y = 5x$
28. $y = -3x$
29. $2x - y = 6$
30. $4x - y = 8$
31. $x + 7y = 21$
32. $x - 3y = 18$
33. $2x + 4y = 12$
34. $4x + 5y = 20$
35. $3x - 6y = 18$
36. $7x - 2y = 14$
37. $x = 5$
38. $y = 4$
39. $y = -7$
40. $x = -1$

For each of the following, find another point on the line that passes through the given point and has the given slope. Then graph each line.

41. $(3, 7)$ $m = \frac{2}{3}$
42. $(8, 5)$ $m = \frac{1}{2}$
43. $(-1, 4)$ $m = -\frac{1}{4}$
44. $(3, -2)$ $m = -\frac{3}{5}$
45. $(5, 1)$ $m = 3$
46. $(4, 2)$ $m = 5$
47. $(6, -3)$ $m = -4$
48. $(-5, 1)$ $m = -2$
49. $(-5, -2)$ $m = 0$
50. $(-4, -3)$ m is undefined.

Graph each of the following pairs of linear equations on the same set of axes and find the slope of each line. Indicate which equations represent parallel lines.

51. $y = 8x + 8$
$y = 8x - 4$

52. $y = -2x + 5$
$y = 2x + 5$

53. $x - 3y = 6$
$x + 3y = 9$

54. $x - 4y = 8$
$x - 4y = 12$

55. $y = -7$
$x = -7$

56. $x = 5$
$y = -5$

57. $9x - 3y = 18$
$-6x + 2y = 12$

58. $2x + 4y = 16$
$5x + 10y = 10$

14.5 Forms of Linear Equations

In this section, we will discuss two different forms of linear equations: *point-slope form* and *slope-intercept form.* If a linear equation is written in one of these forms, we can immediately determine some of its properties without having to graph it. When a linear equation is written in *point-slope form,* we can easily determine a point on the line and the slope of the line. When a linear equation is written in *slope-intercept form,* we can determine the slope and y-intercept of the line.

To derive the formula for point-slope form, we begin with the equation for the slope of a line that passes through the point (x_1, y_1). If we let (x, y) be any other point on that line, the equation for its slope is as follows:

$$m = \frac{y - y_1}{x - x_1},\ x \neq x_1$$

We now multiply both sides of the equation by $x - x_1$ to obtain the following:

$$(x - x_1)m = \left(\frac{y - y_1}{x - x_1}\right)(x - x_1)$$

$$(x - x_1)m = y - y_1$$

$$y - y_1 = m(x - x_1)$$

The equation in the preceding line is the point-slope form for the equation of a line.

Point-Slope Form

The equation of a line that passes through the point (x_1, y_1) and has a slope of m is given by:

$$y - y_1 = m(x - x_1)$$

example 1

Find the equation of a line that passes through the point $(8, -3)$ and has a slope of $-\frac{2}{5}$.

solution Substitute $(x_1, y_1) = (8, -3)$ and $m = -\frac{2}{5}$ in the equation for point-slope form:

$$y - y_1 = m(x - x_1)$$

$$y - (-3) = -\frac{2}{5}(x - 8)$$

$$y + 3 = -\frac{2}{5}(x - 8)$$

●

example 2

Find the equation of a line that passes through the points $(-5, 1)$ and $(3, 7)$.

solution To use point-slope form, we first need to find the slope of the line. Let $(x_1, y_1) = (-5, 1)$ and $(x_2, y_2) = (3, 7)$.

$$m = \frac{y_2 - y_1}{x_2 - x_1} = \frac{7 - 1}{3 - (-5)} = \frac{6}{8} = \frac{3}{4}$$

Now substitute in $(x_1, y_1) = (-5, 1)$ and $m = \frac{3}{4}$ into the equation for point-slope form.

$$y - y_1 = m(x - x_1)$$

$$y - 1 = \frac{3}{4}(x - (-5))$$

$$y - 1 = \frac{3}{4}(x + 5)$$

●

Another form for a linear equation, slope-intercept form, represents the equation of a line that has a slope of m and a y-intercept of b. Since the y-intercept is b, this line passes through the point $(0, b)$, and we can substitute $(x_1, y_1) = (0, b)$ into point-slope form to obtain the following:

$$y - y_1 = m(x - x_1)$$
$$y - b = m(x - 0)$$
$$y - b = mx$$
$$y = mx + b$$

The preceding line is slope-intercept form for the equation of a line.

Slope Intercept Form

The equation of a line that has a slope of m and a y-intercept of b is given by:

$$y = mx + b$$

example 3

Find the equation of a line that has a slope of $\frac{1}{2}$ and a y-intercept of -7.

solution

Substitute $m = \frac{1}{2}$ and $b = -7$ into slope-intercept form.

$$y = mx + b$$

$$y = \frac{1}{2}x - 7$$

example 4

Find the slope and y-intercept of the equation $9x + 3y = 1$.

solution

First rewrite the equation in slope-intercept form.

$$9x + 3y = 1$$
$$3y = -9x + 1$$
$$y = \underset{\substack{\uparrow \\ m}}{-3}x + \underset{\substack{\uparrow \\ b}}{\frac{1}{3}}$$

Since $m = -3$ and $b = \frac{1}{3}$, the slope is -3 and the y-intercept is $\frac{1}{3}$.

example 5

Find the equation of a line that passes through the points $(1, -3)$ and $(3, 7)$. What is the y-intercept of this line?

solution

First find the slope of this line and use point-slope form to write its equation. Let $(x_1, y_1) = (1, -3)$ and $(x_2, y_2) = (3, 7)$.

$$m = \frac{y_2 - y_1}{x_2 - x_1} = \frac{7 - (-3)}{3 - 1} = \frac{10}{2} = 5$$

$$y - y_1 = m(x - x_1)$$
$$y - (-3) = 5(x - 1)$$
$$y + 3 = 5(x - 1)$$

Now rewrite this equation in slope-intercept form to find the y-intercept of the line.

$$y + 3 = 5(x - 1)$$
$$y + 3 = 5x - 5$$
$$y = 5x - 8$$

Since $b = -8$, the y-intercept is -8.

example 6

Find the equation of the line that passes through the point $(4, -7)$ and is parallel to $y = -2x + 5$.

solution

The slope of the line $y = -2x + 5$ is -2.
To find the equation of the line we can substitute the values $(x_1, y_1) = (4, -7)$ and $m = -2$ into point-slope form.

$$y - y_1 = m(x - x_1)$$
$$y - (-7) = -2(x - 4)$$
$$y + 7 = -2(x - 4)$$

A line that is parallel to it also has a slope of -2.

QUICK QUIZ

1. Find the equation of a line that passes through the point (3, 8) and has a slope of $-\frac{5}{6}$.
2. Find the equation of a line that passes through the points (0, 7) and (9, -2).
3. Find the slope and y-intercept of the line $8x - 3y = 6$.

ANSWERS

1. $y - 8 = -\frac{5}{6}(x - 3)$
2. $y - 7 = -x$
3. $m = \frac{8}{3}$, $b = -2$

14.5 Exercises

For each of the following, find the equation of the line given a point on the line and its slope.

1. (3, 5) $m = 3$
2. (7, 4) $m = 2$
3. (2, 0) $m = 5$
4. (0, 6) $m = 1$
5. (-4, 6) $m = -2$
6. (3, -2) $m = -3$
7. (1, -5) $m = -1$
8. (-5, 6) $m = -4$
9. (-4, -3) $m = \frac{1}{2}$
10. (-6, -2) $m = \frac{2}{3}$
11. (9, -2) $m = -\frac{3}{5}$
12. (-8, 1) $m = -\frac{1}{4}$
13. $\left(\frac{1}{2}, 3\right)$ $m = 2$
14. $\left(1, \frac{2}{3}\right)$ $m = -2$
15. (3, 7) $m = 0$
16. (5, 2) m is undefined.

For each of the following, find the equation of the line that contains the given points.

17. (6, 3) and (7, 5)
18. (3, 2) and (5, 6)
19. (1, 8) and (3, 2)
20. (5, 7) and (4, 9)
21. (-5, 2) and (4, -7)
22. (3, -8) and (4, -1)
23. (0, -4) and (-3, 8)
24. (-9, 1) and (-4, -2)
25. (-2, -3) and (-6, 5)
26. (-1, -2) and (2, 7)
27. (-5, 7) and (-3, 1)
28. (-7, 2) and (-6, -3)
29. (2, -6) and (9, 8)
30. (-3, 1) and (5, 0)
31. (3, 7) and (3, -2)
32. (5, 4) and (-7, 4)

For each of the following, find the equation of the line given its slope and y-intercept.

33. $m = 2$ $b = 5$
34. $m = 1$ $b = 7$
35. $m = -1$ $b = 4$
36. $m = -3$ $b = 4$
37. $m = -4$ $b = -3$
38. $m = -2$ $b = -5$
39. $m = \frac{1}{2}$ $b = -5$
40. $m = \frac{2}{3}$ $b = 9$

41. $m = -\frac{3}{4}$ $b = 7$ **42.** $m = -\frac{1}{5}$ $b = -2$ **43.** $m = -\frac{1}{3}$ $b = -1$ **44.** $m = \frac{3}{5}$ $b = -8$

For each of the following linear equations, find the slope and y-intercept.

45. $y = 8x + 3$ **46.** $y = 3x - 9$ **47.** $y = -5x - 7$ **48.** $y = -7x + 2$

49. $y = \frac{3}{4}x + 9$ **50.** $y = -\frac{1}{2}x - 4$ **51.** $y = -4x$ **52.** $y = 7x$

53. $x + y = 9$ **54.** $x - y = -6$ **55.** $6x + 3y = 12$ **56.** $2x - 4y = 16$

57. $2x - 5y = 7$ **58.** $6x - 4y = 9$ **59.** $8x + 2y = 16$ **60.** $3x + 7y = 21$

Answer the following questions.

61. Find the equation of the line that passes through the points (6, 5) and (7, 2). What is the y-intercept of this line?

62. Find the equation of the line that passes through the points (5, 2) and (3, 4). What is the y-intercept of this line?

63. Find the equation of the line that passes through the points (2, −2) and (4, −1). What is the y-intercept of this line?

64. Find the equation of the line that passes through the points (−2, 7) and (−5, 8). What is the y-intercept of this line?

65. Find the equation of the line that passes through the point (3, 2) and is parallel to $y = 5x + 7$.

66. Find the equation of the line that passes through the point (−5, 1) and is parallel to $y = -3x - 9$.

67. Find the equation of the line that passes through the point (−8, −3) and is parallel to $6x - 3y = 11$.

68. Find the equation of the line that passes through the point (7, −6) and is parallel to $2x + 5y = 8$.

14.6 Graphing Linear Inequalities in Two Variables

The graph of a linear inequality having two variables is a portion of the x-y plane that is bounded by the corresponding linear equation. Any point that lies in this region will satisfy the linear inequality if its coordinates are substituted for x and y. Consider the following example.

example 1

Graph the inequality $3x - 5y \geqslant 15$.

solution

To find the boundary of the region represented by the inequality, we first graph the corresponding linear equation $3x - 5y = 15$.

When $x = 0$, $y = -3$: plot (0, −3)

When $y = 0$, $x = 5$: plot (5, 0)

Draw a line through these points.

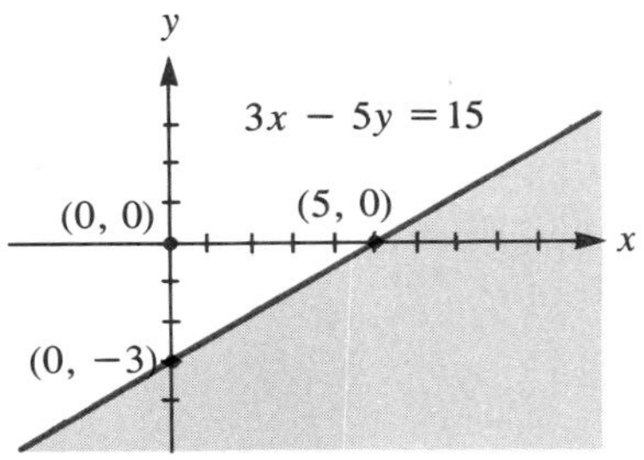

To determine which side of the line is the region that includes the solution set to the inequality, we pick any point on either side of the line and substitute its coordinates for x and y in the inequality.

$$3x - 5y \geq 15 \qquad \text{Choose the point (0, 0).}$$

$$(0, 0): \quad 3(0) - 5(0) \geq 15$$

$$0 \geq 15 \quad \text{is a false statement.}$$

Since (0, 0) does not satisfy the inequality, it cannot be in the solution set. Therefore, the solution set includes the points that lie on the other side of the line. Shade the side of the line that does not include (0, 0). The points in the shaded region, together with the points in the boundary line, represent the solution set for the inequality $3x - 5y \geq 15$. Note that the points on the boundary line represent ordered pairs that are solutions for $3x - 5y = 15$, and the points in the shaded region represent ordered pairs that are solutions for $3x - 5y > 15$. ●

We use a solid line to graph linear equations that correspond to $\geq$ or $\leq$ inequalities to indicate that points on the line are included in the solution set. We use a dotted line to graph linear equations that correspond to $>$ or $<$ inequalities to indicate that points on the line are not included in the solution set.

We will now summarize the procedure for graphing linear inequalities.

To Graph a Linear Inequality

1. Graph the corresponding linear equation to find the boundary of the region that represents the solution set. Use a solid line for $\geq$ or $\leq$ inequalities. Use a dotted line for $>$ or $<$ inequalities.
2. Pick any point on either side of the line and substitute its coordinates for x and y in the inequality. If the resulting statement is true, shade the side of the line that contains that point. If the resulting statement is false, shade the side of the line that does not contain that point.

example 2 Graph the inequality $y < -\frac{1}{4}x + 7$.

solution First graph the equation $y = -\frac{1}{4}x + 7$ using a dotted line for the graph.

When $x = 0$, $y = 7$: plot (0, 7)

When $x = 8$, $y = 5$: plot (8, 5)

Draw a dotted line through these points since it is the boundary for a $<$ inequality.

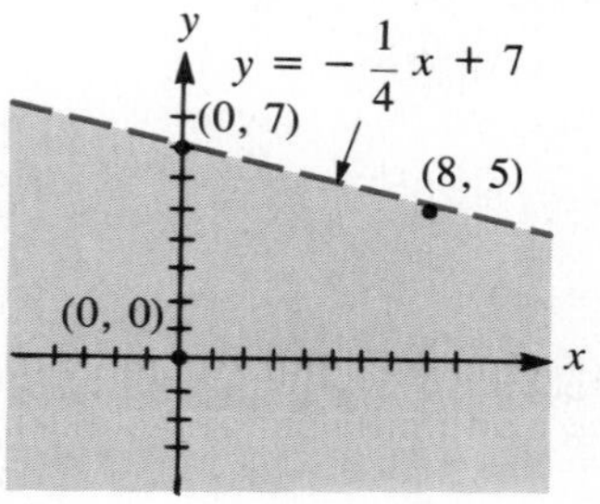

Then pick any point on either side of the line and substitute its coordinates for x and y in the inequality.

$$y < -\frac{1}{4}x + 7 \qquad \text{Choose the point (0, 0).}$$

$$(0, 0): \quad 0 < -\frac{1}{4}(0) + 7$$

$$0 < 7 \quad \text{is a true statement.}$$

Shade the side of the line that contains (0, 0).

example 3

Graph the inequality $x > -2$.

solution

The boundary is the vertical line $x = -2$. Graph this equation using a dotted line. Since the solution set includes all points that have an x-coordinate greater than -2, shade the right-hand side of the line. Note that the points in the boundary line do *not* represent ordered pairs that are solutions of the inequality $x > -2$. Thus, the boundary line is not part of the graph of the solution set, and this is indicated by the *dotted* line.

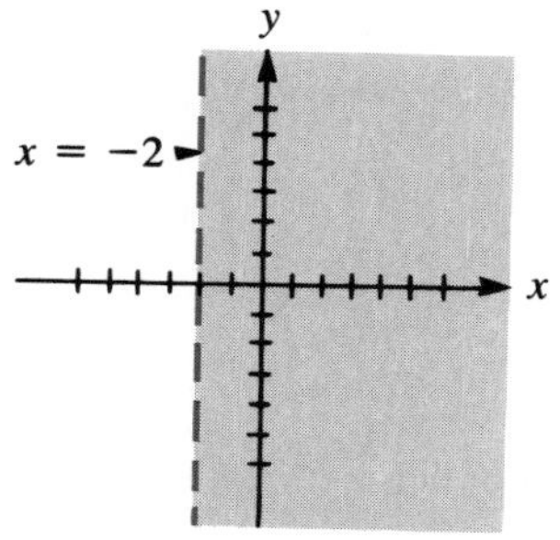

14.6 Exercises

Graph each of the following linear inequalities.

1. $x + y \leqslant 7$
2. $x - y \geqslant -2$
3. $2x - 5y > 10$
4. $3x + 4y < 12$
5. $x + 4y < 4$
6. $x - 2y \leqslant 8$
7. $3x - y \geqslant 9$
8. $5x + y > 15$
9. $x - 6y > 12$
10. $x + 3y \leqslant 18$
11. $y < 5x$
12. $y > -7x$
13. $y \leqslant -2x + 4$
14. $y < 6x + 3$
15. $x > 5y$
16. $x \geqslant 9y$
17. $x < -8y + 4$
18. $x > 2y + 6$
19. $3x + 7y \leqslant 21$
20. $8x - 5y \geqslant 40$
21. $5x - 4y < 20$
22. $2x + 9y > 18$
23. $3y > 9x + 1$
24. $2y \leqslant 8x + 3$
25. $3x + 4y \geqslant 7$
26. $2x + 3y < 5$
27. $y < 5$
28. $x \geqslant 8$
29. $x \geqslant -7$
30. $y < -2$

Summary and Review

Key Terms

[14.2] The **Cartesian Coordinate System** is the two-dimensional plane in which equations in two variables are graphed. It is also called the rectangular coordinate system or x-y plane.

An **ordered pair** represents a point in the x-y plane.

The **abscissa** is the x-coordinate of an ordered pair.

The **ordinate** is the y-coordinate of an ordered pair.

The **slope** of a line is the vertical change divided by the horizontal change.

The **x-intercept** of an equation is the x-coordinate of the point where the graph of the equation crosses the x-axis.

The **y-intercept** of an equation is the y-coordinate of the point where the graph of the equation crosses the y-axis.

Calculations

[14.1] According to the **addition property of inequalities**, if any quantity is added to both sides of an inequality the inequality remains unchanged.

According to the **multiplication property of inequalities**, if both sides of an inequality are multiplied by a positive number, the inequality remains unchanged, and if both sides are multiplied by a negative number, the inequality reverses.

[14.3] **To graph a linear equation**, find at least two ordered pairs that are solutions to the equation, plot the points that correspond to those ordered pairs in the x-y plane, and draw a straight line through these points.

To find the x-intercept of an equation, set $y = 0$ and solve for x.

To find the y-intercept of an equation, set $x = 0$ and solve for y.

[14.4] The **slope** of a line that passes through the points (x_1, y_1) and (x_2, y_2) is given by:

$$\text{slope} = m = \frac{y_2 - y_1}{x_2 - x_1} = \frac{\Delta y}{\Delta x}$$

The equation of a **vertical line** is $x = a$, where a is the x-coordinate of every point on the line. The slope of a vertical line is undefined.

The equation of a **horizontal line** is $y = b$, where b is the y-coordinate of every point on the line. The slope of a horizontal line is zero.

[14.5] **Point-slope form** for the equation of a line that passes through the point (x_1, y_1) and has a slope of m is given by:

$$y - y_1 = m(x - x_1)$$

Slope-intercept form for the equation of a line that has a slope of m and a y-intercept of b is given by:

$$y = mx + b$$

[14.6] **To graph a linear inequality**, graph the corresponding linear equation to find the boundary of the region that represents the solution set using a solid line for $\geqslant$ or $\leqslant$ inequalities and a dotted line for $>$ or $<$ inequalities. Pick any point on either side of the line and substitute its coordinates for x and y in the inequality. If the resulting statement is true, shade the side of the line that contains the point, and if not, shade the other side of the line.

Chapter 14 Review Exercises

Indicate which of the following statements are true and which are false. For those that are false, change the italic expression to make the statement true.

1. If both sides of an inequality are multiplied by a negative number, the inequality *reverses.*
2. The y-coordinate of an ordered pair is called the *abscissa.*

3. The slope of a *vertical* line is undefined.
4. When an equation is written in slope-intercept form, the *x-intercept* can be immediately identified.
5. Lines that run *up* to the right have a positive slope.
6. For any point located in the second quadrant, the abscissa is *less than* the ordinate.

[14.1] Solve each of the following linear inequalities and graph the solution set on a number line.

7. $2x + 9 < 5$
8. $3x - 7 \geqslant 2$
9. $3 - 4t \geqslant 7$
10. $5 - 6t > 7$
11. $4y + 6 > 8y - 3$
12. $2y - 8 < 5y + 4$
13. $7(z - 3) > 6$
14. $9(z + 2) \geqslant 4$
15. $3(2w - 5) - 4w \leqslant 5(4 - w)$
16. $-2(4w - 1) \leqslant 3(3w - 2) + 7w$

[14.3] Graph each of the following linear equations.

17. $x - y = 9$
18. $x + y = 6$
19. $3x + 8y = 24$
20. $5x - 4y = 20$
21. $x - 3y = 9$
22. $6x + y = 12$
23. $y = -8x$
24. $y = 3x$
25. $y = 5x + 2$
26. $y = -2x - 1$
27. $y = 7$
28. $x = -5$

[14.4] Find the slopes of the lines passing through the points given.

29. $(6, 1)$ and $(7, 5)$
30. $(5, 3)$ and $(8, 9)$
31. $(7, -3)$ and $(9, 1)$
32. $(-4, 5)$ and $(-3, -1)$
33. $(-9, 6)$ and $(-3, -2)$
34. $(-3, 4)$ and $(7, -2)$

[14.5] For each of the following, write the equation of the line having the characteristics given.

35. Passes through $(-3, 5)$, slope $= 4$
36. Passes through $(7, -1)$, slope $= -2$
37. Passes through $(0, 8)$, slope $= -\frac{1}{3}$
38. Passes through $(-2, -3)$, slope $= \frac{2}{5}$
39. Passes through $(4, 6)$, slope is undefined.
40. Passes through $(2, -5)$, slope $= 0$
41. Passes through $(4, 2)$ and $(5, 6)$
42. Passes through $(1, 7)$ and $(5, 3)$
43. Passes through $(-5, 8)$ and $(1, -4)$
44. Passes through $(-3, 7)$ and $(2, 6)$
45. Passes through $(8, 5)$ and $(-3, 5)$
46. Passes through $(-2, 7)$ and $(-2, -9)$
47. Slope $= 2$, y-intercept $= -3$
48. Slope $= -1$, y-intercept $= 2$
49. Slope $= -\frac{3}{4}$, y-intercept $= 0$
50. Slope $= \frac{1}{2}$, y-intercept $= \frac{1}{8}$
51. Passes through $(-5, 7)$, parallel to $y = -3x + 8$
52. Passes through $(4, -3)$, parallel to $y = x - 6$

[14.6] Graph each of the following linear inequalities.

53. $y \geqslant 5x$
54. $y > -3x$
55. $x + 3y < 6$
56. $x - 5y \leqslant 15$
57. $4x - y > 12$
58. $6x + y < 18$
59. $8x + 3y \leqslant 24$
60. $9x - 4y \leqslant 36$
61. $y < 4$
62. $x \geqslant -8$

15 SYSTEMS OF LINEAR EQUATIONS

A system of linear equations consists of two or more linear equations that can be graphed on the same set of coordinate axes. A *solution* to a linear system is an ordered pair that is a solution to every linear equation that is included in the system. In this chapter, we learn how to solve systems of linear equations that include two equations in two unknowns. Systems of linear equations that include more than two equations are discussed in more advanced algebra courses.

15.1 Graphical Solutions to Linear Systems

One technique for solving a system of two equations in two unknowns is to graph both equations on the same set of axes and look at the graph to determine the point(s) common to both equations. Since the graph of a linear equation is a straight line, when we graph two linear equations on the same set of axes, there are three cases that describe the different possibilities for the solution set.

1. If both lines intersect in a single point, as shown in Figure 15.1, the solution consists of one ordered pair that is the point of intersection.

figure 15.1

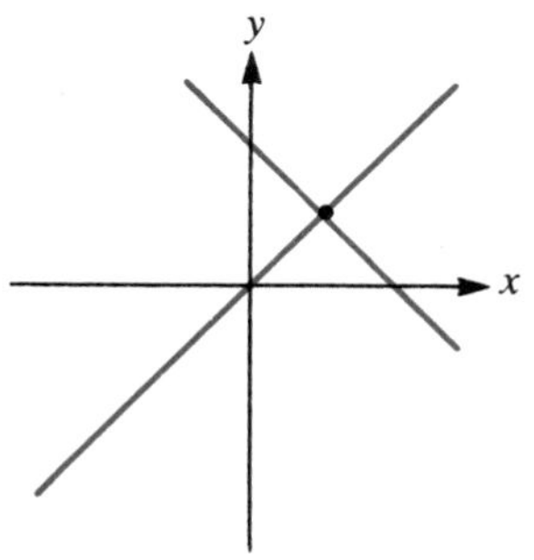

Lines intersect: The solution is the point of intersection.

2. If the two lines are parallel as shown in Figure 15.2, there are no points that are common to both equations. Thus, there is no solution to the system of linear equations.

figure 15.2

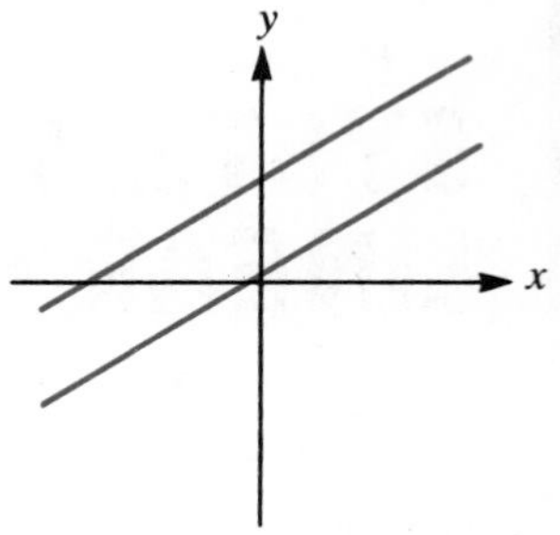

Parallel lines: No solution.

3. If the graph of each equation is represented by the same line, that is, the lines coincide, every point on the line satisfies both equations as shown in Figure 15.3. Thus, an infinite number of ordered pairs represent the solution set to the linear system.

figure 15.3

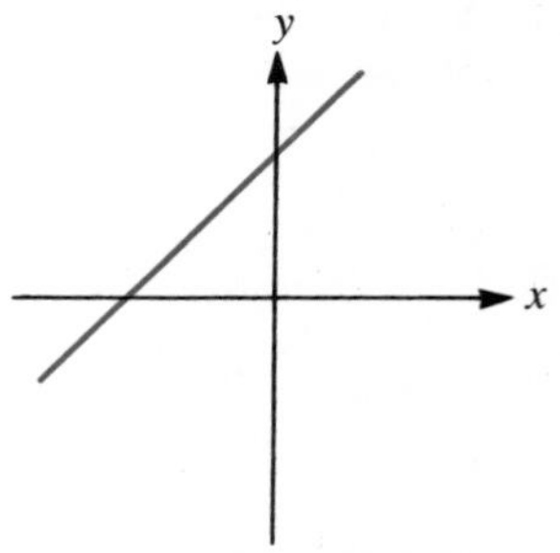

Lines coincide: The solution is every point on the line.

example 1

Solve the following linear system.

$$2x - y = 8$$
$$x + 3y = 18$$

solution

Graph both equations on the same set of axes.

$$2x - y = 8$$

x	y
0	-8
4	0

$$x + 3y = 18$$

x	y
0	6
18	0

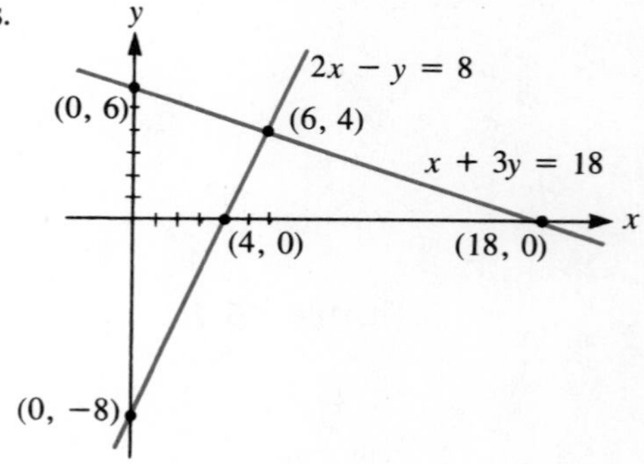

CHECK

The lines intersect at the point (6, 4). Thus, (6, 4) is the solution to the linear system.

$2x - y = 8$	$x + 3y = 18$
$2(6) - 4 \ ? \ 8$	$6 + 3(4) \ ? \ 18$
$12 - 4 \ ? \ 8$	$6 + 12 \ ? \ 18$
$8 = 8 \ \checkmark$	$18 = 18 \ \checkmark$

To check the solution, substitute $x = 6$ and $y = 4$ into each equation.

example 2

Solve the following linear system.

$$3x - 2y = 6$$
$$-6x + 4y = -12$$

solution

Graph both equations on the same set of axes.

$3x - 2y = 6$

x	y
0	-3
2	0

$-6x + 4y = -12$

x	y
0	-3
2	0

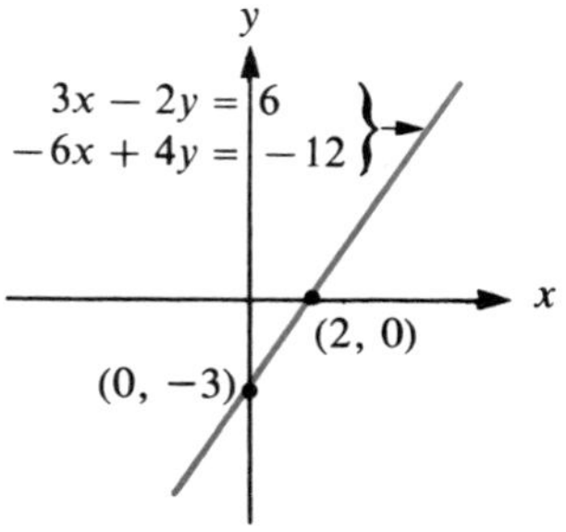

Since both equations are represented by the same line, every point on this line is a solution to the linear system. Thus, the solution set consists of an infinite number of ordered pairs which all lie on the line. ●

example 3

Solve the following linear system.

$$3x + 5y = -15$$
$$3x + 5y = 0$$

solution

Graph both equations on the same set of axes.

$3x + 5y = -15$

x	y
0	-3
-5	0

$3x + 5y = 0$

x	y
0	0
-5	3

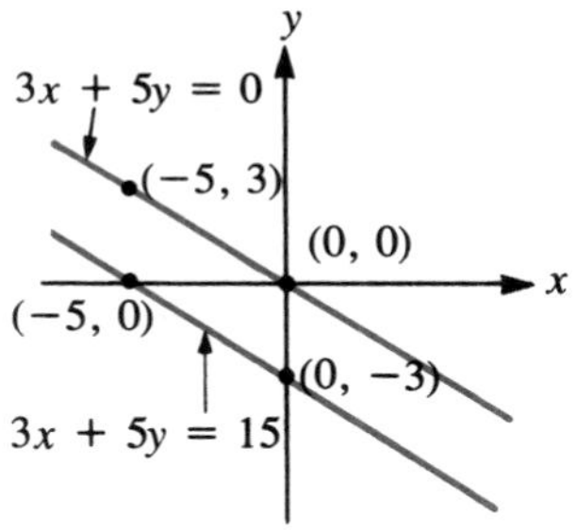

Since the two lines are parallel, there are no points that satisfy both equations. Therefore, this system of linear equations has no solution. ●

example 4

Solve the following system of linear equations.

$$x + 2y = 4$$
$$y = 3$$

solution Graph both equations on the same set of axes.

$x + 2y = 4$

x	y
0	2
4	0

$y = 3$

x	y
0	3
4	3

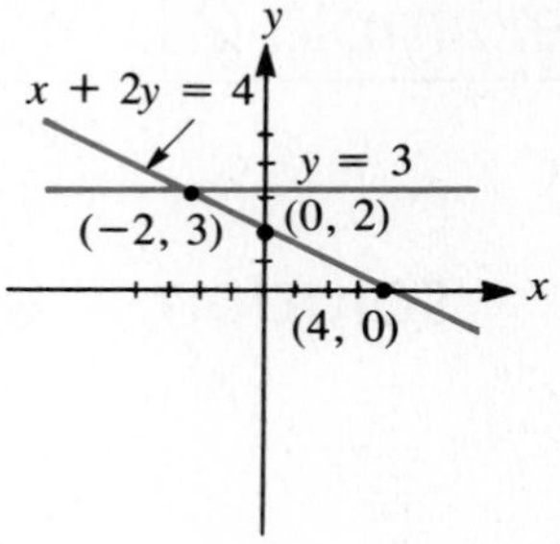

Since the lines intersect at the point $(-2, 3)$, it is the solution to the linear system.

CHECK

$$\begin{aligned} x + 2y &= 4 \\ -2 + 2(3) &\ ?\ 4 \\ -2 + 6 &\ ?\ 4 \\ 4 &= 4 \ \checkmark \end{aligned} \qquad \begin{aligned} y &= 3 \\ 3 &= 3 \ \checkmark \end{aligned}$$

To check the solution substitute $x = -2$ and $y = 3$ into each equation.

15.1 Exercises

Indicate which of the following statements are true and which are false. For those that are false, change the italic expression to make the statement true.

1. If the graph of a system of linear equations consists of two lines that are *parallel*, there is no solution to the linear system.
2. If there are an infinite number of solutions to a system of two linear equations in two unknowns, the lines *intersect*.
3. If two lines *coincide*, the solution set consists of one ordered pair.
4. An ordered pair is a solution to a linear system if it is a solution to *every* equation in the system.

Solve each of the following linear systems by graphing.

5. $x + y = 5$, $x - y = 3$
6. $x + y = -9$, $x - y = 5$
7. $x + y = 6$, $x - 2y = 12$
8. $x - y = 2$, $3x + 5y = 30$
9. $x - y = 5$, $2x + 7y = 28$
10. $x + y = -5$, $x - 2y = 4$
11. $x - y = -8$, $x + 2y = 4$
12. $x - y = 7$, $2x + 3y = -6$
13. $x - 3y = 6$, $x = 3$
14. $5x + 2y = 10$, $y = -5$
15. $2x - y = -8$, $3x + y = 3$
16. $x + 2y = 8$, $2x - y = 6$
17. $-5x + y = 5$, $5x + 2y = -20$
18. $2x + 3y = 6$, $2x - y = -10$
19. $3x + y = 0$, $3x + y = -6$
20. $5x - 2y = 0$, $5x - 2y = -10$
21. $2x - 3y = 18$, $2x + 3y = 6$
22. $-2x + 3y = 30$, $4x + 3y = 12$
23. $-4x + 5y = 40$, $y = 4$
24. $2x + 5y = -10$, $x = 5$
25. $2x + 5y = 10$, $-2x - 5y = 10$
26. $3x + 2y = 12$, $-3x - 2y = 6$
27. $7x - 2y = -14$, $-14x + 4y = 28$
28. $5x + 4y = 20$, $15x + 12y = 60$
29. $-2x + y = 4$, $2x - 3y = -12$
30. $3x - 2y = 18$, $2x + y = -2$

15.2 The Addition-Elimination Technique

In many instances the graphic technique for solving linear systems is not practical. Such cases would include problems in which the coordinates of the point of intersection are not integers and thus cannot be read accurately from the graph. Instead of estimating the solution to these problems, we can use one of two algebraic techniques to find the exact coordinates. These are the addition-elimination technique and the substitution technique. In this section we will discuss the addition-elimination technique, and in Section 15.3 we will discuss the substitution technique.

The addition-elimination technique is based on the addition principle of equality introduced in Section 10.3. According to this principle, if equal quantities are added to both sides of an equation, the solution to that equation remains unchanged. We will illustrate the use of the addition principle by solving the linear system in Example 1.

example 1

Solve the following system of linear equations.

$$3x + y = 4$$
$$2x - y = 11$$

solution

$$\begin{aligned} 3x + y &= 4 \\ \underline{2x - y} &\underline{= 11} \\ 5x \qquad &= 15 \end{aligned}$$

Add both equations together.

Notice that since the coefficients of y add up to zero, the variable y is eliminated.

$$x = 3$$

Solve the resulting equation for x.

$$\begin{aligned} 3x + y &= 4 \\ 3(3) + y &= 4 \\ 9 + y &= 4 \\ y &= -5 \end{aligned}$$

Substitute $x = 3$ into one of the original equations and solve for y.

Thus, the solution to the linear system is the ordered pair $(3, -5)$.

CHECK

$3x + y = 4$	$2x - y = 11$
$3(3) + (-5) \ ? \ 4$	$2(3) - (-5) \ ? \ 11$
$9 + (-5) \ ? \ 4$	$6 + 5 \ ? \ 11$
$4 = 4 \ \checkmark$	$11 = 11 \ \checkmark$

Check the solution by substituting $x = 3$, $y = -5$ into each of the original equations.

When we add two equations using the addition-elimination technique, the coefficients of one of the variables must add up to zero so that one of the variables is eliminated. The sum of the equations consists of one equation in one unknown which can easily be solved for the variable.

Many times, if we simply added the two equations in a linear system, we would find that the coefficients of neither of the variables add up to zero. In these problems we can use the multiplication principle of equality, which states that both sides of a linear equation can be multiplied by the same nonzero quantity without changing the solution to the equation. Therefore, to solve linear systems in which the coefficients of neither of the variables add up to zero, we must first multiply both sides of one or both of the equations by a constant so that the coefficients of one of the variables in the resulting system are opposites. When we then add the two equations, one of the variables will be eliminated. This technique is illustrated in Example 2.

example 2

Solve the following linear system.

$$4x + 5y = 1$$
$$-2x + 3y = -6$$

solution Multiply both sides of the second equation by 2. The first equation remains unchanged:

$$\begin{aligned} 4x + 5y &= 1 & \rightarrow \quad 4x + 5y &= 1 \\ 2(-2x + 3y) &= 2(-6) & \rightarrow \quad -4x + 6y &= -12 \\ & & 11y &= -11 \\ & & y &= -1 \end{aligned}$$

The coefficients of x in the resulting system are 4 and -4.

Add the two equations. The variable x is eliminated. Solve for y.

$$\begin{aligned} 4x + 5y &= 1 \\ 4x + 5(-1) &= 1 \\ 4x - 5 &= 1 \\ 4x &= 6 \\ x &= \frac{3}{2} \end{aligned}$$

Substitute $y = -1$ into the first equation in the original system and solve for x.

Thus the solution to the system is $\left(\frac{3}{2}, -1\right)$.

CHECK

$$4x + 5y = 1 \qquad\qquad -2x + 3y = -6$$
$$4\left(\frac{3}{2}\right) + 5(-1) \ ? \ 1 \qquad\qquad -2\left(\frac{3}{2}\right) + 3(-1) \ ? \ -6$$
$$6 - 5 \ ? \ 1 \qquad\qquad -3 - 3 \ ? \ -6$$
$$1 = 1 \checkmark \qquad\qquad -6 = -6 \checkmark$$

We will now summarize the procedure for solving systems of linear equations using the addition-elimination technique.

The Addition-Elimination Technique

1. If necessary, multiply both sides of one or both of the equations in the linear system by the appropriate constant so that the coefficients of one of the variables in the resulting system are opposites.
2. Add the two equations in the resulting linear system to eliminate one of the variables. The sum will consist of one equation in one unknown.
3. Solve this equation for the variable.
4. Substitute the value obtained for this variable into one of the original equations and solve for the other variable.
5. Check the solution by substituting the values obtained for each of the variables into each of the original equations.

Let us now consider a few more examples that illustrate the addition-elimination technique.

example 3

Solve the following linear system.

$$2x + y = 3$$
$$3x + 2y = 8$$

solution

Multiply both sides of the first equation by -2. The second equation remains unchanged.

$$-2(2x + y) = -2(3) \rightarrow -4x - 2y = -6$$
$$3x + 2y = 8 \rightarrow \underline{3x + 2y = 8}$$

The coefficients y in the resulting system are 2 and -2.

$$-x = 2$$

Add the two equations. The variable y is eliminated. Solve for x.

$$x = -2$$

Substitute $x = -2$ into the first equation in the original system and solve for y.

$$2x + y = 3$$
$$2(-2) + y = 3$$
$$-4 + y = 3$$
$$y = 7$$

Thus, the solution to the system is $(-2, 7)$.

CHECK

$$2x + y = 3 \qquad 3x + 2y = 8$$
$$2(-2) + 7 \ ? \ 3 \qquad 3(-2) + 2(7) \ ? \ 8$$
$$-4 + 7 \ ? \ 3 \qquad -6 + 14 \ ? \ 8$$
$$3 = 3 \ \checkmark \qquad 8 = 8 \ \checkmark$$

example 4

Solve the following linear system.

$$4x - 5y = -2$$
$$3x = 2y - 5$$

solution

Rewrite the second equation so that like terms line up when the second equation is written under the first equation.

$$4x - 5y = 2$$
$$3x - 2y = -5$$

Multiply both sides of the first equation by 3 and both sides of the second equation by -4.

$$3(4x - 5y) = 3(-2) \rightarrow 12x - 15y = -6$$
$$-4(3x - 2y) = -4(-5) \rightarrow \underline{-12x + 8y = 20}$$

The coefficients of x in the resulting system are 12 and -12.

$$-7y = 14$$

Add the two equations. The variable x is eliminated.

$$y = -2$$

Solve for y.

Substitute $y = -2$ into the second equation in the original system and solve for x.

$$3x - 2y = -5$$
$$3x - 2(-2) = -5$$
$$3x + 4 = -5$$
$$3x = -9$$
$$x = -3$$

Thus, the solution is $(-3, -2)$.

CHECK

$$4x - 5y = -2$$
$$4(-3) - 5(-2) \; ? \; -2$$
$$-12 + 10 \; ? \; -2$$
$$-2 = -2 \; \checkmark$$

$$3x - 2y = -5$$
$$3(-3) - 2(-2) \; ? \; -5$$
$$-9 + 4 \; ? \; -5$$
$$-5 = -5 \; \checkmark$$

Thus far, we have only used the addition-elimination technique to solve systems of linear equations in which the solution is a single ordered pair. These systems can be represented graphically by two lines that intersect in a single point. We will now look at what happens when we use the addition-elimination technique to solve a linear system that is represented by two parallel lines. Recall that there is no solution to such a system since parallel lines have no points in common.

example 5

Solve the following linear system.

$$3x - 7y = 6$$
$$3x - 7y = 1$$

Multiply both sides of the first equation by -1. The second equation remains unchanged.

$$-1(3x - 7y) = -1(6) \rightarrow -3x + 7y = -6$$
$$3x - 7y = 1 \rightarrow 3x - 7y = 1$$

Notice that the coefficients of both x and y in the resulting systems are opposites.

$$0 = -5$$

Add the two equations.

$$0 \neq -5$$

The sum on the left is 0, and the sum on the right is -5.

Since the statement $0 = -5$ is always false, it is said to be a *contradiction.* If the sum of the two equations in a linear system is a contradiction, there are no ordered pairs that will satisfy both equations. We can therefore conclude that there is no solution to this linear system, and the lines that represent the equations are parallel.

We now look at an example in which the addition-elimination technique is used to solve a system of linear equations that can be represented by lines that coincide. Recall that the solution to such a system consists of an infinite number of ordered pairs.

example 6

Solve the following linear system.

$$-6x + 2y = 2$$
$$3x - y = -1$$

solution Multiply both sides of the second equation by 2. The first equation remains unchanged.

$$-6x + 2y = 2 \rightarrow -6x + 2y = 2$$
$$2(3x - y) = 2(-1) \rightarrow 6x - 2y = -2$$

The coefficients of both x and y in the resulting system are opposites.

$$0 = 0$$

Add the two equations.

The sum of both sides is 0.

Since the statement $0 = 0$ is always true, it is said to be an *identity*. If the sum of the two equations in a linear system is an identity, any ordered pair that is a solution to one equation in the system is also a solution to the other equation. We can therefore conclude that there are an infinite number of ordered pairs that are solutions to this system, and the lines that represent the equations coincide.

In summary, to use the addition-elimination technique to solve a linear system, we first multiply both sides of one or both of the equations by a constant so that the coefficients of one of the variables in the resulting system are opposites. We then add the two equations, and the sum will be in one of these three forms:

1. If the sum of the equations consists of one equation in one unknown, the solution to the linear system is a single ordered pair, and the lines representing the equations intersect.
2. If the sum of the equations is a contradiction (a statement that is always false), there is no solution to the linear system, and the lines representing the equations are parallel.
3. If the sum of the equations is an identity (a statement that is always true), there are an infinite number of solutions to the linear system, and the lines representing the equations coincide.

QUICK QUIZ

Solve each of the following linear systems.

1. $x + 4y = 7$
   $3x - 4y = 5$
2. $4x - 6y = -1$
   $-2x + 3y = 3$
3. $2x - 6y = -10$
   $-x + 3y = 5$
4. $7x - 4y = 6$
   $3x - 2y = 4$

ANSWERS

1. (3, 1)
2. parallel lines; no solution.
3. lines coincide; infinite number of solutions.
4. $(-2, -5)$

15.2 Exercises

Indicate which of the following statements are true and which are false. For those that are false, change the italic word to make the statement true.

1. If the sum of two linear equations is a statement that is always false, the lines representing these equations *coincide.*
2. When we add two linear equations using the addition-elimination technique, the sum of the coefficients of one of the variables must be *one.*
3. When we multiply both sides of a linear equation by a constant, the solution to the equation is *unchanged.*
4. If the sum of two linear equations using the addition-elimination technique is an identity, the lines representing these equations *intersect.*

Solve each of the following linear systems using the addition-elimination technique.

5. $x - y = 2$
$3x + y = 10$

6. $x + y = 6$
$-x + 4y = 4$

7. $5x + 2y = 4$
$-5x - y = -7$

8. $x - 3y = 9$
$-7x + 3y = 9$

9. $2x = 7y + 17$
$3x + 7y = 8$

10. $4x - 5y = 11$
$-4x = 3y + 13$

11. $2x - 5y = -8$
$2x - y = 8$

12. $3x + 5y = -5$
$x + 5y = 5$

13. $2x + y = 0$
$5x - 2y = -9$

14. $x - 4y = -7$
$-3x + 5y = 0$

15. $4x - 3y = -6$
$x + 5y = -13$

16. $6x - y = 8$
$3x - 2y = -2$

17. $2x - 3y = 10$
$5x + 6y = -2$

18. $-2x - 5y = -1$
$8x + 3y = 21$

19. $3x = 7y + 16$
$9x + 2y = 25$

20. $5x - 2y = 10$
$10y = 3x + 38$

21. $5x - 2y = 8$
$2x + 3y = 7$

22. $4x - 3y = -9$
$-7x + 4y = 7$

23. $7x + 2y = 5$
$7x + 2y = -4$

24. $3x - 9y = -8$
$-3x + 9y = 2$

25. $4x - 5y = 10$
$5x - 3y = -7$

26. $2x + 7y = 8$
$3x - 4y = -17$

27. $-3x + 4y = -7$
$7x - 2y = 20$

28. $4x - 9y = 7$
$5x - 3y = 2$

29. $2x - 3y = 7$
$-6x + 9y = 5$

30. $5x + 2y = -3$
$10x + 4y = 1$

31. $-2x + y = -3$
$8x - 4y = 12$

32. $2x - y = -5$
$-6x + 3y = 15$

33. $5x - 2y = 6$
$-8x + 5y = -9$

34. $3x - 5y = -2$
$-2x + 4y = 1$

15.3 The Substitution Technique

Another algebraic technique for solving systems of linear equations is called the substitution technique. The substitution technique is most convenient for solving systems of linear equations in which the coefficients of at least one of the variables in one of the equations is 1 or -1. To use this technique, we first solve one of the equations for one of the variables in terms of the other variable. We then substitute the resulting expression for that variable in the other equation to obtain one equation in one unknown. After solving for the variable in this equation, we can substitute the result into one of the original equations to find the value of the other variable. We will illustrate the substitution technique with the following examples.

example 1

Solve the following system of linear equations.

$$x = 2y - 4$$
$$x + 3y = 21$$

solution

Since the first equation is already solved for x in terms of y, we can substitute $2y - 4$ for x in the second equation.

$$\underline{x} + 3y = 21$$
$$\underline{2y - 4} + 3y = 21$$

Solve the resulting equation for y.

$$5y - 4 = 21$$
$$5y = 25$$
$$y = 5$$

$x = 2y - 4$
$x = 2(5) - 4$
$x = 10 - 4$
$x = 6$

Substitute $y = 5$ into the first equation and solve for x.

The solution is (6, 5).

CHECK

$x = 2y - 4$	$x + 3y = 21$
$6 \; ? \; 2(5) - 4$	$6 + 3(5) \; ? \; 21$
$6 \; ? \; 10 - 4$	$6 + 15 \; ? \; 21$
$6 = 6 \; \surd$	$21 = 21 \; \surd$

example 2

Solve the following system of linear equations.

$$3x - 5y = 18$$
$$-2x + y = -5$$

solution

Solve the second equation for y in terms of x.

$$-2x + y = -5$$
$$y = 2x - 5$$

Now substitute $2x - 5$ for y in the first equation.

$$3x - 5\underline{y} = 18$$
$$3x - 5\underline{(2x - 5)} = 18$$
$$3x - 10x + 25 = 18$$
$$-7x + 25 = 18$$
$$-7x = -7$$
$$x = 1$$

Solve the resulting equation for x.

$$-2x + y = -5$$
$$-2(1) + y = -5$$
$$-2 + y = -5$$
$$y = -3$$

Substitute $x = 1$ into the second equation and solve for y.

The solution is $(1, -3)$.

CHECK

$3x - 5y = 18$	$-2x + y = -5$
$3(1) - 5(-3) \; ? \; 18$	$-2(1) + (-3) \; ? \; -5$
$3 + 15 \; ? \; 18$	$-2 + (-3) \; ? \; -5$
$18 = 18 \; \surd$	$-5 = -5 \; \surd$

We will now summarize the procedure for solving systems of linear equations using the substitution technique.

The Substitution Technique

1. If necessary, solve one of the equations for one of the variables in terms of the other variable.
2. Substitute the resulting expression for that variable in the other equation to obtain one equation in one unknown.
3. Solve this equation for the variable.
4. Substitute the value obtained for this variable into one of the original equations, and solve for the other variable.
5. Check the solution by substituting the values obtained for each of the variables into each of the original equations.

example 3

Solve the following linear system.

$$x - 4y = 5$$
$$-2x + 8y = 9$$

solution

Solve the first equation for x.

$$x - 4y = 5$$
$$x = 4y + 5$$

Substitute $4y + 5$ for x in the second equation.

$$-2x + 8y = 9$$
$$-2(4y + 5) + 8y = 9 \qquad \text{Solve the resulting equation for } y.$$
$$-8y - 10 + 8y = 9$$
$$-10 = 9 \qquad \text{False.}$$

Notice that the result is a contradiction, a statement that is always false. Therefore, we can conclude that this system has no solution, and the lines representing the two equations are parallel.

example 4

Solve the following linear system.

$$y = -5x + 3$$
$$y = 2x - 11$$

solution

Since both equations are solved for y in terms of x, we can set the two expressions on the right equal to each other. When we do this, we are actually substituting $2x - 11$ for y in the first equation.

$$2x - 11 = -5x + 3 \qquad \text{Solve the resulting equation for } x.$$
$$7x = 14$$
$$x = 2$$

$$y = -5x + 3$$
$$y = -5(2) + 3$$
$$y = -10 + 3$$
$$y = -7$$

Substitute $x = 2$ in the first equation and solve for y.

The solution is $(2, -7)$.

CHECK

$y = -5x + 3$	$y = 2x - 11$
$-7 \ ? \ -5(2) + 3$	$-7 \ ? \ 2(2) - 11$
$-7 \ ? \ -10 + 3$	$-7 \ ? \ 4 - 11$
$-7 = -7 \ \checkmark$	$-7 = -7 \ \checkmark$

example 5

Solve the following linear system.

$$3x - y = -1$$
$$-15x + 5y = 5$$

solution

Solve the first equation for y.

$$3x - y = -1$$
$$-y = -3x - 1$$
$$y = 3x + 1$$

Substitute $3x + 1$ for y in the second equation.

$$-15x + 5\underline{y} = 5$$
$$-15x + 5(\underline{3x + 1}) = 5$$
$$-15x + 15x + 5 = 5$$
$$5 = 5$$

Solve the resulting equation for x.

Notice that the result is an identity, a statement that is always true. We can therefore conclude that any ordered pair that is a solution to the first equation is also a solution to the second equation. Thus, there are an infinite number of solutions to the linear system, and the lines representing the equations coincide.

QUICK QUIZ

Solve each of the following linear systems.

1. $-5x + y = 7$
$7x - 2y = -5$

2. $3x + 6y = -8$
$x + 2y = 3$

3. $x = 5y - 6$
$x = -3y + 10$

ANSWERS

1. $(-3, -8)$

2. parallel lines; no solution.

3. $(4, 2)$

15.3 Exercises

Solve each of the following linear systems using the substitution technique.

1. $y = 4x - 7$
   $5x - 2y = 5$
2. $y = -5x + 6$
   $3x + 4y = -10$
3. $x = -8y + 5$
   $3x + 7y = -2$
4. $x = 2y - 3$
   $3x - 4y = -1$
5. $5x + y = 4$
   $7x + 2y = 2$
6. $7x - 3y = 4$
   $-6x + y = 6$
7. $2x - 5y = 11$
   $x - 9y = -1$
8. $x - 8y = 11$
   $2x + 9y = -3$
9. $5x - 6y = -2$
   $4x - y = -13$
10. $9x - y = -13$
    $-3x + 4y = 19$
11. $-x - 8y = 7$
    $2x + 5y = 8$
12. $9x + 4y = 1$
    $-x - 2y = -11$
13. $-6x + 3y = 8$
    $2x - y = 5$
14. $x - 3y = 7$
    $-3x + 9y = 4$
15. $y = -5x + 3$
    $y = 3x - 13$
16. $y = 4x + 5$
    $y = 2x - 1$
17. $x = 2y - 5$
    $x = -3y + 15$
18. $x = 5y - 7$
    $x = -3y - 5$
19. $x - 3y = 4$
    $-3x + 9y = -12$
20. $-8x - 2y = 10$
    $4x + y = -5$
21. $2x - y = 7$
    $-8x - 3y = 14$
22. $3x - 6y = 13$
    $x + 3y = 6$
23. $8x - 4y = 5$
    $-2x + y = 9$
24. $x - 3y = -8$
    $-2x + 6y = 7$
25. $4x - 7y = 6$
    $x = 8y - 11$
26. $y = 9x - 11$
    $8x - 3y = -5$
27. $8x - 2y = -6$
    $-4x + y = 3$
28. $-x + 3y = 4$
    $5x - 15y = -20$
29. $3x + y = 4$
    $7x - 3y = 8$
30. $x - 7y = -4$
    $-2x + 9y = 5$

15.4 Word Problems Involving Linear Systems

In this section, we will learn how to solve word problems using systems of linear equations. The following procedure outlines the method we will use.

Steps for Solving Word Problems Using Linear Systems

1. Read the problem carefully to determine two unknown quantities that need to be found in order to solve the problem. Define two variables to represent these unknown quantities.
2. Use the information in the problem to write two linear equations that relate the variables.
3. Solve the system of linear equations for the variables.
4. Answer the question asked in the problem. This may or may not be the same as the solution to the linear system.
5. Check your answer by determining whether or not it satisfies the conditions of the problem.

We now use this procedure to solve the following problems.

example 1

Find two numbers whose sum is 5 such that twice the first is equal to 19 more than the second.

solution

1. Define the variables.

   Let x = first number

   y = second number

2. Write two linear equations that relate the variables.

$x + y = 5$	The sum of the numbers is 5.
$2x = y + 19$	Twice the first equals 19 more than the second.

3. Solve this system using the addition-elimination technique.

$x + y = 5$ $2x - y = 19$ $3x = 24$	Rewrite the second equation so that the variable terms are on the left and the constant is on the right. Add the two equations.
$x = 8$	Solve for x.
$x + y = 5$ $8 + y = 5$ $y = -3$	Substitute $x = 8$ into the first equation and solve for y.

4. The problem asks us to find the two numbers. The two numbers are 8 and -3.
5. Check the answer.

$x + y = 5$ $8 + (-3) \ ? \ 5$ $5 = 5 \checkmark$	The sum of the two numbers is 5.
$2x = y + 19$ $2(8) \ ? \ -3 + 19$ $16 = 16 \checkmark$	Twice the first is equal to 19 more than the second.

●

example 2

A piggy bank contains only dimes and quarters which amount to \$5.45. If there are a total of 35 coins in the bank, how many are dimes and how many are quarters?

solution

1. Define the variables.

   Let d = number of dimes

   q = number of quarters

2. Write two equations that relate the variables.
   The first equation involves the number of coins.

$$\text{number of dimes} + \text{number of quarters} = \text{total number of coins}$$

$$d + q = 35$$

The second equation involves the value of the coins.

value of dimes + value of quarters = total value

$$0.10d + 0.25q = 5.45$$

Multiply both sides of this equation by 100 to eliminate decimal places.

$$10d + 25q = 545$$

3. Solve this system using the substitution technique.

$$d + q = 35$$
$$10d + 25q = 545$$

$$d = 35 - q$$ Solve the first equation for d.

$$10d + 25q = 545$$ Substitute $35 - q$ for d in the second equation.
$$10(35 - q) + 25q = 545$$

$$350 - 10q + 25q = 545$$ Solve the resulting equation for q.
$$350 + 15q = 545$$
$$15q = 195$$
$$q = 13 \text{ quarters}$$

$$d + q = 35$$ Substitute $q = 13$ into the first equation and solve for d.
$$d + 13 = 35$$
$$d = 22 \text{ dimes}$$

4. The problem asks for the number of dimes and the number of quarters. There are 22 dimes and 13 quarters.
5. Check the answer.

$$d + q = 35$$ The total number of coins is 35.
$$22 + 13 \ ? \ 35$$
$$35 = 35 \ \checkmark$$

$$0.10d + 0.25q = 5.45$$ The coins amount to \$5.45.
$$0.10(22) + 0.25(13) \ ? \ 5.45$$
$$2.20 + 3.25 \ ? \ 5.45$$
$$5.45 = 5.45 \ \checkmark$$

example 3

A mixture of walnuts and pecans that weighs 75 pounds sells for \$0.98 per pound. If walnuts cost \$1.10 per pound and pecans cost \$0.90 per pound, how many pounds of walnuts and how many pounds of pecans are included in the mixture?

solution

1. Define the variables.

Let w = number of pounds of walnuts
p = number of pounds of pecans

2. Write two equations that relate the variables.
The first equation involves the number of pounds of nuts in the mixture.

number of pounds of walnuts + number of pounds of pecans = total number of pounds

$$w + p = 75$$

The second equation involves the cost of the nuts in the mixture.

cost of walnuts + cost of pecans = total cost

$$1.1w + 0.9p = 0.98(75)$$

3. Solve this system using the substitution technique.

$$p = 75 - w$$ Solve the first equation for p.

$$1.1w + 0.9p = 73.5$$ Substitute $75 - w$ for p in the second equation.

$$1.1w + 0.9(75 - w) = 73.5$$

$$1.1w + 67.5 - 0.9w = 73.5$$ Solve the resulting equation for w.

$$0.2w = 6$$

$$w = 30 \text{ pounds of walnuts}$$

$$w + p = 75$$ Substitute $w = 30$ into the first equation and solve for p.

$$30 + p = 75$$

$$p = 45 \text{ pounds of pecans}$$

4. The problem asks for the number of pounds of walnuts and the number of pounds of pecans. There are 30 pounds of walnuts and 45 pounds of pecans.
5. Check the answer.

$$w + p = 75$$ The mixture weighs 75 pounds.

$$30 + 45 \ ? \ 75$$

$$75 = 75 \ \checkmark$$

$$1.1w + 0.9p = 0.98(75)$$ The mixture sells for \$0.98 per pound.

$$1.1(30) + 0.9(45) \ ? \ 0.98(75)$$

$$33.0 + 40.5 \ ? \ 73.5$$

$$73.5 = 73.5 \ \checkmark$$

example 4

A chemist needs to mix a 10% acid solution with a 35% acid solution to make 500 ml of a solution that is 25% acid. How much of each solution does she need?

solution

1. Define the variables.

x = number ml of 10% acid solution

y = number ml of 35% acid solution

2. Write two equations that relate the variables. For percent mixture problems it is often easiest to use a table to display the information given in the problem.

	ml OF SOLUTION	% PURE ACID	ml OF PURE ACID
1st solution	x	.10	$.10x$
2nd solution	y	.35	$.35y$
Mixture	500	.25	.25(500)

The first equation is obtained by adding x ml of the first solution to y ml of the second solution to obtain 500 ml of the mixture.

$$x + y = 500$$

The second equation is obtained by adding the amount of pure acid in x ml of the 10% solution, $.10x$, to the amount of pure acid in y ml of the 35% solution, $.35y$, to obtain the amount of pure acid in 500 ml of the mixture, which is 25% acid, $.25(500)$.

$$.10x + .35y = .25(500)$$

3. Solve this system using the substitution technique.

$$x + y = 500$$
$$.10x + .35y = 125$$
$$x = 500 - y \qquad \text{Solve the first equation for } x.$$
$$.10(500 - y) + .35y = 125 \qquad \text{Substitute } 500 - y \text{ for } x \text{ in the second equation.}$$
$$50 - .10y + .35y = 125 \qquad \text{Solve the resulting equation for } y.$$
$$50 + .25y = 125$$
$$.25y = 75$$
$$y = 300 \text{ ml of 35\% solution}$$
$$x + y = 500 \qquad \text{Substitute } y = 300 \text{ into the first equation and solve for } x.$$
$$x + 300 = 500$$
$$x = 200 \text{ ml of 10\% solution}$$

4. The problem asks for the amount of each solution required to make the mixture. 300 ml of the 35% acid solution and 200 ml of the 10% acid solution are needed.

5. Check the answers.

$$x + y = 500 \qquad \text{The mixture contains 500 ml.}$$
$$200 + 300 \ ? \ 500$$
$$500 = 500 \ \checkmark$$
$$.10x + .35y = 125 \qquad \text{The mixture contains 125 ml of pure acid.}$$
$$.10(200) + .35(300) \ ? \ 125$$
$$20 + 105 \ ? \ 125$$
$$125 = 125 \ \checkmark$$

example 5

A woman has a total inheritance of \$75,000 which she invested in two accounts. One account has an annual interest rate of 8% and the other has an annual interest rate of $9\frac{1}{2}\%$. If the total amount of interest she earns the first year is \$6,600, how much did she invest at each rate?

solution

1. Define the variables.

Let x = amount invested at 8%

y = amount invested at $9\frac{1}{2}\%$

2. Write two equations that relate the variables.
The first equation involves the value of the inheritance.

$$\text{amount invested at } 8\% + \text{amount invested at } 9\tfrac{1}{2}\% = \text{total inheritance}$$
$$x + y = 75{,}000$$

The second equation involves the amount of total interest.

$$\text{interest from } 8\% \text{ account} + \text{interest from } 9\tfrac{1}{2}\% \text{ account} = \text{total interest}$$
$$0.08x + 0.095y = 6{,}600$$

3. Solve the system using the substitution technique.

$x = 75{,}000 - y$ — Solve the first equation for x.

$0.08(75{,}000 - y) + 0.095y = 6{,}600$ — Substitute $75{,}000 - y$ for x in the second equation and solve for y.

$6{,}000 - .08y + 0.095y = 6{,}600$

$0.015y = 600$

$y = 40{,}000$ dollars at $9\frac{1}{2}\%$

$x + y = 75{,}000$ — Substitute $y = 40{,}000$ into the first equation and solve for x.

$x + 40{,}000 = 75{,}000$

$x = 35{,}000$ dollars at 8%

4. The problem asks for the amount invested at each rate. \$35,000 was invested at 8% and \$40,000 was invested at $9\frac{1}{2}\%$.
5. Check the answer.

$x + y = 75{,}000$ — The total investment is \$75,000.

$35{,}000 + 40{,}000 \ ? \ 75{,}000$

$75{,}000 = 75{,}000 \ \checkmark$

$0.08x + 0.095y \ ? \ 6{,}600$ — The total interest is \$6,600.

$0.08(35{,}000) + 0.095(40{,}000) \ ? \ 6{,}600$

$2{,}800 + 3{,}800 \ ? \ 6{,}600$

$6{,}600 = 6{,}600 \ \checkmark$

●

example 6

The sum of the digits in a two-digit number is 12. If 36 is added to the number, the order of the digits is reversed. Find the number.

solution

1. Define the variables.

Let t = tens digit

u = ones digit

2. Write two equations relating the variables.
The sum of the digits is 12.

$t + u = 12$

$10t + u$ is the original number.
$10u + t$ is the number obtained when the digits are reversed.
36 more than the number equals the number with the digits reversed.

$$10t + u + 36 = 10u + t$$

$$10t + u - 10u - t = -36$$
$$9t - 9u = -36$$
$$t - u = -4$$

Simplify the equation by collecting variable terms on the left and putting the constant on the right.

3. Solve the system using the addition-elimination technique.

$$t + u = 12$$
$$t - u = -4$$
$$2t = 8$$

Add the two equations.

$$t = 4 \text{ tens digit}$$

Solve for t.

$$t + u = 12$$
$$4 + u = 12$$
$$u = 8 \text{ ones digit}$$

Substitute $t = 4$ into the first equation and solve for u.

4. The problem asks for the original number.
$10t + u = 10(4) + 8 = 48$ is the number.

5. Check the answer.

$$t + u = 12$$
$$4 + 8 \ ? \ 12$$
$$12 = 12 \ \checkmark$$

The sum of the digits is 12.

$$10t + u + 36 = 10u + t$$
$$10(4) + 8 + 36 \ ? \ 10(8) + 4$$
$$40 + 8 + 36 \ ? \ 80 + 4$$
$$48 + 36 \ ? \ 84$$
$$84 = 84 \ \checkmark$$

36 more than the number equals the number with the digits reversed.

15.4 Exercises

Solve each of the following word problems.

1. Two numbers have a sum of 9. If three times the first is equal to 5 less than the second, find the two numbers.
2. Two numbers have a sum of 4. If twice the first is equal to 11 more than the second, find the two numbers.
3. The difference between two numbers is 6. If the larger number is 3 less than four times the smaller, find the two numbers.
4. The difference between two numbers is 8. If the smaller number is 14 less than twice the larger, find the two numbers.
5. Find two numbers such that the first is 3 less than five times the second and that 11 more than three times the second is equal to 3 more than twice the first.
6. Find two numbers such that the first is 7 more than four times the second and that 12 less than three times the first is equal to 5 more than eight times the second.
7. A collection of coins that consists of only quarters and nickels is worth $3.65. If a total of 29 coins are in the collection, how many are quarters, and how many are nickels?
8. A piggy bank contains 57 coins which consist of only dimes and nickels. If the coins amount to $4.40, how many are dimes, and how many are nickels?
9. A wallet contains only $5 and $10 bills which amount to $145. If the number of $5 bills is 3 less than twice the number of $10 bills, how many bills of each type are in the wallet?

10. A purse contains only dimes and quarters which amount to \$4.75. If the number of dimes is 2 less than three times the number of quarters, how many coins of each type are in the purse?
11. A lawyer invested a total of \$12,500 in two types of bonds, one yielding 11% a year and the other yielding 12% a year. If her annual income from the two investments is \$1,430, how much money did she invest in each type of bond?
12. A journalist invested a total of \$6,500 in two accounts, one having an annual interest rate of 9% and the other having an annual interest rate of $7\frac{1}{2}\%$. If the total interest earned for one year is \$525, find the amount invested at each rate.
13. A physician invested a sum of money in two different money market accounts, one having an annual interest rate of $12\frac{1}{2}\%$ and the other having an annual interest rate of 10%. The amount invested at $12\frac{1}{2}\%$ is \$500 more than the amount invested at 10%. If her annual income from these two investments is \$2,020, find the amount invested at each rate.
14. A linguist invested an inheritance in two types of bonds, one yielding 10% a year and the other yielding 12% a year. His annual income from these two investments is \$3,050. If the amount invested at 10% is \$2,500 less than the amount invested at 12%, find the amount invested at each rate.
15. A candy store sells marshmallow-filled chocolates for \$3.25 a pound and cream-filled chocolates for \$2.75 a pound. How many pounds of each type of candy should be used to make a 20-pound assortment that sells for \$2.95 a pound?
16. A bakery sells chocolate chip cookies for \$2.25 a pound and sugar cookies for \$1.75 a pound. How many pounds of each type should be used to make 25 pounds of a mixture that sells for \$1.99 a pound?
17. A garage mechanic needs to make 7 l of a 45% antifreeze solution by mixing a 25% antifreeze solution with a 60% antifreeze solution. How much of each solution should he use?
18. How much milk containing 2% butterfat should be mixed with milk containing 5% butterfat to make 8 pints of milk containing 3.5% butterfat?
19. How much pure acid (100% acid solution) should be mixed with a 25% acid solution to make 300 ml of a 40% acid solution?
20. How much pure alcohol should be added to a 40% alcohol solution to make 600 ml of a 75% alcohol solution?
21. A grocer wants to mix coffee selling for \$3.75 a pound with coffee selling for \$4.50 a pound to make 50 pounds of a mixture that sells for \$4.23 a pound. How many pounds of each type of coffee should he use?
22. A store sells hard candies for \$0.75 a pound and carmel candies for \$0.90 a pound. How many pounds of each type should be used to make 60 pounds of a mixture that sells for \$0.84 a pound?
23. A movie theater charges an admission price of \$4.25 for adults and \$2.50 for children. If the theater earned a total of \$662.50 for selling 188 tickets to the 8 P.M. show, how many adult tickets and how many children's tickets were sold?
24. A fast-food restaurant sells hamburgers for \$0.90 a piece and fish sandwiches for \$1.25 a piece. If they made a total of \$184.55 by selling 179 of these sandwiches to the lunch-time crowd, how many hamburgers and how many fish sandwiches were sold?
25. The sum of the digits in a two-digit number is 13. If the ones digit is 8 less than twice the tens digit, find the number.
26. The sum of the digits in a two-digit number is 8. If 36 is subtracted from the number, the order of the digits is reversed. Find the number.
27. The sum of the digits in a two-digit number is 9. If 45 is added to the number, the order of the digits is reversed. Find the number.
28. The ones digit of a two-digit number is 6 more than the tens digit. The number is 9 less than four times the sum of the digits. Find the number.

15.5 Graphical Solutions to Systems of Linear Inequalities

In Section 14.6 we learned that the solution set for a linear inequality in two variables is the portion of the x-y plane that is bounded by the corresponding linear equation and contains all the points that satisfy the inequality. The solution set to a system of linear inequalities is a region that is bounded by the graphs of the corresponding linear equations in the system. This region is the intersection of the solution sets for all the linear inequalities in the system. It contains the set of points that satisfy all of the inequalities in the system. The following example illustrates the procedure for graphing the solution set to a system of linear inequalities.

example 1

Graph the solution set to the following system.

$$x \leqslant 3$$
$$y > -2$$

solution

1. Graph the solution set to the inequality $x \leqslant 3$. We graph the equation $x = 3$ using a solid line because it corresponds to a $\leqslant$ inequality, and we shade the region to the left of this line because it contains all the points that satisfy the inequality.
2. Graph the solution set to the inequality $y > -2$ on the same set of axes. We graph the equation $y = -2$ using a dotted line because it corresponds to a $>$ inequality and we then shade the region above the line because it contains all the points that satisfy the inequality.
3. The region that is shaded twice satisfies both inequalities because it represents the intersection of the regions that satisfy the inequalities $x \leqslant 3$ and $y > -2$. ●

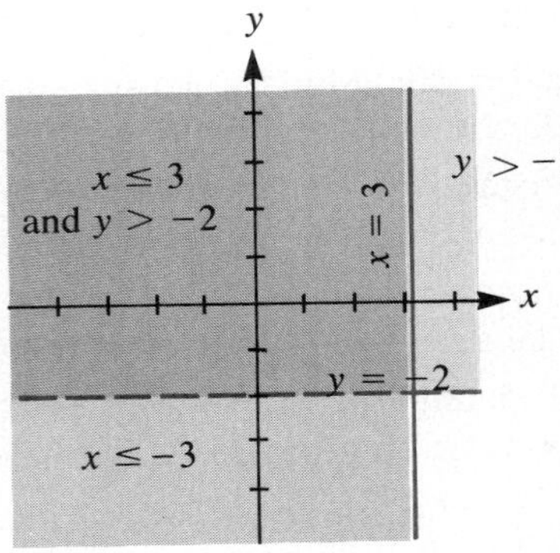

We now summarize this procedure for graphing systems of linear inequalities:

To Solve a System of Linear Inequalities

1. For each inequality, graph the corresponding linear equation to find a boundary of the region that represents the solution set. Use a solid line for $\geqslant$ or $\leqslant$ inequalities and a dotted line for $>$ or $<$ inequalities. Shade the side of the line that includes the points that satisfy the inequality.
2. Graph the solution sets for all inequalities in the system on the same set of axes as described in step 1.
3. The region that is the intersection of all the shaded regions is the solution set to the system of linear inequalities. It contains the set of points that satisfy all the inequalities in the system.

example 2

Graph the solution set to the following system.

$$5x + 3y \leqslant 15$$
$$y \geqslant 2x$$

solution

1. Graph the solution set to the inequality $5x + 3y \leqslant 15$. We graph the equation $5x + 3y = 15$ using a solid line and shade the region below this line.
2. Graph the solution set to the inequality $y \geqslant 2x$. We graph the equation $y = 2x$ using a solid line and shade the region above this line.
3. The region that is shaded twice satisfies both inequalities because it represents the intersection of the regions that satisfy the inequalities $5x + 3y \leqslant 15$ and $y \geqslant 2x$.

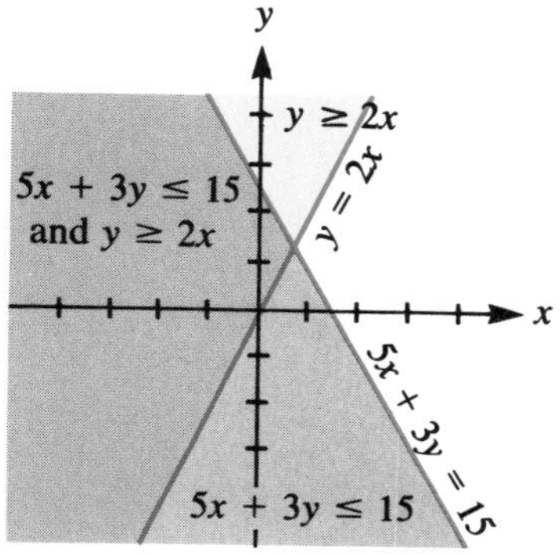

example 3

Graph the solution set to the following system.

$$3x + 4y \geqslant 12$$
$$x - 2y > -6$$
$$x < 7$$

solution

1. Graph the solution set to the inequality $3x + 4y \geqslant 12$. We graph the equation $3x + 4y = 12$ using a solid line and shade the region above this line.

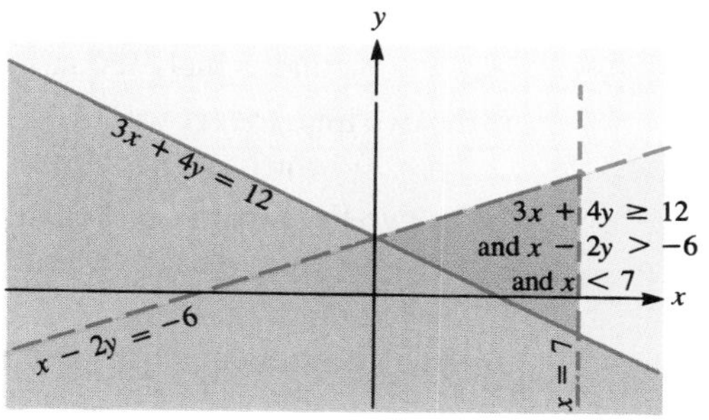

2. Graph the solution set to the inequality $x - 2y > -6$. We graph the equation $x - 2y = -6$ using a dotted line and shade the region below this line.
3. Graph the solution set to the inequality $x < 7$. We graph the equation $x = 7$ using a dotted line and shade the region to the left of this line.
4. The region that is shaded three times satisfies all three inequalities because it represents the intersection of the regions that satisfy the inequalities $3x + 4y \geqslant 12$, $x - 2y > -6$, and $x < 7$. ●

15.5 Exercises

Graph the solution set to each of the following systems of linear inequalities.

1. $x \leqslant 5$, $y > 1$

2. $x > 2$, $y \leqslant 9$

3. $x \geqslant -4$, $y \geqslant 0$

4. $x < 0$, $y \geqslant -5$

5. $x - 3y > 6$, $x < 3$

6. $2x + y \leqslant 8$, $y \geqslant 2$

7. $2x + 5y \leqslant 10$, $y > -2$

8. $3x - 4y > 12$, $x \leqslant 3$

9. $5x + 3y > -15$, $y > 3x$

10. $5x + 6y < 30$, $y \geqslant 4x$

11. $4x - 3y \leqslant 12$, $y < -2x$

12. $x - 2y > -6$, $y < -3x$

13. $2x + 3y \geqslant 6$, $x - y \geqslant -4$

14. $-5x + 2y \leqslant 10$, $x - y < 5$

15. $2x + 5y \leqslant 10$, $2x - y > -6$

16. $-3x + 5y > 15$, $2x + y \geqslant 6$

17. $2x - 3y \geqslant -6$, $7x + 3y < 21$

18. $5x - 2y < 10$, $7x + 4y \geqslant 28$

19. $-4x + 3y \leqslant 12$, $5x + 3y > -15$

20. $-5x + 4y \geqslant 20$, $3x + 7y \geqslant -21$

21. $4x + 5y \leqslant 20$, $x > 1$, $y \geqslant -2$

22. $4x + 3y > -12$, $x \leqslant 5$, $y < 4$

23. $-5x + 2y < 10$, $2x + 3y > 6$, $x < 2$

24. $3x + 2y \geqslant -12$, $2x - 5y \leqslant 10$, $y \leqslant 1$

25. $3x - 4y < -12$, $5x + 3y \leqslant 15$, $x \geqslant -2$

26. $2x - 3y < 6$, $7x + 2y < 14$, $x > -4$

Summary and Review

Key Terms

[15.1] A **system of linear equations** consists of two or more linear equations which are graphed on the same set of coordinate axes.

A **solution** to a linear system is an ordered pair that is a solution to every linear equation that is included in the system.

If the two lines that constitute a linear system **intersect**, the solution is a single ordered pair which is the point of intersection.

If the two lines that constitute a linear system are **parallel**, there are no points common to both equations and there is no solution to the system.

If the two lines that constitute a linear system **coincide**, every point on the line satisfies both equations and there are an infinite number of solutions to the system.

[15.2, 15.3] If the result obtained using either the addition-elimination technique or the substitution technique is a **contradiction**, there is no solution to the linear system.

If the result obtained using either the addition-elimination technique or the substitution technique is an **identity**, there are an infinite number of solutions to the linear system.

Calculations

[15.1] **To solve a system of linear equations by graphing**, graph both equations on the same set of axes and look at the graph to determine the point(s) common to both equations.

[15.2] **To solve a system of linear equations algebraically**, use either the addition-elimination technique or the substitution technique.

[15.2] **To solve a system of linear equations by the addition-elimination technique:**

1. If necessary, multiply both sides of one or both of the equations in the linear system by the appropriate constant so that the coefficients of one of the variables in the resulting system are opposites.
2. Add the two equations in the resulting linear system to eliminate one of the variables.
3. Solve this equation for the variable.
4. Substitute the value obtained for this variable into one of the original equations and solve for the other variable.
5. Check the solution.

[15.3] **To solve a system of linear equations by the substitution technique:**

1. If necessary, solve one of the equations for one of the variables in terms of the other variable.
2. Substitute the resulting expression for that variable into the other equation to obtain one equation in one unknown.
3. Solve this equation for the variable.
4. Substitute the value obtained for this variable into one of the original equations and solve for the other variable.
5. Check the solution.

[15.4] **To solve word problems using linear systems:**

1. Define two variables to represent two unknown quantities that need to be found in order to solve the problem.
2. Use the information in the problem to write two linear equations that relate the variables.
3. Solve the system of linear equations for the variables.
4. Answer the question asked in the problem.
5. Check your answer by determining whether or not it satisfies the conditions of the problem.

[15.5] **To solve a system of linear inequalities**, graph the corresponding linear equation for each inequality and shade the side of the line that includes the points that satisfy the inequality. The region that is the intersection of all the shaded regions is the solution set to the system of linear inequalities.

Chapter 15 Review Exercises

[15.1] Solve each of the following linear systems by graphing.

1. $x + 2y = -2$
$3x - 4y = 24$

2. $x + 2y = 2$
$-2x + y = 6$

3. $x - 2y = -4$
$-3x + 6y = 12$

4. $x + 3y = 9$
$2x - y = 4$

5. $3x - y = 3$
$-3x + y = 6$

6. $x - y = -3$
$-x + y = 6$

[15.2] Solve each of the following linear systems using the addition-elimination technique.

7. $5x - 2y = -5$
$2x - 4y = 14$

8. $3x + 5y = -2$
$6x + 7y = 8$

9. $8x + 5y = 3$
$3x + 2y = -1$

10. $-5x + 3y = 10$
$7x - 2y = -14$

[15.3] Solve each of the following linear systems using the substitution technique.

11. $5x + y = 19$
$3x - 4y = 16$

12. $x - 4y = 18$
$-6x + 3y = -3$

13. $-x + 5y = -15$
$8x - 3y = 9$

14. $6x - y = 11$
$7x - 2y = 17$

[15.2, 15.3] Solve each of the following linear systems using any technique.

15. $5x - 2y = -6$
$10x - 3y = -4$

16. $4x - 3y = -5$
$-5x + 9y = 1$

17. $7x - y = 25$
$5x + 3y = 3$

18. $x + 4y = 18$
$6x + 5y = 13$

19. $-4x - 3y = 10$
$8x + 9y = -2$

20. $5x - 2y = 27$
$-x + 6y = 17$

[15.4] Solve each of the following word problems.

21. Two numbers have a difference of 9. If the larger is 7 less than twice the smaller, find the two numbers.

22. A collection of nickels and quarters amounts to \$3.85. If the number of nickels is 5 more than three times the number of quarters, how many of each type of coin are in the collection?

23. A filmmaker invests a \$151,000 inheritance in two kinds of money market accounts, one having an annual interest rate of 12% and the other having an annual interest rate of 11%. If the interest earned from these investments the first year is \$17,450, how much money was invested at each rate?

24. A grocer sells pistachios for \$6.79 a pound and peanuts for \$2.89 a pound. How many pounds of each must he use to make a 10-pound mixture that would sell for \$4.45 a pound?

25. A chemist needs to make 6 l of a 60% alcohol solution. If the lab has a 35% alcohol solution and a 75% alcohol solution in stock, how much of each solution should be used?

26. The sum of the digits in a two digit number is 12. If 18 is subtracted from the number, the order of the digits is reversed. Find the number.

[15.5] Solve each of the following systems of linear inequalities.

27. $4x - 3y \leqslant 12$
$y > -3$

28. $2x + 7y > 14$
$x < 5$

29. $5x + 2y > 10$
$3x - 7y < 21$

30. $9x + 2y \leqslant 18$
$-3x + 4y \leqslant 12$

31. $4x - 5y \geqslant -20$
$x + 2y > -2$
$x < 2$

32. $-2x + 7y < 14$
$5x - 2y < 10$
$x \geqslant -1$

16 RATIONAL EXPRESSIONS

16.1 Reducing to Lowest Terms

In Section 7.5 we defined a rational number to be any number that can be written in the form $\frac{a}{b}$, where a and b are integers and $b \neq 0$. In this chapter, we will be working with algebraic expressions called *rational expressions.*

definition

A *rational expression* is any expression that can be written in the form $\frac{P}{Q}$, where P and Q are polynomials and $Q \neq 0$.

Some examples of rational expressions are:

$$\frac{3}{x-7}, \quad \frac{8x-5}{x^2+9}, \quad \text{and} \quad \frac{x+2}{x^2+5x+6}.$$

According to this definition, the denominator of a rational expression cannot be equal to zero. This is because a rational expression represents the operation division, and division by zero is undefined. Thus, a rational expression is undefined if its denominator is equal to zero. We can determine the value(s) of an unknown variable that will make a rational expression undefined by setting the denominator equal to zero and solving for the unknown variable. This technique is illustrated in the following example.

example 1

Find the value(s) of x that will make each of the following rational expressions undefined.

solution

(a) $\frac{5}{x+8}$

$x + 8 = 0$ Set the denominator equal to 0.

$x = -8$ Solve for x.

The expression $\frac{5}{x+8}$ is undefined when $x = -8$. Note that the expression $\frac{5}{x+8}$ is defined for all values of x except -8.

(b) $\dfrac{2x - 1}{x^2 + 2x - 15}$

$x^2 + 2x - 15 = 0$	Set the denominator equal to 0.
$(x + 5)(x - 3) = 0$	Factor.
$x + 5 = 0 \qquad x - 3 = 0$	Set each factor equal to 0.
$x = -5 \qquad x = 3$	Solve for x.

The expression $\dfrac{2x - 1}{x^2 + 2x - 15}$ is undefined when $x = -5$ and when $x = 3$.

Note that the expression $\dfrac{2x - 1}{x^2 + 2x - 15}$ is defined for all values of x except -5 and 3. ●

We will now learn a technique for reducing a rational expression to lowest terms. Recall from Section 2.2 that to reduce a fraction to lowest terms, we divide both the numerator and denominator by the greatest common factor. For example, to reduce $\frac{14}{63}$ to lowest terms, we divide both the numerator and denominator by the greatest common factor of 14 and 63, which is 7:

$$\frac{14}{63} = \frac{14 \div 7}{63 \div 7} = \frac{2}{9}$$

Another way of showing this procedure is to factor both numerator and denominator completely and cancel out the common factors:

$$\frac{14}{63} = \frac{2 \cdot 7}{3 \cdot 3 \cdot 7} = \frac{2 \cdot \cancel{7}}{3 \cdot 3 \cdot \cancel{7}} = \frac{2}{9}$$

Keep in mind that whenever we cancel a common factor that appears in the numerator and denominator, we are actually dividing both numerator and denominator by that factor. Also remember that whenever all factors are canceled in either the numerator or denominator, the remaining factor is 1 and not 0.

$$\frac{5}{35} = \frac{\cancel{5}}{\cancel{5} \cdot 7} = \frac{1}{7}$$

We will use this same procedure to reduce a rational expression to lowest terms.

To Reduce a Rational Expression to Lowest Terms

1. Factor the numerator and denominator completely.
2. Divide the numerator and denominator by the factors that appear in both. We indicate this by canceling the common factors in the numerator and denominator.

Let us now look at some examples that illustrate this procedure. In all expressions, the values of the variables are such that the denominator is not equal to zero.

example 2

Reduce $\dfrac{35}{5x-15}$ to lowest terms.

solution

$$\frac{35}{5x-15}=\frac{5\cdot 7}{5(x-3)}$$

Factor numerator and denominator completely.

$$=\frac{\cancel{5}\cdot 7}{\cancel{5}(x-3)}$$

Divide numerator and denominator by 5.

$$=\frac{7}{x-3}$$

example 3

Reduce $\dfrac{y^2-16}{y^2+9y+20}$ to lowest terms.

solution

$$\frac{y^2-16}{y^2+9y+20}=\frac{(y+4)(y-4)}{(y+4)(y+5)}$$

Factor numerator and denominator completely.

$$=\frac{\cancel{(y+4)}(y-4)}{\cancel{(y+4)}(y+5)}$$

Divide numerator and denominator by $y+4$.

$$=\frac{y-4}{y+5}$$

example 4

Reduce $\dfrac{3t+6}{3t^2-21t-54}$ to lowest terms.

solution

$$\frac{3t+6}{3t^2-21t-54}=\frac{3(t+2)}{3(t^2-7t-18)}$$

Factor out greatest common monomial factor.

$$=\frac{3(t+2)}{3(t+2)(t-9)}$$

Factor the trinomial in the denominator.

$$=\frac{\cancel{3(t+2)}}{\cancel{3(t+2)}(t-9)}$$

Divide numerator and denominator by $3(t+2)$.

$$=\frac{1}{t-9}$$

Recall that when all factors are canceled, the remaining factor is 1 and not 0.

example 5

Reduce $\dfrac{x^3-3x^2-28x}{x^3-13x^2+42x}$ to lowest terms.

$$\frac{x^3-3x^2-28x}{x^3-13x^2+42x}=\frac{x(x^2-3x-28)}{x(x^2-13x+42)}$$

Factor out the greatest common monomial factors.

$$=\frac{x(x-7)(x+4)}{x(x-7)(x-6)}$$

Factor the trinomials in the numerator and denominator.

$$=\frac{\cancel{x(x-7)}(x+4)}{\cancel{x(x-7)}(x-6)}$$

Divide the numerator and denominator by $x(x-7)$.

$$=\frac{x+4}{x-6}$$

A common algebra mistake is to attempt to cancel common factors of the individual terms in the numerator and denominator before factoring. For example,

$$\frac{2x-3}{4x^2-9} \neq \frac{\cancel{2x}-\cancel{3}}{\underset{2x}{\cancel{4x^2}}-\underset{3}{\cancel{9}}} \neq \frac{1}{2x-3}$$

To do this problem correctly, we must remember the following:

FACTOR FIRST, THEN CANCEL.

Thus,

$$\frac{2x-3}{4x^2-9} = \frac{2x-3}{(2x+3)(2x-3)} = \frac{\cancel{2x-3}}{(2x+3)\cancel{(2x-3)}} = \frac{1}{2x+3}$$

QUICK QUIZ

Reduce each of the following to lowest terms.

1. $\dfrac{8x}{x^2-3x}$
2. $\dfrac{y+5}{y^2+14y+45}$
3. $\dfrac{5x^2-20}{5x^2+5x-30}$

ANSWERS

1. $\dfrac{8}{x-3}$
2. $\dfrac{1}{y+9}$
3. $\dfrac{x+2}{x+3}$

We are frequently asked to reduce fractions that appear in the form $\dfrac{a-b}{b-a}$. At first, there seems to be no way to reduce this fraction because it is difficult to see any common factors for $a-b$ and $b-a$. The technique for reducing fractions of this form is to factor out -1 from each term in the numerator or denominator, as shown in Example 6.

example 6

Reduce $\dfrac{x-7}{7-x}$ to lowest terms.

solution

$$\frac{x-7}{7-x} = \frac{x-7}{-1(-7+x)}$$

Factor out -1 from each term in the denominator.

$$= \frac{x-7}{-1(x-7)}$$

In the denominator, $-7+x = x-7$, by the commutative property.

$$= \frac{\cancel{x-7}}{-1\cancel{(x-7)}}$$

Divide the numerator and denominator by $x-7$.

$$= \frac{1}{-1}$$

Remaining factor in numerator is 1.

$$= -1$$

In general, any fraction of the form

$$\frac{a-b}{b-a},\ a \neq b, \text{ is equal to } -1.$$

example 7

Reduce $\dfrac{6-3x}{x^2-4}$ to lowest terms.

solution

$$\frac{6-3x}{x^2-4} = \frac{3(2-x)}{(x+2)(x-2)}$$ Factor the numerator and denominator.

$$= \frac{3}{(x+2)} \cdot \frac{(2-x)}{(x-2)}$$ Rewrite as product of two fractions.

$$= \frac{3}{x+2} \cdot (-1)$$ Since $\dfrac{2-x}{x-2} = -1$.

$$= \frac{-3}{x+2}$$ Multiply numerators.

A common algebra mistake is to attempt to reduce fractions to -1 that are not opposite in sign. For example,

$$\frac{x+8}{x-8} \neq -1$$

This fraction cannot be reduced because there are no common factors in the numerator and denominator. If we factor out -1 from each term in the denominator, we obtain the following:

$$\frac{x+8}{x-8} = \frac{x+8}{-1(-x+8)} = \frac{x+8}{-1(8-x)}$$

Therefore, this fraction cannot be reduced.

In general, fractions of the form

$$\frac{a+b}{a-b} \quad \text{cannot be reduced.}$$

QUICK QUIZ

Reduce each of the following to lowest terms.

1. $\dfrac{x-9}{9-x}$
2. $\dfrac{-5+y}{y-5}$
3. $\dfrac{4x-12}{9-x^2}$

ANSWERS

1. -1
2. 1
3. $\dfrac{-4}{3+x}$

16.1 Exercises

Find the value(s) of x that will make each of the following rational expressions undefined.

1. $\dfrac{5}{x-2}$
2. $\dfrac{8}{x+7}$
3. $\dfrac{2x}{x+9}$
4. $\dfrac{5x}{x-4}$
5. $\dfrac{-3}{x^2-16}$
6. $\dfrac{-8}{x^2-25}$
7. $\dfrac{x+7}{x^2-3x-18}$
8. $\dfrac{x-5}{x^2-9x+14}$
9. $\dfrac{3x-2}{x^3+13x^2+36x}$
10. $\dfrac{2x+1}{x^3+14x^2+48x}$

Reduce each of the following to lowest terms.

11. $\dfrac{6}{3x-9}$
12. $\dfrac{10}{5x+15}$
13. $\dfrac{4}{8x+12}$
14. $\dfrac{2}{6x-4}$
15. $\dfrac{-8}{4y-6}$
16. $\dfrac{-9}{6y+18}$
17. $\dfrac{18x}{9x+3}$
18. $\dfrac{12x}{8x-4}$
19. $\dfrac{7a}{a^2+5a}$
20. $\dfrac{5a}{a^2-4a}$
21. $\dfrac{5x+20}{4x+16}$
22. $\dfrac{8x-24}{5x-15}$
23. $\dfrac{x+5}{x^2-25}$
24. $\dfrac{x-3}{x^2-9}$
25. $\dfrac{2x-14}{x^2-49}$
26. $\dfrac{3x+27}{x^2-81}$
27. $\dfrac{y+4}{2y^2-32}$
28. $\dfrac{y-2}{3y^2-12}$
29. $\dfrac{3z+9}{z^2+8z+15}$
30. $\dfrac{4z+16}{z^2+11z+28}$
31. $\dfrac{x+8}{x^2-x-72}$
32. $\dfrac{x+7}{x^2+15x+56}$
33. $\dfrac{x^2+3x-40}{2x+16}$
34. $\dfrac{x^2-2x-63}{3x+21}$
35. $\dfrac{t^2-9}{t^2+11t+24}$
36. $\dfrac{t^2-49}{t^2+2t-35}$
37. $\dfrac{2x+14}{2x^2+10x-28}$
38. $\dfrac{3x-12}{3x^2-6x-24}$
39. $\dfrac{3y^2+24y+36}{3y^2-12}$
40. $\dfrac{4y^2-36}{4y^2+8y-60}$
41. $\dfrac{x^2-10x+21}{x^2-x-42}$
42. $\dfrac{x^2+14x+45}{x^2-2x-35}$
43. $\dfrac{2a^2+4a-48}{2a^2-22a+56}$
44. $\dfrac{3a^2+9a+6}{3a^2-6a-24}$
45. $\dfrac{z^3-64z}{z^3+2z^2-48z}$
46. $\dfrac{z^3-49z}{z^3-z^2-56z}$
47. $\dfrac{x^3-6x^2-27x}{x^3-4x^2-45x}$
48. $\dfrac{x^3+x^2-30x}{x^3+15x^2+54x}$
49. $\dfrac{9x+27}{x^4-81}$
50. $\dfrac{7x-14}{x^4-16}$
51. $\dfrac{x-8}{8-x}$
52. $\dfrac{3-x}{x-3}$
53. $\dfrac{-2+y}{y-2}$
54. $\dfrac{y-7}{-7+y}$
55. $\dfrac{5-5x}{x-1}$
56. $\dfrac{x-4}{8-2x}$
57. $\dfrac{6-x}{x^2-36}$
58. $\dfrac{x^2-81}{9-x}$
59. $\dfrac{y^2-49}{21-3y}$
60. $\dfrac{16-2y}{y^2-64}$

16.2 Multiplication and Division of Rational Expressions

Recall from Section 2.6 that in order to multiply fractions, we multiply the numerators and place the result over the product of the denominators. To avoid obtaining an answer that needs to be reduced to lowest terms, we can factor the numerators and denominators before multiplying them. We then divide the numerator and denominator of the product by the common factors.

$$\frac{7}{9}\cdot\frac{6}{35} = \frac{7}{3\cdot 3}\cdot\frac{2\cdot 3}{5\cdot 7}$$ Factor numerators and denominators.

$$= \frac{\cancel{7}\cdot 2\cdot \cancel{3}}{3\cdot\cancel{3}\cdot 5\cdot\cancel{7}}$$ Multiply numerators and denominators.

$$= \frac{2}{15}$$ Divide numerator and denominator by common factors.

We will use this same procedure to multiply rational expressions.

1. **Factor the numerator and denominator of each rational expression completely.**
2. **Multiply the numerators and denominators.**
3. **Divide the numerator and denominator of the product by the common factors that appear in both.**

We now illustrate this procedure with the following example.

example 1

Multiply $\frac{3x-6}{10}\cdot\frac{5x}{x^2-4}$

solution

$$\frac{3x-6}{10}\cdot\frac{5x}{x^2-4} = \frac{3(x-2)}{2\cdot 5}\cdot\frac{5\cdot x}{(x+2)(x-2)}$$ Factor numerators and denominators.

$$= \frac{3\cancel{(x-2)}\cdot\cancel{5}\cdot x}{2\cdot\cancel{5}(x+2)\cancel{(x-2)}}$$ Multiply numerators and denominators.

$$= \frac{3x}{2(x+2)}$$ Divide numerators and denominators by common factors.

In Section 2.7 we learned that in order to divide fractions, we multiply the first fraction by the reciprocal of the second. For example,

$$\frac{2}{9}\div\frac{4}{3} = \frac{2}{9}\cdot\frac{3}{4}$$ Multiply the first fraction by the reciprocal of the second.

$$= \frac{2}{3\cdot 3}\cdot\frac{3}{2\cdot 2}$$ Factor numerators and denominators.

$$= \frac{\not{2} \cdot \not{3}}{\not{3} \cdot 3 \cdot \not{2} \cdot 2}$$ Multiply numerators and denominators.

$$= \frac{1}{6}$$ Divide numerators and denominators by common factors.

We use this same procedure to divide rational expressions.

To Divide Rational Expressions

1. Multiply the first expression by the reciprocal of the second.
2. Simplify using the procedure for multiplying rational expressions.

We now illustrate this procedure with the following example.

example 2

Divide $\frac{4x + 6}{x^2 + 2x - 15} \div \frac{2x^2 + 3x}{x + 5}$.

solution

$$\frac{4x + 6}{x^2 + 2x - 15} \div \frac{2x^2 + 3x}{x + 5}$$

$$= \frac{4x + 6}{x^2 + 2x - 15} \cdot \frac{x + 5}{2x^2 + 3x}$$ Multiply the first expression by the reciprocal of the second.

$$= \frac{2(2x + 3)}{(x + 5)(x - 3)} \cdot \frac{x + 5}{x(2x + 3)}$$ Factor numerators and denominators.

$$= \frac{2\cancel{(2x + 3)}\cancel{(x + 5)}}{\cancel{(x + 5)}(x - 3)x\cancel{(2x + 3)}}$$ Multiply numerators and denominators.

$$= \frac{2}{x(x - 3)}$$ Divide numerators and denominators by common factors. Leave the denominator in factored form. ●

Let us now look at some additional examples of multiplying and dividing rational expressions.

example 3

Multiply $\frac{y + 7}{y^2 - 5y + 6} \cdot \frac{y^2 - 9}{y^2 + 10y + 21}$.

solution

$$\frac{y + 7}{y^2 - 5y + 6} \cdot \frac{y^2 - 9}{y^2 + 10y + 21}$$

$$= \frac{y + 7}{(y - 2)(y - 3)} \cdot \frac{(y + 3)(y - 3)}{(y + 3)(y + 7)}$$ Factor numerators and denominators.

$$= \frac{\cancel{(y + 7)}\cancel{(y + 3)}\cancel{(y - 3)}}{(y - 2)\cancel{(y - 3)}\cancel{(y + 3)}\cancel{(y + 7)}}$$ Multiply numerators and denominators.

$$= \frac{1}{y - 2}$$ Divide numerators and denominators by common factors. ●

Recall that when all the factors in the numerator are canceled, the numerator becomes 1 and not 0.

example 4

Divide $\dfrac{10a^3}{a^2 - 2a - 24} \div \dfrac{5a^3 + 5a^2}{3a^2 - 3a - 60}$.

solution

$$\frac{10a^3}{a^2 - 2a - 24} \div \frac{5a^3 + 5a^2}{3a^2 - 3a - 60}$$

$$= \frac{10a^3}{a^2 - 2a - 24} \cdot \frac{3a^2 - 3a - 60}{5a^3 + 5a^2}$$

Multiply the first expression by the reciprocal of the second.

$$= \frac{2 \cdot 5 \cdot a \cdot a \cdot a}{(a-6)(a+4)} \cdot \frac{3(a^2 - a - 20)}{5a^2(a+1)}$$

Factor numerators and denominators.

$$= \frac{2 \cdot 5 \cdot a \cdot a \cdot a}{(a-6)(a+4)} \cdot \frac{3(a-5)(a+4)}{5 \cdot a \cdot a(a+1)}$$

$$= \frac{2 \cdot \cancel{5} \cdot \cancel{a} \cdot \cancel{a} \cdot a \cdot 3(a-5)\cancel{(a+4)}}{(a-6)\cancel{(a+4)}\cancel{5} \cdot \cancel{a} \cdot \cancel{a}(a+1)}$$

Multiply numerators and denominators.

$$= \frac{6a(a-5)}{(a-6)(a+1)}$$

Divide numerator and denominator by common factors. ●

The next example illustrates how this procedure can be applied to multiply and/or divide more than two rational expressions.

example 5

Simplify $\dfrac{2x+6}{x^2-25} \cdot \dfrac{x^2+5x}{x^2-x-12} \div \dfrac{6x-30}{x^2-3x-4}$.

solution

$$\frac{2x+6}{x^2-25} \cdot \frac{x^2+5x}{x^2-x-12} \div \frac{6x-30}{x^2-3x-4} = \frac{2x+6}{x^2-25} \cdot \frac{x^2+5x}{x^2-x-12} \cdot \frac{x^2-3x-4}{6x-30}$$

$$= \frac{2(x+3)}{(x+5)(x-5)} \cdot \frac{x(x+5)}{(x-4)(x+3)} \cdot \frac{(x-4)(x+1)}{6(x-5)}$$

$$= \frac{\cancel{2}\cancel{(x+3)}x\cancel{(x+5)}\cancel{(x-4)}(x+1)}{\cancel{(x+5)}(x-5)\cancel{(x-4)}\cancel{(x+3)}\cancel{2} \cdot 3(x-5)}$$

$$= \frac{x(x+1)}{3(x-5)(x-5)} = \frac{x(x+1)}{3(x-5)^2}$$ ●

QUICK QUIZ

Perform each of the operations indicated.

1. $\dfrac{8x^3}{x^2+6x+5} \cdot \dfrac{x+5}{2x}$
2. $\dfrac{y^2-16}{y^2-6y+8} \div \dfrac{y^2+7y+12}{5y-10}$
3. $\dfrac{3a+15}{a^2+2a-8} \div \dfrac{6a^2-6}{a-2} \cdot \dfrac{a^2-6a-7}{a+5}$

ANSWERS

1. $\dfrac{4x^2}{x+1}$
2. $\dfrac{5}{y+3}$
3. $\dfrac{a-7}{2(a-1)(a+4)}$

16.2 Exercises

Perform each of the operations indicated.

1. $\dfrac{2x-16}{6x} \cdot \dfrac{x^2}{x-8}$

2. $\dfrac{x+5}{x^3} \cdot \dfrac{9x}{3x+15}$

3. $\dfrac{2x^3}{3x+12} \cdot \dfrac{5x+20}{10x^2}$

4. $\dfrac{x}{2x-4} \cdot \dfrac{4x-8}{12}$

5. $\dfrac{y-5}{6y} \div \dfrac{3y-15}{2y^2}$

6. $\dfrac{2y+14}{4y^2} \div \dfrac{y+7}{8y}$

7. $\dfrac{12}{2y-1} \div \dfrac{9y}{6y-3}$

8. $\dfrac{16y^2}{3y+2} \div \dfrac{4y}{6y+4}$

9. $\dfrac{3x^3}{x^2-4} \cdot \dfrac{5x+10}{6x^2}$

10. $\dfrac{4x-12}{8x} \cdot \dfrac{2x}{x^2-9}$

11. $\dfrac{8a}{2a-12} \div \dfrac{a^3}{a^2-36}$

12. $\dfrac{a^2-25}{a} \div \dfrac{3a+15}{9a^2}$

13. $\dfrac{x^2+5x-14}{3x-6} \cdot \dfrac{12x}{x^2-49}$

14. $\dfrac{10x+30}{x^2+8x+15} \cdot \dfrac{x^2-25}{5x}$

15. $\dfrac{4x^2}{x^2-16} \div \dfrac{2x^2+12x}{x^2+10x+24}$

16. $\dfrac{8x^2+72x}{x^2+6x-27} \div \dfrac{2x^2}{x^2-9}$

17. $\dfrac{x^2+2x-15}{12x^2} \cdot \dfrac{3x^2-12x}{x^2+x-20}$

18. $\dfrac{x^2-12x+35}{8x^2-56x} \cdot \dfrac{20x^2}{x^2-3x-10}$

19. $\dfrac{x^2-9x}{4x^2+28x} \div \dfrac{x^2-7x-18}{x^2+9x+14}$

20. $\dfrac{x^2-x-12}{x^2-10x+24} \div \dfrac{5x^2+15x}{x^2-6x}$

21. $\dfrac{t^2+10t+21}{t+5} \cdot \dfrac{t^2+t-20}{t^2+3t-28}$

22. $\dfrac{t^2+5t-14}{t^2+10t+21} \cdot \dfrac{3t-18}{t^2-8t+12}$

23. $\dfrac{2z^2+7z+3}{z^2-z-12} \cdot \dfrac{7z-28}{2z^2-z-1}$

24. $\dfrac{z^2+8z+7}{3z^2+5z+2} \cdot \dfrac{15z+10}{z^2+5z-14}$

25. $\dfrac{x^2-5x-24}{x^2-12x+32} \div \dfrac{7x+21}{x^2-x-12}$

26. $\dfrac{x^2-11x+30}{x^2+3x-40} \div \dfrac{x^2-4x-12}{x+8}$

27. $\dfrac{2y^2+4y-30}{y^2+3y-18} \div \dfrac{4y^2+20y}{y^2+5y-6}$

28. $\dfrac{9y^2-18y}{y^2+7y+12} \div \dfrac{3y^2-24y+36}{y^2-3y-18}$

29. $\dfrac{10x^3-40x}{x^2+8x-9} \cdot \dfrac{x^2+7x-18}{5x^2+10x}$

30. $\dfrac{2x^2+16x+30}{x^2+3x-10} \cdot \dfrac{x^2-2x}{4x^3-36x}$

31. $\dfrac{3x-21}{x^2-x-2} \cdot \dfrac{x^2-1}{x^2-x-42} \cdot \dfrac{x^2+4x-12}{6x-6}$

32. $\dfrac{x^2-6x+27}{x+2} \cdot \dfrac{x-5}{7x-63} \cdot \dfrac{14x+28}{x^2-2x-15}$

33. $\dfrac{x^2+2x-63}{3x^2+6x} \cdot \dfrac{x^2-4}{x^2+12x+27} \div \dfrac{x-7}{6x+18}$

34. $\dfrac{x+8}{12x-12} \cdot \dfrac{4x+20}{x^2+4x-32} \div \dfrac{x^2+13x+40}{x^2-5x+4}$

35. $\dfrac{2x^3+4x^2}{x^2+x-20} \div \dfrac{10x^3-90x}{x^2-2x-35} \div \dfrac{x^2-5x-14}{x^2-x-12}$

36. $\dfrac{x^2-3x-28}{3x^3+3x^2} \div \dfrac{x^2-5x-36}{x-9} \div \dfrac{x^2-10x+21}{12x^3-12x}$

16.3 The Lowest Common Denominator

In Chapter 2, we learned that in order to add or subtract fractions having different denominators, first we must change each fraction to an equivalent fraction having the same denominator. Our calculations were simplest if we used the LCD or lowest common denominator. The LCD is the smallest number that is divisible by the denominators of all fractions being considered.

Recall that in order to find the LCD, we completely factor each denominator and write each factorization in exponential notation. The LCD is the product of all unique factors, each raised to the highest exponent appearing in any of the factorizations. For example, we can find the LCD of $\frac{3}{8}$ and $\frac{5}{12}$ as follows:

$8 = 2 \cdot 2 \cdot 2 = 2^3$ — Completely factor each denominator.

$12 = 2 \cdot 2 \cdot 3 = 2^2 \cdot 3$

The highest exponent 2 is raised to is 3: 2^3. — The unique prime factors are 2 and 3.

The highest exponent 3 is raised to is 1: 3^1.

$\text{LCD} = 2^3 \cdot 3 = 8 \cdot 3 = 24$ — The LCD is the product $2^3 \cdot 3$.

We use the same procedure to find the LCD of two rational expressions as follows.

To Find the LCD of Two or More Rational Expressions

1. Completely factor all denominators and write each factorization in exponential notation.
2. List every unique factor and raise each one to the highest exponent that appears in any of the factorizations.
3. The LCD is the product of all the expressions found in step 2.

We will illustrate this procedure with the following examples.

example 1

Find the LCD of $\frac{5}{9x^4}$ and $\frac{2}{15x^3}$.

$9x^4 = 3 \cdot 3 \cdot x^4 = 3^2 \cdot x^4$ — Factor each denominator completely.

$15x^3 = 3 \cdot 5 \cdot x^3$

The highest exponent 3 is raised to is $2: 3^2$. — The unique factors are 3, 5, and x.

The highest exponent 5 is raised to is $1: 5$.

The highest exponent x is raised to is $4: x^4$.

$\text{LCD} = 3^2 \cdot 5 \cdot x^4 = 45x^4$ — The LCD is the product $3^2 \cdot 5 \cdot x^4$. ●

example 2

Find the LCD for $\dfrac{4x+1}{x^2-36}$ and $\dfrac{3x}{2x+12}$.

solution

$x^2 - 36 = (x+6)(x-6)$
$2x + 12 = 2(x+6)$
The unique factors are 2, $(x+6)$, and $(x-6)$.
The highest exponent each factor is raised to is 1.
$\text{LCD} = 2(x+6)(x-6)$ ●

example 3

Find the LCD for $\dfrac{9y}{y^2+4y-21}$ and $\dfrac{2y-5}{y^2-9}$.

solution

$y^2 + 4y - 21 = (y+7)(y-3)$ — Factor each denominator.
$y^2 - 9 = (y+3)(y-3)$
The unique factors are $(y+7)$, $(y-3)$, and $(y+3)$.
The highest exponent each factor is raised to is 1.
$\text{LCD} = (y+7)(y-3)(y+3)$ ●

Once we have found the LCD, it is important to be able to determine what factor(s) each of the original denominators must be multiplied by in order to obtain the LCD. We do this by comparing the factorizations of each denominator with that of the LCD and crossing out the factors common to both. For example, let us consider the fractions $\frac{3}{8}$ and $\frac{5}{12}$ which we found to have an LCD of 24:

$$\left.\begin{aligned} 8 &= \cancel{2}\cdot\cancel{2}\cdot\cancel{2} \\ \text{LCD} &= \cancel{2}\cdot\cancel{2}\cdot\cancel{2}\cdot 3 \end{aligned}\right\} \quad \text{Multiply by 3.}$$

$$\left.\begin{aligned} 12 &= \cancel{2}\cdot\cancel{2}\cdot\cancel{3} \\ \text{LCD} &= \cancel{2}\cdot\cancel{2}\cdot 2\cdot\cancel{3} \end{aligned}\right\} \quad \text{Multiply by 2.}$$

Therefore, to obtain the LCD we must multiply 8 by 3 and 12 by 2.

example 4

Find the LCD of $\dfrac{4}{5x^2-15x}$ and $\dfrac{3}{10x}$.

Then determine what factor(s) each denominator must be multiplied by to obtain the LCD.

solution

$15x^2 - 15x = 5x(x-3)$
$10x = 2\cdot 5\cdot x$
The unique factors are 5, x, $(x-3)$, and 2.
The highest exponent each factor is raised to is 1.
$\text{LCD} = 5\cdot x\cdot(x-3)\cdot 2 = 10x(x-3)$

Now compare the factorizations of each denominator with that of the LCD:

$$\left.\begin{aligned} 5x^2 - 15x &= \cancel{5}\cdot\cancel{x}\cdot\cancel{(x-3)} \\ \text{LCD} &= 2\cdot\cancel{5}\cdot\cancel{x}\cdot\cancel{(x-3)} \end{aligned}\right\} \quad \text{Multiply by 2.}$$

$$\left.\begin{aligned} 10x &= \cancel{2}\cdot\cancel{5}\cdot\cancel{x} \\ \text{LCD} &= \cancel{2}\cdot\cancel{5}\cdot\cancel{x}\cdot(x-3) \end{aligned}\right\}$$ Multiply by $(x - 3)$.

Thus, to obtain the LCD we must multiply $5x^2 - 15x$ by 2 and $10x$ by $(x - 3)$.

example 5

Find the LCD for $\dfrac{5}{z^3 - 9z}$ and $\dfrac{2z - 1}{z^2 + 7z + 12}$.

Then determine what factor(s) each denominator must be multiplied by to obtain the LCD.

solution

$z^3 - 9z = z(z^2 - 9) = z(z + 3)(z - 3)$

$z^2 + 7z + 12 = (z + 3)(z + 4)$

The unique factors are z, $(z + 3)$, $(z - 3)$, and $(z + 4)$.

The highest exponent each factor is raised to is 1.

LCD $= z(z + 3)(z - 3)(z + 4)$

Now compare the factorization of each denominator with that of the LCD:

$$\left.\begin{aligned} z^3 - 9z &= \cancel{z}\cdot\cancel{(z+3)}\cdot\cancel{(z-3)} \\ \text{LCD} &= \cancel{z}\cdot\cancel{(z+3)}\cdot\cancel{(z-3)}\cdot(z+4) \end{aligned}\right\}$$ Multiply by $(z + 4)$.

$$\left.\begin{aligned} z^2 + 7z + 12 &= \cancel{(z+3)}\cdot\cancel{(z+4)} \\ \text{LCD} &= z\cdot\cancel{(z+3)}\cdot(z-3)\cdot\cancel{(z+4)} \end{aligned}\right\}$$ Multiply by $z(z - 3)$.

Thus, to obtain the LCD we must multiply $z^3 - 9z$ by $(z + 4)$ and $z^2 + 7z + 12$ by $z(z - 3)$.

QUICK QUIZ

Find the LCD for each of the following pairs of fractions.

1. $\dfrac{5}{4k^3}$ and $\dfrac{7}{18k^2}$
2. $\dfrac{2y}{y^2 - 36}$ and $\dfrac{4}{5y + 30}$
3. $\dfrac{x - 3}{x^2 - 3x - 4}$ and $\dfrac{2x + 5}{x^2 - 9x + 20}$

ANSWERS

1. $36k^3$
2. $5(y + 6)(y - 6)$
3. $(x + 1)(x - 4)(x - 5)$

16.3 Exercises

Find the LCD for each of the following pairs of fractions.

1. $\dfrac{4}{9k}$ and $\dfrac{5}{12k}$
2. $\dfrac{2}{15k}$ and $\dfrac{7}{10k}$
3. $\dfrac{2}{5x}$ and $\dfrac{4}{x^3}$
4. $\dfrac{5}{x^4}$ and $\dfrac{3}{2x^2}$

5. $\frac{5}{2y^4}$ and $\frac{3}{6y^2}$

6. $\frac{1}{12y^3}$ and $\frac{4}{3y^5}$

7. $\frac{1}{a^2b^3}$ and $\frac{6}{ab^4}$

8. $\frac{3}{a^5b^2}$ and $\frac{8}{a^3b^3}$

9. $\frac{3}{2x - 6}$ and $\frac{1}{4x - 12}$

10. $\frac{1}{3x + 15}$ and $\frac{5}{2x + 10}$

11. $\frac{8}{3t + 21}$ and $\frac{2}{t^2 + 7t}$

12. $\frac{7}{4t - 24}$ and $\frac{4}{t^2 - 6t}$

13. $\frac{5}{k - 9}$ and $\frac{4}{9 - k}$

14. $\frac{6}{3 - k}$ and $\frac{2}{k - 3}$

15. $\frac{3}{6y + 30}$ and $\frac{4}{9y + 45}$

16. $\frac{5}{8y - 16}$ and $\frac{3}{7y - 14}$

17. $\frac{5x}{x + 7}$ and $\frac{2x}{x^2 - 49}$

18. $\frac{4x}{x - 9}$ and $\frac{3x}{x^2 - 81}$

19. $\frac{y}{y^2 - 4}$ and $\frac{3}{5y - 10}$

20. $\frac{5}{y^2 - 9}$ and $\frac{y}{6y + 18}$

21. $\frac{x - 3}{x^2 + 3x - 40}$ and $\frac{x + 5}{x^2 - 64}$

22. $\frac{x + 2}{x^2 + 8x - 9}$ and $\frac{x - 4}{x^2 - 1}$

23. $\frac{2z}{z^2 - 25}$ and $\frac{7z}{z^2 - 12z + 35}$

24. $\frac{5z}{z^2 - 16}$ and $\frac{3z}{z^2 + 9z + 20}$

25. $\frac{x - 1}{x^3 - 16x}$ and $\frac{3x + 4}{x^2 + 3x - 28}$

26. $\frac{x + 8}{x^3 - 25x}$ and $\frac{2x - 7}{x^2 - 13x + 40}$

27. $\frac{3x}{x^2 + 6x - 7}$ and $\frac{5x}{x^2 - 5x + 4}$

28. $\frac{2x}{x^2 + 4x - 21}$ and $\frac{4x}{x^2 + 11x + 28}$

29. $\frac{x + 4}{x^3 - 9x^2 + 18x}$ and $\frac{2x - 5}{x^2 - x - 30}$

30. $\frac{x - 7}{x^3 + 7x^2 + 10x}$ and $\frac{3x + 2}{x^2 - 4x - 45}$

Find the LCD for each of the following pairs of fractions. Then determine what factor(s) each denominator must be multiplied by to obtain the LCD.

31. $\frac{4}{5x^3}$ and $\frac{8}{15x}$

32. $\frac{3}{16x^4}$ and $\frac{7}{24x^2}$

33. $\frac{5}{6y - 18}$ and $\frac{3}{2y - 6}$

34. $\frac{8}{3y + 21}$ and $\frac{5}{2y + 14}$

35. $\frac{4}{x^2 + 5x}$ and $\frac{x}{4x + 20}$

36. $\frac{3}{x^2 - 8x}$ and $\frac{9x}{5x - 40}$

37. $\frac{y}{3y + 27}$ and $\frac{7}{y^2 - 81}$

38. $\frac{3}{y^2 - 25}$ and $\frac{y}{3y - 15}$

39. $\frac{3x}{x^2 - 9}$ and $\frac{4x}{x^2 - 5x - 24}$

40. $\frac{2x}{x^2 - 6x + 8}$ and $\frac{9x}{x^2 - 16}$

41. $\frac{3x - 1}{x^2 + 3x - 10}$ and $\frac{x + 5}{x^2 - 9x + 14}$

42. $\frac{x - 7}{x^2 + 2x - 15}$ and $\frac{2x + 3}{x^2 + 8x + 15}$

16.4 Addition and Subtraction of Rational Expressions

In Section 2.3 we learned that to add two fractions having the same denominator, we add the numerators, put the result over the common denominator, and, if possible, reduce to lowest terms. For example,

$$\frac{3}{7} + \frac{2}{7} = \frac{3+2}{7} = \frac{5}{7}$$

We will use this same procedure to add two rational expressions having the same denominator as shown in the following example.

example 1

Add $\frac{5}{y} + \frac{8}{y}$.

solution

$$\frac{5}{y} + \frac{8}{y} = \frac{5+8}{y}$$

Add the numerators and put the result over the common denominator.

$$= \frac{13}{y}$$

●

To subtract two rational expressions having the same denominator, we subtract the numerators, put the result over the common denominator and, if possible, reduce to lowest terms, as shown in the following example.

example 2

Subtract $\frac{x}{x^2-4} - \frac{2}{x^2-4}$.

solution

$$\frac{x}{x^2-4} - \frac{2}{x^2-4} = \frac{x-2}{x^2-4}$$

Subtract the numerators and put the result over the common denominator.

$$= \frac{\cancel{x-2}}{(x+2)\cancel{(x-2)}}$$

To reduce to lowest terms, factor the denominator and divide numerator and denominator by the common factor.

$$= \frac{1}{x+2}$$

●

Recall that if we want to add or subtract two fractions that do not have the same denominator, we must find the LCD or lowest common denominator of the two fractions. The LCD is the smallest number that is divisible by the denominators of all fractions being considered. We then change each fraction to an equivalent fraction having the same LCD, and then perform the addition or subtraction. For example, to subtract the fractions $\frac{5}{6} - \frac{2}{9}$, we must first find the LCD by factoring the two denominators:

$$6 = 2 \cdot 3$$
$$9 = 3 \cdot 3$$
$$\text{LCD} = 2 \cdot 3 \cdot 3 = 18$$

To obtain the LCD, we must multiply the denominator 6 by 3 and the denominator 9 by 2.

$$\frac{5}{6}-\frac{2}{9}=\frac{5\cdot 3}{6\cdot 3}-\frac{2\cdot 2}{9\cdot 2}$$

Change each fraction to an equivalent fraction with a denominator of 18.

$$=\frac{15}{18}-\frac{4}{18}$$

Multiply numerator and denominator of the first by 3 and numerator and denominator of the second by 2.

$$=\frac{15-4}{18}$$

Subtract the numerators and put the result over the LCD.

$$=\frac{11}{18}$$

We use this same procedure to add or subtract rational expressions.

To Add or Subtract Rational Expressions

1. **Factor** the denominators of the expressions to be added or subtracted and find the LCD. The LCD is the smallest expression that is divisible by the denominators of all the rational expressions being considered.
2. **Rewrite** each of the original rational expressions as an equivalent expression having a denominator equal to the LCD. This is done by multiplying numerator and denominator of each expression by the appropriate factor(s) of the LCD.
3. **Add or subtract** the numerators and put the result over the LCD.
4. **If necessary, reduce** to lowest terms by dividing numerator and denominator by the common factors that appear in both.

We now illustrate this procedure with the following examples.

example 3

Add $\frac{7}{x}+\frac{4}{3x}$.

solution

The LCD for x and $3x$ is $3x$.

To obtain the LCD, we must multiply the first denominator by 3.

$$\frac{7}{x}+\frac{4}{3x}=\frac{7\cdot 3}{x\cdot 3}+\frac{4}{3x}$$

Change the first fraction to an equivalent fraction with a denominator of $3x$ by multiplying numerator and denominator by 3.

$$=\frac{21}{3x}+\frac{4}{3x}$$

$$=\frac{21+4}{3x}$$

Add the numerators and place the result over the common denominator.

$$=\frac{25}{3x}$$

example 4

Subtract $\frac{9}{2x+4}-\frac{x}{3x+6}$.

solution

$$2x + 4 = 2(x + 2)$$
$$3x + 6 = 3(x + 2)$$
$$\text{LCD} = 6(x + 2)$$

Factor the denominators. The LCD is $6(x + 2)$.

To obtain the LCD, we must multiply the first denominator by 3 and the second denominator by 2.

$$\frac{9}{2x+4} - \frac{x}{3x+6} = \frac{9}{2(x+2)} - \frac{x}{3(x+2)}$$
$$= \frac{3 \cdot 9}{3 \cdot 2(x+2)} - \frac{2 \cdot x}{2 \cdot 3(x+2)}$$
$$= \frac{27}{6(x+2)} - \frac{2x}{6(x+2)}$$

Change each fraction to an equivalent fraction with a denominator of $6(x + 2)$ by multiplying numerator and denominator of the first by 3 and numerator and denominator of the second by 2.

$$= \frac{27 - 2x}{6(x+2)}$$

Subtract the numerators and put the result over the LCD.

example 5

Add $\dfrac{x}{x^2 - 9} + \dfrac{3}{4x + 12}$.

solution

$$x^2 - 9 = (x + 3)(x - 3)$$
$$4x + 12 = 4(x + 3)$$
$$\text{LCD} = 4(x + 3)(x - 3)$$

Factor the two denominators. The LCD is $4(x + 3)(x - 3)$.

To obtain the LCD, we must multiply the first denominator by 4 and the second denominator by $x - 3$.

$$\frac{x}{x^2-9} + \frac{3}{4x+12}$$
$$= \frac{x}{(x+3)(x-3)} + \frac{3}{4(x+3)}$$
$$= \frac{4 \cdot x}{4 \cdot (x+3)(x-3)} + \frac{3 \cdot (x-3)}{4(x+3) \cdot (x-3)}$$
$$= \frac{4x}{4(x+3)(x-3)} + \frac{3x-9}{4(x+3)(x-3)}$$

Change each fraction to an equivalent fraction having a denominator of $4(x + 3)(x - 3)$ by multiplying numerator and denominator of the first by 4 and numerator and denominator of the second by $x - 3$.

$$= \frac{4x + 3x - 9}{4(x+3)(x-3)}$$
$$= \frac{7x - 9}{4(x+3)(x-3)}$$

Add the numerators and put the result over the LCD.

example 6

Subtract $\dfrac{2x}{x^2 + 5x - 14} - \dfrac{5}{x^2 - 49}$.

solution

$$x^2 + 5x - 14 = (x + 7)(x - 2)$$
$$x^2 - 49 = (x + 7)(x - 7)$$
$$\text{LCD} = (x + 7)(x - 7)(x - 2)$$

Factor the denominators. The LCD is $(x + 7)(x - 7)(x - 2)$.

To obtain the LCD, we must multiply the first denominator by $x - 7$ and the second denominator by $x - 2$.

$$\frac{2x}{x^2 + 5x - 14} - \frac{5}{x^2 - 49}$$

$$= \frac{2x}{(x+7)(x-2)} - \frac{5}{(x+7)(x-7)}$$

Change each fraction to an equivalent fraction having a denominator of $(x + 7)(x - 7)(x - 2)$ by multiplying numerator and denominator of the first by $x - 7$ and numerator and denominator of the second by $x - 2$.

$$= \frac{2x \cdot (x-7)}{(x+7)(x-2) \cdot (x-7)} - \frac{5 \cdot (x-2)}{(x+7)(x-7) \cdot (x-2)}$$

$$= \frac{2x^2 - 14x}{(x+7)(x-7)(x-2)} - \frac{5x - 10}{(x+7)(x-7)(x-2)}$$

$$= \frac{2x^2 - 14x - (5x - 10)}{(x+7)(x-7)(x-2)}$$

Subtract the numerators and put the result over the LCD. Be certain to use parentheses since the numerator being subtracted is a binomial

$$= \frac{2x^2 - 19x + 10}{(x+7)(x-7)(x-2)}$$

Since the numerator cannot be factored, the expression is in lowest terms. ●

QUICK QUIZ

Perform each of the operations indicated.

1. $\dfrac{2}{x+5} + \dfrac{6}{x+5}$
2. $\dfrac{y}{3y-6} - \dfrac{4}{y-2}$
3. $\dfrac{3}{x^2 + 5x + 4} + \dfrac{2x}{x^2 - 16}$

ANSWERS

1. $\dfrac{8}{x+5}$
2. $\dfrac{y - 12}{3(y-2)}$
3. $\dfrac{2x^2 + 5x - 12}{(x+4)(x-4)(x+1)}$

16.4 Exercises

Perform each of the operations indicated.

1. $\dfrac{5}{x} + \dfrac{8}{x}$
2. $\dfrac{9}{x} + \dfrac{7}{x}$
3. $\dfrac{9}{a} - \dfrac{5}{a}$
4. $\dfrac{4}{a} - \dfrac{8}{a}$
5. $\dfrac{x}{x+3} + \dfrac{4}{x+3}$
6. $\dfrac{x}{x-5} + \dfrac{7x}{x-5}$
7. $\dfrac{2x}{x-8} - \dfrac{9x}{x-8}$
8. $\dfrac{11}{x+2} - \dfrac{3}{x+2}$
9. $\dfrac{y}{y^2-16} + \dfrac{4}{y^2-16}$
10. $\dfrac{3y}{3y+5} + \dfrac{5}{3y+5}$

11. $\dfrac{9x}{9x-2} - \dfrac{2}{9x-2}$

12. $\dfrac{x}{x^2-25} - \dfrac{5}{x^2-25}$

13. $\dfrac{4}{x} + \dfrac{x}{5}$

14. $\dfrac{x}{8} + \dfrac{7}{x}$

15. $\dfrac{t}{7} - \dfrac{3}{t}$

16. $\dfrac{4}{x} - \dfrac{x}{9}$

17. $\dfrac{4}{5x} + \dfrac{7}{2x}$

18. $\dfrac{2}{9x} + \dfrac{5}{6x}$

19. $\dfrac{7}{8z} - \dfrac{5}{12z}$

20. $\dfrac{11}{3z} - \dfrac{7}{8z}$

21. $\dfrac{x}{x+5} + \dfrac{7}{3x+15}$

22. $\dfrac{9}{4x-12} + \dfrac{x}{x-3}$

23. $\dfrac{3}{2x-14} - \dfrac{x}{x-7}$

24. $\dfrac{x}{x+4} - \dfrac{1}{5x+20}$

25. $\dfrac{2x}{5x+15} + \dfrac{5x}{2x+6}$

26. $\dfrac{x}{6x-30} + \dfrac{7}{2x-10}$

27. $\dfrac{4}{3y-18} - \dfrac{y}{2y-12}$

28. $\dfrac{9y}{3y+6} - \dfrac{2y}{5y+10}$

29. $\dfrac{5}{x} + \dfrac{3}{x^2-7x}$

30. $\dfrac{9}{x^2+5x} + \dfrac{2}{x}$

31. $\dfrac{3}{x^2+9x} - \dfrac{8}{x}$

32. $\dfrac{7}{x} - \dfrac{5}{x^2-3x}$

33. $\dfrac{5z}{z^2-25} + \dfrac{3z}{2z-10}$

34. $\dfrac{3}{4z+8} + \dfrac{z}{z^2-4}$

35. $\dfrac{x}{x^2-36} + \dfrac{5}{3x+18}$

36. $\dfrac{8}{3x-21} - \dfrac{x}{x^2-49}$

37. $\dfrac{3}{x^2-9} + \dfrac{1}{x^2+2x-15}$

38. $\dfrac{x}{x^2+11x+18} + \dfrac{2}{x^2-81}$

39. $\dfrac{a}{a^2+9a+8} - \dfrac{7}{a^2-64}$

40. $\dfrac{9}{a^2-13a+42} - \dfrac{4}{a^2-36}$

41. $\dfrac{5}{x^2+5x+6} + \dfrac{3x}{x^2+x-2}$

42. $\dfrac{1}{x^2+6x+8} + \dfrac{4x}{x^2-5x-36}$

43. $\dfrac{3x}{x^2+5x-14} - \dfrac{2}{x^2-7x+10}$

44. $\dfrac{8}{x^2-x-30} - \dfrac{2x}{x^2+6x+5}$

16.5 Solving Equations with Rational Expressions

In this section, we will learn how to solve equations that contain rational expressions for an unknown variable. The procedure for solving these equations, which we will refer to as *rational equations*, is outlined in the following steps.

To Solve a Rational Equation

1. Find the LCD of the denominators of all rational expressions in the equation.
2. Multiply both sides of the equation by the LCD to eliminate all denominators. This step is sometimes called "clearing the fractions."
3. Solve the resulting equation for the unknown variable.
4. Check each solution by substituting it for the variable in the original equation. Reject all solutions that would make an expression in the original equation undefined.

We now illustrate this procedure with the following examples.

example 1 Solve $\frac{x}{4} + \frac{5}{12} = \frac{2}{3}$.

solution

$$\frac{x}{4} + \frac{5}{12} = \frac{2}{3}$$ The LCD of 4, 12, and 3 is 12.

$$12\left(\frac{x}{4} + \frac{5}{12}\right) = 12\left(\frac{2}{3}\right)$$ Multiply both sides of the equation by 12.

$$12\left(\frac{x}{4}\right) + 12\left(\frac{5}{12}\right) = 12\left(\frac{2}{3}\right)$$ By distributive property.

$$\overset{3}{\cancel{12}}\left(\frac{x}{\cancel{4}}\right) + \cancel{12}\left(\frac{5}{\cancel{12}}\right) = \overset{4}{\cancel{12}}\left(\frac{2}{\cancel{3}}\right)$$ Divide numerators and denominators by common factors.

$$3x + 5 = 8$$ Notice that all denominators have been eliminated.

$$3x = 3$$

$$x = 1$$ Solve for x.

CHECK

$$\frac{x}{4} + \frac{5}{12} = \frac{2}{3}$$ Check the solution.

$$\frac{1}{4} + \frac{5}{12} \overset{?}{=} \frac{2}{3}$$

$$\frac{3}{12} + \frac{5}{12} \overset{?}{=} \frac{8}{12}$$

$$\frac{8}{12} = \frac{8}{12} \quad \checkmark$$

Thus, the solution is $x = 1$.

example 2 Solve $\frac{3}{y+2} = \frac{9}{y}$ for y.

solution

$$\frac{3}{y+2} = \frac{9}{y}$$ The LCD of $y + 2$ and y is $y(y + 2)$.

$$y(y+2)\left(\frac{3}{y+2}\right) = y(y+2)\left(\frac{9}{y}\right)$$ Multiply both sides of the equation by $y(y + 2)$.

$$\frac{y(y+2)3}{y+2} = \frac{y(y+2)9}{y}$$

Divide numerators and denominators by common factors.

$$3y = 9(y+2)$$

Notice that all denominators have been eliminated.

$$3y = 9y + 18$$
$$-6y = 18$$
$$y = -3$$

CHECK

$$\frac{3}{y+2} = \frac{9}{y}$$

Check the solution.

$$\frac{3}{-3+2} \; ? \; \frac{9}{-3}$$
$$\frac{3}{-1} \; ? \; -3$$
$$-3 = -3 \; \checkmark$$

Thus, the solution is $y = -3$.

example 3

Solve $\dfrac{4x}{x-5} - 4 = \dfrac{5}{x}$.

solution

$$\frac{4x}{x-5} - 4 = \frac{5}{x}$$

The LCD of $x - 5$ and x is $x(x - 5)$.

$$x(x-5)\left(\frac{4x}{x-5} - 4\right) = x(x-5)\left(\frac{5}{x}\right)$$

Multiply both sides of the equation by $x(x - 5)$.

$$\frac{x(x-5)4x}{x-5} - x(x-5)4 = \frac{x(x-5)5}{x}$$

By distributive property.

$$\frac{x(x-5)4x}{x-5} - x(x-5)4 = \frac{x(x-5)5}{x}$$

Divide numerators and denominators by common factors.

$$4x^2 - 4x(x-5) = 5(x-5)$$
$$4x^2 - 4x^2 + 20x = 5x - 25$$
$$20x = 5x - 25$$
$$15x = -25$$
$$x = -\frac{25}{15}$$
$$x = -\frac{5}{3}$$

CHECK

$$\frac{4x}{x-5} - 4 = \frac{5}{x}$$

Check the solution.

$$\frac{4\left(-\frac{5}{3}\right)}{-\frac{5}{3} - 5} - 4 \; ? \; \frac{5}{-\frac{5}{3}}$$

$$\frac{-\frac{20}{3}}{-\frac{20}{3}} - 4 \ ? \ \frac{-15}{5}$$

$$1 - 4 \ ? \ -3$$

$$-3 = -3 \ \checkmark$$

Thus, the solution is $x = -\frac{5}{3}$.

example 4

Solve $\dfrac{x}{x-2} + \dfrac{3}{7} = \dfrac{2}{x-2}$ for x.

solution

$$\frac{x}{x-2} + \frac{3}{7} = \frac{2}{x-2}$$

The LCD of $x - 2$ and 7 is $7(x - 2)$.

$$7(x-2)\left(\frac{x}{x-2} + \frac{3}{7}\right) = 7(x-2)\left(\frac{2}{x-2}\right)$$

Multiply both sides of the equation by $7(x - 2)$.

$$\frac{7(x-2)x}{x-2} + \frac{7(x-2)3}{7} = \frac{7(x-2)2}{x-2}$$

By distributive property.

$$\frac{7\cancel{(x-2)}x}{\cancel{x-2}} + \frac{\cancel{7}(x-2)3}{\cancel{7}} = \frac{7\cancel{(x-2)}2}{\cancel{x-2}}$$

Divide numerators and denominators by common factors.

$$7x + 3(x-2) = 7 \cdot 2$$

$$7x + 3x - 6 = 14$$

$$10x = 20$$

$$x = 2$$

CHECK

$$\frac{x}{x-2} + \frac{3}{7} = \frac{x}{x-2}$$

Check the solution.

$$\frac{2}{2-2} + \frac{3}{7} \ ? \ \frac{2}{2-2}$$

$$\frac{2}{0} + \frac{3}{7} \neq \frac{2}{0} \ \checkmark$$

Notice that if 2 is substituted for x in the original equation, two rational expressions are undefined because these denominators are equal to 0. Therefore, $x = 2$ cannot be a solution to this equation, and the equation has no solution.

example 5

Solve $\dfrac{-4}{y^2 - 9} = \dfrac{y+1}{y^2 + 3y}$ for y.

solution

$$y^2 - 9 = (y+3)(y-3)$$

Factor the two denominators.

$$y^2 + 3y = y(y+3)$$

The LCD is $y(y + 3)(y - 3)$.

$$\text{LCD} = y(y+3)(y-3)$$

$$y(y+3)(y-3)\left(\frac{-4}{(y+3)(y-3)}\right) = y(y+3)(y-3)\left(\frac{y+1}{y(y+3)}\right)$$

Multiply both sides by $y(y+3)(y-3)$.

$$y\cancel{(y+3)}\cancel{(y-3)}\left(\frac{-4}{\cancel{(y+3)}\cancel{(y-3)}}\right) = \cancel{y}\cancel{(y+3)}(y-3)\left(\frac{y+1}{\cancel{y}\cancel{(y+3)}}\right)$$

Divide numerators and denominators by common factors.

$$-4y = (y-3)(y+1)$$
$$-4y = y^2 - 2y - 3$$
$$0 = y^2 + 2y - 3$$

Add $4y$ to both sides.

$$y^2 + 2y - 3 = 0$$
$$(y+3)(y-1) = 0$$

Solve the resulting quadratic equation by factoring.

$$y + 3 = 0 \quad \text{or} \quad y - 1 = 0$$

Set each factor equal to 0.

$$y = -3 \quad \text{or} \quad y = 1$$

Solve each equation for y.

CHECK

$$\frac{-4}{y^2-9} = \frac{y+1}{y^2+3y}$$

Check each solution in the original equation.

$$y = -3: \quad \frac{-4}{(-3)^2 - 9} \stackrel{?}{=} \frac{-3+1}{(-3)^2 + 3(-3)}$$
$$\frac{-4}{0} \neq \frac{-2}{0}$$

The rational expressions are undefined when $y = -3$.

Thus, $y = -3$ is not a solution to the equation.

$$y = 1: \quad \frac{-4}{1^2 - 9} \stackrel{?}{=} \frac{1+1}{1^2 + 3(1)}$$
$$\frac{-4}{-8} \stackrel{?}{=} \frac{2}{4}$$
$$\frac{1}{2} = \frac{1}{2} \ \checkmark$$

The solution $y = 1$ checks.

Thus, the solution is $y = 1$. ●

QUICK QUIZ

Solve each of the following equations.

1. $2 - \frac{x}{6} = \frac{1}{3}$
2. $\frac{a}{a+4} + \frac{2}{5} = \frac{10}{a+4}$
3. $\frac{6}{y^2 - 16} = \frac{5}{y^2 + y - 12}$

ANSWERS

1. $x = 10$
2. $a = 6$
3. $y = -2$

16.5 Exercises

Solve each of the following equations. Be sure to check each solution in the original equation.

1. $\frac{1}{2} + \frac{x}{10} = \frac{2}{5}$

2. $\frac{5}{6} + \frac{x}{12} = \frac{1}{4}$

3. $\frac{3}{4} - \frac{x}{8} = \frac{5}{8}$

4. $\frac{x}{9} - \frac{1}{3} = \frac{5}{9}$

5. $\frac{1}{x} + 3 = \frac{7}{x}$

6. $\frac{x}{4} + \frac{9}{x} = 5$

7. $\frac{10}{x} - \frac{x}{2} = 4$

8. $2 - \frac{1}{x} = \frac{5}{x}$

9. $\frac{5}{6} + \frac{x-3}{2} = \frac{x}{3}$

10. $\frac{x}{4} + \frac{1}{2} = \frac{x+2}{4}$

11. $\frac{1}{2} - \frac{x}{10} = \frac{x-2}{5}$

12. $\frac{x}{3} - \frac{x+3}{12} = \frac{1}{4}$

13. $\frac{1}{x} = \frac{4}{x+6}$

14. $\frac{2}{x-3} = \frac{4}{x}$

15. $\frac{5}{a} = \frac{9}{a+4}$

16. $\frac{7}{a} = \frac{2}{a-5}$

17. $\frac{4}{y+5} = \frac{3}{y-3}$

18. $\frac{7}{y-2} = \frac{6}{y+1}$

19. $1 + \frac{5}{x} = \frac{6}{x^2}$

20. $1 - \frac{3}{x} = \frac{10}{x^2}$

21. $\frac{1}{x} + \frac{x-3}{x+4} = \frac{x-1}{x+4}$

22. $\frac{2}{x} + \frac{1}{4x} = \frac{x}{x+1}$

23. $\frac{3x}{x+2} - 3 = \frac{2}{x}$

24. $\frac{5x}{x-3} - 5 = \frac{9}{x}$

25. $\frac{y}{y-5} + \frac{3}{4} = \frac{5}{y-5}$

26. $\frac{x}{x+3} - \frac{2}{5} = \frac{-3}{x+3}$

27. $\frac{5}{a^2-49} = \frac{2}{a^2-6a-7}$

28. $\frac{2}{a^2-9} = \frac{4}{a^2+2a-15}$

29. $\frac{-2}{y^2-16} = \frac{y+3}{y^2+4y}$

30. $\frac{8}{y^2-36} = \frac{y-2}{y^2-6y}$

31. $\frac{4}{x-5} + \frac{5}{x^2-25} = \frac{7}{x+5}$

32. $\frac{2}{x+3} + \frac{6}{x^2-9} = \frac{5}{x-3}$

33. $\frac{3}{x+2} + \frac{2}{x-4} = \frac{6}{x^2-2x-8}$

34. $\frac{5}{x-1} + \frac{3}{x+5} = \frac{-2}{x^2+4x-5}$

35. $\frac{4}{x+3} + \frac{12}{x^2-3x-18} = \frac{7}{x-6}$

36. $\frac{3}{x+2} - \frac{9}{x^2+7x+10} = \frac{8}{x+5}$

16.6 Word Problems Involving Rational Expressions

We will now look at some word problems that can be solved using equations that contain rational expressions. Some of the most common applications of rational expressions include word problems involving fractions, motion, and work. Consider the following examples.

example 1

The denominator of a fraction is 4 less than three times the numerator. If the value of the fraction is equivalent to $\frac{3}{8}$, find the fraction.

solution

The problem asks us to find an unknown fraction.

Let x = numerator of the fraction. — Define the variables.

$3x - 4$ = denominator of the fraction.

$$\frac{x}{3x-4} = \frac{3}{8}$$

Since the unknown fraction is equivalent to $\frac{3}{8}$, we can set up an equation to find x.

$$8(3x-4)\left(\frac{x}{3x-4}\right) = 8(3x-4)\left(\frac{3}{8}\right)$$

Multiply both sides by the LCD, which is $8(3x - 4)$.

$$\frac{8\cancel{(3x-4)}x}{\cancel{3x-4}} = \frac{\cancel{8}(3x-4)3}{\cancel{8}}$$

Divide numerators and denominators by common factors.

$$8x = 3(3x-4)$$

Solve the resulting equation for x.

$$8x = 9x - 12$$

$$-x = -12$$

$x = 12$ is the numerator

$3x - 4 = 3(12) - 4 = 32$ is the denominator

CHECK

The fraction is $\frac{12}{32}$.

This fraction should **not** be reduced to lowest terms because the reduced fraction will not satisfy the conditions of the problem.

$$\frac{12}{32} = \frac{12 \div 4}{32 \div 4} = \frac{3}{8} \ \checkmark$$

The next two examples are motion problems which involve the relationship between distance D, rate R, and time T. These problems can be solved using the distance formula, which can be expressed in any of the following ways:

$$R \times T = D \qquad \frac{D}{T} = R \qquad \frac{D}{R} = T$$

For example, if a car travels at a rate of 50 miles per hour for 3 hours, the total distance traveled would be 150 miles. If any of these three quantities (rate, time, or distance) is unknown, it can be determined if the other two are known, using a form of the distance formula:

$$50 \text{ mi/hr} \times 3 \text{ hr} = 150 \text{ mi} \qquad \frac{150 \text{ mi}}{3 \text{ hr}} = 50 \text{ mi/hr} \qquad \frac{150 \text{ mi}}{50 \text{ mi/hr}} = 3 \text{ hr}$$

Examples 2 and 3 illustrate applications of the distance formula.

example 2

A sportscar is traveling along a highway at a rate that is 10 miles per hour faster than a pick-up truck. If the sportscar travels 110 miles in the same amount of time that the pick-up truck travels 90 miles, find the rate at which each is traveling.

solution The problem asks us to find the rate of the sportscar and the rate of the pick-up truck.

Let r = rate of pick-up truck — Define the variables.

$r + 10$ = rate of sportscar

Since the sportscar travels 110 miles in the same amount of time that the pick-up truck travels 90 miles, we can use the third form of the distance formula, $\frac{D}{R} = T$, to express the time for each and set the times equal.

$$T(\text{sportscar}) = T(\text{pick-up truck})$$

$$\frac{D(\text{sportscar})}{R(\text{sportscar})} = \frac{D(\text{pick-up truck})}{R(\text{pick-up truck})}$$

$$\frac{110}{r + 10} = \frac{90}{r}$$

$$(r + 10)r\left(\frac{110}{r + 10}\right) = (r + 10)r\left(\frac{90}{r}\right)$$

Multiply both sides by the LCD, which is $(r + 10)r$.

$$\frac{\cancel{(r + 10)}r(110)}{\cancel{r + 10}} = \frac{(r + 10)\cancel{r}\,90}{\cancel{r}}$$

Divide numerators and denominators by common factors.

$$110r = 90(r + 10)$$
$$110r = 90r + 900$$
$$20r = 900$$
$$r = 45 \text{ mi/hr is the rate of pick-up truck}$$
$$r + 10 = 45 + 10 = 55 \text{ mi/hr is the rate of sportscar}$$

The sportscar travels 55 mi/hr and the pick-up truck travels 45 mi/hr.

CHECK

$$\frac{110 \text{ mi}}{55 \text{ mi/hr}} = 2 \text{ hr}$$

$$\frac{90 \text{ mi}}{45 \text{ mi/hr}} = 2 \text{ hr} \checkmark$$

The two times are equal. ●

example 3

A boat can travel 12 mi/hr in still water. If it can travel 35 miles downstream in the same amount of time that it can travel 21 miles upstream, what is the rate of the current?

solution

The problem asks us to find the rate of the current.

Let c = rate of the current — Define the variables.

$12 + c$ = rate of boat traveling downstream

$12 - c$ = rate of boat traveling upstream

Since the boat travels 35 miles downstream in the same amount of time it takes to travel 21 miles upstream, we can set the two times equal.

$$T(\text{downstream}) = T(\text{upstream})$$

$$\frac{D(\text{downstream})}{R(\text{downstream})} = \frac{D(\text{upstream})}{R(\text{upstream})}$$

$$\frac{35}{12 + c} = \frac{21}{12 - c}$$

$$(12 + c)(12 - c)\left(\frac{35}{12 + c}\right) = (12 + c)(12 - c)\left(\frac{21}{12 - c}\right)$$

$$\frac{\cancel{(12 + c)}(12 - c)(35)}{\cancel{12 + c}} = \frac{(12 + c)\cancel{(12 - c)}(21)}{\cancel{12 - c}}$$

$$(12 - c)35 = (12 + c)21$$

$$420 - 35c = 252 + 21c$$

$$-56c = -168$$

$$c = 3 \text{ miles/hour is rate of current.}$$

The rate of the current is 3 miles/hour.

CHECK

$$\frac{35 \text{ mi}}{12 + 3 \text{ mi/hr}} = \frac{35}{15} \text{ hr} = 2\frac{1}{3} \text{ hr}$$

$$\frac{21 \text{ mi}}{12 - 3 \text{ mi/hr}} = \frac{21}{9} \text{ hr} = 2\frac{1}{3} \text{ hr} \checkmark$$

The two times are equal. ●

Example 4 illustrates the use of an equation involving rational expressions to solve a work problem.

example 4

A math professor can paint a classroom in 5 hours and one of her students can paint the same room in 4 hours. How long would it take them to paint the classroom if they worked together?

solution

The problem asks us to find the amount of time needed for the student and professor to paint the room working together.

Let t = number of hours needed working together

In 1 hour, the professor can paint $\frac{1}{5}$ of the room working alone, and the student can paint $\frac{1}{4}$ of the room working alone. Thus, in t hours the professor can paint a portion equal to $t \cdot \frac{1}{5}$ or $\frac{t}{5}$ and the student can paint a portion equal to $t \cdot \frac{1}{4}$ or $\frac{t}{4}$.

Let $\frac{t}{5}$ = portion painted by the professor in t hours

$\frac{t}{4}$ = portion painted by the student in t hours

Since the job consists of painting 1 classroom, we can add the two portions together and set them equal 1.

$$\frac{t}{5} + \frac{t}{4} = 1$$

$$20\left(\frac{t}{5} + \frac{t}{4}\right) = 20(1)$$

$$20\left(\frac{t}{5}\right) + 20\left(\frac{t}{4}\right) = 20$$

$$4t + 5t = 20$$

$$9t = 20$$

$$t = \frac{20}{9}$$

$$t = 2\frac{2}{9} \text{ hours.}$$

They can finish the job in $2\frac{2}{9}$ hours working together.

CHECK

$$\frac{2\frac{2}{9}}{5} + \frac{2\frac{2}{9}}{4} = \frac{\frac{20}{9}}{5} + \frac{\frac{20}{9}}{4}$$

$$= \frac{\frac{20}{9} \cdot 4}{5 \cdot 4} + \frac{\frac{20}{9} \cdot 5}{4 \cdot 5}$$

$$= \frac{\frac{80}{9}}{20} + \frac{\frac{100}{9}}{20} = \frac{\frac{180}{9}}{20}$$

$$= \frac{20}{20} = 1 \checkmark$$

16.6 Exercises

1. The numerator of a fraction is 5 less than four times the denominator. If the value of the fraction is equivalent to seven-halves, find the fraction.
2. The denominator of a fraction is 3 less than twice the numerator. If the value of the fraction is equivalent to three-fifths, find the fraction.

3. What number must be added to both the numerator and denominator of the fraction one-thirteenth so that the resulting fraction is equivalent to five-ninths?
4. What number must be added to both the numerator and denominator of the fraction five-elevenths so that the resulting fraction is equivalent to two-thirds?
5. The numerator of a fraction is 8 less than the denominator. If 2 is subtracted from the numerator and 5 is added to the denominator, the resulting fraction is equivalent to one-fourth. Find the original fraction.
6. The denominator of a fraction is 7 more than the numerator. If 3 is added to the numerator and 2 is subtracted from the denominator, the resulting fraction is equivalent to four-fifths. Find the original fraction.
7. A student received grades of 88, 75, 91, and 83 on four math quizzes. What score must he obtain on the fifth quiz so that his average grade on the five quizzes is 82?
8. The combined weight of three football players is 672 pounds. What is the weight of a fourth football player if the average weight of the four men is 220 pounds?
9. The rate of a passenger train is 15 miles per hour faster than that of a freight train. If the passenger train can travel 165 miles in the same time that it takes the freight train to travel 132 miles, what is the rate of each train?
10. A jogger can run 0.5 miles per hour faster than her husband. If she can run 17 miles in the same time it takes her husband to run 16 miles, what is the rate of each?
11. A woman rode a distance of 30 miles on a bicycle and returned on horseback. The rate of the horse was twice as fast as her rate via bicycle. If the entire trip took $3\frac{3}{4}$ hours, what was her rate on the bicycle and the rate of the horse?
12. A man rowed a distance of 15 miles to an island and returned in a motorboat. The rate of the motorboat was three times as fast as his rate rowing. If the entire trip took 2 hours, what was his rate rowing and the rate of the motorboat?
13. A plane can travel 450 miles per hour in still air. While flying with the wind the plane can travel 960 miles in the same amount of time it requires to fly 840 miles against the wind. Find the rate of the wind.
14. A boat travels 20 miles per hour in still water. If it can travel 75 miles downstream in the same time it takes to travel 45 miles upstream, what is the rate of the current?
15. A boat traveling upstream can go 70 miles in the same amount of time it would take for it to go 110 miles traveling downstream. If the speed of the current is 10 miles per hour, what would be the rate of the boat traveling in still water?
16. An airplane can go 540 miles flying with the wind in the same amount of time it would take to go 420 miles flying against the wind. If the wind speed is 40 miles per hour, what would be the rate of the airplane in still air?
17. It takes a father 40 minutes to shovel the driveway, and it takes his daughter 60 minutes to shovel the same driveway. How long would it take them if they worked together?
18. A printing press takes 2 hours to print the morning edition of a daily newspaper, and another press can do the job in 3 hours. How long would it take both presses to do the job working together?
19. A swimming pool can be filled by one hose in 3 hours and emptied by another hose in 6 hours. How long would it take to fill the pool if both hoses are used?
20. An inlet pipe can fill a tank in 45 minutes, and an outlet pipe can empty a tank in 15 minutes. If the tank is full and both pipes are opened, how long would it take to empty the tank?

16.7 Simplifying Complex Fractions

In Section 2.7 we defined a complex fraction to be a fraction whose numerator and/or denominator contains fractions. Recall that there are two techniques to simplify complex fractions. The first is to simplify numerator and denominator separately and multiply the numerator by the reciprocal of the denominator. For example, the complex fraction

$$\frac{\frac{3}{4}-\frac{2}{4}}{\frac{2}{3}}$$

may be simplified as follows:

$$\frac{\frac{3}{4}-\frac{1}{2}}{\frac{2}{3}}=\frac{\frac{3}{4}-\frac{2}{4}}{\frac{2}{3}}$$

$$=\frac{\frac{1}{4}}{\frac{2}{3}}$$ Simplify the numerator.

$$=\frac{1}{4}\cdot\frac{3}{2}$$ Multiply the numerator by the reciprocal of the denominator.

$$=\frac{3}{8}$$

The second technique is to find the LCD of all fractions that appear in the numerator and denominator and multiply the numerator and denominator by the LCD. To simplify the previous example using the second technique we obtain the following:

$$\frac{\frac{3}{4}-\frac{1}{2}}{\frac{2}{3}}=\frac{\left(\frac{3}{4}-\frac{1}{2}\right)12}{\left(\frac{2}{3}\right)\cdot 12}$$ The LCD is 12. Multiply numerator and denominator by 12.

$$=\frac{\frac{3}{4}\cdot 12-\frac{1}{2}\cdot 12}{\frac{2}{3}\cdot 12}$$ By distributive property.

$$=\frac{9-6}{8}=\frac{3}{8}$$

Recall that the second technique is much easier to use when simplifying complicated complex fractions because it immediately eliminates all fractions in the numerator and denominator.

We will now use this technique to simplify rational expressions that appear in the form of a complex fraction, that is, rational expressions that contain fractions in the numerator and/or denominator.

> **To simplify a complex fraction, multiply both numerator and denominator by the LCD of all rational expressions that appear in the numerator and denominator.**

example 1

Simplify $\dfrac{\dfrac{3x}{y^2}}{\dfrac{6x}{y^3}}$.

solution The LCD of $\dfrac{3x}{y^2}$ and $\dfrac{6x}{y^3}$ is y^3.

$$\frac{\frac{3x}{y^2}}{\frac{6x}{y^3}} = \frac{\frac{3x}{y^2} \cdot y^3}{\frac{6x}{y^3} \cdot y^3}$$

Multiply numerator and denominator by the LCD which is y^3.

$$= \frac{3xy}{6x}$$

Simplify. Notice that the fractions in the numerator and denominator have been eliminated.

$$= \frac{y}{2}$$

Reduce to lowest terms. ●

example 2

Simplify $\dfrac{\dfrac{1}{a} + \dfrac{1}{ab}}{1 + \dfrac{1}{b}}$.

solution The LCD of $\dfrac{1}{a}$, $\dfrac{1}{ab}$, and $\dfrac{1}{b}$ is ab.

$$\frac{\frac{1}{a} + \frac{1}{ab}}{1 + \frac{1}{b}} = \frac{\left(\frac{1}{a} + \frac{1}{ab}\right) \cdot ab}{\left(1 + \frac{1}{b}\right) \cdot ab}$$

Multiply numerator and denominator by the LCD which is ab.

$$= \frac{\left(\frac{1}{a} \cdot ab\right) + \left(\frac{1}{ab} \cdot ab\right)}{\left(1 \cdot ab\right) + \left(\frac{1}{b} \cdot ab\right)}$$

By distributive property.

$$= \frac{b + 1}{ab + a}$$

Simplify. Notice that all fractions in the numerator and denominator have been eliminated.

$$= \frac{\cancel{b + 1}}{a(\cancel{b + 1})}$$

Factor denominator.

$$= \frac{1}{a}$$

Reduce to lowest terms. ●

example 3

Simplify $\dfrac{1 + \dfrac{5}{x} + \dfrac{6}{x^2}}{1 - \dfrac{4}{x^2}}$.

solution

The LCD of $\dfrac{5}{x}$, $\dfrac{6}{x^2}$, and $\dfrac{4}{x^2}$ is x^2.

$$\frac{1 + \frac{5}{x} + \frac{6}{x^2}}{1 - \frac{4}{x^2}} = \frac{\left(1 + \frac{5}{x} + \frac{6}{x^2}\right) \cdot x^2}{\left(1 - \frac{4}{x^2}\right) \cdot x^2}$$

Multiply numerator and denominator by the LCD which is x^2.

$$= \frac{(1 \cdot x^2) + \left(\frac{5}{x} \cdot x^2\right) + \left(\frac{6}{x^2} \cdot x^2\right)}{(1 \cdot x^2) - \left(\frac{4}{x^2} \cdot x^2\right)}$$

By distributive property.

$$= \frac{x^2 + 5x + 6}{x^2 - 4}$$

Simplify.

$$= \frac{\cancel{(x+2)}(x + 3)}{\cancel{(x+2)}(x - 2)}$$

Factor numerator and denominator.

$$= \frac{x + 3}{x - 2}$$

Reduce to lowest terms. ●

example 4

Simplify $\dfrac{\dfrac{5}{x - 5} + 1}{\dfrac{5}{x + 5} - 1}$.

solution

The LCD of $\dfrac{5}{x - 5}$ and $\dfrac{5}{x + 5}$ is $(x - 5)(x + 5)$.

$$\frac{\frac{5}{x-5} + 1}{\frac{5}{x+5} - 1} = \frac{\left(\frac{5}{x-5} + 1\right) \cdot (x - 5)(x + 5)}{\left(\frac{5}{x+5} - 1\right) \cdot (x - 5)(x + 5)}$$

$$= \frac{\left(\frac{5}{\cancel{x-5}}\right)\cancel{(x-5)}(x + 5) + 1 \cdot (x - 5)(x + 5)}{\left(\frac{5}{\cancel{x+5}}\right)(x - 5)\cancel{(x+5)} - 1 \cdot (x - 5)(x + 5)}$$

$$= \frac{5(x + 5) + (x - 5)(x + 5)}{5(x - 5) - (x - 5)(x + 5)}$$

$$= \frac{5x + 25 + x^2 - 25}{5x - 25 - (x^2 - 25)}$$

$$= \frac{5x + x^2}{5x - x^2} = \frac{\cancel{x}(5 + x)}{\cancel{x}(5 - x)} = \frac{5 + x}{5 - x}$$ ●

QUICK QUIZ

Simplify each of the following complex fractions.

1. $\dfrac{\frac{2}{3x^2}}{\frac{5}{6x}}$

2. $\dfrac{\frac{1}{x}+\frac{1}{y}}{\frac{1}{xy}}$

3. $\dfrac{1-\frac{1}{a^2}}{1-\frac{8}{a}+\frac{7}{a^2}}$

ANSWERS

1. $\dfrac{4}{5x}$

2. $y + x$

3. $\dfrac{a+1}{a-7}$

16.7 Exercises

Simplify each of the following complex fractions.

1. $\dfrac{\frac{8}{3x^2}}{\frac{2}{9x^3}}$

2. $\dfrac{\frac{3}{5x^4}}{\frac{6}{10x}}$

3. $\dfrac{\frac{4x}{y^3}}{\frac{8x}{y^2}}$

4. $\dfrac{\frac{6x}{y^3}}{\frac{2x}{y^5}}$

5. $\dfrac{\frac{9a^2}{b^2}}{\frac{3a}{b^5}}$

6. $\dfrac{\frac{3a^4}{b^4}}{\frac{12a}{b^3}}$

7. $\dfrac{\frac{1}{x}-2}{\frac{1}{x}+5}$

8. $\dfrac{1+\frac{1}{x}}{5-\frac{1}{x}}$

9. $\dfrac{1+\frac{3}{y}}{y-\frac{9}{y}}$

10. $\dfrac{1-\frac{4}{y}}{y-\frac{16}{y}}$

11. $\dfrac{3+\frac{3}{4a}}{2+\frac{1}{2a}}$

12. $\dfrac{5-\frac{5}{6a}}{3-\frac{1}{2a}}$

13. $\dfrac{1-\frac{25}{x^2}}{\frac{1}{x}-\frac{5}{x^2}}$

14. $\dfrac{\frac{1}{x}+\frac{3}{x^2}}{1-\frac{9}{x^2}}$

15. $\dfrac{\frac{1}{ab}+\frac{1}{b}}{\frac{1}{a^2b}}$

16. $\dfrac{\frac{1}{ab^2}}{\frac{1}{a}-\frac{1}{ab}}$

17. $\dfrac{\frac{1}{a+2}}{\frac{1}{a^2+7a+10}}$

18. $\dfrac{\frac{1}{a-5}}{\frac{1}{a^2-8a+15}}$

19. $\dfrac{1-\frac{5}{x}+\frac{4}{x^2}}{1-\frac{16}{x^2}}$

20. $\dfrac{1-\frac{36}{x^2}}{1+\frac{7}{x}+\frac{6}{x^2}}$

21. $\dfrac{1+\frac{8}{a}+\frac{15}{a^2}}{1+\frac{1}{a}-\frac{6}{a^2}}$

22. $\dfrac{1+\frac{4}{a}-\frac{12}{a^2}}{1+\frac{5}{a}-\frac{6}{a^2}}$

23. $\dfrac{\frac{2}{x+2}-1}{\frac{2}{x-2}+1}$

24. $\dfrac{\frac{3}{x-3}+1}{\frac{3}{x-3}-1}$

Summary and Review

Key Terms

[16.1] A **rational expression** is any expression that can be written in the form $\frac{P}{Q}$, where P and Q are polynomials and $Q \neq 0$.

A rational expression is **undefined** if its denominator is equal to zero.

[16.7] A rational expression that appears in the form of a **complex fraction** is a rational expression that contains fractions in the numerator and/or denominator.

Calculations

[16.1] **To reduce a rational expression to lowest terms**, factor the numerator and denominator completely and divide the numerator and denominator by the common factors that appear in both.

[16.2] **To multiply rational expressions**, factor the numerator and denominator of each rational expression completely, multiply the numerators and denominators, and divide the numerator and denominator by the product of the common factors that appear in both.

To divide rational expressions, multiply the first expression by the reciprocal of the second and simplify using the procedure for multiplying rational expressions.

[16.3] **To find the LCD of two or more rational expressions**, factor all denominators and write each factorization in exponential notation. The LCD is the product of all unique factors, each one raised to the highest exponent that appears in any of the factorizations.

[16.4] **To add or subtract rational expressions**, find the LCD and rewrite each expression as an equivalent rational expression having a denominator equal to the LCD. Add or subtract the numerators, put the result over the LCD, and if necessary, reduce to lowest terms.

[16.5] **To solve a rational equation**, find the LCD of the denominators of all the rational expressions in the equation. Multiply both sides of the equation by the LCD to eliminate all denominators and solve the resulting equation for the unknown variable. Check each solution by substituting it for the variable in the original equation and reject all solutions that would make an expression in the original equation undefined.

[16.7] **To simplify a complex fraction**, multiply both numerator and denominator by the LCD of all rational expressions that appear in the numerator and denominator.

Chapter 16 Review Exercises

[16.1] Find the value(s) of x that will make each of the following rational expressions undefined and then reduce to lowest terms.

1. $\frac{8}{4x + 16}$ **2.** $\frac{15}{5x - 20}$ **3.** $\frac{3x + 21}{x^2 - 2x - 35}$ **4.** $\frac{2x + 12}{x^2 + 10x + 24}$

5. $\frac{x^2 - 1}{x^2 + 9x + 8}$ **6.** $\frac{x^2 - 4}{x^2 - 9x + 14}$ **7.** $\frac{3x^2 + 9x - 30}{3x^2 + 24x + 45}$ **8.** $\frac{2x^2 - 10x + 8}{2x^2 + 2x - 40}$

[16.2–16.4] Perform each of the operations indicated.

9. $\dfrac{x-4}{x} \cdot \dfrac{6x^2}{3x-12}$

10. $\dfrac{x^3}{x+4} \div \dfrac{2x}{4x+16}$

11. $\dfrac{2y+6}{y^2-16} \div \dfrac{y^2+5y+6}{2y+8}$

12. $\dfrac{y^2+3y-4}{3y-15} \cdot \dfrac{y^2-25}{6y-6}$

13. $\dfrac{5}{x} + \dfrac{7}{3x}$

14. $\dfrac{2}{5x} - \dfrac{6}{x}$

15. $\dfrac{t}{t+6} - \dfrac{2}{3t+18}$

16. $\dfrac{3}{4t-12} + \dfrac{t}{t-3}$

17. $\dfrac{x^2+9x+20}{3x^2} \cdot \dfrac{6x^2-18x}{x^2+2x-15}$

18. $\dfrac{8x^3}{x^2-9x+14} \div \dfrac{2x^2+8x}{x^2-3x-28}$

19. $\dfrac{x^2-9x+20}{4x^2+32x+48} \cdot \dfrac{2x^2-10x-28}{x^2-11x+28}$

20. $\dfrac{3x^2+27x+54}{x^2-5x-24} \cdot \dfrac{x^2-6x-16}{6x^2+42x+36}$

21. $\dfrac{3y}{2y+14} + \dfrac{y}{y^2-49}$

22. $\dfrac{5y}{y^2-36} - \dfrac{2y}{3y-18}$

23. $\dfrac{x}{x^2-3x-10} - \dfrac{5}{x^2+8x+12}$

24. $\dfrac{6}{x^2-2x-3} + \dfrac{x}{x^2-11x+24}$

[16.5] Solve each of the following equations. Be sure to check each solution in the original equation.

25. $\dfrac{3}{x} + \dfrac{x}{6} = \dfrac{3}{2}$

26. $\dfrac{x}{4} - \dfrac{12}{x} = -2$

27. $\dfrac{a}{a-5} - \dfrac{3}{2} = \dfrac{4}{a-5}$

28. $\dfrac{a}{a+3} + \dfrac{4}{5} = \dfrac{6}{a+3}$

29. $\dfrac{-6}{y^2-16} = \dfrac{y+1}{y^2+4y}$

30. $\dfrac{y+2}{y^2-5y} = \dfrac{14}{y^2-25}$

[16.6] Solve each of the following word problems.

31. The denominator of a fraction is 9 more than twice the numerator. If the value of the fraction is equivalent to three-sevenths, find the fraction.

32. A commercial jet can fly 150 miles per hour faster than a private plane. If the jet can fly 1,100 miles in the same amount of time it takes the private plane to fly 800 miles, find the rate of each.

33. A boat travels 25 miles per hour in still water. If it can travel 77 miles downstream in the same amount of time it takes to travel 33 miles upstream, find the rate of the current.

34. It takes a bricklayer 50 minutes to build a wall and 75 minutes for his assistant to complete the same job. How long would it take them to build the wall if they worked together?

[16.7] Simplify each of the following complex fractions.

35. $\dfrac{\dfrac{10x}{x^2}}{\dfrac{5x}{x^4}}$

36. $\dfrac{\dfrac{4x}{x^5}}{\dfrac{12x}{x^2}}$

37. $\dfrac{\dfrac{1}{ab^2} - \dfrac{1}{a^2b}}{\dfrac{1}{ab}}$

38. $\dfrac{\dfrac{1}{a}}{\dfrac{1}{a^2b} + \dfrac{1}{ab}}$

39. $\dfrac{1 - \dfrac{49}{x^2}}{1 + \dfrac{10}{x} + \dfrac{21}{x^2}}$

40. $\dfrac{1 - \dfrac{11}{x} + \dfrac{24}{x^2}}{1 - \dfrac{64}{x^2}}$

17 RADICAL EXPRESSIONS

17.1 Simplifying Radicals

In Chapter 7 we defined the nth root of a number to be some number whose nth power is equal to the given number.

definition

The expression $\sqrt[n]{a}$ is called a *radical* and it indicates the nth root of a. The number a is called the *radicand*, the number n is called the *index*, and the symbol $\sqrt{\ }$ is called a *radical sign*.

For example, the expression $\sqrt[5]{32}$ is a radical which indicates the 5th root of 32. The number 32 is the radicand and the number 5 is the index. It is equivalent to some number that, when raised to the 5th power, is equal to 32.

$$\sqrt[5]{32} = 2 \quad \text{since} \quad 2^5 = 2 \cdot 2 \cdot 2 \cdot 2 \cdot 2 = 32$$

In this chapter, we will learn how to perform algebraic operations on *radical expressions.*

definition

A *radical expression* indicates a root of an algebraic quantity.

For example, $\sqrt{4x^2}$ indicates the square root of $4x^2$. It is equivalent to some quantity, that when squared, is equal to $4x^2$.

$$\sqrt{4x^2} = 2|x| \quad \text{since} \quad (2x)^2 = 2^2 \cdot x^2 = 4x^2$$

Recall from Section 7.1 that the square roots of negative numbers are imaginary numbers, which are not a part of the real number system. When working with higher roots of numbers, we learned that even roots of negative numbers simplify to imaginary numbers, whereas odd roots of negative numbers simplify to negative real numbers.

In this chapter, we concern ourselves with finding even roots of only positive numbers. Thus, any variable expression that appears under a radical sign is assumed to be positive unless otherwise indicated.

example 1

Simplify each of the following radical expressions.

(a) $\sqrt{49x^4}$

This is equivalent to some quantity that, when squared, is equal to $49x^4$.

$$\sqrt{49x^4} = 7x^2 \quad \text{since} \quad (7x^2)^2 = 7^2(x^2)^2 = 49x^4$$

(b) $\sqrt[3]{27x^3y^3}$

This is equivalent to some quantity that, when cubed, is equal to $27x^3y^3$.

$$\sqrt[3]{27x^3y^3} = 3xy \quad \text{since} \quad (3xy)^3 = 3^3x^3y^3 = 27x^3y^3$$

●

Thus far, we have learned how to simplify only radical expressions in which the quantity under the radical sign, or radicand, was a perfect square or perfect cube. In these problems, the simplified form contained no radicals. We now learn how to simplify radical expressions in which at least one of the factors of the radicand is a perfect square, but the entire radicand is not. For these problems, we use the first property of radicals introduced in Section 7.2.

For any positive numbers, a and b,

$$\sqrt{a \cdot b} = \sqrt{a} \cdot \sqrt{b}$$

The square root of a product is the product of the square roots of the factors.

For example $\sqrt{8}$ may be simplified as follows:

$$\sqrt{8} = \sqrt{4 \cdot 2} = \sqrt{4} \cdot \sqrt{2}$$
$$= 2\sqrt{2}$$

Recall that $2\sqrt{2}$ is read as "two times the square root of two."

This procedure for simplifying the square root of a product is summarized as follows.

To Simplify the Square Root of a Product

1. Rewrite the radicand as the product of two factors, so that the first is the largest perfect square that is a factor of the radicand.
2. Express the square root of the product as the product of the square roots of the factors.
3. Find the square root of the factor that is a perfect square.

We now illustrate this procedure with the following examples:

example 2

Simplify each of the following radical expressions.

(a) $\sqrt{xy^3}$

$$\sqrt{xy^3} = \sqrt{y^2 \cdot xy}$$
$$= \sqrt{y^2} \cdot \sqrt{xy} = y\sqrt{xy}$$

y^2 is the largest factor of xy^3 that is a perfect square.

(b) $\sqrt{12x^3y^2}$

$$\begin{aligned}\sqrt{12x^3y^2} &= \sqrt{(4x^2y^2)(3x)} \\ &= \sqrt{4x^2y^2} \cdot \sqrt{3x} \\ &= 2xy\sqrt{3x}\end{aligned}$$

$4x^2y^2$ is the largest factor of $12x^3y^2$ that is a perfect square.

(c) $-2\sqrt{18a^7}$

$$\begin{aligned}-2\sqrt{18a^7} &= -2\sqrt{(9a^6)(2a)} \\ &= -2\sqrt{9a^6} \cdot \sqrt{2a} \\ &= -2(3a^3)\sqrt{2a} \\ &= -6a^3\sqrt{2a}\end{aligned}$$

(d) $\sqrt{75x^5y^4z^3}$

$$\begin{aligned}\sqrt{75x^5y^4z^3} &= \sqrt{(25x^4y^4z^2)(3xz)} \\ &= \sqrt{25x^4y^4z^2} \cdot \sqrt{3xz} \\ &= 5x^2y^2z\sqrt{3xz}\end{aligned}$$

●

We now learn how to simplify radical expressions in which the radicand is a fraction. To do this, we use the second property of radicals introduced in Section 7.2.

For any positive numbers a and b,

$$\sqrt{\frac{a}{b}} = \frac{\sqrt{a}}{\sqrt{b}}$$

The square root of a quotient is the square root of the numerator divided by the square root of the denominator.

For example, $\sqrt{\frac{9}{25}} = \frac{\sqrt{9}}{\sqrt{25}} = \frac{3}{5}$.

To simplify square roots of fractions in which the denominator is not a perfect square, we must eliminate the square root in the denominator, which is an irrational number. Since whenever we multiply the square root of a number by itself the result is that number, we can do this by multiplying numerator and denominator by the square root in the denominator. This procedure is called *rationalizing the denominator.* For example,

$$\begin{aligned}\sqrt{\frac{1}{3}} = \frac{\sqrt{1}}{\sqrt{3}} &= \frac{1}{\sqrt{3}} \\ &= \frac{1}{\sqrt{3}} \cdot \frac{\sqrt{3}}{\sqrt{3}} \\ &= \frac{\sqrt{3}}{3}\end{aligned}$$

Multiply numerator and denominator by $\sqrt{3}$.

The denominator is now a rational number.

Notice that when we rationalize the denominator using this technique, we are using the following property regarding square roots.

For any number a,

$$\sqrt{a} \cdot \sqrt{a} = a$$

We now summarize the procedure for simplifying the square root of a fraction (or quotient).

To Simplify the Square Root of a Fraction (or Quotient)

1. Express the square root of the fraction as the square root of the numerator divided by the square root of the denominator.
2. Simplify the square roots of the numerator and denominator separately, using the procedure for simplifying the square root of a product, if necessary.
3. If there is a square root in the denominator, rationalize the denominator by multiplying the numerator and denominator by the square root in the denominator.

We now illustrate this procedure with the following examples.

example 3

Simplify each of the following radical expressions.

(a) $\sqrt{\frac{4}{5}}$

$$\sqrt{\frac{4}{5}} = \frac{\sqrt{4}}{\sqrt{5}}$$ Rewrite the square root of the fraction.

$$= \frac{2}{\sqrt{5}}$$ Simplify the square root in the numerator.

$$= \frac{2}{\sqrt{5}} \cdot \frac{\sqrt{5}}{\sqrt{5}}$$ Multiply the numerator and denominator by $\sqrt{5}$ to rationalize the denominator.

$$= \frac{2\sqrt{5}}{5}$$

(b) $\sqrt{\frac{25x^3}{4y^2}}$

$$\sqrt{\frac{25x^3}{4y^2}} = \frac{\sqrt{25x^3}}{\sqrt{4y^2}}$$ Rewrite the square root of the fraction.

$$= \frac{\sqrt{25x^2}\sqrt{x}}{2y}$$ Simplify the square roots in the numerator and denominator.

$$= \frac{5x\sqrt{x}}{2y}$$

(c) $5\sqrt{\frac{t^2}{2}}$

$$5\sqrt{\frac{t^2}{2}} = \frac{5\sqrt{t^2}}{\sqrt{2}}$$

Rewrite the square root of the fraction.

$$= \frac{5t}{\sqrt{2}} \cdot \frac{\sqrt{2}}{\sqrt{2}}$$

Multiply the numerator and denominator by $\sqrt{2}$ to rationalize the denominator.

$$= \frac{5t\sqrt{2}}{2}$$

In each of the examples presented in this section, we have expressed all of our answers in *simplified form.*

A Radical Expression Is in Simplified Form if:

1. There are no perfect squares that are factors of the quantity under the square root sign.
2. There are no radicals in the denominator.

QUICK QUIZ

Simplify each of the following radical expressions.

1. $\sqrt[3]{8a^3}$
2. $\sqrt{27xy^2}$
3. $\sqrt{\frac{25x^2}{y}}$

ANSWERS

1. $2a$
2. $3y\sqrt{3x}$
3. $\frac{5x\sqrt{y}}{y}$

17.1 Exercises

Simplify each of the following radical expressions.

1. $\sqrt{9a^2}$
2. $\sqrt{36b^2}$
3. $\sqrt{49t^6}$
4. $\sqrt{16w^4}$
5. $\sqrt{x^2y^4}$
6. $\sqrt{x^6y^2}$
7. $\sqrt[3]{27x^3}$
8. $\sqrt[3]{8y^6}$
9. $\sqrt[3]{a^3b^6}$
10. $\sqrt[3]{a^9b^3}$
11. $\sqrt{144x^4y^2z^{10}}$
12. $\sqrt{64x^6y^8z^2}$

Use the first property of radicals to simplify the square roots of each of the following products.

13. $\sqrt{20x^2}$
14. $\sqrt{50y^2}$
15. $\sqrt{a^3b^2}$
16. $\sqrt{ab^2}$
17. $2\sqrt{x^2y^5}$
18. $3\sqrt{x^3y^4}$
19. $\sqrt{32ab^4}$
20. $\sqrt{72a^5b}$
21. $-\sqrt{12w^6}$
22. $-\sqrt{27t^4}$
23. $\sqrt{16x^3y^5z^2}$
24. $\sqrt{9x^4y^5z^2}$
25. $5\sqrt{54x^5y^6}$
26. $6\sqrt{24x^4y^7}$
27. $-3\sqrt{5a^4b^7c^2}$
28. $-4\sqrt{3a^3b^5c^4}$

29. $\sqrt{128x^3y^6z^9}$ **30.** $\sqrt{48x^8y^5z^2}$ **31.** $\frac{1}{5}\sqrt{75a^8b^5c^4}$ **32.** $\frac{1}{3}\sqrt{18a^3b^9c^6}$

Use the second property of radicals to simplify the square roots of each of the following quotients.

33. $\sqrt{\frac{2}{3}}$ **34.** $\sqrt{\frac{5}{6}}$ **35.** $\sqrt{\frac{a}{2}}$ **36.** $\sqrt{\frac{b}{5}}$ **37.** $\sqrt{\frac{25x^2}{9}}$

38. $\sqrt{\frac{49y}{4}}$ **39.** $\sqrt{\frac{64a^4}{81b^2}}$ **40.** $\sqrt{\frac{16a^2}{25b^4}}$ **41.** $\sqrt{\frac{24x^6}{5y^6}}$ **42.** $\sqrt{\frac{32x^4}{3y^2}}$

43. $3\sqrt{\frac{x^4y^5}{9}}$ **44.** $5\sqrt{\frac{x^2y^3}{25}}$ **45.** $-\sqrt{\frac{ab^6}{c^8}}$ **46.** $-\sqrt{\frac{a^7b}{c^6}}$ **47.** $\sqrt{\frac{8x^3y^6}{27}}$

48. $\sqrt{\frac{18x^4y^5}{50}}$ **49.** $\sqrt{\frac{a^3b^7}{2c}}$ **50.** $\sqrt{\frac{a^8b^3}{3c}}$ **51.** $\sqrt{\frac{27x^3y^4}{z^3}}$ **52.** $\sqrt{\frac{32x^4y^5}{z^3}}$

17.2 Addition and Subtraction of Radical Expressions

In this section we learn how to add and subtract radical expressions. The technique we use involves the distributive property, and it is similar to that used for combining like terms to simplify algebraic expressions. Only *like radicals*, which are radical expressions that have the same index and radicand, can be combined using this technique. Consider the following example.

example 1

Combine the radicals in the expression $8\sqrt{3} + 5\sqrt{3} - 2\sqrt{3}$.

solution

We first factor out $\sqrt{3}$ from each term using the distributive property and then add and subtract the coefficients of $\sqrt{3}$.

$$\begin{aligned} 8\sqrt{3} + 5\sqrt{3} - 2\sqrt{3} &= (8 + 5 - 2)\sqrt{3} \\ &= 11\sqrt{3} \end{aligned}$$

●

Remember that the answer is read as "eleven times the square root of three." Whenever we express the product of a whole number and a radical, the whole number appears first.

If all the radical factors of an algebraic expression we wish to simplify do not have the same index and radicand, the simplified form may contain more than one term. We express the result as the sum or difference of terms containing unlike radicals. Consider this example.

example 2

Simplify $4\sqrt{5} + 7\sqrt{2} - 3\sqrt{5} + \sqrt{2}$.

solution

$$\begin{aligned} &4\sqrt{5} + 7\sqrt{2} - 3\sqrt{5} + \sqrt{2} \\ &= 4\sqrt{5} - 3\sqrt{5} + 7\sqrt{2} + \sqrt{2} && \text{By commutative property.} \\ &= (4 - 3)\sqrt{5} + (7 + 1)\sqrt{2} && \text{By distributive property.} \\ &= 1 \cdot \sqrt{5} + 8\sqrt{2} \\ &= \sqrt{5} + 8\sqrt{2} \end{aligned}$$

●

Often we must first rewrite the radicals in an algebraic expression in simplified form before we can combine like radicals. This technique is illustrated in the following example.

example 3

Add $4\sqrt{3} + 5\sqrt{12}$.

solution

The $\sqrt{12}$ must be written in simplified form before we can add.

$$
\begin{aligned}
4\sqrt{3} + 5\sqrt{12} &= 4\sqrt{3} + 5\sqrt{4}\sqrt{3} && \text{Rewrite } \sqrt{12} = \sqrt{4}\sqrt{3}. \\
&= 4\sqrt{3} + 5 \cdot 2\sqrt{3} && \text{Simplify } \sqrt{4} = 2. \\
&= 4\sqrt{3} + 10\sqrt{3} && \text{Multiply } 5 \cdot 2 = 10. \\
&= (4 + 10)\sqrt{3} && \text{By distributive property.} \\
&= 14\sqrt{3}
\end{aligned}
$$

We now summarize the procedure we have used to add or subtract radical expressions.

To Add or Subtract Radical Expressions

1. Simplify all radicals that are not expressed in simplified form.
2. Add or subtract terms containing like radicals using the distributive property.

example 4

Simplify $2\sqrt{45} - 9\sqrt{5} + \sqrt{20}$.

solution

$$
\begin{aligned}
&2\sqrt{45} - 9\sqrt{5} + \sqrt{20} && \text{Rewrite } \sqrt{45} = \sqrt{9}\sqrt{5};\ \sqrt{20} = \sqrt{4}\sqrt{5}. \\
&\quad = 2\sqrt{9}\sqrt{5} - 9\sqrt{5} + \sqrt{4}\sqrt{5} \\
&\quad = 2 \cdot 3\sqrt{5} - 9\sqrt{5} + 2\sqrt{5} && \text{Simplify } \sqrt{9} = 3;\ \sqrt{4} = 2. \\
&\quad = 6\sqrt{5} - 9\sqrt{5} + 2\sqrt{5} \\
&\quad = (6 - 9 + 2)\sqrt{5} && \text{By distributive property.} \\
&\quad = (-1)\sqrt{5} \\
&\quad = -\sqrt{5}
\end{aligned}
$$

We will now look at some examples in which the radicands contain variables.

example 5

Subtract $7\sqrt{4x^3} - 3\sqrt{9x}$.

solution

$$
\begin{aligned}
&7\sqrt{4x^3} - 3\sqrt{9x} \\
&\quad = 7\sqrt{4x^2}\sqrt{x} - 3\sqrt{9}\sqrt{x} && \text{Simplify radical expressions.} \\
&\quad = 7(2x)\sqrt{x} - 3(3)\sqrt{x} \\
&\quad = 14x\sqrt{x} - 9\sqrt{x} && \text{Multiply factors of radicals.} \\
&\quad = (14x - 9)\sqrt{x} && \text{By distributive property.}
\end{aligned}
$$

example 6 Simplify $5\sqrt{18a^2b^3} + \sqrt{2a^4b} - 3a\sqrt{8b^3}$.

solution

$$\begin{aligned} 5\sqrt{18a^2b^3} + \sqrt{2a^4b} - 3a\sqrt{8b^3} &= 5\sqrt{9a^2b^2}\,\sqrt{2b} + \sqrt{a^4}\,\sqrt{2b} - 3a\sqrt{4b^2}\,\sqrt{2b} \\ &= 5(3ab)\sqrt{2b} + a^2\sqrt{2b} - 3a(2b)\sqrt{2b} \\ &= 15ab\sqrt{2b} + a^2\sqrt{2b} - 6ab\sqrt{2b} \\ &= (15ab + a^2 - 6ab)\sqrt{2b} \\ &= (a^2 + 9ab)\sqrt{2b} \end{aligned}$$

QUICK QUIZ

Simplify and combine like radicals in each of the following.

1. $3\sqrt{7} - 8\sqrt{7} + 2\sqrt{7}$
2. $5\sqrt{27} - 2\sqrt{48}$
3. $\sqrt{5x^3} + 2\sqrt{18x^2} + 3x\sqrt{20x}$

ANSWERS

1. $-3\sqrt{7}$
2. $7\sqrt{3}$
3. $7x\sqrt{5x} + 6x\sqrt{2}$

17.2 Exercises

Simplify and combine like radicals in each of the following expressions.

1. $8\sqrt{6} + 3\sqrt{6}$
2. $2\sqrt{5} + 4\sqrt{5}$
3. $5\sqrt{2} - \sqrt{2}$
4. $7\sqrt{3} - 5\sqrt{3}$
5. $2\sqrt{5} - 9\sqrt{5} + \sqrt{5}$
6. $\sqrt{7} - 8\sqrt{7} + 3\sqrt{7}$
7. $-8\sqrt{11} - 4\sqrt{11}$
8. $-6\sqrt{10} - 5\sqrt{10}$
9. $5\sqrt{3} - 7\sqrt{2} + 4\sqrt{3} + 2\sqrt{2}$
10. $8\sqrt{5} + 7\sqrt{3} - 2\sqrt{5} - 8\sqrt{3}$
11. $5\sqrt{27} + 2\sqrt{3}$
12. $5\sqrt{8} + 9\sqrt{2}$
13. $6\sqrt{5} - 3\sqrt{45}$
14. $8\sqrt{7} - 2\sqrt{28}$
15. $8\sqrt{2} - 9\sqrt{32}$
16. $3\sqrt{3} - 7\sqrt{75}$
17. $3\sqrt{12} + 7\sqrt{48}$
18. $5\sqrt{18} + 4\sqrt{50}$
19. $\sqrt{80} + 5\sqrt{20}$
20. $\sqrt{24} + 2\sqrt{54}$
21. $7\sqrt{18} - \sqrt{72}$
22. $3\sqrt{63} - \sqrt{28}$
23. $5\sqrt{27} + 9\sqrt{3} - 2\sqrt{48}$
24. $5\sqrt{18} + 9\sqrt{2} - 3\sqrt{32}$
25. $8\sqrt{45} - 4\sqrt{5} + 3\sqrt{20}$
26. $2\sqrt{54} - 6\sqrt{6} + \sqrt{96}$
27. $7\sqrt{7} + 6\sqrt{63} + \sqrt{28}$
28. $5\sqrt{10} + \sqrt{40} + 2\sqrt{90}$
29. $\sqrt{50} - 3\sqrt{98} - 2\sqrt{72}$
30. $2\sqrt{27} - 5\sqrt{75} - \sqrt{48}$
31. $7\sqrt{54} + \sqrt{24} - \sqrt{96}$
32. $\sqrt{125} + \sqrt{20} - 4\sqrt{45}$
33. $4\sqrt{18} - 2\sqrt{12} + 8\sqrt{32}$
34. $8\sqrt{27} - 3\sqrt{8} + 2\sqrt{48}$
35. $5\sqrt{x} + 3\sqrt{x}$
36. $2\sqrt{y} - 3\sqrt{y}$
37. $4\sqrt{a^3} - 7\sqrt{a}$
38. $6\sqrt{b} + 2\sqrt{b^3}$
39. $2x\sqrt{4x^2} + \sqrt{25x^4}$
40. $\sqrt{16x^4} + 3x\sqrt{9x^2}$
41. $5y\sqrt{18y} - 4\sqrt{8y^3}$
42. $2y\sqrt{27y} - 6\sqrt{12y^3}$
43. $2\sqrt{xy^2} + 5\sqrt{x^3}$
44. $3\sqrt{x^2y^3} + 7\sqrt{y}$
45. $\sqrt{12x^2y^3} - 7\sqrt{3x^4y}$
46. $8\sqrt{2xy^4} - \sqrt{18x^3y^2}$
47. $5a\sqrt{ab^2} - 3\sqrt{a^2b^2} + 2\sqrt{a^3}$
48. $4\sqrt{a^2b^2} + 7\sqrt{b^3} - 3b\sqrt{a^2b}$
49. $x\sqrt{18xy^2} + 2\sqrt{xy} - 2\sqrt{50x^3}$
50. $\sqrt{48y^3} - 5\sqrt{3xy} + y\sqrt{27x^2y}$

17.3 Multiplication and Division of Radical Expressions

In this section, we learn how to multiply and divide radical expressions. Recall that according to the first property of radicals, the square root of the product of two positive numbers is the product of the square roots of the factors. Thus, to multiply two radicals, we can simply find the square root of the product of the radicands.

For any positive numbers a and b,

$$\sqrt{a} \cdot \sqrt{b} = \sqrt{a \cdot b}$$

For example,

$$\sqrt{5} \cdot \sqrt{3} = \sqrt{5 \cdot 3} = \sqrt{15}$$

If the radicals are multiplied by factors that are rational numbers, we multiply the coefficients of the radicals separately. A summary of this procedure follows.

To Multiply Two Square Roots

1. Multiply the coefficients of the radicals.
2. Multiply the radicands and take the square root of the product.
3. Write the results of steps 1 and 2 as a product and simplify if necessary.

example 1

Multiply each of the following.

(a) $3\sqrt{2} \cdot 5\sqrt{6}$

$3\sqrt{2} \cdot 5\sqrt{6} = (3 \cdot 5)(\sqrt{2}\sqrt{6})$ — By commutative and associative properties.

$= 15\sqrt{2 \cdot 6}$ — Multiply coefficients and radicands.

$= 15\sqrt{12}$

$= 15\sqrt{4}\sqrt{3}$ — Simplify the result.

$= 15(2)\sqrt{3} = 30\sqrt{3}$

(b) $-3\sqrt{2x} \cdot 8\sqrt{2x}$

$-3\sqrt{2x} \cdot 8\sqrt{2x} = (-3 \cdot 8)(\sqrt{2x} \cdot \sqrt{2x})$ — By commutative and associative properties.

$= -24(\sqrt{2x} \cdot \sqrt{2x})$ — Multiply coefficients of radicals.

$= -24(2x)$ — Multiply radicals.

$= -48x$

●

To multiply radical expressions that contain more than one term, we use the distributive property as shown in these examples.

example 2

Multiply each of the following.

(a) $\sqrt{3}(\sqrt{7}+\sqrt{3})$

$\sqrt{3}(\sqrt{7}+\sqrt{3}) = \sqrt{3}\cdot\sqrt{7}+\sqrt{3}\cdot\sqrt{3}$ — By distributive property.

$= \sqrt{21}+3$ — Multiply radicals.

(b) $2\sqrt{2}(4\sqrt{5}-5\sqrt{6}+\sqrt{10})$

$2\sqrt{2}(4\sqrt{5}-5\sqrt{6}+\sqrt{10})$

$= 2\sqrt{2}\cdot 4\sqrt{5}-2\sqrt{2}\cdot 5\sqrt{6}+2\sqrt{2}\cdot\sqrt{10}$ — By distributive property.

$= (2)(4)\sqrt{2}\sqrt{5}-(2)(5)\sqrt{2}\sqrt{6}+2\sqrt{2}\sqrt{10}$ — Multiply coefficients of radicals.

$= 8\sqrt{10}-10\sqrt{12}+2\sqrt{20}$ — Multiply radicals.

$= 8\sqrt{10}-10\sqrt{4}\sqrt{3}+2\sqrt{4}\sqrt{5}$ — Simplify radicals.

$= 8\sqrt{10}-10(2)\sqrt{3}+2(2)\sqrt{5}$

$= 8\sqrt{10}-20\sqrt{3}+4\sqrt{5}$ ●

To multiply two binomial radical expressions we use the FOIL method as shown in the following examples.

example 3

Multiply each of the following.

(a) $(7+\sqrt{2})(5+\sqrt{2})$

$(7+\sqrt{2})(5+\sqrt{2})$

$= 7\cdot 5 + 7\cdot\sqrt{2} + 5\cdot\sqrt{2} + \sqrt{2}\cdot\sqrt{2}$

FIRST   OUTER   INNER   LAST

$= 35 + (7+5)\sqrt{2} + 2$ — Combine like radicals.

$= 35 + 12\sqrt{2} + 2$

$= 37 + 12\sqrt{2}$ — Combine whole numbers.

(b) $(\sqrt{5}-\sqrt{2})^2$

$(\sqrt{5}-\sqrt{2})^2$

$= (\sqrt{5}-\sqrt{2})(\sqrt{5}-\sqrt{2})$ — Rewrite as multiplication.

$= \sqrt{5}\sqrt{5}-\sqrt{5}\sqrt{2}-\sqrt{5}\sqrt{2}$ — Using FOIL method.

$+\sqrt{2}\sqrt{2}$

$= 5-\sqrt{10}-\sqrt{10}+2$

$= 7-2\sqrt{10}$ — Combine like terms.

(c) $(\sqrt{7}+4)(\sqrt{7}-4)$

$(\sqrt{7}+4)(\sqrt{7}-4)$

$= \sqrt{7}\sqrt{7}-4\sqrt{7}+4\sqrt{7}-16$ — Using FOIL method.

$= 7-16$ — The middle terms add up to zero.

$= -9$ ●

In Example 3(c) the expressions $(\sqrt{7}+4)$ and $(\sqrt{7}-4)$ are said to be *conjugates* of one another. Notice that whenever we multiply a radical expression by its conjugate, the middle terms add up to zero and the result contains no radicals.

To divide radicals, we use the second property of radicals which states that the square root of a quotient is equal to the square root of the numerator divided by the square root of the denominator.

For any positive numbers a and b,

$$\frac{\sqrt{a}}{\sqrt{b}} = \sqrt{\frac{a}{b}}$$

For example,

$$\frac{\sqrt{50}}{\sqrt{2}} = \sqrt{\frac{50}{2}} = \sqrt{25} = 5$$

If the radicals are multiplied by rational coefficients, we divide the coefficients and the radicands separately. This procedure is summarized as follows.

To Divide Two Square Roots

1. Divide the coefficients of the radicals.
2. Divide the radicands and take the square root of the quotients.
3. Write the results of steps 1 and 2 as a product and simplify if necessary.

example 4

Divide each of the following radicals.

(a) $\dfrac{21\sqrt{12}}{7\sqrt{3}}$

$$\frac{21\sqrt{12}}{7\sqrt{3}} = \frac{21}{7}\sqrt{\frac{12}{3}}$$ Divide coefficients and radicands.

$$= 3\sqrt{4}$$ Simplify the result.

$$= 3 \cdot 2$$

$$= 6$$

(b) $\dfrac{\sqrt{15x^3}}{\sqrt{3x}}$

$$\frac{\sqrt{15x^3}}{\sqrt{3x}} = \sqrt{\frac{15x^3}{3x}}$$ Divide radicals.

$$= \sqrt{5x^2}$$

$$= \sqrt{5}\sqrt{x^2}$$ Simplify the result.

$$= x\sqrt{5}$$

(c) $\dfrac{3\sqrt{8}+\sqrt{18}}{\sqrt{2}}$

$$\frac{3\sqrt{8}+\sqrt{18}}{\sqrt{2}} = \frac{3\sqrt{8}}{\sqrt{2}} + \frac{\sqrt{18}}{\sqrt{2}}$$ Each term in the numerator must be divided by $\sqrt{2}$.

$$= (3)\sqrt{\frac{8}{2}} + \sqrt{\frac{18}{2}}$$ Divide radicals.

$$= 3\sqrt{4} + \sqrt{9}$$ Simplify the result.

$$= 3(2) + 3$$
$$= 6 + 3 = 9$$

●

If the radical in the denominator is not a factor of the numerator, we can simplify the expression by rationalizing the denominator as shown in the following examples.

example 5

Rationalize the denominator in each of the following expressions.

(a) $\dfrac{8}{\sqrt{2}}$

$$\frac{8}{\sqrt{2}} = \frac{8}{\sqrt{2}} \cdot \frac{\sqrt{2}}{\sqrt{2}}$$ Multiply the numerator and denominator by $\sqrt{2}$ to rationalize the denominator.

$$= \frac{8\sqrt{2}}{2}$$

$$= 4\sqrt{2}$$ Divide whole numbers.

(b) $\dfrac{9\sqrt{5}}{\sqrt{12}}$

$$\frac{9\sqrt{5}}{\sqrt{12}} = \frac{9\sqrt{5}}{\sqrt{4}\sqrt{3}}$$ First simplify radical in the denominator.

$$= \frac{9\sqrt{5}}{2\sqrt{3}} \cdot \frac{\sqrt{3}}{\sqrt{3}}$$ Then rationalize the denominator.

$$= \frac{9\sqrt{5 \cdot 3}}{2 \cdot 3}$$ Multiply radicals.

$$= \frac{3\sqrt{15}}{2}$$ Divide the numerator and denominator by 3. Result is now in simplified form.

(c) $\dfrac{4x}{\sqrt{x}}$

$$\frac{4x}{\sqrt{x}} = \frac{4x}{\sqrt{x}} \cdot \frac{\sqrt{x}}{\sqrt{x}}$$ Rationalize the denominator.

$$= \frac{4x\sqrt{x}}{x}$$ Multiply radicals.

$$= 4\sqrt{x}$$ Divide the numerator and denominator by x.

●

If the expression in the denominator consists of a sum or difference that includes a square root, we rationalize the denominator by multiplying the numerator and denominator by the *conjugate* of the denominator.

definition

For any positive numbers a and b, the expressions $\sqrt{a} + \sqrt{b}$ and $\sqrt{a} - \sqrt{b}$ are said to be conjugates of one another.

Whenever we multiply an expression by its conjugate, we obtain a product that contains no radicals because the middle terms add up to zero.

$$(\sqrt{a} + \sqrt{b})(\sqrt{a} - \sqrt{b}) = a - b$$

Notice that this is the same formula for the difference of two perfect squares written in a different form

$$(a + b)(a - b) = a^2 - b^2$$

We now have two techniques for rationalizing a denominator. Before using either technique, make sure that all radicals are in simplified form.

To Rationalize a Denominator

1. If the denominator contains a single term that includes a square root, multiply the numerator and denominator by that square root.
2. If the denominator expresses the sum or difference of two terms, such that at least one term includes a square root, multiply the numerator and denominator by the conjugate of the denominator.

Let us now look at some examples that require us to rationalize a denominator by multiplying numerator and denominator by the conjugate of the denominator.

example 6

Rationalize the denominator in each of the following expressions.

(a) $\dfrac{2}{5 - \sqrt{3}}$

$$\frac{2}{5 - \sqrt{3}} = \frac{2}{(5 - \sqrt{3})} \frac{(5 + \sqrt{3})}{(5 + \sqrt{3})}$$

Multiply the numerator and denominator by the conjugate, which is $5 + \sqrt{3}$.

$$= \frac{2(5 + \sqrt{3})}{5^2 - 3}$$

Multiply out the denominator. The product contains no radicals.

$$= \frac{2(5 + \sqrt{3})}{25 - 3}$$

Leave the numerator in factored form.

$$= \frac{2(5 + \sqrt{3})}{22}$$

$$= \frac{5 + \sqrt{3}}{11}$$

Divide the numerator and denominator by 2.

(b) $\dfrac{20}{\sqrt{2}+\sqrt{7}}$

$\dfrac{20}{\sqrt{2}+\sqrt{7}} = \dfrac{20}{(\sqrt{2}+\sqrt{7})}\dfrac{(\sqrt{2}-\sqrt{7})}{(\sqrt{2}-\sqrt{7})}$ Multiply the numerator and denominator by the conjugate, which is $\sqrt{2}-\sqrt{7}$.

$= \dfrac{20(\sqrt{2}-\sqrt{7})}{2-7}$ Multiply out the denominator. The product contains no radicals.

$= \dfrac{20(\sqrt{2}-\sqrt{7})}{-5}$

$= -4(\sqrt{2}-\sqrt{7})$ Divide the numerator and denominator by -5. ●

QUICK QUIZ

Perform each of the operations indicated.

1. $3\sqrt{3}\cdot 5\sqrt{6}$
2. $(1+\sqrt{3})(8-\sqrt{3})$
3. $\dfrac{9\sqrt{20}}{3\sqrt{5}}$
4. $\dfrac{4}{\sqrt{7}-2}$

ANSWERS

1. $45\sqrt{2}$
2. $5+7\sqrt{3}$
3. 6
4. $\dfrac{4(\sqrt{7}+2)}{3}$

17.3 Exercises

Multiply each of the following radical expressions. Express all answers in simplified form.

1. $\sqrt{7}\cdot\sqrt{3}$ **2.** $\sqrt{2}\cdot\sqrt{5}$ **3.** $4\sqrt{6}\cdot 3\sqrt{8}$

4. $5\sqrt{12}\cdot 2\sqrt{3}$ **5.** $7\sqrt{2}\cdot 2\sqrt{6}$ **6.** $3\sqrt{6}\cdot 4\sqrt{7}$

7. $-8\sqrt{x}\cdot 4\sqrt{x}$ **8.** $-9\sqrt{y}\cdot 5\sqrt{y}$ **9.** $\sqrt{12a}\cdot\sqrt{3a}$

10. $\sqrt{2a}\cdot\sqrt{8a}$ **11.** $-2\sqrt{5t}\cdot 5\sqrt{5t}$ **12.** $-3\sqrt{3z}\cdot 6\sqrt{3z}$

13. $\sqrt{xy^2}\cdot\sqrt{9x}$ **14.** $\sqrt{16y}\cdot\sqrt{x^2y}$ **15.** $\sqrt{25ab}\cdot\sqrt{b}$

16. $\sqrt{49ab^2}\cdot\sqrt{a}$ **17.** $\sqrt{3}(9+2\sqrt{3})$ **18.** $\sqrt{7}(4+3\sqrt{7})$

19. $-\sqrt{6}(4+5\sqrt{2})$ **20.** $-\sqrt{2}(5\sqrt{6}+8)$ **21.** $\sqrt{5}(3\sqrt{5}-\sqrt{3})$

22. $\sqrt{6}(\sqrt{2}-7\sqrt{6})$ **23.** $\sqrt{10}(3\sqrt{2}+\sqrt{5}-9)$ **24.** $\sqrt{8}(7+5\sqrt{2}-4\sqrt{3})$

25. $3\sqrt{2}(8\sqrt{6}+7\sqrt{2})$ **26.** $4\sqrt{3}(2\sqrt{15}-5\sqrt{3})$ **27.** $2\sqrt{7}(5\sqrt{2}-\sqrt{14})$

28. $3\sqrt{5}(4\sqrt{3}+\sqrt{10})$ **29.** $(8+\sqrt{2})(7+\sqrt{2})$ **30.** $(5+\sqrt{3})(6+\sqrt{3})$

31. $(4+\sqrt{5})(9-\sqrt{5})$ **32.** $(8+\sqrt{7})(3-\sqrt{7})$ **33.** $(8+\sqrt{3})^2$

34. $(7-\sqrt{2})^2$ **35.** $(2-\sqrt{7})^2$ **36.** $(4+\sqrt{6})^2$

37. $(5+\sqrt{3})(5-\sqrt{3})$ **38.** $(9+\sqrt{2})(9-\sqrt{2})$ **39.** $(5-\sqrt{2})(9-\sqrt{2})$

40. $(4-\sqrt{2})(\sqrt{7}-\sqrt{2})$ **41.** $(\sqrt{3}+2\sqrt{6})(\sqrt{3}-\sqrt{6})$ **42.** $(\sqrt{5}+\sqrt{3})(3\sqrt{5}-\sqrt{3})$

43. $(2\sqrt{5}-\sqrt{3})(\sqrt{7}+4\sqrt{3})$ **44.** $(7\sqrt{2}+\sqrt{5})(\sqrt{6}-3\sqrt{5})$ **45.** $(\sqrt{x}-\sqrt{y})(\sqrt{x}+\sqrt{y})$

46. $(\sqrt{p}+\sqrt{q})(\sqrt{p}-\sqrt{q})$ **47.** $(\sqrt{a}+4)(\sqrt{a}-4)$ **48.** $(\sqrt{b}-6)(\sqrt{b}+6)$
49. $(\sqrt{x}-7)(\sqrt{x}+3)$ **50.** $(\sqrt{x}+5)(\sqrt{x}-9)$

Divide each of the following radical expressions. Express all answers in simplified form.

51. $\frac{\sqrt{98}}{\sqrt{2}}$ **52.** $\frac{\sqrt{48}}{\sqrt{3}}$ **53.** $\frac{5\sqrt{28}}{\sqrt{7}}$ **54.** $\frac{7\sqrt{20}}{\sqrt{5}}$

55. $\frac{12\sqrt{24}}{3\sqrt{3}}$ **56.** $\frac{28\sqrt{42}}{4\sqrt{7}}$ **57.** $\frac{6\sqrt{45}}{2\sqrt{5}}$ **58.** $\frac{15\sqrt{54}}{3\sqrt{3}}$

59. $\frac{\sqrt{75}+\sqrt{48}}{\sqrt{3}}$ **60.** $\frac{\sqrt{32}-\sqrt{72}}{\sqrt{2}}$ **61.** $\frac{10\sqrt{6}-5\sqrt{50}}{5\sqrt{2}}$ **62.** $\frac{4\sqrt{27}+20\sqrt{6}}{4\sqrt{3}}$

63. $\frac{\sqrt{42}-\sqrt{28}}{\sqrt{7}}$ **64.** $\frac{\sqrt{20}-\sqrt{35}}{\sqrt{5}}$ **65.** $\frac{\sqrt{27x^3}}{\sqrt{3x}}$ **66.** $\frac{\sqrt{54y^3}}{\sqrt{6y}}$

67. $\frac{8\sqrt{10a^2}}{4\sqrt{2a}}$ **68.** $\frac{6\sqrt{15b^2}}{2\sqrt{5b}}$ **69.** $\frac{\sqrt{9x^3y}}{\sqrt{xy}}$ **70.** $\frac{\sqrt{25xy^3}}{\sqrt{xy}}$

Rationalize the denominator in each of the following expressions. Express all answers in simplified form.

71. $\frac{4}{\sqrt{5}}$ **72.** $\frac{2}{\sqrt{7}}$ **73.** $\frac{7}{\sqrt{6}}$ **74.** $\frac{8}{\sqrt{3}}$ **75.** $\frac{8}{\sqrt{12}}$

76. $\frac{6}{\sqrt{18}}$ **77.** $\frac{5\sqrt{2}}{\sqrt{50}}$ **78.** $\frac{4\sqrt{3}}{\sqrt{48}}$ **79.** $\frac{9a}{\sqrt{a}}$ **80.** $\frac{8a}{\sqrt{a}}$

81. $\frac{7x}{\sqrt{16x^3}}$ **82.** $\frac{5x}{\sqrt{9x^3}}$ **83.** $\frac{8}{5+\sqrt{2}}$ **84.** $\frac{6}{4+\sqrt{3}}$ **85.** $\frac{2}{7-\sqrt{5}}$

86. $\frac{3}{2-\sqrt{7}}$ **87.** $\frac{\sqrt{3}}{\sqrt{3}-5}$ **88.** $\frac{\sqrt{2}}{\sqrt{2}+9}$ **89.** $\frac{2\sqrt{6}}{\sqrt{6}+\sqrt{3}}$ **90.** $\frac{4\sqrt{5}}{\sqrt{3}-\sqrt{5}}$

91. $\frac{3\sqrt{5}-7}{\sqrt{5}+\sqrt{2}}$ **92.** $\frac{2\sqrt{3}+5}{\sqrt{3}-\sqrt{7}}$ **93.** $\frac{5\sqrt{3}+2\sqrt{5}}{4\sqrt{3}-\sqrt{5}}$ **94.** $\frac{3\sqrt{2}-5\sqrt{3}}{\sqrt{2}+2\sqrt{3}}$

17.4 Solving Radical Equations

In this section we learn how to solve *radical equations.* A *radical equation* is an equation in which a variable appears under a radical sign. To solve radical equations, we use the following property which allows us to square both sides of an equation.

For any numbers a and b,

if $a = b$

then $a^2 = b^2$

In other words, if two quantities are equal, the squares of these quantities are equal.

Sometimes, when we square both sides of an equation, the resulting equation has a different solution set than the original equation. This is because even if $a^2 = b^2$, it is not necessarily true that $a = b$. For example, even though

$$(7)^2 = (-7)^2$$
$$7 \neq -7$$

Therefore, when solving radical equations, we must be careful to check all solutions in the original equation. We now summarize the procedure for solving radical equations in which the radical expression is a square root.

To Solve a Radical Equation

1. Rewrite the equation so that the radical expression is isolated on the left.
2. Square both sides of the equation.
3. Solve the resulting equation for the unknown.
4. Check all solutions in the original equation.

To solve a radical equation that involves a higher root, we must raise both sides of the equation to the corresponding power. For example, to solve a radical equation involving a cube root, we must raise both sides of the equation to the third power. We will not be working with these kinds of equations in this book, but you will learn how to solve them in more advanced algebra courses.

We now illustrate the procedures for solving radical equations with the following examples.

example 1

Solve $\sqrt{x + 5} = 3$ for x.

solution

$$\sqrt{x+5} = 3$$
$$(\sqrt{x+5})^2 = (3)^2 \qquad \text{Square both sides.}$$
$$x + 5 = 9$$
$$x = 4 \qquad \text{Solve for } x.$$

CHECK

$$\sqrt{x+5} = 3$$
$$\sqrt{4+5} \ ? \ 3$$
$$\sqrt{9} \ ? \ 3$$
$$3 = 3 \ \checkmark \qquad \text{The solution } x = 4 \text{ checks.}$$

Thus, the solution is $x = 4$.

example 2

Solve $\sqrt{3a + 7} = -2$ for a.

solution

$$\sqrt{3a+7} = -2$$
$$(\sqrt{3a+7})^2 = (-2)^2 \qquad \text{Square both sides.}$$
$$3a + 7 = 4$$
$$3a = -3$$
$$a = -1 \qquad \text{Solve for } a.$$

CHECK

$$\sqrt{3a+7} = -2$$
$$\sqrt{3(-1)+7} \ ? \ -2$$
$$\sqrt{4} \ ? \ -2$$
$$2 \neq -2$$

The solution $a = -1$ does not check.

Thus, the equation has no solution.

example 3

Solve $2\sqrt{5y} - 7 = 13$ for y.

solution

$$2\sqrt{5y} - 7 = 13$$ Isolate the radical on the left:

$$2\sqrt{5y} = 20$$ Add 7 to both sides.

$$\sqrt{5y} = 10$$ Multiply both sides by $\frac{1}{2}$.

$$(\sqrt{5y})^2 = (10)^2$$ Square both sides.

$$5y = 100$$

$$y = 20$$ Solve for y.

CHECK

$$2\sqrt{5y} - 7 = 13$$
$$2\sqrt{5 \cdot 20} - 7 \ ? \ 13$$
$$2\sqrt{100} - 7 \ ? \ 13$$
$$2(10) - 7 \ ? \ 13$$
$$20 - 7 \ ? \ 13$$
$$13 = 13 \ \surd$$

The solution $y = 20$ checks.

Thus, the solution is $y = 20$.

example 4

Solve $\sqrt{2x+11} = x + 4$ for x.

solution

$$\sqrt{2x+11} = x + 4$$

$$(\sqrt{2x+11})^2 = (x+4)^2$$ Square both sides.

$$2x + 11 = x^2 + 8x + 16$$ Solve the resulting quadratic equation.

$$0 = x^2 + 6x + 5$$

$$x^2 + 6x + 5 = 0$$ Rewrite in standard form.

$$(x+5)(x+1) = 0$$ Factor.

$$x + 5 = 0 \quad \text{or} \quad x + 1 = 0$$ Set each factor equal to zero.

$$x = -5 \quad \text{or} \quad x = -1$$ Solve for x.

CHECK

$$\sqrt{2x+11} = x + 4$$ Check both solutions in the original equation.

$$x = -5: \quad \sqrt{2(-5)+11} \ ? \ -5 + 4$$
$$\sqrt{1} \ ? \ -1$$
$$1 \neq -1$$

The solution $x = -5$ does not check.

Thus, $x = -5$ is not a solution to the equation.

$$x = -1: \quad \sqrt{2(-1) + 11} \ ? \ -1 + 4$$
$$\sqrt{9} \ ? \ 3$$
$$3 = 3 \ \checkmark$$

The solution $x = -1$ checks.

Thus, the only solution is $x = -1$.

WARNING

A common mistake in attempting to solve radical equations is to break up a sum or difference that appears under a radical sign

$$\sqrt{a + b} \neq \sqrt{a} + \sqrt{b}$$

For example,

$$\sqrt{4 + 9} \neq \sqrt{4} + \sqrt{9}$$
$$\sqrt{13} \neq 2 + 3$$
$$\sqrt{13} \neq 5$$

Therefore,

$$\sqrt{x + 16} \neq \sqrt{x} + \sqrt{16} = \sqrt{x} + 4$$

Even though we have established properties for the square root of a product and quotient, there are no similar properties for the square root of a sum or difference.

17.4 Exercises

Solve each of the following radical equations.

1. $\sqrt{3x} = 6$
2. $\sqrt{2x} = 4$
3. $4\sqrt{x} = 20$
4. $5\sqrt{x} = 15$
5. $\sqrt{x + 2} = 3$
6. $\sqrt{x + 8} = 5$
7. $\sqrt{a - 4} = 7$
8. $\sqrt{a - 1} = 6$
9. $\sqrt{x + 9} = -4$
10. $\sqrt{x + 4} = -3$
11. $\sqrt{y - 3} = -5$
12. $\sqrt{y - 6} = -1$
13. $\sqrt{2x + 15} = 3$
14. $\sqrt{3x + 7} = 2$
15. $\sqrt{5a - 4} = 4$
16. $\sqrt{2a - 1} = 3$
17. $3\sqrt{x} + 11 = 2$
18. $5\sqrt{x} + 18 = 8$
19. $4\sqrt{x} - 7 = 1$
20. $2\sqrt{x} - 5 = 9$
21. $\sqrt{x - 2} + 9 = 14$
22. $\sqrt{x + 5} + 4 = 10$
23. $\sqrt{3y + 19} - 11 = -7$
24. $\sqrt{2y + 17} - 8 = -5$
25. $9 + \sqrt{5x + 1} = 15$
26. $3 - \sqrt{6x - 5} = 6$
27. $\sqrt{x - 1} = x - 3$
28. $\sqrt{x - 3} = x - 5$
29. $\sqrt{x + 18} = x + 6$
30. $\sqrt{x + 9} = x + 7$
31. $\sqrt{3y + 13} = y + 5$
32. $\sqrt{7y + 2} = y + 2$
33. $\sqrt{5a - 16} = a - 2$
34. $\sqrt{2a - 9} = a - 6$
35. $\sqrt{7x + 15} = x + 3$
36. $\sqrt{2x - 5} = x - 4$

Summary and Review

Key Terms

[17.1] The expression $\sqrt[n]{a}$ is called a **radical** and it indicates the nth root of a. The number a is called the **radicand**, the number n is called the **index**, and the symbol $\sqrt{\ }$ is called a **radical sign**.

A **radical expression** is in **simplified form** if there are no perfect squares that are factors of the quantity under the square root sign, and there are no radicals in the denominator.

[17.3] For any positive numbers a and b, the expressions $\sqrt{a} + \sqrt{b}$ and $\sqrt{a} - \sqrt{b}$ are said to be **conjugates** of one another.

Properties

[17.1] 1. For any positive numbers a and b,

$$\sqrt{a \cdot b} = \sqrt{a} \cdot \sqrt{b}$$

2. For any positive numbers a and b,

$$\sqrt{\frac{a}{b}} = \frac{\sqrt{a}}{\sqrt{b}}$$

3. For any number a,

$$\sqrt{a} \cdot \sqrt{a} = a$$

[17.4] 4. For any numbers a and b,

if $a = b$, then $a^2 = b^2$.

Calculations

[17.1] **To simplify the square root of a product**, rewrite the radicand as the product of two factors, so that the first is the largest perfect square that is a factor of the radicand. Express the square root of the product as the product of the square roots of the factors, and find the square root of the factor that is a perfect square.

To simplify the square root of a fraction (or quotient), rewrite the expression as the square root of the numerator divided by the square root of the denominator, and simplify the numerator and denominator separately. If there is a square root in the denominator, rationalize the denominator by multiplying the numerator and denominator by the square root in the denominator.

[17.2] **To add or subtract radical expressions**, simplify all radicals not in simplified form and add or subtract terms containing like radicals using the distributive property.

[17.3] **To multiply two square roots**, multiply the coefficients of the radicals and take the square root of the product of the radicands.

To multiply radical expressions that contain more than one term, use the distributive property or the FOIL method.

To divide two square roots, divide the coefficients of the radicals and take the square root of the quotient of the radicands.

To rationalize a denominator containing a single term that includes a square root, multiply numerator and denominator by that square root. If the denominator expresses the sum or difference of two terms such that at least one term includes a square root, multiply numerator and denominator by the conjugate of the denominator.

[17.4] **To solve a radical equation**, rewrite the equation so that the radical expression is isolated on the left, and square both sides of the equation. Solve the resulting equation for the unknown and check all solutions in the original equation.

Chapter 17 Review Exercises

[17.1] Simplify each of the following radical expressions.

1. $\sqrt{64x^2y^6}$
2. $\sqrt{81x^6y^4}$
3. $\sqrt[3]{27y^3}$
4. $\sqrt[3]{8y^7}$
5. $\sqrt{50w^5}$
6. $\sqrt{45t^3}$
7. $\sqrt{72a^3b^4}$
8. $\sqrt{20ab^5}$
9. $\sqrt{\frac{9x^4}{64}}$
10. $\sqrt{\frac{4x^8}{49}}$
11. $\sqrt{\frac{xy^3}{18}}$
12. $\sqrt{\frac{x^2y^5}{32}}$

[17.2, 17.3] Perform each of the operations indicated. Express all answers in simplified form.

13. $5\sqrt{7} + 4\sqrt{7} - \sqrt{7}$

14. $9\sqrt{3} - \sqrt{3} + 8\sqrt{3}$

15. $3\sqrt{72} + 3\sqrt{80}$

16. $5\sqrt{75} - 2\sqrt{27}$

17. $3y\sqrt{x^4y} - x\sqrt{x^2y^3}$

18. $8\sqrt{x^3y^6} + xy^2\sqrt{xy^2}$

19. $2\sqrt{3} \cdot 4\sqrt{6}$

20. $3\sqrt{10} \cdot 9\sqrt{2}$

21. $5\sqrt{2}(3\sqrt{2} + \sqrt{8})$

22. $\sqrt{3}(5\sqrt{6} - 2\sqrt{3})$

23. $(8 - 7\sqrt{5})(3 + 2\sqrt{5})$

24. $(6 + 3\sqrt{7})(5 + 4\sqrt{7})$

25. $\dfrac{\sqrt{72}}{\sqrt{2}}$

26. $\dfrac{\sqrt{45}}{\sqrt{5}}$

27. $\dfrac{\sqrt{27} - \sqrt{12}}{\sqrt{3}}$

28. $\dfrac{\sqrt{50} + \sqrt{18}}{\sqrt{2}}$

29. $\dfrac{5\sqrt{6}}{\sqrt{54}}$

30. $\dfrac{4\sqrt{3}}{\sqrt{48}}$

31. $\dfrac{4}{7 + \sqrt{3}}$

32. $\dfrac{3}{8 - \sqrt{2}}$

33. $\dfrac{8\sqrt{2}}{\sqrt{7} - \sqrt{2}}$

34. $\dfrac{4\sqrt{3}}{\sqrt{3} + \sqrt{5}}$

[17.4] Solve each of the following radical equations.

35. $\sqrt{3x - 2} = 4$

36. $\sqrt{2x + 3} = 5$

37. $2\sqrt{y} - 11 = 5$

38. $3\sqrt{y} - 8 = 7$

39. $\sqrt{5x + 6} = x + 2$

40. $\sqrt{2x - 3} = x - 3$

18 QUADRATIC EQUATIONS

In this chapter, we learn some additional techniques for solving quadratic equations. As we have seen, a quadratic equation is an equation that can be written in the form $ax^2 + bx + c = 0$ where a, b, and c are constants and $a \neq 0$.

In Section 13.4 we learned how to solve a quadratic equation by factoring the left-hand side and setting each factor equal to zero. Though factoring is one of the simplest techniques for solving a quadratic equation, it is possible to use this method only when the equation can be factored. To solve quadratic equations that cannot be factored, we must learn a more general technique called *completing the square*. This technique is then used to derive a formula called the *quadratic formula*, which can also be used to solve any quadratic equation, regardless of whether or not it can be factored.

18.1 Solving Quadratic Equations Involving Perfect Squares

In this section, we begin to lay the groundwork for learning the technique of completing the square by learning a technique for solving quadratic equations involving perfect squares. To solve a quadratic equation in the form $x^2 = k$, where the left-hand side is a perfect square and the right-hand side is a constant, we can simply take the square root of both sides of the equation to obtain the following:

$$x^2 = k$$
$$\sqrt{x^2} = \pm\sqrt{k}$$
$$x = \pm\sqrt{k}$$

Note that the equation $x^2 = k$ has two solutions, a positive and a negative square root: $x = \sqrt{k}$ and $x = -\sqrt{k}$. We indicate this by writing $x = \pm\sqrt{k}$. In all of the problems we will be solving in this chapter, the constant k will be a positive number. If k were a negative number, the solution would be the square root of a negative number, which is an imaginary number. You will learn how to solve such equations in more advanced algebra courses.

We now illustrate this technique with the following examples.

example 1

Solve $x^2 = 49$ for x.

$$x^2 = 49$$
$$\sqrt{x^2} = \pm\sqrt{49} \qquad \text{Take the square root of both sides.}$$
$$x = \pm 7 \qquad \text{There are two solutions: } x = 7 \text{ or } x = -7.$$

CHECK $x^2 = 49$ — Check each solution in the original equation.

$x = 7$: $7^2 \ ?\ 49$

$49 = 49$ ✓ — The solution $x = 7$ checks.

$x = -7$: $(-7)^2 \ ?\ 49$

$49 = 49$ ✓ — The solution $x = -7$ checks.

The solution set is $\{7, -7\}$. ●

example 2 Solve $5a^2 + 3 = 18$ for a.

solution

$5a^2 + 3 = 18$ — First rewrite the equation in the form $a^2 = k$:

$5a^2 = 15$ — Add -3 to both sides.

$a^2 = 3$ — Multiply both sides by $\frac{1}{5}$.

$\sqrt{a^2} = \pm\sqrt{3}$ — Take the square root of both sides.

$a = \pm\sqrt{3}$ — There are two solutions: $a = \sqrt{3}$ or $a = -\sqrt{3}$.

CHECK $5a^2 + 3 = 18$ — Check each solution in the original equation.

$a = \sqrt{3}$: $(5\sqrt{3}) + 3 \ ?\ 18$

$5(3) + 3 \ ?\ 18$

$18 = 18$ ✓ — The solution $a = \sqrt{3}$ checks.

$a = -\sqrt{3}$: $5(-\sqrt{3})^2 + 3 \ ?\ 18$

$5(3) + 3 \ ?\ 18$

$18 = 18$ ✓ — The solution $a = -\sqrt{3}$ checks.

The solution set is $\{\sqrt{3}, -\sqrt{3}\}$. ●

We now look at how this technique can be applied to quadratic equations in which the square of a binomial is on the left and a constant is on the right.

example 3 Solve $(t - 3)^2 = 25$ for t.

solution

$(t - 3)^2 = 25$

$\sqrt{(t - 3)^2} = \pm\sqrt{25}$ — Take the square root of both sides.

$t - 3 = \pm 5$

$t = 3 \pm 5$ — Add 3 to both sides.

$t = 3 + 5$ or $t = 3 - 5$ — Rewrite the two solutions separately and solve for t.

$t = 8$ or $t = -2$

CHECK $(t - 3)^2 = 25$ — Check each solution in the original equation.

$t = 8$: $(8 - 3)^2 \ ?\ 25$

$5^2 \ ?\ 25$

$25 = 25$ ✓ — The solution $t = 8$ checks.

$t = -2$: $(-2-3)^2 \ ? \ 25$
$(-5)^2 \ ? \ 25$
$25 = 25 \ \checkmark$ — The solution $t = -2$ checks.

The solution set is $\{8, -2\}$.

example 4

solution

Solve $(5y+2)^2 = 9$ for y.

$(5y+2)^2 = 9$

$\sqrt{(5y+2)^2} = \pm\sqrt{9}$ — Take the square root of both sides.

$5y + 2 = \pm 3$

$5y = -2 \pm 3$ — Add -2 to both sides.

$y = \dfrac{-2 \pm 3}{5}$ — Multiply both sides by $\frac{1}{5}$.

$y = \dfrac{-2+3}{5}$ or $y = \dfrac{-2-3}{5}$ — Rewrite the two solutions separately and solve for y.

$y = \dfrac{1}{5}$ or $y = -\dfrac{5}{5}$

$y = \dfrac{1}{5}$ or $y = -1$

CHECK $(5y+2)^2 = 9$ — Check each solution in the original equation.

$y = \frac{1}{5}$: $\left[5\left(\frac{1}{5}\right) + 2\right]^2 \ ? \ 9$

$(1+2)^2 \ ? \ 9$

$3^2 \ ? \ 9$

$9 = 9 \ \checkmark$ — The solution $y = \frac{1}{5}$ checks.

$y = -1$: $[5(-1)+2]^2 \ ? \ 9$

$(-5+2)^2 \ ? \ 9$

$(-3)^2 \ ? \ 9$

$9 = 9 \ \checkmark$ — The solution $y = -1$ checks.

The solution set is $\left\{\frac{1}{5}, -1\right\}$.

example 5

solution

Solve $(2x-7)^2 - 5 = 3$ for x.

$(2x-7)^2 - 5 = 3$ — Rewrite so that the square of the binomial is isolated on the left.

$(2x-7)^2 = 8$ — Add 5 to both sides.

$\sqrt{(2x-7)^2} = \pm\sqrt{8}$ — Take the square root of both sides.

$2x - 7 = \pm 2\sqrt{2}$ — Simplify: $\sqrt{8} = \sqrt{4}\sqrt{2} = 2\sqrt{2}$.

$2x = 7 \pm 2\sqrt{2}$ — Add 7 to both.

$x = \dfrac{7 \pm 2\sqrt{2}}{2}$ — Multiply both sides by $\frac{1}{2}$.

$$x = \frac{7 + 2\sqrt{2}}{2} \quad \text{or} \quad x = \frac{7 - 2\sqrt{2}}{2}$$

CHECK $(2x - 7)^2 - 5 = 3$

Check each solution in the original equation.

$$x = \frac{7 + 2\sqrt{2}}{2}: \quad \left[2\left(\frac{7 + 2\sqrt{2}}{2}\right) - 7\right]^2 - 5 \ ? \ 3$$

$$(7 + 2\sqrt{2} - 7)^2 - 5 \ ? \ 3$$

$$(2\sqrt{2})^2 - 5 \ ? \ 3$$

$$2^2(2) - 5 \ ? \ 3$$

$$8 - 5 \ ? \ 3$$

$$3 = 3 \ \surd$$

The solution $x = \frac{7 + 2\sqrt{2}}{2}$ checks.

$$x = \frac{7 - 2\sqrt{2}}{2}: \quad \left[2\left(\frac{7 - 2\sqrt{2}}{2}\right) - 7\right]^2 - 5 \ ? \ 3$$

$$(7 - 2\sqrt{2} - 7)^2 - 5 \ ? \ 3$$

$$(-2\sqrt{2})^2 - 5 \ ? \ 3$$

$$(-2)^2(2) - 5 \ ? \ 3$$

$$8 - 5 \ ? \ 3$$

$$3 = 3 \ \surd$$

The solution $x = \frac{7 - 2\sqrt{2}}{2}$ checks.

The solution set is $\left\{\frac{7 + 2\sqrt{2}}{2}, \frac{7 - 2\sqrt{2}}{2}\right\}$. ●

WARNING

A common mistake in using this technique to solve quadratic equations is to forget the negative square root when taking the square root of both sides of the equation. If $x^2 = k$, $x = \sqrt{k}$ or $x = -\sqrt{k}$. Whenever you take the square root of both sides of an equation, remember that there are two roots, a positive and a negative one.

QUICK QUIZ

Solve each of the following quadratic equations by the square root method.

1. $2x^2 - 7 = 3$
2. $(a + 6)^2 = 12$
3. $(3t - 2)^2 = 4$

ANSWERS

1. $x = \pm\sqrt{5}$
2. $a = -6 \pm 2\sqrt{3}$
3. $t = 0$ or $t = \frac{4}{3}$

18.1 Exercises

Solve each of the following quadratic equations by the square root method and check your solutions.

1. $x^2 = 4$ **2.** $x^2 = 36$ **3.** $z^2 = 63$ **4.** $z^2 = 20$

5. $7y^2 = 28$ **6.** $5y^2 = 45$ **7.** $a^2 + 5 = 86$ **8.** $a^2 + 3 = 19$

9. $t^2 - 6 = 19$
10. $t^2 - 8 = 41$
11. $x^2 + 3 = 21$
12. $x^2 - 4 = 68$
13. $3a^2 - 2 = 10$
14. $2a^2 + 5 = 37$
15. $4t^2 + 7 = 87$
16. $5t^2 - 6 = 34$
17. $(x - 3)^2 = 36$
18. $(x - 7)^2 = 81$
19. $(y + 9)^2 = 49$
20. $(y + 2)^2 = 64$
21. $(z + 6)^2 = 27$
22. $(z + 8)^2 = 48$
23. $(t - 1)^2 = 50$
24. $(t - 5)^2 = 28$
25. $(2x - 5)^2 = 1$
26. $(3x + 8)^2 = 16$
27. $(7y + 2)^2 = 25$
28. $(5y - 1)^2 = 36$
29. $(3t - 4)^2 = 64$
30. $(7z + 4)^2 = 100$
31. $(4a + 1)^2 = 75$
32. $(2a + 5)^2 = 72$
33. $(8t - 3)^2 = 48$
34. $(6t - 1)^2 = 27$
35. $(a - 7)^2 - 3 = 13$
36. $(a + 8)^2 - 9 = 40$
37. $(5t + 4)^2 + 7 = 88$
38. $(4t + 1)^2 + 3 = 28$
39. $(2x - 3)^2 - 9 = 23$
40. $(3x - 5)^2 - 8 = 46$

18.2 Completing the Square

In this section, we learn a technique which will enable us to solve any quadratic equation, even those that cannot be factored. This technique, called *completing the square*, requires us to rewrite a quadratic equation so that the square of a binomial appears on the left and a constant appears on the right. We will then use the technique presented in Section 18.1 to solve the resulting equation.

Many trinomials can also be expressed as the square of a binomial. These trinomials are called perfect squares. Consider the following examples:

$$x^2 + 6x + 9 = (x + 3)^2$$
$$x^2 - 10x + 25 = (x - 5)^2$$
$$x^2 + \frac{4}{3}x + \frac{4}{9} = \left(x + \frac{2}{3}\right)^2$$

Notice that in each of these trinomials, the coefficient of x^2 is 1. Also notice that in each case, the constant term is equal to the square of one-half of the coefficient of x:

$$x^2 + 6x + 9: \quad \left[\frac{1}{2}(6)\right]^2 = 3^2 = 9$$
$$x^2 - 10x + 25: \quad \left[\frac{1}{2}(-10)\right]^2 = (-5)^2 = 25$$
$$x^2 + \frac{4}{3}x + \frac{4}{9}: \quad \left[\frac{1}{2}\left(\frac{4}{3}\right)\right]^2 = \left(\frac{2}{3}\right)^2 = \frac{4}{9}$$

If we are given a binomial in the form $x^2 + bx$ where the coefficient of x^2 is 1, we can add the square of one-half the coefficient of x to the expression and the resulting trinomial will be a perfect square. This process is called *completing the square.*

To complete the square given a binomial in the form $x^2 + bx$, add to the binomial $\left(\frac{1}{2} \cdot b\right)^2$, which is the square of one-half the coefficient of x.

example 1

Add the appropriate number to each of the following binomials to complete the square. Then express the resulting trinomial as the square of a binomial.

(a) $x^2 + 14x +$ ____

$$\left[\frac{1}{2}(14)\right]^2 = 7^2 = 49$$ The coefficient of x is 14.

$$x^2 + 14x + 49 = (x + 7)^2$$ Add 49 to the binomial.

(b) $x^2 - 18x +$ ____

$$\left[\frac{1}{2}(-18)\right]^2 = (-9)^2 = 81$$ The coefficient of x is -18.

$$x^2 - 18x + 81 = (x - 9)^2$$ Add 81 to the binomial.

(c) $x^2 + x +$ ____

$$\left[\frac{1}{2}(1)\right]^2 = \left(\frac{1}{2}\right)^2 = \frac{1}{4}$$ The coefficient of x is 1.

$$x^2 + x + \frac{1}{4} = \left(x + \frac{1}{2}\right)^2$$ Add $\frac{1}{4}$ to the binomial. ●

We now describe the procedure of completing the square that can be used to solve quadratic equations.

To Solve a Quadratic Equation by Completing the Square

1. Rewrite the equation in the form $ax^2 + bx = c$ so that the variable terms are on the left and the constant term is on the right.
2. If the coefficient of x^2 is not equal to 1, divide both sides of the equation by that number.
3. Add the square of one-half the resulting coefficient of x, $\left(\frac{1}{2}\frac{b}{a}\right)^2$, to both sides of the equation to complete the square.
4. Rewrite the trinomial on the left as the square of a binomial.
5. Take the square root of both sides of the equation and solve for x.

We now illustrate this procedure with the following examples.

example 2

Solve $x^2 - 6x - 16 = 0$ by completing the square.

solution

$$x^2 - 6x - 16 = 0$$

$$x^2 - 6x = 16$$ Add 16 to both sides.

$$x^2 - 6x + 9 = 16 + 9$$ Add $\left[\frac{1}{2}(-6)\right]^2 = (-3)^2 = 9$ to both sides.

$(x-3)^2 = 25$ — Rewrite the left as the square of a binomial.

$\sqrt{(x-3)^2} = \pm\sqrt{25}$ — Take the square root of both sides.

$x - 3 = \pm 5$

$x = 3 \pm 5$ — Add 3 to both sides.

$x = 3 + 5$ or $x = 3 - 5$

$x = 8$ or $x = -2$

CHECK $x^2 - 6x - 16 = 0$ — Check each solution in the original equation.

$x = 8$: $8^2 - 6(8) - 16 \ ?\ 0$

$64 - 48 - 16 \ ?\ 0$

$0 = 0 \ \checkmark$ — The solution $x = 8$ checks.

$x = -2$: $(-2)^2 - 6(-2) - 16 \ ?\ 0$

$4 + 12 - 16 \ ?\ 0$

$0 = 0 \ \checkmark$ — The solution $x = -2$ checks.

The solution set is $\{8, -2\}$.

example 3

Solve $4x^2 + 4x - 3 = 0$ by completing the square.

solution

$4x^2 + 4x - 3 = 0$

$4x^2 + 4x = 3$ — Add 3 to both sides.

$x^2 + x = \frac{3}{4}$ — Divide both sides by 4.

$x^2 + x + \frac{1}{4} = \frac{3}{4} + \frac{1}{4}$ — Add $\left[\frac{1}{2}(1)\right]^2 = \frac{1}{4}$ to both sides.

$\left(x + \frac{1}{2}\right)^2 = 1$ — Rewrite the left as the square of a binomial.

$\sqrt{\left(x + \frac{1}{2}\right)^2} = \pm\sqrt{1}$ — Take the square root of both sides.

$x + \frac{1}{2} = \pm 1$

$x = -\frac{1}{2} \pm 1$ — Add $-\frac{1}{2}$ to both sides.

$x = -\frac{1}{2} + 1$ or $x = -\frac{1}{2} - 1$

$x = \frac{1}{2}$ or $x = -\frac{3}{2}$

CHECK $4x^2 + 4x - 3 = 0$ — Check each solution in the original equation.

$$x = \frac{1}{2}: \quad 4\left(\frac{1}{2}\right)^2 + 4\left(\frac{1}{2}\right) - 3 \ ? \ 0$$

$$4\left(\frac{1}{4}\right) + 2 - 3 \ ? \ 0$$

$$1 + 2 - 3 \ ? \ 0$$

$$0 = 0 \ \checkmark$$

The solution $x = \frac{1}{2}$ checks.

$$x = -\frac{3}{2}: \quad 4\left(-\frac{3}{2}\right)^2 + 4\left(-\frac{3}{2}\right) - 3 \ ? \ 0$$

$$4\left(\frac{9}{4}\right) + \left(-\frac{12}{2}\right) - 3 \ ? \ 0$$

$$9 - 6 - 3 \ ? \ 0$$

$$0 = 0 \ \checkmark$$

The solution $x = -\frac{3}{2}$ checks.

The solution set is $\left\{\frac{1}{2}, -\frac{3}{2}\right\}$.

example 4 Solve $3x^2 - 2x - 6 = 0$ by completing the square.

solution

$$3x^2 - 2x - 6 = 0$$

Add 6 to both sides.

$$3x^2 - 2x = 6$$

$$x^2 - \frac{2}{3}x = 2$$

Divide both sides by 3.

$$x^2 - \frac{2}{3}x + \frac{1}{9} = 2 + \frac{1}{9}$$

Add $\left[\frac{1}{2}\left(-\frac{2}{3}\right)\right]^2 = \left(-\frac{1}{3}\right) = \frac{1}{9}$ to both sides.

$$\left(x - \frac{1}{3}\right)^2 = \frac{18}{9} + \frac{1}{9}$$

Rewrite the left as the square of a binomial.

$$\left(x - \frac{1}{3}\right)^2 = \frac{19}{9}$$

$$\sqrt{\left(x - \frac{1}{3}\right)^2} = \pm\sqrt{\frac{19}{9}}$$

Take the square root of both sides.

$$x - \frac{1}{3} = \pm\frac{\sqrt{19}}{3}$$

$$x = \frac{1}{3} \pm \frac{\sqrt{19}}{3}$$

Add $\frac{1}{3}$ to both sides.

$$x = \frac{1}{3} + \frac{\sqrt{19}}{3} \quad \text{or} \quad x = \frac{1}{3} - \frac{\sqrt{19}}{3}$$

$$x = \frac{1 + \sqrt{19}}{3} \quad \text{or} \quad x = \frac{1 - \sqrt{19}}{3}$$

CHECK $3x^2 - 2x - 6 = 0$

$x = \frac{1 + \sqrt{19}}{3}$: $\quad 3\left(\frac{1 + \sqrt{19}}{3}\right)^2 - 2\left(\frac{1 + \sqrt{19}}{3}\right) - 6 \ ? \ 0$

$$3\left(\frac{1 + 2\sqrt{19} + 19}{9}\right) - \frac{2 + 2\sqrt{19}}{3} - 6 \ ? \ 0$$

$$\frac{1 + 2\sqrt{19} + 19}{3} - \frac{2 + 2\sqrt{19}}{3} - \frac{18}{3} \ ? \ 0$$

$$\frac{1 + 2\sqrt{19} + 19 - 2 - 2\sqrt{19} - 18}{3} \ ? \ 0$$

$$\frac{0}{3} \ ? \ 0$$

$$0 = 0 \ \checkmark$$

$x = \frac{1 - \sqrt{19}}{3}$: $\quad 3\left(\frac{1 - \sqrt{19}}{3}\right)^2 - 2\left(\frac{1 - \sqrt{19}}{3}\right) - 6 \ ? \ 0$

$$3\left(\frac{1 - 2\sqrt{19} + 19}{9}\right) - \frac{2 - 2\sqrt{19}}{3} - 6 \ ? \ 0$$

$$\frac{1 - 2\sqrt{19} + 19}{3} - \frac{2 - 2\sqrt{19}}{3} - \frac{18}{3} \ ? \ 0$$

$$\frac{1 - 2\sqrt{19} + 19 - 2 + 2\sqrt{19} - 18}{3} \ ? \ 0$$

$$\frac{0}{3} \ ? \ 0$$

The solution set is $\left\{\frac{1 + \sqrt{19}}{3}, \frac{1 - \sqrt{19}}{3}\right\}$. $\qquad 0 = 0 \ \checkmark$

QUICK QUIZ

Solve each of the following quadratic equations by completing the square.

1. $x^2 + 4x - 21 = 0$
2. $9y^2 - 3y - 2 = 0$
3. $2a^2 + 4a - 7 = 0$

ANSWERS

1. $x = 3, -7$
2. $y = \frac{2}{3}, -\frac{1}{3}$
3. $a = \frac{-2 \pm 3\sqrt{2}}{2}$

18.2 Exercises

Add the appropriate number to each of the following binomials to complete the square. Then express the resulting trinomial as the square of a binomial.

1. $x^2 + 8x +$ ____ **2.** $x^2 + 6x +$ ____ **3.** $x^2 - 14x +$ ____ **4.** $x^2 - 2x +$ ____

5. $y^2 + 10y +$ ____ **6.** $y^2 + 18y +$ ____ **7.** $y^2 - 4y +$ ____ **8.** $y^2 - 16y +$ ____

9. $a^2 - 12a + \_\_\_$ **10.** $a^2 + 20a + \_\_\_$ **11.** $t^2 + 5t + \_\_\_$ **12.** $t^2 + t + \_\_\_$

13. $x^2 - 3x + \_\_\_$ **14.** $x^2 - 9x + \_\_\_$ **15.** $a^2 + \frac{1}{3}a + \_\_\_$ **16.** $a^2 - \frac{1}{4}a + \_\_\_$

17. $y^2 - \frac{1}{2}y + \_\_\_$ **18.** $y^2 + \frac{2}{5}y + \_\_\_$ **19.** $x^2 + \frac{3}{4}x + \_\_\_$ **20.** $x^2 - \frac{2}{3}x + \_\_\_$

Solve each of the following quadratic equations by completing the square.

21. $x^2 + 12x + 35 = 0$ **22.** $x^2 + 13x + 36 = 0$ **23.** $y^2 - 5y - 14 = 0$ **24.** $y^2 - 2y - 15 = 0$
25. $a^2 + 6a - 27 = 0$ **26.** $a^2 + 7a - 8 = 0$ **27.** $t^2 - 8t + 15 = 0$ **28.** $t^2 - 11t + 18 = 0$
29. $x^2 + 15x + 56 = 0$ **30.** $x^2 + 16x + 63 = 0$ **31.** $y^2 - 3y - 10 = 0$ **32.** $y^2 - 4y - 45 = 0$
33. $a^2 + 4a - 21 = 0$ **34.** $a^2 + 3a - 54 = 0$ **35.** $a^2 - 13a + 40 = 0$ **36.** $a^2 - 15a + 56 = 0$
37. $2x^2 - 9x - 5 = 0$ **38.** $3x^2 - 10x - 8 = 0$ **39.** $5y^2 - 11y + 2 = 0$ **40.** $4y^2 + y - 3 = 0$
41. $6t^2 + 7t - 3 = 0$ **42.** $6t^2 + 11t - 10 = 0$ **43.** $2x^2 + 2x - 5 = 0$ **44.** $3x^2 + 3x - 4 = 0$
45. $3x^2 - 6x + 2 = 0$ **46.** $4x^2 - 8x + 3 = 0$ **47.** $4x^2 + 2x - 3 = 0$ **48.** $9x^2 + 3x - 1 = 0$
49. $2x^2 + 3x - 4 = 0$ **50.** $4x^2 + 3x - 4 = 0$

18.3 The Quadratic Formula

Since we can use the technique of completing the square to solve any quadratic equation, we can use this method to solve the general quadratic equation, $ax^2 + bx + c = 0$. The result of this process is called the **quadratic formula.**

> To solve a quadratic equation in the form $ax^2 + bx + c = 0$, where a, b, and c are constants and $a \neq 0$, using the *quadratic formula*, substitute a, b, and c into the following equation and solve for x:
>
> $$x = \frac{-b \pm \sqrt{b^2 - 4ac}}{2a}$$

Notice that the result of the quadratic formula produces two values for x:

$$x = \frac{-b + \sqrt{b^2 - 4ac}}{2a} \quad \text{or} \quad x = \frac{-b - \sqrt{b^2 - 4ac}}{2a}$$

Once we have memorized the quadratic formula, it is no longer necessary to go through the process of completing the square to solve a quadratic equation.

We now illustrate how the quadratic formula is derived by solving the equation $ax^2 + bx + c = 0$ using the method of completing the square.

$$ax^2 + bx + c = 0$$

$$ax^2 + bx = -c \qquad \text{Add } -c \text{ to both sides.}$$

$$x^2 + \frac{b}{a}x = -\frac{c}{a} \qquad \text{Divide both sides by } a.$$

$$x^2 + \frac{b}{a}x + \frac{b^2}{4a^2} = \frac{-c}{a} + \frac{b^2}{4a^2} \qquad \text{Add } \left(\frac{1}{2}\cdot\frac{b}{a}\right)^2 = \frac{b^2}{4a^2} \text{ to both sides.}$$

$$\left(x + \frac{b}{2a}\right)^2 = \frac{-c}{a} \cdot \frac{4a}{4a} + \frac{b^2}{4a^2}$$

Rewrite the left as the square of a binomial. The LCD of $\frac{-c}{a}$ and $\frac{b^2}{4a^2}$ is $4a^2$.

$$\left(x + \frac{b}{2a}\right)^2 = \frac{-4ac + b^2}{4a^2}$$

Add the terms on the right.

$$\sqrt{\left(x + \frac{b}{2a}\right)^2} = \pm\sqrt{\frac{b^2 - 4ac}{4a^2}}$$

Take the square root of both sides.

$$x + \frac{b}{2a} = \pm\frac{\sqrt{b^2 - 4ac}}{\sqrt{4a^2}}$$

Rewrite the square root of the quotient on the right.

$$x = \frac{-b}{2a} \pm \frac{\sqrt{b^2 - 4ac}}{2a}$$

Add $-\frac{b}{2a}$ to both sides. $\sqrt{4a^2} = 2a$.

$$x = \frac{-b \pm \sqrt{b^2 - 4ac}}{2a}$$

Combine the terms on the right.

The last step is the quadratic formula. ●

Let us now look at an example which illustrates how the quadratic formula is used to solve a quadratic equation.

example 1 Solve $2x^2 - 7x - 4 = 0$ for x using the quadratic formula.

solution

$$2x^2 - 7x - 4 = 0$$

The equation is in standard form: $ax^2 + bx + c = 0$.

$$a = 2 \qquad b = -7 \qquad c = -4$$

Identify a, b, and c.

$$x = \frac{-b \pm \sqrt{b^2 - 4ac}}{2a}$$

Write down the quadratic formula.

$$x = \frac{-(-7) \pm \sqrt{(-7)^2 - 4(2)(-4)}}{2(2)}$$

Substitute in $a = 2$, $b = -7$, and $c = -4$.

$$x = \frac{7 \pm \sqrt{49 + 32}}{4}$$

$$x = \frac{7 \pm \sqrt{81}}{4}$$

$$x = \frac{7 \pm 9}{4}$$

$$x = \frac{7 + 9}{4} \quad \text{or} \quad x = \frac{7 - 9}{4}$$

$$x = \frac{16}{4} \quad \text{or} \quad x = -\frac{2}{4}$$

$$x = 4 \quad \text{or} \quad x = -\frac{1}{2}$$

We leave checking the solution to the student. The solution set is $\left\{4, -\frac{1}{2}\right\}$. ●

In the quadratic formula, the quantity under the radical sign, $b^2 - 4ac$, is called the *discriminant*.

$$x = \frac{-b \pm \sqrt{b^2 - 4ac}}{2a} \longleftarrow \text{discriminant}$$

The value of the discriminant indicates the kind of solution to a quadratic equation.

1. If the discriminant equals zero, the quadratic equation has one solution which is a real number.
2. If the discriminant is positive, the quadratic equation has two solutions which are real numbers.
3. If the discriminant is negative, the quadratic equation has two solutions which are not real numbers. This is because each solution will contain the square root of a negative number.

In this book, we deal only with quadratic equations of the first two types. You will work with examples of the third type in more advanced algebra courses.

Let us now look at some additional examples which illustrate the use of the quadratic formula.

example 2

Solve $3t^2 = 1 - 5t$ for t using the quadratic formula.

solution

$3t^2 = 1 - 5t$

$3t^2 + 5t - 1 = 0$ — Rewrite in standard form: $at^2 + bt + c = 0$.

$a = 3 \qquad b = 5 \qquad c = -1$ — Identify a, b, and c.

$$t = \frac{-b \pm \sqrt{b^2 - 4ac}}{2a}$$ Write the quadratic formula.

$$t = \frac{-5 \pm \sqrt{(5)^2 - 4(3)(-1)}}{2(3)}$$ Substitute in the values for a, b, and c.

$$t = \frac{-5 \pm \sqrt{25 + 12}}{6}$$

$$t = \frac{-5 \pm \sqrt{37}}{6}$$

The solution set is $\left\{\frac{-5 + \sqrt{37}}{6}, \frac{-5 - \sqrt{37}}{6}\right\}$. ●

example 3

Find the roots of the equation $2x^2 = 3(2x - 1)$ using the quadratic formula.

solution

$2x^2 = 3(2x - 1)$

$2x^2 = 6x - 3$ — Rewrite in standard form: $ax^2 + bx + c = 0$.

$2x^2 - 6x + 3 = 0$

$a = 2 \qquad b = -6 \qquad c = 3$ — Identify a, b, and c.

$$x = \frac{-b \pm \sqrt{b^2 - 4ac}}{2a}$$ Write the quadratic formula.

$$x = \frac{-(-6) \pm \sqrt{(-6)^2 - 4(2)(3)}}{2(2)}$$ Substitute in the values $a = 2$, $b = -6$, and $c = 3$.

$$x = \frac{6 \pm \sqrt{36 - 24}}{4}$$

$$x = \frac{6 \pm \sqrt{12}}{4}$$

$$x = \frac{6 \pm 2\sqrt{3}}{4}$$ Simplify $\sqrt{12} = \sqrt{4}\sqrt{3} = 2\sqrt{3}$.

$$x = \frac{3 \pm \sqrt{3}}{2}$$ Divide numerator and denominator by 2.

The roots of the equation are $x = \dfrac{3 + \sqrt{3}}{2}$ and $x = \dfrac{3 - \sqrt{3}}{2}$. ●

QUICK QUIZ

Solve each of the following quadratic equations using the quadratic formula.

1. $x^2 - 7x - 18 = 0$
2. $2y^2 + 7y = 15$
3. $4x(2 - x) = 1$

ANSWERS

1. $x = -2, 9$
2. $y = \frac{3}{2}, -5$
3. $x = \dfrac{2 \pm \sqrt{3}}{2}$

18.3 Exercises

Solve each of the following quadratic equations using the quadratic formula.

1. $x^2 + 7x + 12 = 0$ **2.** $x^2 + 8x + 15 = 0$ **3.** $y^2 + 4y - 45 = 0$ **4.** $y^2 + 4y - 21 = 0$
5. $x^2 = 8x - 15$ **6.** $x^2 = 10x - 24$ **7.** $y^2 = 6y + 27$ **8.** $y^2 = 3y + 40$
9. $a^2 = 4(3 - a)$ **10.** $a^2 = 3(6 - a)$ **11.** $x^2 = 10x - 25$ **12.** $x^2 = 4x - 4$
13. $2y^2 + 3y - 5 = 0$ **14.** $3y^2 - 2y - 8 = 0$ **15.** $3x^2 + 10x - 8 = 0$ **16.** $2x^2 - 3x - 9 = 0$
17. $2a^2 - 13a = 7$ **18.** $3a^2 - 10a = 8$ **19.** $x^2 + 5x + 5 = 0$ **20.** $x^2 + 3x + 1 = 0$
21. $3x^2 - x = 4$ **22.** $2x^2 + x = 3$ **23.** $2y^2 + 3y = 5$ **24.** $3y^2 + 5y = 1$
25. $5t^2 = 2t + 3$ **26.** $4t^2 = 5t + 6$ **27.** $6a^2 = 2(5a - 1)$ **28.** $5a^2 = 3(3a - 1)$
29. $5(x^2 + x) = 1$ **30.** $3(x^2 + 2x) = 1$

18.4 Graphing Quadratic Equations

In this section, we learn how to graph quadratic equations, which are equations of the form $y = ax^2 + bx + c$. The method we use initially is to set up a table which includes values of x, chosen arbitrarily and the corresponding values of y obtained by substituting each x value into the equation and solving for y. We will then plot each pair of coordinates (x, y) on the x-y plane and connect them with a smooth curve. The graphs of all quadratic equations have the same general shape, which is called a *parabola*. Examples 1 and 2 illustrate this method for graphing parabolas.

example 1

Graph $y = x^2 + 2$.

solution

x	y	
-2	6	$y = (-2)^2 + 2 = 4 + 2 = 6$
-1	3	$y = (-1)^2 + 2 = 1 + 2 = 3$
0	2	$y = (0)^2 + 2 = 0 + 2 = 2$
1	3	$y = (1)^2 + 2 = 1 + 2 = 3$
2	6	$y = (2)^2 + 2 = 4 + 2 = 6$

Choose values for x and substitute into the equation to find the corresponding values of y.

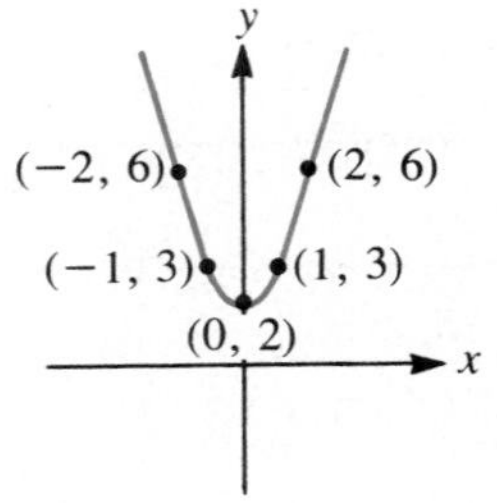

The lowest point on this parabola (0, 2) is called the *vertex*. Notice that this parabola opens upward and that the left-hand side and the right-hand side are identical. ●

example 2

Graph $y = -x^2 + 6x - 7$.

solution

x	y	
1	-2	$y = -(1)^2 + 6(1) - 7 = -1 + 6 - 7 = -2$
2	1	$y = -(2)^2 + 6(2) - 7 = -4 + 12 - 7 = 1$
3	2	$y = -(3)^2 + 6(3) - 7 = -9 + 18 - 7 = 2$
4	1	$y = -(4)^2 + 6(4) - 7 = -16 + 24 - 7 = 1$
5	-2	$y = -(5)^2 + 6(5) - 7 = -25 + 30 - 7 = -2$

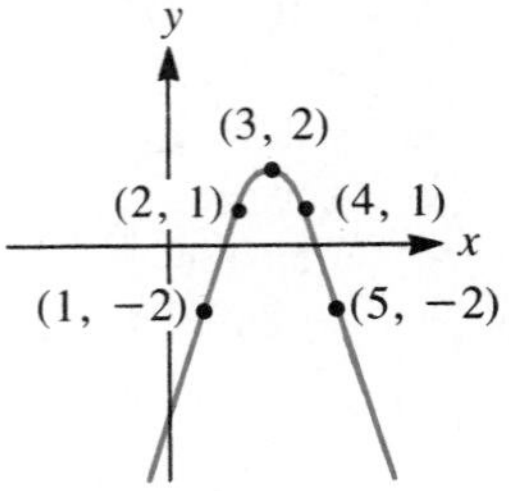

Notice that the parabola opens downward and that its vertex is the point (3, 2) the highest point on the parabola. ●

Now that we can recognize the shape of a parabola, we can save ourselves some work in graphing a quadratic equation if we can determine the vertex and whether or not the parabola opens upward or downward. The following procedure outlines a method for graphing a parabola.

To Graph a Quadratic Equation

1. If necessary, rewrite the equation in the form $y = ax^2 + bx + c$.
2. Find the x-coordinate of the vertex by substituting the values for b and a into the equation $x = -\frac{b}{2a}$.
3. Find the y-coordinate of the vertex by substituting the value found for x in step 2 into the original equation $y = ax^2 + bx + c$.
4. Determine the direction in which the parabola opens:

   If $a > 0$, the parabola opens upward.
   If $a < 0$, the parabola opens downward.
5. Plot as many additional points as are necessary to get a reasonable picture.

We now illustrate this procedure with the following examples.

example 3

Graph $y - 5 = -3x^2$.

$y - 5 = -3x^2$

$y = -3x^2 + 5$

Rewrite the equation in the form: $y = ax^2 + bx + c$.

$a = -3 \quad b = 0 \quad c = 5$

Identify a, b, and c.

$x = \frac{-b}{2a} = \frac{-0}{2(-3)} = 0$

Find the x-coordinate of the vertex.

$y = -3x^2 + 5$

$y = -3(0)^2 + 5 = 5$

Find the y-coordinate of the vertex.

The vertex is (0, 5).

The parabola opens downward.

Since $a = -3 < 0$.

$x = 2$: $y = -3(2)^2 + 5 = -7$

$x = -2$: $y = -3(-2)^2 + 5 = -7$

Find two other points on the parabola.

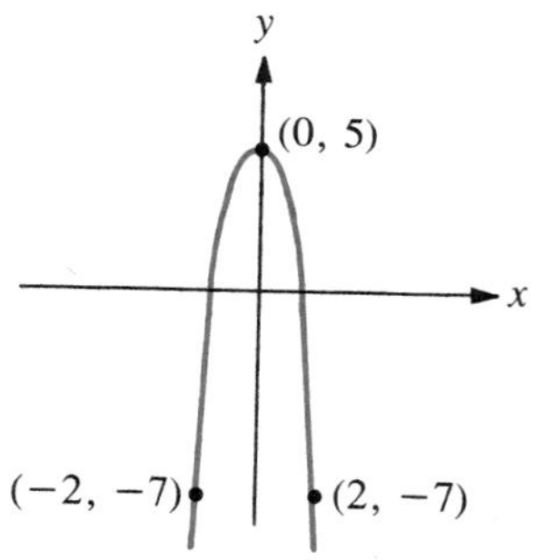

example 4 Graph $y = 2(x - 1)^2 + 3$.

solution

$y = 2(x - 1)^2 + 3$

$y = 2(x^2 - 2x + 1) + 3$

$y = 2x^2 - 4x + 2 + 3$

$y = 2x^2 - 4x + 5$

Rewrite the equation in the form: $y = ax^2 + bx + c$.

$a = 2 \qquad b = -4 \qquad c = 5$

Identify a, b, and c.

$$x = \frac{-b}{2a} = \frac{-(-4)}{2(2)} = \frac{4}{4} = 1$$

Find the x-coordinate of the vertex.

$y = 2x^2 - 4x + 5$

$y = 2(1)^2 - 4(1) + 5 = 3$

Find the y-coordinate of the vertex.

The vertex is (1, 3).

The parabola opens upward.

$x = 2$: $y = 2(2)^2 - 4(2) + 5 = 5$

$x = 0$: $y = 2(0)^2 - 4(0) + 5 = 5$

Find two other points on the parabola.

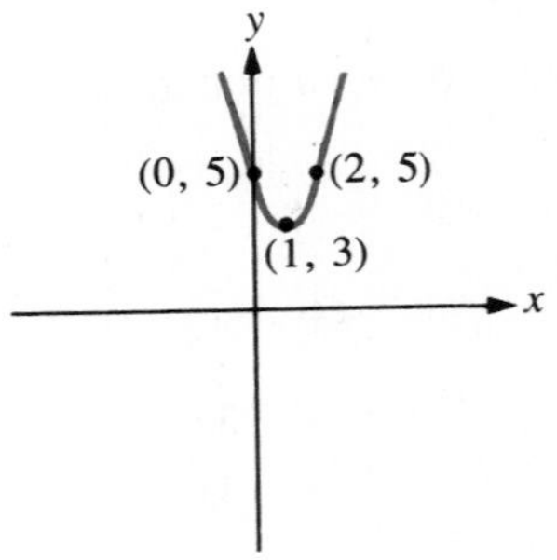

18.4 Exercises

Graph each of the following quadratic equations. Be sure to indicate the vertex of each parabola.

1. $y = x^2$
2. $y = -x^2$
3. $y = -5x^2 + 2$
4. $y = 3x^2 - 1$
5. $y = x^2 - 4x + 7$
6. $y = x^2 - 8x + 17$
7. $y = -x^2 - 6x + 16$
8. $y = -x^2 + 2x + 6$
9. $y = x^2 - 10x + 23$
10. $y = x^2 - 6x + 3$
11. $y - 48 = x^2 + 14x$
12. $y - 30 = x^2 + 12x$
13. $y - 8 = -x^2 - 2x$
14. $y + 21 = -x^2 - 10x$
15. $y = (x + 4)^2$
16. $y = (x - 2)^2$
17. $y + 3 = (x - 1)^2$
18. $y - 8 = (x + 4)^2$
19. $y - 5 = -2(x + 3)^2$
20. $y - 1 = -3(x - 2)^2$

Summary and Review

Key Terms

[18.1] A **quadratic equation** is an equation that can be written in the form $ax^2 + bx + c = 0$, where a, b, and c are constants and $a \neq 0$.

[18.4] A **parabola** is the graph of a quadratic equation.

The **vertex** is the highest or lowest point on a parabola.

Calculations

To solve a quadratic equation in the form $x^2 = k$, where $k > 0$, take the square root of both sides to obtain $x = \pm\sqrt{k}$.

[18.2] **To solve a quadratic equation by completing the square**:

1. Rewrite the equation in the form $ax^2 + bx = c$ so that the variable terms are on the left and the constant term is on the right.
2. If the coefficient of x^2 is not equal to 1, divide both sides of the equation by that number.
3. Add the square of one-half the resulting coefficient of x, $\left(\frac{1}{2}\frac{b}{a}\right)^2$, to both sides of the equation to complete the square.
4. Rewrite the trinomial on the left as the square of a binomial.
5. Take the square root of both sides of the equation and solve for x.

[18.3] **To solve a quadratic equation** in the form $ax^2 + bx + c = 0$, where a, b, and c are constants and $a \neq 0$ using the **quadratic formula**, substitute a, b, and c into this equation and solve for x:

$$x = \frac{-b \pm \sqrt{b^2 - 4ac}}{2a}$$

The **discriminant** is the quantity $b^2 - 4ac$ in the quadratic formula. If the discriminant equals zero, the quadratic equation has one solution which is a real number. If the discriminant is positive, the quadratic equation has two solutions which are real numbers. If the discriminant is negative, the quadratic equation has two solutions which are not real numbers.

[18.4] **To graph a quadratic equation** in the form $y = ax^2 + bx + c$, find the x-coordinate of the vertex, which is $\frac{-b}{2a}$, and substitute into the equation to find the y-coordinate of the vertex. The parabola opens upward if $a > 0$ and it opens downward if $a < 0$.

Chapter 18 Review Exercises

[18.1] Solve each of the following quadratic equations by the square root method.

1. $x^2 + 3 = 67$ **2.** $x^2 - 5 = 44$ **3.** $(a - 5)^2 = 36$ **4.** $(a + 2)^2 = 16$
5. $(3x + 1)^2 + 2 = 27$ **6.** $(2x - 5)^2 - 4 = 5$

[18.2] Solve each of the following quadratic equations by completing the square.

7. $x^2 - 5x - 36 = 0$ **8.** $x^2 + x - 42 = 0$ **9.** $2y^2 - 11y + 5 = 0$ **10.** $3y^2 + 14y + 8 = 0$
11. $6x^2 + 3x - 2 = 0$ **12.** $4x^2 - 2x - 5 = 0$

[18.3] Solve each of the following quadratic equations using the quadratic formula.

13. $x^2 + 3x - 10 = 0$ **14.** $x^2 - 5x - 24 = 0$ **15.** $3t^2 - 17t - 6 = 0$ **16.** $2t^2 - 7t + 5 = 0$
17. $5x^2 = x + 4$ **18.** $3x^2 = 7x - 3$

[18.4] Graph each of the following quadratic equations.

19. $y = 2x^2 + 3$ **20.** $y = 4x^2 - 1$ **21.** $y = -x^2 + 8x - 15$
22. $y = x^2 - 6x + 2$ **23.** $y + 4 = (x - 3)^2$ **24.** $y - 6 = -(x + 5)^2$

APPENDIX: HOW TO USE A CALCULATOR

There are basically two kinds of calculators—those that use algebraic logic, and those that use reverse Polish notation (RPN). Generally, a calculator with an equal sign [=] will use algebraic logic, and one with an enter key [ENTER] will use RPN. The best way to learn how to use your calculator properly is to read the instruction manual that accompanies it.

We will briefly discuss how a few arithmetic operations are performed on a calculator that uses algebraic logic. Some important keys you will use are the ten numerical keys, the four basic operations keys [+] [−] [×] [÷], the decimal point key [.], the square root key [√], and the clear key [C], which should always be pressed before beginning a new calculation in order to clear the memory. The examples in the following table illustrate the use of these keys.

OPERATION	EXAMPLE	KEYS TO PRESS	ANSWER IN DISPLAY WINDOW
Addition	$53 + 84$	[5] [3] [+] [8] [4] [=]	137.
Subtraction	$95 - 27$	[9] [5] [−] [2] [7] [=]	68.
Multiplication	5.7×8	[5] [.] [7] [×] [8] [=]	45.6
Division	$-265 \div 5$	[2] [6] [5] [+/−] [÷] [5] [=]	−53.
Square Root	$\sqrt{361}$	[3] [6] [1] [√]	19.

When more than one operation is performed in sequence, some calculators will automatically do multiplication and division before addition and subtraction, according to the order of operations rules. Others will simply perform the operations in the order in which they are entered into the calculator. Thus, pressing the sequence of keys

[5] [+] [3] [×] [4] [=]

will yield one of two possible answers, depending on your calculator. If the answer that appears in the display window is 17, your calculator automatically performed multiplication before addition.

$$5 + (3 \times 4) = 17$$

If the answer that appears in the display window is 32, your calculator performed the addition first.

$$(5 + 3) \times 4 = 32$$

Regardless of the kind of calculator you have, you can dictate the order in which you want the operations to be performed by entering the operation you want performed first and pressing the equal sign [=] before proceeding on to the next operation. For example, to calculate $(34 - 8) \div 2$, press this series of buttons:

[3] [4] [−] [8] [=] [÷] [2] [=]

The result in the display window should be 13.

ANSWERS TO ODD-NUMBERED EXERCISES

Chapter 1

1.1 Exercises, page 7

1. false; does **3.** false; 7,008 **5.** tens; fifty-eight **7.** hundreds; eight hundred six
9. thousands; four thousand thirty **11.** ten thousands; twenty-one thousand, nine
13. hundred thousands; five hundred thousand, one
15. hundred thousands; one million, four hundred ninety-six thousand, six hundred twenty-eight
17. millions; thirty-eight million, two thousand, five hundred forty
19. hundred millions; seven hundred fifteen million, one hundred fifty-one thousand, five hundred fifteen
21. billions; eighty-nine billion, six hundred forty-one million, three hundred seventy-two thousand, fifty-five
23. trillions; seven trillion, four hundred eighty-six thousand, five hundred thirty-two **25.** 47 **27.** 602 **29.** 1,066
31. 8,008 **33.** 50,600 **35.** 944,026 **37.** 7,000,008 **39.** 11,064,218 **41.** 2,053,103,481
43. 50,009,300,500 **45.** 5,705,006,474,302 **47.** $(3 \times 10) + (3 \times 1)$ **49.** $(6 \times 100) + (5 \times 1)$
51. $(4 \times 1{,}000) + (2 \times 1)$ **53.** $(1 \times 10{,}000) + (4 \times 10)$
55. $(6 \times 100{,}000) + (3 \times 10{,}000) + (8 \times 1{,}000) + (5 \times 100) + (1 \times 10) + (8 \times 1)$
57. $(3 \times 10{,}000{,}000) + (4 \times 10{,}000) + (1 \times 1{,}000) + (3 \times 1)$
59. $(9 \times 100{,}000{,}000) + (9 \times 10{,}000{,}000) + (9 \times 1{,}000{,}000) + (9 \times 100{,}000) + (9 \times 10{,}000) + (9 \times 1{,}000) + (9 \times 100) + (9 \times 10) + (9 \times 1)$
61. $(5 \times 1{,}000{,}000{,}000) + (3 \times 10{,}000{,}000) + (8 \times 1{,}000{,}000) + (6 \times 100{,}000) + (2 \times 1{,}000) + (1 \times 100) + (2 \times 10) + (4 \times 1)$
63. $(3 \times 100{,}000{,}000{,}000) + (2 \times 1{,}000{,}000) + (4 \times 100{,}000) + (4 \times 1{,}000) + (1 \times 100)$
65. $(4 \times 10{,}000{,}000{,}000{,}000) + (4 \times 1{,}000{,}000{,}000{,}000) + (2 \times 100{,}000{,}000{,}000) + (9 \times 10{,}000{,}000{,}000) + (5 \times 10{,}000{,}000) + (3 \times 1{,}000{,}000) + (7 \times 100{,}000) + (7 \times 10{,}000) + (7 \times 1{,}000) + (1 \times 100) + (1 \times 10) + (9 \times 1)$
67. Multiply 7 and 3. **69.** Divide 18 by 2. **71.** a is equal to 10. **73.** w is greater than t. **75.** $<$ **77.** $<$
79. $>$ **81.** $=$ **83.** $\neq$; $<$ **85.** $=$ **87.** $=$ **89.** $\neq$; $<$ **91.** $=$ **93.** 64,000,000
95. five trillion, eight hundred eighty billion **97.** hundred millions
99. six million, eight hundred eight thousand, three hundred seventy

1.2 Exercises, page 15

1. true **3.** false; rightmost **5.** false; less **7.** 4 **9.** 14 **11.** 6 **13.** 13 **15.** 17 **17.** 15
19. 17 **21.** 16 **23.** 18 **25.** 19 **27.** commutative **29.** associative **31.** associative
33. commutative **35.** commutative and associative **37.** 70; 72 **39.** 600; 648 **41.** 130; 134 **43.** 900; 901
45. 1,400; 1,426 **47.** 49 **49.** 133 **51.** 1,371 **53.** 785 **55.** 146 **57.** 725 **59.** 13,682
61. 492 **63.** 98,588 **65.** 9,361 **67.** 7,200,158 **69.** 44,037 **71.** $4 + 9 = 13$ **73.** $3 + 8$
75. $5 + x = 9$ **77.** $x + y$ **79.** 1,037 miles **81.** \$10,150 **83.** 437 tickets **85.** 9,075,129 cars

1.3 Exercises, page 23

1. true **3.** true **5.** 9 **7.** 8 **9.** 7 **11.** 8 **13.** 5 **15.** 9 **17.** 3 **19.** 9 **21.** 10 **23.** 500; 536 **25.** 500; 523 **27.** 4,000; 4,224 **29.** 19,000; 18,732 **31.** 45 **33.** 13 **35.** 46 **37.** 645 **39.** 117 **41.** 226 **43.** 4,619 **45.** 1,884 **47.** 4,166 **49.** 27,728 **51.** 5,607,775 **53.** 31,677,314 **55.** $12 - 5$ **57.** $15 - 7 = 8$ **59.** $10 - 3 = 7$ **61.** $x - y$ **63.** \$361 **65.** 181,618,000 **67.** \$3,700 **69.** \$162,417,320,000

1.4 Exercises, page 33

1. false; addition **3.** true **5.** 42 **7.** 49 **9.** 45 **11.** 63 **13.** 81 **15.** 21 **17.** 72 **19.** 63 **21.** associative **23.** distributive **25.** distributive **27.** distributive **29.** $3(6) + 3(2) = 18 + 6 = 24$ **31.** $8(7 + 3) = 8(10) = 80$ **33.** $5(7) - 5(2) = 35 - 10 = 25$ **35.** $9(3) + 8(3) = 27 + 24 = 51$ **37.** $6(12 + 8) = 6(20) = 120$ **39.** $7(9 - 4) = 7(5) = 35$ **41.** $8(40) + 8(5) = 320 + 40 = 360$ **43.** $5(73 + 7) = 5(80) = 400$ **45.** 318 **47.** 567 **49.** 294 **51.** 1,312 **53.** 2,904 **55.** 10,011 **57.** 58,926 **59.** 144,666 **61.** 2,608,320 **63.** 29,925,372 **65.** 1,500 **67.** 30,000 **69.** 74,958 **71.** 640,000 **73.** 434,654 **75.** 4,800,000 **77.** 4,602,070 **79.** 140,000,000 **81.** 21,794,736 **83.** 1,500; 1,377 **85.** 12,000; 11,603 **87.** 560,000; 577,163 **89.** 1,500,000; 1,531,242 **91.** $5 \cdot 4 = 20$ **93.** $3a$ **95.** 504 miles **97.** \$3,910 **99.** 6,697,000 cars **101.** 525,600 minutes

1.5 Exercises, page 43

1. false; dividend **3.** false; subtraction **5.** false; less **7.** 9 **9.** 0 **11.** 8 **13.** 3 **15.** 7 **17.** 8 **19.** 4 **21.** undefined **23.** 3 **25.** 1 **27.** 7 R3 **29.** 5 R4 **31.** 6 R1 **33.** 18 **35.** 52 **37.** 41 **39.** 37 R2 **41.** 153 R2 **43.** 1,460 R1 **45.** 1,276 R2 **47.** 586 **49.** 6,351 R5 **51.** 16,434 R1 **53.** 2,156,831 R3 **55.** 1,333,333 R1 **57.** 4 R2 **59.** 1 R24 **61.** 11 R11 **63.** 5 R57 **65.** 185 R8 **67.** 373 R20 **69.** 25 R346 **71.** 25 R279 **73.** 73 R197 **75.** 92 R570 **77.** 35 **79.** 239 **81.** $36 \div 9$ **83.** $x \div y$ **85.** \$860 **87.** 17 cards; 1 card **89.** 4 pounds **91.** \$645 **93.** 14 seats **95.** 13 hours

1.6 Exercises, page 49

1. false; base **3.** true **5.** $3^3 \cdot 8^2$ **7.** $2^6 \cdot 6^1 \cdot 7^2$ **9.** $5^3 \cdot 8^2 + 5^2 \cdot 7^3$ **11.** 2^2 **13.** 5^2 **15.** 2^5 **17.** 3^4 **19.** 5^4 **21.** 100 **23.** 9 **25.** 216 **27.** 720 **29.** 49,000 **31.** 160,000 **33.** 33,872,256 **35.** $(5 \times 10^2) + (3 \times 10^1) + (8 \times 10^0)$ **37.** $(6 \times 10^2) + (2 \times 10^0)$ **39.** $(7 \times 10^3) + (4 \times 10^2) + (3 \times 10^1) + (6 \times 10^0)$ **41.** $(4 \times 10^4) + (1 \times 10^3) + (7 \times 10^2) + (8 \times 10^1) + (3 \times 10^0)$ **43.** $(7 \times 10^4) + (4 \times 10^3) + (6 \times 10^1) + (2 \times 10^0)$ **45.** $(3 \times 10^5) + (4 \times 10^4) + (5 \times 10^0)$ **47.** $(6 \times 10^6) + (9 \times 10^5) + (4 \times 10^4) + (7 \times 10^3) + (8 \times 10^1) + (9 \times 10^0)$ **49.** $(5 \times 10^9) + (2 \times 10^8) + (1 \times 10^6) + (8 \times 10^5) + (4 \times 10^4) + (6 \times 10^2) + (9 \times 10^1) + (3 \times 10^0)$ **51.** $7 \times 10^1 \times 4 \times 10^2 = 28 \times 10^3$ **53.** $6 \times 10^2 \times 8 \times 10^3 = 48 \times 10^5$ **55.** $11 \times 10^2 \times 7 \times 10^3 = 77 \times 10^5$ **57.** $4 \times 10^4 \times 5 \times 10^4 = 20 \times 10^8 = 2 \times 10^9$ **59.** $7 \times 10^5 \times 3 \times 10^5 = 21 \times 10^{10}$ **61.** $25 \times 10^4 \times 2 \times 10^5 = 50 \times 10^9 = 5 \times 10^{10}$ **63.** 3^9 **65.** 4^6 **67.** 2^{14} **69.** 5^{11} **71.** 3^{12} **73.** 4^{10} **75.** 7^{10} **77.** 5^{11} **79.** $8^6 \cdot 6^7$ **81.** $2^9 \cdot 10^8$ **83.** $7^{13} \cdot 8^8$ **85.** 6^{20} **87.** 4^{38}

1.7 Exercises, page 52

1. false; first **3.** true **5.** 17 **7.** 100 **9.** 13 **11.** 16 **13.** 77 **15.** 16 **17.** 40 **19.** 35 **21.** 9 **23.** 2 **25.** 135 **27.** 5 **29.** 3,850 **31.** 8,888 **33.** 17 **35.** 3 **37.** 23 **39.** 2 **41.** 32 **43.** $7^5 + 7^7$ **45.** $5^{10} + 5^6$ **47.** 2,003 **49.** 116,447 **51.** multiply first; 14 **53.** divide first; 5 **55.** raise to power first; 18 **57.** subtract first; 4 **59.** add numbers in parentheses first; 81 **61.** subtract numbers in parentheses first; 6 **63.** $2(6 + 2) = 16$ **65.** $5 \cdot 4 - 2 = 18$ **67.** $2^3 + 6 = 14$ **69.** $(3 \cdot 4) + (10 \div 2) = 17$ **71.** $(2 + 5)(6 + 3) = 63$

1.8 Exercises, page 55

1. true **3.** false; less than or equal to **5.** 46 **7.** 344 **9.** 41,576 **11.** 29 minutes **13.** 62 words per minute **15.** 214 miles **17.** 8 **19.** \$1,000 **21.** 140 deaths per day **23.** 142,845 square miles

1.9 Exercises, page 62

1. true **3.** false; 3 **5.** 109 **7.** 131 R2 **9.** 672 **11.** 464 R7 **13.** 24,613 R2 **15.** 7,449 **17.** 143,350 R6 **19.** 2 **21.** 2, 3, 4, 5, 6, 8, 10 **23.** 2, 3, 4, 6, 8 **25.** 3 **27.** 3, 5, 9 **29.** 2, 3, 4, 6, 9 **31.** 3, 9 **33.** 2, 4, 8 **35.** 2 **37.** 2, 4 **39.** 2, 3, 4, 6, 8 **41.** 2, 3, 4, 6, 8 **43.** 3, 5, 9 **45.** 2, 3, 6 **47.** 2, 4, 5, 8, 10 **49.** 0, 2, 4, 6, 8 **51.** 0, 4, 8 **53.** 4 **55.** 1 **57.** 0, 2, 4, 6, 8 **59.** 2, 6 **61.** 4 **63.** 1 **65.** yes **67.** yes **69.** no **71.** no **73.** no **75.** no **77.** no **79.** yes **81.** yes **83.** no **85.** no **87.** yes

1.10 Exercises, page 66

1. false; prime **3.** true **5.** composite **7.** prime **9.** prime **11.** composite **13.** prime **15.** composite **17.** composite **19.** composite **21.** prime **23.** prime **25.** composite **27.** prime **29.** composite **31.** composite **33.** composite **35.** 0, 4, 8, 12, 16 **37.** 0, 7, 14, 21, 28 **39.** 0, 10, 20, 30, 40 **41.** 0, 15, 30, 45, 60 **43.** 0, 20, 40, 60, 80 **45.** 2^4 **47.** $2 \cdot 3^3$ **49.** $2 \cdot 3 \cdot 7$ **51.** 2^5 **53.** $2 \cdot 3 \cdot 13$ **55.** $2^3 \cdot 7$ **57.** 2^6 **59.** $2^2 \cdot 3 \cdot 5$ **61.** $2^2 \cdot 3^3$ **63.** $3^2 \cdot 5^2$ **65.** $2 \cdot 3 \cdot 5^2$ **67.** $2 \cdot 89$ **69.** $2^3 \cdot 3^2 \cdot 5$ **71.** $2 \cdot 3 \cdot 5 \cdot 23$ **73.** $2^2 \cdot 5^2 \cdot 7$ **75.** $2 \cdot 5^2 \cdot 19$ **77.** $2^2 \cdot 5^4$ **79.** $2 \cdot 5^5$ **81.** $2^4 \cdot 5^4$

1.11 Exercises, page 69

1. false; product **3.** false; less than **5.** 6 **7.** 8 **9.** 12 **11.** 5 **13.** 12 **15.** 2 **17.** 8 **19.** 13 **21.** 12 **23.** 14 **25.** 8 **27.** 15 **29.** 12 **31.** 30 **33.** 40 **35.** 108 **37.** 96 **39.** 168 **41.** 24 **43.** 200 **45.** 126 **47.** 720 **49.** 2,940 **51.** 288 **53.** GCF, 3; LCM, 252 **55.** GCF, 6; LCM, 270 **57.** GCF, 7; LCM, 294 **59.** GCF, 12; LCM, 1,512 **61.** GCF, 4; LCM, 1,008

Chapter 1 Review Exercises, page 72

1. false; does not **3.** false; division **5.** false; less than **7.** nine hundred seven; $(9 \times 100) + (7 \times 1)$ **9.** thirteen thousand, sixty-nine; $(1 \times 10{,}000) + (3 \times 1{,}000) + (6 \times 10) + (9 \times 1)$ **11.** four million, seventy-nine thousand, five hundred eighty-one; $(4 \times 1{,}000{,}000) + (7 \times 10{,}000) + (9 \times 1{,}000) + (5 \times 100) + (8 \times 10) + (1 \times 1)$ **13.** $>$ **15.** $=$ **17.** $>$ **19.** commutative property of multiplication **21.** distributive property **23.** commutative property of multiplication **25.** distributive property **27.** commutative property of multiplication **29.** commutative property of multiplication **31.** 1,259 **33.** 2,914 **35.** 33 R4 **37.** 247 **39.** 351 **41.** 157 R7 **43.** 95,238 **45.** 6,058 **47.** 905,148 **49.** 526,183 **51.** 24,187,113 **53.** 47,163,214 **55.** $5 - 2$ **57.** $33 \div 3$ **59.** $2 \cdot 8 = 16$ **61.** 5^9 **63.** 2^8 **65.** 7^{20} **67.** 2^{27} **69.** 44 **71.** 648 **73.** 36 **75.** 16 **77.** 15 **79.** 3 **81.** 12 **83.** 24 **85.** 23 **87.** 3 **89.** 135 **91.** 59,031 **93.** 2, 3, 4, 6 **95.** none **97.** 3, 5 **99.** composite **101.** prime **103.** composite **105.** $2^2 \cdot 17$ **107.** $2^3 \cdot 3 \cdot 5$ **109.** $2^2 \cdot 3 \cdot 5^3$ **111.** GCF, 8; LCM, 224 **113.** GCF, 18; LCM, 432 **115.** GCF, 4; LCM, 96 **117.** GCF, 7; LCM, 168 **119.** $6 \cdot 9 - 8 = 46$ **121.** $(3 \cdot 9) + (20 \div 4)$ **123.** $63 \div 7 - 3 = 6$ **125.** 132,497 immigrants **127.** \$135 **129.** 15,580 miles **131.** 11,793,000 people **133.** 32 points **135.** \$23 **137.** 72 **139.** 82 people per square mile **141.** 131,075 passengers

Chapter 2

2.1 Exercises, page 82

1. true **3.** true **5.** false; improper **7.** false; $\frac{20}{5}$ **9.** true **11.** $\frac{2}{5}$ **13.** $\frac{39}{8}$ **15.** $\frac{1}{11}$ **17.** $\frac{6}{1}$ **19.** $\frac{13}{52}$ **21.** improper **23.** improper **25.** proper **27.** improper **29.** improper **31.** $54 \div 6 = 9$ **33.** $0 \div 6 = 0$ **35.** $21 \div 7 = 3$ **37.** $\frac{2}{3}$ **39.** $1\frac{2}{3} = \frac{5}{3}$ **41.** $1\frac{3}{4}$ **43.** $6\frac{1}{3}$ **45.** $8\frac{1}{5}$ **47.** $27\frac{1}{3}$

49. $14\frac{3}{7}$ **51.** $35\frac{5}{9}$ **53.** $77\frac{4}{7}$ **55.** $9\frac{7}{12}$ **57.** $40\frac{13}{21}$ **59.** $\frac{16}{5}$ **61.** $\frac{74}{9}$ **63.** $\frac{31}{7}$ **65.** $\frac{85}{8}$
67. $\frac{99}{8}$ **69.** $\frac{131}{6}$ **71.** $\frac{151}{20}$ **73.** $\frac{359}{10}$ **75.** $\frac{680}{13}$ **77.** $\frac{201}{101}$ **79.** $\frac{6,559}{94}$ **81.** $\frac{3,168}{23}$ **83.** $\frac{3}{32}$
85. $\frac{13}{52}$ **87.** $\frac{7}{12}$ **89.** $\frac{7}{11}$ (men); $\frac{4}{11}$ (women) **91.** $7\frac{21}{60} = \frac{441}{60}$ minutes **93.** $1\frac{5}{16} = \frac{21}{16}$ pounds

2.2 Exercises, page 89

1. false; dividing **3.** false; division **5.** $\frac{18}{42}$ **7.** $\frac{3}{7}$ **9.** $\frac{16}{26}$ **11.** $\frac{5}{9}$ **13.** $\frac{35}{28}$ **15.** $\frac{6}{7}$ **17.** $\frac{1}{2}$
19. $\frac{21}{4}$ **21.** $\frac{11}{14}$ **23.** $\frac{6}{5}$ **25.** $\frac{15}{8}$ **27.** $\frac{7,497}{10,682}$ **29.** $\frac{13}{81}$ **31.** 1 **33.** 40 **35.** 15 **37.** 12
39. 48 **41.** 4 **43.** 3 **45.** 1 **47.** 14 **49.** 5 **51.** 108 **53.** 1,122 **55.** 2 **57.** 5
59. 150 **61.** 9 **63.** 3,355 **65.** 3 **67.** $\frac{2}{5}$ **69.** $\frac{1}{3}$ **71.** $\frac{3}{4}$ **73.** $\frac{3}{4}$ **75.** 6 **77.** $\frac{4}{7}$
79. $\frac{1}{6}$ **81.** $\frac{5}{2}$ **83.** $\frac{1}{2}$ **85.** $\frac{2}{7}$ **87.** $\frac{7}{9}$ **89.** $\frac{21}{44}$ **91.** $\frac{11}{18}$ **93.** $\frac{1}{6}$ **95.** $\frac{4}{9}$ **97.** $\frac{25}{7}$ **99.** $\frac{11}{12}$
101. $\frac{3}{16}$ **103.** $\frac{64}{3}$ **105.** $\frac{1}{6}$ **107.** $\frac{11}{25}$ **109.** $\frac{8}{9}$ **111.** $\frac{1}{21}$ week

2.3 Exercises, page 97

1. true **3.** true **5.** $\frac{4}{7}$ **7.** 1 **9.** $\frac{4}{5}$ **11.** $\frac{15}{17}$ **13.** $\frac{23}{20}$ **15.** $\frac{13}{12}$ **17.** $\frac{5}{4}$ **19.** $\frac{25}{33}$
21. $\frac{41}{50}$ **23.** $\frac{5}{28}$ **25.** $\frac{123}{35}$ **27.** $\frac{127}{100}$ **29.** $\frac{35}{144}$ **31.** $\frac{49}{150}$ **33.** $\frac{23}{54}$ **35.** $\frac{17}{63}$ **37.** $\frac{91}{132}$
39. $\frac{183}{20}$ **41.** $\frac{93}{10}$ **43.** $\frac{85}{98}$ **45.** $\frac{31}{30}$ **47.** $\frac{1,091}{1,000}$ **49.** $\frac{34}{75}$ **51.** $\frac{25}{24}$ **53.** $\frac{117}{112}$ **55.** $\frac{104}{213}$
57. $\frac{1,460}{1,421}$ **59.** $\frac{16}{7}$ **61.** $\frac{52}{17}$ **63.** $\frac{111}{8}$ **65.** $1\frac{13}{30}$ hours **67.** $\frac{59}{60}$ **69.** $1\frac{1}{8}$ pounds

2.4 Exercises, page 104

1. true **3.** true **5.** $<$ **7.** $<$ **9.** $<$ **11.** $<$ **13.** $<$ **15.** $>$ **17.** $<$
19. $>$ **21.** $>$ **23.** $<$ **25.** $\frac{4}{7}$ **27.** $\frac{1}{9}$ **29.** $\frac{11}{72}$ **31.** $\frac{13}{18}$ **33.** $\frac{31}{90}$ **35.** $\frac{7}{72}$ **37.** $\frac{22}{39}$
39. $\frac{23}{48}$ **41.** $\frac{84}{275}$ **43.** $\frac{25}{168}$ **45.** $\frac{25}{6}$ **47.** $\frac{79}{12}$ **49.** $\frac{3}{28}$ **51.** $\frac{138}{2,183}$ **53.** $\frac{37}{72}$ **55.** $\frac{5}{6}$ **57.** $\frac{29}{42}$
59. $\frac{13}{60}$ **61.** $\frac{79}{144}$ **63.** $\frac{1}{12}$ cup **65.** The teammate swam $\frac{1}{40}$ mile farther. **67.** Democrats

2.5 Exercises, page 110

1. $7\frac{3}{4}$ **3.** $7\frac{5}{8}$ **5.** $10\frac{5}{8}$ **7.** $135\frac{7}{8}$ **9.** $47\frac{1}{16}$ **11.** $16\frac{7}{12}$ **13.** $17\frac{17}{24}$ **15.** $16\frac{51}{80}$ **17.** $26\frac{29}{36}$
19. $19\frac{5}{56}$ **21.** $140\frac{19}{25}$ **23.** $10\frac{7}{180}$ **25.** $17\frac{43}{144}$ **27.** $84\frac{13}{27}$ **29.** $14\frac{3}{5}$ **31.** $46\frac{1}{3}$ **33.** $15\frac{31}{54}$

35. $9\frac{11}{12}$ **37.** $11\frac{1}{48}$ **39.** $18\frac{13}{16}$ **41.** $252\frac{358}{817}$ **43.** $993\frac{187}{910}$ **45.** $6\frac{1}{2}$ **47.** $5\frac{5}{8}$ **49.** $2\frac{5}{12}$ **51.** $9\frac{11}{21}$ **53.** $27\frac{7}{15}$ **55.** $3\frac{2}{3}$ **57.** $1\frac{1}{3}$ **59.** $3\frac{1}{2}$ **61.** $15\frac{16}{27}$ **63.** $13\frac{1}{18}$ **65.** $\frac{13}{24}$ **67.** $2\frac{4}{7}$ **69.** $7\frac{13}{18}$ **71.** $24\frac{3}{8}$ **73.** $3\frac{16}{21}$ **75.** $6\frac{27}{56}$ **77.** $2\frac{19}{144}$ **79.** $2\frac{5}{12}$ **81.** $7\frac{43}{60}$ **83.** $7\frac{111}{160}$ **85.** $9\frac{26}{75}$ **87.** $9\frac{269}{288}$ **89.** $5\frac{173}{378}$ **91.** 8 **93.** $15\frac{1}{20}$ **95.** $47\frac{1}{6}$ **97.** $13\frac{361}{1{,}000}$ **99.** $5\frac{5}{96}$ **101.** $4\frac{46}{55}$ miles **103.** $16\frac{19}{60}$ gallons **105.** $2\frac{3}{8}$ inches **107.** Saturday: $5\frac{1}{3}$ hours; Sunday: $5\frac{7}{12}$ hours; Total: $10\frac{11}{12}$ hours **109.** $6\frac{29}{48}$ feet

2.6 Exercises, page 118

1. false; multiplied by **3.** false; $\frac{15}{8}$ **5.** $\frac{2}{21}$ **7.** $\frac{5}{32}$ **9.** $\frac{12}{77}$ **11.** $\frac{10}{27}$ **13.** $\frac{2}{3}$ **15.** $\frac{1}{56}$ **17.** $\frac{25}{4}$ **19.** $\frac{5}{66}$ **21.** $\frac{4}{21}$ **23.** $\frac{5}{24}$ **25.** $\frac{15}{224}$ **27.** $\frac{14}{9}$ **29.** $\frac{2}{121}$ **31.** 7 **33.** $1\frac{5}{7}$ **35.** $34\frac{2}{5}$ **37.** $1\frac{1}{16}$ **39.** $52\frac{1}{2}$ **41.** $7\frac{7}{12}$ **43.** $32\frac{1}{2}$ **45.** $33\frac{1}{3}$ **47.** $38\frac{1}{2}$ **49.** $16\frac{13}{36}$ **51.** $10\frac{5}{7}$ **53.** $73\frac{1}{3}$ **55.** $\frac{185}{324}$ **57.** $1{,}615\frac{59}{899}$ **59.** $\frac{5}{24}$ **61.** $\frac{71}{80}$ **63.** $\frac{7}{20}$ **65.** $\frac{1}{108}$ **67.** $\frac{1{,}311}{1{,}220}$ **69.** 31 **71.** $\frac{9}{5}$ **73.** $\frac{5}{21}$ **75.** $\frac{11}{10}$ **77.** $\frac{8}{35}$ **79.** $133\frac{1}{3}$ yards **81.** $21\frac{7}{8}$ feet

2.7 Exercises, page 125

1. true **3.** false; 9 **5.** $\frac{8}{3}$ **7.** 7 **9.** $\frac{1}{11}$ **11.** 1 **13.** $\frac{4}{31}$ **15.** $\frac{2}{27}$ **17.** $\frac{13}{28}$ **19.** 6 **21.** $\frac{14}{15}$ **23.** $\frac{10}{21}$ **25.** $\frac{6}{5}$ **27.** $\frac{5}{14}$ **29.** $\frac{6}{5}$ **31.** $\frac{4}{3}$ **33.** $\frac{8}{21}$ **35.** $\frac{35}{4}$ **37.** 14 **39.** $\frac{1}{12}$ **41.** 9 **43.** $\frac{2}{5}$ **45.** $6\frac{2}{3}$ **47.** $\frac{2}{5}$ **49.** $1\frac{47}{88}$ **51.** $2\frac{14}{15}$ **53.** $\frac{5}{9}$ **55.** $3\frac{1}{2}$ **57.** $\frac{110}{279}$ **59.** $\frac{1{,}961}{1{,}944}$ **61.** $2\frac{6{,}412}{20{,}925}$ **63.** $\frac{2}{9}$ **65.** $\frac{2}{7}$ **67.** $\frac{31}{7}$ **69.** $\frac{13}{6}$ **71.** $\frac{73{,}358}{19{,}095}$ **73.** $5\frac{1}{3}$ laps **75.** $\frac{5}{16}$ pound **77.** $4\frac{1}{3}$ boxes **79.** $4\frac{1}{2}$ times

2.8 Exercises, page 129

1. $\frac{9}{20}$ **3.** $\frac{16}{5}$ **5.** $\frac{39}{20}$ **7.** 12 **9.** $\frac{19}{21}$ **11.** $\frac{7}{5}$ **13.** $\frac{26}{63}$ **15.** $\frac{9}{43}$ **17.** $\frac{9}{38}$ **19.** $21\frac{3}{4}$ **21.** $\frac{9}{200}$ **23.** $6\frac{33}{40}$ **25.** $1\frac{54}{83}$ **27.** $\frac{37}{249}$ **29.** $\frac{139}{143}$ **31.** $\frac{74}{129}$

Chapter 2 Review Exercises, page 131

1. false; sum **3.** true **5.** false; equivalent **7.** incorrect; $\frac{15}{8}$ **9.** incorrect; 0 **11.** correct

13. incorrect; greater **15.** $\frac{5}{8}$ **17.** $\frac{7}{24}$ **19.** $\frac{40}{33}$ **21.** $\frac{35}{4}$ **23.** $\frac{19}{54}$ **25.** $1\frac{1}{2}$ **27.** 21 **29.** $\frac{31}{84}$
31. $5\frac{1}{3}$ **33.** $11\frac{19}{24}$ **35.** $4\frac{23}{24}$ **37.** $\frac{1}{16}$ **39.** $26\frac{2}{5}$ **41.** $2\frac{28}{33}$ **43.** $\frac{1}{2}$ **45.** $9\frac{35}{54}$ **47.** $\frac{19}{6}$ **49.** $\frac{8}{7}$
51. $\frac{19}{36}$ **53.** $\frac{13}{32}$ **55.** $\frac{15}{7}$ **57.** $\frac{7}{16}$ **59.** $\frac{58}{33}$ **61.** $\frac{9}{10}$ **63.** 1 **65.** $7\frac{7}{33}$ **67.** $\frac{41}{85}$ **69.** $3\frac{1}{8}$
71. $\frac{25}{192}$ **73.** $5\frac{1}{3}$ cups sugar; 7 cups flour **75.** Republicans; $\frac{5}{36}$ **77.** 18 bottles **79.** 104 people

Chapter 3

3.1 Exercises, page 142

1. false; hundredths **3.** false; denominator **5.** 0.3 **7.** 0.08 **9.** 0.005 **11.** 0.529 **13.** 0.0021
15. 0.00003 **17.** 0.00127 **19.** 0.006589 **21.** 7.3 **23.** 5.19 **25.** 2.837 **27.** 7.0803
29. tenths; three tenths **31.** hundredths; fifty-three hundredths **33.** tenths; five and nine tenths
35. tenths; forty-seven and eight tenths **37.** thousandths; thirty-eight and nine hundred seventy-four thousandths
39. hundred thousandths; eight hundred and sixty-five thousand, three hundred twenty-seven hundred thousandths
41. tenths; three thousand, six hundred three and forty-six hundredths
43. ten-thousandths; five hundred and four thousand, six hundred eighty-nine ten-thousands
45. ten-thousandths; eight thousand, sixty-seven and four ten-thousandths
47. millionths; three hundred sixty-two thousand, six and two hundred thousand, eight millionths
49. 0.4 **51.** 0.006 **53.** 0.0004 **55.** 0.0386 **57.** 0.000006 **59.** 0.002059 **61.** 2.6 **63.** 308.12
65. 11,000.00058 **67.** $(5 \times 10^0) + \left(2 \times \frac{1}{10^1}\right)$ **69.** $(3 \times 10^1) + (6 \times 10^0) + \left(4 \times \frac{1}{10^1}\right) + \left(6 \times \frac{1}{10^2}\right)$
71. $(2 \times 10^2) + (3 \times 10^1) + \left(5 \times \frac{1}{10^3}\right)$ **73.** $(3 \times 10^4) + (5 \times 10^0) + \left(6 \times \frac{1}{10^1}\right) + \left(2 \times \frac{1}{10^2}\right) + \left(7 \times \frac{1}{10^3}\right) + \left(8 \times \frac{1}{10^4}\right)$
75. $(5 \times 10^7) + (8 \times 10^4) + (3 \times 10^3) + (7 \times 10^1) + \left(4 \times \frac{1}{10^2}\right) + \left(6 \times \frac{1}{10^4}\right)$ **77.** \$0.16; zero and 16/100 dollars
79. \$3.74; three and 74/100 dollars **81.** \$46.48; forty-six and 48/100 dollars
83. \$805.67; eight hundred five and 67/100 dollars **85.** \$7.14; seven and 14/100 dollars **87.** thousandths **89.** 27.32

3.2 Exercises, page 146

1. true **3.** false; thousandth **5.** 3.3; 3.28; 3.279 **7.** 14.0; 13.96; 13.963 **9.** 7.0; 7.01; 7.006
11. 0.5; 0.50; 0.502 **13.** 62.8; 62.83; 62.830 **15.** 38.1; 38.07; 38.071 **17.** 4.8; 4.80; 4.798 **19.** 68.0; 68.00; 68.000
21. 6,800; 6,850; 6,847 **23.** 7,000; 7,050; 7,046 **25.** 700; 710; 709 **27.** 700; 700; 696 **29.** 7,100; 7,080; 7,080
31. 500; 500; 500 **33.** 1,000; 990; 986 **35.** 60,700; 60,710; 60,710 **37.** 330,000 **39.** 6.0504 **41.** 2.857410
43. 70.0 **45.** 10,000.0 **47.** \$368 **49.** \$32.85 **51.** \$88.6 **53.** \$5.031 **55.** \$6.8340 **57.** 51 miles
59. 53.7 seconds **61.** \$500

3.3 Exercises, page 151

1. false; vertically **3.** false; right to left **5.** 12.1 **7.** 46.95 **9.** 15.564 **11.** \$46,75 **13.** 154.2783
15. 26.187 **17.** 1,309.7149 **19.** \$863.58 **21.** 41.15147 **23.** 4.6 **25.** \$8.12 **27.** 45.668
29. \$159.06 **31.** 5.4376 **33.** 2.7226 **35.** 410.066 **37.** 631.8205 **39.** 6.4 **41.** 424.519
43. \$76.45 **45.** 8,644.6516 **47.** \$3,151.87 **49.** 33.751 **51.** 170; 170.14 **53.** 350; 349.2
55. 820; 820.9 **57.** 5400; 5392.4 **59.** 20.8 miles **61.** \$15.14 **63.** \$5.70 **65.** \$185.99

3.4 Exercises, page 157

1. true **3.** false; product **5.** 0.014 **7.** 0.00224 **9.** 3.078 **11.** 1.4418 **13.** $46.472 **15.** 0.0002698 **17.** 45.39084 **19.** 0.3023172 **21.** 512.144448 **23.** $35,999.88 **25.** 1.371816 **27.** 185,353.38 **29.** 80,000 **31.** 0.31 **33.** 3,602.005 **35.** 570,000,000 **37.** 280,300 **39.** 14,400 **41.** 0.0282 **43.** 22,740 **45.** 270 **47.** 140 **49.** 4,400,000 **51.** 0.089 **53.** 0.064 **55.** 1,750 **57.** 91 **59.** 0.00002 **61.** 4.8 **63.** 7.68 **65.** 289 **67.** 30 **69.** 0.081 **71.** 1,062.99 inches **73.** 5,151,600 **75.** $648.31 **77.** $9,205 **79.** $2,280.61

3.5 Exercises, page 164

1. false; left **3.** true **5.** 9 **7.** 0.007 **9.** 13,000 **11.** 0.013 **13.** 0.14 **15.** 23,000 **17.** 0.52 **19.** 56 **21.** 470,000 **23.** 4.6667 **25.** 186.7 **27.** 693.333 **29.** $2.98 **31.** 2,551.85 **33.** 3,052 **35.** $14.2 **37.** 24,831 **39.** 0.06818 **41.** 0.000783 **43.** 0.0006005 **45.** 0.90009 **47.** 0.0002817 **49.** 0.005386 **51.** 5,000 **53.** 3.2 **55.** 40 **57.** 0.00288 **59.** 0.0000128 **61.** 10 **63.** 6.12 **65.** 20 **67.** 0.0816 **69.** 2,000 **71.** $711.54 **73.** 6.3 pounds **75.** 5.793 persons per square mile **77.** 2.4 **79.** 160.9 miles per hour

3.6 Exercises, page 169

1. false; repeating **3.** true **5.** $\frac{2}{25}$ **7.** $\frac{4}{25}$ **9.** $\frac{3}{125}$ **11.** $\frac{3}{400}$ **13.** $\frac{5}{16}$ **15.** $\frac{5}{2}$ **17.** $\frac{64}{5}$ **19.** $\frac{309}{20}$ **21.** $\frac{803}{20}$ **23.** $\frac{323}{80}$ **25.** 0.25 **27.** 0.4 **29.** $0.\overline{45}$ **31.** 0.15625 **33.** $0.2\overline{6}$ **35.** $0.\overline{4}$ **37.** $0.3\overline{8}$ **39.** 0.75 **41.** $0.\overline{2}$ **43.** $2.\overline{6}$ **45.** $1.\overline{4}$ **47.** 1.1875 **49.** $30.58\overline{3}$ **51.** $242.\overline{3}$ **53.** $0.0\overline{95238}$ **55.** $0.\overline{678}$ **57.** 0.158 **59.** 0.23 **61.** 0.8667 **63.** 0.54545 **65.** 0.286 **67.** 2.57 **69.** 40.89 **71.** 0.763

Chapter 3 Review Exercises, page 171

1. false; right **3.** false; division **5.** false; subtracting **7.** eight and seven tenths; $(8 \times 10^0) + \left(7 \times \frac{1}{10^1}\right)$

9. five and eighty-seven hundredths; $(5 \times 10^0) + \left(8 \times \frac{1}{10^1}\right) + \left(7 \times \frac{1}{10^2}\right)$

11. forty-three and six hundred twenty-eight thousandths; $(4 \times 10^1) + (3 \times 10^0) + \left(6 \times \frac{1}{10^1}\right) + \left(2 \times \frac{1}{10^2}\right) + \left(8 \times \frac{1}{10^3}\right)$

13. eight hundred six and five thousand, six hundred fourteen ten-thousandths; $(8 \times 10^2) + (6 \times 10^0) + \left(5 \times \frac{1}{10^1}\right) + \left(6 \times \frac{1}{10^2}\right) + \left(1 \times \frac{1}{10^3}\right) + \left(4 \times \frac{1}{10^4}\right)$

15. 4.5 **17.** 18 **19.** 60 **21.** 1,000 **23.** $5.048 **25.** 628.432 **27.** 737.47 **29.** 0.1926 **31.** 2,380 **33.** 9.3607 **35.** 3.24292 **37.** 8.37 **39.** $3,661.49 **41.** $2,519.72 **43.** 88.0589 **45.** 1.48304 **47.** 7 **49.** 2,880,000 **51.** 190 **53.** 0.296 **55.** 0.000021 **57.** 3.6 **59.** 12 **61.** 64 **63.** 627.8428 **65.** 1.4567886 **67.** 5.3 **69.** 0.000000078 **71.** 16.894 **73.** 9,209,000 **75.** $\frac{3}{25}$ **77.** $\frac{1}{40}$ **79.** $\frac{1}{6,250}$ **81.** $\frac{9}{4,000}$ **83.** $\frac{22}{5}$ **85.** $\frac{409}{20}$ **87.** $0.\overline{6}$ **89.** 0.4375 **91.** $0.\overline{72}$ **93.** 0.714 **95.** 2.4 **97.** $4.\overline{72}$ **99.** 0.246 **101.** first day; 983.3 miles **103.** 2.5 **105.** $18.25 **107.** $1.98

Chapter 4

4.1 Exercises, page 182

1. false; numerator **3.** true **5.** $\frac{3}{5}$ **7.** $\frac{4}{1}$ **9.** $\frac{5}{8}$ **11.** $\frac{22}{1}$ **13.** $\frac{1}{14}$ **15.** $\frac{1}{12}$ **17.** $\frac{15}{16}$ **19.** $\frac{14}{33}$

21. $\frac{3}{1}$ **23.** 12 feet **25.** 4 quarters **27.** 10 pints **29.** $1\frac{1}{2}$ ft **31.** 10 ounces **33.** $\frac{7}{6}$ **35.** $\frac{19}{25}$

37. $\frac{4}{3}$ **39.** $\frac{7}{8}$ **41.** $\frac{8}{5}$ **43.** $\frac{7}{6}$ **45.** $\frac{3}{8}$ **47.** $\frac{3}{8}$ **49.** $\frac{3}{2}$ **51.** $\frac{7}{25}$ **53.** $\frac{17}{15}$ **55.** $\frac{4}{3}$ **57.** $\frac{6}{1}$

59. $\frac{2}{5}$ **61.** **a.** 2:7; **b.** 7:1; **c.** $\frac{1}{5}$; **d.** $\frac{3}{10}$; **e.** $\frac{9}{10}$ **63.** **a.** 1:9; **b.** 9:1; **c.** 9:1,000; **d.** $\frac{100}{501}$; **e.** $\frac{400}{501}$ **65.** $\frac{1}{4}$ **67.** the sedan

69. 5.5¢ per ounce **71.** 1:19 **73.** 19:50 **75.** 2.375¢ per fluid ounce; 2.75¢ per fluid ounce; the frozen orange juice

77. $\frac{9}{8}$; Monday **79.** 62 miles per hour **81.** 125 persons per square mile

83. 8.75¢ per egg (dozen); 8.5¢ per egg (eight); 8-egg carton **85.** $\frac{8}{9}$; red ribbon **87.** 6.19 persons per square mile

89. 5.47 miles per hour

4.2 Exercises, page 190

1. true **3.** false; less **5.** extremes, 6 and 3; means, 9 and 2; 6 is to 9 as 2 is to 3
7. extremes, 5 and 8; means, 2 and 20; 5 is to 2 as 20 is to 8

9. extremes, $\frac{1}{2}$ and 22; means, 11 and 1; $\frac{1}{2}$ is to 11 as 1 is to 22 **11.** extremes, $2\frac{1}{3}$ and 24; means, 8 and 7; $2\frac{1}{3}$ is to 8 as 7 is to 24

13. false **15.** true **17.** false **19.** false **21.** false **23.** true **25.** true **27.** false **29.** true
31. true **33.** false **35.** false **37.** $<$ **39.** $>$ **41.** $<$ **43.** $<$ **45.** $<$ **47.** $>$
49. $>$ **51.** $>$ **53.** $>$ **55.** the White Sox player **57.** deodorant soap **59.** U.S. skater
61. 18.5-ounce bottle **63.** Alabama

4.3 Exercises, page 196

1. true **3.** false; cross multiply **5.** variable, x; coefficient, 5 **7.** variable, t; coefficient, 17

9. variable, u; coefficient, $\frac{5}{8}$ **11.** variable, a; coefficient, 1.4 **13.** variable, x; coefficient, 18 **15.** $x = 15$

17. $y = 14$ **19.** $x = \frac{40}{13}$ **21.** $y = 14$ **23.** $m = 18$ **25.** $q = 78$ **27.** $y = \frac{23}{6}$ **29.** $x = \frac{147}{5}$

31. $y = \frac{24}{7}$ **33.** $z = 243$ **35.** $y = \frac{192}{5}$ **37.** $q = 90$ **39.** $x = 2$ **41.** $s = \frac{18}{55}$ **43.** $x = \frac{13}{30}$

45. $w = \frac{3}{4}$ **47.** $v = 2\frac{5}{6}$ **49.** $x = 0.08$ **51.** $y = 0.0025$ **53.** $w = 0.18$ **55.** $t = 0.025$ **57.** $y = 6$

59. $z = 0.0000005$ **61.** $t = 0.06$ **63.** $x = 144$ **65.** $w = 672$ **67.** $x = 0.190$ **69.** $w = 22$

4.4 Exercises, page 204

1. true **3.** true **5.** $\frac{7}{2} = \frac{x}{12}; \frac{7}{x} = \frac{2}{12}; \frac{2}{7} = \frac{12}{x}; \frac{x}{7} = \frac{12}{2}$; $x = 42$ passes

7. $\frac{20}{4} = \frac{45}{x}; \frac{20}{45} = \frac{4}{x}; \frac{4}{20} = \frac{x}{45}; \frac{45}{20} = \frac{x}{4}$; $x = 9$ pages

9. $\frac{10}{4} = \frac{6}{x}; \frac{10}{6} = \frac{4}{x}; \frac{4}{10} = \frac{x}{6}; \frac{6}{10} = \frac{x}{4}; x = 2.4$ hours **11.** 8 feet **13.** 15 hits **15.** 240 dentists **17.** 14 car sales **19.** \$1.70 **21.** 2,463 students **23.** 435 balls **25.** 77 people **27.** 2 feet **29.** 42 students **31.** 1 gram **33.** 50 people **35.** 315 miles **37.** 165 calories **39.** \$1.05 **41.** 1 cup **43.** \$8.73 **45.** 14 minutes **47. a.** 20 nickels; **b.** 30 quarters; **c.** 62 coins **49. a.** 24 juniors; **b.** 18 sophomores; **c.** 42 seniors **51.** 6,146 women **53.** 144,000 square miles

Chapter 4 Review Exercises, page 207

1. true **3.** false; multiplied by **5.** true **7.** $\frac{2}{3}$ **9.** $\frac{8}{1}$ **11.** $\frac{5}{6}$ **13.** $\frac{6}{5}$ **15.** $\frac{28}{9}$ **17.** $\frac{22}{5}$ **19.** $>$ **21.** $<$ **23.** $>$ **25.** $>$ **27.** $<$ **29.** $<$ **31.** $<$ **33.** variable, x; coefficient, 8 **35.** variable, z; coefficient, 1 **37.** variable, t; coefficient, $\frac{3}{5}$ **39.** variable, b; coefficient, 15 **41.** $x = 24$ **43.** $x = 14$ **45.** $t = \frac{9}{2}$ **47.** $n = 24$ **49.** $x = \frac{1}{14}$ **51.** $y = 3\frac{3}{4}$ **53.** $t = 6$ **55.** $n = 24$ **57.** $x = 0.0432$ **59.** $\frac{1}{13}$ **61.** \$0.258 per pound **63.** $\frac{21}{20}$; math **65.** the shortstop **67.** \$246 **69.** 681 mothers **71.** 8 teaspoons **73.** \$12,950 **75.** 4.77 persons per square mile **77.** \$123.1 billion

Chapter 5

5.1 Exercises, page 216

1. true **3.** false; less **5.** $\frac{3}{4}$ **7.** $\frac{1}{50}$ **9.** $\frac{57}{100}$ **11.** $\frac{16}{25}$ **13.** $\frac{6}{5}$ **15.** 5 **17.** $\frac{11}{250}$ **19.** $\frac{9}{2{,}000}$ **21.** $\frac{307}{10{,}000}$ **23.** $\frac{173}{1{,}000}$ **25.** $\frac{3}{16}$ **27.** $\frac{15}{16}$ **29.** $\frac{1}{200}$ **31.** $\frac{5}{16}$ **33.** $\frac{351}{500}$ **35.** $\frac{2}{3}$ **37.** 13% **39.** 1% **41.** 3.2% **43.** $5\frac{1}{3}\%$ **45.** 75% **47.** 40% **49.** 28% **51.** 65% **53.** 80% **55.** 2.7% **57.** 3.18% **59.** 17.5% **61.** 62.5% **63.** $11\frac{1}{9}\%$ **65.** $46\frac{2}{3}\%$ **67.** 60% **69.** $\frac{7}{20}$

5.2 Exercises, page 220

1. false; left **3.** true **5.** 0.71 **7.** 0.05 **9.** 0.248 **11.** 0.0409 **13.** 0.0028 **15.** 2.25 **17.** 36.84 **19.** 0.05529 **21.** 0.1803 **23.** 0.085 **25.** 0.456 **27.** 1.30125 **29.** 0.0214 **31.** 37% **33.** 4% **35.** 66% **37.** 61.3% **39.** 0.9% **41.** 60% **43.** 6.75% **45.** 450% **47.** 300% **49.** 1,320% **51.** 81.3% **53.** 507% **55.** 6.05% **57.** 50% **59.** 87.5% **61.** $11\frac{1}{9}\%$ **63.** 18.75% **65.** 67.5% **67.** $41\frac{2}{3}\%$ **69.** $13\frac{1}{3}\%$ **71.** $28\frac{4}{7}\%$ **73.** $38\frac{6}{13}\%$ **75.** 360% **77.** 550% **79.** 275% **81.** 150% **83.** 13.0% **85.** 60.4% **87.** 82.5% **89.** 37.5% **91.** 48%

5.3 Exercises, page 226

1. true **3.** false; base **5.** 52 **7.** 75 **9.** 25% **11.** 23.75 **13.** 0.34% **15.** 8.75% **17.** 100 **19.** 15.5 **21.** 340% **23.** 81.25 **25.** 3,000 **27.** 1.875% **29.** 21 **31.** 250% **33.** 62.5

35. $18\frac{3}{4}$ **37.** $77\frac{7}{9}\%$ **39.** 18 **41.** 0.8619204 **43.** 397.15 **45.** 18 students **47.** 480 students **49.** 600 people **51.** 144 students **53.** 16.25% **55.** 350 people **57.** 75% **59.** 2% **61.** 31.25% **63.** 1,105 workers **65.** 60 grams **67.** 7 people **69.** 2.1% **71.** 62.1%

5.4 Exercises, page 236

1. true **3.** false; 15% **5.** $13,529 **7.** $24,076 **9.** $27,824 **11.** $33,744.18 **13.** $2.25 **15.** $0.60 **17.** $9.00 **19.** $26.25 **21.** $8.078.50 **23.** $305.50 **25.** $36,980 **27.** $1.11 **29.** $405 **31.** 7% **33.** $1,200 **35.** $17,000 salary plus 8% commission **37.** $12,504.50 **39.** 5.5% **41.** $636 **43.** $20,000 **45.** $3.30 **47.** $160 **49.** $18.30 **51.** $68.43 **53.** $2,547.88

5.5 Exercises, page 240

1. false; sum of **3.** true **5.** 30% **7.** $6,296.50 **9.** $6,839.10 **11.** $307.34 **13.** $18.45 **15.** $424.15 **17.** 62.5% **19.** 140 patients **21.** $4.92 **23.** $11,243.75 **25.** $241.50 **27.** $166.140 **29.** 5% **31.** $30 **33.** 638 students **35.** $53.57 **37.** $13,123,000 **39.** $9,359.35 **41.** $3,433.83 **43.** 4.94%

5.6 Exercises, page 247

1. a. third grade; 200 students; **b.** 800 students; **c.** 12.5%
3. a. Sault Ste. Marie, MI; 115 inches; **b.** Cheyenne, WY; **c.** 65 inches **5. a.** 1981; **b.** 11%; **c.** 14.7%
7. a. 14–24: 553,401; 25–44: 1,605,948; 45–64: 3,624,234; 65 and over: 5,067,417; **b.** 1,443,183; **c.** 85.2%; 9,245,052 people
9. a. 3 years; **b.** 6.875%; **c.** 5 million; **d.** 2.4 times greater
11. a. 1978 and 1980; 11 million bales; **b.** 5 million bales; **c.** 1979; **d.** 12 million bales
13. a. 1981; **b.** 1979 and 1980; 33¢; **c.** 37%
15. a. transportation: $250; miscellaneous expenses: $350; books and supplies: $400; room and board: $3,500; tuition: $8,000; **b.** $4,500; **c.** 8%; $1,000

Chapter 5 Review Exercises, page 256

1. true **3.** false; salesperson **5.** false; denominator **7.** $\frac{57}{100}$ **9.** $\frac{41}{500}$ **11.** $\frac{9}{5}$ **13.** $\frac{3}{400}$ **15.** $\frac{5}{6}$ **17.** 9% **19.** 80% **21.** 250% **23.** 43.75% **25.** $38\frac{6}{13}\%$ **27.** 18.6% **29.** 0.18 **31.** 0.02 **33.** 1.23 **35.** 0.0009 **37.** 0.182 **39.** 28% **41.** 31.9% **43.** 80% **45.** 0.43% **47.** 720% **49.** 92.4 **51.** 112.5 **53.** 16% **55.** 15.5 **57.** 3,000% **59.** 1.35% **61.** 60% **63.** 110 people **65.** $39.69 **67.** 396 students **69.** $44\frac{4}{9}\%$ **71.** $611.50 **73.** $31.96 **75.** 548 people **77.** 14.35% **79. a.** 15,018,450 families; **b.** 458,190 families

Chapter 6

6.1 Exercises, page 264

1. true **3.** false; greater than **5.** < **7.** = **9.** > **11.** < **13.** 7; positive **15.** 51; positive **17.** −5.8; negative **19.** $-\frac{1}{5}$; negative **21.** $3\frac{1}{2}$; positive **23.** −0.23; negative **25.** −6; negative

27. $-a$; positive **29.** 3 **31.** 51 **33.** $\frac{3}{5}$ **35.** 3.9 **37.** −62 **39.** $-\frac{5}{9}$ **41.** 82 **43.** −52
45. $-8\frac{7}{9}$ **47.** $-9\frac{3}{5}$ **49.** −5.4 **51.** −1.85 **53.** −5.48 **55.** −28 **57.** = **59.** > **61.** >
63. = **65.** > **67.** < **69.** > **71.** = **73.** >

6.2 Exercises, page 270

1. false; negative **3.** true **5.** 2 **7.** −11 **9.** 1 **11.** −14 **13.** 11 **15.** −41 **17.** 21
19. −25 **21.** 44 **23.** −82 **25.** −2.5 **27.** 2.9 **29.** 6.6 **31.** −0.12 **33.** 222 **35.** $\frac{2}{9}$
37. $\frac{1}{8}$ **39.** $-\frac{1}{4}$ **41.** $1\frac{1}{4}$ **43.** −5 **45.** 1 **47.** 6 **49.** −535 **51.** $\frac{2}{3}$ **53.** 34,335,564
55. −1.71271 **57.** −28 **59.** $\frac{3}{5}$ **61.** −0.675 **63.** 534,987 **65.** −7°F **67.** 14,494 feet
69. 15-yard line

6.3 Exercises, page 274

1. true **3.** false; greater than **5.** −2 **7.** −6 **9.** −7 **11.** 14 **13.** −1 **15.** −23 **17.** 36
19. −1.8 **21.** 14.1 **23.** −751 **25.** −0.334 **27.** −0.525 **29.** −2,266 **31.** $-\frac{1}{4}$ **33.** $-\frac{3}{2}$
35. $-8\frac{5}{8}$ **37.** −2,637,949 **39.** −12.16497 **41.** 2 **43.** 0 **45.** −5 **47.** 7 **49.** −3.9
51. −50 **53.** $\frac{1}{8}$ **55.** $4\frac{1}{2}$ **57.** 2,164,054 **59.** 11°F **61.** $44,920

6.4 Exercises, page 279

1. true **3.** true **5.** −21 **7.** 30 **9.** −0.21 **11.** −24.6 **13.** −1.68 **15.** −0.66 **17.** $\frac{3}{10}$
19. $\frac{1}{10}$ **21.** 3 **23.** −9 **25.** 45 **27.** 0 **29.** −54 **31.** −24 **33.** 48 **35.** $\frac{1}{12}$ **37.** $\frac{1}{12}$
39. −420 **41.** 240 **43.** −144 **45.** 76.93443 **47.** 146,730.24 **49.** 81 **51.** 1 **53.** −20
55. −35 **57.** −300 **59.** 37 **61.** −30 **63.** −62 **65.** 73 **67.** 34 **69.** −2,409 **71.** −307,695

6.5 Exercises, page 284

1. false; positive **3.** true **5.** −8 **7.** 8 **9.** −5 **11.** 12 **13.** 3 **15.** −0.09 **17.** −0.8
19. $-\frac{1}{6}$ **21.** $-\frac{1}{10}$ **23.** $-1\frac{41}{55}$ **25.** 12 **27.** $\frac{3}{8}$ **29.** $-\frac{3}{4}$ **31.** $-\frac{1}{3}$ **33.** $\frac{10}{3}$ **35.** 1
37. $-\frac{3}{4}$ **39.** $-\frac{9}{7}$ **41.** $-\frac{1}{4}$ **43.** −5 **45.** 1 **47.** 9 **49.** 4 **51.** −29 **53.** −3
55. −7 **57.** $-\frac{2}{3}$ **59.** −8 **61.** $\frac{6}{7}$ **63.** $\frac{2}{3}$

Chapter 6 Review Exercises, page 286

1. incorrect; 2 **3.** incorrect; 63 **5.** correct **7.** correct **9.** incorrect; -4 **11.** $<$ **13** $=$ **15.** $<$ **17.** $>$ **19.** $=$ **21.** $<$ **23.** $<$ **25.** 2 **27.** -5 **29.** 42 **31.** -3 **33.** -5 **35.** -69 **37.** -0.56 **39.** -40 **41.** $\frac{1}{6}$ **43.** $\frac{13}{30}$ **45.** $-\frac{1}{6}$ **47.** $\frac{1}{6}$ **49.** -1 **51.** -10 **53.** 42 **55.** -180 **57.** $\frac{8}{3}$ **59.** $-\frac{9}{7}$ **61.** -21 **63.** 26 **65.** -80 **67.** $-4{,}700$ **69.** -28 **71.** -9 **73.** $\frac{4}{3}$ **75.** $-84{,}260$ **77.** $-10{,}610{,}873$

Chapter 7

7.1 Exercises, page 292

1. false; perfect square **3.** false; positive **5.** 1 **7.** 7 **9.** -9 **11.** ± 8 **13.** 12 **15.** -3 **17.** 18 **19.** 15 **21.** 17 **23.** 13 **25.** 30 **27.** 1.41 **29.** 2.24 **31.** 3.87 **33.** 9.22 **35.** 7.87 **37.** 7.94 **39.** 24 **41.** 27 **43.** 32 **45.** 64 **47.** 71 **49.** 62 **51.** 95 **53.** 2.32 **55.** 7.84 **57.** 13.71 **59.** 27.60 **61.** 23.54 **63.** 25.27 **65.** 73.13 **67.** 168.60 **69.** 0.66 **71.** 0.18 **73.** 0.05 **75.** 2,713.89

7.2 Exercises, page 296

1. false; product **3.** true **5.** 20 **7.** 50 **9.** 90 **11.** 110 **13.** 800 **15.** $\frac{1}{5}$ **17.** $\frac{2}{9}$ **19.** $\frac{6}{11}$ **21.** $\frac{8}{13}$ **23.** 0.5 **25.** 0.7 **27.** 1.1 **29.** 0.3 **31.** 0.08 **33.** 0.03 **35.** $4\sqrt{3}$ **37.** $2\sqrt{5}$ **39.** $5\sqrt{2}$ **41.** $4\sqrt{5}$ **43.** $7\sqrt{2}$ **45.** $8\sqrt{2}$ **47.** $6\sqrt{10}$ **49.** $12\sqrt{2}$

7.3 Exercises, page 301

1. 5.2 **3.** 6.7 **5.** 31.4 **7.** 36.1 **9.** 357 **11.** 794 **13.** 520 **15.** 0.62 **17.** 0.29 **19.** 0.694 **21.** 0.068 **23.** 1,200 **25.** 2.4 **27.** 4.2 **29.** 8.24 **31.** 30.4 **33.** 19.30 **35.** 65.7 **37.** 242 **39.** 0.85 **41.** 0.52 **43.** 0.114 **45.** 0.21 **47.** 0.07 **49.** 2,596

7.4 Exercises, page 304

1. 3 **3.** 2 **5.** 4 **7.** 5 **9.** 10 **11.** $\frac{1}{2}$ **13.** $\frac{1}{10}$ **15.** 0.2 **17.** 0.4 **19.** -2 **21.** -8 **23.** $-\frac{1}{4}$ **25.** $2\sqrt[3]{2}$ **27.** $2\sqrt[3]{3}$ **29.** $5\sqrt[3]{2}$ **31.** $3\sqrt[3]{10}$ **33.** $10\sqrt[4]{2}$ **35.** $10\sqrt[5]{5}$

7.5 Exercises, page 307

1. true **3.** true **5.** false; a rational number **7.** false; real **9.** integers, rational numbers, real numbers **11.** irrational numbers, real numbers **13.** rational numbers, real numbers **15.** irrational numbers, real numbers **17.** rational numbers, real numbers **19.** rational numbers, real numbers

21. counting numbers, whole numbers, integers, rational numbers, real numbers **23.** rational numbers, real numbers **25.** irrational numbers, real numbers **27.** whole numbers, integers, rational numbers, real numbers **29.** $<$ **31.** $=$ **33.** $>$ **35.** $=$ **37.** $<$ **39.** $<$ **41.** $<$ **43.** $>$ **45.** $>$ **47.** $=$

Chapter 7 Review Exercises, page 309

1. 7 **3.** $\pm\frac{3}{5}$ **5.** -0.6 **7.** -70 **9.** -0.03 **11.** 100 **13.** 3.7 **15.** 18 **17.** 127 **19.** 0.28 **21.** 6,580 **23.** 0.365 **25.** 2.43 **27.** 25.2 **29.** 197 **31.** 0.530 **33.** 7,331 **35.** 0.797 **37.** $10\sqrt{2}$ **39.** $2\sqrt{2}$ **41.** $2\sqrt{6}$ **43.** 2 **45.** 3 **47.** -1 **49.** irrational numbers, real numbers **51.** rational numbers, real numbers **53.** rational numbers, real numbers **55.** rational numbers, real numbers **57.** counting numbers, whole numbers, integers, rational numbers, real numbers **59.** irrational numbers, real numbers **61.** $>$ **63.** $=$ **65.** $=$ **67.** $>$ **69.** $>$ **71.** $>$

Chapter 8

8.1 Exercises, page 315

1. true **3.** false; centimeters **5.** 96 **7.** $2\frac{1}{3}$ **9.** 17 **11.** 86 **13.** $7\frac{1}{3}$ **15.** 3,168 **17.** 282 **19.** 9 **21.** 880 **23.** 9.5 **25.** 15 **27.** 500 **29.** 19,000 **31.** 3,500 **33.** 38 **35.** 5.3 **37.** 700 **39.** 5.9 **41.** 20,000 **43.** 85.7 **45.** 72 **47.** 81 **49.** 700,000 **51.** 9 cm **53.** 91 m **55.** 68.5 cm **57.** 3.2 m **59.** 15 cm **61.** 12,755 km **63.** 14; 13.78 **65.** 150; 152.4 **67.** 2; 1.829 **69.** 720; 724.1 **71.** 450; 457.2 **73.** Minneapolis **75.** 26 tiles **77.** 66.6 ft **79.** \$15.94

8.2 Exercises, page 320

1. false; decreases **3.** false; pounds **5.** 2 **7.** 1,000 **9.** 124 **11.** 137 **13.** 106 **15.** 86 **17.** 0.125 **19.** 5 **21.** 6,800 **23.** 2.4 **25.** 5,000 **27.** 3.5 **29.** 5,000 **31.** 0.0067 **33.** 0.75 **35.** 347,000 **37.** 4,180 **39.** 3,000,000 **41.** 0.088 **43.** 0.0095 **45.** 625 **47.** 70,000 **49.** 454 g **51.** 300 g **53.** 5 g **55.** 1 t **57.** 15; 13.61 **59.** 8.8; 8.82 **61.** 7.2; 6.532 **63.** 21,000; 19,845 **65.** 96; 105.8 **67.** 50 g **69.** chocolate bar without nuts **71.** 4.5 kg **73.** 420-g package

8.3 Exercises, page 324

1. false; more than **3.** 100 **5.** 144 **7.** $2\frac{1}{2}$ **9.** 1.75 **11.** 21 **13.** 12 **15.** 11 **17.** 74 **19.** $1\frac{5}{8}$ **21.** $1\frac{1}{8}$ **23.** 3 **25.** 0.85 **27.** 0.062 **29.** 57,600 **31.** 8.8 **33.** 897 **35.** 7,500 **37.** 5,750 **39.** 3.62 **41.** 69 **43.** 0.92 **45.** 2,100 mL **47.** 3.8 L **49.** 70 L **51.** 0.4 kL **53.** 9; 8.517 **55.** 5; 5.073 **57.** 32; 30.43 **59.** 20; 20.29 **61.** 100; 101.4 **63.** milk **65.** 24 doses **67.** 1.893 L **69.** \$0.50

8.4 Exercises, page 329

1. false; C **3.** true **5.** 38°C **7.** 190°C **9.** 30°C **11.** 20 **13.** -5 **15.** 95 **17.** 239 **19.** -20 **21.** 23 **23.** 153 **25.** 75 **27.** 11 **29.** 54 **31.** 28 **33.** 22; 25 **35.** 142; 149 **37.** 14; 15 **39.** 192; 194 **41.** -36; -40 **43.** -2; $-2\frac{2}{9}$ **45.** 294; $285\frac{4}{5}$

8.5 Exercises, page 332

1. 16 ft 1 in. **3.** 6 lb 4 oz **5.** 12 wk 1 da **7.** 7 hr 1 min **9.** 3 ft 8 in. **11.** 5 lb 11 oz **13.** 4 da **15.** 5 hr 52 min 33 sec **17.** 2 hr 25 min **19.** 5 lb 2 oz **21.** 5 in. **23.** 2 lb 9 oz

Chapter 8 Review Exercises, page 333

1. false; hundredth **3.** true **5.** 23.8 m **7.** 232 g **9.** 2 L **11.** 6 cm **13.** 37°C **15.** $\frac{2}{3}$ **17.** 2.25 **19.** 3 **21.** 0.3 **23.** 8.2 **25.** 53,000 **27.** 66 **29.** 10,500 **31.** $1\frac{2}{3}$ **33.** 920 **35.** 230 **37.** 860 **39.** 10 **41.** 185 **43.** 75 **45.** 18,000 **47.** 0.063 **49.** $-6\frac{1}{9}$ **51.** 60; 59.06 **53.** 20; 19.96 **55.** 12; 11.36 **57.** 25; $22\frac{2}{9}$ **59.** 16 sandwiches **61.** 8,914 km **63.** 237 mL bottle **65.** 102.2°F **67.** 2 hr 15 min

Chapter 9

9.1 Exercises, page 337

1. false; perpendicular **3.** false; two **5.** false; 90° **7.** false; five **9.** false; 90° **11.** true **13.** 45° **15.** 180° **17.** 30° **19.** parallelogram **21.** right triangle **23.** quadrilateral **25.** pentagon

9.2 Exercises, page 341

1. true **3.** false; sum **5.** 17 m **7.** 14 mi **9.** 33 mm **11.** 26 ft **13.** 28 in. **15.** 24 km **17.** 4.8 m **19.** $17\frac{1}{2}$ yd **21.** 18 cm **23.** 60 mm **25.** 45 dm **27.** 66 in. **29.** 36 ft **31.** 156 cm **33.** 13 ft **35.** 21.84 m

9.3 Exercises, page 351

1. false; square **3.** true **5.** 108 ft^2 **7.** 56.7 cm^2 **9.** $11\frac{7}{8}$ mi^2 **11.** 120 ft^2 **13.** 1.69 ft^2 **15.** 1,936 cm^2 **17.** $21\frac{7}{9}$ yd^2 **19.** 7.56 dm^2 **21.** 10 yd^2 **23.** 3 yd^2 **25.** $3\frac{15}{16}$ m^2 **27.** 3 ft^2 **29.** 45 ft^2 **31.** 16.43 cm^2 **33.** 50 in.^{2} **35.** 2 **37.** 288 **39.** 3 **41.** 0.75 **43.** 700 **45.** 0.25 **47.** 32,670 **49.** 0.0008 **51.** 26.25 ft^2 **53.** 486 in.^{2} **55.** 875 tiles **57.** 1,500 ha **59.** $526.80 **61.** 12 in. **63.** $1\frac{3}{5}$ cm **65.** 3 in. **67.** 9 dm **69.** $60\frac{1}{8}$ ft^2 **71.** 25 ft^2

9.4 Exercises, page 357

1. 10 in. **3.** 20 m **5.** 12 yd **7.** 1.5 cm **9.** 1,500 mm **11.** 21.7 cm **13.** 40.3 yd **15.** no **17.** yes **19.** yes **21.** yes **23.** 13 ft **25.** 1,000 mi **27.** 16 ft **29.** 13.5 m^2
31. $a = 12$ mm; $b = 9$ mm; perimeter = 42 mm; area = 84 mm^2

9.5 Exercises, page 363

1. false; radius **3.** false; diameter **5.** 31π in. **7.** 14.4π cm **9.** 198 in. **11.** 11 km **13.** 21.98 ft **15.** $27\frac{19}{40}$ m **17.** 5.28 hm **19.** 1,701.88 ft **21.** 361π dm^2 **23.** 6.76π km^2 **25.** 15,400 mm^2 **27.** $21\frac{21}{32}$ yd^2 **29.** 282,600 cm^2 **31.** $55\frac{11}{25}$ hm^2 **33.** 28.26 ft^2 **35.** 6,443.5626 mi^2 **37.** 69.08 in. **39.** $9\frac{5}{8}$ ft^2 **41.** 81.64 in. **43.** 37 cm **45.** 14,826 ft^2 **47.** 487.485 m^2

9.6 Exercises, page 372

1. false; cubic **3.** false; feet **5.** 96 in.^{3} **7.** $116\frac{2}{3}$ yd^3 **9.** $13\frac{1}{2}$ ft^3 **11.** 64 cm^3 **13.** 1.728 in.^{3} **15.** $11\frac{25}{64}$ dm^3 **17.** 36π cm^3 **19.** 15.7 ft^3 **21.** 192π dm^3 **23.** $457\frac{1}{3}\pi$ ft^3 **25.** 113.04 cm^3 **27.** 562.5π in.^{3} **29.** 1,526.04 ft^3 **31.** 3,140 ft^3 **33.** 200.96 in.^{3} **35.** $\frac{1}{27}\pi$ ft^3 **37.** 312.9 cm^3 **39.** 5,184 **41.** $115\frac{1}{2}$ **43.** 9,000 **45.** 7 **47.** 19 **49.** 60 **51.** 0.8 **53.** 600,000,000 **55.** 523,600 gal **57.** 4,851 in.^{3} **59.** 462 mL **61.** 376.8 ft^3 **63.** 1,582.56 in.^{3}

Chapter 9 Review Exercises, page 375

1. false; never **3.** false; area **5.** false; perimeter **7.** 1,280 **9.** 4 **11.** 4 **13.** 0.025 **15.** 345.6 **17.** 0.1 **19.** 14 **21.** 5.4 **23.** 22 cm **25.** $17\frac{3}{4}$ in. **27.** 14π mm **29.** 12 ft **31.** 63 ft^2 **33.** 57 m^2 **35.** 30 ft^2 **37.** 24 m^2 **39.** $81\pi/4$ **41.** 27 dm^2 **43.** 512 in.^{3} **45.** 36π in.^{3} **47.** 8π m^3 **49.** 4,050 cm^3 **51.** 6.1 m **53.** 5 ft **55.** 420 ft^2 **57.** 37.5 cm **59.** 60 ft **61.** 7.1 in. **63.** 62 m **65.** 4 cm **67.** 24,891 mi **69.** 16.7 L

Chapter 10

10.1 Exercises, page 381

1. true **3.** true **5.** terms, 1; variable, w; coefficient, 3 **7.** terms, 2; variable, x; coefficient, 2; variable, y; coefficient, 5 **9.** terms, 3; variable, p; coefficient, 3; variable, q; coefficient, 9; variable, r; coefficient, 1 **11.** terms, 2; variable, c; coefficient, $\frac{1}{2}$; variable, d; coefficient, -1 **13.** 4 **15.** 27 **17.** -4 **19.** -4 **21.** 6 **23.** -90 **25.** 170 **27.** -24 **29.** $x - 5$ **31.** $-2t$ **33.** $\frac{2}{3}z$ **35.** $x - 35$ **37.** $7a + 5$ **39.** $-8(a + b)$ **41.** $(14 - x) \div 2$ **43.** $2(a + 5)$ **45.** $\frac{5}{8}(u + 7)$ **47.** Four less than x **49.** The product of six and a **51.** Nine more than t **53.** w divided by eleven **55.** Eight more than the product of two and y **57.** Two less than the product of three and n **59.** Five times the quantity z minus seven **61.** The quantity four minus u, divided by three **63.** Three-eighths the quantity x plus five

10.2 Exercises, page 384

1. false; variable **3.** true **5.** $17x$ **7.** $2a$ **9.** $-8t$ **11.** $\frac{1}{2}a$ **13.** $-53u$ **15.** $2.3p$ **17.** $15y$
19. $14z$ **21.** $43c$ **23.** $-14.2u$ **25.** $5y + 7$ **27.** $43 - 8n$ **29.** $1 - 9w$ **31.** $5 - 5p$
33. $-3x + 20$ **35.** $-11s + 6$ **37.** -12 **39.** $68z - 64$ **41.** $-22t + 40$ **43.** $-62k - 9$

10.3 Exercises, page 389

1. false; solution **3.** true **5.** yes **7.** no **9.** no **11.** no **13.** yes **15.** no **17.** $x = 8$
19. $y = -3$ **21.** $w = -2$ **23.** $x = -2$ **25.** $y = 8$ **27.** $a = 45$ **29.** $z = 81$ **31.** $x = 4.8$
33. $x = -5$ **35.** $y = 4$ **37.** $z = 3$ **39.** $t = -4$ **41.** $t = 4$ **43.** $p = -13$ **45.** $w = -12$
47. $x = -11$ **49.** $x = -2$ **51.** $x = -9$ **53.** $c = -8$ **55.** $d = -11$ **57.** $x = 7$ **59.** $w = 2$
61. $a = 6$ **63.** $b = -9$ **65.** $x = 40$ **67.** $x = -24$ **69.** $x = -44$ **71.** $x = 13$ **73.** $x = 30$
75. $t = -8$ **77.** $x = 16$ **79.** $x = 11$

10.4 Exercises, page 393

1. $x = 7$ **3.** $x = -9$ **5.** $x = 8$ **7.** $x = -9$ **9.** $z = 5$ **11.** $p = \frac{19}{7}$ **13.** $a = -6$ **15.** $x = 0$
17. $w = 15$ **19.** $x = 15$ **21.** $x = -72$ **23.** $x = 28$ **25.** $t = -4$ **27.** $y = \frac{35}{18}$ **29.** $z = \frac{27}{28}$
31. $m = 8$ **33.** $y = -5$ **35.** $x = 6$ **37.** $x = -9$ **39.** $t = 9$ **41.** $w = 4$ **43.** $z = -5$ **45.** $y = 8$
47. $r = -3$

10.5 Exercises, page 398

1. true **3.** false; distributive **5.** $y = 3$ **7.** $x = -9$ **9.** $z = -10$ **11.** $t = \frac{1}{3}$ **13.** $b = \frac{9}{5}$
15. $n = 2$ **17.** $x = 3$ **19.** $x = 1$ **21.** $x = \frac{9}{7}$ **23.** $a = \frac{3}{2}$ **25.** $u = -3$ **27.** $t = \frac{3}{7}$ **29.** $s = 8$
31. $p = 2$ **33.** $x = -4$ **35.** $x = -4$ **37.** $y = 4$ **39.** $y = -1$ **41.** $t = \frac{13}{4}$ **43.** $y = 5$
45. $y = 18$ **47.** $p = 3$ **49.** $z = 2$ **51.** $z = \frac{8}{3}$ **53.** $z = -20$ **55.** $t = 6$ **57.** $s = 2$ **59.** $x = -10$
61. $x = -2$

10.6 Exercises, page 401

1. 72 **3.** 6 **5.** 9 **7.** 8 **9.** 140 **11.** 3 **13.** 64π **15.** 15π **17.** 30 **19.** 16
21. $m = \frac{F}{a}$ **23.** $r = \frac{d}{t}$ **25.** $t = \frac{i}{pr}$ **27.** $m = \frac{E}{c^2}$ **29.** $w = \frac{V}{lh}$ **31.** $h = \frac{E}{mg}$ **33.** $r = \frac{A - P}{P}$
35. $r^3 = \frac{3V}{4\pi}$ **37.** $r^2 = \frac{A}{\pi}$ **39.** $r = \frac{s - a}{s}$ **41.** $r^2 = \frac{V}{\pi h}$ **43.** $h = \frac{2A}{B + b}$ **45.** 28 in.^{2} **47.** 14 ft^2
49. 8 ft

10.7 Exercises, page 405

1. 18 years old **3.** 3 and 29 **5.** 27 and 28 **7.** 18 men **9.** 660 miles **11.** length, 8 ft; width, 3 ft
13. $22,500 **15.** candy bar, 35¢; pen, 49¢; notebook, 98¢ **17.** 2 in., 6 in., 6 in. **19.** 3 lb

Chapter 10 Review Exercises, page 407

1. 11 **3.** -56 **5.** -1 **7.** $x - 7$ **9.** $5(z + 6)$ **11.** $(7 + t) \div 3$ **13.** The product of negative five and y **15.** Three times the quantity x minus two **17.** Two less than the product of six and w **19.** $-5x$ **21.** $-3y - 8$ **23.** $8x - 20$ **25.** $-2a - 29$ **27.** $x = 8$ **29.** $u = 3$ **31.** $z = -8$ **33.** $p = 5$ **35.** $a = \frac{1}{2}$ **37.** $m = -15$ **39.** $k = -30$ **41.** $x = \frac{9}{35}$ **43.** 7 **45.** 35 **47.** 4 **49.** $t = \frac{d}{r}$ **51.** $d = \frac{C}{\pi}$ **53.** $h = \frac{3V}{\pi r^2}$ **55.** 33 and 35 **57.** $3\frac{1}{2}$ hr **59.** length, 12 cm; width, 4 cm **61.** 750 ft

Chapter 11

11.1 Exercises, page 413

1. false; one **3.** false; the opposite of y squared **5.** 16 **7.** -4 **9.** 64 **11.** x^9 **13.** t^3 **15.** n^{22} **17.** y^{16} **19.** u^{40} **21.** z^{13} **23.** y^{14} **25.** a^7 **27.** q^{36} **29.** x^3y^6 **31.** $9x^6$ **33.** $125t^3$ **35.** x^6y^{21} **37.** $m^{30}n^{12}$ **39.** $-a^{10}b^5$ **41.** t^6 **43.** $27v^3$ **45.** $16y^{20}$ **47.** $-64x^6$ **49.** $\frac{27}{64}t^3$ **51.** $25x^8y^2$ **53.** $-a^9b^{12}$ **55.** $\frac{1}{9}a^2b^4$ **57.** $x^{12}y^{20}z^4$ **59.** $32a^{10}b^5c^{20}$ **61.** $(x + 2)^8$ **63.** z^{3+n} **65.** a^{3c} **67.** w^{5a} **69.** t^{9k}

11.2 Exercises, page 418

1. false; subtract **3.** true **5.** x^7 **7.** n^8 **9.** t^9 **11.** a^6 **13.** 1 **15.** c^7 **17.** x^{4n} **19.** a^{7-t} **21.** $(x - 3)^5$ **23.** $\frac{1}{64}$ **25.** $\frac{1}{81}$ **27.** 9 **29.** $\frac{2}{x^7}$ **31.** $\frac{b^5}{a^3}$ **33.** $\frac{-r^5}{st^8}$ **35.** x^5 **37.** $\frac{1}{p^5q^2}$ **39.** $\frac{yz^2}{x^3}$ **41.** $\frac{5bd^7}{a^2c^3}$ **43.** $\frac{1}{9}$ **45.** $\frac{3}{16}$ **47.** $\frac{-20}{9}$ **49.** $-\frac{2}{27}$ **51.** $-\frac{3}{4}$ **53.** $\frac{1}{a^5}$ **55.** $\frac{1}{t^{11}}$ **57.** $\frac{1}{27}$ **59.** 125 **61.** $-y^8$ **63.** n^8 **65.** $\frac{1}{k^6}$ **67.** x^3 **69.** 27 **71.** $\frac{1}{2^4}$ or $\frac{1}{16}$ **73.** $\frac{1}{m^7}$ **75.** $\frac{1}{r^{18}}$ **77.** $\frac{1}{x}$ **79.** $\frac{1}{9a^{10}}$ **81.** $\frac{1}{49}t^4$ **83.** $\frac{x^6}{y^{21}}$ **85.** $\frac{s^{15}}{r^5t^{20}}$ **87.** p^6 **89.** $\frac{1}{64}x^2y^6$ **91.** $\frac{1}{w^{22}}$ **93.** $\frac{16}{9}$ **95.** $\frac{8}{125}$ **97.** $\frac{a^{10}}{64}$ **99.** r^5s^2 **101.** $49x^6y^4$ **103.** $\frac{1}{t^{14}}$ **105.** $\frac{1}{8}k^6$ **107.** $\frac{16y^2z^2}{x^4}$

11.3 Exercises, page 423

1. $-16x^8$ **3.** $27a^8$ **5.** $5t^{11}$ **7.** $-80u^{12}$ **9.** $-15x^5y^5$ **11.** $6a^8b^9$ **13.** $21x^5y^7z^2$ **15.** $-10a^9b^7c^{13}$ **17.** $-30x^6y^5$ **19.** $40a^7b^9$ **21.** $8x^{19}y^{26}$ **23.** $324x^{10}y^{12}$ **25.** $4w^3$ **27.** $\frac{-x}{2}$ **29.** $\frac{8}{3t^8}$ **31.** $\frac{-4y^2}{x}$ **33.** $\frac{1}{3x^2y^6}$ **35.** $\frac{7b^9}{a^3}$ **37.** $\frac{y^3}{4x^3}$ **39.** $-\frac{1}{4x^3yz^5}$ **41.** $\frac{25}{27a^8b^{19}}$ **43.** xz^{11}

45. $\frac{4b^4}{a^2}$ **47.** $\frac{3t^4}{2}$ **49.** $-4a^3$ **51.** $4xy$ **53.** $-3m^4$ **55.** $12r^2s^2$ **57.** $7abc$ **59.** $-3x^2 + 8y^2$
61. $4b^5 - 3b^4$ **63.** $11xy - 5yz$ **65.** $-3ab^3 + 7a^3b^3$ **67.** $-4x^3y^2 + 7x^2y^3$ **69.** $-4t^2 - 5st + 4s^2$
71. $6w^5z^3 - 9w^3z^5$

11.4 Exercises, page 427

1. 3 **3.** 12 **5.** 2 **7.** 5 **9.** -9 **11.** $\frac{2}{3}$ **13.** -3 **15.** 8 **17.** 8 **19.** 100,000
21. -8 **23.** -32 **25.** 16 **27.** $\frac{1}{32}$ **29.** $\frac{27}{64}$ **31.** $-\frac{1}{7}$ **33.** 11 **35.** 3 **37.** 80 **39.** $\frac{1}{8}$
41. $\frac{1}{9}$ **43.** $\frac{1}{11}$ **45.** $\frac{1}{4}$ **47.** 7 **49.** $\frac{27}{8}$ **51.** $\frac{1}{10}$ **53.** $\frac{1}{2}$ **55.** $x^{3/5}$ **57.** $a^{4/3}$ **59.** w
61. $t^{1/2}$ **63.** $k^{1/2}$ **65.** $\frac{1}{p^{1/3}}$ **67.** $x^{4/3}$ **69.** m^3 **71.** $\frac{1}{a^2}$ **73.** $2x^{2/3}$ **75.** xy^4 **77.** $9a^6b^8$
79. $-2x^{1/3}y$ **81.** $7x^3y^{1/2}z^2$ **83.** $8ab^2$ **85.** $\frac{25y^2}{xz^4}$ **87.** $\frac{y}{16x^{2/3}}$

11.5 Exercises, page 430

1. 6.8×10^4 **3.** 5×10^{-5} **5.** 3.56×10^2 **7.** 4.8×10^9 **9.** 1.4×10^{-6} **11.** 5.07×10^5
13. 3.2564×10^{-2} **15.** 5,400 **17.** 0.000038 **19.** 32.7 **21.** 0.000008 **23.** 12,750,000 **25.** 0.00604
27. 536,281.4 **29.** 2.1×10^{10} **31.** 3.6×10^{-10} **33.** 4.5×10^4 **35.** 4.2×10^{12} **37.** 1.4×10^{-9}
39. 6.3×10^{-1} **41.** $1.2 \times 10^0 = 1.2$ **43.** 7×10^{12} **45.** 6.4×10^{-9} **47.** 5.4×10^{-2} **49.** 7×10^2
51. 9×10^1 **53.** 5×10^{-12} **55.** 9×10^4 **57.** 1.2×10^{-2} **59.** 7×10^6 **61.** 8×10^{10} **63.** 4×10^{-7}
65. 3×10^8 **67.** 1×10^{20} **69.** 2.39×10^{-4} **71.** 300,000,000 **73.** 9.3×10^7

Chapter 11 Review Exercises, page 432

1. -64 **3.** 8 **5.** $\frac{1}{4}$ **7.** $-\frac{1}{4}$ **9.** $\frac{1}{4}$ **11.** 2 **13.** p^{13} **15.** t^{21} **17.** a^6b^{15} **19.** $64x^6y^2$
21. $\frac{1}{25b^6}$ **23.** $\frac{y^{18}}{x^{15}}$ **25.** q^9 **27.** $-\frac{1}{z^{21}}$ **29.** $\frac{36}{x^6y^{10}z^8}$ **31.** 32 **33.** $\frac{1}{a}$ **35.** $\frac{y^3}{7x^{3/2}z^4}$
37. $-15a^{11}$ **39.** $\frac{3}{4t^5}$ **41.** $14x^8y^7$ **43.** $72x^{21}y^{14}$ **45.** $\frac{-8x^3}{y^5}$ **47.** $10a^3b^5$ **49.** $12wz$ **51.** $2x^5$
53. $-3pq^4 + p^4q$ **55.** 3.5×10^{10} **57.** 7×10^5 **59.** 1.2×10^{-2} **61.** 7×10^{-7}
63. 8.4×10^7 **65.** 3×10^3

Chapter 12

12.1 Exercises, page 439

1. false; binomial **3.** false; descending **5.** $-4x^2 + 7$; second degree **7.** $4a^3 + 2a - 9$; third degree
9. $6t^5 + 2t^3 - 3t^2 - 4t$; fifth degree **11.** $3a + 2b$ **13.** $x^2 - x - 7$ **15.** $-4y^3 + 2y^2 - 13y - 7$ **17.** $-2x - 4y$
19. $-3t^2 + 3t + 3$ **21.** $-9a - 2b - c$ **23.** $-6z^2 + 10z + 8$ **25.** $8r + 6s - 7t$ **27.** $3x^3 - 8x^2 + 7x - 1$
29. $8w^2 - 7w + 3$ **31.** $-2x - 7y$ **33.** $6t^2 - 16t + 11$ **35.** $-8z^3 - 5z^2 + 11z - 4$ **37.** $5r + 8s$
39. $7a^2 - 11a + 11$ **41.** $10x^2 + 4xy + 4y^2$ **43.** $-9x + 8y + 12z$ **45.** $-7y^2 - 7y + 2$ **47.** $7z^3 - 7z^2 - z + 5$

49. $-4p^3 - 10p^2 + 15p + 13$ **51.** $3a - 8b$ **53.** $11x - 19$ **55.** $3x^2 - 8x + 4$ **57.** $2x - 10y$ **59.** $-5t - 1$ **61.** $-3w^2 + 17w - 14$ **63.** $4a + 8b$

12.2 Exercises, page 443

1. $21a - 6ab$ **3.** $-16w^2 + 18w$ **5.** $21p - 12q$ **7.** $-2a^3 + 12a^2 + 6a$ **9.** $-15x^2y + 12xy^2$ **11.** $2a^3b^2 - 8ab^3$ **13.** $20xy^2 - 35xy^3$ **15.** $-5t^3 + 20t^2 - 40t$ **17.** $-15a^3b^2 - 21a^2b^3 + 15a^5b^2$ **19.** $2a^3b^2c - 14ab^2c^3 - 6a^2bc^2$ **21.** $17x^2 - 21x$ **23.** $-23s^2t + 3s^3$ **25.** $26x + 32y$ **27.** $-n^2 - 13n$ **29.** $2w^3 + 14w^2 - 10w - 28$ **31.** $-22x^3y + 21x^2y^2 - 10xy^2$ **33.** $-95z^2 - 175z$ **35.** $-x^2y^2 + 2xy^2 - 7xy$ **37.** $14a^3 + 267a^2$ **39.** $x^2 + 2x - 8$ **41.** $3ab - 24a - 4b + 32$ **43.** $x^2 - 64$ **45.** $10y^2 - 3y - 27$ **47.** $-15x^2 + 11xy + 14y^2$ **49.** $49t^2 - 42t + 9$ **51.** $x^3 - x^2 - 2x + 8$ **53.** $3z^3 + 8z^2 - 39z - 20$ **55.** $40m^3 + 17m^2 - 2m + 8$ **57.** $4x^4 - 26x^3 + 19x^2 + 13x - 3$ **59.** $-72w^3 + 180w^2$ **61.** $-6x^3 + 39x^2 + 21x$ **63.** $y^3 + 6y^2 + 12y + 8$ **65.** $6p^3 - 23p^2 - 118p - 105$ **67.** $4x^3 - 24x^2 + 36x$ **69.** $-5x^3 - 20x^2 - 20x$

12.3 Exercises, page 447

1. $x^2 + 12x + 35$ **3.** $x^2 - 6x - 27$ **5.** $x^2 - 6x + 8$ **7.** $a^2 - a - 42$ **9.** $t^2 - 9t + 8$ **11.** $z^2 + 11z + 18$ **13.** $-x^2 - x + 72$ **15.** $3x^2 + 25x + 28$ **17.** $2y^2 - y - 45$ **19.** $12x^2 - 40x - 7$ **21.** $3y^2 + 27y + 42$ **23.** $21y^2 - 17y + 2$ **25.** $14w^2 + 29w + 12$ **27.** $-15x^2 + 4x + 32$ **29.** $9x^2 - 35xy - 4y^2$ **31.** $16x^2 + 62xy + 21y^2$ **33.** $a^4 + 2a^2 - 15$ **35.** $5x^3 - 40x^2 - 2x + 16$ **37.** $x^2 + 8x + 16$ **39.** $y^2 - 16y + 64$ **41.** $z^2 + 12z + 36$ **43.** $y^2 + 6y + 9$ **45.** $p^2 - 18p + 81$ **47.** $16x^2 + 8x + 1$ **49.** $64x^2 - 48x + 9$ **51.** $4x^2 + 28x + 49$ **53.** $64y^2 - 80y + 25$ **55.** $x^2 + 6xy + 9y^2$ **57.** $25x^2 - 10xy + y^2$ **59.** $9x^2 + 12xy + 4y^2$ **61.** $81x^2 - 72xy + 16y^2$ **63.** $25a^2 + 80ab + 64b^2$ **65.** $x^4 + 6x^2 + 9$ **67.** $4x^4 - 4x^2 + 1$ **69.** $x^2 - 64$ **71.** $x^2 - 1$ **73.** $9 - y^2$ **75.** $16x^2 - 9$ **77.** $81x^2 - 4$ **79.** $9 - 25a^2$ **81.** $x^2 - 49y^2$ **83.** $16x^2 - 9y^2$ **85.** $x^4 - 49$

12.4 Exercises, page 449

1. $6x + 4y$ **3.** $-2a - 4b$ **5.** $2p + 4q - r$ **7.** $-5x - 6y$ **9.** $-5t + 2$ **11.** $4x + 2$ **13.** $5a^2 - 3a - 2$ **15.** $-4z^3 + 7z^2 - 8z$ **17.** $\frac{p}{3} - 3 + \frac{2}{p}$ **19.** $6c^2 - 3c - 9$ **21.** $9t^3 - 5t + 4$ **23.** $-7y^3 + 3y + \frac{2}{y}$ **25.** $\frac{x^3}{2} - x - \frac{2}{x}$ **27.** $-\frac{2a^2}{3} + 3 + \frac{2}{a}$ **29.** $x^2y - x$ **31.** $3b - \frac{4}{ab}$ **33.** $a^4b^3 - a^2 - ab^2$ **35.** $-5x + 3y^2 + 2y$ **37.** $4x^2y - 5x + 3y$ **39.** $2a^2 - \frac{1}{2b} + \frac{3}{2a^2}$

12.5 Exercises, page 453

1. $8x + 5$ **3.** $9x + 4$ **5.** $9y + 2$ **7.** $4t^2 - 5t + 2$ **9.** $7x^2 + 5x + 4 - \frac{3}{x + 5}$ **11.** $3a^2 - 6a - 2$ **13.** $x^2 + 4x - 3 + \frac{4}{9x - 1}$ **15.** $\frac{1}{3}x^2 - x + 2$ **17.** $x^2 - 4x + 16$ **19.** $x - 6 + \frac{72}{x + 6}$ **21.** $3x^2 + 6x + 5 + \frac{15}{x - 2}$ **23.** $z^2 - 2z + 4 - \frac{16}{z + 2}$ **25.** $5x - 6$ **27.** $a^3 + 3a + 9 + \frac{9a + 22}{a^2 - 3}$ **29.** $x^2 - 7xy + 2y^2$

Chapter 12 Review Exercises, page 455

1. false; not equal **3.** true **5.** $3x + 10y$ **7.** $4a + b$ **9.** $8x^2 - 10x + 11$ **11.** $-3z^2 - 15z - 7$ **13.** $28x^2 - 56x$ **15.** $-15a^3b + 6ab^3$ **17.** $12x^5 - 30x^4y + 42x^3y^2$ **19.** $-38x^2 - 26x$ **21.** $-24n^3 + 84n^2$ **23.** $-12x - 1$ **25.** $a^2 + 2a - 15$ **27.** $z^2 + 12z + 32$ **29.** $10x^2 - 39x + 35$ **31.** $-28y^2 + 15y + 27$ **33.** $10a^2 + 47ab + 9b^2$ **35.** $t^2 + 14t + 49$ **37.** $9n^2 - 48n + 64$ **39.** $49x^2 + 56xy + 16y^2$ **41.** $a^2 - 49$ **43.** $4x^2 - 25$ **45.** $16a^2 - 25b^2$ **47.** $2x + 3y$ **49.** $3x - 5$ **51.** $-6a^2 - 2a + 7$ **53.** $7t^3 - 5t + 3$ **55.** $5x^2 + \frac{6x}{y} + 9y$ **57.** $2x + 9 + \frac{2}{x - 3}$ **59.** $9x + 7$ **61.** $x^2 + 8x - 7 - \frac{4}{x + 2}$ **63.** $x^2 + x + 1$

Chapter 13

13.1 Exercises, page 461

1. 5 **3.** $3x$ **5.** $4z^2$ **7.** a^2b^2 **9.** $6xy^2$ **11.** $a + b$ **13.** $3p^2(p + 2)$ **15.** $3(3x + y)$
17. $8(p - 3)$ **19.** $9(1 + 3t)$ **21.** $x(a + b)$ **23.** $8(x^2 + y^2)$ **25.** $5a(3a - 1)$ **27.** $n^3(n^2 + 1)$
29. $-3x(x + 2)$ **31.** $7x^2(2x + 3)$ **33.** $8p^3(5p - 1)$ **35.** $5y(3y^2 + 6y - 4)$ **37.** $18t(4t^7 + 2t^4 + 3)$
39. $x^3y(x^2y + 2)$ **41.** $3x^2y^3(xy - 2)$ **43.** $9ab^3(7a + 3b^2)$ **45.** $-13x^2y(xy^4 + 2)$ **47.** $x^3y^2(4x^2y + x + 3)$
49. $6ab(4ab - 8 + 3b^2)$ **51.** $9x^2y(7y^4 - 3xy - 2)$ **53.** $6x^2yz^2(7xy + 3z^2 - 6y^2z^3)$ **55.** $(a - b)(3x + y)$
57. $5(x + y)(x + 2)$ **59.** $3a(a - 5)(3a + 2)$ **61.** $(x + 7)(x + 2)$ **63.** $(x + 5)(x - 8)$ **65.** $(x - 4)(x - 7)$
67. $(3x + 4)(x + 5)$ **69.** $(4x - 3)(x - 7)$ **71.** $(3x^2 + 8)(x + 4)$ **73.** correct **75.** incorrect; $7x(3x - 1)$
77. incorrect; $5y(5y^2 + 6y + 1)$ **79.** incorrect; $-3z(z + 2)$ **81.** incorrect; $4xy(x^2y - 2xy^4 + 3)$ **83.** correct

13.2 Exercises, page 466

1. $(x + 3)(x - 3)$ **3.** $(a + 10)(a - 10)$ **5.** $(t + 1)(t - 1)$ **7.** $(7 + n)(7 - n)$ **9.** $(8 + p)(8 - p)$
11. $(2x + 9)(2x - 9)$ **13.** $(5z + 12)(5z - 12)$ **15.** $(5 + 6y)(5 - 6y)$ **17.** $(4x + 11y)(4x - 11y)$ **19.** $(x + 7)^2$
21. $(y - 6)^2$ **23.** $(a + 2)^2$ **25.** $(z - 1)^2$ **27.** $(t + 10)^2$ **29.** $(p - 8)^2$ **31.** $(y + 13)^2$ **33.** $(3a + 5)^2$
35. $(4x - 3)^2$ **37.** $(7t - 2)^2$ **39.** $(6x + y)^2$ **41.** $(a - 5b)^2$ **43.** $(3x + 4y)^2$ **45.** $(t - 5)(t^2 + 5t + 25)$
47. $(4 - w)(16 + 4w + w^2)$ **49.** $(x + 10)(x^2 - 10x + 100)$ **51.** $(6 + y)(36 - 6y + y^2)$ **53.** $(a + 2b)(a^2 - 2ab + 4b^2)$
55. $(3a - b)(9a^2 + 3ab + b^2)$ **57.** $(4x + 5y)(16x^2 - 20xy + 25y^2)$ **59.** $(6a - 5b)(36a^2 + 30ab + 25b^2)$
61. $3(x + 5)(x - 5)$ **63.** $2(y + 4)(y - 4)$ **65.** $2(6 + t)(6 - t)$ **67.** $3(x + 3)^2$ **69.** $5(b - 4)^2$ **71.** $2(t + 6)^2$
73. $x(x - 7)^2$ **75.** $2t(t + 8)^2$ **77.** $2x(x + 6)(x - 6)$ **79.** $(x^2 + 4)(x + 2)(x - 2)$ **81.** $ab(a + 5b)(a - 5b)$
83. $2y(x + 7y)(x - 7y)$ **85.** incorrect; $x^2 - 9$ **87.** correct **89.** correct **91.** incorrect; $3(w - 5)(w - 5)$
93. correct

13.3 Exercises, page 471

1. $(x + 5)(x + 4)$ **3.** $(w + 7)(w - 2)$ **5.** $(a + 6)(a - 4)$ **7.** $(t - 9)(t - 8)$ **9.** $(x - 9)(x + 2)$
11. $(y - 5)(y + 4)$ **13.** $(w - 8)(w + 2)$ **15.** $(n + 6)(n + 5)$ **17.** $(u - 9)(u - 6)$ **19.** $(r - 5)(r + 2)$
21. $(x + 3)(x - 2)$ **23.** $(z + 7)(z + 5)$ **25.** $(s - 6)(s + 3)$ **27.** $(x + 5y)(x + 3y)$ **29.** $(a + 9b)(a - 7b)$
31. $(s - 8t)(s - 3t)$ **33.** $(x + 9y)(x - y)$ **35.** $(a + 7b)(a + 3b)$ **37.** $(s - 9t)(s - 4t)$ **39.** $(x - 8y)(x + 5y)$
41. $(2x + 5)(x + 2)$ **43.** $(3y + 8)(y - 9)$ **45.** $(4a - 1)(a - 2)$ **47.** $(7t - 1)(t + 5)$ **49.** $(3s + 8)(s - 4)$
51. $(2x - 5)(2x - 3)$ **53.** $(5y - 1)(y + 6)$ **55.** $(2a - 7)(3a + 5)$ **57.** $(4y + 1)(3y + 7)$ **59.** $(3t + 2)(3t - 4)$
61. $(5x - 8)(4x - 7)$ **63.** $(2x - 5y)(x + y)$ **65.** $(5a + 3b)(a + 6b)$ **67.** $(2x + 9y)(2x - 5y)$
69. $(2a - 3b)(3a - 8b)$ **71.** $2(x + 7)(x + 5)$ **73.** $5(w - 8)(w - 2)$ **75.** $6(p - 7)(p + 3)$ **77.** $-(x - 5)(x + 1)$
79. $-2(z - 7)(z + 5)$ **81.** $x(x + 9)(x + 6)$ **83.** $a^2(a + 7)(a + 4)$ **85.** $3x(x - 8)(x - 2)$ **87.** $3y(x - 8)(x + 2)$
89. $a(a - 5b)(a - 2b)$ **91.** $2(2x - 3)(x + 5)$ **93.** $4(4a - 3)(a - 1)$ **95.** $5(2w + 1)(3w - 1)$
97. $-(2x + 5)(2x + 3)$ **99.** $-5(3y + 5)(y - 2)$ **101.** $x(2x + 3)(2x - 5)$ **103.** $a^3(3a - 2)(a - 4)$
105. $3t(5t + 6)(t + 1)$ **107.** $4z(4z + 3)(z + 2)$ **109.** $x(2x + y)(x + 4y)$

13.4 Exercises, page 474

1. $2(x + 7)(x - 4)$ **3.** $4x(x + 5)(x - 5)$ **5.** $x(x - 8)^2$ **7.** $(x + 9)(x + 7)$ **9.** $3x^2(x + 4)(x - 4)$
11. $(2x + 7)(x + 3)$ **13.** $(x^2 + 9)(x + 3)(x - 3)$ **15.** $x^3(x + 7)(x + 6)$ **17.** $9x(x - 2)^2$ **19.** $(3x - 7)(x + 2)$
21. $-x(x - 8)(x + 6)$ **23.** $8xy(x + 3)(x + 1)$ **25.** $xy^2(x + 6)(x - 6)$ **27.** $5x^2(x + 3)^2$ **29.** $4x(x - 2)(x - 7)$
31. $y(3x + 1)(x - 4)$ **33.** $(5x^2 + 3)(x - 8)$ **35.** $(x + 1)(x - 1)(x - 7)$ **37.** $5xy(x + 4)(x - 3)$
39. $xy^2(x + 6)(x - 8)$ **41.** $5y(x + 2)(x^2 - 2x + 4)$ **43.** $b^2(a - 3b)(a^2 + 3ab + 9b^2)$

13.5 Exercises, page 477

1. $x = -5, -8$ **3.** $z = 4, 6$ **5.** $t = 3, -6$ **7.** $m = 9, -1$ **9.** $x = -6, -7$ **11.** $y = -7$

13. $t = 0, -5$ **15.** $y = 7, -7$ **17.** $y = 5, -5$ **19.** $x = -\frac{5}{2}, -4$ **21.** $z = \frac{6}{5}, 2$ **23.** $a = \frac{5}{7}, -2$
25. $x = 0, 8$ **27.** $y = 0, -1$ **29.** $x = \frac{5}{4}, -\frac{5}{4}$ **31.** $a = -3, -7$ **33.** $w = -3$ **35.** $t = -\frac{7}{3}, -\frac{3}{2}$
37. $x = 0, \frac{3}{7}$ **39.** $x = \frac{7}{5}$ **41.** $x = 0, 5$ **43.** $y = 0, 8, -3$ **45.** $a = 0, 3, 4$ **47.** $z = 0, -6, -\frac{7}{2}$
49. $t = 0, -\frac{1}{5}, 3$ **51.** $x = 0, -\frac{7}{6}, 2$ **53.** $w = 0, -\frac{3}{7}, \frac{1}{3}$ **55.** $a = 0, \frac{2}{5}, -\frac{2}{5}$ **57.** $x = 0, -\frac{3}{7}, -\frac{3}{2}$

13.6 Exercises, page 481

1. 6 and 7 or −7 and −6 **3.** 7 and 9 or −9 and −7 **5.** 5 and 13 **7.** father: 32 years old; daughter: 10 years old
9. length: 11 inches; width: 3 inches **11.** base: 8 m; height: 3 m **13.** length: 8 yards; width: 7 yards
15. 12 cm and 16 cm **17.** 5 feet **19.** base: 15 cm; height: 5 cm

Chapter 13 Review Exercises, page 483

1. false; first **3.** false; second **5.** true **7.** $6(x - 9)$ **9.** $5a(3a + 1)$ **11.** $3w^2(4w^4 + 3w - 8)$
13. $-4x^2y^2(y - 5x^3)$ **15.** $2(a + b)(x + 4y)$ **17.** $(x + 4)(x + 8)$ **19.** $(y + 8)(y - 8)$ **21.** $(2x + 5)(2x - 5)$
23. $(t + 9)^2$ **25.** $(p - 11)^2$ **27.** $(a + 3b)^2$ **29.** $(w + 5)(w^2 - 5w + 25)$ **31.** $(x + 4)(x + 3)$ **33.** $(w - 6)(w + 4)$
35. $(t + 7)(t + 9)$ **37.** $(x + 5y)(x - 3y)$ **39.** $(2x + 5)(x + 6)$ **41.** $(x - 5)(3x - 2)$ **43.** $(5w - 3)(w + 6)$
45. $(3a - 4)(3a - 2)$ **47.** $(4x + 3)(x - 7)$ **49.** $(2x + 3y)(x - 4y)$ **51.** $(x^2 + 8)(x - 7)$ **53.** $2(x + 5)(x - 5)$
55. $4(y + 3)^2$ **57.** $7(z - 2)^2$ **59.** $xy(x + 7y)(x - 7y)$ **61.** $x(x - 6)(x - 7)$ **63.** $4(x - 5)(x + 3)$
65. $-(y + 6)(y + 4)$ **67.** $a(a - 5)(a - 3)$ **69.** $5t(t + 6)(t - 3)$ **71.** $6(2x + 1)(x + 5)$ **73.** $-(5y - 4)(y - 3)$
75. $2p(2p + 7)(p - 6)$ **77.** $2(x^2 - 3)(x + 9)$ **79.** $x(x - 5y)(x + 3y)$ **81.** $x = 7, -5$ **83.** $a = 8$
85. $w = 0, 4$ **87.** $x = -\frac{7}{2}, -4$ **89.** $p = 6, -2$ **91.** $x = 0, -8, -9$ **93.** $a = 0, 7, -7$ **95.** $t = 0, \frac{6}{5}, 3$
97. 3 and 5 **99.** length: 9 feet; width: 3 feet

Chapter 14

14.1 Exercises, page 491

1. $x > 7$

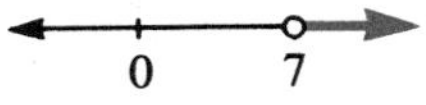

3. $y \leqslant -2$

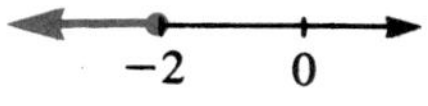

5. $t \geqslant \frac{1}{3}$

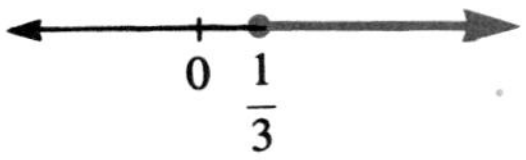

7. $p < -4.9$

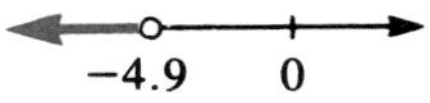

9. $x \leqslant 9$

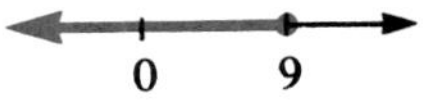

11. $x > 6$

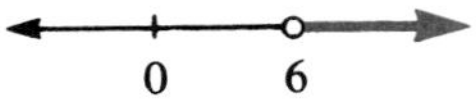

13. $x \leqslant 12$

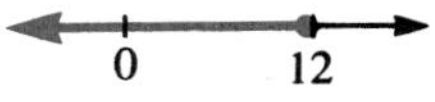

15. $y \geqslant 9$

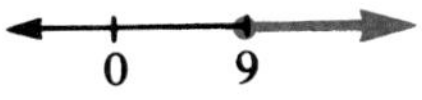

17. $t > 3$

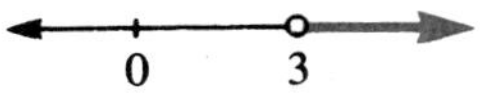

19. $t \leqslant 10$

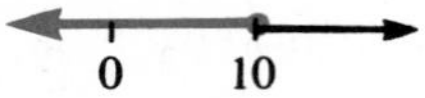

21. $t > -7$

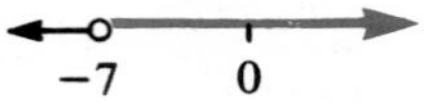

23. $x \leqslant 4$

25. $t > -3$

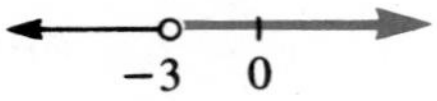

27. $y > 32$

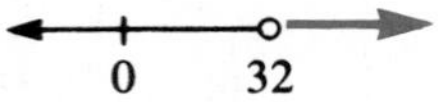

29. $x \leqslant -24$

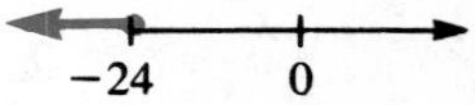

31. $p > 3$

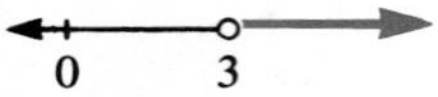

33. $z \geqslant \frac{5}{2}$

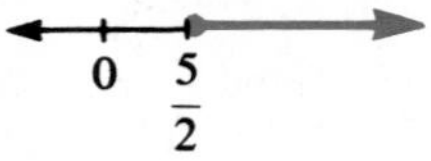

35. $x \geqslant 2$

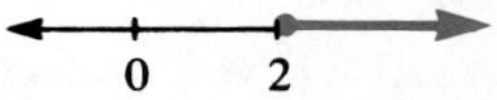

37. $x > -\frac{8}{7}$

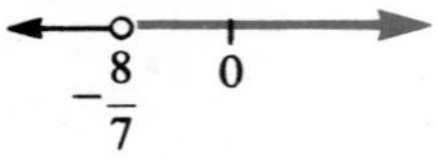

39. $y \geqslant \frac{3}{8}$

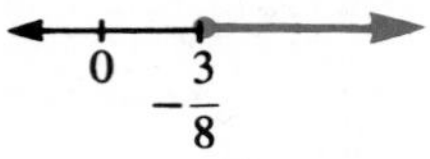

41. $t > \frac{11}{3}$

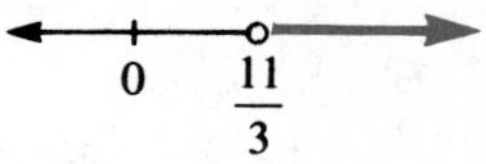

43. $t \leqslant 7$

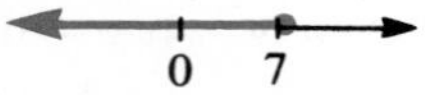

45. $y < -\frac{19}{6}$

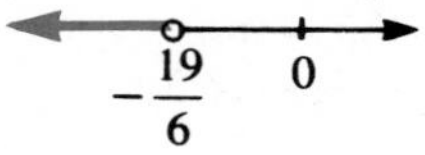

47. $x < 5$

49. $y > \frac{6}{5}$

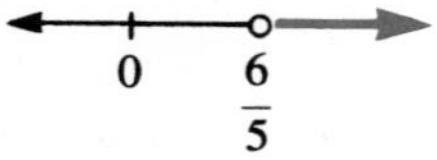

51. $t \leqslant \frac{-43}{2}$

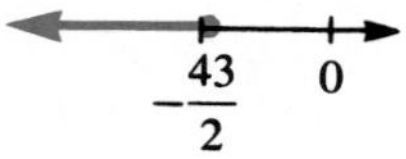

53. $w \leqslant 9$

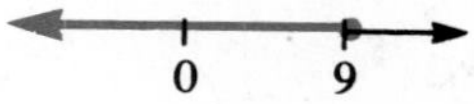

55. $x < -5$

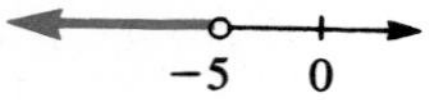

57. $x > 2$

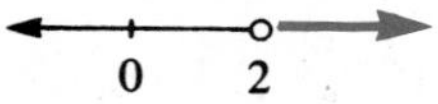

59. $x \leqslant 7$

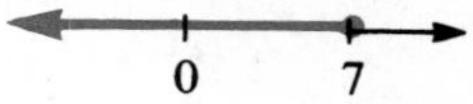

14.2 Exercises, page 494

1. true **3.** false; abscissa **5.** false; less **7.** false; *x*-coordinate

9–27.

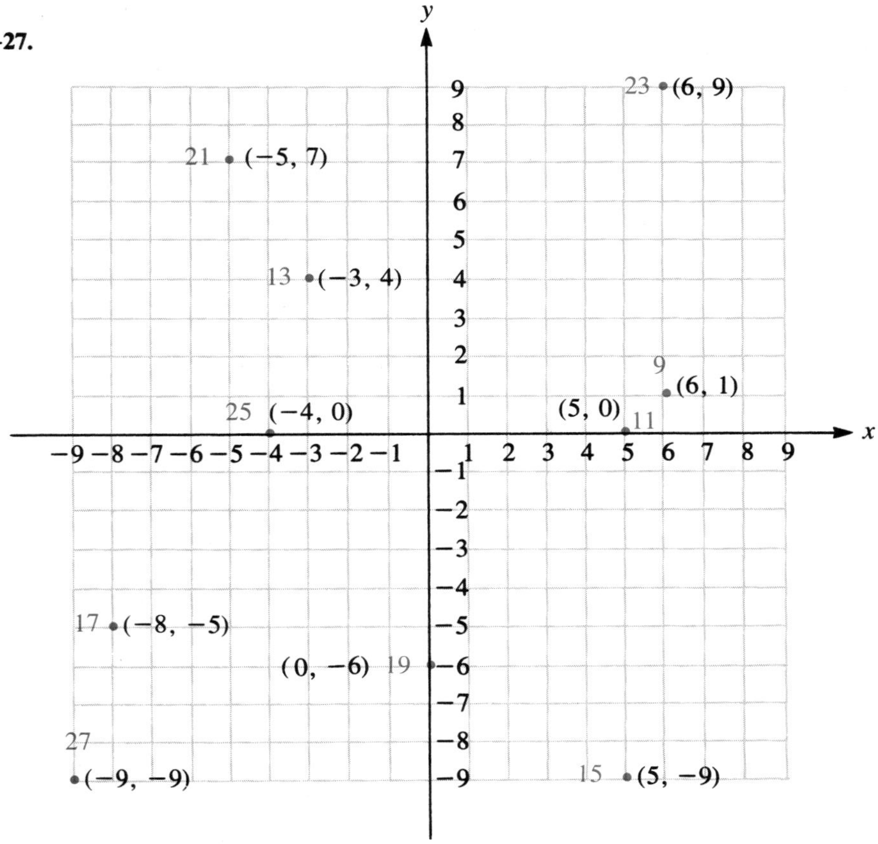

29. (5, 4) **31.** (−3, 6) **33.** (4, 0) **35.** (−7, 0) **37.** (−5, −9)

14.3 Exercises, page 500

1. true **3.** true **5.** yes **7.** no **9.** yes **11.** yes **13.** no **15.** −2; 0 **17.** 2; 5 **19.** −5; −2 **21.** 17; −8 **23.** 2; 35 **25.** 0; 4 **27.** 0; 6 **29.** 7; any real number **31.** (8, 0); (0, −8) **33.** (0, 0) **35.** $\left(-\frac{2}{3}, 0\right)$; (0, 2) **37.** (8, 0); (0, 2) **39.** (5, 0); (0, −2) **41.** (7, 0); (0, 3)

43.

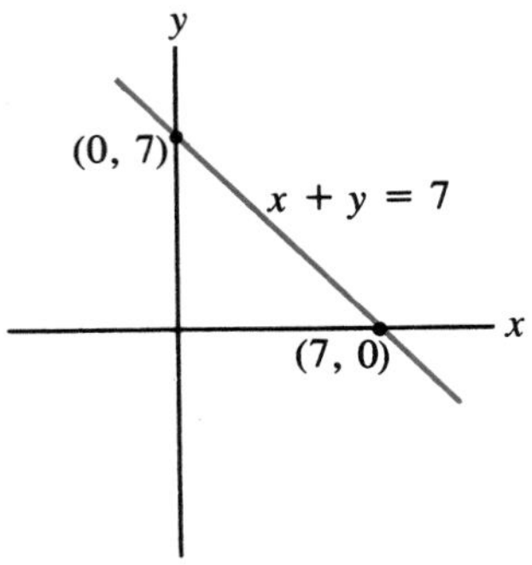

45.

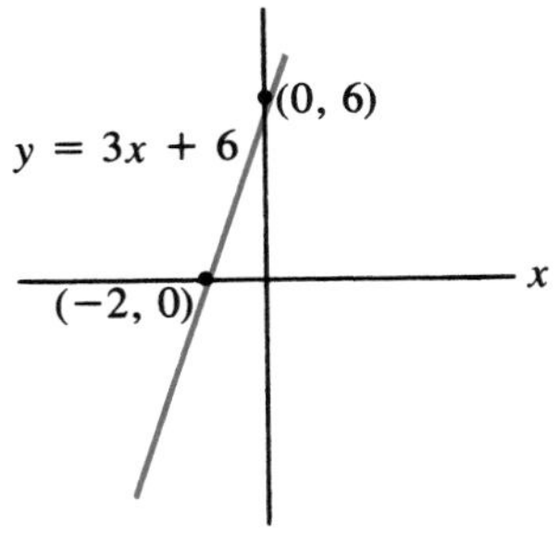

47.

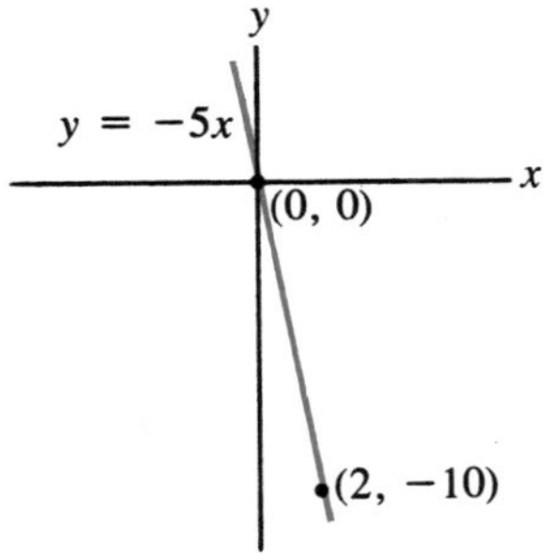

49.

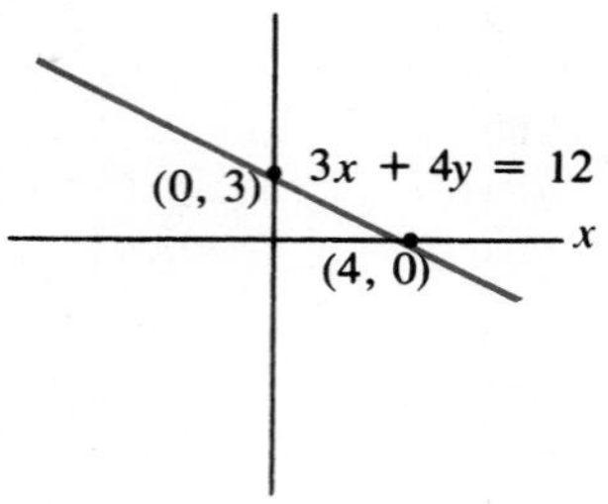

51.

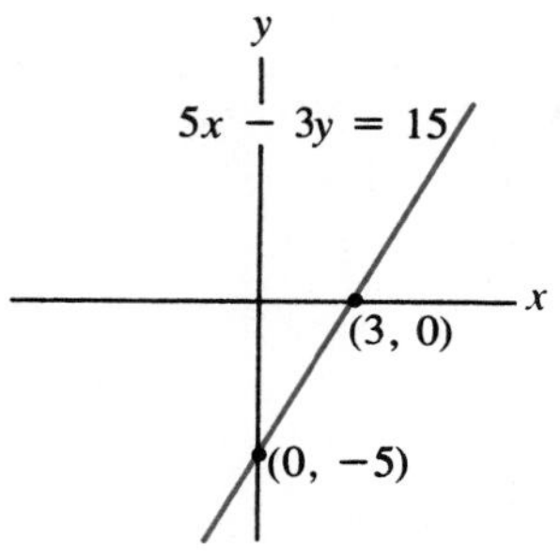

53.

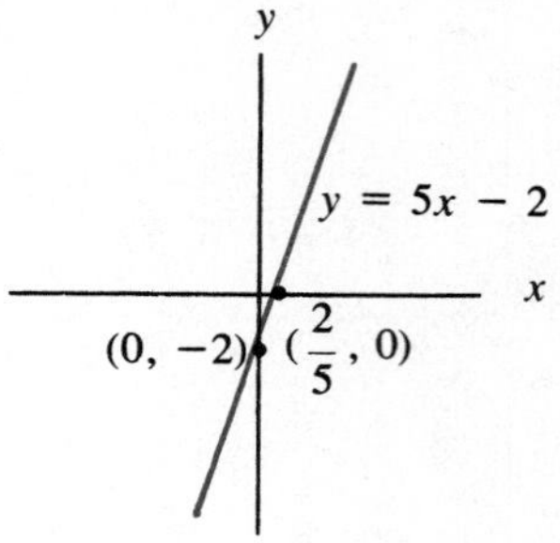

55.

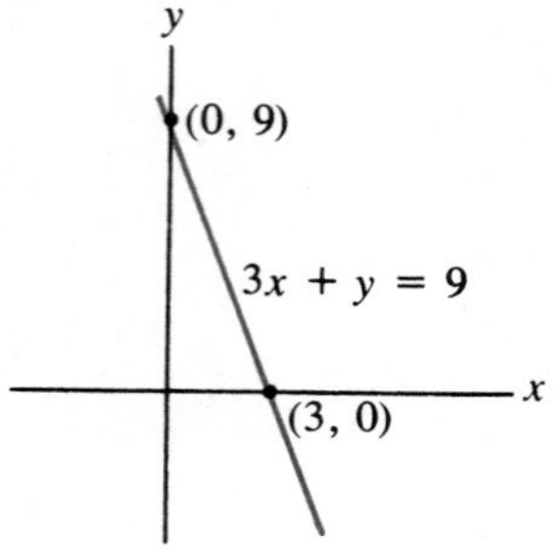

57.

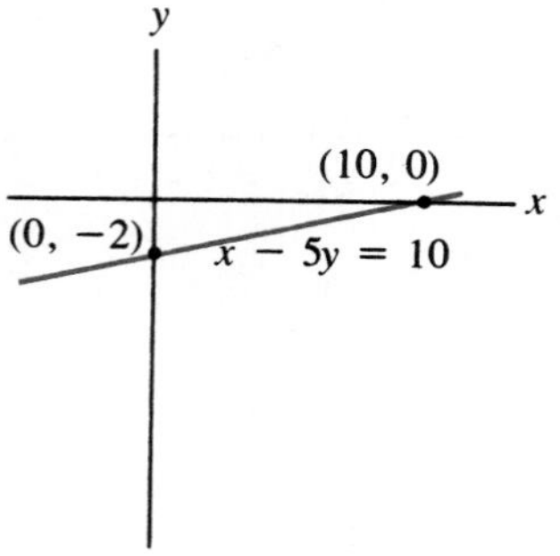

59.

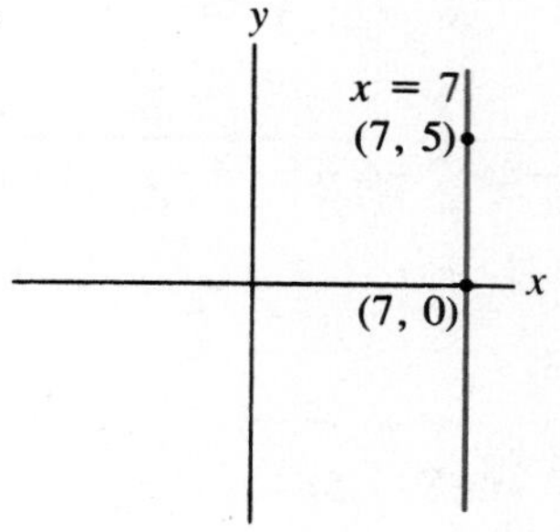

61.

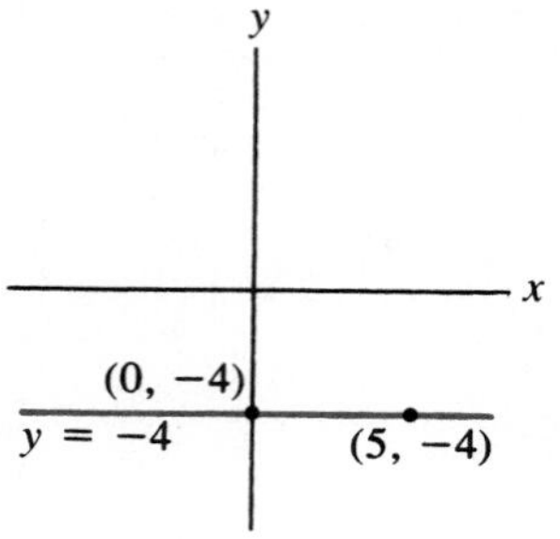

63. $x = 2$

65. $y = \frac{2}{3}$

67. 0; $y = 0$

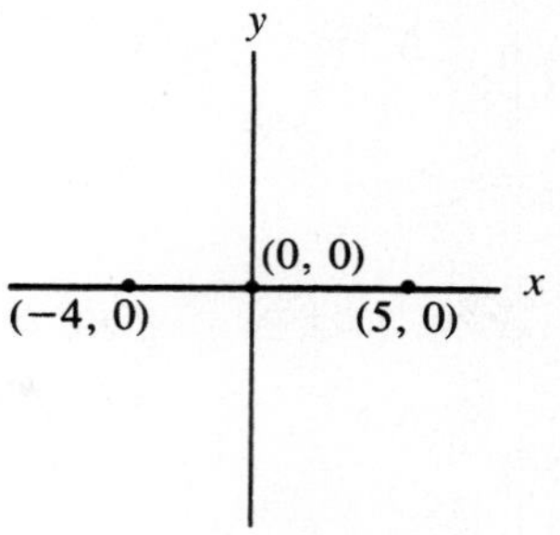

69. The lines are parallel.

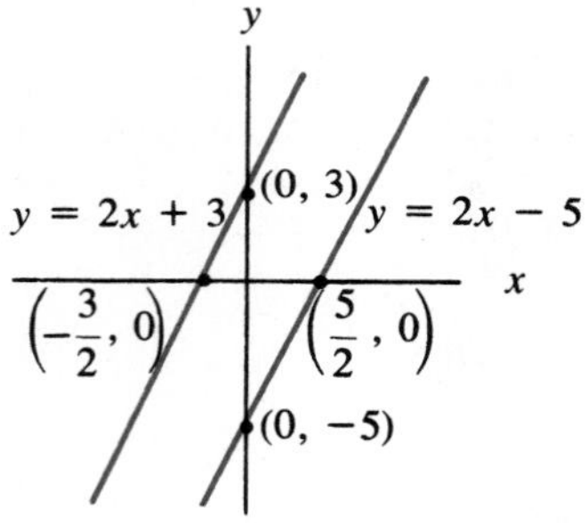

14.4 Exercises, page 506

1. false; positive **3.** false; zero **5.** 1 **7.** -4 **9.** $\frac{1}{5}$ **11.** $-\frac{1}{2}$ **13.** 2 **15.** $\frac{2}{3}$ **17.** -5
19. undefined **21.** $\frac{1}{3}$ **23.** 0 **25.** -1 **27.** 5 **29.** 2 **31.** $-\frac{1}{7}$ **33.** $-\frac{1}{2}$ **35.** $\frac{1}{2}$
37. undefined **39.** 0

41.

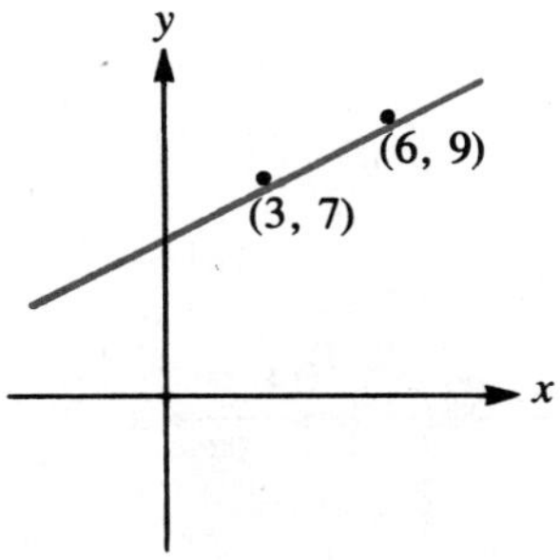

43.

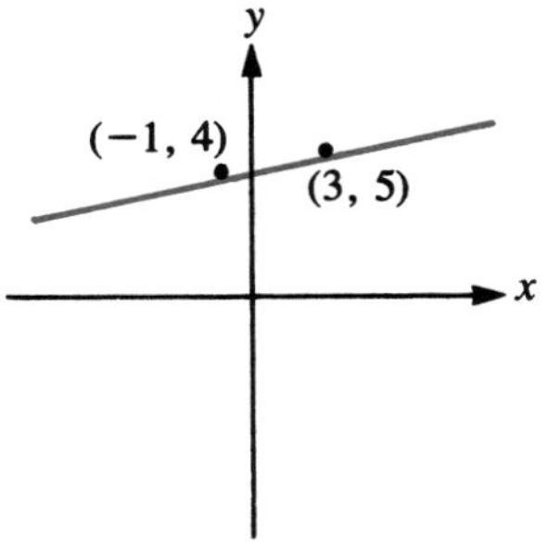

45.

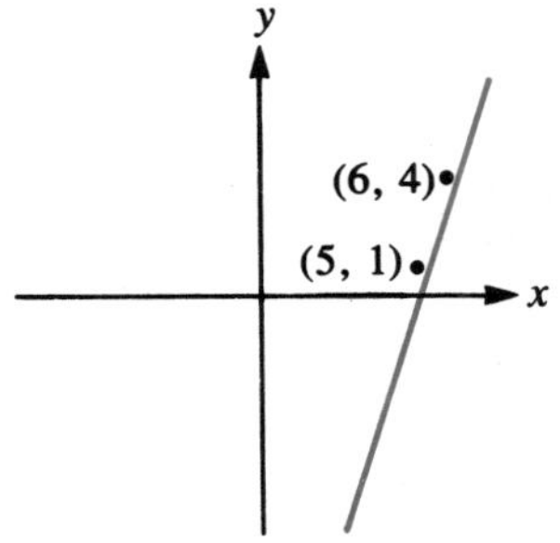

47.

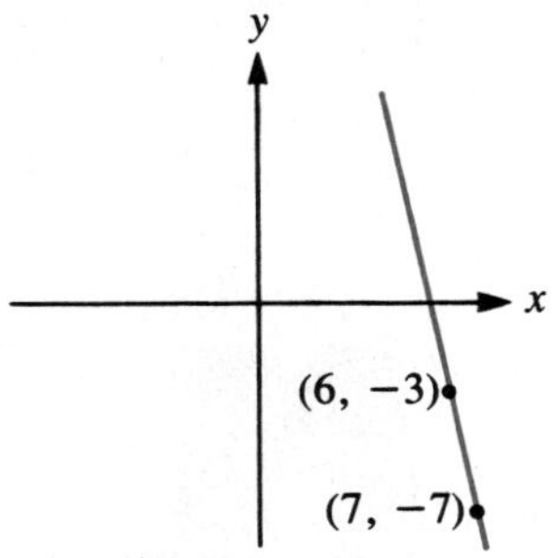

49.

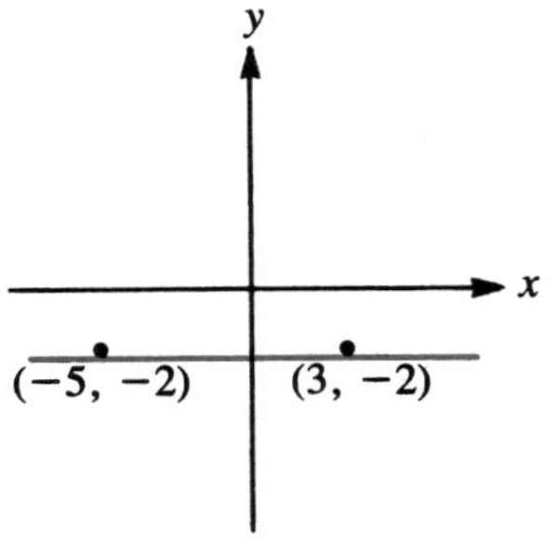

51. These lines are parallel.

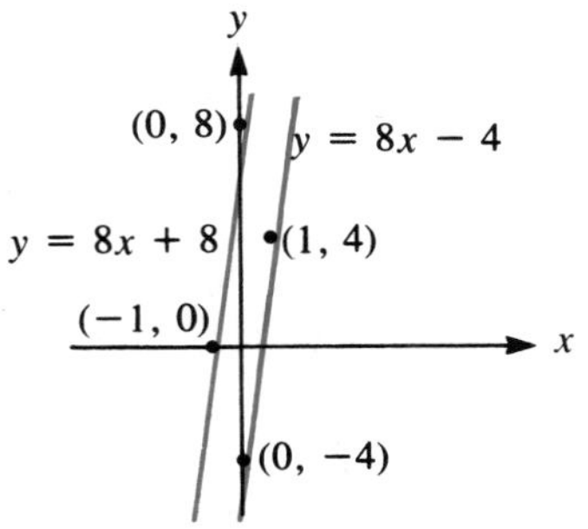

53. These lines are not parallel.

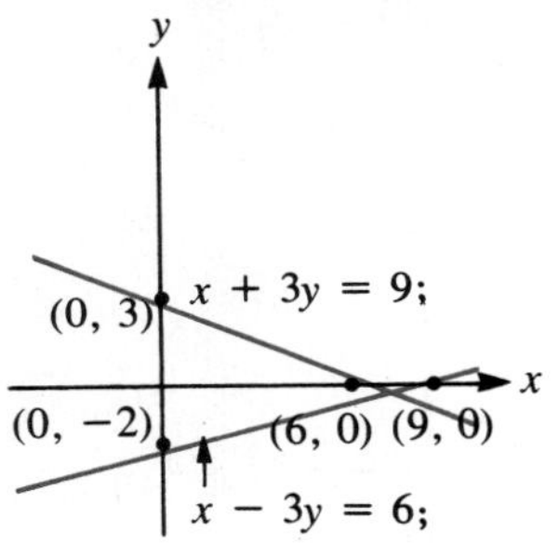

55. These lines are not parallel.

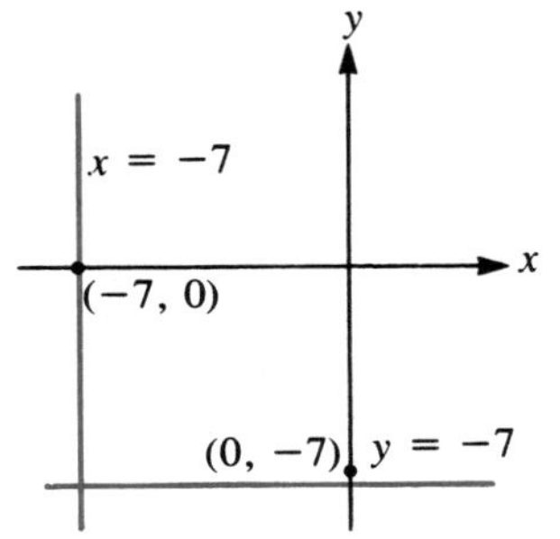

57. These lines are parallel.

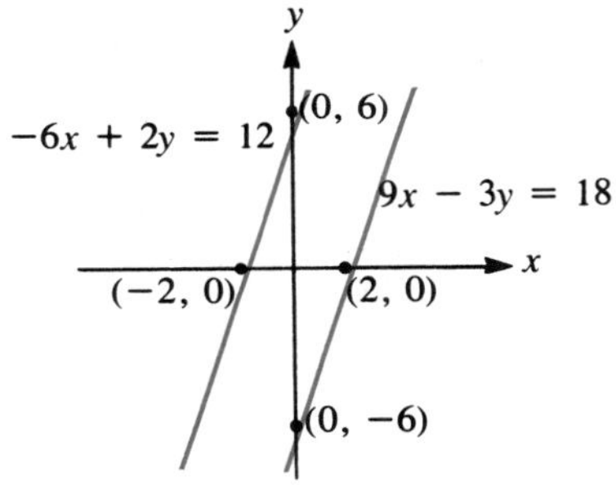

14.5 Exercises, page 510

1. $y - 5 = 3(x - 3)$ **3.** $y = 5(x - 2)$ **5.** $y - 6 = -2(x + 4)$ **7.** $y + 5 = -(x - 1)$ **9.** $y + 3 = \frac{1}{2}(x + 4)$

11. $y + 2 = -\frac{3}{5}(x - 9)$ **13.** $y - 3 = 2\left(x - \frac{1}{2}\right)$ **15.** $y = 7$ **17.** $y - 3 = 2(x - 6)$ **19.** $y - 8 = -3(x - 1)$

21. $y - 2 = -(x + 5)$ **23.** $y + 4 = -4x$ **25.** $y + 3 = -2(x + 2)$ **27.** $y - 7 = -3(x + 5)$

29. $y + 6 = 2(x - 2)$ **31.** $x = 3$ **33.** $y = 2x + 5$ **35.** $y = -x + 4$ **37.** $y = -4x - 3$ **39.** $y = \frac{1}{2}x - 5$

41. $y = -\frac{3}{4}x + 7$ **43.** $y = -\frac{1}{3}x - 1$ **45.** $m = 8; b = 3$ **47.** $m = -5; b = -7$ **49.** $m = \frac{3}{4}; b = 9$

51. $m = -4; b = 0$ **53.** $m = -1; b = 9$ **55.** $m = -2; b = 4$ **57.** $m = \frac{2}{5}; b = \frac{-7}{5}$ **59.** $m = -4; b = 8$

61. $y - 5 = -3(x - 6)$ or $y = -3x + 23; b = 23$ **63.** $y + 2 = \frac{1}{2}(x - 2)$ or $y = \frac{1}{2}x - 3; b = -3$ **65.** $y - 2 = 5(x - 3)$

67. $y + 3 = 2(x + 8)$

14.6 Exercises, page 513

1.

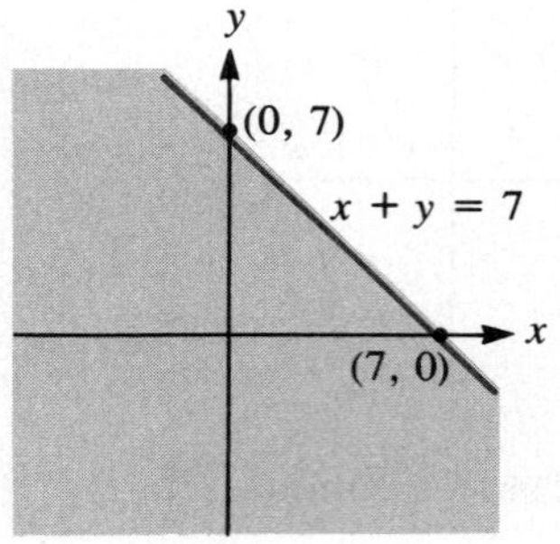

3.

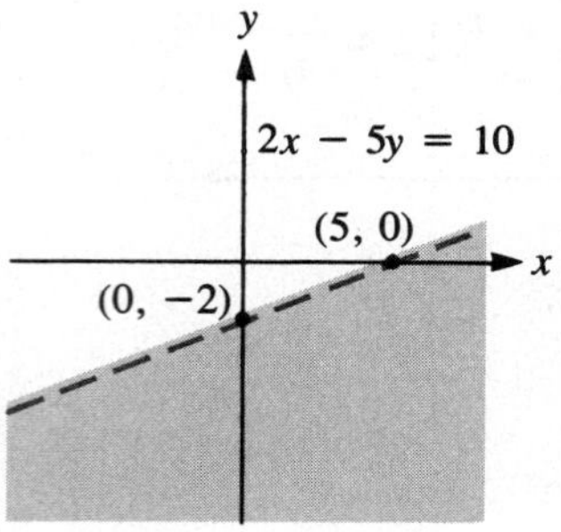

5.

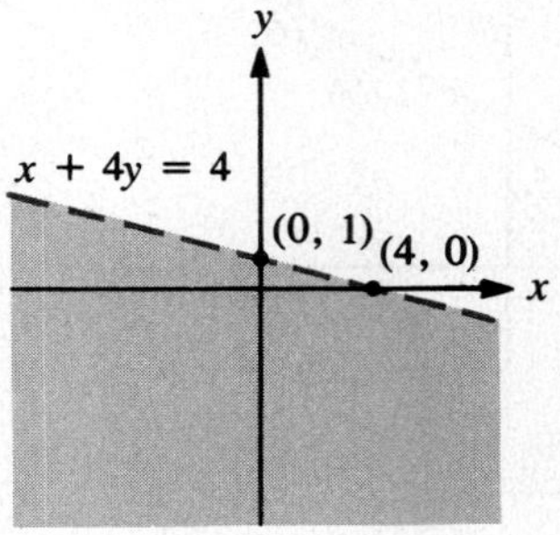

7.

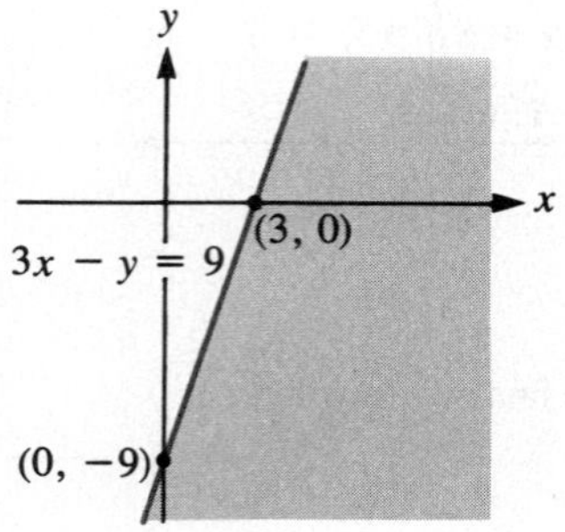

9.

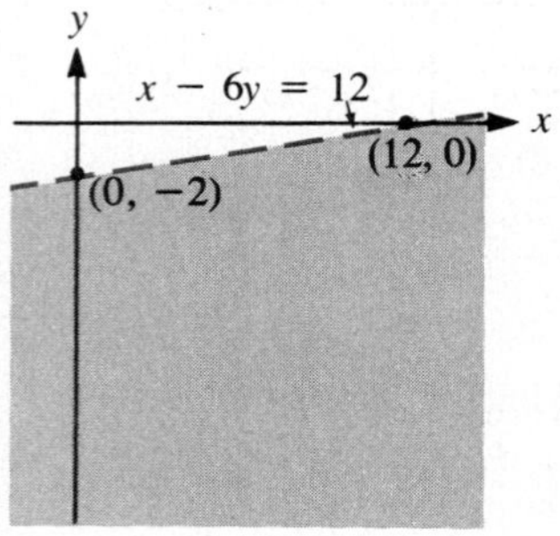

11.

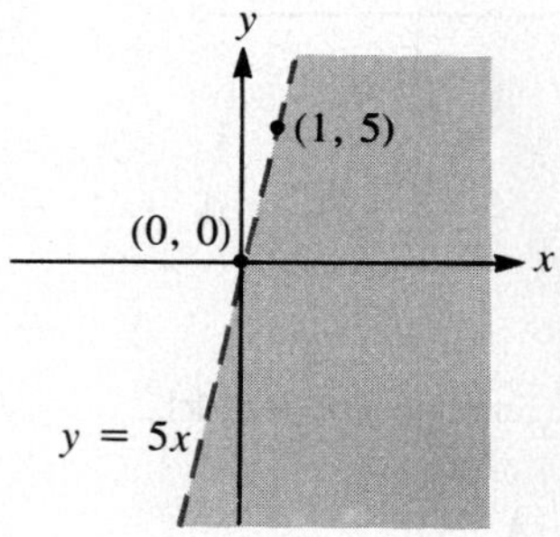

13.

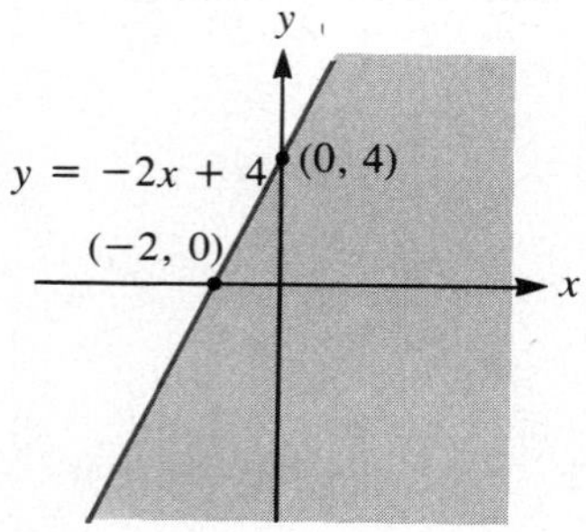

15.

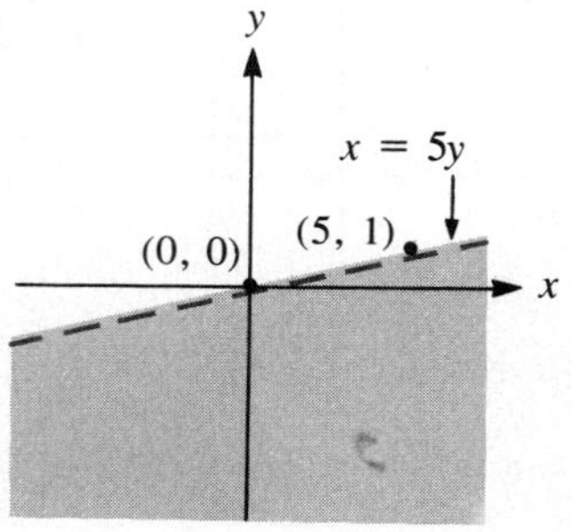

17.

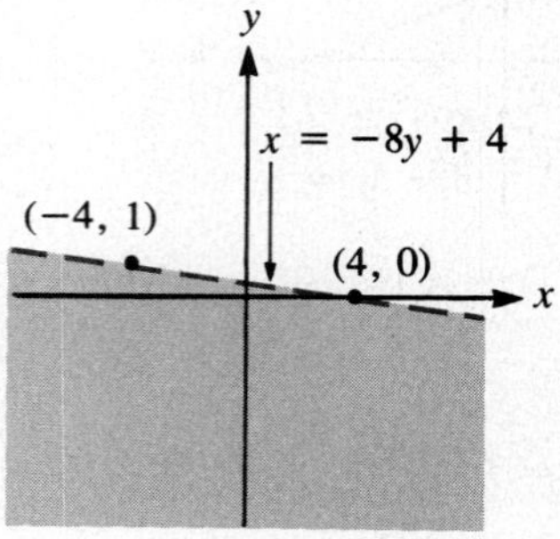

19.

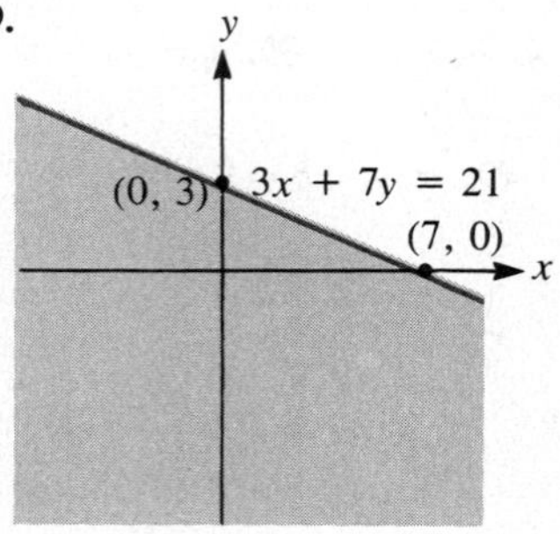

21.

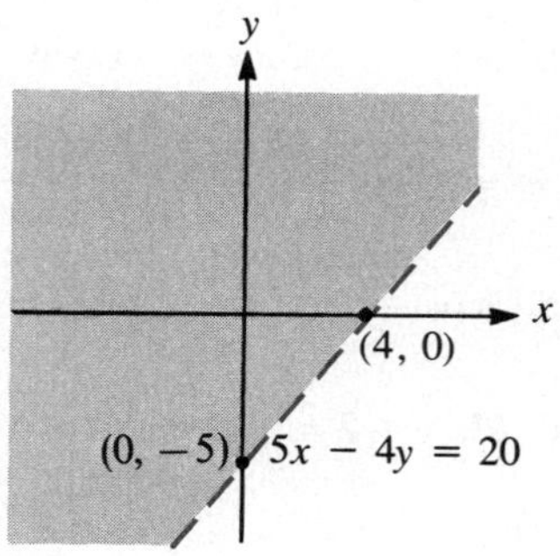

23.

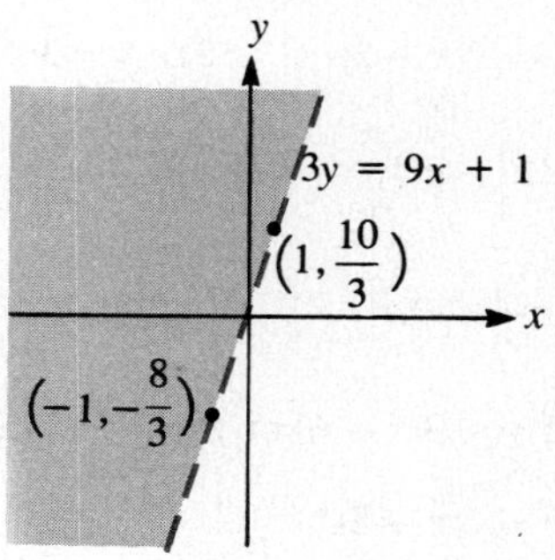

25.

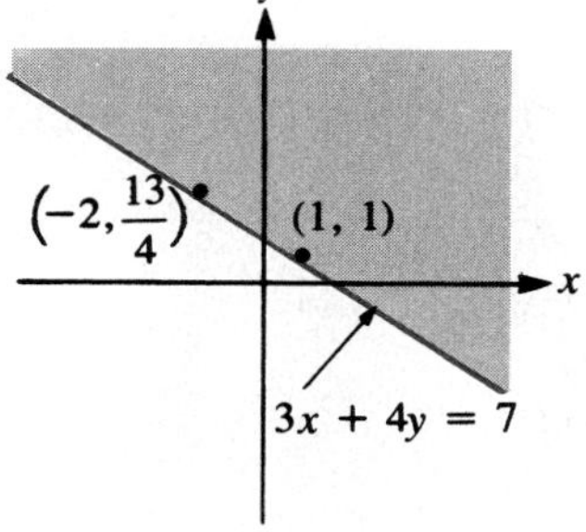

27.

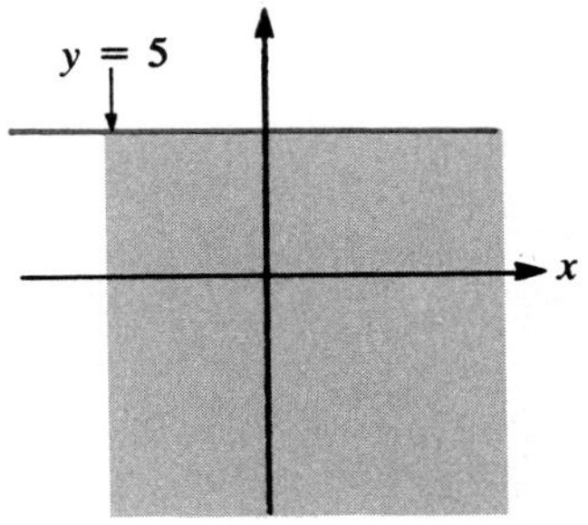

29.

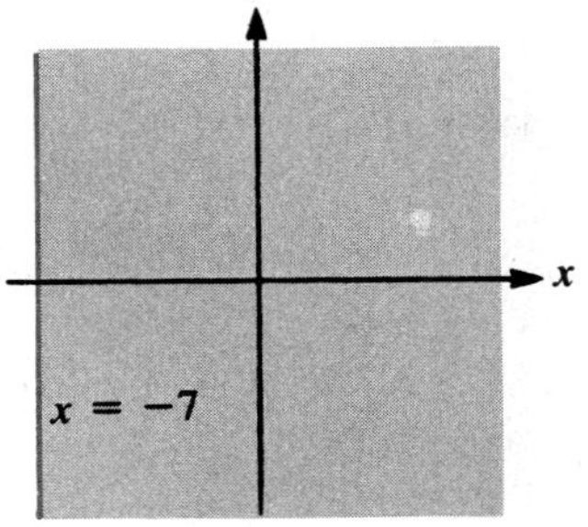

Chapter 14 Review Exercises, page 514

1. true **3.** true **5.** true

7. $x < -2$

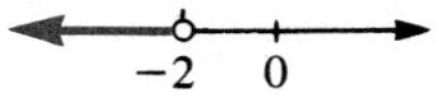

9. $t \leqslant -1$

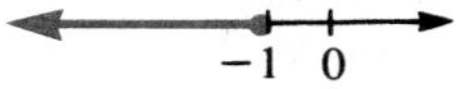

11. $y < \frac{9}{4}$

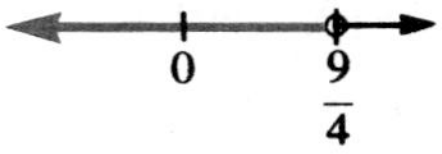

13. $z > \frac{27}{7}$

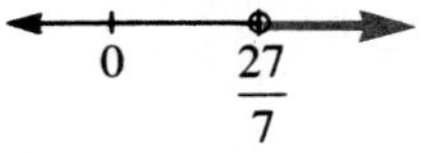

15. $w \leqslant 5$

17.

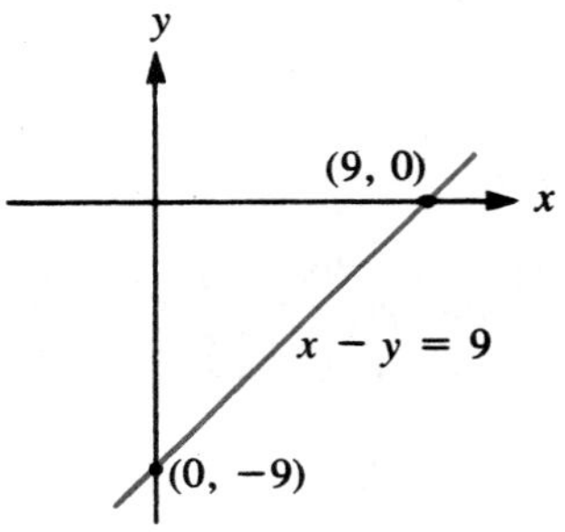

19.

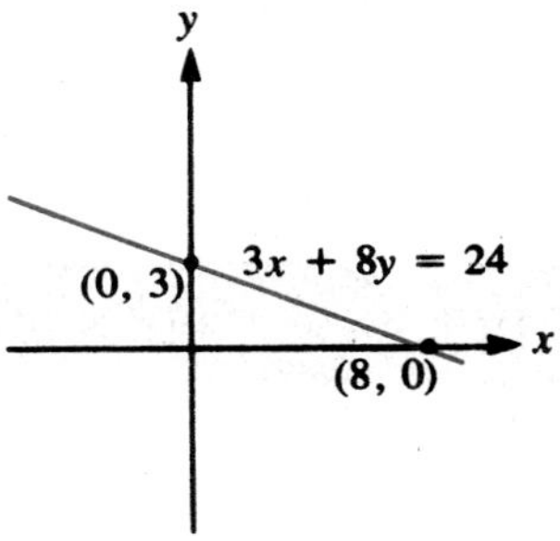

21.

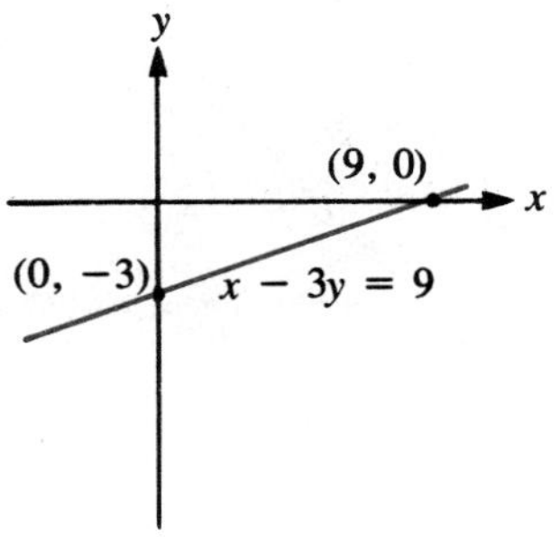

23.

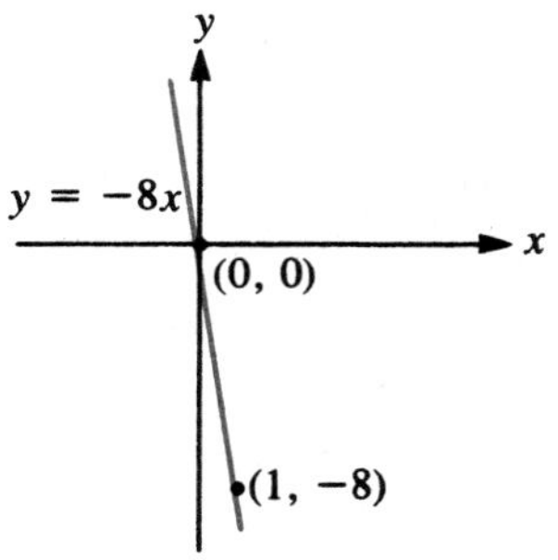

25.

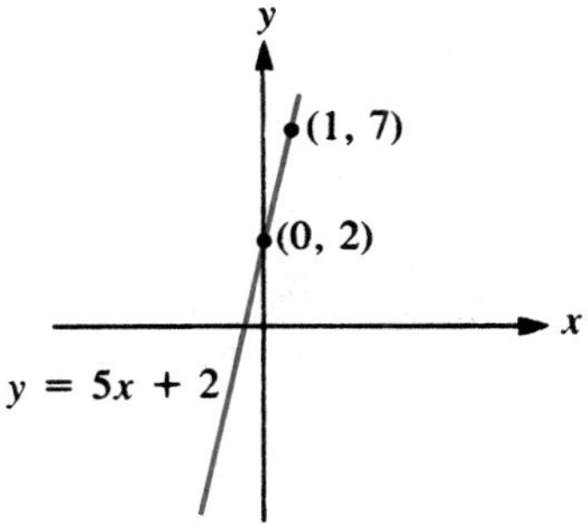

27.

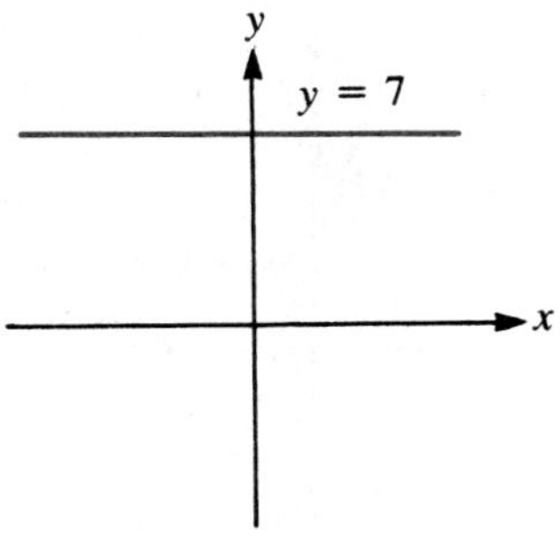

29. 4 **31.** 2 **33.** $-\frac{4}{3}$ **35.** $y - 5 = 4(x + 3)$ **37.** $y - 8 = -\frac{1}{3}x$

39. $x = 4$ **41.** $y - 2 = 4(x - 4)$ **43.** $y - 8 = -2(x + 5)$ **45.** $y = 5$

47. $y = 2x - 3$ **49.** $y = -\frac{3}{4}x$ **51.** $y - 7 = -3(x + 5)$

53.

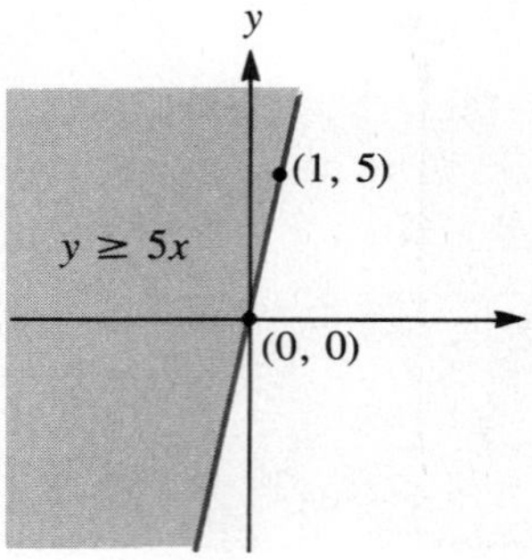

55.

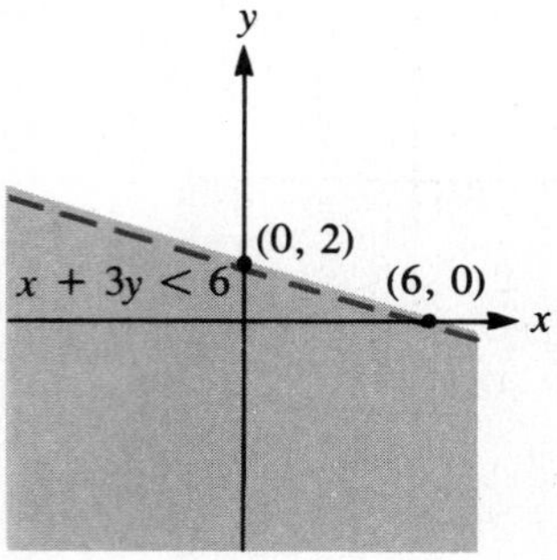

57.

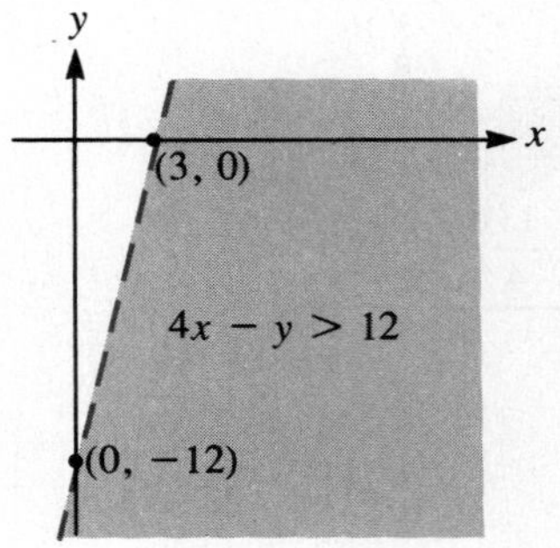

59.

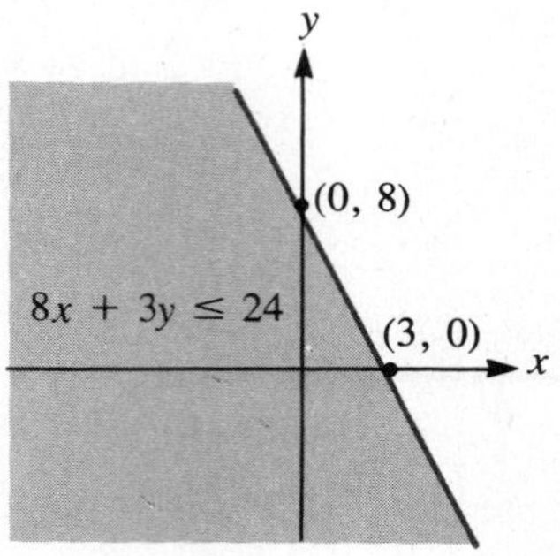

61.

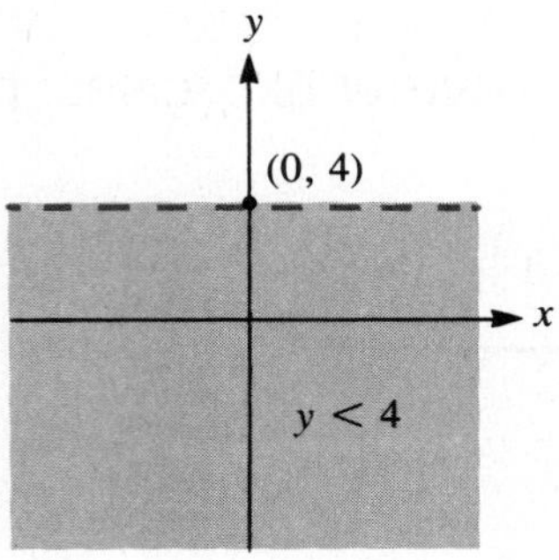

Chapter 15

15.1 Exercises, page 520

1. true **3.** false; intersect

5.

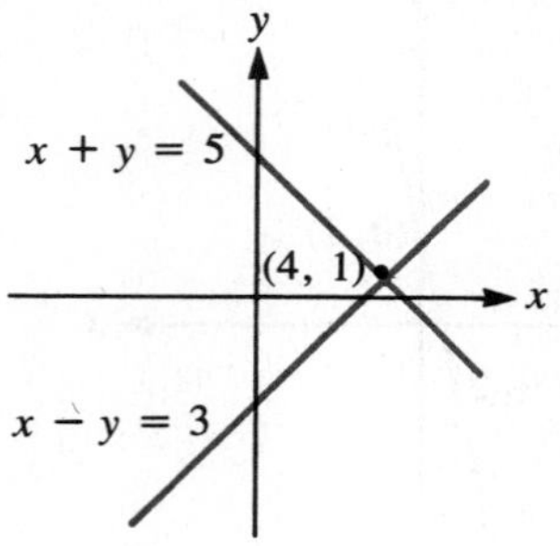

7.

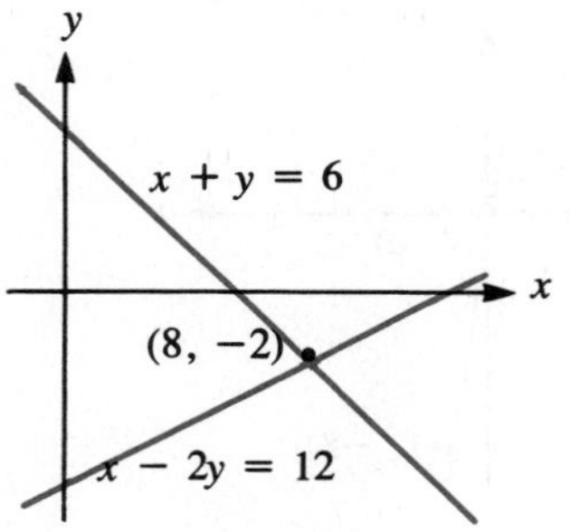

9.

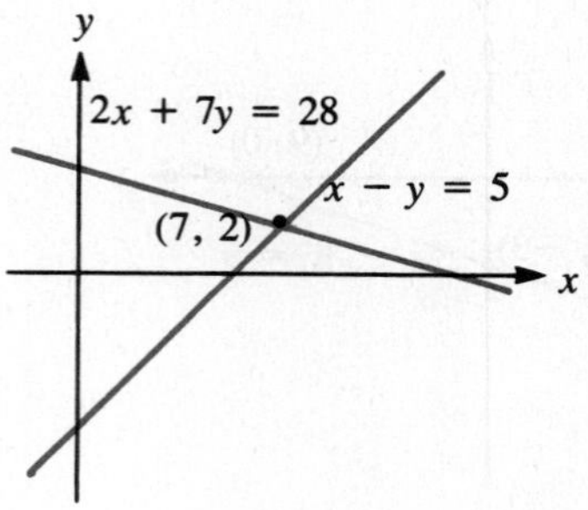

11.

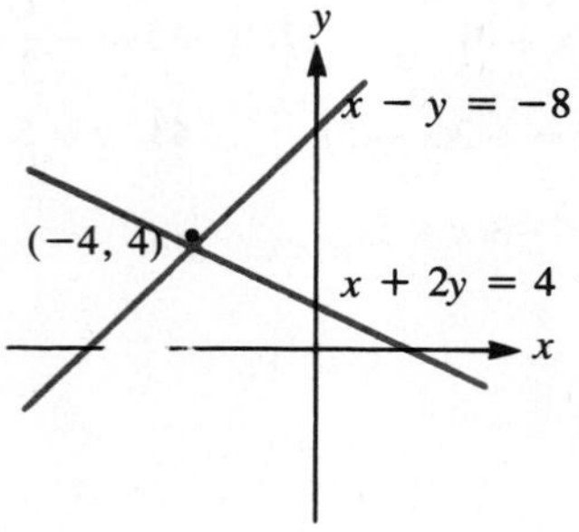

13.

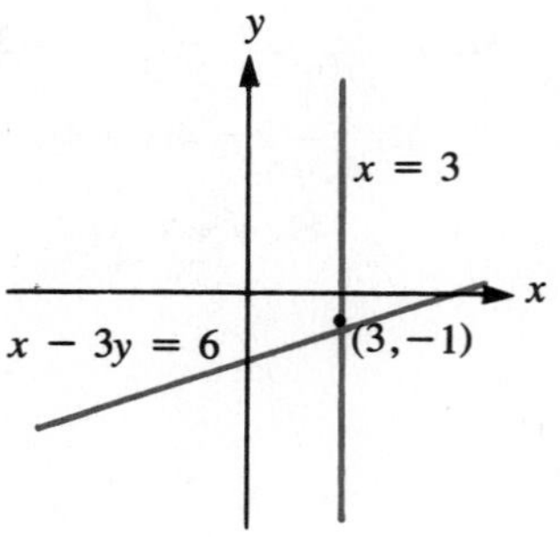

15.

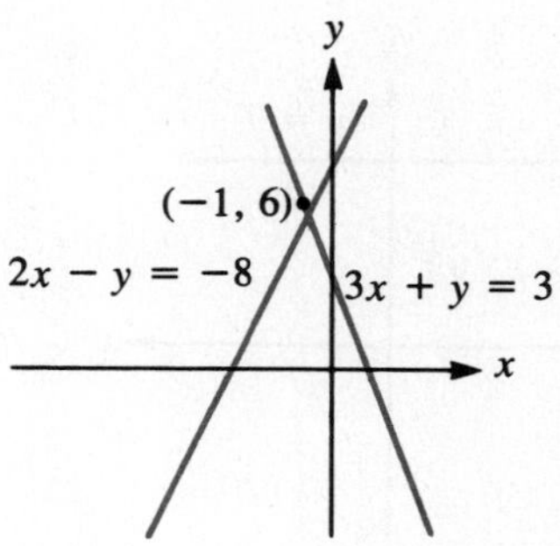

17.

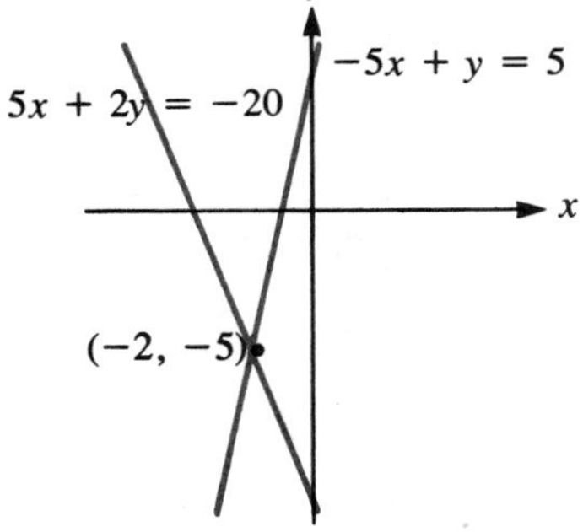

19. parallel lines; no solution

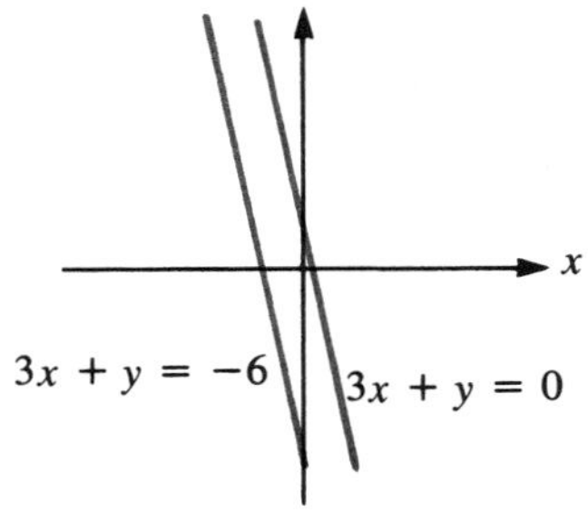

21.

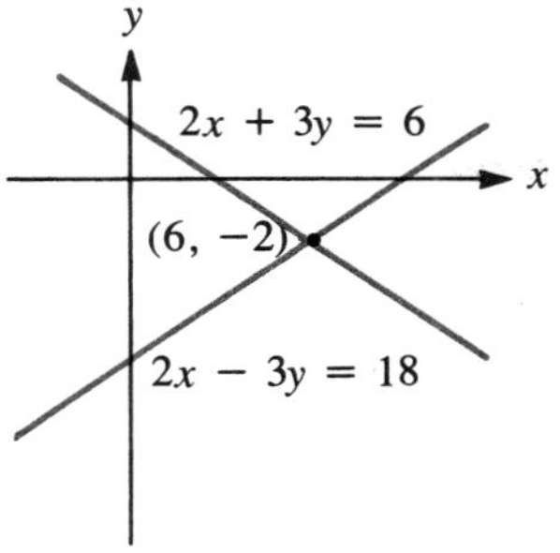

23.

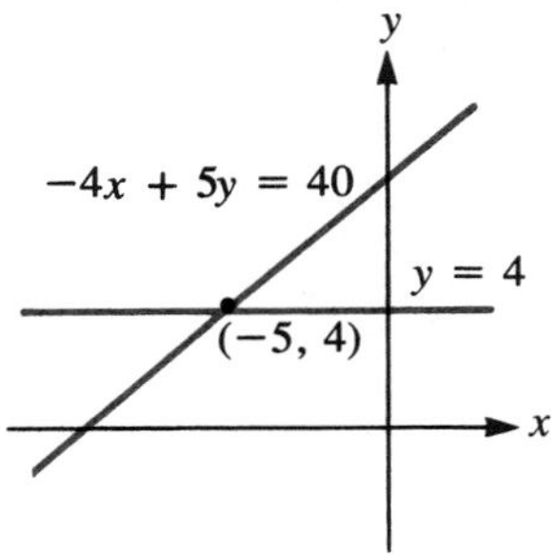

25. parallel lines; no solution

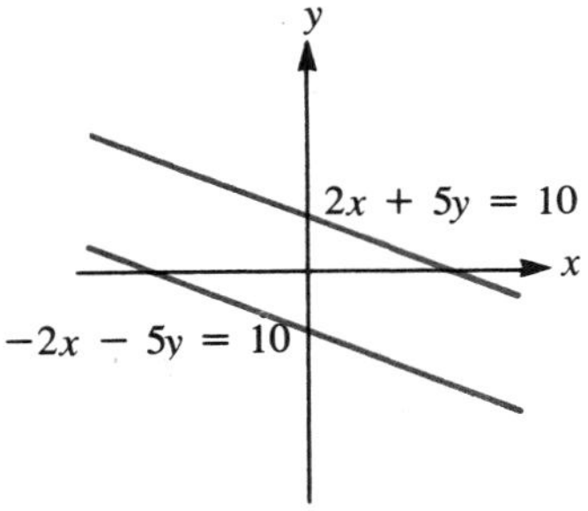

27. lines coincide; infinite number of solutions

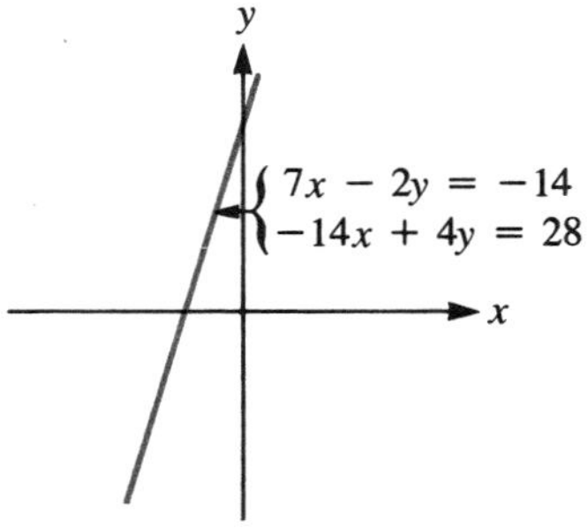

29.

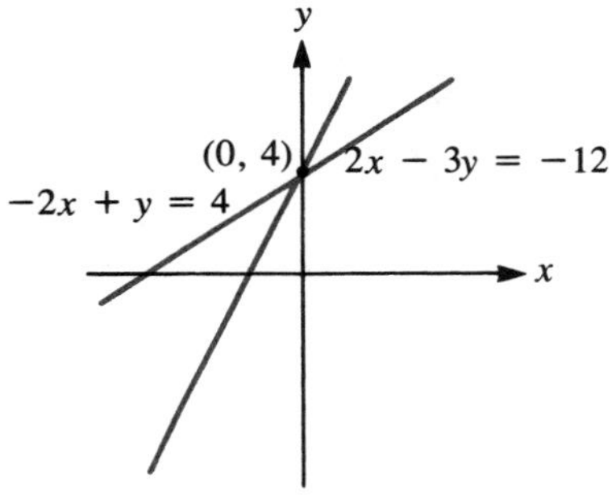

15.2 Exercises, page 525

1. false; are parallel **3.** true **5.** (3, 1) **7.** (2, −3) **9.** (5, −1) **11.** (6, 4) **13.** (−1, 2) **15.** (−3, −2) **17.** (2, −2) **19.** (3, −1) **21.** (2, 1) **23.** parallel lines; no solution **25.** (−5, −6) **27.** $\left(3, \frac{1}{2}\right)$ **29.** parallel lines; no solution **31.** lines coincide; infinite number of solutions **33.** $\left(\frac{4}{3}, \frac{1}{3}\right)$

15.3 Exercises, page 530

1. (3, 5) **3.** (−3, 1) **5.** (2, −6) **7.** (8, 1) **9.** (−4, −3) **11.** (9, −2) **13.** parallel lines; no solution **15.** (2, −7) **17.** (3, 4) **19.** lines coincide; infinite number of solutions **21.** $\left(\frac{1}{2}, -6\right)$ **23.** parallel lines; no solution **25.** (5, 2) **27.** lines coincide; infinite number of solutions **29.** $\left(\frac{5}{4}, \frac{1}{4}\right)$

15.4 Exercises, page 536

1. 1 and 8 **3.** 9 and 3 **5.** 7 and 2 **7.** 11 quarters; 18 nickels **9.** 13 \$5 bills; 8 \$10 bills

11. \$7,000 at 11%; \$5,500 at 12% **13.** \$9,200 at $12\frac{1}{2}$%; \$8,700 at 10%

15. 8 pounds marshmallow-filled; 12 pounds cream-filled **17.** 3 l 25% solution; 4 l 60% solution **19.** 60 ml.
21. 18 pounds at \$3.75 per lb; 32 pounds at \$4.50 per lb **23.** 110 adult; 78 child **25.** 76 **27.** 27

15.5 Exercises, page 540

1.

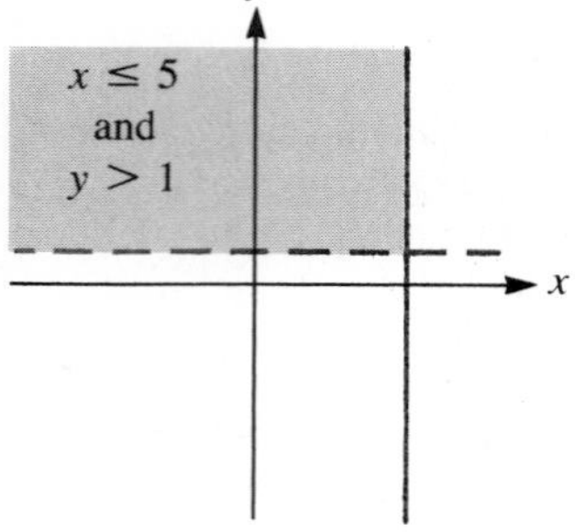

3.

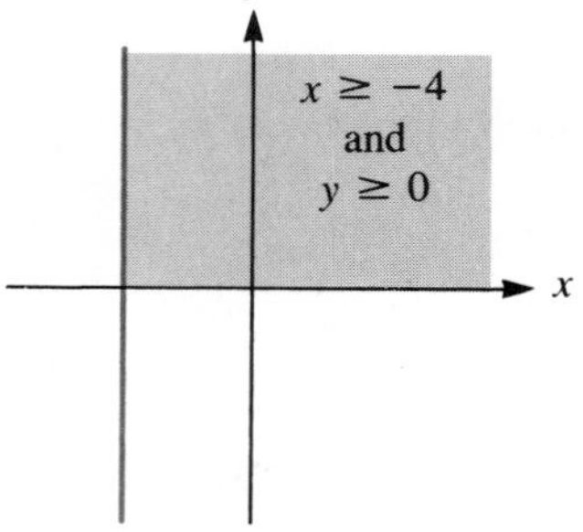

5.

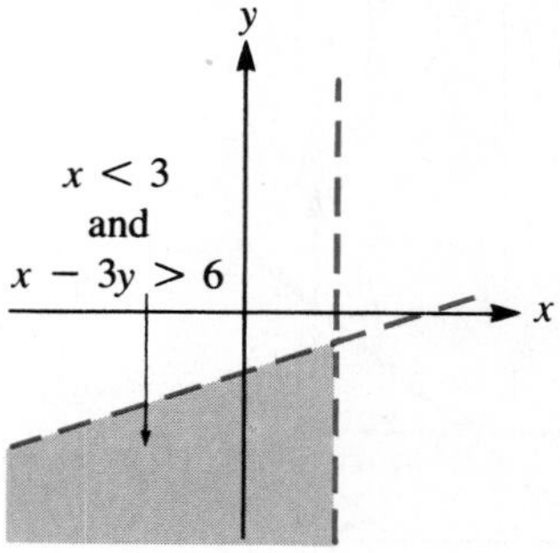

7.

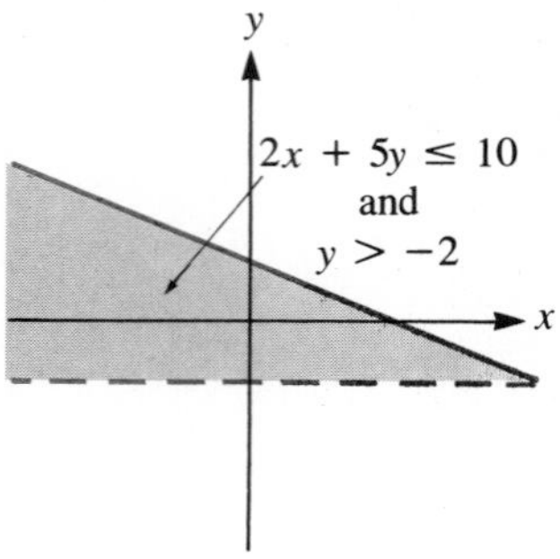

9.

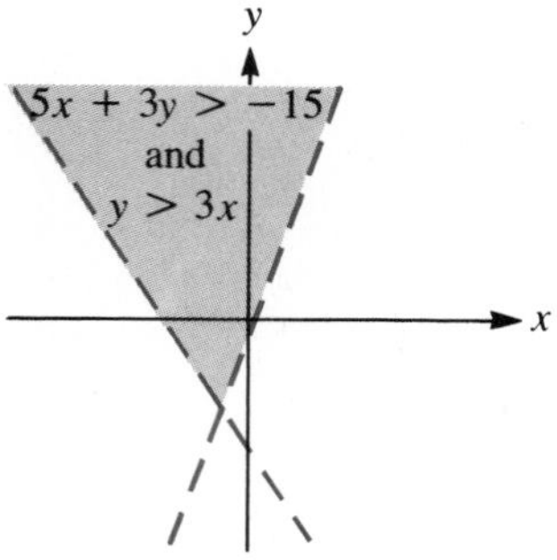

11.

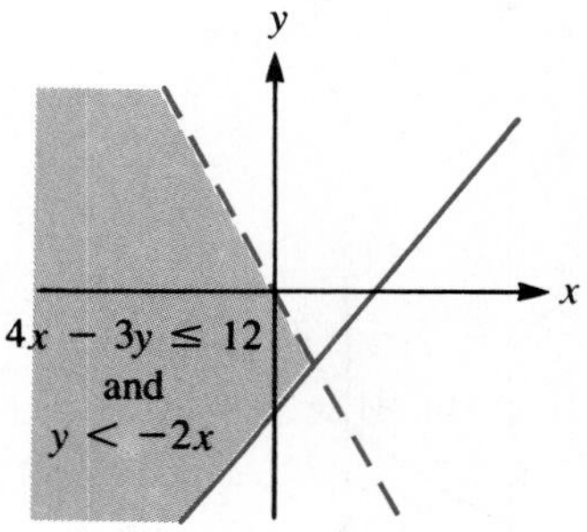

13.

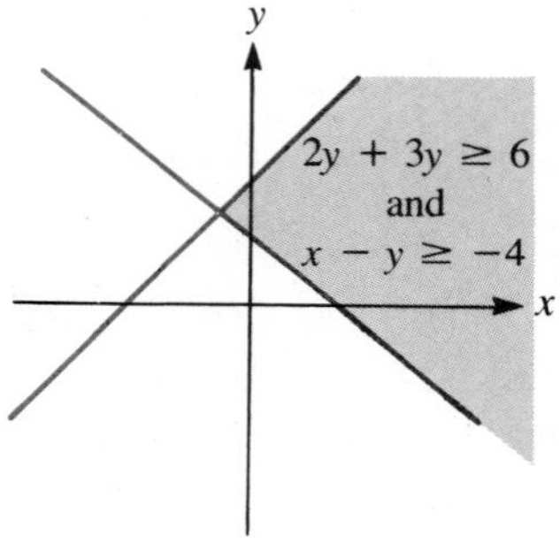

15.

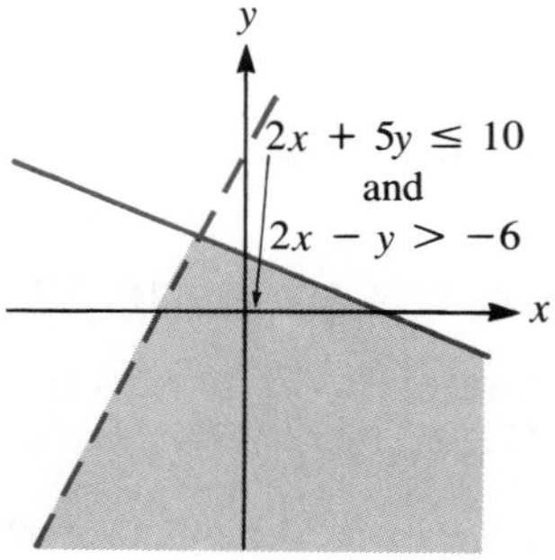

17.

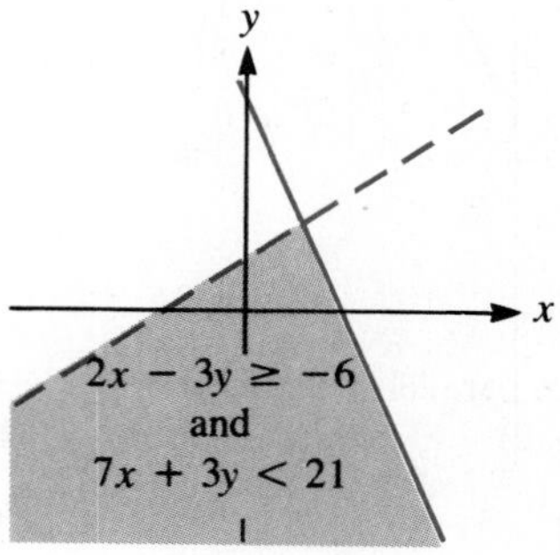

19.

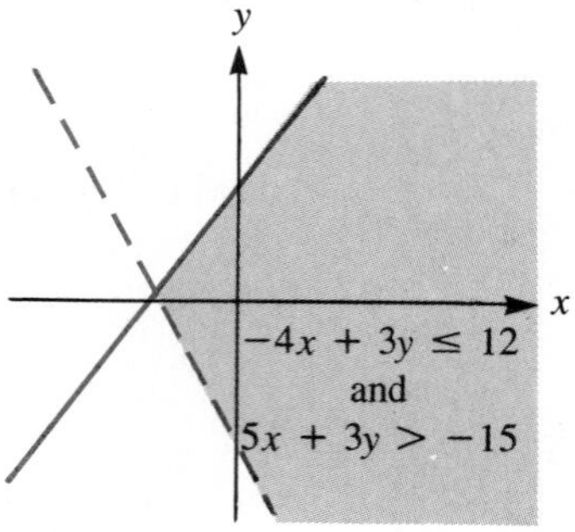

21.

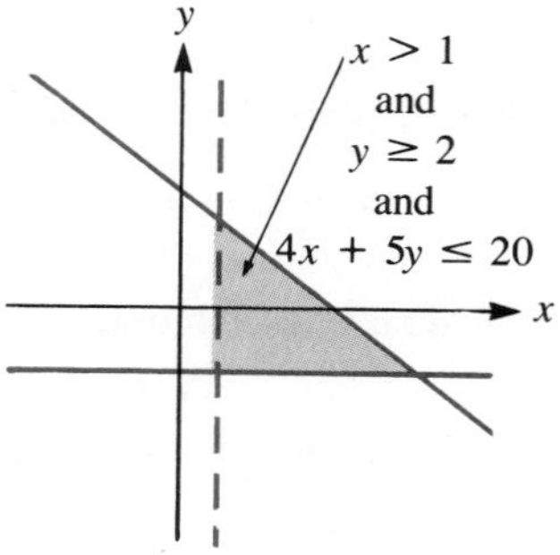

23.

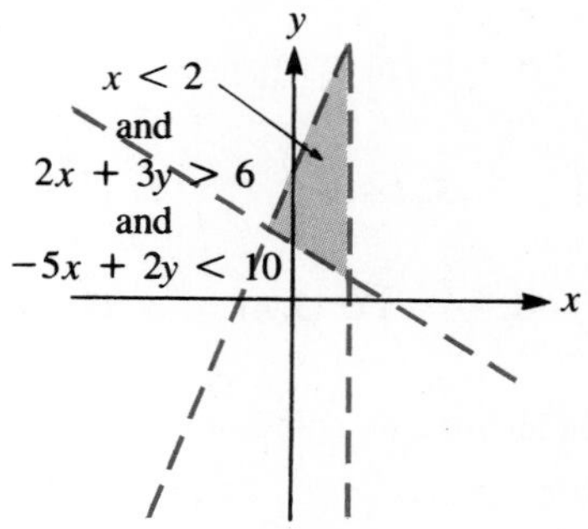

25.
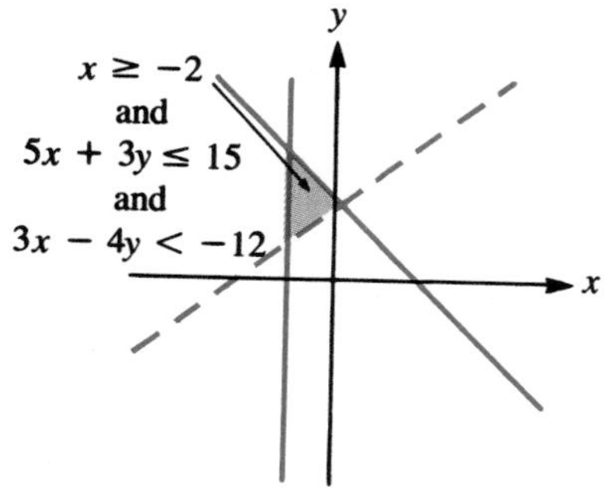

Chapter 15 Review Exercises, page 542

1.
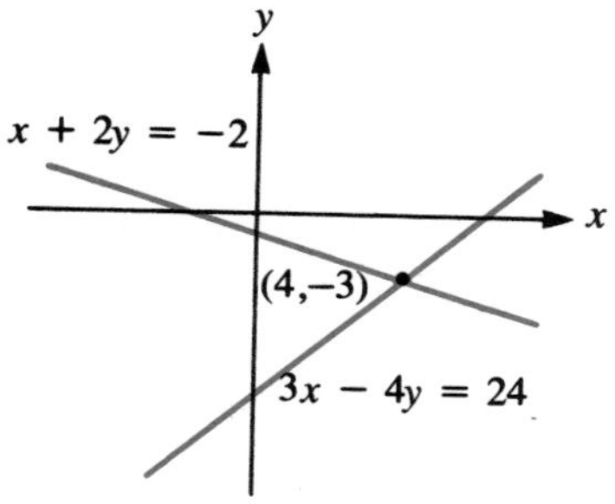

3. lines coincide; infinite number of solutions
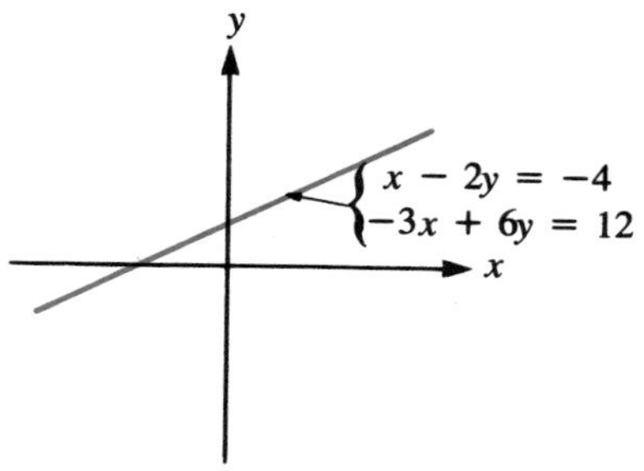

5. parallel lines; no solution
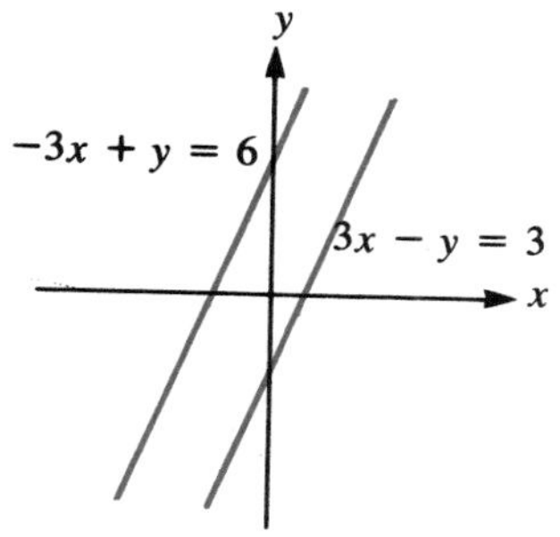

7. $(-3, -5)$ 9. $(11, -17)$ 11. $(4, -1)$ 13. $(0, -3)$ 15. $(2, 8)$ 17. $(3, -4)$ 19. $(-7, 6)$
21. 25 and 16 23. \$84,000 at 12%; \$67,000 at 11% 25. 2.25 l 60% solution; 3.75 l 75% solution

27.
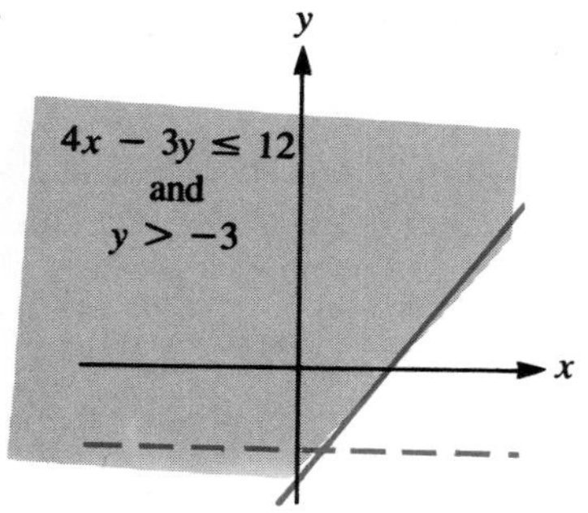

29.
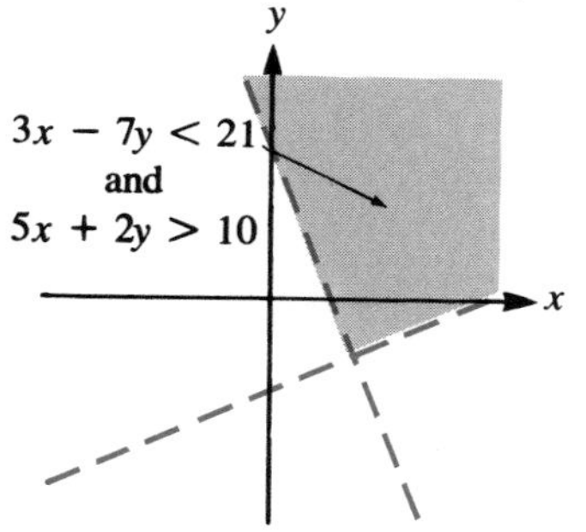

31.
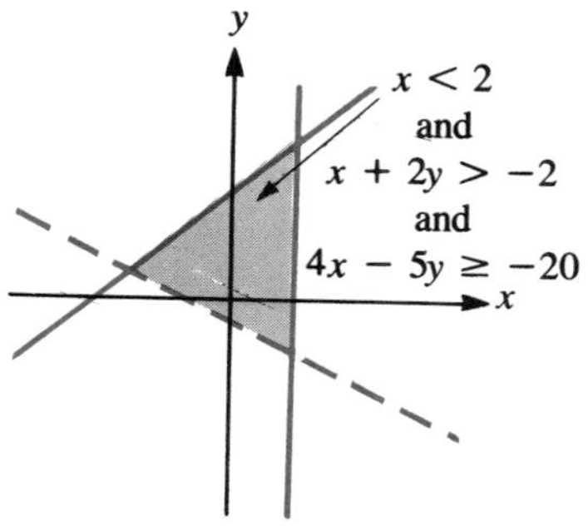

Chapter 16

16.1 Exercises, page 548

1. $x \neq 2$ 3. $x \neq -9$ 5. $x \neq 4, -4$ 7. $x \neq -3, 6$ 9. $x \neq 0, -4, -9$ 11. $\frac{2}{x-3}$ 13. $\frac{1}{2x+3}$

15. $\frac{-4}{2y-3}$ 17. $\frac{6x}{3x+1}$ 19. $\frac{7}{a+5}$ 21. $\frac{5}{4}$ 23. $\frac{1}{x-5}$ 25. $\frac{2}{x+7}$ 27. $\frac{1}{2(y-4)}$ 29. $\frac{3}{z+5}$

31. $\frac{1}{x-9}$ **33.** $\frac{x-5}{2}$ **35.** $\frac{t-3}{t+8}$ **37.** $\frac{1}{x-2}$ **39.** $\frac{y+6}{y-2}$ **41.** $\frac{x-3}{x+6}$ **43.** $\frac{a+6}{a-7}$ **45.** $\frac{z-8}{z-6}$
47. $\frac{x+3}{x+5}$ **49.** $\frac{9}{(x-3)(x^2+9)}$ **51.** -1 **53.** 1 **55.** -5 **57.** $\frac{-1}{x+6}$ **59.** $\frac{-(y+7)}{3}$

16.2 Exercises, page 552

1. $\frac{x}{3}$ **3.** $\frac{x}{3}$ **5.** $\frac{y}{9}$ **7.** $\frac{4}{y}$ **9.** $\frac{5x}{2(x-2)}$ **11.** $\frac{4(a+6)}{a^2}$ **13.** $\frac{4x}{x-7}$ **15.** $\frac{2x}{x-4}$ **17.** $\frac{x-3}{4x}$
19. $\frac{1}{4}$ **21.** $t+3$ **23.** $\frac{7}{z-1}$ **25.** $\frac{x+3}{7}$ **27.** $\frac{y-1}{2y}$ **29.** $\frac{2(x-2)^2}{x-1}$ **31.** $\frac{1}{2}$ **33.** $\frac{2(x-2)}{x}$
35. $\frac{x}{5(x-3)}$

16.3 Exercises, page 555

1. $36k$ **3.** $5x^3$ **5.** $6y^4$ **7.** a^2b^4 **9.** $4(x-3)$ **11.** $3t(t+7)$ **13.** $k-9$ **15.** $18(y+5)$
17. x^2-49 **19.** $5(y+2)(y-2)$ **21.** $(x+8)(x-5)(x-8)$ **23.** $(z-7)(z-5)(z+5)$
25. $x(x+4)(x-4)(x+7)$ **27.** $(x+7)(x-4)(x-1)$ **29.** $x(x-6)(x-3)(x+5)$
31. LCD $= 15x^3$; multiply $5x^3$ by 3; multiply $15x$ by x^2 **33.** LCD $= 6(y-3)$; multiply $(6y-18)$ by 1; multiply $(2y-6)$ by 3
35. LCD $= 4x(x+5)$; multiply (x^2+5x) by 4; multiply $(4x+20)$ by x
37. LCD $= 3(y+9)(y-9)$; multiply $(3y+27)$ by $(y-9)$; multiply (y^2-81) by 3
39. LCD $= (x-3)(x+3)(x-8)$; multiply (x^2-9) by $(x-8)$; multiply $(x^2-5x-24)$ by $(x-3)$
41. LCD $= (x+5)(x-2)(x-7)$; multiply $(x^2+3x-10)$ by $(x-7)$ multiply $(x^2-9x+14)$ by $(x+5)$

16.4 Exercises, page 560

1. $\frac{13}{x}$ **3.** $\frac{4}{a}$ **5.** $\frac{x+4}{x+3}$ **7.** $\frac{-7x}{x-8}$ **9.** $\frac{1}{y-4}$ **11.** 1 **13.** $\frac{20+x^2}{5x}$ **15.** $\frac{t^2-21}{7t}$ **17.** $\frac{43}{10x}$
19. $\frac{11}{24x}$ **21.** $\frac{3x+7}{3(x+5)}$ **23.** $\frac{3-2x}{2(x-7)}$ **25.** $\frac{29x}{10(x+3)}$ **27.** $\frac{8-3y}{6(y-6)}$ **29.** $\frac{5x-32}{x(x-7)}$ **31.** $\frac{-8x-69}{x(x+9)}$
33. $\frac{z(3z+25)}{2(z+5)(z-5)}$ **35.** $\frac{2(4x-15)}{3(x+6)(x-6)}$ **37.** $\frac{2(2x+9)}{(x+5)(x+3)(x-3)}$ **39.** $\frac{a^2-15a-7}{(a+8)(a-8)(a+1)}$
41. $\frac{3x^2+14x-5}{(x+3)(x+2)(x-1)}$ **43.** $\frac{3x^2-17x-14}{(x+7)(x-2)(x-5)}$

16.5 Exercises, page 566

1. $x=-1$ **3.** $x=1$ **5.** $x=2$ **7.** $x=2, -10$ **9.** $x=4$ **11.** $x=3$ **13.** $x=2$ **15.** $a=5$
17. $y=27$ **19.** $x=1, -6$ **21.** $x=4$ **23.** $x=-\frac{1}{2}$ **25.** no solution **27.** $a=3$ **29.** $y=3$
31. $x=20$ **33.** $x=\frac{14}{5}$ **35.** $x=-11$

16.6 Exercises, page 570

1. $\frac{35}{10}$ **3.** 14 **5.** $\frac{7}{15}$ **7.** 73 **9.** passenger train: 75 mph; freight train: 60 mph
11. bicycle: 12 mph; horse: 24 mph **13.** 30 mph **15.** 45 mph **17.** 24 minutes **19.** 6 hours

16.7 Exercises, page 575

1. $12x$ **3.** $\frac{1}{2y}$ **5.** $3ab^3$ **7.** $\frac{1-2x}{1+5x}$ **9.** $\frac{1}{y-3}$ **11.** $\frac{3}{2}$ **13.** $x+5$ **15.** $a(1+a)$ **17.** $a+5$ **19.** $\frac{x-1}{x+4}$ **21.** $\frac{a+5}{a-2}$ **23.** $\frac{2-x}{2+x}$

Chapter 16 Review Exercises, page 576

1. $x \neq -4$ **3.** $x \neq 7, -5$ **5.** $x \neq -1, -8$ **7.** $x \neq -3, -5$ **9.** $2x$ **11.** $\frac{4}{(y-4)(y+2)}$ **13.** $\frac{22}{3x}$ **15.** $\frac{3t-2}{3(t+6)}$ **17.** $\frac{2(x+4)}{x}$ **19.** $\frac{x-5}{2(x+6)}$ **21.** $\frac{y(3y-19)}{2(y+7)(y-7)}$ **23.** $\frac{x^2+x+25}{(x-5)(x+2)(x+6)}$ **25.** $x = 3, 6$ **27.** $a = 7$ **29.** $y = 1, -4$ **31.** $\frac{27}{63}$ **33.** 10 mph **35.** $2x^2$ **37.** $\frac{a-b}{ab}$ **39.** $\frac{x-7}{x+3}$

Chapter 17

17.1 Exercises, page 583

1. $3a$ **3.** $7t^3$ **5.** xy^2 **7.** $3x$ **9.** ab^2 **11.** $12x^2yz^5$ **13.** $2x\sqrt{5}$ **15.** $ab\sqrt{a}$ **17.** $2xy^2\sqrt{y}$ **19.** $4b^2\sqrt{2a}$ **21.** $-2w^3\sqrt{3}$ **23.** $4xy^2z\sqrt{xy}$ **25.** $15x^2y^3\sqrt{6x}$ **27.** $-3a^2b^3c\sqrt{5b}$ **29.** $8xy^3z^4\sqrt{2xz}$ **31.** $a^4b^2c^2\sqrt{3b}$ **33.** $\frac{\sqrt{6}}{3}$ **35.** $\frac{\sqrt{2a}}{2}$ **37.** $\frac{5x}{3}$ **39.** $\frac{8a^2}{9b}$ **41.** $\frac{2x^3}{5y^3}\sqrt{30}$ **43.** $x^2y^2\sqrt{y}$ **45.** $\frac{-b^3\sqrt{a}}{c^4}$ **47.** $\frac{2xy^3\sqrt{6x}}{9}$ **49.** $\frac{ab^3\sqrt{2abc}}{2c}$ **51.** $\frac{3xy^2\sqrt{3xz}}{z^2}$

17.2 Exercises, page 586

1. $11\sqrt{6}$ **3.** $4\sqrt{2}$ **5.** $-6\sqrt{5}$ **7.** $-12\sqrt{11}$ **9.** $9\sqrt{3}-5\sqrt{2}$ **11.** $17\sqrt{3}$ **13.** $-3\sqrt{5}$ **15.** $-28\sqrt{2}$ **17.** $34\sqrt{3}$ **19.** $14\sqrt{5}$ **21.** $15\sqrt{2}$ **23.** $16\sqrt{3}$ **25.** $26\sqrt{5}$ **27.** $27\sqrt{7}$ **29.** $-28\sqrt{2}$ **31.** $19\sqrt{6}$ **33.** $44\sqrt{2}-4\sqrt{3}$ **35.** $8\sqrt{x}$ **37.** $(4a-7)\sqrt{a}$ **39.** $9x^2$ **41.** $7y\sqrt{2y}$ **43.** $(2y+5x)\sqrt{x}$ **45.** $(2y-7x)x\sqrt{3y}$ **47.** $(5b+2)a\sqrt{a}-3ab$ **49.** $(3y-10)x\sqrt{2x}+2\sqrt{xy}$

17.3 Exercises, page 592

1. $\sqrt{21}$ **3.** $48\sqrt{3}$ **5.** $28\sqrt{3}$ **7.** $-32x$ **9.** $6a$ **11.** $-50t$ **13.** $3xy$ **15.** $5b\sqrt{a}$ **17.** $9\sqrt{3}+6$ **19.** $-4\sqrt{6}-10\sqrt{3}$ **21.** $15-\sqrt{15}$ **23.** $6\sqrt{5}+5\sqrt{2}-9\sqrt{10}$ **25.** $48\sqrt{3}+42$ **27.** $10\sqrt{14}-14\sqrt{2}$ **29.** $58+15\sqrt{2}$ **31.** $31+5\sqrt{5}$ **33.** $67+16\sqrt{3}$ **35.** $11-4\sqrt{7}$ **37.** 22 **39.** $47-14\sqrt{2}$ **41.** $-9+3\sqrt{2}$ **43.** $2\sqrt{35}+8\sqrt{15}-\sqrt{21}-12$ **45.** $x-y$ **47.** $a-16$ **49.** $x-4\sqrt{x}-21$ **51.** 7 **53.** 10 **55.** $8\sqrt{2}$ **57.** 9 **59.** 9 **61.** $2\sqrt{3}-5$ **63.** $\sqrt{6}-2$ **65.** $3x$ **67.** $2\sqrt{5a}$ **69.** $3x$ **71.** $\frac{4\sqrt{5}}{5}$ **73.** $\frac{7\sqrt{6}}{6}$ **75.** $\frac{4\sqrt{3}}{3}$ **77.** 1 **79.** $9\sqrt{a}$ **81.** $\frac{7\sqrt{x}}{4x}$ **83.** $\frac{40-8\sqrt{2}}{23}$ **85.** $\frac{7+\sqrt{5}}{22}$ **87.** $\frac{3+5\sqrt{3}}{-22}$ **89.** $4-2\sqrt{2}$ **91.** $\frac{15-7\sqrt{5}-3\sqrt{10}+7\sqrt{2}}{3}$ **93.** $\frac{70+13\sqrt{15}}{43}$

17.4 Exercises, page 596

1. $x = 12$ **3.** $x = 25$ **5.** $x = 7$ **7.** $a = 53$ **9.** no solution **11.** no solution **13.** $x = -3$ **15.** $a = 4$ **17.** no solution **19.** $x = 4$ **21.** $x = 27$ **23.** $y = -1$ **25.** $x = 7$ **27.** $x = 5$ **29.** $x = -2$ **31.** $y = -3, -4$ **33.** $a = 4, 5$ **35.** $x = 3, -2$

Chapter 17 Review Exercises, page 597

1. $8xy^3$ **3.** $3y$ **5.** $5w^2\sqrt{2w}$ **7.** $6ab^2\sqrt{2a}$ **9.** $\frac{3x^2}{8}$ **11.** $\frac{y\sqrt{2xy}}{6}$ **13.** $8\sqrt{7}$ **15.** $18\sqrt{2} + 12\sqrt{5}$ **17.** $2x^2y\sqrt{y}$ **19.** $24\sqrt{2}$ **21.** 50 **23.** $-46 - 5\sqrt{5}$ **25.** 6 **27.** 1 **29.** $\frac{5}{3}$ **31.** $\frac{14 - 2\sqrt{3}}{23}$ **33.** $\frac{8\sqrt{14} + 16}{5}$ **35.** $x = 6$ **37.** $y = 64$ **39.** $x = 2, -1$

Chapter 18

18.1 Exercises, page 602

1. $x = \pm 2$ **3.** $z = \pm 3\sqrt{7}$ **5.** $y = \pm 2$ **7.** $a = \pm 9$ **9.** $t = \pm 5$ **11.** $x = \pm 3\sqrt{2}$ **13.** $a = \pm 2$ **15.** $t = \pm 2\sqrt{5}$ **17.** $x = 9, -3$ **19.** $y = -2, -16$ **21.** $z = -6 \pm 3\sqrt{3}$ **23.** $t = 1 \pm 5\sqrt{2}$ **25.** $x = 2, 3$ **27.** $y = -1, \frac{3}{7}$ **29.** $t = 4, -\frac{4}{3}$ **31.** $a = \frac{-1 \pm 5\sqrt{3}}{4}$ **33.** $t = \frac{3 \pm 4\sqrt{3}}{8}$ **35.** $a = 11, 3$ **37.** $t = 1, -\frac{13}{5}$ **39.** $x = \frac{3 \pm 4\sqrt{2}}{2}$

18.2 Exercises, page 607

1. 16 **3.** 49 **5.** 25 **7.** 4 **9.** 36 **11.** $\frac{25}{4}$ **13.** $\frac{9}{4}$ **15.** $\frac{1}{36}$ **17.** $\frac{1}{16}$ **19.** $\frac{9}{64}$ **21.** $x = -5, -7$ **23.** $x = 7, -2$ **25.** $a = 3, -9$ **27.** $t = 3, 5$ **29.** $x = -7, -8$ **31.** $y = 5, -2$ **33.** $a = 3, -7$ **35.** $a = 5, 8$ **37.** $x = -\frac{1}{2}, 5$ **39.** $y = \frac{1}{5}, 2$ **41.** $t = -\frac{3}{2}, \frac{1}{3}$ **43.** $x = \frac{-1 \pm \sqrt{11}}{2}$ **45.** $x = \frac{3 \pm \sqrt{3}}{3}$ **47.** $x = \frac{-1 \pm \sqrt{13}}{4}$ **49.** $x = \frac{-3 \pm \sqrt{41}}{4}$

18.3 Exercises, page 611

1. $x = -3, -4$ **3.** $y = 5, -9$ **5.** $x = 3, 5$ **7.** $y = 9, -3$ **9.** $a = 2, -6$ **11.** $x = 5$ **13.** $y = -\frac{5}{2}, 1$ **15.** $x = \frac{2}{3}, -4$ **17.** $a = 7, -\frac{1}{2}$ **19.** $x = \frac{-5 \pm \sqrt{5}}{2}$ **21.** $x = \frac{4}{3}, -1$ **23.** $y = 1, -\frac{5}{2}$ **25.** $y = 1, -\frac{3}{5}$ **27.** $a = \frac{5 \pm \sqrt{13}}{6}$ **29.** $x = \frac{-5 \pm 3\sqrt{5}}{10}$

18.4 Exercises, page 614

1. $y = x^2$

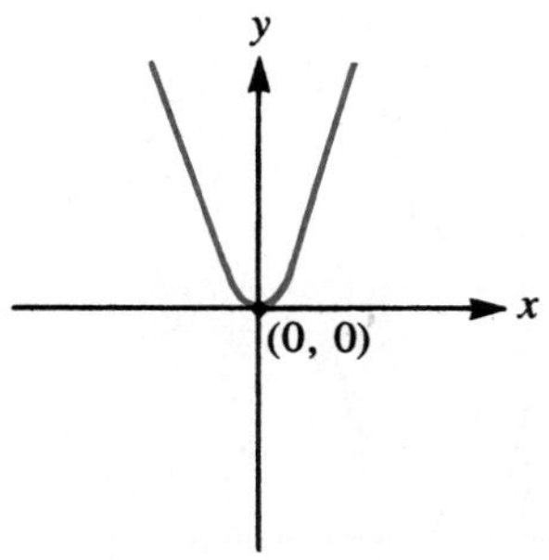

3. $y = -5x^2 + 2$

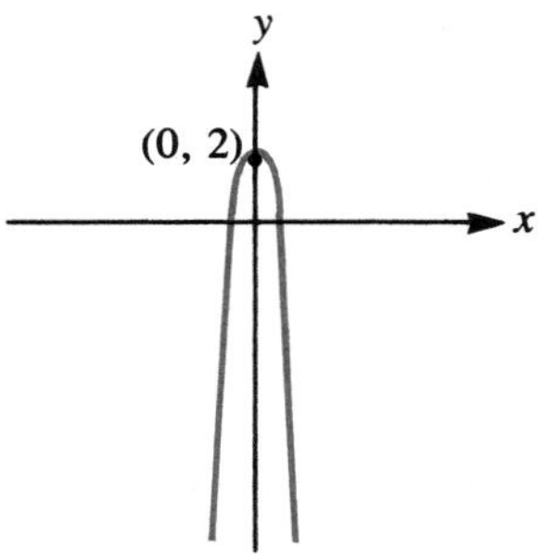

5. $y = x^2 - 4x + 7$

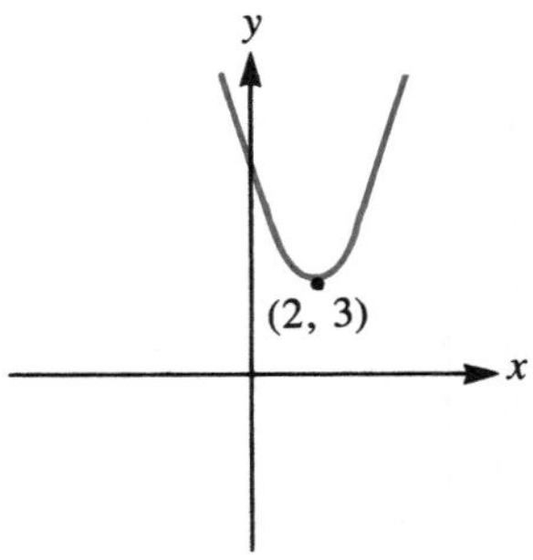

7. $y = -x^2 - 6x + 16$

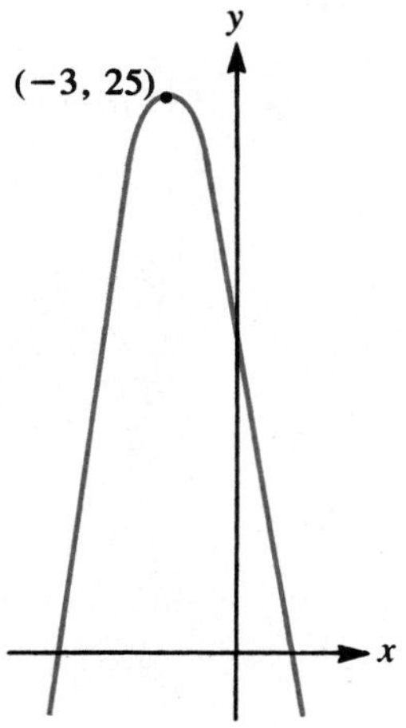

9. $y = x^2 - 10x + 23$

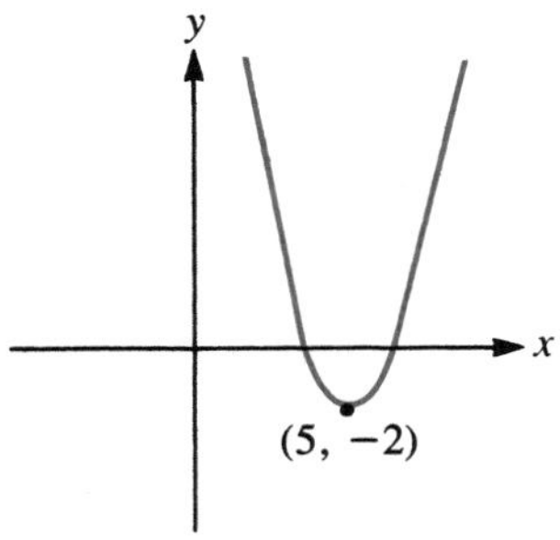

11. $y - 48 = x^2 + 14x$

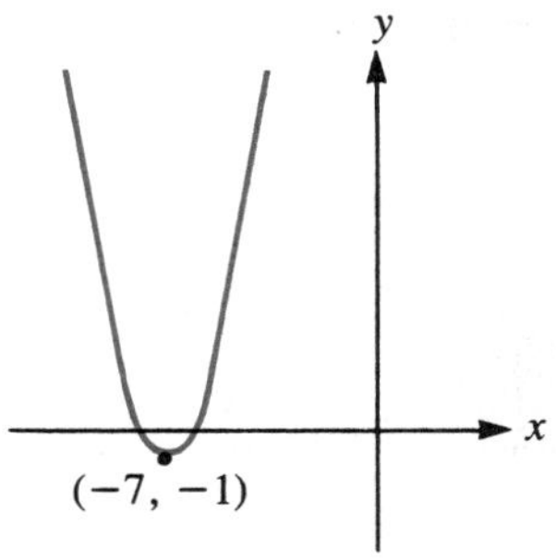

13. $y - 8 = -x^2 - 2x$

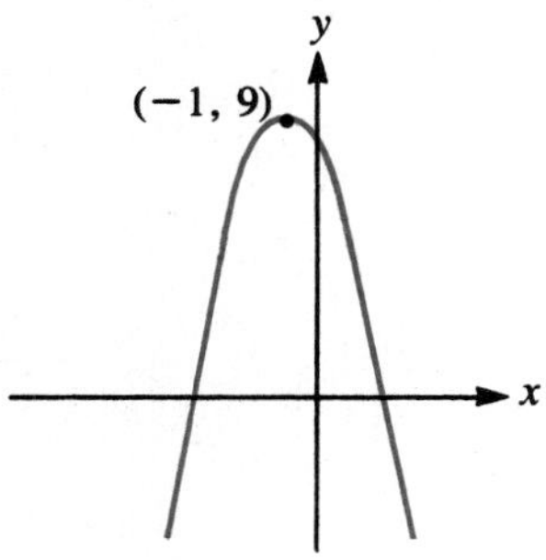

15. $y = (x + 4)^2$

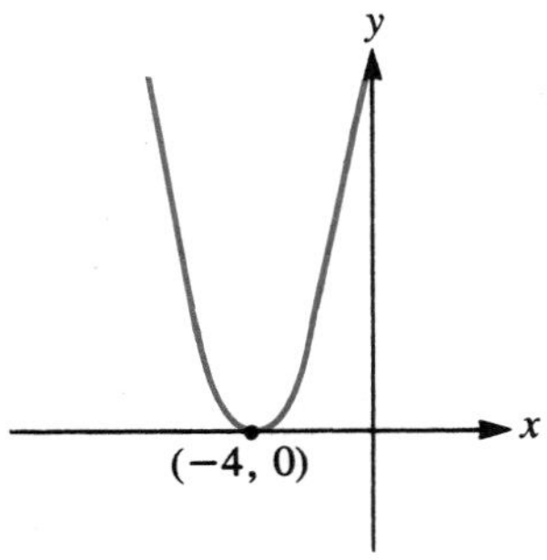

17. $y + 3 = (x - 1)^2$

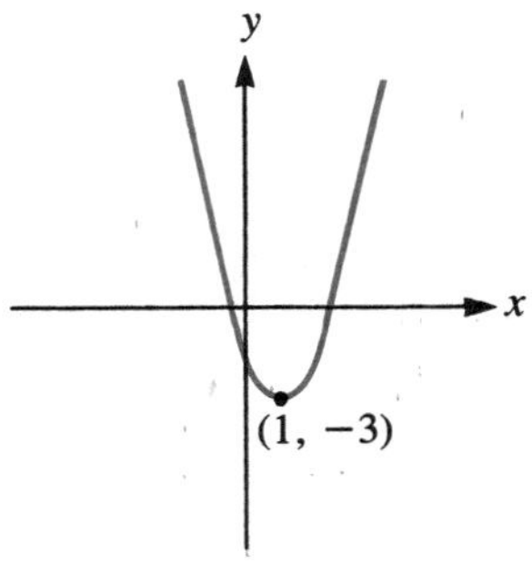

19. $y - 5 = -2(x + 3)^2$

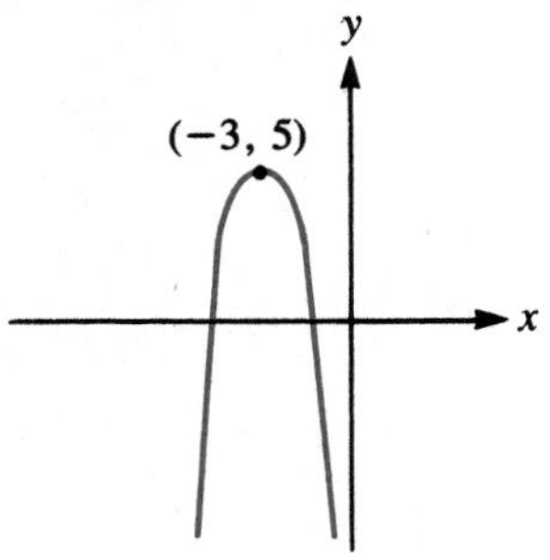

Chapter 18 Review Exercises, page 615

1. $x = \pm 8$ **3.** $a = 11, -1$ **5.** $x = \frac{4}{3}, -2$ **7.** $x = 9, -4$ **9.** $y = \frac{1}{2}, 5$ **11.** $x = \frac{-3 \pm \sqrt{57}}{12}$

13. $x = 2, -5$ **15.** $t = 6, -\frac{1}{3}$ **17.** $x = 1, -\frac{4}{5}$

19. $y = 2x^2 + 3$

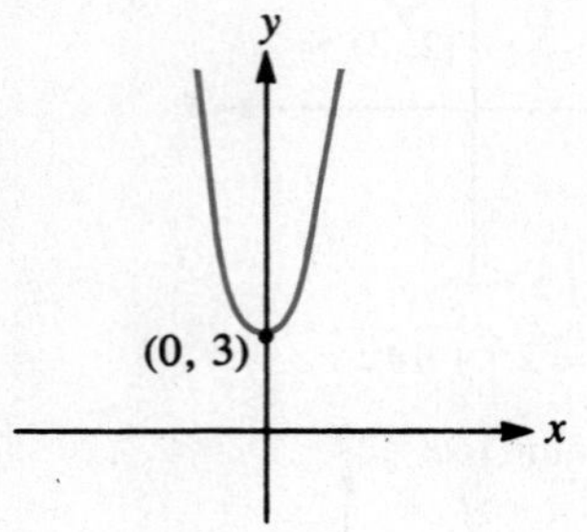

21. $y = -x^2 + 8x - 15$

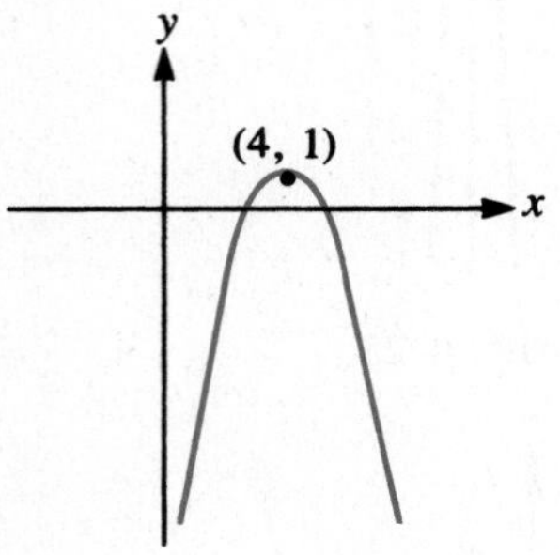

23. $y + 4 = (x - 3)^2$

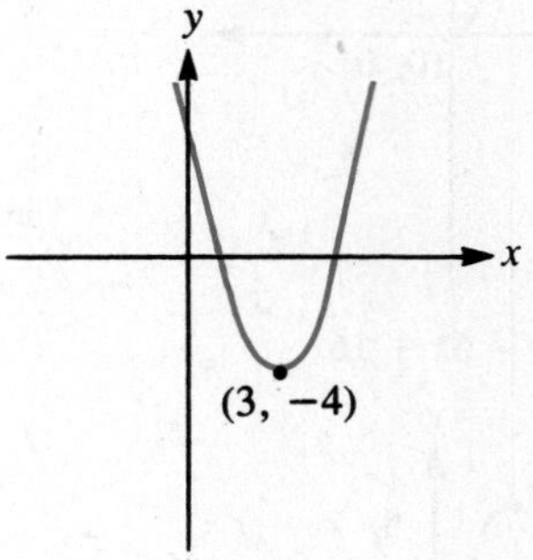

INDEX